Encyclopedia of Earth Sciences

Editorial Board

ENCYCLOPEDIA OF
EARTH SCIENCES

E. Julius Dasch

Editor in Chief

Volume 1

MACMILLAN REFERENCE USA
Simon & Schuster Macmillan
NEW YORK

Simon & Schuster and Prentice Hall International
LONDON   MEXICO CITY   NEW DELHI   SINGAPORE   SYDNEY   TORONTO

Simon & Schuster Macmillan
1633 Broadway
New York, NY 10019

Library of Congress Catalog Card Number: 96-11302

PRINTED IN THE UNITED STATES OF AMERICA

Printing Number

1 2 3 4 5 6 7 8 9 10

LIBRARY OF CONGRESS CATALOGING-IN-PUBLICATION DATA
Encyclopedia of Earth Sciences / E. Julius Dasch, editor in
    chief
        p. cm.
    Includes bibliographical references and index.
    ISBN 0-02-883000-8 (set). — ISBN 0-02-897112-4
    (v. 1). — ISBN 0-02-897114-0 (v. 2)
    1. Earth Sciences — Encyclopedias.   I. Dasch, E. Julius.
QE5.E5137   1996   96-11302
550′ .3—dc20   CIP

This paper meets the requirements of ANSI-NISO Z39.48-1992
(Permanence of Paper)

CONTENTS

Preface   vii

Geologic Timescale   x

Measurements and Their Conversion in the Earth Sciences
(includes conversion tables)   xi

Estimates of the Bulk Composition of Different Portions of
the Earth and Primitive Meteorites
(includes tables of elements)   xix

List of Articles   xxvii

List of Contributors   xxxvii

Acronyms and Standard Abbreviations   xlv

Encyclopedia of Earth Sciences   **1–1192**

Index   1193

Editorial and Production Staff

Hélène G. Potter
Project Editor

Jonathan G. Aretakis
Copy Editor

Dorothy Bauhoff Kachouh
Illustrations Editor

Patricia Brecht
Ellen P. Raspitha
Greg Teague
Helen Wallace
Proofreaders

AEIOU, Inc.
Indexer

Maureen Frantino
Production Manager

Debra Alpern
Alison Avery
Jessica Brent
Editorial Assistants

MACMILLAN REFERENCE
Elly Dickason, *Publisher*
Paul Bernabeo, *Editor in Chief*

PREFACE

A hybrid of various classical disciplines, the earth sciences in the most general sense encompass not only the sciences of the solid Earth, the oceanographic and atmospheric sciences, and the biological sciences, but also the study of our solar system and its place in the galaxy and the universe. No area in the natural sciences has expanded so rapidly in the past two decades as the earth sciences. Owing to widened usage of the term "earth sciences," and to the great expansion of the topic, we have titled this work the *Encyclopedia of Earth Sciences*, avoiding the less inclusive terms "geology" or "geological sciences."

The earth sciences, in the strictest sense, comprise the collection of fundamental sciences—physics, chemistry, biology, and mathematics—applied to the study of Earth. Time is a major factor that sets the earth sciences apart from most other areas of science. Because Earth's history is long—about 4.5 billion years (Ga), roughly one-third the age of the universe—fundamental principles operating over such an immensity of time may be viewed in a much more evolutionary, geologic sense. Thus, stratigraphy, the study of layered rocks and the fossils they may contain, is a quintessential earth science discipline—a time-calibrated, integrated study involving principles of erosion and weathering, transportation, deposition, incorporation of fossil life forms, and finally, chemical and physical reorganization in new rocks.

Our understanding of Earth and its environment has undergone several major shifts in the past two centuries, the period during which the earth sciences became a separate and maturing scientific discipline. Before the late eighteenth century, geological theory was intertwined with religion, natural history, and especially the extraction

and utilization of resources such as coal, metals, and building stone. The realization at the beginning of the nineteenth century that Earth's history is vastly longer than that described in the biblical record constituted one of the great intellectual watersheds of western science. The Darwinian revolution of the mid-nineteenth century, another of the great "revolutions" of western science, created a firm scientific basis for the understanding of paleontology and the way life has developed on Earth, thereby creating an indelible departure point for the study of geology as well as for that of biology.

Major changes in the study of the earth sciences during the second half of the twentieth century have transformed the field from a descriptive, qualitative endeavor to a highly rigorous and quantitative discipline. Perhaps the most important of this century's milestones in the geological sciences occurred in the 1940s and 1950s, when geologists, physicists, and chemists began to make fundamental analytical and theoretical observations about Earth processes such as the crystallization of minerals from molten rock and the nature of earthquake waves. Laboratory analysis and experimentation, combined with testable theoretical considerations, became essential to a more complete understanding of Earth. Although careful observation and description were, and continued to be, critical tools (for example, in the description and mapping of rocks, soils, and mineral resources), increasingly field and laboratory data have become augmented by sophisticated analytical and experimental techniques. A notable example has been the development of the mass spectrometer and its application to the determination of geologic time through the measurement of radioactive isotopes and their daughter products in minerals and rocks. Isotopic dating thus provided a quantitative calibration of earth science's most important cornerstone, the geological timescale, which in the past had been constructed from qualitative studies of faunal succession.

Perhaps this century's most dramatic breakthrough in the earth sciences occurred in the 1960s and 1970s. The development of a quantitative basis for the theory of plate tectonics, which holds that outer, rigid concentric shells of Earth's lithosphere (known as plates) move about on Earth's surface relative to one another, provided a suitable mechanism for explaining the long-suspected drifting of the continents, plate divergence (seafloor spreading), and plate collisions. Almost at once disparate geological topics such as earthquakes, volcanism, the formation of mountains and mountain chains, and ore genesis and emplacement, became tractable as facets of a grand, unified model for Earth processes.

Another paradigmatic shift in the earth sciences began in the late 1960s with the Apollo and Luna missions to the Moon, robotic exploration of many other bodies in our solar system, the first photographs of the whole Earth, and the coordinated, direct analysis of the first extraterrestrial rocks other than meteorites. The first photograph of Earth from outer space (see color plates in Vol. 2), appropriately taken by a geologist from the surface of the Moon, revealed to a wider public not only the wonderful colors of "Spaceship Earth" but the planet's perilously thin atmosphere. For the first time in history, human inhabitants of Earth observed the whole planet's unique but fragile environment through a camera lens. Other photographs from the Apollo program, and later spaceflights, especially those of the space shuttle, returned clear evidence of this fragility—soil erosion and deforestation from clear-cutting, and the burning of large segments of the world's rainforests; a widening hole in the protective ozone layer at the polar regions; and the resulting alteration of ecosystems through a contamination of air and water.

Discoveries of the space age have dramatically affected our view of Earth as a planet. Missions to all of the planets except Pluto and its large moon Charon, as well as to a large number of the satellites of the outer gaseous planets, have documented striking and largely unforeseen differ-

ences between Earth and the other major bodies in the solar system. Before the intense geological study of Mars and Venus, scientists thought that comparative planetology alone could provide critical insight into Earth genesis and evolution, a better understanding of the development of life, and address such topics as ozone budgets in Earth's atmosphere and the effect of greenhouse gases and global climatic changes. While these questions have been partly answered, the space age has documented extreme and unsuspected variability among the planets and their satellites. Understanding the reasons for this variability will enable scientists to unlock questions concerning the evolution of our own remarkable planet.

At the advent of the twenty-first century, the increasingly complex and interdisciplinary earth sciences are benefitting from rapidly developing, innovative technologies for collecting, organizing, and manipulating large databases such as those assembled from the remote sensing of our environment. Enhanced computational capabilities are proving invaluable in the modeling and testing of energy and material transfers among Earth's lithosphere, hydrosphere, atmosphere, and biosphere.

A major change in employment patterns in the earth sciences, especially in the United States and in other industrial countries, has been the marked reduction in jobs related to resource extraction and a shift to enterprises concerned more with Earth stewardship and planning. In the United States, for instance, petroleum and natural gas companies provided vast employment opportunities for geologists until the 1980s, when much lower fuel prices had a negative impact on that job market. Fortunately, emerging opportunities in environmental and water-related industries have lessened the overall impact of job reduction, but there remain fewer jobs for traditionally trained geologists. As a result, university enrollment in geology and related earth science departments has dropped precipitously; statistics in the mid-1990s indicate a possible slowing of these trends, and some new and modified programs are growing remarkably.

Because the earth sciences cover such a broad range of topics, this encyclopedia presented a challenge in organization and editorial philosophy. The result is a two-volume work consisting of about 360 entries of various lengths, which, taken together, cover the main aspects of the earth sciences. The *Encyclopedia of Earth Sciences* is designed to serve both the advanced high-school student and the university undergraduate student. Additional targeted audiences include decision-makers, such as politicians, municipal leaders, and community activists, along with journalists, educators, and the interested public.

Although the entries contained in this work appear in alphabetical order, the many subjects can be grouped into five organizational categories. The first two categories are Solid Earth Processes and Surficial Earth Processes. The third section, Earth Resources and Stewardship, includes studies of resources and their use, and takes into account concerns for environmental awareness. A group of entries, arranged under the section title Earth Sciences in the Public Eye, addresses the increasing importance of topics concerning natural hazards, public health, and land-rights and land-use questions. The most far-reaching group of entries is the fifth group, Earth in Space. The addition of space-related topics was deemed necessary by the continuing advance in our understanding of Earth processes, a result of rapidly accumulating knowledge about the solar system and processes affecting the other planets.

The five organizational sections carry roughly equal weight in terms of numbers of entries. Added to the mix of topical articles is a series of entries on the history of the more fundamental subdisciplines; a similar series on employment opportunities in these major fields; and, finally, about fifty short biographical sketches of notable earth scientists of the past. Prominent scientists and scholars were asked to write the individual entries; these authors are recognized leaders in their fields. The authors include members of academia, government, industry, and not-for-profit institutions, among other sources. The overall integrity and quality of the encyclopedia directly result from their collective participation.

Numerous illustrations and photographs, including two color inserts in Vols. 1 and 2, enhance discussion of the highly visual earth sciences. Each entry contains references for further study, and cross references and blind entries lead the reader to other relevant topics. Essential materials, such as the geologic timescale, structure of the world's oceans and atmosphere, concentrations of the chemical elements in major Earth reservoirs, scientific units of measurement, and explanation of acronyms and abbreviations, and an index, may be found within the appropriate article or as part of the encyclopedia's supporting documentation in the front and backmatter.

The editors wish to thank the members of the Macmillan Library Reference staff, especially Hélène Potter and Elly Dickason, for their vision and professionalism. Dr. David K. Ekroth conceived the idea for this work. Jonathan Aretakis provided painstaking and creative editing.

E. JULIUS DASCH

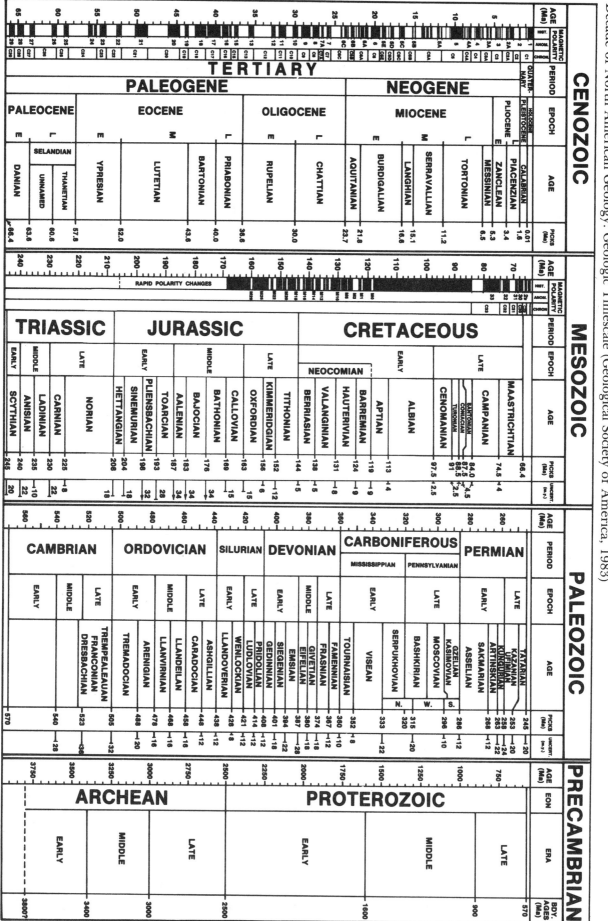

Decade of North American Geology: Geologic Timescale (Geological Society of America, 1983)

MEASUREMENTS AND THEIR CONVERSION IN THE EARTH SCIENCES

The many disciplines comprising this encyclopedia—in the basic sciences, derivative studies, and applied disciplines—require an understanding of standard units of international measurement, units derived from the standard units, and "customary" (national or regional) units. The most commonly used of these "weights and measures," and their conversions, are discussed and listed below.

Use of the terms "units" and "standards" should be distinct. A *unit* is a special quantity in terms of which other quantities are expressed; it is fixed by definition (meter, gram, gallon, etc.) and is independent of other conditions such as temperature. A *standard* is a physical representation of a unit, not independent of other conditions (e.g., a meter standard has a length of 1 meter at some definite temperature and supported in a certain manner).

The metric systems of units began their evolution in France in 1791. The use of these systems has spread, first to non-English-speaking countries, and more recently to English-speaking countries. The first system was based on the centimeter, gram, and second (cgs), units especially of interest in science and technology; later, the more practicable meter, kilogram, and second (mks) system was used. The present (modernized) metric system is the International System of Units—Le Système International d'Unités, or SI, adopted in 1960—based on mks, with the addition of units for temperature (thermodynamic), electricity, radiation, and the quantity of substance that enters into chemical reactions (seven "base" units). The modern SI also presently includes two "supplementary" units for plane angle and solid angle, and many "derived" units for all other quantities (work, force, power), expressed in terms of the seven base and supplementary units, and other derived units. The SI metric system is now either obligatory or permissible throughout the world.

SI Base Unit Definitions

Meter. The unit of length of path traveled by light in a vacuum during a time interval of 1/299,792,458th of a second.

Kilogram. The unit of mass equal to the mass of the international prototype of the kilogram, a platinum-iridium standard kept at the International Bureau of Weights and Measures (General Conference of Weights and Measures, or CGPM).

Second. The unit of duration 9,192,631,770 periods of the radiation corresponding to the transition between the two hyperfine levels of the ground state of the cesium-133 atom (^{133}Cs) (CGPM).

Ampere. The unit of constant current that, if maintained in two straight parallel conductors of infinite length, of negligible circular cross section, and placed 1 meter apart in a vacuum, would produce between these conductors a force equal to 2×10^{-7} newton (force that, acting for 1 second on a mass of 1 kilogram, gives it a velocity of 1 meter per second) per meter of length (CGPM).

Kelvin. The unit of thermodynamic temperature that is 1/273.15th of the thermodynamic temperature of the triple point of water (CGPM).

Mole. The unit of amount of substance of a system that contains as many elementary entities as there are atoms in 0.012 kilograms of carbon-12 (^{12}C) (CGPM).

Candela. The unit of luminous intensity, in a given direction, of a source that emits monochromatic radiation of frequency 540×10^{12} hertz (see "Units Derived from SI" for a definition of hertz) and that has a radiant intensity in that direction of 1/683 watt per steradian (see "SI Supplementary Units" for a definition of steradian) (CGPM).

SI Base Unit Names and Symbols

Physical Quantity	Name	Symbol
Length	meter	m
Mass	kilogram	kg
Time	second	s
Electric current	ampere	A
Thermodynamic temperature	kelvin	K
Amount of substance	mole	mol
Luminous intensity	candela	cd

SI Supplementary Unit Names and Symbols

SI supplementary units are: (1) the radian (rad), the plane angle between two radii of a circle that cut off on the circumference an arc equal in length to the radius; and (2) the steradian (sr), the solid angle that has its vertex in the center of a sphere equal to that of a square with sides of length equal to the radius of the sphere.

The radian and steradian are currently interpreted as dimensionless derived units equal to 1. These units are thus included in the table of units derived from SI, below, rather than in a separate table.

Units Derived from SI, with Special Names and Symbols

Derived Quantity	Name of SI Unit	Symbol for SI Unit	Expression in Terms of SI Base Units	
Plane angle	radian	rad	$(m)\,(m^{-1}) = 1$	
Solid angle	steradian	sr	$(m^2)\,(m^{-2}) = 1$	
Frequency	hertz	Hz	s^{-1}	
Force	newton	N	$m\ kg\ s^{-2}$	
Pressure, stress	pascal	Pa	$N\ m^{-2}$	$= m^{-1}\ kg\ s^{-2}$
Energy, work, heat	joule	J	$N\ m$	$= m^2\ kg\ s^{-2}$
Power, radiant flux	watt	W	$J\ s^{-1}$	$= m^2\ kg\ s^{-3}$
Electric charge	coulomb	C	$A\ s$	
Electric potential, electromotive force	volt	V	$J\ C^{-1}$	$= m^{-2}\ kg\ s^{-3}\ A^{-1}$
Electric resistance	ohm	Ω	$V\ A^{-1}$	$= m^2\ kg\ s^{-3}\ A^{-2}$
Electric conductance	siemens	S	Ω^{-1}	$= m^{-2}\ kg^{-1}\ s^3\ A^2$
Electric capacitance	farad	F	$C\ V^{-1}$	$= m^{-2}\ kg^{-1}\ s^4\ A^2$
Magnetic flux density	tesla	T	$V\ s\ m^{-2}$	$= kg\ s^{-2}\ A^{-1}$
Magnetic flux	weber	Wb	$V\ s$	$= m^2\ kg\ s^{-2}\ A^{-1}$
Inductance	henry	H	$V\ A^{-1}\ s$	$= m^2\ kg\ s^{-2}\ A^{-2}$
Celsius temperature	degree Celsius	°C	K	
Luminous flux	lumen	lm	$cd\ sr$	
Illuminance	lux	lx	$cd\ sr\ m^{-2}$	
Activity	becquerel	Bq	s^{-1}	

Units Used with SI, with Name, Symbol, and Values in SI Units

The following units, not part of the SI, will continue to be used in appropriate contexts (e.g., angstrom):

Physical Quantity	Name of Unit	Symbol for Unit	Value in SI Units
Time	minute	min	60 s
Time	hour	h	3,600 s
Time	day	d	86,400 s
Plane angle	degree	°	$(\pi/180)$ rad
Plane angle	minute	′	$(\pi/10\ 800)$ rad
Plane angle	second	″	$(\pi/648\ 000)$ rad
Length	angström	Å	10^{-10} m
Area	barn	b	10^{-28} m^2
Volume	liter	l, L	dm^3 = 10^{-3} m^3
Mass	ton	t	Mg = 10^3 kg
Pressure	bar	bar	10^5 Pa = 10^5 N m^{-2}
Energy	electronvolt	eV ($= e \times$ V)	$\approx 1.60218 \times 10^{-19}$ J
Mass	unified atomic mass unit	u ($= m_a(^{12}C)/12$)	$\approx 1.66054 \times 10^{-27}$ kg

In the SI system (and, where appropriate, in SI-derived units and units used together with the SI), designations of multiples and subdivisions of any unit may be arrived at by combining the name of the unit with the prefixes *deka*, *hecto*, and *kilo* (10, 100, and 1,000) and *deci*, *centi*, and *milli* (1/10, 1/100, 1/1,000), etc. These generally recognized prefixes are given below:

yotta	10^{24}	deci	10^{-1}
zetta	10^{21}	centi	10^{-2}
exa	10^{18}	milli	10^{-3}
peta	10^{15}	micro	10^{-6}
tera	10^{12}	nano	10^{-9}
giga	10^{9}	pico	10^{-12}
mega	10^{6}	femto	10^{-15}
kilo	10^{3}	atto	10^{-18}
hecto	10^{2}	zepto	10^{-21}
deka	10^{1}	yocto	10^{-24}

Thus, a kilometer is 1,000 meters, a millimeter is 0.001 meter.

Astronomical Units

The more common astrophysical and astronomical quantities used in this encyclopedia are described below. For more complete information on these and other astronomical quantities, see Allen (1973):

Astronomical unit
of distance = AU = 1.495979×10^{13} cm
mean Sun–Earth distance = semimajor axis of Earth orbit

Parsec (= 206264.806
AU)　　　　pc = 3.085678×10^{18} cm
　　　　　= 3.261633 light-years

Light-year = 9.460530×10^{17} cm

Light time for 1 AU = 499.00479 s = 0.005775 52 d

Solar mass　$\mathcal{M}_\odot$ = 1.989×10^{33} g

Solar radius　$\mathcal{R}_\odot$ = 6.9599×10^{10} cm

Solar radiation　$\mathcal{L}_\odot$ = 3.826×10^{33} erg/s

Earth mass　$\mathcal{M}_\oplus$ = 5.976×10^{27} g

Earth mean density　$\bar{\rho}_\oplus$ = 5.517 g cm^{-3}

Earth equatorial
radius　　　= 6378.164 km

Atomic Units

For certain atomic and electronic uses, some fundamental constants may be regarded as constants. Conversion to SI units is not exact:

Physical Quantity	Name of Unit	Symbol for Unit	Definition and Value of Unit in SI
Mass	electron rest mass	m_e	$m_e \approx 9.1095 \times 10^{-31}$ kg
Charge	elementary charge	e	$e \approx 1.6022 \times 10^{-19}$ C
Action	Planck constant/2π	$\hbar$	$\hbar = h/2\pi \approx 1.0546 \times 10^{-34}$ J s
Length	Bohr	a_0	$4\pi\varepsilon_0\hbar^2/m_e e^2 \approx 5.2918 \times 10^{-11}$ m
Energy	hartree	$E_\hbar$	$\hbar^2/m_e a_0^2 \approx 4.3598 \times 10^{-18}$ J

Conversions for Standard, Derived, and Customary Measurements

LENGTH

angstrom (Å)	0.1 nanometer (exactly)
	0.000 1 micrometer (exactly)
	0.000 000 1 millimeter (exactly)
	0.000 000 004 inch
1 centimeter (cm)	0.393 7 inch
1 decimeter (dm)	3.937 inches
1 fathom	1.828 8 meters
1 foot (ft)	0.304 8 meter (exactly)
1 inch (in)	2.54 centimeters (exactly)
1 kilometer (km)	0.621 mile
1 meter (m)	39.37 inches
	1.094 yards
1 micrometer (μm)	0.001 millimeter (exactly)
	0.000 039 37 inch
1 mil	0.001 inch (exactly)
	0.025 4 millimeter (exactly)
1 mile (mi) (U.S. statute)	5 280 feet (exactly)
	1.609 kilometers
1 mile (mi) (international)	5 280 feet (exactly)
1 millimeter (mm)	0.039 37 inch
1 nanometer (nm)	0.001 micrometer (exactly)
	0.000 000 039 37 inch
1 rod (rd), pole, or perch	16 1/2 feet (exactly)
	5.029 2 meters
1 yard (yd)	0.914 4 meter (exactly)

AREA

1 acre	43 560 square feet (exactly)
	0.405 hectare
1 are	119.599 square yards
	0.025 acre
1 hectare	2.471 acres
1 square centimeter (cm²)	0.155 square inch
1 square foot (ft²)	929.030 square centimeters
1 square inch (in²)	6.451 6 square centimeters (exactly)
1 square kilometer (km²)	247.104 acres
	0.386 square mile
1 square meter (m²)	1 196 square yards
	10.764 square feet
1 square mile (mi²)	258.999 hectares
1 square millimeter (mm²)	0.002 square inch
1 square rod (rd²), square pole, or square perch	25.293 square meters
1 square yard (yd²)	0.836 square meter

VOLUME

1 barrel (bbl), liquid*	31 to 42 gallons
1 cubic centimeter (cm³)	0.061 cubic inch
1 cubic decimeter (dm³)	61.024 cubic inches
1 cubic foot (ft³)	7.481 gallons 28.316 cubic decimeters
1 cubic inch (in.³)	0.554 fluid ounce 4.433 fluid drams 16.387 cubic centimeters
1 cubic meter (m³)	1.308 cubic yards
1 cubic yard (yd³)	0.765 cubic meter
1 dekaliter (daL)	2.642 gallons 1.135 pecks
1 dram, fluid (or liquid)	1/8 fluid ounce (exactly) 0.226 cubic inch 3.697 milliliters
1 gallon (gal) (U.S.)	231 cubic inches (exactly) 3.785 liters 128 U.S. fluid ounces (exactly)
1 gallon (gal) (British Imperial)	277.42 cubic inches 1.201 U.S. gallons 4.546 liters
1 gill (gi)	7.219 cubic inches 4 fluid ounces (exactly) 0.118 liter
1 hectoliter (hL)	26.418 gallons 2.838 bushels
1 liter (1 cubic decimeter exactly)	1.057 liquid quarts 0.908 dry quart 61.025 cubic inches
1 milliliter (mL)	0.271 fluid dram 16.231 minims 0.061 cubic inch

* There are a variety of "barrels" established by law or usage. For example, U.S. federal taxes on fermented liquors are based on a barrel of 31 gallons (141 liters); many state laws fix the "barrel for liquids" as 31½ gallons; one state fixes a 36-gallon (160.5 liters) barrel for cistern measurement; federal law recognizes a 40-gallon (178 liters) barrel for "proof spirits"; by custom, 42 gallons comprise a barrel of crude oil or petroleum products for statistical purposes, and this equivalent is recognized "for liquids" by four states.

1 ounce, fluid (or liquid)	1.805 cubic inches 29.573 milliliters 0.961 U.S. fluid ounce
1 ounce, fluid (fl oz) (British)	1.734 cubic inches 28.412 milliliters
1 pint (pt), liquid	28.875 cubic inches (exactly) 0.473 liter
1 quart (qt), dry (U.S.)	67.201 cubic inches 1.101 liters
1 quart (qt), liquid (U.S.)	57.75 cubic inches (exactly) 0.946 liter

UNITS OF MASS

1 assay ton (AT)	29.167 grams
1 carat (ct)	200 milligrams (exactly) 3.086 grains
1 dram, avoirdupois (dr avdp)	27 11/32 (= 27.344) grains 1.777 grams
1 gamma (γ)	1 microgram (exactly)
1 grain	64.798 91 milligrams (exactly)
1 gram (g)	15.432 grains 0.035 ounce, avoirdupois
1 kilogram (kg)	2.205 pounds
1 microgram (μg)	0.000 001 gram (exactly)
1 milligram (mg)	0.015 grain
1 ounce, avoirdupois (oz avdp)	437.5 grains (exactly) 0.911 troy or apothecaries ounce 28.350 grams
1 pound, avoirdupois (lb avdp)	7 000 grains (exactly) 1.215 troy or apothecaries pounds 453.592 37 grams (exactly)
1 pound, troy or apothecaries (lb t or lb ap)	5 760 grains (exactly) 0.823 avoirdupois pound 373.242 grams
1 ton, gross or long	2 240 pounds (exactly) 1.12 net tons (exactly) 1.016 metric tons
1 ton, metric (t)	2 204.623 pounds 0.984 gross ton 1.102 net tons
1 ton, net or short	2 000 pounds (exactly) 0.893 gross ton 0.907 metric ton

PRESSURE

1 pascal (Pa)	1 newton/square meter (N/m^2)
1 kilogram/ square centi- meter (kg/cm^2)	0.96784 atmosphere (atm) = 14.2233 pounds/square inch (lb/in.^{2}) = 0.98067 bar
1 bar	0.98692 atmosphere (atm) = 10^5 pascals (Pa) = 1.02 kilograms/ square centimeter (kg/cm^2)

ENERGY AND POWER

Energy

1 joule (J)	1 newton meter (N · m) 2.390×10^{-1} calories (cal) 9.47×10^{-4} British ther- mal units (Btu) 2.78×10^{-7} kilowatt hours (kWh)
1 calorie (cal)	4.184 joule (J) 3.968×10^{-3} British ther- mal units (Btu) 1.16×10^{-6} kilowatt hours (kWh)
1 British thermal unit (Btu)	1055.87 joules (J) 252.19 calories (cal) 2.928×10^{-4} kilowatt hours (kWh)
1 kilowatt hour (kWh)	3.6×10^6 joules (J) 8.60×10^5 calories (cal) 3.41×10^3 British thermal units (Btu)

Power (energy per unit time)

1 watt (W)	1 joule per second (J/s) 3.4129 Btu/h 1.341×10^{-3} horsepower (hp) 14.34 calories per minute (cal/min)
1 horsepower (hp)	550 ft lb/s 7.46×10^2 watts (W)

Temperature

Temperature is the measure of the internal kinetic energy (expressed as movement) of the atoms and molecules in a body. In the SI system, temperature is expressed on the Kelvin scale (K), although temperature intervals are the same as on the more familiar Celsius (°C) scale. In the following equation, Celsius temperature (t) is defined in terms of thermodynamic temperature (T) by t = T − T$_0$, where T$_0$ = 273.15 K by definition. On the Celsius scale, 100 degrees is selected as the temperature at which water boils at sea level (= 212°C). On the K scale, 0 degrees is absolute zero, the temperature at which all atomic and molecular motion ceases.

Scientists commonly use the Celsius system. Although not recommended for scientific and technical use, earth scientists also use the familiar Fahrenheit temperature scale (°F). 1°F = 1.8°C or K. The triple point of H$_2$O, where gas, liquid, and solid water coexist, is 32°F.

TEMPERATURE CONVERSIONS

• To change from Fahrenheit (F) to Celsius (C)

$$°C = \frac{(°F - 32°)}{1.8}$$

• To change from Celsius (C) to Fahrenheit (F)

$$°F = (°C \times 1.8) + 32°$$

• To change from Celsius (C) to Kelvin (K)

$$K - °C + 273.15$$

• To change from Fahrenheit (F) to Kelvin (K)

$$K = \frac{(°F - 32°)}{1.8} + 273.15$$

Bibliography

ALLEN, C. W. *Astrophysical Quantities*, 3d ed. London, 1973.

AMERICAN SOCIETY FOR TESTING AND MATERIALS. *Standard Practice for the Use of International System of Units (SI) (The Modernized Metric System)*. Special Publication E 380-93. Philadelphia, PA, 1993.

ANSI/IEEE. *American National Standard for Metric Practice*. Std 268-1992. New York, 1992.

JUDSON, L. V. *Weights and Measures Standards of the United States: A Brief History*. National Bureau of Standards (now National Institute of Standards and Technology), Special Publication 447. Gaithersburg, MD, 1976.

LIDE, D. R., ed. *CRC Handbook of Chemistry and Physics, 1990–91*. Boston, MA, 1991.

NATIONAL INSTITUTE OF STANDARDS AND TECHNOLOGY (NIST). *The International System of Units (SI)*. Special Publication 330. Gaithersburg, MD, 1991.

TAYLOR, B. N. *Guide for the Use of the International System of Units (SI)*. National Institute of Standards and Technology (NIST), Special Publication 811. Gaithersburg, MD, 1995.

E. JULIUS DASCH

ESTIMATES OF THE BULK COMPOSITIONS OF DIFFERENT PORTIONS OF THE EARTH AND PRIMITIVE METEORITES

Earth scientists have found it useful to make estimates of the bulk compositions of various discrete reservoirs of Earth, and even of the whole or bulk Earth. These estimates are used to infer the geochemical behavior of elements during geological processes, refine models of the heat budget of Earth, and model the evolution of our atmosphere, to name just a few earth science applications. Some estimates are much more easily made than others. The atmosphere and oceans are readily sampled and are fairly well mixed by currents, winds, and storms, so the compositions of these reservoirs are undoubtedly the best constrained. The crust is also easily sampled, but is not well mixed (note the difference between estimated upper crust and lower crust compositions), making estimates of its composition less well constrained. The mantle is mixed by convection to some extent, but it is not easily sampled. The core is almost completely unknown; it cannot be sampled, and its composition can only be estimated by starting with estimates for the bulk Earth and the primitive mantle and by using mass balance for the elements to guess the concentrations of siderophile elements in the core (geophysical information helps to constrain these results).

As an example of the uncertainties involved in these estimates, McDonough and Sun (1995) provided their "subjective judgment" of the uncertainties for each element in their primitive mantle composition. The uncertainties range from ± 10 percent for elements such as magnesium (Mg), which are easily and often measured and whose geochemical behavior is well understood, to ± a factor of 4 (between 0.25 times and 4 times the estimate) for elements such as Mercury (Hg), which are infrequently measured in rocks and whose geochemistry is less well understood.

The estimates given below are not unique; various earth scientists have made estimates of geologic reservoirs of interest to them, with varying results. Two tables are provided. Table 1 is arranged in order of atomic number. This table has the advantage that the units of concentration of neighboring elements tend to be the same because of generally decreasing abundance with increasing atomic number as a result of nucleosynthesis of the elements, and that geochemically similar elements tend to be grouped. Table 2 is arranged in alphabetical order of element names.

DAVID W. MITTLEFEHLDT

Table 1. Elements Arranged by Atomic Number

Name	Symbol	Atomic Number	Units[1]	CI Chondrites[2]	Bulk Earth[3]	Core[4]	Primitive Mantle[5]	Oceanic Crust[6]	Continental Crust[7]			Oceans[8]	Atmosphere[9]
									Bulk	Lower	Upper		
Hydrogen	H	1	mg/g	20.2	0.078	‡						108	0.21
Helium	He	2	ng/g	10	1.3							0.005	723
Lithium	Li	3	μg/g	1.50	2.7		1.6		13	11	20	0.17	
Beryllium	Be	4	ng/g	24.9	56		68		1,500	1,000	3,000	0.0005	
Boron	B	5	μg/g	0.87	0.47		0.30		10	8.3	15	4.6	
Carbon	C	6	mg/g	34.5	0.35	‡	0.12					0.025	0.131
Nitrogen	N	7	μg/g	3,180	9.1		2					17	7.538×10^5
Oxygen	O	8	mg/g	464	285.0	‡						857	233.0
Fluorine	F	9	μg/g	60.7	53		25					1	
Neon	Ne	10	pg/g	201	25							100	1.27×10^7
Sodium	Na	11	mg/g	5.00	1.58		2.67	19.9	23.0	20.8	28.9	10.8	
Magnesium	Mg	12	mg/g	98.9	132.1		228	45.7	32.0	38.0	13.3	1.28	
Aluminum	Al	13	mg/g	8.68	17.7		23.5	15.3	84.1	85.2	80.4	0.0008	
Silicon	Si	14	mg/g	106.4	143.4		210	236	268	254	308	5.6	
Phosphorus	P	15	μg/g	1,220	2,150	5,900	90				700	120	
Sulfur	S	16	mg/g	62.5	18.4	‡	0.25					0.905	
Chlorine	Cl	17	μg/g	704	25		17					19,400	
Argon	Ar	18	ng/g	397	70.4							630	1.29×10^7
Potassium	K	19	μg/g	558	170		240	884	9,100	2,800	28,000	400	
Calcium	Ca	20	mg/g	9.28	19.3		25.3	80.8	52.9	60.7	30.0	0.412	
Scandium	Sc	21	μg/g	5.82	12.1		16.2	41.4	30	36	11	0.00004	
Titanium	Ti	22	mg/g	0.436	1.03		1.205	9.68	5.4	6.0	3.0	2×10^{-8}	
Vanadium	V	23	μg/g	56.5	103		82		230	285	60	0.002	
Chromium	Cr	24	mg/g	2.66	4.78		2.625		0.185	0.235	0.035	2×10^{-7}	
Manganese	Mn	25	mg/g	1.99	0.59		1.045		1.4	1.7	0.6	4×10^{-6}	
Iron	Fe	26	mg/g	190.4	358.7	891	62.6	81.0	70.7	82.4	35.0	0.00006	
Cobalt	Co	27	μg/g	502	940	2,450	105	47.1	29	35	10	0.0019	
Nickel	Ni	28	mg/g	11.0	20.4	53.7	1.96	0.150	0.105	0.135	0.02	0.00047	
Copper	Cu	29	μg/g	126	57	100	30	74.4	75	90	25	0.25	
Zinc	Zn	30	μg/g	312	93	160	55		80	83	71	0.39	
Gallium	Ga	31	μg/g	10.0	5.5	7.8	4.0		18	18	17	0.0002	
Germanium	Ge	32	μg/g	32.7	13.8	37	1.1		1.6	1.6	1.6	0.00007	
Arsenic	As	33	μg/g	1.86	3.6	10	0.05		1.0	0.8	1.5	0.003	
Selenium	Se	34	μg/g	18.6	6.1	17	0.075		50	50	50	0.005	
Bromine	Br	35	ng/g	3,570	134		50					67,400	
Krypton	Kr	36	pg/g	57	4.3							300	3.23×10^6
Rubidium	Rb	37	μg/g	2.30	0.58		0.600	1.26	32	5.3	112	0.12	

Table 1. Continued

Name	Symbol	Atomic Number	Units[1]	CI Chondrites[2]	Bulk Earth[3]	Core[4]	Primitive Mantle[5]	Oceanic Crust[6]	Continental Crust[7]			Oceans[8]	Atmosphere[9]
									Bulk	Lower	Upper		
Strontium	Sr	38	µg/g	7.80	18.2		19.9	113	260	230	350	8	
Yttrium	Y	39	µg/g	1.56	3.29		4.30	35.8	20	19	22	0.0003	
Zirconium	Zr	40	µg/g	3.94	7.42		10.5	104	100	70	190		
Niobium	Nb	41	µg/g	0.246	0.501		0.658	3.51	11	6	25	0.00002	
Molybdenum	Mo	42	µg/g	0.928	2.96	8.2	0.050		1.0	0.8	1.5	0.008	
Technetium	Tc	43	*										
Ruthenium	Ru	44	µg/g	0.712	1.48	4.2	0.0050					0.14	
Rhodium	Rh	45	µg/g	0.134	0.32	0.90	0.0009				0.5	0.11	
Palladium	Pd	46	ng/g	560	1,000	2,800	3.9		1	1			
Silver	Ag	47	ng/g	199	80	210	8		80	90	50	0.0003	
Cadmium	Cd	48	ng/g	686	21		40		98	98	98	0.11	
Indium	In	49	ng/g	80	2.7		11		50	50	50		
Tin	Sn	50	µg/g	1.72	0.71	1.8	0.13	1.38	2.5	1.5	5.5	0.0003	
Antimony	Sb	51	ng/g	142	64	170	5.5		200	200	200	0.5	
Tellurium	Te	52	ng/g	2,320	940	2,600	12						
Iodine	I	53	ng/g	433	17		10					60	
Xenon	Xe	54	pg/g	190	10							100	394
Cesium	Cs	55	ng/g	187	59		21	14.1	1,000	100	3,700	0.5	
Barium	Ba	56	µg/g	2.34	5.1		6.60	13.9	250	150	550	0.04	
Lanthanum	La	57	µg/g	0.2347	0.48		0.648	3.90	16	11	30	0.0003	
Cerium	Ce	58	µg/g	0.6032	1.28		1.675	12.0	33	23	64	0.0004	
Praseodymium	Pr	59	µg/g	0.0891	0.162		0.254	2.07	3.9	2.8	7.1		
Neodymium	Nd	60	µg/g	0.4524	0.87		1.25	11.2	16	12.7	26		
Promethium	Pm	61	*										
Samarium	Sm	62	µg/g	0.1471	0.26		0.406	3.75	3.5	3.17	4.5		
Europium	Eu	63	µg/g	0.056	0.100		0.154	1.34	1.1	1.17	0.88		
Gadolinium	Gd	64	µg/g	0.1966	0.37		0.544	5.08	3.3	3.13	3.8		
Terbium	Tb	65	µg/g	0.0363	0.067		0.099	0.885	0.60	0.59	0.64		
Dysprosium	Dy	66	µg/g	0.2427	0.45		0.674	6.30	3.7	3.6	3.5		
Holmium	Ho	67	µg/g	0.0556	0.101		0.149	1.34	0.78	0.77	0.80		
Erbium	Er	68	µg/g	0.1589	0.29		0.438	4.14	2.2	2.2	2.3		
Thulium	Tm	69	µg/g	0.0242	0.044		0.068	0.621	0.32	0.32	0.33		
Ytterbium	Yb	70	µg/g	0.1625	0.29		0.441	3.90	2.2	2.2	2.2		
Lutetium	Lu	71	µg/g	0.0243	0.049		0.0675	0.589	0.30	0.29	0.32		
Hafnium	Hf	72	µg/g	0.104	0.29		0.283	2.97	3.0	2.1	5.8		
Tantalum	Ta	73	ng/g	14.2	29		37	192	1,000	600	2,200		
Tungsten	W	74	ng/g	92.6	250	650	29		1,000	700	2,000		
Rhenium	Re	75	ng/g	36.5	76	210	0.28		0.4	0.4	0.4	0.12	
Osmium	Os	76	ng/g	486	1,110	3,100	3.4		0.05	0.05	0.05		

Table 1. Continued

Name	Symbol	Atomic Number	Units[1]	CI Chondrites[2]	Bulk Earth[3]	Core[4]	Primitive Mantle[5]	Oceanic Crust[6]	Continental Crust[7] Bulk	Continental Crust[7] Lower	Continental Crust[7] Upper	Oceans[8]	Atmosphere[9]
Iridium	Ir	77	ng/g	481	1,060	3,000	3.2		0.1	0.13	0.02		
Platinum	Pt	78	ng/g	990	2,100	5,900	7.1						
Gold	Au	79	ng/g	140	290	820	1.0		3.0	3.4	1.8	0.2	
Mercury	Hg	80	ng/g	258	9.9	8.7	10					0.2	
Thallium	Tl	81	ng/g	142	4.9		3.5		360	230	750		
Lead	Pb	82	ng/g	2,470	141		150	489	8,000	4,000	20,000	1	
Bismuth	Bi	83	ng/g	114	3.7		2.5		60	38	127	0.02	
Polonium	Po	84	†										
Astatine	At	85	†										
Radon	Rn	86	†										
Francium	Fr	87	†										
Radium	Ra	88	†										
Actinium	Ac	89	†										
Thorium	Th	90	ng/g	29.4	65		79.5	187	3,500	1,060	10,700	0.05	
Protactinium	Pa	91	†										
Uranium	U	92	ng/g	8.1	18		20.3	71.1	910	280	2,800	3	

[1] The units are SI units rather than the more familiar parts per million (ppm), parts per billion (ppb), etc. These tables use SI units in part because they are the standard, but more important because "billion" does not have a single definition; it means 10^9 in American usage, but 10^{12} in British usage. See "Measurements and Their Conversion in the Earth Sciences" in the frontmatter.

[2] CI Chondrites: primitive meteorite composition, modified from E. Anders and N. Grevese, *Geochimica et Cosmochimica Acta* 53 (1989): 197–214. *See also* entry METEORITES.

[3] Bulk Earth: modified from E. Anders, *Philosophical Transactions of the Royal Society of London A* 285 (1977):23–40. *See also* the entry EARTH, COMPOSITION OF.

[4] Core: estimated by mass balance from the bulk Earth and primitive mantle compositions.

[5] Primitive Mantle: from W. F. McDonough and S.-S. Sun, *Chemical Geology* 120 (1995):223–253.

[6] Oceanic Crust: average normal mid-ocean ridge basalt of A. W. Hofmann, *Earth and Planetary Science Letters* 90 (1988):297–314. *See also* the entry OCEANIC CRUST, STRUCTURE OF.

[7] Continental Crust—Bulk, Lower and Upper: from S. R. Taylor and S. M. McLennan, *Reviews of Geophysics* 33 (1995):241–265.

[8] Oceans: modified from the entry OCEANOGRAPHY, CHEMICAL; and D. A. Ross, *Introduction to Oceanography*, 4th ed. (1988).

[9] Atmosphere: modified from the entry EARTH'S ATMOSPHERE, CHEMICAL COMPOSITION OF, and M. Ozima and F. A. Podosek, *Nobel Gas Geochemistry* (1983).

* No long-lived radioactive (>10^8 years) or stable nuclei of these elements exist.

† No long-lived radioactive (>10^8 years) or stable nuclei of these elements exist. Their abundances in the earth are steady-state concentrations in the ^{232}Th, ^{235}U, and ^{238}U decay schemes.

‡ Earth's core contains about 10% of a light element, of which these are the most likely candidates. See the entry CORE, COMPOSITION OF.

Table 2. Elements Arranged Alphabetically by Name

Name	Symbol	Atomic Number	Units[1]	Cl Chondrites[2]	Bulk Earth[3]	Core[4]	Primitive Mantle[5]	Oceanic Crust[6]	Continental Crust[7]			Oceans[8]	Atmosphere[9]
									Bulk	Lower	Upper		
Actinium	Ac	89	†										
Aluminum	Al	13	mg/g	8.68	17.7		23.5	15.3	84.1	85.2	80.4	0.0008	
Antimony	Sb	51	ng/g	142	64	170	5.5		200	200	200	0.5	
Argon	Ar	18	ng/g	397	70.4							630	1.29×10^7
Arsenic	As	33	µg/g	1.86	3.6	10	0.05	13.9	1.0	0.8	1.5	0.0003	
Astatine	At	85	†										
Barium	Ba	56	µg/g	2.34	5.1		6.60		250	150	550	0.04	
Beryllium	Be	4	ng/g	24.9	56		68		1,500	1,000	3,000	0.0005	
Bismuth	Bi	83	ng/g	114	3.7		2.5		60	38	127	0.02	
Boron	B	5	µg/g	0.87	0.47		0.30		10	8.3	15	4.6	
Bromine	Br	35	ng/g	3,570	134		50					67,400	
Cadmium	Cd	48	ng/g	686	21		40		98	98	98	0.11	
Calcium	Ca	20	mg/g	9.28	19.3		25.3	80.8	52.9	60.7	30.0	0.412	
Carbon	C	6	mg/g	34.5	0.35	‡	0.12					0.025	0.131
Cerium	Ce	58	µg/g	0.6032	1.28		1.675	12.0	33	23	64	0.0004	
Cesium	Cs	55	µg/g	187	59		21	14.1	1,000	100	3,700	0.5	
Chlorine	Cl	17	µg/g	704	25		17					19,400	
Chromium	Cr	24	mg/g	2.66	4.78	2,450	2.625		0.185	0.235	0.035	2×10^{-7}	
Cobalt	Co	27	µg/g	502	940	100	105	47.1	29	35	10	0.0019	
Copper	Cu	29	µg/g	126	57		30	74.4	75	90	25	0.25	
Dysprosium	Dy	66	µg/g	0.2427	0.45		0.674	6.30	3.7	3.6	3.5		
Erbium	Er	68	µg/g	0.1589	0.29		0.438	4.14	2.2	2.2	2.3		
Europium	Eu	63	µg/g	0.056	0.100		0.154	1.34	1.1	1.17	0.88		
Fluorine	F	9	µg/g	60.7	53		25					1	
Francium	Fr	87	†										
Gadolinium	Gd	64	µg/g	0.1966	0.37		0.544	5.08	3.3	3.13	3.8		
Gallium	Ga	31	µg/g	10.0	5.5	7.8	4.0		18	18	17	0.0002	
Germanium	Ge	32	µg/g	32.7	13.8	37	1.1		1.6	1.6	1.6	0.00007	
Gold	Au	79	ng/g	140	290	820	1.0		3.0	3.4	1.8	0.2	
Hafnium	Hf	72	µg/g	0.104	0.29		0.283	2.97	3.0	2.1	5.8		
Helium	He	2	ng/g	10	1.3							0.005	723
Holmium	Ho	67	µg/g	0.0556	0.101		0.149	1.34	0.78	0.77	0.80		
Hydrogen	H	1	mg/g	20.2	0.078	‡	11					108	
Indium	In	49	ng/g	80	2.7				50	50	50		
Iodine	I	53	ng/g	443	17		10					60	0.21
Iridium	Ir	77	ng/g	481	1,060	3,000	3.2		0.1	0.13	0.02		
Iron	Fe	26	mg/g	190.4	358.7	891	62.6	81.0	70.7	82.4	35.0	0.00006	

Table 2. Continued

Name	Symbol	Atomic Number	Units[1]	Cl Chondrites[2]	Bulk Earth[3]	Core[4]	Primitive Mantle[5]	Oceanic Crust[6]	Continental Crust[7]			Oceans[8]	Atmosphere[9]
									Bulk	Lower	Upper		
Krypton	Kr	36	pg/g	57	4.3							300	3.23×10^8
Lanthanum	La	57	μg/g	0.2347	0.48		0.648	3.90	16	11	30	0.0003	
Lead	Pb	82	ng/g	2,470	141		150	489	8,000	4,000	20,000	1	
Lithium	Li	3	μg/g	1.50	2.7		1.6		13	11	20	0.17	
Lutetium	Lu	71	μg/g	0.0243	0.049		0.0675	0.589	0.30	0.29	0.32		
Magnesium	Mg	12	mg/g	98.8	132.1		228	45.7	32.0	38.0	13.3	1.28	
Manganese	Mn	25	mg/g	1.99	0.59		1.045		1.4	1.7	0.6	4×10^{-6}	
Mercury	Hg	80	ng/g	258	9.9	8.7	10					0.2	
Molybdenum	Mo	42	μg/g	0.928	2.96	8.2	0.050		1.0	0.8	1.5	0.008	
Neodymium	Nd	60	μg/g	0.4524	0.87		1.25	11.2	16	12.7	26		
Neon	Ne	10	pg/g	201	25							100	1.27×10^7
Nickel	Ni	28	mg/g	11.0	20.4	53.7	1.96	0.150	0.105	0.135	0.02	0.00047	
Niobium	Nb	41	μg/g	0.246	0.501		0.658	3.51	11	6	25	0.00002	
Nitrogen	N	7	μg/g	3,180	9.1		2					17	7.538×10^5
Osmium	Os	76	ng/g	486	1,100	3,100	3.4		0.05	0.05	0.05		
Oxygen	O	8	mg/g	464	285.0	‡						857	233.0
Palladium	Pd	46	ng/g	560	1,000	2,800	3.9		1	1	0.5		
Phosphorus	P	15	μg/g	1,220	2,150	5,900	90				700	120	
Platinum	Pt	78	ng/g	900	2,100	5,900	7.1						
Polonium	Po	84	†										
Potassium	K	19	μg/g	558	170		240	884	9,100	2,800	28,000	400	
Praseodymium	Pr	59	μg/g	0.0891	0.162		0.254	2.07	3.9	2.8	7.1		
Promethium	Pm	61	*										
Protactinium	Pa	91	†										
Radium	Ra	88	†										
Radon	Rn	86	†										
Rhenium	Re	75	ng/g	36.5	76	210	0.28		0.4	0.4	0.4		
Rhodium	Rh	45	μg/g	0.134	0.32	0.90	0.0009						
Rubidium	Rb	37	μg/g	2.30	0.58		0.600	1.26	32	5.3	112	0.12	
Ruthenium	Ru	44	μg/g	0.712	1.48	4.2	0.0050						
Samarium	Sm	62	μg/g	0.1471	0.26		0.406	3.75	3.5	3.17	4.5		
Scandium	Sc	21	μg/g	5.82	12.1		16.2	41.4	30	36	11	0.00004	
Selenium	Se	34	μg/g	18.6	6.1	17	0.075		50	50	50	0.005	
Silicon	Si	14	mg/g	106.4	143.4		210	236	268	254	308	5.6	
Silver	Ag	47	ng/g	199	80	210	8		80	90	50	0.14	
Sodium	Na	11	mg/g	5.00	1.58		2.67	19.9	23.0	20.8	28.9	10.8	

Table 2. Continued

Name	Symbol	Atomic Number	Units[1]	Cl Chondrites[2]	Bulk Earth[3]	Core[4]	Primitive Mantle[5]	Oceanic Crust[6]	Continental Crust[7] Bulk	Continental Crust[7] Lower	Continental Crust[7] Upper	Oceans[8]	Atmosphere[9]
Strontium	Sr	38	μg/g	7.80	18.2		19.9	113	260	230	350	8	
Sulfur	S	16	mg/g	62.5	18.4	‡	0.25					0.905	
Tantalum	Ta	73	ng/g	14.2	29		37	192	1,000	600	2,200		
Technetium	Tc	43	*										
Tellurium	Te	52	ng/g	2,320	940	2,600	12	0.885					
Terbium	Tb	65	μg/g	0.0363	0.067		0.099		0.60	0.59	0.64		
Thallium	Tl	81	ng/g	142	4.9		3.5		360	230	750	0.05	
Thorium	Th	90	ng/g	29.4	65		79.5	187	3,500	1,060	10,700	0.0003	
Thulium	Tm	69	μg/g	0.0242	0.044		0.068	0.621	0.32	0.32	0.33	2×10^{-8}	
Tin	Sn	50	μg/g	1.72	0.71	1.8	0.13	1.38	2.5	1.5	5.5		
Titanium	Ti	22	mg/g	0.436	1.03		1.205	9.68	5.4	6.0	3.0	0.12	
Tungsten	W	74	ng/g	92.6	250	650	29		1,000	700	2,000		
Uranium	U	92	ng/g	8.1	18		20.3	71.1	910	280	2,800	3	
Vanadium	V	23	μg/g	56.5	103		82		230	285	60	0.002	
Xenon	Xe	54	pg/g	190	10							100	394
Ytterbium	Yb	70	μg/g	0.1625	0.29		0.441	3.90	2.2	2.2	2.2		
Yttrium	Y	39	μg/g	1.56	3.29		4.30	35.8	20	19	22	0.0003	
Zinc	Zn	30	μg/g	312	93	160	55		80	83	71		
Zirconium	Zr	40	μg/g	3.94	7.42		10.5	104	100	70	190	0.39	

[1] The units are SI units rather than the more familiar parts per million (ppm), parts per billion (ppb), etc. These tables use SI units in part because they are the standard, but more important because "billion" does not have a single definition; it means 10^9 in American usage, but 10^{12} in British usage. See "Measurements and Their Conversion in the Earth Sciences" in the frontmatter.

[2] Cl Chondrites: primitive meteorite composition, modified from E. Anders and N. Grevese, *Geochimica et Cosmochimica Acta* 53 (1989): 197–214. *See also* entry METEORITES.

[3] Bulk Earth: modified from E. Anders, *Philosophical Transactions of the Royal Society of London* A 285 (1977):23–40. *See also* the entry EARTH, COMPOSITION OF.

[4] Core: estimated by mass balance from the bulk Earth and primitive mantle compositions.

[5] Primitive Mantle: from W. F. McDonough and S.-S. Sun, *Chemical Geology* 120 (1995):223–253.

[6] Oceanic Crust: average normal mid-ocean ridge basalt of A. W. Hofmann, *Earth and Planetary Science Letters* 90 (1988):297–314. *See also* the entry OCEANIC CRUST, STRUCTURE OF.

[7] Continental Crust—Bulk, Lower and Upper: from S. R. Taylor and S. M. McLennan, *Reviews of Geophysics* 33 (1995):241–265.

[8] Oceans: modified from the entry OCEANOGRAPHY, CHEMICAL, and D. A. Ross, *Introduction to Oceanography*, 4th ed. (1988).

[9] Atmosphere: modified from the entry EARTH'S ATMOSPHERE, CHEMICAL COMPOSITION OF, and M. Ozima and F. A. Podosek, *Nobel Gas Geochemistry* (1983).

* No long-lived radioactive ($>10^8$ years) or stable nuclei of these elements exist.

† No long-lived radioactive ($>10^8$ years) or stable nuclei of these elements exist. Their abundances in the earth are steady-state concentrations in the ^{232}Th, ^{235}U, and ^{238}U decay schemes.

‡ Earth's core contains about 10% of a light element, of which these are the most likely candidates. See the entry CORE, COMPOSITION OF.

LIST OF ARTICLES

A

Abrasive Materials
Richard V. Dietrich

Agassiz, Jean Louis Rodolphe
Robert Schoch

Alloy Metals
Philip A. Candela

Aluminum Resources
Brian J. Skinner

Amphibians and Reptiles
Robert L. Carroll

Angiosperms
David Taylor

Apollo Astronauts
Harrison Schmitt

Archeological Geoscience
George (Rip) Rapp

Arthropods
Loren Babcock

Asteroids
Clark R. Chapman

Atmospheres, Planetary
Andrew Ingersoll

Auroras
Neal Brown

B

Bagnold, Ralph Alger
James R. Underwood

Bioerosion
Richard G. Bromley

Biosphere
Carole Hickman

Birds
John Ruben
Dave Shutler

Bjerknes, J. A. B. and V. F. K.
Keith R. Benson

Bowen, Norman L.
Hatten S. Yoder

Brachiopods
Danita Brandt

Bragg, William Lawrence and William Henry
Peter Heaney

Brongniart, Alexander
Kennard Bork

C

Careers in the Earth Sciences: Atmospheric Sciences
Keith Seitter
Stephanie Kehoe

Careers in the Earth Sciences: Engineering Geology and Geologic Engineering
Allen Hatheway

Careers in the Earth Sciences: Environmental Science
Donald Coates

Careers in the Earth Sciences: Exploration and Mining
Spencer R. Titley

Careers in the Earth Sciences: Hydrology
John E. Moore

Careers in the Earth Sciences: Oceanography
M. Grant Gross

Careers in the Earth Sciences: Oil and Gas
William L. Fisher

Careers in the Earth Sciences: Planetary
Sciences
Ted A. Maxwell

Careers in the Earth Sciences: Teaching
James R. Underwood

Careers in the Earth Sciences: Writing, Photography, and Filmmaking
Lawrence Engel

Carson, Rachel
Patricia Dasch

Caves and Karst Topography
William White

Ceramic Materials
Haydn H. Murray

Chemistry of Earth Materials
Gunter Faure

Chromium
Brian J. Skinner

Climatology
Susan Marshall

Coal
Heinz Damberger

Coal Mining
Heinz Damberger

Coastal Processes
Paul Komar

Colonial Invertebrate Fossils
Danita Brandt

Comets
Humberto Campins

Construction Materials
Richard V. Dietrich

Continental Crust, Structure of
Robert A. Phinney

Continental Margins
Warren Manspeizer

Continents, Evolution of
Paul F. Hoffman

Core, Structure of
Raymond Jeanloz

Cosmology, Big Bang, and the Standard Model
David Talent

Cuvier, Georges
Robert Schoch

D

Daly, Reginald A.
Eugene C. Robertson

Dana, James D.
E. Julius Dasch

Darwin, Charles
Ellis Yochelson

Davis, William Morris
Mark L. Hineline

Debris in Earth's Orbit
Phillip Anz-Meador

Desalination
Frederick C. Eubank

Diagenesis
Gerald M. Friedman

Dinosaurs
Alan J. Charig

Drilling for Scientific Research
Arthur Barber

Du Toit, Alexander
Brian J. Skinner

E

Earth, Composition of
David W. Mittlefehldt

Earth, Harmonic Motions of
Michael G. Rochester

Earth, Motions of
Thomas A. Herring

Earth, Origin of
Horton E. Newsom

Earth, Shape of
Kurt Lambeck

Earth, Structure of
Bradford H. Hager

Earth as a Dynamic System
W. Richard Peltier

Earth-Moon System, Earliest History of
William K. Hartmann

Earth Observing Satellites
Ghassem Asrar

Earthquakes
Charles K. Scharnberger

Earthquakes and Seismicity
Mary Lou Zoback

Earth's Atmosphere
James C. G. Walker

Earth's Atmosphere, Chemical Composition of
James C. G. Walker

Earth Science Information
James O'Donnell
Dena Hanson

Earth's Atmosphere, Structure of
Kirk A. Maasch

Earth's Crust, History of
Dallas H. Abbott

Earth's Freshwaters
M. Gordon Wolman

Earth's Glaciers and Frozen Waters
Stephanie S. Shipp

Earth's Hydrosphere
Jeff Raffensperger

Earth's Magnetic Field
Catherine Constable

Echinoderms
Danita Brandt

Eclipses and Occultations
James L. Elliot

Economic Geology, History of
Cyrus W. Field

Ekman, Vagn Valfrid
Artur Svansson

Elsasser, Walter M.
Peter Olson

Energy, Geothermal
James B. Koenig

Energy from Streams and Oceans
Robert Livingston

Energy from the Atom
Robert Livingston

Energy from the Biomass
Lynn Wright

Energy from the Sun
Robert Livingston

Energy from the Wind
Robert Livingston

Energy Use Around the World
Robert Livingston

Engineering Aspects of Earth Sciences
Christopher C. Mathewson

Engineering Geology, History of
Allen Hatheway

Environment and Earth Science
Donald Coates

Eskola, Pentti
Eric J. Essene

Ewing, William Maurice
Manik Talwani

Extinctions
Arthur Boucot

Extraterrestrial Life, Search for
Bob Arnold

F

Famous Controversies in Geology
Anthony Hallam

Fertilizer Raw Materials
Joseph L. Graf

Fish Evolution
John A. Long

Floods
Peter Patton

Fossil Fuel Use, History of
James R. Craig

Fossil Fuels
Ronald W. Stanton

Fossil Record of Human Evolution
John R. Lukacs

Fossil Soils
  Greg Retallack

Fossilization and the Fossil Record
  Loren Babcock

G

Galaxies
  Deidre Hunter

Galilean Satellites
  Deborah Domingue

Garrels, Robert M.
  E. Julius Dasch

Gems and Gem Minerals
  Cornelis Klein

Geochemical Techniques
  Phillip Dean Ihinger

Geochemistry, History of
  Brian Mason

Geographic Information Systems
  Kenneth C. McGwire

Geologic Time
  E. Julius Dasch

Geologic Time, Measurement of
  Kenneth A. Foland

Geologic Work by Streams
  Ellen Wohl

Geologic Work by Wind
  Nicholas Lancaster

Geological Surveys
  D. Chris Findlay

Geology, History of
  Ellen Tan Drake

Geology and Public Policy
  Elisabeth G. Newton

Geomorphology, History of
  John D. Vitek

Geophysical Techniques
  George H. Shaw

Geophysics, History of
  Rufus D. Catchings

Geothermometry and Geobarometry
  Frank S. Spear

Gilbert, Grove Karl
  Ellis Yochelson

Glacial Ages Through Geologic Time
  Harold W. Borns

Glaessner, Martin F.
  Brian McGowran

Glass
  Richard V. Dietrich

Global Climatic Changes, Human Intervention
  Eric Barron

Global Environmental Changes, Natural
  Richard S. Williams, Jr.

Global Positioning Satellites, Geographical Information Systems, and Automated Mapping
  Geoffrey Blewitt

Gold
  James R. Craig

Goldschmidt, Victor Moritz
  E. Julius Dasch

Groundwater
  Jean M. Bahr

Gutenberg, Beno
  Don Anderson

Gymnosperms
  Gene Mapes

H

Hazardous Waste Disposal
  Konrad Krauskopf

Heat Budget of the Earth
  Frank D. Stacey

Herschel Family
  Nadine G. Barlow

Hess, Harry
  Alan Allwardt

Higher Life Forms, Earliest Evidence of
  S. Conway Morris

Holmes, Arthur
  Gordon Craig

Hubbert, M. King
  Larry Drew

Hubble, Edwin
Ron Brashear

Hurricanes
Robert W. Burpee

Hutton, James
Paul Farber

Hydrothermal Alteration and Hydrothermal Mineral Deposits
Naomi Oreskes

I

Igneous Processes
Bruce Marsh

Igneous Rocks
James S. Beard

Impact Cratering
Mark Cintala

Impact Cratering and the Bombardment Record
Nadine G. Barlow

Industrial Minerals
Joseph L. Graf

Insects, History of
Loren Babcock

Interplanetary Medium, Cosmic Dust, and Micrometeorites
Michael E. Zolensky

Intrusive Rocks and Intrusions
James S. Beard

Invertebrates
Stefan Bengtson

Iron Deposits
Gene L. LaBerge

Isostasy
Marcia McNutt

Isotope Tracers, Radiogenic
David W. Mittlefehldt

Isotope Tracers, Stable
Chris S. Romanek

J

Jupiter and Saturn
Gordon Bjoraker

K

King, Clarence Rivers
Ellis Yochelson

Kuiper, Gerard
Dale P. Cruikshank

L

Lakes
Pamela Gore

Lamarck, Jean-Baptiste
Robert Schoch

Land Degradation and Desertification
Farouk El-Baz

Landscape Evolution
Arthur Bloom

Landslides and Rockfalls
Allen Hatheway

Lehmann, Inge
Erik Hjortenberg

Libby, Willard
James Arnold

Life, Evolution of
Laurel Collins
Tim Collins

Life, Origin of
Ralph E. Taggart

Lindgren, Waldemar
Brian J. Skinner

Lyell, Charles
Leonard Wilson

M

Magma
Allan Treiman

Mammals
Robert Schoch

Manganese Deposits
Gene L. LaBerge

Mantle, Structure of
Bradford H. Hager

Mantle Convection and Plumes
W. Richard Peltier

Mapping, Geologic
Scott Southworth

Mapping, Planetary
Raymond Batson

Mars
Nadine G. Barlow
Joseph Boyce

Meinzer, Oscar E.
William Back

Mercury
Nadine G. Barlow

Metallic Mineral Deposits, Formation of
Brian J. Skinner

Metallic Mineral Deposits Formed by Sedimentary Processes
J. Barry Maynard

Metallic Mineral Deposits Formed by Weathering
Gene L. LaBerge

Metallic Mineral Resources from the Sea
Mark D. Hannington
Peter M. Herzig

Metallogenic Provinces
Ian M. Lange

Metals, Geochemically Abundant and Geochemically Scarce
Brian J. Skinner

Metamorphic Processes
Jane Selverstone

Metamorphic Rocks
Jill J. Schneiderman

Meteorites
David W. Mittlefehldt

Meteorites from the Moon and Mars
Allan Treiman

Meteorology, History of
James R. Fleming

Mineral Deposits, Exploration for
Willard C. Lacy

Mineral Deposits, Formation Through Geologic Time
Spencer R. Titley

Mineral Deposits, Igneous
Ian M. Lange

Mineral Deposits, Metamorphic
Naomi Oreskes

Mineral Physics
Raymond Jeanloz

Mineral Structure and Crystal Chemistry
David R. Veblen

Mineralized Microfossils
Jere H. Lipps

Mineralogy, History of
Michael E. Zolensky

Minerals, Nonsilicates
Michael E. Zolensky

Minerals, Silicates
Michael E. Zolensky

Minerals and Their Study
Michael E. Zolensky

Miners, History of
Willard C. Lacy

Mining Techniques, Past and Present
Willard C. Lacy

Minorities in the Earth Sciences
Marilyn J. Suiter

Mitchell, Maria
E. Dorrit Hoffleit

Modern Ore Deposits
Mark D. Hannington
Peter M. Herzig

Mollusks
Loren Babcock

Moon
B. Ray Hawke

Mountains and Mountain Building
Robert J. Lillie

Murchison, Roderick
Doug Erwin

N

Nansen, Fridtjof
Alexander R. McBirney

Nier, Alfred O. C.
  Lawrence E. Nyquist
Nonferrous Metals
  P. Geoffrey Feiss
Nucleosynthesis and the Origin of the Elements
  Charles R. Harper

O

Observational Astronomy
  David Levy
Ocean Basins, Evolution of
  John Mutter
Oceanic Crust, Structure of
  Don Elthon
Oceanographic Expeditions
  Ellen Kappel
Oceanography, Biological
  Linda E. Duguay
Oceanography, Chemical
  Martha Scott
Oceanography, Geological
  Constance Sancetta
Oceanography, History of
  M. Grant Gross
Oceanography, Physical
  Lynne Talley
Oil and Gas, Improved Recovery of
  Faruk Civan
  Anuj Gupta
Oil and Gas, Physical and Chemical Characteristics of
  Jack A. Williams
Oil and Gas, Prospecting for
  Raymon L. Brown
Oil and Gas, Reserves and Resources of
  Ronald D. Evans
  Michael L. Wiggins
Oil Shale
  John R. Dyni
Oldest Rocks in the Solar System
  E. Julius Dasch

Ore
  Half Zantop
Organic Geochemistry
  Keith A. Kvenvolden
Owen, Sir Richard
  Phillip Sloan

P

Paleobotany
  Patricia Gensel
Paleoclimatology
  Lisa Sloan
Paleoclimatology, History of
  Thomas Crowley
Paleoecology
  Arthur Boucot
  Jane Gray
Paleogeography
  Christopher R. Scotese
Paleontology
  Danita Brandt
Paleontology, History of
  Ronald Rainger
Palynology
  Jane Gray
Permanently Frozen Ground
  Harold W. Borns
Petroleum
  Jack A. Williams
Petrologic Techniques
  Amy J. G. Jurewicz
  Stephen R. Jurewicz
Petrology, History of
  Hatten S. Yoder
Physical Properties of Magmas
  Donald B. Dingwell
Physical Properties of Rocks
  Scott King
  Nikolas Christensen
Physics of the Earth
  Frank D. Stacey

Placer Deposits
  Eric R. Force

Planetary Geoscience, History of
  Ursula B. Marvin

Planetary Magnetic Fields
  Janet Luhmann

Planetary Missions
  Louis Friedman

Planetary Rings
  Nadine G. Barlow

Planetary Systems, Other
  David C. Black

Planetology, Comparative
  Jayne Aubele

Plate Tectonics
  Christopher R. Scotese

Platinum-Group Elements
  Louis J. Cabri

Playfair, John
  Virginia C. Gulick

Pluto and Charon
  Robert L. Marcialis

Polar Earth Science
  Winifred Reuning

Pollution of Atmosphere
  Vickie S. Connors

Pollution of Groundwater Aquifers
  Jean M. Bahr

Pollution of Lakes and Streams
  Dennis O. Nelson

Powell, John Wesley
  Clifford M. Nelson

Prior, George Thurland
  Robert Hutchison

Pteridophytes
  Judy Skog

Public Health and Earth Science
  H. Catherine W. Skinner

R

Refractory Materials
  Haydn H. Murray

Research in the Earth Sciences
  Jonathan G. Price
  Thomas M. Usselman

Resource Use, History of
  James R. Craig

Resources, Renewable and Nonrenewable
  Kula C. Misra

Richter, Charles
  Hiroo Kanamori

Rifting of the Crust
  Dale Sawyer

Ringwood, A. E. (Ted)
  David H. Green

Rivers, Geomorphology of
  Ellen Wohl

Rocks and Their Study
  Allan Treiman

Rossby, Carl-Gustaf A.
  John M. Lewis

Rubey, William W.
  W. Gary Ernst

S

Satellite Laser Ranging and Very Long Baseline
  Interferometry
  John Degnan

Satellites, Midsized
  Bonnie Buratti

Satellites, Small
  Peter Thomas

Satellites, Solar Power
  Frederick Koomanoff

Schuchert, Charles
  Naomi Oreskes

Scientific Creationism
  Paul Roberts

Seawater, Physical and Chemical Characteristics
  of
  Nathaniel Ostrom
  David Long

Sedgwick, Adam
  Robert Schoch

Sedimentary Structures
Elana Leithold

Sedimentology
Sam Boggs

Sedimentology, History of
Gerald M. Friedman

Sediments and Sedimentary Rocks, Chemical and Organic
Gerald M. Friedman

Sediments and Sedimentary Rocks, Terrigenous (Clastic)
Kitty Milliken

Seismic Lines, COCORP, and Related Geophysical Techniques
Jack E. Oliver

Seismic Tomography
Adam M. Dziewonski

Shepard, Francis P.
G. Ross Heath

Silver
Brian J. Skinner

Simpson, George Gaylord
Robert Schoch

Soil Degradation
Les McFadden

Soil Types
Garth Voigt

Soil Types and Land Use
Garth Voigt

Soils, Formation of
Garth Voigt

Solar System
Renu Malhotra

Solid Earth-Hydrosphere-Atmosphere-Biosphere Interface
William S. Fyfe

Sorby, Henry C.
Robert L. Folk

Space, Future Exploration of
Geoffrey Briggs

Special Metals
P. Geoffrey Feiss

Stars
Stephen Becker

Stille, Hans
Gerhard Oertel

Stommel, Henry
James Luyten

Strategic Mineral Resources and Stockpiles
Half Zantop

Stratigraphy
William B. N. Berry

Stratigraphy, History of
William B. N. Berry

Structural Geology
Gerhard Oertel

Structural Geology, History of
Terry Engelder

Suess, Eduard
Eugen F. Stumpfl

Sun
Raymond N. Smartt

Supergene Enrichment
Spencer R. Titley

Surface Water, Storage and Distribution
Scott Curry

Sverdrup, Harald U.
Robert Marc Friedman

T

Taphonomy
Carlton E. Brett

Tars, Tar Sands, and Extra Heavy Oils
Richard F. Meyer

Tectonic Blocks
George W. Moore

Tectonism, Active
Robert S. Yeats

Tectonism, Planetary
Matthew P. Golombek

Tektites
Christian Koeberl

Telescopes
Nadine G. Barlow

Terrestrialization
Jane Gray

Thermodynamics and Kinetics
Virginia M. Oversby

Titan
Cindy Cunningham

Titanium Deposits
Eric R. Force

Trace Elements
Denis M. Shaw

Trace Fossils
Richard G. Bromley

Triton
Candice Hansen

U

Uranium and Thorium
Richard Sanford
Gary R. Winkler

Uranus and Neptune
Mark Marley

Urban Planning and Land Use
Allen Hatheway

Urey, Harold
Jacob Bigeleisen

Useful Mineral Substances
Kula C. Misra

V

Venus
Larry S. Crumpler

Vernadsky, Vladimir I.
Brian Mason

Views of the Earth
Patricia Dasch

Volcanic Eruptions
Stanley N. Williams

Volcanism
Stephen Self

Volcanism, Planetary
Rosaly Lopes-Gautier

W

Walcott, Charles
Ellis Yochelson

Waste Disposal, Municipal
Christopher VanCantfort
Clyde Frank

Water Quality
Dennis O. Nelson

Water Supply and Management
Michael Whiteley

Water Use
Kurt Putnam

Weathering and Erosion
Michael Velbel

Weathering and Erosion, Planetary
Steven H. Williams

Wegener, Alfred
Mott Greene

Werner, Abraham Gottlob
Wilfried Schroder

Wilson, J. Tuzo
W. Richard Peltier

Women in the Earth Sciences
H. Catherine W. Skinner

LIST OF CONTRIBUTORS

Dallas H. Abbott
Lamont Doherty Earth Observatory

Alan Allwardt
Santa Cruz, CA

Don Anderson
California Institute of Technology, Pasadena

Phillip Anz-Meador
Viking Science and Technology, Inc.

Bob Arnold
SETI Institute, Mountain View, CA

James Arnold
Cal Space, University of California, San Diego

Ghassem Asrar
NASA, Washington, DC

Jayne Aubele
Brown University

Loren Babcock
Ohio State University

William Back
U.S. Geological Survey

Jean M. Bahr
University of Wisconsin, Madison

Arthur Barber
Littleton, CO

Nadine G. Barlow
University of Central Florida

Eric Barron
Pennsylvania State University

Raymond Batson
Hurricane, VT

James S. Beard
Virginia Museum of Natural History

Stephen Becker
Los Alamos National Laboratory

Stefan Bengtson
Swedish Museum of Natural History, Stockholm

Keith R. Benson
University of Washington

William B. N. Berry
University of California, Berkeley

Jacob Bigeleisen
St. James, NY

Gordon Bjoraker
NASA Goddard Space Flight Center, Greenbelt, MD

David C. Black
Lunar and Planetary Institute, Houston, TX

Geoffrey Blewitt
University of Newcastle upon Tyne, England

Arthur Bloom
Cornell University

Sam Boggs
University of Oregon, Eugene

Kennard Bork
Denison University

Harold W. Borns
University of Maine, Orono

Arthur Boucot
Oregon State University

Joseph Boyce
NASA Headquarters, Washington, DC

Danita Brandt
Michigan State University

Ron Brashear
Huntington Library

Carlton E. Brett
University of Rochester

Geoffrey Briggs
NASA Ames Research Center, Moffett Field, CA

Richard G. Bromley
University of Copenhagen, Denmark

Neal Brown
University of Alaska, Fairbanks

Raymon L. Brown
University of Oklahoma, Norman

Bonnie Buratti
Jet Propulsion Laboratory, Pasadena, CA

Robert W. Burpee
U.S. Department of Commerce, National Oceanic and Atmospheric Administration

Louis J. Cabri
Canada Centre for Mineral and Energy Technology

Humberto Campins
University of Florida, Gainesville

Philip A. Candela
University of Maryland

Robert L. Carroll
Redpath Museum, McGill University, Canada

Rufus D. Catchings
U.S. Geological Survey, Menlo Park, CA

Clark R. Chapman
Planetary Science Institute, Tucson, AZ

Alan J. Charig
Hurst Green, Oxted, Surrey, England

Nikolas Christensen
Purdue University

Mark Cintala
NASA Johnson Space Center

Faruk Civan
University of Oklahoma

Donald Coates
Tucson, AZ

Laurel Collins
Florida International University

Tim Collins
Florida International University

Vickie S. Connors
NASA Headquarters, Washington, DC

Catherine Constable
University of California, San Diego

Gordon Craig
Lasswade, Lothians, Scotland

James R. Craig
Virginia Polytechnic Institute and State University

Thomas Crowley
Texas A&M University

Dale P. Cruikshank
NASA Ames Research Center, Moffett Field, CA

Larry S. Crumpler
Brown University

Cindy Cunningham
York University, Canada

Scott Curry
Oregon Health Division, Drinking Water Program

Heinz Damberger
Illinois Geological Survey

E. Julius Dasch
NASA Headquarters, Washington, DC

Patricia Dasch
National Space Society, Washington, DC

John Degnan
Laboratory for Terrestrial Physics, NASA Goddard Space Flight Center

Richard V. Dietrich
Mt. Pleasant, MI

Donald B. Dingwell
Bayerisches Geoinstitut, Universitat Bayreuth, Bayreuth, Germany

Deborah Domingue
Lunar and Planetary Institute, Houston, TX

Ellen Tan Drake
Oregon State University

Larry Drew
U.S. Geological Survey, Reston, VA

Linda E. Duguay
Center for Environment and Estuarine Studies, University of Maryland System

John R. Dyni
U.S. Geological Survey, Denver, CO

Adam M. Dziewonski
Harvard University

Farouk El-Baz
Boston University

James L. Elliot
Massachusetts Institute of Technology

Donald Elthon
University of Houston

Lawrence Engel
Columbia University

Terry Engelder
Pennsylvania State University

W. Gary Ernst
Stanford University

Doug Erwin
National Museum of Natural History, Smithsonian Institution

Eric J. Essene
University of Michigan

Frederick C. Eubank
U.S. Army Corps of Engineers, Washington, DC

Ronald D. Evans
University of Oklahoma

Paul Farber
Oregon State University

Gunter Faure
Ohio State University

P. Geoffrey Feiss
University of North Carolina, Chapel Hill

Cyrus W. Field
Oregon State University

D. Chris Findlay
Geological Survey of Canada

William L. Fisher
University of Texas

James R. Fleming
Colby College

Kenneth A. Foland
Ohio State University

Robert L. Folk
University of Texas

Eric R. Force
University of Arizona

Clyde Frank
U.S. Department of Energy, Washington, DC

Gerald M. Friedman
Brooklyn College and Graduate School of the City University of New York (CUNY) and Northeastern Science Foundation Affiliated with CUNY

Louis Friedman
The Planetary Society, Pasadena, CA

Robert Marc Friedman
University of California, San Diego

William S. Fyfe
University of Western Ontario, Canada

Patricia Gensel
University of North Carolina, Chapel Hill

Matthew P. Golombek
Jet Propulsion Laboratory, Pasadena, CA

Pamela Gore
Dekalb College

Joseph L. Graf
Southern Oregon State College, Ashland

Jane Gray
University of Oregon

David H. Green
Australian National University, Canberra

Mott Greene
University of Puget Sound

M. Grant Gross
Baltimore, MD

Virginia C. Gulick
NASA Ames Research Center, Moffett Field, CA

Anuj Gupta
University of Oklahoma, Norman

Bradford H. Hager
Massachusetts Institute of Technology

Anthony Hallam
The University of Birmingham School, Edgbaston, Birmingham

Mark D. Hannington
Geological Survey of Canada, Ottawa, Ontario

Candice Hansen
Jet Propulsion Laboratory, Pasadena, CA

Dena Hanson
Pasadena, CA

Charles R. Harper
Harvard University

William K. Hartmann
Planetary Science Institute, Tucson, AZ

Allen Hatheway
University of Missouri, Rolla

B. Ray Hawke
University of Hawaii

Peter Heaney
Princeton University

G. Ross Heath
University of Washington

Thomas A. Herring
Massachusetts Institute of Technology

Peter M. Herzig
Institut für Mineralogie und Lagerstattenlehre, Aachen, Germany

Carole Hickman
University of California, Berkeley

Mark L. Hineline
University of California, San Diego

Erik Hjortenberg
Office of Seismology, Denmark

E. Dorrit Hoffleit
Yale University

Paul F. Hoffman
Harvard University

Deidre Hunter
Lowell Observatory

Robert Hutchison
British Museum, England

Philip Dean Ihinger
Yale University

Andrew Ingersoll
California Institute of Technology

Raymond Jeanloz
University of California, Berkeley

Amy J. G. Jurewicz
Lockheed Engineering and Science Company, NASA

Stephen R. Jurewicz
NASA Johnson Space Center

Hiroo Kanamori
California Institute of Technology, Pasadena, CA

Ellen Kappel
Washington, DC

Stephanie Kehoe
Boston, MA

Scott King
Purdue University

Cornelis Klein
University of New Mexico

Christian Koeberl
Institute of Geochemistry, University of Vienna

James B. Koenig
GeothermEx Inc.

Paul Komar
Oregon State University

Frederick Koomanoff
U.S. Department of Energy, Washington, DC

Konrad Krauskopf
Stanford University

Keith A. Kvenvolden
U.S. Geological Survey

Gene L. LaBerge
University of Wisconsin, Oshkosh

Willard C. Lacy
Green Valley, AZ

Kurt Lambeck
Australian National University

Nicholas Lancaster
Desert Research Institute, Reno, NV

Ian M. Lange
University of Montana

Elana Leithold
North Carolina State University

David Levy
Tucson, AZ

John M. Lewis
National Severe Storms Laboratory, Norman, OK

Robert J. Lillie
Oregon State University

Jere H. Lipps
University of California, Berkeley

Robert Livingston
Silver Springs, MD

David Long
Michigan State University

John A. Long
Western Australian Museum, Perth, Australia

Rosaly Lopes-Gautier
Jet Propulsion Laboratory, Pasadena, CA

Janet Luhmann
University of California, Los Angeles

John R. Lukacs
University of Oregon

James Luyten
Woods Hole Oceanographic Institution, Woods Hole, MA

Kirk A. Maasch
University of Maine, Orono

Renu Malhotra
Lunar and Planetary Institute, Houston, TX

Warren Manspeizer
Rutgers University, Newark, NJ

Gene Mapes
Ohio University

Robert L. Marcialis
Tucson, AZ

Mark Marley
New Mexico State University

Bruce Marsh
The Johns Hopkins University

Susan Marshall
University of North Carolina, Charlotte

Ursula B. Marvin
Smithsonian, Cambridge, MA

Brian Mason
Smithsonian Institution

Christopher C. Mathewson
Texas A&M University

Ted A. Maxwell
Smithsonian Institution

J. Barry Maynard
University of Cinncinnati

Alexander R. McBirney
University of Oregon

Les McFadden
University of New Mexico

Brian McGowran
University of Adelaide, Australia

Kenneth C. McGwire
Desert Research Institute, Reno, NV

Marcia McNutt
Massachusetts Institute of Technology

Richard F. Meyer
U.S. Geological Survey, Reston, VA

Kitty Milliken
University of Texas

Kula C. Misra
The University of Tennessee

David W. Mittlefehldt
Lockheed Engineering and Sciences Company, NASA

George W. Moore
Oregon State University

John E. Moore
Denver, CO

S. Conway Morris
Sedgwick Museum, University of Cambridge, England

Haydn H. Murray
Indiana University

John Mutter
Lamont and Doherty Earth Observatory, Palisades, NY

Clifford M. Nelson
U.S. Geological Survey, Reston, VA

Dennis O. Nelson
Oregon Health Division, Drinking Water Program

Horton E. Newsom
University of New Mexico

Elisabeth G. Newton
E. G. Newton & Associates, Inc., Washington, DC

Lawrence E. Nyquist
NASA Johnson Space Center, Houston, TX

James O'Donnell
California Institute of Technology

Gerhard Oertel
University of California, Los Angeles

Jack E. Oliver
Cornell University

Peter Olson
The Johns Hopkins University

Naomi Oreskes
Dartmouth College

Nathaniel Ostrom
Michigan State University

Virginia M. Oversby
Lawrence Livermore National Laboratory, Livermore, CA

Peter Patton
Wesleyan University

W. Richard Peltier
University of Toronto

Robert A. Phinney
Princeton University

Jonathan G. Price
University of Nevada, Reno

Kurt Putnam
Oregon Health Division, Drinking Water Program

Jeff Raffensperger
University of Virginia, Charlottesville

Ronald Rainger
Texas Tech University

George (Rip) Rapp
University of Minnesota

Greg Retallack
University of Oregon

Winifred Reuning
Arlington, VA

Paul Roberts
Oregon State University

Eugene C. Robertson
U.S. Geological Survey, Reston, VA

Michael G. Rochester
Memorial University, St. John's, Newfoundland

Chris S. Romanek
University of Georgia

John Ruben
Oregon State University

Constance Sancetta
National Science Foundation

Richard Sanford
Chicago, IL

Dale Sawyer
Rice University

Charles K. Scharnberger
Millersville University

Harrison Schmitt
Albuquerque, NM

Jill S. Schneiderman
Pomona College

Robert Schoch
Boston University

Wilfried Schroder
Geophysical Station, Bremen-Roennenbeck, Germany

Christopher R. Scotese
University of Texas

Martha Scott
Texas A&M University

Keith Seitter
American Meteorological Society

Stephen Self
University of Hawaii

Jane Selverstone
University of New Mexico

Denis M. Shaw
McMaster University, Canada

George H. Shaw
Union College

Stephanie S. Shipp
Rice University

Dave Shutler
Canadian Wildlife Service, Saskatoon, Canada

Brian J. Skinner
Yale University

H. Catherine W. Skinner
Yale University

Judy Skog
George Mason University

Lisa Sloan
University of California, Santa Cruz

Phillip Sloan
University of Notre Dame

Raymond N. Smartt
National Solar Observatory, Sunspot, NM

Scott Southworth
U.S. Geological Survey, Reston, VA

Frank S. Spear
Rennsaeler Polytechnic Institute

Frank D. Stacey
The University of Queensland, Australia

Ronald W. Stanton
U.S. Geological Survey, Reston, VA

Eugen F. Stumpfl
Institut für Geowissenschaften, Austria

Marilyn J. Suiter
American Geological Institute, Alexandria, VA

Artur Svansson
Göteborg University, Sweden

Ralph E. Taggart
Michigan State University

David Talent
University of Houston

Lynne Talley
Scripps Institution of Oceanography, La Jolla, CA

Manik Talwani
Houston, TX

David Taylor
Indiana University Southeast

Peter Thomas
Cornell University

Spencer R. Titley
University of Arizona

Allan Treiman
Lunar and Planetary Institute, Houston, TX

James R. Underwood
Kansas State University

Thomas M. Usselman
National Research Council, Washington, DC

Christopher VanCantfort
Waste Policy Institute, Blacksburg, VA

David R. Veblen
The Johns Hopkins University

Michael Velbel
Michigan State University

John D. Vitek
Oklahoma State University

Garth Voigt
Boseman, MT

James C. G. Walker
University of Michigan

William White
Pennsylvania State University

Michael Whiteley
Oregon Health Division, Drinking Water Program

Michael L. Wiggins
University of Oklahoma

Jack A. Williams
Tulsa, OK

Stanley N. Williams
Arizona State University

Richard S. Williams, Jr.
U.S. Department of the Interior, Woods Hole, MA

Steven H. Williams
University of North Dakota

Leonard Wilson
University of Minnesota

Gary R. Winkler
U.S. Geological Survey, Denver, CO

Ellen Wohl
Colorado State University

M. Gordon Wolman
The Johns Hopkins University

Lynn Wright
Oak Ridge National Laboratory, Oak Ridge, TN

Robert S. Yeats
Oregon State University

Ellis Yochelson
Bowie, MD

Hatten S. Yoder
Geophysical Laboratory, Washington, DC

Half Zantop
Dartmouth College

Mary Lou Zoback
U.S. Geological Survey, Menlo Park, CA

Michael E. Zolensky
NASA Johnson Space Center

ACRONYMS AND STANDARD ABBREVIATIONS

A/E   architectural/engineering

AABW   Antarctic Bottom Water

AAG   Association of American Geographers

AAIW   Antarctic Intermediate Water

ABET   Accreditation Board for Engineers and Technologists

ACCP   Atlantic Climate Change Program

ADEOS   Advanced Earth Observing System

AEG   Association of Engineering Geologists

AGI   American Geological Institute

API   American Petroleum Institute

ASCE   American Society of Civil Engineers

ATLAS   Atmospheric Laboratory for Applications and Science

ATOC   Acoustic Thermometry for Ocean Climate

AU   astronomical units

b.y.   billion years

B.A.   bachelor of arts

B.C.E.   before common era

B.S.   bachelor of sciences

Bbo   barrels of oil

BBO   billion barrels of oil

BMP   best management practices

BOD   biological oxygen demand

BP   before present

C.E.   common era

ca.   circa

CalTech   California Institute of Technology

CD-ROM   compact disc read-only memory

CMB   core-mantle boundary

CME   coronal mass ejection

COCORP   Consortium for Continental Reflection Profiling

CofM   center of mass

ct   carat

CZCS   Coastal Zone Color Scanner

D/H   deuterium-to-hydrogen ratio

DOD   Department of Defense

DOE   Department of Energy

DSDP   Deep Sea Drilling Project

DTA   differential thermal analysis

ED   electrodialysis

EDR   electrodialysis-reversal

EDX (or EDS)   energy dispersive X-ray spectroscopy

EELS   electron energy loss spectroscopy

EGs   engineering geologists

EOS   Earth Observing System

EOSDIS   EOS Data and Information System

EPA   Environmental Protection Agency

ERBE   Earth Radiation Budget Experiment

ESA   European Space Agency

FAMOUS   French-American Mid-Ocean Undersea Study

Ga   billion years

GAC   granular activated charcoal

GCM   General Circulation Model

GEs   geological engineers

GIS   geographic information system

GISP   Greenland Ice Sheet Project

GPS   Global Positioning System

GSA   Geological Society of America

GWP   Global Warming Potential

ICB   inner core boundary

ICBM   intercontinental ballistic missile

ICES   International Council for the Exploration of the Sea

IGY   International Geophysical Year

IPCC   Intergovernmental Panel on Climate Change

JOI   Joint Oceanographic Institutions

JOIDES   Joint Oceanographic Institutions for Deep Earth Sampling

Ka   thousand years
kya   thousands of years ago
LAGEOS   Laser Geodynamics Satellite
LANDSAT Land Remote-Sensing Satellite
LDG   Libyan Desert Glass
LIL   large-ion-lithophile
LITE   Lidar In-Space Technology Experiment
LVZ   lower velocity zone
M.A.   master of arts
Ma   million years
MCL   maximum containment level
MED   multiple effect distillation
MESUR   Mars Environmental Survey Mission
MHD   magnetohydrodynamics
MM   Modified Mercalli
MORB   mid-ocean ridge basalt
MSF   multistage flash
MSW   municipal solid waste
MTPE   Mission to Planet Earth
mya   millions of years ago
NADW   North Atlantic Deep Water
NAPL   non-aqueous phase liquid
NAS   National Academy of Sciences
NASA   National Aeronautics and Space
   Administration
NEOs   near-Earth objects
NMR   nuclear magnetic resonance
NOAA   National Oceanic and Atmospheric
   Administration
NPC   National Petroleum Council
NPDES   National Pollutant Discharge
   Elimination System
NSCAT   NASA Scatterometer
NSF   National Science Foundation
NTIS   National Technical Information Service
NWS   National Weather Service
ODP   Ocean Drilling Project
PCE   pyrometric cone equivalent
PDR   precision depth recorder
PGE   platinum-group elements
PGM   platinum-group minerals
PICs   products of incomplete combustion
PIXE   proton-induced X-ray emission
ppb   parts per billion
ppm   parts per million
ppt   parts per thousand

PREM   Preliminary Reference Earth Model
PSU   practical salinity unit
PV   photovoltaic
RDF   refuse-derived fuel
REE   rare-earth elements
RO   reverse osmosis
ROV   remotely operated vehicle
RPM   regulatory program manager
SAR   Synthetic Aperture Radar
SDI   Strategic Defense Initiative
SeaWiFS   Sea-Viewing Wide Field Sensor
SEM   scanning electron microscope
SETI   Search for Extraterrestrial Intelligence
SIMS   secondary ion mass spectrometry
SLR   satellite laser ranging
SNC   shergottite, nakhlite, and chassignite
SOC   synthetic organic chemical
SPS   solar power satellite
SST   sea surface topography
STEM   scanning transmission electron
   microscope
SYNROC   synthetic rock
TDS   total dissolved solids
TEM   transmission electron microscope
TGA   thermal gravimetric analysis
TMDL   total maximum daily load
TOF   time of flight
TOGA   Tropical Ocean–Global Atmosphere
TOMS   Total Ozone Mapping Spectrometer
TOPEX   Ocean Topography Experiment
TRMM   Tropical Rainfall Measuring Mission
TVA   Tennessee Valley Authority
UARS   Upper Atmosphere Research Satellite
UNEP   United Nations Environment Program
UNESCO   United Nations Educational, Scientific,
   and Cultural Organization
USBR   U.S. Bureau of Reclamation
USGCRP   U.S. Global Change Research Program
USGS   U.S. Geological Survey
VC   vapor compression
VLBI   Very Long Baseline Interferometry
VOC   volatile organic chemical
WOCE   World Ocean Circulation Experiment
WWSSN   World Wid Standarized Seismograph
   Network
WWW   World Wide Web

A

ABRASIVE MATERIALS

Abrasive materials are used to smooth, roughen, or scour surfaces by removing fragments and protuberances from those surfaces. The processes are accomplished by rubbing or blasting the surfaces with the abrasives or abrasive-charged media.

Uses

A large percentage of abrasive grit is used for working steel and other metals. Among other uses of loose abrasive grit or grit-charged media are: finishing wood, leather, felt, hard rubber, and plastic; grinding and polishing glass, including lenses; core drilling for prospecting or recovering ore or oil; sharpening tools and fishhooks; cutting and polishing dimension stones; rubbing concrete; sandblasting; buffing varnished, painted, or lacquered surfaces; lapping semiconductors that require high-quality scratch-free surfaces; grinding pulpwood; milling grain; tumbling, faceting, and polishing gemstones; erasing pencil and ink marks; and, as aggregate, producing scratching surfaces for safety matches and wear-resistant, nonskid surfaces for ramps and stair treads, for example.

Procedures

For most purposes, abrasive materials are crushed and sized by sieving or, if extremely fine particles are required, by elutriation. The resulting grit is used loose, coated on cloth or paper, bonded to form solid blocks, or set or impregnated into metals. A few rocks such as sandstone and novaculite can be cut into abrasive stones such as pulpstones, grindstones, mill wheels, and hones.

Loose cobble-sized rock fragments are used as charges in ball mills; sand-sized grains are used as charges for wire and flat gangsaws and for air- or hydro-blasting; smaller grains are incorporated into slurries (i.e., suspensions of water or oil) or pastes for tumbling, lapping, and buffing. Coated abrasives are used for such purposes as dressing and polishing wood. Bonded grains are molded or cut into shapes for use as grindstones, knife sharpeners, cutting tools, hones, and so on. Diamonds are set in rock drills, and abrasive-impregnated metals are used as knives, saws, files, and lapidary wheels (Figure 1).

Abrasive Characteristics

Hardness, grain shape, toughness, density, grainsize, and purity dictate the applications of diverse abrasion materials.

1

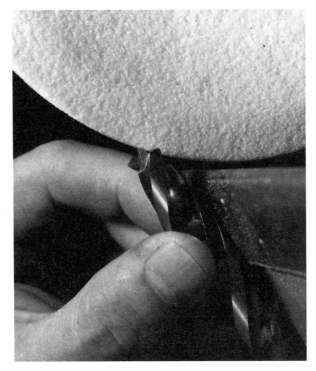

Figure 1. Off-hand grinding of brad-point drill with 60-grit, white aluminum oxide wheel (photograph courtesy of Emmett Mason).

Hardness, as determined by scratching, has been used to help identify minerals for almost two centuries. Mohs' hardness scale is the most widely used base (Table 1). On this scale, each mineral can be scratched by any mineral having a higher hardness number. Thus, for example, quartz (H-7) will scratch feldspar (H-6) but can be scratched by topaz (H-8); and similarly a substance that can be scratched by quartz but will scratch feldspar is usually said to have a hardness of $6^{1/2}$.

Shapewise, blocky grains with sharp points and chisel-like edges are preferred for most abrasion procedures. Some abrasion materials (e.g., garnet, both natural and manufactured glass, and silicon carbide) have the additional advantage of being "self-sharpening"; that is, during use, they rebreak to yield similarly sharp, chisel-edged, albeit smaller, particles.

As for toughness, the best abrasive particles are those that survive over long periods of use. Brittle materials that break down easily to powder soon have their efficiencies reduced. In general, poly-crystalline-grained abrasives are tougher than those consisting of monocrystalline grains.

High-density abrasive grains are most suitable for air- and hydro-blasting whereby surfaces are abraded as the result of the force of impact. If the impelled fragments have sharp points or edges, the density factor is even greater because of a so-called spike-heel effect.

Both grain size and purity must be controlled carefully. This is because the presence of any relatively large particle within a fine-grained grit or of a harder grain within a "mild abrasive" may lead to undesirable scratches on finished surfaces.

Abrasive Materials

Several natural and manufactured substances are used as abrasives; the most widely used natural substances are diamond, emery, garnet, quartz (including diverse cherts and tripoli), staurolite, natural glass (e.g., pumice), and diverse sands, sandstones, and pebble conglomerates. Manufactured substances include synthetic diamonds, carborundum (SiC), fused Al_2O_3 (aluminum oxide), glass, diverse "rouges" (e.g., fine-grained red iron oxide), metallic abrasives (e.g., steel wool and chilled and annealed steel shot), and several less well known substances such as alumina-zirconia oxide, boron carbide, boron nitride, cerium oxide, chromium boride, tin oxide, and zirconium oxide.

The following statements about a few of these materials illustrate diverse aspects of abrasive materials:

- Industrial diamond is, so far as hardness is concerned, by far the most effective abrasive (Table 1). Particularly in the past, many natural diamonds used as abrasives were referred to as carbonado or bort. Synthetic diamonds are often preferred over natural diamonds because their properties and shapes can be tailor-made for certain applications.
- Carborundum and fused Al_2O_3, both widely used synthetic abrasives, are produced by heating high-silica sand along with high-grade, low-ash petroleum or pitch coke and by fusing calcined, abrasive-grade bauxite, respectively.
- Emery is composed largely of corundum plus noteworthy amounts of iron oxides (commonly magnetite, less commonly hematite), plus or minus spinel. It is produced by crushing "emery" rock.
- Quartz is the major constituent of most sands and rocks such as sandstones, quartz pebble con-

Table 1. Mohs Scale and the Hardness of Widely Used Abrasives[1]

Mohs Scale[2]	Other Substances	Approximate Knoop Hardness[3] (kgf/mm^2)
10 +DIAMOND		7950
	*CUBIC BORON NITRIDE	4690
	*CARBORUNDUM	2450
	*FUSED-Al_2O_3	2140
9 +CORUNDUM		2040
8 Topaz		1330
	GARNET	1330
	STAUROLITE	
7 +(*?)QUARTZ (i.e., special silica stone)		810
	*HIGH CARBON STEEL	740
	*PIG IRON	
	TRIPOLI	
6 FELDSPAR		
	+GLASS (e.g., PUMICE)	
	*ROUGE	
5 Apatite		—
4 Flourite		—
3 Calcite		—
2 Gypsum		—
1 Talc		—

[1] Names of abrasives are in uppercase letters; those preceded with a plus sign (+) are used in both their natural and synthetic forms; those preceded with an asterisk (*) are synthetic.

[2] Mohs numbers indicate relative hardness. Although hardness differences between superjacent minerals up to corundum (H.9) are more or less equivalent, that between corundum (9) and diamond (10) is greater than that between talc (1) and corundum (9).

[3] Values have been recalculated from data compiled by Rue (1991) except for the 740 value given for high carbon steel, which is from Hight (1983). All are semiquantitative in that they are based on measurements of indentations made when rhombus-shaped diamonds are forced into a given material under certain static loads. So far as considerations of abrasives, these values may be less applicable than Mohs numbers, which are based on scratch tests.

glomerates, quartzites, and chert (e.g., flint and novaculite). It is also common in veins and pegmatite masses. In the abrasive industry, all of these are referred to widely under the group name special silica stone.

- Staurolite, a common by-product of heavy-mineral concentrates recovered from beach sands, for example, has recently found increased use as an abrasive—particularly for sandblasting and impact finishing of metals. Much staurolite supplants the former use of quartz in response to stringent regulations limiting amounts of free silica in air-blasting media.

- Tripoli is the name widely applied to partially decomposed, friable, microcrystalline silica derived from silica-rich rocks that have undergone noteworthy chemical weathering. A common precursor of these "rotten stones" is fine-grained, siliceous (i.e., cherty) limestone. The crushed abrasive product is usually marketed as a "mild abrasive" because its grains have no distinct corners or edges. Tripoli and several other "mild" or "soft" abrasives (Table 2) are used for such things as scouring powders and abrasive soaps, dental powders and pastes, and for buffing soft metals, ceramic ware, pearls, leather, and rice grains.

Table 2. Fairly Common "Mild" and "Soft" Abrasives and Their Typical Sources

---

Bauxite, a rock made up largely of aluminum oxide minerals such as gibbsite, diaspore, and boehmite

Calcite, the major constituent of rocks such as limestone and chalk

Diatomite and diatomaceous earth, siliceous sediments consisting largely of the microscopic, opaline frustules of single-celled plants

Feldspar, a common rock-forming mineral, which for use as an abrasive is usually recovered from pegmatite masses

Fuller's earth, primarily montmorillonite or attapulgite, which are clay minerals

Lampblack, a residue of burning oil

Rouge, diverse materials are designated as rouges (for example, "jeweler's rouge," which is a fine, red iron), i.e., ferric oxide

Talc, a relatively common mineral comprising metamorphic rocks such as talc schist and soapstone

Whiting, chalk, a limestone made up largely of calcareous tests of microorganisms such as foraminifers

---

Metallic abrasives are used chiefly as charges for gang saws to cut relatively soft rocks such as soapstone and verd antique.

Production

Natural abrasives and the raw materials for both synthetic and manufactured abrasion materials are derived from several different rocks and unconsolidated rock materials. The many uses of abrasives support a large industry (Table 3).

Natural Abrasion

Natural abrasion, which accounts for features such as potholes, glacial grooves and pavements, and ventifacts, is also noteworthy here. These erosional features reflect impacts on surfaces by natural abrasion "tools"—for example, boulders, cobbles, pebbles, sand grains, clay-size particles—carried by streams or other water currents, glacial ice, or the wind. In most cases, the "tools" also undergo abrasion—usually termed attrition—that is manifested by, for example, the rounded stones that occur in streambeds and along some beaches.

Table 3. Summary of Statistics for Abrasion Materials of the United States of America for 1989[1]

Production	$227,761,000+*
Exports	260,363,000
Reexports	33,771,000
Imports for consumption	420,982,000

---

[1] Reported by G. T. Austin (1991).

* This figure does not include values for emery and staurolite, both of which were "withheld to avoid disclosing company proprietary data."

Bibliography

AUSTIN, G. T. "Abrasive Materials." In *Minerals Yearbook*. Washington, DC, 1989. U.S. Bureau of Mines (1991): 77–96.

BOLEN, W. P. "Pumice and Pumicite." In *Minerals Yearbook*. Washington, DC, 1989. U.S. Bureau of Mines (1991): 817–820.

DAVIS, L. L., and V. V. TEPORDEI. "Sand and Gravel." In *Mineral Facts and Problems*. Washington, DC, 1989. U.S. Bureau of Mines, Bulletin 675 (1986): 689–703.

EARDLEY-WILMOT, V. L. *Abrasives* (Part I). Canadian Department of Mines, Publication 673. Ottawa, 1927.

HIGHT, R. P. "Abrasives." In *Industrial Minerals and Rocks*, 5th ed. New York, 1983, pp. 11–32.

McCAWLEY, F. X., and L. H. BAUMGARDNER. "Aluminum." In *Mineral Facts and Problems*. Washington, DC, 1985. U.S. Bureau of Mines, Bulletin 675 (1986): 9–31.

POTTER, M. J. "Feldspar." In *Mineral Facts and Problems*. Washington, DC, 1985. U.S. Bureau of Mines, Bulletin 675 (1986): 255–263.

RUE, C. V. "Abrasives." In vol. 1 of *Encyclopedia of Chemical Technology*, eds. R. E. Kirk and D. F. Othmer, 4th ed. New York, 1991, pp. 17–37.

SMOAK, J. F. "Corundum and Emery." In *Mineral Facts and Problems*. Washington, DC, 1985. U.S. Bureau of Mines, Bulletin 675 (1986): 223–233.

——. "Diamond, Industrial." In *Mineral Facts and Problems*. Washington, DC, 1985. U.S. Bureau of Mines, Bulletin 675 (1986): 233–247.

——. "Garnet." In *Mineral Facts and Problems*. Washington, DC, 1985. U.S. Bureau of Mines, Bulletin 675 (1986): 297–304.

RICHARD V. DIETRICH

ACID RAIN

See Pollution of Atmosphere; Pollution of Lakes and Streams

AGASSIZ, JEAN LOUIS RODOLPHE

A descendant of a Huguenot family that had moved to Switzerland, Agassiz was born in Motiers, Canton de Fribourg, on 28 May 1807. He studied at Bienne, Lausanne, Zurich, Heidelberg, and Munich, receiving a degree in medicine from the University of Heidelberg in 1830. From early childhood Louis Agassiz had a strong interest in natural history, and after receiving his medical degree he went to Paris to study under the great comparative anatomist and founder of vertebrate paleontology, GEORGES CUVIER. In 1832 Agassiz was appointed a professor of natural history at the College of Neuchâtel, where he remained until 1846. In that year he was invited to give a series of lectures at the Lowell Institute in Boston. The success of these lectures, plus Agassiz's interest in the natural history of the United States, resulted in a permanent move to America. In 1848 he was made professor of natural history at the Lawrence Scientific School at Harvard University. Though he traveled extensively during the latter part of his career, he remained at Harvard until his death. This lifelong residence was made despite the fact that he was offered, among other honors, the professorship of paleontology at the Museum of Natural History in Paris in 1859. At Harvard in 1859–1860 Agassiz founded the Museum of Comparative Zoology and also served as its first director. In 1863 Agassiz helped to found the National Academy of Sciences in Washington, DC.

Agassiz is perhaps best known for his work on fossil fishes; his work on glaciers, which resulted in the concept of the "Ice Age;" his contributions to general natural history (particularly of North America); and his opposition to Darwin's theory of evolution by natural selection (*see* DARWIN, CHARLES).

Relatively early in his career Agassiz established his reputation with a massive, five-volume study of fossil fishes, *Recherches sur les poissons fossiles* (1833–1844); in addition he published *Monographie des poissons fossiles du vieux grès rouge ou système dévonien des Iles Britanniques et de Russie* (1844). Agassiz also had a general interest in zoological nomenclature and classification, publishing *Nomenclator Zoologicus* (1842–1846).

While pursuing the above-mentioned studies Agassiz also became interested in problems of glacial geology, building on the work of earlier geologists (especially Johann H. Charpentier [1786–1855]). Beginning about 1836 he began investigating the modern glaciers of the Swiss Alps. Agassiz demonstrated that glaciers move or flow. One demonstration of this characteristic was that a cabin built on a glacier in 1827 moved about a mile from its original site in a dozen years. Agassiz also performed experiments on glaciers, such as driving a straight line of stakes across a glacier and then returning several years later to find that the middle stakes had moved downhill, forming a "U" shape. Agassiz also documented the physical effects of glacial movement on the land's surface. He noted grooves and scourings on rock surfaces over which glaciers had moved, as well as various rock accumulations that formed on the sides and ends of glaciers (lateral and terminal moraines, erratic boulders, and so forth; *see* EARTH'S GLACIERS AND FROZEN WATERS). Finding similar features in northern Europe and North America, Agassiz established that glaciers had once (in the relatively recent geological past) covered a much more extensive area than they do today. This fact established the concept of an Ice Age. Among Agassiz's important works on glacial geology are *Études sur les glaciers* (1840), *Système glaciaire* (1846), and *Nouvelles études et expériences sur les glaciers actuels* (1847).

In North America, Agassiz continued his studies of natural history, publishing such works as *Contributions to the Natural History of the United States* (1857–1862), a work that was never completed although four volumes were published, *The Structure of Animal Life* (1862), *Methods of Study in Natural History* (1863), *A Journal in Brazil* (1868), based on his travels to that country, and *Geological Sketches* (two volumes, 1866–1876). During this time he also traveled widely, including journeys to Florida and the western United States, Europe, Brazil, and a trip around Cape Horn to the Pacific.

Throughout his life Agassiz opposed the concept of evolution, both as formulated by Darwin and by Lamarck (*see* LAMARCK, JEAN-BAPTISTE) and his followers. Agassiz acknowledged (indeed, helped to elucidate) the progression of fossil forms that outlined the history of life through geological time. But, like Cuvier, he did not interpret this history as being the result of evolution (see CUVIER, GEORGES). Rather, Agassiz approached the fossil record from what has been considered an "idealist" and "embryological" point of view. Agassiz viewed the development of a human embryo as being a process in which the organism goes through a progressive hierarchy of levels of organization—fish, amphibian, reptile, mammal, and so forth—as it approaches a certain perfection of being in the final adult form. Embryological development is a goal-directed process. Likewise, according to Agassiz, the history of different species and biotas through geological time is also a developmental series representing progress toward the highest organic beings, namely humans (Homo sapiens). These concepts of Agassiz and of like-minded thinkers of the nineteenth century (such as the Scottish geologist Hugh Miller [1802–1856]) might seem to approach evolutionary thinking, but they were actually far from it. Agassiz was convinced that the history of life on Earth was a series of discontinuous stages, and each stage of fauna and flora had been catastrophically destroyed and replaced with a new stage. The driving force behind the history of life was a supernatural Deity who regularly intervened to miraculously produce new suites of species. Agassiz believed that any particular species, once created, was fixed and incapable of change or evolution (see LIFE, EVOLUTION OF).

Despite his opposition to evolution, Agassiz is generally acknowledged as one of the great naturalists of the nineteenth century. He was known as an enthusiastic and eloquent teacher, and his lectures were well attended. His son, Alexander Agassiz, was also a well-known naturalist and director of the Museum of Comparative Zoology (1874–1910) after his father's death, in Cambridge, Massachusetts, on 14 December 1873. In 1879 a large Pleistocene lake, which covered parts of North Dakota, Minnesota, and Manitoba, was named after Louis Agassiz. Louis Agassiz was elected to the Hall of Fame for Great Americans in 1915.

Bibliography

AGASSIZ, L. Methods of Study in Natural History. New York, 1863.
———. Essay on Classification. E. Lurie, ed. Cambridge, MA, 1962.
———. Studies on Glaciers. A. V. Carozzi, transl. New York, 1967.
BOWLER, P. Evolution: The History of an Idea. Berkeley, CA, 1984.
MERRILL, G. P. The First One Hundred Years of American Geology. New Haven, 1924.
TELLER, J. D. Louis Agassiz, Scientist and Teacher. Columbus, OH, 1947.
ZITTEL, K. A. VON. History of Geology and Paleontology to the End of the Nineteenth Century. M. M. Ogilvie-Gordon, transl. London, 1901.

ROBERT M. SCHOCH

ALLOY METALS

An alloy is a solid mixture of metallic chemical elements. Common examples of alloys are brass, steel, and pewter. Metals, the constituents of alloys, possess a high reflectivity of light (metallic luster) and are good conductors of heat and electricity. Many metals are malleable (e.g., they can be hammered into sheets) and ductile (e.g., they can be drawn into wire). Alloys can be fabricated so that they have a hardness, strength, or resistance to corrosion that differs significantly from the properties of the constituent metals. Alloys are classified as either ferrous alloys, those that contain significant iron, or as nonferrous alloys, those that consist dominantly or entirely of elements other than iron. This entry addresses the origin of mineral deposits that are mined for metals used principally for their alloying properties with iron.

An alloy may be a crystalline, chemical mixture of metals (an alloy solid solution), such as a copper-nickel alloy; a compound that possesses the atomic regularity of a simple chemical compound, such as beta-brass ($CuZn$); or a multiphase alloy that is composed of at least two physically separable compounds (different types of crystals, or phases) such as carbon steel, which is composed of iron metal and cementite (Fe_3C).

A number of metals, including chromium (Cr), molybdenum (Mo), tungsten (W), and also vanadium (V), nickel (Ni), cobalt (Co), tantalum (Ta), niobium (Nb), and manganese (Mn) are added to steel to produce certain desirable physical and chemical properties. Stainless steel, by definition, has at least 11 percent Cr. A thin coating of Cr_2O_3 forms on the surface and "passivates" or protects the bulk of the alloy from further corrosion. Addition of small amounts of Mo, W, or other metals to the "300 series," or austenitic stainless steels (soft steels), increases their strength and corrosion resistance. The "400 series" steels (martensitic, or hard steels) are harder than the 300 series steels, but are less corrosion resistant. Some of the metals that are added to steels are also used in nonferrous alloys. For example, stellite, an alloy of Cr with Co and W, is used to make lathing tools, and pressure-containment vessels for high pressure–high temperature experiments, because it maintains its strength at high temperature.

Three alloy metals, Cr, Mo, and W, occur in group VIB (the "chromium group") of the periodic table of the chemical elements, and represent the first-, second-, and third-row transition elements in the group, respectively. Membership in the chromium group of metals confers a highest (group) oxidation state of 6+ (the oxidation state represents the idealized charge on the metal when it loses electrons to a nonmetal, such as sulfur or oxygen, during chemical bonding); therefore their highest (simple) oxides have the formula RO_3, where R = Cr, Mo, or W. Cr with this oxidation state is found only in deposits formed by the near-surface alteration of Cr-bearing rocks. The mineral crocoite ($PbCrO_4$) contains Cr^{6+} and is the mineral from which the metal Cr was first isolated.

Cr^{3+} is the most common oxidation state of Cr in Earth's crust and occurs in the only important ore mineral of chromium, chromite (Fe, Mg) (Cr, $Al)_2O_4$, which is in the spinel group of oxide minerals. Mo^{6+} and W^{6+} have wider stability ranges than Cr^{6+}, in part because the outermost electrons in Mo and W are farther from the nucleus, and therefore all six valence electrons are more easily lost. The most common ore minerals of W are wolframite, (Fe, $Mn)WO_4$, and scheelite, $Ca(W, Mo)O_4$; both contain W^{6+}. The most important ore mineral for molybdenum is molybdenite, MoS_2, which contains Mo^{4+}, although some molybdenum is present in scheelite, where its oxidation state is 6+.

All the important ores of the group VIB metals form by magma-related processes. Further, these metals also have some affinity for metallic iron, and a significant proportion of these metals are found in the iron-nickel phase in meteorites. We expect, therefore, that some significant proportion of Earth's budget of these metals is present in the core, and is, therefore, inaccessible. Mo and W are good examples of elements that seem to prefer the earth's crust over the mantle; that is, they are lithophile or (crustal) rock-loving elements. During the partial melting processes that formed the crust from the mantle, a portion of these elements tended to concentrate in the melt phase (i.e., they are crystal-incompatible elements), and were thereby concentrated upward in the mantle-crust system by virtue of the buoyancy of hot, gas-charged magma. These magmas cool, forming primitive continental crust; later remelting of these crustal rocks produced granite magmas with Mo and W again favoring the molten rock. When granite magmas crystallize, they eventually saturate with respect to a gaseous, bubble-forming, water-rich (aqueous) phase, into which Mo and W partition from the crystallizing melt. This highly buoyant, low-viscosity, aqueous fluid moves into fractures in the preexisting crust, and in the already crystallized portion of the granite. The decrease in temperature and pressure, and the reaction of the fluid with cooler rocks, causes tungsten and molybdenum minerals to be deposited from the flowing aqueous fluid. Furthermore, the reaction of the hydrothermal fluid with the enclosing rock produces hydrothermal alteration (e.g., reaction of feldspar to a fine-grained mixture of mica and quartz); this is common in mineral deposits, and serves as a ready guide to ore during mineral exploration. Quartz, pyrite, pyrrhotite, chalcopyrite, and other minerals can also precipitate from the flowing aqueous solutions. If conditions are right, molybdenite, wolframite, or scheelite may be part of the precipitating assemblage of vein minerals. Veins may be many centimeters across, and are traceable in some places for hundreds of meters, or they may be small hairline fractures that cut across large masses of hydrothermally altered rock.

Most of the geologists who have studied tungsten and molybdenum deposits consider the aqueous fluids that deposited them were derived dominantly from a crystallizing magma, by virtue of the spatial, temporal, and chemical relationships between granites and the veins. However, some proportion of the aqueous fluid may also have been derived from deeply circulating surficial waters, or even water released from the heating of hydrous rock upon intrusion of the magmatic body. The relative proportions of these nonmagmatic aqueous fluids in any deposit may vary from place to place and from time to time. In some deposits, metals may have been leached from the surrounding rocks by surficial water that is heated and circulates due to the presence of the body of magma.

Tungsten (W) and molybdenum (Mo) deposits may be described as vein-type, skarn (where ore and other minerals replace preexisting limestone), or as porphyry/disseminated/stockwork type (or simply, porphyry-type). In porphyry-type deposits, the associated granites are usually porphyritic (i.e., the granite has larger crystals set in a matrix of smaller crystals, indicative, most likely, of a two-stage cooling history). Further, the ore minerals are disseminated and are found in many small, intersecting fractures that host the mineralization. The world's largest Mo deposits are of the porphyry type (commonly called porphyry molybdenum deposits), and many are located in western North America. The most famous of these deposits is the Climax deposit of Colorado. One of the world's greatest concentrations of W deposits is in the Jiangxi Province of China, where both vein-type and porphyry deposits occur.

Varying proportions of W and Mo, along with associated tin (Sn) and copper (Cu), and other metals, may occur in these deposits. Research on the controls of the ratios of these metals in these high-temperature, magmatic-hydrothermal deposits continues, but some workers think that details of magmatic crystallization, such as the types of igneous minerals crystallizing from the "boiling" magma, or how oxidized the magma is, may be determining factors.

Cr behaves differently from Mo and W during rock melting or magma crystallization; it is a crystal-compatible element. During melting of Earth's mantle, chromium either remains behind during melting, or its host mineral (chromite) is among the first minerals to crystallize when the resulting mantle-derived melt begins to cool. Because chromite is dense and crystallizes while the magma is mostly liquid, it sinks to the bottom of the chamber of magma where it can accumulate in layers. Most of the world's chromite resources are in large, shallow, funnel- or sill-shaped, layered mafic igneous intrusions that are found in very old, stable continental areas. The world's largest such body, the Precambrian Bushveld complex of South Africa, fills a volume of 100,000 km^3, and represents solid-ified basaltic magma; this intrusion contains enormous reserves of chromite. During cooling of a large mass of basaltic magma, a number of minerals including plagioclase feldspar, pyroxenes, olivine, and chromite crystallize as the mostly liquid mush turns to stone. One of the early formed minerals, chromite, is also the most dense. Therefore, chromite settles out (by the action of gravity), accumulating in massive layers that, in the case of the Bushveld and the related Great Dyke of Zimbabwe, are from a few millimeters to over a meter thick and are traceable for tens of kilometers. Some spinel-rich layers in these deposits are Cr-poor, but are rich in V-bearing magnetite; in fact, large minable deposits of V occur in the Bushveld intrusion.

A significant proportion of the world's chromite has been produced from what are called podiform chromite deposits. The host rocks are, like those of the Bushveld, cumulates from the crystallization of basaltic magma, but they are highly deformed and variably altered. These bodies are irregularly shaped and occur in geological complexes that are believed to represent transported fragments of oceanic crust and uppermost mantle. Some of the larger deposits of this type occur in the Ural region of Russia. In contrast to the shallow, stable continental environment that supported the crystallization of Bushveld-type chromite deposits, the podiform chromites crystallized in the earth's upper mantle and were transported tectonically into the crust.

Some researchers think that mixing of slightly different magmas is required to produce the almost pure chromite layers that crystallized from the mantle-derived basaltic lava that cooled and crystallized during the formation of magmatic chromite deposits.

Bibliography

EDWARDS, R., and K. ATKINSON. *Ore Deposit Geology.* New York, 1986.

EVANS, A. M. *Ore Geology and Industrial Minerals: An Introduction,* 3rd ed. Oxford, Eng., 1993.

SKINNER, B. J. *Earth Resources,* 2nd ed. Englewood Cliffs, NJ, 1976.

WHITNEY, J. A., and A. J. NALDRETT, eds. *Ore Deposition Associated with Magmas.* El Paso, TX, 1989.

PHILIP A. CANDELA

ALUMINUM RESOURCES

Aluminum is a silvery white metal, with the atomic number 13, an atomic weight of 26.98, and a density of 2.7 g/cm^3. The name and chemical symbol (Al) are derived from *alumen*, the Latin name for alum, a potassium-bearing sulfate salt of aluminum. Because aluminum is an efficient conductor of electricity, is strong but light, is resistant to oxidation, and an intrinsically abundant element, it has become one of the most important industrial metals. Aluminum is used in the construction industry, in the manufacture of automobiles and aircraft, in electrical equipment, in a wide variety of fabricated metal products, in machinery, beverage cans, containers of all kinds, and in many home consumer products.

Geochemical Abundance

Aluminum is the most abundant metal and the third most abundant chemical element in the earth's crust, accounting for 8.0 percent of the mass of the crust by weight.

Because of its abundance, aluminum is present as a major constituent in many common minerals such as the feldspars, micas, and clays. Despite its abundance, the difficulty of separating aluminum from the elements with which it combines in nature is so great that the metal was first prepared in a pure form only in 1827; the separation process was so expensive that aluminum could at first only be used for expensive jewelery and items of fashion. Indeed, its cost and rarity were so great that in the mid-nineteenth century French nobility used aluminum instead of gold for utensils at banquets.

Production of Aluminum

Aluminum metal first became available in large amounts and at reasonable prices in 1886, following development of the Hall-Heroult electrolytic method of separation.

The separation process used today is a two-stage one, the second of which is the basic Hall-Heroult process. In the first stage, the Bayer process, aluminum is leached from minerals with caustic soda (NaOH) and precipitated as $Al_2O_3 \cdot 3H_2O$. The precipitate is then heated to drive off the H_2O, leaving anhydrous alumina (Al_2O_3). In the Hall-Heroult stage, the alumina is dissolved in a vat of molten cryolite (Na_3AlF_6) at about 950°C and an electric current is passed through the solution, reducing the aluminum to molten metal. The molten metal is then cast into ingots.

Aluminum production requires a very great input of energy. Not only are the heat requirements high, the amount of electrical energy used is enormous. For this reason, aluminum smelting is carried out near inexpensive sources of electricity, such as hydroelectric power plants in northern British Columbia and Norway, and natural gas-fired plants in the Middle East. An indication of the comparative energy demands of aluminum and iron ore smelting is illustrative: 25 million Btu's of energy are needed to produce a ton of steel while 150 million Btu's are required to produce a ton of aluminum.

Aluminum Ores

Despite the large number of minerals that contain aluminum, only three—the aluminum hydroxides diaspore ($Al_2O_3 \cdot H_2O$) and its dimorphs, boehmite ($Al_2O_3 \cdot H_2O$), and gibbsite ($Al_2O_3 \cdot 3H_2O$)—are desirable ore minerals. They are desirable because they are more readily treated by the Bayer process than any other aluminum-bearing mineral. The hydroxides are rarely found pure in nature. Most commonly they are mixed in varying proportions and also mixed with clays and other minerals such as quartz (SiO_2) and goethite ($Fe_2O_3 \cdot H_2O$). Such mixtures are called "bauxites" after the town of Les Baux in southern France where the material was first identified.

Bauxites are aluminous laterites and they form by intense weathering in which the most soluble portion of a rock is dissolved away, leaving a residue of aluminum and iron hydroxides. When a laterite contains between 40 and 60 percent Al_2O_3 by weight, it is called bauxite. The conditions under which bauxites form are high rainfall and high temperature together with the right drainage conditions; they can be met only where tropical or subtropical conditions prevail. The world's major bauxite deposits are found in northern Australia, West Africa, northern South America, and southeast Asia. Paleobauxites, formed in prior geological ages and moved to their present positions as a result of continental drift, are found in Europe, northern Asia, and in the United States in Arkansas.

Because bauxites are restricted geographically and the deposits are limited in size, extensive studies have been carried out to find ways of recovering aluminum from non-bauxitic sources. It is certainly possible to do so, but the cost is considerably greater than production from bauxites. Materials that have been successfully tested are clays (especially kaolinite, $Al_2Si_2O_5(OH)_4$), the calcium feldspar, anorthite ($CaAl_2Si_2O_8$), nepheline ($NaAlSiO_4$), sillimanite and kyanite (polymorphs of Al_2SiO_5), and alunite ($KAl_3(SO_4)_2(OH)_6$). Known resources of bauxite exceed 50 billion tons and the annual world production of aluminum is about 20 million tons. Bauxite supplies are clearly adequate for current needs and it will be well into the twenty-first century before aluminum will have to be produced from sources other than bauxite.

Bibliography

GUILBERT, J. M., and C. F. PARK, JR. *The Geology of Ore Deposits.* New York, 1985.

KESLER, S. E. *Mineral Resources, Economics and the Environment.* New York, 1994.

BRIAN J. SKINNER

AMPHIBIANS AND REPTILES

Amphibians

Modern amphibians, exemplified by frogs, salamanders, and the tropical, limbless caecilians, live at the interface between water and land. Many amphibians live at least part of their life on land, but most lay their eggs in the water and have an aquatic larval stage. Their way of life seems intermediate between fully aquatic fish and fully terrestrial reptiles and mammals.

Remains of primitive amphibians are known from the Upper Devonian into the Mesozoic (Figure 1). Many features of their skeletons are intermediate between one group of sarcopterygian (lobe-finned) fish and primitive reptiles. These primitive amphibians are collectively known as stegocephalians (roofed skulls) because they had a solid bony covering over most of the head, except for openings for the eyes and nose, in contrast with the very open skulls of modern frogs and salamanders. For 70 million years (Ma), until the diversification of reptiles, these archaic amphibians were the dominant land vertebrates.

A gap of approximately 30 Ma separates the well-known Devonian amphibians and animals from the Late Mississippian. It was during this time that the major diversification of this group occurred, and it is difficult to establish relationships among the many, highly specialized lineages of the later Carboniferous. Two major groups of Paleozoic amphibians have long been recognized, the labyrinthodonts and the lepospondyls. The labyrinthodonts resemble more closely their ancestors among the rhipidistians fish and the early Devonian amphibians in being of large size—most were at least one meter in length—and in having their vertebrae formed of several pieces of bone; the base of their teeth was enfolded like the paths of a labyrinth. This assemblage probably includes the ancestors of all later amphibians as well as the amniotes (the group including reptiles, birds, and mammals).

The labyrinthodonts occupied a wide range of habitats, and their physical attributes included large semiaquatic forms resembling crocodiles and fully terrestrial animals. Two major groups are recognized: the Reptiliomorpha, which included the ancestors of amniotes, and the Batrachomorpha, which may include the ancestors of some of the living amphibians. The labyrinthodonts were the dominant land vertebrates until the end of the Carboniferous, when the early amniotes began to diversify. A second diversification among the batrachomorphs in the late Paleozoic led to a diversity of primarily aquatic groups with large flat heads and very short limbs, some of which persisted into the Cretaceous, where they are found with dinosaurs.

The lepospondyls were generally of small size, had spool-shaped vertebrae (as do modern amphibians and amniotes), and their teeth were not enfolded. Lepospondyls are thought to have evolved from labyrinthodonts, but it is not known whether they shared a single common ancestry or evolved separately from several different lineages. They were extremely diverse in body form, from the extremely elongate, snakelike aïstopods to the newtlike nectrideans. Many greatly reduced or completely lost their limbs. In contrast to the cosmopolitan labyrinthodonts, the lepospondyls were

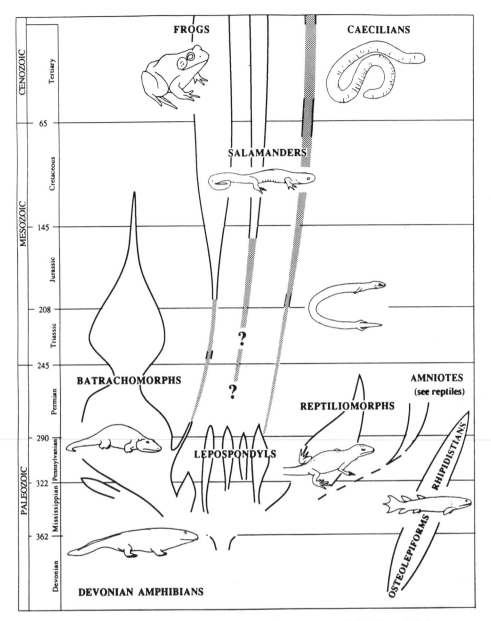

Figure 1. Geological distribution of the major groups of fossil and living amphibians. Numerical scale on left is in millions of years.

limited primarily to what are now the northern continents, and did not survive into the Mesozoic.

The earliest fossils of the modern amphibian orders are known from the Jurassic. Jurassic frogs and salamanders closely resemble their living counterparts, but the early Jurassic caecilian still retained tiny limbs, although its highly specialized jaw structure resembles that of living relatives. Frogs, salamanders, and caecilians resemble one another in their small body size, damp skin, and

particularly in the structure of their teeth, in which the crown is attached to the base by fibrous tissue. They are commonly grouped together as the Lissamphibia, but no fossils are known that share specializations of all three groups. Only the frogs can be linked to any particular group of Paleozoic amphibians. They share with dissosophid labyrinthodonts a specialized ear structure capable of detecting high-frequency airborne vibrations. Caecilians share specializations of the skull and lower jaw with

11

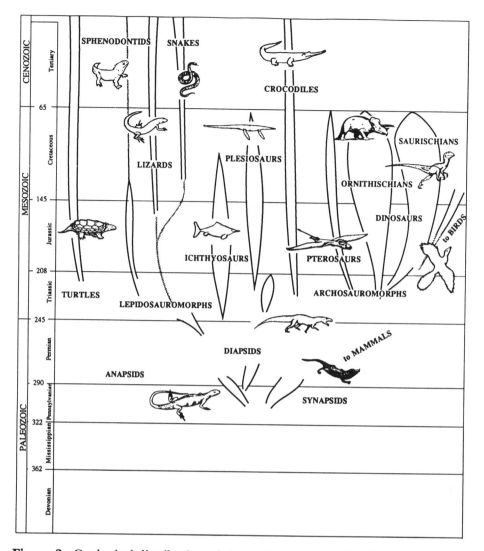

Figure 2. Geological distribution of the major groups of fossil and living reptiles. Numerical scale on left is in millions of years.

microsaurs among the lepospondyls. Salamanders may be related to either labyrinthodonts or lepospondyls, but no intermediate forms are known that support either relationship.

Only a small number of caecilian fossils are known between the Lower Jurassic and the present. Frogs and salamanders have a very poor fossil record in the Mesozoic. All of the modern families had evolved by the Early Cenozoic.

Reptiles

Reptiles are among the most spectacular fossils. Dinosaurs, flying reptiles, and giant aquatic reptiles dominated the land, sea, and air during the Mesozoic. The more modest reptile groups that survive today—lizards, snakes, crocodiles, and turtles—have an extensive fossil record going back to the Triassic period (Figure 2).

Reptiles are one of three types of animals making up the vertebrate group Amniota, which also includes birds and mammals. The amniotes are distinguished from fish and amphibians by the presence of membranes, including the amnion, that surround the embryo and allow the eggs to be laid on land or retained in the body of the mother. The young hatch or are born at an advanced stage of development, in contrast with having an aquatic larval stage as in most amphibians.

Amniote fossils first appear in rocks of Pennsylvanian age. At that time, two major lineages can be recognized: one, the Synapsida, led eventually to mammals; a second was ancestral to modern reptiles and birds. Later in the Carboniferous period, the second group divided again to give rise in one direction to turtles and in the other to an assemblage called the Diapsida. (The term "synapsid" refers to the presence of a single pair of openings in the side of the skull, behind the eyes; the term "diapsid" refers to the presence of two pairs of openings. The most primitive amniotes have no openings behind the eyes—the anapsid condition.)

Turtles (Chelonia) are the only living reptiles retaining the primitive anapsid condition. Even the earliest turtles, from the Upper Triassic, appear modern in the elaboration of a heavy bony covering, consisting of an integrated carapace and plastron (the dorsal and ventral elements of the shell). The marginal teeth had already been replaced by a horny beak, but the neck was not yet capable of retraction. No fossils are known that illustrate the transition between more primitive reptiles and turtles, and their specific ancestry is currently much debated.

The diapsid assemblage radiated extensively in the Late Paleozoic. One major subgroup, the lepidosauromorphs, gave rise in the late Permian to the ancestors of lizards, snakes, and Sphenodon, a genus that is now isolated on small islands off the coast of New Zealand. The modern lizard groups differentiated late in the Jurassic and achieved nearly their modern form by the end of the Cretaceous. The most spectacular fossil lizards were the mosasaurs, which were the dominant marine carnivores in the Late Cretaceous, reaching a length of 12 m. Snakes are known as early as the Lower Cretaceous and became dominant in the Cenozoic as predators on mammals.

The second major group of diapsids are the archosauromorphs, which include the dinosaurs, pterosaurs, crocodiles, and birds. Early archosauromorphs, which first appeared in the Upper Permian, are distinguished from lepidosauromorphs by changes in the rear leg and foot that made it possible for their descendants to draw the legs more directly under the body, and enabled the dinosaurs to assume a bipedal posture. Several groups of primitive archosauromorphs are known only from the Triassic. The term "thecodont" is applied to a diverse assemblage of carnivorous forms, known first in the Upper Permian, that includes the ancestors of crocodiles, pterosaurs, and dinosaurs.

Crocodiles appear at the end of the Triassic, distinguished from other archosaurs by their solidly built but relatively flat skulls. The earliest known members of this group were small, with very long limbs, indicating that they were agile, terrestrial predators. As early as the Lower Jurassic, several crocodile lineages became adapted to an aquatic way of life.

Flying reptiles (pterosaurs) and dinosaurs both appeared by the end of the Triassic. Upper Triassic pterosaurs had already evolved large wings for powered flight, and fossils are known that illustrate their origin from terrestrial archosaurs.

The oldest known dinosaurs, Herrerasaurs and Eoraptor, were fully bipedal carnivorous animals, presaging the giant Tyrannosaurus of the Late Cretaceous. Birds evolved from small, bipedal, carnivorous dinosaurs early in the Mesozoic. Two groups of herbivorous dinosaurs had also appeared by the end of the Triassic, the gigantic sauropods, culminating in Diplodocus and Apatosaurus [Brontosaurus], and the initially bipedal ornithischians. Ornithischians include the primarily bipedal heterodontosaurids, iguanidontids, hadrosaurs, and pachycephalosaurids, the armored stegosaurs and ankylosaurs, and the typically horned and quadrapedal ceratopsians.

The large size of most dinosaurs would have allowed them to maintain a high, constant body temperature. Their upright posture implies a high level of activity, as in warm-blooded mammals. Pterosaurs would have had to be warm-blooded to produce the continuous power necessary for flight.

Accompanying the dinosaurs in the Mesozoic were several groups of marine reptiles that appear to have diverged from the diapsid stock at about the time of the split between lepidosauromorphs and archosauromorphs. The ichthyosaurs were the most highly specialized for rapid swimming. The Jurassic and Cretaceous genera have a body form resembling that of the fastest swimming modern fish, the tuna, with a spindle-shaped body, and a high, lunate tail. The earliest ichthyosaurs, from the Lower Triassic, were already highly specialized for aquatic feeding and locomotion, and their specific ancestry has not been established.

The placodonts, nothosaurs, and plesiosaurs had a more rigid trunk, slender or shortened tail, and used the limbs for propulsion. Placodonts and nothosaurs are restricted to the Triassic. Some placodonts evolved a massive bony covering, analogous with that of turtles, and lost their teeth. The nothosaurs and plesiosaurs represent successive radiations within a single group in which the limbs evolved into gigantic paddles that propelled the body somewhat as in the modern sea lion. Cretaceous plesiosaurs reached lengths of more than 12 m.

Dinosaurs, pterosaurs, mosasaurs, and plesiosaurs were all numerous and diverse until the very end of the Cretaceous period, at which time they suddenly became extinct.

The concept of reptiles was once extended to the pelycosaurs and therapsids (together termed the Synapsida, or mammal–like reptiles) that link the most primitive amniotes and mammals. Most paleontologists would now agree that these groups should be considered with mammals rather than with reptiles.

Bibliography

Amphibians

BENTON, M. J., ed. *The Phylogeny and Classification of the Tetrapods.* Vol. 1, *Amphibians, Reptiles, Birds.* Oxford, Eng., 1988.
CARROLL, R. L. *Vertebrate Paleontology and Evolution.* New York, 1987.
———. "The Primary Radiation of Terrestrial Vertebrates." *Annual Review of Earth and Planetary Science* 20 (1992):45–84.

Reptiles

McGOWAN, G. *Dinosaurs, Spitfires, and Sea Dragons.* Cambridge, MA, 1991.
WEISHAMPEL, D. B.; P. DODSON; and H. OSMOLSKA, eds., *The Dinosauria.* Berkeley and Los Angeles, CA, 1993.

ROBERT L. CAROLL

ANGIOSPERMS

Today there are around 300,000 species of flowering plants, or angiosperms, and these are the dominant plant life-forms on Earth. Yet only 135 million years ago (Ma) during the early Cretaceous period angiosperms were so rare that only fossil pollen grains have been discovered. The reasons flowering plants have been successful have long been of interest to botanists and paleobotanists.

Angiosperm paleobotany has played an important role in the understanding of angiosperm origin and evolution. These fossils range from dispersed pollen in sediment to compressions of leaves, stems, fruits, seeds, and flowers, to three dimensionally preserved wood and reproductive organs (Friis et al., 1987). Like fossils in other groups, the study of these fossils has led to an understanding of the minimum date for the origin of groups and specific characters, and the identification of evolutionary mosaics (fossil species with combination of characters restricted to different groups).

The earliest undoubted remains of angiosperms are pollen grains from the Valanginian stage of the Cretaceous period (138 Ma). Leaves have been discovered from the Barremian (127 Ma) and flowers and possible wood from the Aptian (120 Ma). The earliest fossil flowers are unusual in their preservation as they are grouped on spikes that are still attached to an axis and subtended by leaves (Taylor and Hickey, 1990). The fossil is unassignable to any group of living angiosperms as the leaf characters are in a combination not found in any living angiosperm leaf. Other fossils of interest are the Triassic plant Sanmiguelia and angiosperm-like pollen grains from the Triassic and Jurassic periods. Continued work will find whether these intriguing fossils are the remains of angiosperms or angiosperm relatives.

Although angiosperm fossils are rare and quite similar during the Valanginian, they rapidly diversify in form and number during the early Cretaceous. The fossils are assignable to the living groups of angiosperms. The two major groups of angiosperms are the class Magnoliopsida (the dicotyledons) and the class Liliopsida (the monocotyledons); the dicotyledons have two seed leaves and flowers with floral parts in fours or fives (e.g., balloon flowers) while monocotyledons have a single seed leaf and parts in threes (e.g., lilies). Magnoliopsida includes the most ancestral angiosperms that are placed within the subclass Magnoliidae. Yet most of Magnoliopsida species are members of the eudicots, the group with tricolpate-type pollen (Doyle and Donoghue, 1989).

Many early angiosperm fossils are similar in morphology (or form) to the living members of the subclass Magnoliidae, which are perennial herbs usually with small flowers. These herbaceous species are informally called the eoangiosperms. Yet other members of the subclass, from shrubby groups such as Magnoliales, Laurales, and winteroids, quickly appear. Other early appearing groups include Liliopsida and eudicots, specifically the subclasses Hamamelidae and Rosidae (Friis et al., 1987). By the end of the Cretaceous, species from most subclasses have evolved and many fossils are assignable to living families (Figure 1; Muller, 1981).

The early Cretaceous radiation (or increase in number of species) of flowering plants dramatically changed the earth's terrestrial vegetation. The fossil record shows that angiosperms initially appeared in the equatorial regions. From here they radiated and spread toward the poles. The increase was so rapid that by the earliest late Cretaceous at least 50 percent of the species in the fossil floras were flowering plants (Figure 1; Lidgard and Crane, 1988). This radiation and diversifica-

tion were paralleled by extinction of several seed plant groups, although conifers and ferns maintained a near constant number of species (Friis et al., 1987).

Angiosperm fossils have also aided comprehension of the ecology of the early members. Studies of the paleoecology of early localities show that angiosperms are initially found in disturbed sites along river margins. Flowering plants appear first in these habitats and are found mixed with fern fossils. Later they appear in the backswamp margins and channel deposits, and last in the river terraces. Studies of the paleoecology of the late Cretaceous show that most angiosperms were small and although high in the number of species, they were not dominant.

Molecular fossils are a new source of data on angiosperm evolution. Fossil DNA has been reported from Miocene age (12 Ma) fossils and allows the fossil species to be included in molecular

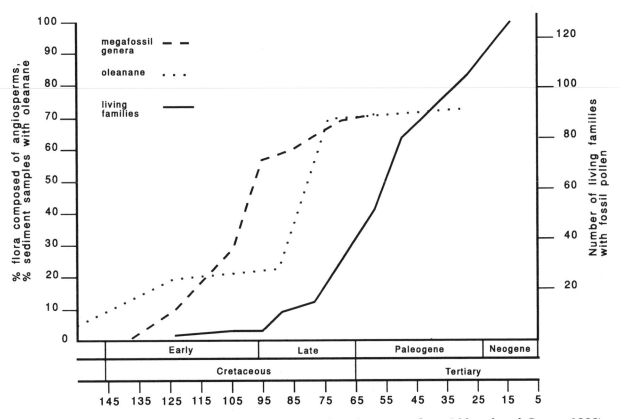

Figure 1. A comparison of the increased percentage of angiosperms (from Lidgard and Crane, 1988), increased percentage of sediments with the molecular fossil oleanane (from Moldowan et al., 1994), and the number of living families represented by fossil pollen through the Cretaceous and Tertiary (from Muller, 1981).

phylogenies. Another molecular fossil is oleanane, a triterpanoid chemical compound (Moldowan et al., 1994). Oleanane is predominantly produced by flowering plants where it appears to be used for plant defense. Examination of the frequency of occurrence of oleanane in sediments shows that it is rare before the Cretaceous (Figure 1). The frequency increases during the early Cretaceous with the major increase occurring during the latest Cretaceous. This curve is similar to that found for increases in angiosperm species in fossil floras. Yet the oleanane increase and the increase in the amount of fossil angiosperm wood in the record are delayed compared to the floristic increase. This delay may be due to increased use of oleanane-type defense compounds or reflect increases in the amount of angiosperm biomass as opposed to the number of species.

Like all fossils, fossil angiosperms have their limitations. The fossil record is incomplete with gaps in morphology and time. Some important characters used in understanding living flowering plant evolution and relationships are not preserved in fossils (see FOSSILIZATION AND THE FOSSIL RECORD). In general the fossils are single organs unlike the groups of vegetative and fertile characters used to understand living plant evolution. Finally, most work on living angiosperm relationships is based on reproductive organs like flowers. Although fossil flowers are now more commonly found, they are still rare compared to pollen and leaves.

The research on living plants has been important for understanding the evolution of characters and for understanding the relationship of major groups of flowering plants. Recent phylogenies based on morphological characters or DNA sequences clearly show that the closest living relatives to angiosperms are from the seed plant group called Gnetophyta. This understanding has led to new interpretations of the evolution of the angiosperm fruit and seed. The evolution of pollen grains and of leaves has also been possible through the study of living angiosperms.

The phylogenetic analyses also have resulted in two hypotheses of the ancestral angiosperm (Taylor and Hickey, 1992). The first suggests that the ancestral angiosperm was a woody shrub with large, multiparted flowers. These plants are members of the Magnoliales, winteroids, and Laurales (Figure 2). The more recent and less well accepted hypothesis indicates that the ancestral angiosperm

was a perennial herb with small, simple flowers. Each of these hypotheses has different implications for the early evolution of flowering plants.

Nevertheless, the fossil record has had a supporting role in these other aspects of angiosperm origin and evolution. The affinities of the earliest fossils, the oleanane record, and the infrequent occurrence of wood until the late Cretaceous provide supporting evidence for the herbaceous origin hypothesis (Taylor and Hickey, 1992). The paleoecology data also have been interpreted as supporting a herbaceous origin. The ancestral angiosperm is suggested to be the first seed plant with a herbaceous habit. This habit allowed rapid growth and was adapted to these unstable streamside habitats. From these sites the early angiosperms radiated into the channel and pond aquatic habitats, and to the margins of the backswamp. Only later did they evolve into woody shrubs and begin competing with existing tree species in the more stable areas of the river terraces. Yet these data are not definitive as species related to currently woody groups are found soon after in the fossil record.

The early appearance of certain characters in the fossil record has influenced our understanding of several angiosperm characters. The fossil record has documented the early occurrence of primitive vessels in wood. It has also shown that earliest leaf fossils have less organized venation patterns than those found in the Magnoliidae (Taylor and Hickey, 1990). Finally the knowledge of pollen evolution has been supplemented by the discovery of early pollen grains with a morphology unlike that in living flowering plants.

The discovery of fossil flowers has advanced the understanding of pollination biology in angiosperms (Friis et al., 1987). Flowers are frequently adapted in morphology for specific pollinators such as bees, butterflies, or hummingbirds. The record of fossil flowers clearly shows a wide range of morphologies from wind-pollinated forms to those that appear adapted to insect pollination during the mid-Cretaceous. Younger fossil flowers have morphologies that are related to advanced pollination systems. Although insect pollination may have not been important for the origin of angiosperms, as it is found in close relatives of angiosperms, it may have been important for the later diversification.

Angiosperm fossils are also important for our understanding of Cretaceous and Tertiary geol-

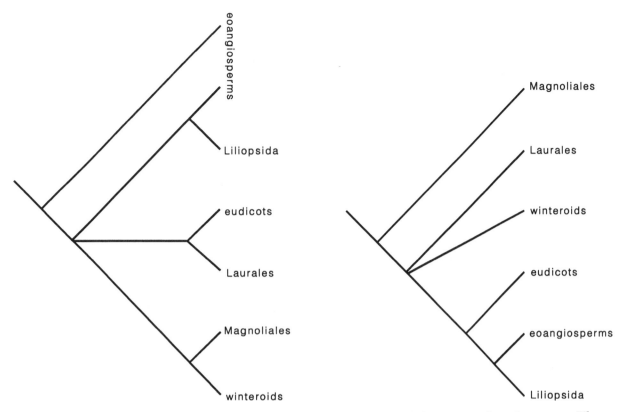

Figure 2. The two phylogenetic trees showing the two hypotheses of the ancestral angiosperms. The top shows the shrubby, large flowered groups basally placed (from Donoghue and Doyle, 1989), while the bottom shows the herbaceous, small flowered groups at the base (Taylor and Hickey, 1992).

ogy. Fossils, particularly pollen grains, are important for biostratigraphic correlations. Both pollen and leaf fossils have added considerable data on the understanding of the Cretaceous-Tertiary extinction. Both fossil types show higher extinction rates at the boundary than are found earlier in the Maestrichtian (69 Ma). Angiosperm fossils, usually leaves, have also been used to interpret paleoclimates. Another area is the use of angiosperm fossils to test hypotheses of former distributions of organisms. The record has shown strong support for a widespread Laurasian tropical flora (the Boreotropical paleoflora) extending from Mexico to Europe and along the margins of the Tethys seaway to east Asia. Thus, in southern North America a tropical flora exists. This flora apparently exchanged elements with the South American flora during the late Cretaceous and more recently in the late Tertiary.

Bibliography

DONOGHUE, M. J., and J. A. DOYLE. "Phylogenetic Analysis of Angiosperms and the Relationships of Hamamelidae." In vol. 1 of *Evolution, Systematics, and Fossil History of the Hamamelidae*, eds. P. R. Crane and S. Blackmore. Oxford, Eng., 1989.

FRIIS, E. M., W. G. CHALONER, and P. R. CRANE, eds. *The Origins of Angiosperms and Their Biological Consequences.* Cambridge, Eng., 1987.

LIDGARD, S., and P. R. CRANE. "Quantitative Analyses of the Early Angiosperm Radiation." *Nature* 331 (1988): 344–346.

MOLDOWAN, J. M.; J. DAHL; B. J. HUIZINGA; F. J. FAGO; L. J. HICKEY; T. M. PEAKMAN; and D. W. TAYLOR. "The Molecular Fossil Record of Oleanane and Its Relationship to Angiosperms." *Science* 265 (1994): 768–771.

MULLER, J. "Fossil Pollen Records of Extant Angiosperms." *Botanical Review* 47 (1981):1–146.

TAYLOR, D. W., and L. J. HICKEY. "An Aptian Plant with

Attached Leaves and Flowers: Implications for Angiosperm Origin." *Science* 247 (1990):702–704.
———. "Phylogenetic Evidence for the Herbaceous Origin of Angiosperms." *Plant Systematics and Evolution*
180 (1992):137–156.

DAVID WINSHIP TAYLOR

APOLLO ASTRONAUTS

On 25 May 1961, President John F. Kennedy challenged the United States to land "a man on the
moon and return him safely to Earth." This statement defined a need for astronauts who would
have the knowledge and experience to help develop and test space systems and then to use those
systems to land on and explore the Moon (Gilruth,
1975).

Astronaut Selection

The thirty pilots that made up the first three
groups of Apollo astronauts, the first seven of
whom had been selected originally for Project
Mercury in 1959, had received their professional
flying training in the military. All but two came to
the National Aeronautics and Space Administration (NASA) following recommendations from
their parent services. NASA chose the six scientists
of the fourth group in 1965 from among sixteen
recommended by a committee of the National
Academy of Sciences. The fifth group consisted of
nineteen more pilots; eleven scientists and engineers comprised the sixth group; and seven astronauts from the U.S. Air Force's Manned Orbiting
Laboratory Program transferred to NASA in 1969
as the seventh and last group to be involved directly with Apollo. If not qualified as a jet pilot,
each scientist spent his first year in Air Force flight
training (Sherrod, 1975).

Apollo Missions

Only twenty-four of the seventy-three men selected for Apollo actually flew to the Moon, twelve
of whom landed there. All of the astronauts, however, contributed their expertise and insights to
the success of Apollo through assignments to oversee engineering projects or to participate in mission support crews. Table 1 summarizes the assignments and history of the astronauts who flew
specific Apollo missions.

Astronaut Training

Lunar science training concentrated on honing the
observational, sampling, photographic, and communicative skills of the astronauts. Geological field
training included the planning, execution, and revision of traverses across appropriate geological
features in the southwestern United States and Hawaii, emphasizing as many of the mission constraints of time, equipment, and communications
as feasible.

Mission Results

Only one of the seven attempts to land on the
Moon during the Apollo program, *Apollo XIII*,
proved unsuccessful. In addition to providing humankind with its first steps into the solar system,
the galaxy, and the universe, these missions gave
science a first-order understanding of the evolution of the Moon, a planet similar to Earth except
in size and in the absence of water and air. Each of
the six landings contributed unique insights into
this understanding as well as strengthening the
conclusions reached as a result of preceding missions.

For example, the 3.7-billion-year-old volcanic
basalts (called "mare" basalts after Galileo's first
impression that the Moon had seas, or "maria")
sampled at Tranquillity Base by *Apollo XI*, established the very old age of the Moon's features, the
absence of water, and that gases from solar wind
had accumulated in the several meters of surface
materials pulverized by continued impacting by
meteorites and asteroids. *Apollo XII*'s samples from
Mare Insularum included not only a broad sampling of other basalts but also types of unsuspected
rock rich in potassium, rare-earth elements, and
phosphorus. Samples from Fra Mauro collected by
Apollo XIV suggested that such unusual rocks as
well as other very old crystalline rocks formed
from a lunarwide magma ocean soon after the
Moon was formed 4.6 billion years (Ga) ago. Fra
Mauro samples also dated the last of the great basin forming events showing that it occurred just
before the first major eruption of mare basalt.

Table 1. Apollo Missions

Flights	Astronauts	Mission
APOLLO I 1/27/57	Virgil I. Grissom Edward H. White II Roger B. Chaffee	Crew killed during test on launch pad
APOLLO II–III		Designations not used (no mission flown)
APOLLO IV–VI 11/9/67 1/22/68 4/4/68		Unmanned test flights
APOLLO VII 10/11/68	Walter M. Schirra R. Walter Cunningham Donn F. Eisele	Earth orbit test flight of the Command and Service Module
APOLLO VIII 12/21/68	Frank Borman James A. Lovell William A. Anders	Lunar orbit test flight of the Command and Service Module
APOLLO IX 3/3/69	James A. McDivitt David R. Scott Russell L. Schweikart	Earth orbit test flight of the Command and Service Module and Lunar Module
APOLLO X 5/18/69	Thomas P. Stafford John W. Young Eugene A. Cernan	Lunar orbit test flight of the Command and Service Module and Lunar Module
APOLLO XI 7/16/69	*Neil A. Armstrong Michael Collins *Buzz Aldrin	First lunar landing, Mare Tranquillitatis **Science: Mare basalt and regolith
APOLLO XII 11/14/69	*Charles Conrad, Jr. Richard F. Gordon, Jr. *Alan L. Bean	Second lunar landing, Mare Insularum **Science: Mare basalt, regolith, and Copernicus ray material
APOLLO XIII 4/11/70	James A. Lovell John L. Sweigert Fred W. Haise	Aborted on the way to Fra Mauro on the Moon due to explosion in Service Module
APOLLO XIV 1/31/71	*Alan B. Shepard Stuart Roosa *Edgar D. Mitchell	Third lunar landing, Fra Mauro area **Science: Ejecta from the Imbrium Basin
APOLLO XV 7/26/71	*David R. Scott Alfred M. Worden *James B. Irwin	Fourth lunar landing, Hadley-Appennines area **Science: Mare basalt, ancient crust, and Hadley Rille
APOLLO XVI 4/16/72	*John W. Young Thomas K. Mattingly *Charles M. Duke	Fifth lunar landing, Descartes area **Science: Lunar Highlands material
APOLLO XVII 12/7/72	*Eugene A. Cernan Ronald E. Evans *Harrison H. Schmitt	Sixth lunar landing, Taurus-Littrow Valley **Science: Young volcanics, basin rim material, and mare basalt

* Indicates an astronaut who landed on the Moon.
** Each landing crew also deployed an array of remotely operated geophysical and space physics experiments.

Apollo XV's observations and samples at Hadley-Apennines provided additional insights into the crystallization of the magma ocean, established a true network of geophysical stations from which the internal structure of the Moon became apparent, and began the geochemical mapping of the Moon's surface from orbit. The rocks and soils returned by *Apollo XVI* from Descartes accented the puzzles of the extremely complex lunar highlands. *Apollo XVII* explored the varied features of the valley of Taurus-Littrow, where, with the help of the first geologist-astronaut, studies tested most of the developing theories on the evolution of the Moon and discovered the "orange soil," new volcanic material from the lunar deep interior (Schmitt, 1975, 1991; Wilhelms, 1987).

Future Missions

The Apollo program established the psychological, technological, and scientific basis for future missions to the Moon and subsequently to Mars. The rationale for returning to the Moon will include scientific investigations, particularly astronomy, and the potential for lunar helium-3 (He^3) to serve as a terrestrial alternative to fossil fuels. Mars as well as lunar settlements offer potential enclaves for the expansion of the human species in addition to providing new scientific insights on the origin of life (Schmitt, 1992).

Bibliography

CORTRIGHT, E. M. *Apollo Expeditions to the Moon.* Washington, DC, 1975.

GILRUTH, R. R. "I Believe We Should Go to the Moon." In *Apollo Expeditions to the Moon,* ed. E. M. Cortright. Washington, DC, 1975.

HEIKEN, G. H.; D. T. VANIMAN; and B. M. FRENCH, eds. *Lunar Sourcebook: A User's Guide to the Moon.* Cambridge, Eng., 1991.

LOGSDON, J. M. *The Decision to Go to the Moon.* Cambridge, Mass., 1970.

SCHMITT, H. H. "Evolution of the Moon: The 1974 Model." *Space Science Reviews* 18 (1975):259–279.

———. "Evolution of the Moon: Apollo Model." *American Mineralogist* 76 (1991):773–784.

———. "Moon, Mars by Third Millennium." *Space News,* October 5–11, 1992.

SHERROD, R. "Men for the Moon." In *Apollo Expeditions to the Moon,* ed. E. M. Cortright. Washington, DC, 1975.

WILHELMS, D. E. *The Geological History of the Moon.* Washington, DC, 1987.

HARRISON H. SCHMITT

ARCHAEOLOGICAL GEOSCIENCE

Both geology and archaeology were born in the intellectual ferment of the first half of the nineteenth century. From approximately 1830 until well into the twentieth century these young disciplines were often united in a common goal—a study of the evidence for the "antiquity of man." The Englishman Sir CHARLES LYELL, one of the founders of geology, in his book *The Geological Evidence for the Antiquity of Man* (1870), established clearly the role of geology in archaeologic inquiry. In the United States, JOHN WESLEY POWELL was instrumental in the birth and early development of both the United States Geological Survey (USGS) and the Bureau of American Ethnology and became director of both.

Many geologic methods and concepts that are applied to studying the natural forces that shape the landscape and affect the nature of sedimentary deposits are also applicable to the solution of archaeologic problems. Indeed, archaeology has borrowed heavily from geology in developing archaeologic stratigraphy and chronology. The reasons are not hard to find. Both the field geologist and the excavation archaeologist:

1. have a sedimentary sequence to interpret;
2. deal with problems of chronology—potsherds in archaeologic strata are analogous to index fossils in geologic strata—and rely on radiometric dating;
3. must deal with an incomplete record—any initial depositional sequence is subject to later erosion, intrusion, and alteration;
4. are faced with a wide variety of rock and mineral materials that contain information needed to reconstruct the origin and subsequent history of the deposit;
5. use similar maps, cross sections, and geographic information systems to handle the spatial aspects of the depositional record and the changing landscape;
6. may use core drilling, geophysical prospecting, and other remote sensing techniques to gather data.

Every archaeologic site and survey region has a distinct geologic setting that governs the availability of raw materials, water, shelter, agricultural land, the stability of the landscape, many environmental hazards, and the energy costs of most human activities. Without understanding these environmental conditions and limitations it is not possible to reconstruct the lifeways of ancient societies.

Archaeologic geoscience uses geomorphic and sedimentologic techniques to reconstruct past landscapes. Such reconstructions are especially useful in areas of rapid landscape change—along major river systems and adjacent to oceans and inland seas—because these environments have been the foci of human settlement and activity. The floodplains of three great river systems, the Nile, the Tigris/Euphrates, and the Indus were the birthplaces of urban civilizations and their irrigation agriculture. Such floodplain environments are marked by dynamic change—swift erosion and deposition, lateral migration of the river channel, and construction of deltas at river mouths or the infill of estuaries.

For example, the present Nile River has only two main branches in the delta region. In the fifth century B.C.E. the Greek historian Herodotus reported five active branches; Ptolemy, in the second century C.E. recorded eight. The deposition at the mouth of the Tigris/Euphrates river system pushed the shoreline south into the Persian Gulf more than 200 km since the great Mesopotamian city of Ur was in its heyday. Historians long believed that the combat at ancient Troy recorded in Homer's *Iliad* took place on the plain of Troy. However, geologic core drilling that allowed dating of the sediments on the plain has shown that in Homeric times the plain was still under water in an estuary. Whether a stretch of shoreline advances seaward or retreats landward depends on a mix of three geologic processes: worldwide rise or fall of sea level, vertical tectonic movements of the regional bedrock, and deposition or erosion of coastal sedimentary deposits.

Three distinct methodologies have been used to investigate coastal change in an archaeologic context. Geomorphic field surveys can detail the evolution of the relevant landforms, river systems, and land-sea relations. In a second approach archaeologic geoscientists survey the vertical position of datable coastal archaeologic structures, then conduct mathematical analyses to reestablish the original relationship between the structures and their contemporaneous coastline. The third method can give the most precise paleogeographic reconstruc-

tions of ancient coastlines in depositional areas. In this method intensive core drilling is undertaken to recover the vertical sequence of coastal geologic environments. Core drilling commonly recovers material that can be radiocarbon dated, also allowing the establishment of an absolute chronology for the sequence of changes.

Two examples illustrate results that have been obtained using the core-drilling method. One of the western world's most famous battles was the one between the Spartans and the Persians at Thermopylae, Greece, in 480 B.C.E. The combat was on a narrow strip of land between the mountains and the sea. The area is one of extensive tectonic activity, sea level fluctuations, and sediment deposition. The modern topography is a broad agricultural plain. Using core drilling, Kraft and others (1987) were able to determine that in 480 B.C.E. a coastal pass of perhaps 100 m existed, with waves possibly lapping at the foot of the mountains at the narrowest point. But it is critical to understand that the land surface of that time is currently buried under up to 20 m of terrigenous clastic sediment and hot spring travertine deposits. This points up the extreme difficulty encountered by those who use only observations of the current landscape to attempt a detailed reconstruction of earlier landscapes in depositional areas.

The location of Punic Carthage (North Africa) has never been in doubt. Scholars since the early nineteenth century have disagreed, however, on the origin, shape, and number of its harbors. Using two types of corers, Gifford and others (1992) were able to unravel the coastal sedimentary sequences and thereby make paleogeographic reconstructions of the coastline and place constraints on the nature and location of harbor installations predating the visible remains of the third century B.C.E. They also proposed a geologically and historically plausible reconstruction of the famous Roman siege and capture of the city in 146 B.C.E.

The sedimentary record of an archaeologic site provides critical environmental, stratigraphic, and cultural information because the sediments are a mix of geologic, biogenic, and anthropogenic components. The composition, structure, texture, spatial contexts, and pedology all contain unique information on the human element in the formation of these deposits. Yet only in the last few decades have sedimentologic studies become a significant part of excavation efforts. Archaeologic geoscientists now use a framework for studying sedimentation on habitation sites based on the traditional geologic concepts of sediment source, transportation and deposition agents, deposition locations, and post-deposition alterations, as well as the concept of facies to better explain the spatial distribution of deposits (see STRATIGRAPHY).

The silt fraction of archaeologic sediments often contains phytoliths (mineral matter with distinct shapes found in plants) that can indicate the paleoecology of the area as well as be the only material remains of agricultural plants. Related chemical studies to identify anthrosols (soils derived from human activity) provide diagnostic indicators of abandoned habitation sites and agricultural fields. Using petrographic techniques and the polarizing microscope on archaeologic sediments and soils, earth and soil scientists can reconstruct many minute details in the often long history of reciprocal impact of humans with the environment.

Provenance studies involve a statistical comparison between the trace element, isotope, or other pattern in an object of interest, and the known pattern in potential source deposits. Such "fingerprinting" has proven successful in determining the geographic/geologic source of artifacts made of ceramic clays, copper, chert, obsidian, marble, and other lithic materials. Such studies have yielded some startling results. On the Plain of Thebes in Upper Egypt sit two larger-than-life stone figures representing a pharaoh of the Eighteenth Dynasty. The very hard quartzite stone could have come from only a half dozen or so quarries along the Nile. Trace-element analyses showed that the quartzite came from a quarry more than 400 km downstream, near Cairo, rather than from quarries upriver and much nearer Thebes.

Distinctive variations in the isotopic composition of carbon and oxygen in marble frequently allow archaeologic geoscientists to assign ancient Mediterranean area marble statues and monuments to the quarry from which the raw marble came (see ISOTOPE TRACERS, STABLE). In like fashion geochemists have analyzed isotopically ancient marble inscriptions to verify that inscription fragments were in fact once part of the same object. Stable isotope studies are also proving successful in sourcing lead, silver, and copper. Stable isotopes of carbon and nitrogen also provide insights into ancient human and animal diets.

Some of the most fruitful and challenging research opportunities often lie at the interface be-

tween disciplines. A technique illustrating this is archaeomagnetic dating. Over periods of decades or centuries Earth's magnetic poles are not fixed but are in constant motion. To chart the wandering poles, samples of materials that became magnetic during ancient human activity and which have since remained fixed in space are used to determine the location of the poles for specific times in the past. This technique allows the construction of an archaeomagnetic dating system.

In many parts of the world violent volcanic eruptions sporadically interrupt daily life, sometimes causing catastrophic destruction. A massive volcanic eruption in Central America resulting in major human migration may have contributed to the origin of Mayan civilization. During the middle of the second millennium B.C.E. another such eruption blew away most of the Aegean island of Thera in Greece, blanketing the inhabitants in the surrounding region with volcanic ash. Much of the eastern Mediterranean region as well as Central America and parts of western South America lie in volcanic belts. Since all volcanic rocks contain radioactive elements that can be used to date eruptions, archaeologic geoscientists now use these ash falls, which can cover an area of thousands of square kilometers, to date archaeologic strata.

History is also replete with accounts of earthquake events of tragic cultural consequence. Archaeologists define chronologic horizons by stratigraphic and cultural (e.g., changes in ceramic styles) discontinuities. In those parts of the world subject to severe earth tremors, stratigraphic discontinuities are often defined by "destruction" layers that archaeologists attribute to earthquakes. From the geophysical view, one must be cautious in attributing such damage to seismic violence. Other geologic and cultural phenomena (e.g., warfare) can damage structures—so that "the walls came tumbling down."

Finally, earth scientists trying to reconstruct the landscapes of the past make ever-increasing use of modern high technology. Remote-sensing instruments in orbiting satellites measure radar, infrared, and visible light radiation. Each part of the radiation spectrum records different kinds of information. In 1981 the space shuttle Columbia, using a Shuttle Imaging Radar System, recorded images that showed river channels now buried under the sands of the Western Desert in Egypt and the Sudan. These images helped locate new archaeologic sites in the desert.

Bibliography

GIFFORD, J. A.; G. RAPP, JR.; and V. VITALI. "Paleogeography of Carthage (Tunisia): Coastal Change During the First Millennium BC." *Journal of Archaeological Science* 19 (1992):575–596.

KRAFT, J. C.; G. RAPP, JR.; G. J. SZEMLER; C. TZIAVOS; and E. W. KASE. "The Pass at Thermopylae, Greece." *Journal of Field Archaeology* 14 (1987):181–198.

LYELL, C. *The Geological Evidence for the Antiquity of Man*, 2nd ed. Philadelphia, 1870.

RAPP, G, JR. and J. A. GIFFORD, eds. *Archaeological Geology*. New Haven, 1985.

WATERS, M. R. *Principles of Geoarchaeology: A North American Perspective*. Tucson, 1992.

GEORGE (RIP) RAPP, JR.

ARTHROPODS

The phylum Arthropoda comprises an enormous array of invertebrates (Figure 1) whose fossil record extends through virtually all of the Phanerozoic eon. Living representatives of the Arthropoda include insects (*see* INSECTS, HISTORY OF), spiders, ticks, scorpions, millipedes, centipedes, pill bugs, crayfishes, shrimps, crabs, lobsters, barnacles, ostracodes, and horseshoe crabs. Other familiar, extinct representatives of the group are trilobites and eurypterids.

The term arthropod is derived from Greek words meaning "jointed leg." Jointed appendages, which are present in all members of the phylum, have resulted from the covering of the body by a chitinous cuticle. This cuticle functions as an external skeleton (or exoskeleton) that provides support for muscles and protection for soft body parts, including protection against desiccation. In a few groups such as the ostracodes, trilobites, crabs, lobsters, and barnacles, some parts of the cuticle have been reinforced by the addition of calcium carbonate or calcium phosphate.

Besides a chitinous exoskeleton and jointed appendages, all members of the phylum Arthropoda share a bilaterally symmetrical, segmented body, a complete digestive tract, and a dorsal brain in the head followed by a ventral nerve cord. Most species reproduce sexually, although some insect species can reproduce asexually from females. There is a common tendency for the body segments to be combined into functional groups, called tagmata. The head, for example, is made up of fused segments that have nervous, sensory, and other functions. Other tagmata are the thorax and abdomen or tail region. The appendages are commonly specialized for a division of labor; antennae have a sensory function, whereas other appendages may be specialized for feeding, walking, swimming, respiring, or mating. Eyesight is nearly universal among the arthropods. The capability for sight ranges from simple light-sensitive eyes, which are present in many species, to well-developed compound eyes present in the trilobites, horseshoe crabs, eurypterids, insects, and others. Many arthropods undergo metamorphosis, or change, from a larval form to an adult form. Larvae are commonly quite distinct in morphology from adults of the same species.

Arthropods have achieved an extraordinary level of evolutionary success. The number of living insects alone has been estimated at one to five million species, which is more than all other animal species combined. Among groups represented commonly in the fossil record, trilobite diversity has been estimated to be in the range of 250,000 to 2,000,000 species, and the diversity of living and fossil ostracode crustaceans probably exceeds that of trilobites.

Arthropods have radiated into virtually every conceivable habitat on Earth and are an integral part of many food chains. They are found in abyssal depths of oceans, through shallow marine water, brackish water, fresh water, and all terrestrial environments. Aquatic species respire directly through the cuticle or by means of gills, and terrestrial species respire directly through the cuticle or by means of tracheae (air tubes) or book lungs. Styles of locomotion represented in the phylum are swimming, walking, crawling, and flying. Arthropods may be herbivores, carnivores (both active predators and scavengers), filter-feeders, or parasites. Arthropods, especially insects and crustaceans, serve as a major source of food for other animals. Many living arthropods are of considerable economic interest to humans. Commercial harvesting of shrimps, crabs, lobsters, and crayfishes is commonplace in coastal areas of the world. In inland areas, certain insects are important as pollinators of plants, whereas others are notorious as crop-destroying or disease-spreading pests.

Figure 1. Representative fossil arthropods, illustrating the great morphological diversity of the phylum. Specimens in figures F, H, and I are of animals that had calcified exoskeletons; other specimens had unmineralized exoskeletons, and are from deposits of exceptional preservation (Lagerstätten).
A. A horseshoe crab, *Paleolimulus signatus*, showing well-preserved appendages under the head shield; from the Permian of Kansas; ×5.
B. A spider, *Geraphrynus carbonarius*; from the Pennsylvanian of Illinois; ×2.
C. An eurypterid, *Eurypterus remipes*; from the Silurian of New York; ×2.
D. An aggregation of trilobites, *Homotelus bromidensis*, preserved variously in dorsal side up and down orientations; from the Ordovician of Oklahoma; ×0.75.

Figure 1. Continued

E. A lobster, *Eryon arctiformis*; from the Jurassic of Germany; ×0.75.

F. An unusually large ostracode, *Herrmania baltica*, from the Silurian of Sweden; ×2. The specimen is viewed from the left side, and, because the two valves are slightly asymmetrical, the edge of the right valve overhangs the left valve, forming a small rim near the bottom of the specimen.

G. A millipede, *Dasyodontus*? sp., preserved in amber from the Oligocene of the Dominican Republic; ×8.

H. Barnacles, *Balanus concavus*, encrusting a scallop shell; from the Miocene of Maryland; ×0.75.

I. A dipteran insect; from the Eocene of Colorado; ×2.

All photos by Loren E. Babcock

Classification and Evolution of Arthropods

There are about six superclasses of arthropods that are commonly recognized (Table 1). Superclasses are distinguished primarily on the basis of the number of appendages on the head in front of the mouth and the nature of the mouthparts. Some extinct groups of animals whose relationships to the Arthropoda are disputed, such as the arthropod-like *Anomalocaris* and its close relatives from Cambrian rocks, are not covered in this entry.

Arthropods have enjoyed inordinate evolutionary success on Earth, although their fossil record documents only a small fraction of that success. Those groups that are most commonly preserved as fossils are ostracode crustaceans and trilobites. Trilobites, which lived during the Paleozoic era, are characterized by generally having an elongate, oval, calcified exoskeleton with an articulating, multisegmented thorax; the appendages have two branches, one for walking or swimming, and the other for respiring. The name of the group is derived from the lengthwise division of the exoskeleton into three lobes. Ostracodes have a bivalved,

Table 1. Classification of Major Arthropod Groups

Scientific Name	Common Name(s) or Representatives
+Superclass Trilobito-morpha	
+Class Trilobita	trilobites
Superclass Crustacea	crustaceans
Class Branchiopoda	branchiopods or fairy shrimps
Class Branchiura	branchiurans
Class Cephalocarida	cephalocarids
Class Cirripedia	barnacles
Class Copepoda	copepods
+Class Euthycarci-noidea	euthycarcinoids
Class Malacostraca	shrimps, crabs, lobsters, crayfishes
Class Mystacocarida	mystacocarids
Class Ostracoda	ostracodes
Superclass Chelicerata	chelicerates
Class Arachnida	spiders, scorpions, whip scorpions
Class Merostomata	horseshoe crabs, euryp-terids
Superclass Pycnogonida	pycnogonids
Class Pantopoda	pantopods
Superclass Myriapoda	myriapods
+Class Archipolypoda	archipolypods
Class Arthropleurida	arthropleurids
Class Chilopoda	centipedes
Class Diplopoda	millipedes
Class Pauropoda	pauropods
Class Symphyla	garden centipedes
Superclass Hexapoda	hexapods
Class Collembola	springtails
Class Diplura	diplurids
Class Insecta	insects
Class Protura	proturids

+ represents an extinct group.

clamlike exoskeleton that protects the internal soft-parts and appendages. Most ostracodes are smaller than 5 mm. Trilobites and ostracodes tend to be more commonly preserved than other members of the phylum because all trilobites and most ostracodes are marine species, and most sedimentary rocks are of marine origin; because most species had calcified exoskeletons that made them more resistant to postmortem or post-molting destruction; because they (like most arthropods) molted,

or shed their old exoskeletons, as they grew, and each of these old exoskeletons could potentially become fossilized; and because of great initial diversity of species and abundance of individuals. An additional reason that trilobites are common as fossils in Paleozoic rocks is because of their multi-segmented exoskeleton, each piece of which could potentially become a fossil. Ostracodes are also common as fossils because the clamlike shape of the exoskeleton, and its generally minute size, resisted postmortem or post-molting breakage. Trilobites and ostracodes are commonly used in biostratigraphy. Crabs, lobsters, and barnacles, which have calcified exoskeletons, are locally common in some Mesozoic and Cenozoic marine rocks.

Arthropods that lack a calcified exoskeleton generally have a poor and incomplete fossil record, and with the exception of species from some extraordinary deposits, are best known from living representatives. The fossil record is especially incomplete for nonmarine arthropods. Much of our information about uncalcified (or lightly skeletalized) arthropods is derived from deposits of extraordinary fossil preservation (or Lagerstätten) such as the Cambrian-age Burgess Shale of British Columbia. Lagerstätten (*see* FOSSILS AND FOSSILIZATION) preserve the remains of animals that ordinarily would not be preserved, and thus provide us with a series of glimpses of the evolutionary history of uncalcified crustaceans, insects, spiders, scorpions, horseshoe crabs, myriapods, and other groups. An extinct group of scorpion-like chelicerates called the eurypterids, and an extinct group of shrimplike crustaceans called phyllocarids, are mostly known from Lagerstätten of the Paleozoic era.

The evolutionary success of the arthropods is accounted for largely by adaptations made possible by their cuticular exoskeleton, jointed appendages, and efficient sensory organs. The evolutionary success of certain land-dwelling groups has been further enhanced by efficient respiration, complex behavior, and metamorphosis (which results in the development of wings in adults). Initially, arthropods were exclusively marine animals, but today their greatest diversity is on land. Assemblages of Cambrian marine arthropods were dominated by a large number of crustaceans, some of which seem to be bizarre by modern standards, and trilobites. Arthropods were the first animals to develop acute eyesight. Compound eyes were present in several groups by the early Cambrian period, but they were particularly well-developed with calcified lenses in trilobites. Chelicerates had a small number of representatives in Cambrian seas. Myriapods, which seem to have been exclusively non-marine from the middle Paleozoic to the present, began as marine animals in the Cambrian.

Arthropods were the first animals to make the transition from the oceans to land. Some crustaceans, chelicerates, and myriapods had invaded the terrestrial environment by the Silurian period. One major obstacle to surviving in the terrestrial environment is the protection of eggs from desiccation. Terrestrial arthropods must have developed the first desiccation-resistant eggs by at least the late Silurian period. Once arthropods had a firm foothold on land, one group, the insects (*see* INSECTS, HISTORY OF), quickly evolved wings and became the first animals to fly. Insects seem to have become the most numerous animals on land by the end of the Paleozoic era, partly because of their ability to fly. Complex social behavior seems to have been developed in land-dwelling termites in the Cretaceous period, although it may have been present in ants, bees, and wasps as early as the Triassic period.

Extinctions of arthropods during the late Paleozoic era and subsequent evolution in remaining groups led to the development of essentially modern assemblages of arthropods by about the middle of the Mesozoic era. Many arthropod groups suffered decline during a dramatic late Permian extinction interval, an event that also affected a large number of nonarthropod organisms (*see* FOSSILIZATION AND THE FOSSIL RECORD). Trilobites, whose species numbers had been in decline since the Ordovician, seem to have become extinct during or shortly following the late Permian event. Eurypterids also became extinct by the end of the Permian. Crabs, shrimps, and lobsters of modern type date back to at least the Jurassic period.

Bibliography

BARNES, R. D. *Invertebrate Zoology.* Philadelphia, 1987.
GOULD, S. J. *Wonderful Life.* New York, 1989.
MANTON, S. *The Arthropoda: Habits, Functional Morphology, and Evolution.* Oxford, Eng., 1977.
MIKULIC, D. G., ed. *Arthropod Paleobiology.* Knoxville, TN, 1990.
MOORE, R. C., ed. *Treatise on Invertebrate Paleontology, Arthropoda.* Lawrence, KS, 1955–1969. Parts O, P, Q, and R.
ROBISON, R. A., and R. L. KAESLER. "Phylum Arthropoda." In *Fossil Invertebrates,* eds. R. S. Boardman, A. H. Cheetham, and A. J. Rowell. Palo Alto, CA, 1987.
SCHRAM, F. R. *Crustacea.* New York, 1986.

LOREN E. BABCOCK

ASTEROIDS

Asteroids are numerous planetary objects that orbit the Sun primarily between the orbits of Mars and Jupiter. Astronomers in the late eighteenth century determined that the planets seemed to occur at predicted intervals from the Sun, but the planet thought to exist between Mars and Jupiter had not been discovered. The first (and largest) asteroid, Ceres, was discovered in 1801, and soon other asteroids were found in this same area. It became apparent that a belt of small objects, rather than one large planet, existed in this area, probably the result of Jupiter's intense gravitational pull keeping these objects from ever accreting into a planet.

Asteroid diameters range from nearly 1,000 km (Ceres) down to mere boulders and smaller. Most asteroids orbit within the so-called main belt of asteroids, located beyond the orbit of Mars between 2.2 and 3.2 astronomical units from the Sun (1 AU = the Earth's distance from the Sun). Another large population of asteroids, called the Trojans, orbit at Jupiter's distance from the Sun, roughly 60 degrees ahead of and behind Jupiter. A small percentage of asteroids, mainly fragments of main-belt asteroids but also including "dead" (ice-depleted) comets, approach or cross the orbit of Earth. Given their potential for colliding with our planet, these objects are particularly interesting.

Until recently, asteroids have been studied by astronomers, not earth scientists. With telescopes, astronomers could only measure asteroid positions and brightnesses. Positions yield orbits, which help dynamical astronomers understand the physics of orbital changes; clusters of asteroids in similar orbits (called families) are probably pieces of once-whole precursor objects that were smashed and disrupted by inter-asteroidal collisions. Measurements of an asteroid's brightness changes over time reveal its rate of spin and pole orientation, and shed some light on body shape and surface structure. Measurements of an asteroid's brightness at different wavelengths—including reflected sunlight at ultraviolet, visible, and near infrared wavelengths as well as re-emitted (thermal) radiation at longer infrared and radio wavelengths—provide information on body size, the mineralogical surface composition, and physical properties of the near-surface layer.

Origin of Asteroids

It is widely accepted that the planets in our solar system grew from a hierarchy of smaller objects (called planetesimals) that came together in collisions and were held together by the ever-increasing gravity fields of the growing planets. Many planetesimals, ranging in diameters from several kilometers to hundreds of kilometers, remained after the planets had grown to their final sizes, primarily in orbits between (or beyond) the planets. Those in the cold, outer reaches of the solar system retain abundant ices; those that are perturbed and penetrate the inner solar system develop visible tails as the Sun's heat sublimates their ices. Remnant planetesimals of the outer solar system are termed "comets" (see COMETS).

Remaining planetesimals in the inner solar system are evidently lacking in ices (at least surface ices), and fragments launched into orbits closer to the Sun show no atmospheres or tails. These asteroids are warmed sufficiently so that any ices at their surfaces have vanished long ago. Moreover, there is evidence from meteorites (small asteroid fragments that reach Earth's surface) that asteroids were actually heated to much higher temperatures early in the history of the solar system (see METEORITES). Indeed, some—perhaps many—asteroids were heated so much that they melted and differentiated, with dense metal sinking to form cores and buoyant lavas spilling out onto their surfaces. The basalt-covered asteroid Vesta is an example.

Evidence of such ancient processes has been obscured by the dominant process that has affected asteroids ever since: collisions with other asteroids and comets, and cratering by interplanetary debris. At a typical collision speed of 5 km/s, an asteroid just one-tenth the size of another can smash the larger body into fragments. A shattered asteroid may reaccumulate into a "rubble pile." But if the collision is energetic enough, the fragments will disperse, creating an asteroid family and a cloud of smaller debris.

Galileo Flybys of Gaspra and Ida

Until 29 October 1991, when the *Galileo* spacecraft flew past the asteroid 951 Gaspra enroute to Jupiter, scientists had to theorize about how collisional fragmentation and cratering affect the geology and geophysical configurations of asteroids. Pictures of the exceptionally angular Gaspra (it is about 18 × 11 × 9 km in size) reveal a surface surprisingly devoid of large craters; rather, there is an abundance of small, "fresh" craters (i.e., comparatively recent craters, unmodified by subsequent erosion). Indeed, the prevalence of small craters is so great that Gaspra's surface is being effectively sandblasted. Yet its irregular, faceted shape implies that older, large collisions had also shaped its surface. Evidently the latest collision had covered up (or shaken up) much of the pre-existing topography, leaving a nearly clean slate to record the peppering by small impacts during roughly the last 100 million years (Ma).

Slight color differences on Gaspra hinted at some surficial soil layer that has been gradually modified in color, perhaps by a kind of "space-weathering" alteration by micrometeorites and the unshielded radiation environment of interplanetary space. On Gaspra, the soil evidently migrates downhill, leaving ridges slightly blue in color.

The closer photographs of the large asteroid 243 Ida, obtained on 28 August 1993, reveal unique asteroidal processes in sharp detail. Unlike Gaspra, Ida is densely covered with large craters, some exceeding 10 km in diameter. It has apparently been cratered for a billion years or more. Evidently Ida has a deep surface layer of rubbled soil (termed regolith), perhaps 100 to 200 m thick. High-resolution pictures reveal numerous grooves on one part of Ida as well as enormous boulders, up to several hundred meters across (see Plate 1).

Ida is even more oddly shaped than Gaspra, although it is not quite so angular. There is a "neck" that circumscribes about three-quarters of Ida, so it seems to be composed of two major chunks. The most startling discovery of *Galileo*'s flyby of Ida was that the asteroid possesses a small satellite. This egg-shaped moonlet, named Dactyl, is about 1.3

28

km across and has a surprisingly smooth profile (Figure 1). From Dactyl's orbital motion and Kepler's Third Law, scientists have determined that the mass of Ida is less than had been expected; evidently Ida's density is only about 2½ grams per cubic centimeter. Such a density implies not only that there must be little metal within Ida, but probably a large amount of void space. Ida's bifurcated shape and low density seem to imply that it is a rubble pile, with two embedded, coherent pieces.

An especially important discovery about Ida concerns its color. Although Ida's spectrum is generally slightly reddish, some of the most recent craters appear slightly bluish, with comparatively deep spectral absorption bands caused by the minerals orthopyroxene and olivine. The most widespread bluish unit on Ida appears to be ejecta from Azzurra, Ida's most recent large impact crater. It seems as though a space-weathering process, like that inferred for Gaspra, is changing the deep-banded spectrum of pristine minerals on Ida into a redder spectrum with weaker features. An even more prominent space-weathering process operates on the Moon—spectra of lunar soils bear no resemblance to laboratory spectra of moon rocks returned by the Apollo astronauts. But, until the *Galileo* flybys, researchers had doubted that asteroids were subject to space weathering because of the lower velocities of micrometeorites, the greater distance from the Sun, and so on.

Asteroid Mineralogy and Relation to Meteorites

The discovery that surficial processes gradually change asteroidal colors surprises some astronomers, who have been classifying asteroid spectra since the late 1960s. Most asteroids are very dark—low in albedo (percentage of sunlight reflected) with neutral-colored, rather featureless spectra. They resemble laboratory spectra of carbonaceous chondrites, primitive meteorites believed to be the most common rocks in space. Another common group of asteroids, especially prevalent in the in-

Figure 1. Image of Ida and its satellite Dactyl.

ner half of the asteroid belt, are higher in albedo and slightly reddish in color, showing weak silicate absorption features. These so-called S-types were spectral matches for only a very rare type of meteorite—the stony irons.

A problem concerned the common ordinary chondrite meteorites. No main-belt asteroid was found with a reflectance spectrum like these important metamorphosed, but unmelted, meteorites. S-type asteroids were interpreted as the collisionally stripped metallic cores of numerous melted, differentiated precursors, while ordinary chondrite asteroids were assumed to be "hiding" among the population of asteroids that are too faint to be observed, those less than a few kilometers in size. The rarity of the supposed S-type stony irons, contrasted with the abundance of ordinary chondrites from unseen parent asteroids, became difficult to reconcile with the new understanding of how meteorites are derived from the main asteroid belt (collisional fragmentation followed by subsequent chaotic orbital changes that bring fragments into Earth-crossing orbits). Sampling processes should not be so highly selective.

Galileo may have resolved the problem. The S-types are evidently a grab bag of diverse mineral assemblages, including some stony-iron parent bodies, and perhaps also the ordinary chondrites. The kind of space weathering documented on Ida tends to work in the direction that would convert the spectrum of pristine ordinary chondrite material into one with an S-like character. It remains to be proven whether Ida is of ordinary chondritic composition and exactly how the space-weathering process works. More insight may be gained when the Near Earth Asteroid Rendezvous (NEAR) spacecraft, scheduled for launch in early 1996, goes into orbit around another S-type asteroid, 433 Eros, in 1999. NEAR contains, in addition to imaging and spectral instruments, X-ray and gamma-ray sensors that can directly determine chemical abundances.

Earth-Approaching Asteroids

Although less than 10 percent of Earth-approaching asteroids have been discovered, it is believed that nearly two thousand objects larger than 1 km in diameter are in Earth-approaching orbits—that is, asteroids whose orbits bring them close enough that they could one day smash into Earth. Most

Earth-approachers, like NEAR's target 433 Eros—at 20 km long, it is one of the biggest—are simply the largest examples of meteoroids that strike Earth. Collisional fragments of main-belt asteroids, they are converted into elongated, Earth-approaching orbits by the chaotic gravitational influences of Mars, Jupiter, and Saturn. The largest "escape hatch" from the asteroid belt is a so-called Kirkwood gap at 2.5 AU from the Sun, where an asteroid would orbit the Sun exactly three times for every one Jupiter orbit; any asteroid near 2.5 AU is quickly converted into a short-lived, planet-crossing orbit.

Another unknown percentage of Earth-approachers are dead short-period comets; after hundreds of passes near the Sun, they have lost all of their ices and no longer exhibit tails. Radar echoes from near-Earth objects (NEOs) show a range of compositions, ranging from solid metal, to rock, to as-yet-unknown materials. Many of them also have strange shapes, including "contact binary" configurations which suggest that two more nearly spherical objects came together.

NEOs are Earth-approaching asteroids (including dead comets) together with the live comets that penetrate the inner solar system. They play an important role in the earth sciences because they inevitably strike Earth from time to time, as illustrated by the heavily cratered face of the nearby Moon. Earth's atmosphere protects us only from objects smaller than about 100 m in diameter. Comparatively few scars of the larger impacts remain because of rapid erosion and tectonism on our planet. But around 4 billion years ago (4 Ga), when the lunar cratering record reveals there was an exceptionally heavy bombardment, Earth was undoubtedly pummelled by NEOs, including some larger than any known today. Life probably had difficulty gaining a foothold on our planet until the bombardment declined and the last impact large enough to sterilize our planet had happened (*see* IMPACT CRATERING AND BOMBARDMENT RECORD).

Although impacts continue at a slower pace, we are still at risk. There are enough 10-km-sized NEOs that Earth must be hit by one every 100 Ma or so. Indeed, the presence of Chicxulub, the 200-km-diameter fossil crater on the Yucatan Peninsula, and the evidence for a global environmental cataclysm in the paleontological record at the time it formed 65 Ma ago, are a reminder that our fragile ecosphere could be susceptible again (*see* EXTINCTIONS).

Impacts with objects as small as 2 km in diameter, which occur every several hundred thousand years (or one chance in a few thousand during the next century), could produce global climatic effects that might threaten civilization as we know it. The dramatic impact of fragments of Comet Shoemaker-Levy 9 into Jupiter in July 1994 provided ample warning: the larger comet fragments struck Jupiter with energies equivalent to an object of 2 km striking Earth, and the resulting firestorms and subsequent year-long blackened regions of Jupiter's atmosphere were as large as our entire planet.

An international "Spaceguard Survey" has been proposed to find the 90 percent of the NEO population that remains undiscovered and to learn if any particular one happens to be headed our way. A by-product of such a search, besides rich scientific rewards, would be discovery of hundreds of objects that would be easier for an astronaut (or would-be miner) to get to than the Moon. Earth-approaching asteroids may yet become astronauts' stepping stones to more distant worlds, such as Mars.

Bibliography

BINZEL, R. P., T. GEHRELS, and M. S. MATTHEWS, eds. *Asteroids II*. Tucson, AZ, 1989.

CHAPMAN, C. R., and D. MORRISON. *Cosmic Catastrophes*. New York, 1989.

McSWEEN, H. Y., JR. *Meteorites and Their Parent Planets*. Cambridge, Eng., 1987.

CLARK R. CHAPMAN

ASTRONAUTS

See Apollo Astronauts

ATMOSPHERES, PLANETARY

All nine planets, four moons, and most comets have atmospheres. These gaseous envelopes are remarkably diverse, although they are made largely from the same six elements—hydrogen (H), helium (He), oxygen (O), carbon (C), nitrogen (N), and sulfur (S). These are the most abundant elements in the Sun and stars. And with the exception of helium, they are the main elements in living things.

Atmospheres differ in composition, mass, temperature, and degree of interaction with the underlying planet. At one extreme, the atmospheres of the four giant planets merge imperceptibly with the planetary interiors, which are made of the same gaseous mixtures as the atmospheres. Pressures in the interiors of these fluid planets reach millions of bars—millions of times the sea level pressure on Earth. At the other extreme, the atmospheres of the Moon and Mercury consist of individual molecules hopping around the surface, never colliding with each other because the density is too low. The intermediate cases are Venus, Earth, Mars, Pluto, Io (a satellite of Jupiter), Titan (a satellite of Saturn), and Triton (a satellite of Neptune). These atmospheres rest on solid or liquid surfaces and interact chemically with the solid planet on a variety of timescales.

This article is organized around two main themes—chemistry and dynamics. The former involves composition and how it has evolved. The latter involves weather and climate. Table 1 gives the general properties of atmospheres around solar system objects.

Chemical Composition and Evolution

Hydrogen and helium are the most abundant elements in the giant planet atmospheres. Chemically reactive elements like oxygen, carbon, and nitrogen exist mainly in combination with hydrogen in compounds such as water (H_2O), methane (CH_4), and ammonia (NH_3). The latter three compounds are gases in the warm interiors of the giant planets, but they condense in the cooler atmospheres. Temperatures on Jupiter and Saturn are too high to allow methane condensation, and the visible clouds are made of ammonia. On Uranus and Neptune the high-altitude clouds are methane ice; the deeper clouds are ammonia and possibly hydrogen sulfide. The giant planets resemble the Sun in overall elemental abundances, but with varying degrees of enrichment of heavy elements relative to hydrogen and helium (*see* PLANETOLOGY, COMPARATIVE).

Carbon dioxide is the most abundant constituent in the atmospheres of Venus and Mars, and

Table 1. General Properties of the Atmospheres around Solar System Objects

Classes of Atmospheres	Solar System Object	Composition—Main Components	P, Tᵃ at Base or Visible Cloud and T in Thermosphere	Nature of Upper Cloud Deck	Typical Flow Speed	Typical Temperature Contrasts
H_2-He atmospheres of the giant planets	Jupiter	H_2, He, CH_4, NH_3, . . .	0.5 bar, 150 K, 1000 K	NH_3	100 m/s	5 K
	Saturn	H_2, He, CH_4, NH_3, . . .	1 bar, 160 K, 400 K	NH_3	400 m/s	5 K
	Uranus	H_2, He, CH_4^b, . . .	1 bar, 80 K, 870 K	CH_4	200 m/s	2 K
	Neptune	H_2, He, CH_4^b, . . .	1 bar, 80 K, 600 K	CH_4	400 m/s	2 K
Terrestrial CO_2 atmospheres	Venus	CO_2, N_2, SO_2^b, $H_2SO_4^b$, . . .	90 bar, 730 K, 200 K	H_2SO_4	100 m/s	5 K
N_2 atmospheres	Mars	CO_2^b N_2, H_2O^b, . . .	7 mbar, 200 K, 400 K	H_2O	40 m/s	40 K
	Titan	N_2^b, CH_4^b, . . .	1.5 bar, 90 K, 180 K	CH_4	?	5 K
	Earth	N_2, O_2, Ar, H_2O^b, . . .	1 bar, 280 K, 1000 K	H_2O	20 m/s	40 K
	Pluto	N_2^b, CO^b, CH_4^b, . . .	3–600 μbar?, 35–45 K?, ?	N_2	?	<20 K
	Triton	N_2^b, CO^b, CH_4^b, . . .	14 μbar, 38 K, 100 K	N_2	?	?
Volcanic	Io	SO_2^b	0.1–10 nbar,ᶜ 120 K, 600 K	SO_2	200 m/s	?
Exospheres	Moon	Na, K, . . .	?	—	?	?
	Mercury	Na, K, . . .	?	—	?	—
Unknown	Charon	?	—	?	?	?
Comae	Comets	H_2O, CH_4, CO, CO_2, CH_3OH	—	—	—	—
	Chiron	CO?, ?	?	?	?	?

ᵃ P, pressure; T, temperature.
ᵇ A condensing, time-variable constituent.
ᶜ Time variable.
From Table 4.2 of *An Integrated Strategy for the Planetary Sciences: 1995–2010* (1994).

might have been the most abundant constituent in the atmosphere of Earth. On our planet, carbon dioxide from the atmosphere has reacted with the rocks to form calcium carbonate—limestone. This reaction takes place in solution, and therefore requires liquid water. It is estimated that Earth's limestone deposits are equivalent to an amount of carbon dioxide that is 40–100 times the mass of Earth's current atmosphere, including all constituents. Without liquid water, this CO_2 would reside in the atmosphere. Venus, which lacks oceans and liquid water, has a dense CO_2 atmosphere whose surface pressure is ninety times that on Earth. Thus Earth and Venus have roughly the same total inventory of CO_2. The CO_2 inventory on Mars is largely unknown, since large amounts of CO_2 might lie buried in polar ice deposits and carbonate rocks. The atmospheric pressure of Mars is only 0.7 percent that of Earth, and is controlled by equilibrium with frozen CO_2 at the poles.

Molecular nitrogen (N_2) is the most abundant constituent in the atmospheres of Earth, Titan, Triton, and Pluto. As with CO_2 on Mars, the N_2 in the atmosphere on Triton and Pluto is in equilibrium with N_2 frost on the ground. As the frost heats up or cools off in response to seasonal changes, the atmosphere expands or contracts. The surface pressure of Mars varies by 30 percent during the martian year. The thinner atmospheres of Triton and Pluto vary by orders of magnitude. This control by N_2 frost is only possible at the low temperatures of the outer solar system.

Oxygen (O_2) is the second most abundant constituent in Earth's atmosphere, constituting 20 percent of the molecules in dry air. Oxygen is produced during photosynthesis from CO_2 and H_2O, and is consumed during respiration and decay. Without life, our planet would have little or no O_2 in its atmosphere. Ozone (O_3), atomic oxygen (O), and O_2 are trace constituents in the atmospheres of Mars and Venus, where they are produced by photodissociation—the destruction of CO_2 molecules by sunlight.

Water is not the major constituent of any planetary atmosphere, although it is the most abundant volatile compound on Earth. The ocean is 300 times as massive as the atmosphere, but Earth temperatures are too low to vaporize more than 0.002 percent of it. Venus contains almost no water, although traces of hydrogen are found in the clouds, which are concentrated solutions of sulfuric acid (H_2SO_4). The hydrogen on Venus tells an interesting story. The proportion of deuterium, the heavy isotope of hydrogen, is 100 times greater than on Earth. Such enrichment comes about when the lighter isotope of hydrogen escapes, suggesting that Venus lost massive amounts of water. Instead of oceans, Venus may have formed with a massive water vapor atmosphere maintained by its own "runaway greenhouse" and greater energy from the Sun. Once up in the atmosphere, water was destroyed by photodissociation, and the lighter hydrogen was lost to space. These processes occurred in the inner solar system, where temperatures are high. On Mars, water is mostly frozen in polar and subsurface ice. On the surfaces of the satellites of the giant planets, water is as hard as rock.

Hydrocarbons in Titan's atmosphere also tell an interesting story. The satellite is not massive enough to hold onto hydrogen over the age of the solar system, so photodissociation of hydrogen-bearing compounds like CH_4 and NH_3 leads to hydrogen escape. The remaining carbon atoms bind to themselves to form compounds like ethane (C_2H_6), acetylene (C_2H_2), and higher hydrocarbons. The result is a photochemical smog that hides the surface and may have condensed out in sufficient quantities to form lakes or oceans.

Sulfur dioxide (SO_2) is the dominant constituent in the atmosphere of Io. The surface of the satellite is cold enough to freeze out all but trace amounts of the gas, especially on the nightside. But a detectable atmosphere is present above the volcanic vents and above the warm patches of frost on the dayside. Maximum surface pressures never exceed 10^{-7} Earth atmospheres, and the pressure away from the vents is several orders of magnitude less. The winds away from the vents rise to supersonic speeds as the thin atmosphere flows out over the cold surface and disappears into a vacuum. Io's atmosphere is patchy.

Sodium and potassium atoms have been detected above the surfaces of Earth, Io, Mercury, and the Moon. These volatile elements scatter sunlight effectively, and so are easy to detect. On the Moon and Mercury, the density is so low that the molecules collide only with the surface and not with each other. On Earth, the collisionless region where the atmosphere merges with the vacuum of space is termed the "exosphere." On Mercury and the Moon, the exosphere extends down to the surface. Sodium and potassium atoms are released from surface rocks through impact by high-energy particles in a process called sputtering. Both Mercury and the Moon may have ice deposits in the permanently shaded craters at the north and south poles. With no sunlight and no atmosphere to transport heat from the hot equator, these regions are as cold as Triton and Pluto. The ice comes from comets that crash onto the surface: Water molecules from the comet hop randomly around until they are destroyed by sunlight or find a safe haven in one of the permanently shaded cold traps.

The noble gases—helium, neon (Ne), argon (Ar), krypton (Kr), and xenon (Xe)—are rare on the terrestrial planets but are abundant on the Sun and the giant planets. The later objects incorporated both gas and dust from the solar nebula—the cloud of material out of which the solar system formed. The noble gases were incorporated along with hydrogen. Objects in the inner solar system incorporated only dust. Their atmospheres come from outgassing—release of material that was chemically bound to the dust—and from comets, which deliver ice from the outer solar system.

Weather and Climate

Atmospheres move because they are heated, and because the heat sources are in different locations than the heat sinks. Sunlight is the main energy source for most planetary atmospheres. The Sun heats only the dayside and heats the equator more than the poles. Internal heat is comparable to sunlight for the giant planets and Io. Convection currents from the warm tropics and the warm interior carry heat poleward and upward. The energy reaches altitudes where the atmosphere is transparent, at which point the energy is radiated to space. On most planets that altitude is near the 100 mbar level—the altitude where the pressure is 0.1 times Earth's sea-level pressure.

Figure 1 shows the mean temperature profiles for the major planet atmospheres. Temperature falls off rapidly with altitude in the troposphere and less rapidly in the stratosphere. The boundary between the two regions, the tropopause, is near 100 mbar for all planets except Mars and Pluto, whose atmospheres end at the 6 mbar and 3 microbar levels, respectively.

Convection is usually a small-scale process consisting of cool downdrafts and warm updrafts that carry heat from where the sunlight is absorbed and the internal heat is stored to where energy is radiated to space. Shimmering of air over the desert,

cumulus clouds, dust devils, and thunderstorms are all examples of small-scale convection. Convection can be either moist or dry. In moist convection, the condensable vapor rises, releases its latent heat, and then falls out as rain or snow. Tornadoes and hurricanes are intense forms of moist convection on small and intermediate scales, respectively.

Large-scale motions transport heat laterally. On Earth the Hadley circulation—the slow overturning of the tropical troposphere—transports heat from the equator out to 30° latitude. Extratropical cyclones and anticyclones—rotating masses of air thousands of kilometers across—transport heat beyond 30° to the polar regions. These huge vortices develop from waves on the jet stream, which is an eastward-moving current of air at midlatitude in each hemisphere.

These same weather elements are present in other planetary atmospheres, but there are differences. Wind speeds are generally larger than on Earth, even in the outer solar system where sunlight is hundreds of times weaker. Storms last much longer: The Great Red Spot on Jupiter was discovered three hundred years ago shortly after the invention of the telescope. It is a huge vortex whose spiraling winds bring up compounds that form the red clouds. The cloud patterns and wind directions vary widely. Venus, Titan, and the giant planets have banded, axisymmetric features— cloud patterns and jet streams that circle the planet on lines of constant latitude. Io, Triton, and Pluto have winds that always blow toward the dark side. Earth and Mars are intermediate cases in which axisymmetric features—Hadley cells and jet streams—coexist with large-scale cyclones and waves.

Paradoxically, wind speeds do not decrease as one moves out in the solar system. Although the power per unit area emitted by Neptune is only one twentieth that at Jupiter and one four hundredth that at Earth, Neptune's large-scale wind speeds reach 400 m/s, while Jupiter's reach 160 m/s and Earth's only reach 50 m/s. The explanation may lie in the role of small-scale convective motions, which provide dissipation for the large-scale flow. As one moves outward from the Sun in the solar system, the small-scale motions decrease, allowing the large-scale motions to build up. At Neptune these large-scale motions are coasting along with almost no dissipation. On Earth, where small-scale convection is more vigorous, the large-scale winds are weaker. Low dissipation may also

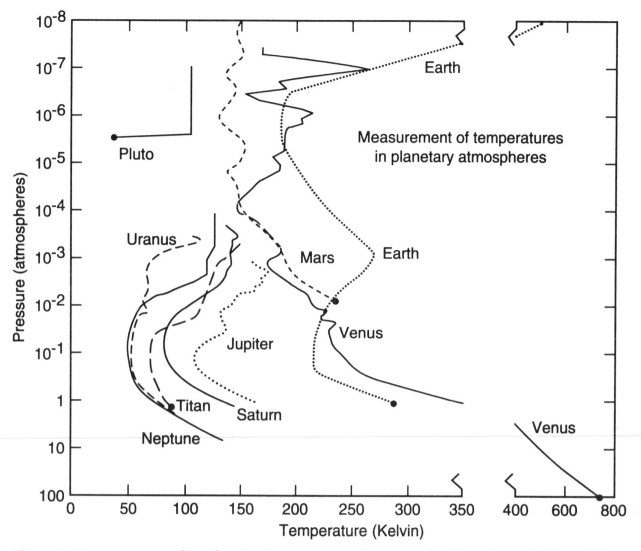

Figure 1. Temperature profiles of major planetary atmospheres. For the giant planets, the base of the profile is the deepest point sampled, so the base pressure has no special significance. For the other planets and Titan, the base pressure is the surface pressure and is indicated by a solid circle. Reproduced from Figure 4.4 of *An Integrated Strategy for the Planetary Sciences: 1995–2010.*

explain the longevity of the large vortices in the giant planet atmospheres. Numerical models show that these vortices arise spontaneously and are stable at the interfaces between the eastward- and westward-flowing jet streams, provided the dissipation is low.

Venus presents a number of questions: Why is Venus so dry? Why is the surface so hot? What makes the atmosphere rotate fifty times faster than the solid planet? As discussed earlier, Venus may have lost an amount of water equal to Earth's

oceans. The high surface temperatures reflect the greenhouse effect of the massive CO_2 atmosphere: Enough sunlight diffuses down to the surface to provide an energy source, but the heat cannot readily escape because the infrared bands are blocked by atmospheric gases. The superrotation of the atmosphere is strange because friction tends to make all layers spin with the solid planet. The explanation involves waves, which can propagate vertically in planetary atmospheres. The waves carry momentum and energy, and could balance the effects of friction. Thus far, however, the waves responsible for the superrotation of Venus have not been identified.

Mars offers interesting examples of climate change. Although the surface now is cold and dry, flowing water carved enormous channels early in the planet's history. The questions are: What caused the climate to change so drastically? And, is the water now frozen in the substrate and at the poles? Layered deposits in the polar regions may provide the answers, but so far the layers have been observed only from orbiting spacecraft and not from the ground. Because Mars is so dry, dust builds up in the atmosphere and creates enormous changes in the weather. The global dust storms are unpredictable and seem to occur every few years. High wind speeds are required, but the mechanism that triggers a global dust storm is unknown.

Earth's climate has also changed drastically over geologic time. The early Earth was warm, despite less output from the early Sun. The present ice sheets in Greenland and Antarctica only appeared a few million years ago. Every hundred thousand years the ice sheets advanced down to about the latitude of New York City and then retreated abruptly. As on Mars, changes in Earth's spin in relation to its orbit may account at least partly for the ice ages—the advances and retreats of the glaciers. The atmosphere was dustier during the cold dry periods. As on Venus, changes in the amount of carbon dioxide may account for the high temperatures of the early Earth. The mystery is how the rather small changes in the orbit could have caused the ice ages, and why the CO_2 should have changed at all. Studying planets does not provide all the answers, but it does reveal some of the extreme possibilities.

Bibliography

ATREYA, S. K., J. B. POLLACK, and M. S. MATTHEWS, eds. *Origin and Evolution of Planetary and Satellite Atmospheres.* Tucson, AZ, 1989.

BEATTY, J. K., and A. CHAIKIN, eds. *The New Solar System,* 3d ed. New York, 1990.

CHAMBERLAIN, J. W., and D. M. HUNTEN. *Theory of Planetary Atmospheres.* Orlando, FL, 1987.

COMMITTEE ON PLANETARY AND LUNAR EXPLORATION 1994. *An Integrated Strategy for the Planetary Sciences: 1995–2010.* Space Studies Board, National Research Council. Washington, DC, 1994.

ANDREW INGERSOLL

AURORAS

At their simplest, auroras appear to be hanging curtains of light in an otherwise dark nighttime sky. The light is created by the discharge of electricity in Earth's upper atmosphere. The electricity is generated as an electromagnetic storm from the Sun as it sweeps past Earth.

In general, auroras occur in narrow oval belts displaced away from, but surrounding, Earth's magnetic poles. Nearly identical auroras occur simultaneously in the Northern and Southern Hemispheres—the Aurora Borealis in the north and the Aurora Australis in the south. Satellite-borne electronic images show Earth's auroras occurring all the time.

The vertical striations that appear in the hanging curtains of light are similar to the streaks of light that often precede sunrise. In 1621 the French scientist Pierre Gassendi was the first to call them "aurora" after the Greek goddess Aurora, who, according to myth, was responsible for the "rosy fingers of dawn." The striations result from the alignment of electrical discharges on the normally invisible lines of Earth's magnetic field. Light-producing electrical discharges start occurring around 250 kilometers above Earth and continue growing brighter until they are extinguished near 115 kilometers. On rare occasions auroras extend thousands of kilometers high. More rarely the sky will be partially covered with a blood red glow that is a high-altitude, low-energy aurora.

The colors of auroras are indicative of the types of elemental gases in Earth's atmosphere where the electrical discharges are taking place. At the altitudes of the auroras, nitrogen and oxygen exist in both their atomic and molecular forms. Electrical discharges in each create a unique discontinuous spread of colors. Electrical discharges in atomic oxygen produce the greenish-white and deep-red colors, while those in singly ionized molecular nitrogen produce blue colors.

A low-level flow of energy from the Sun creates a low level of continuous aurora that is confined to regions near Earth's magnetic poles. Observations of the Sun have revealed that, in general, a high level of aurora activity follows an approximate eleven-year cycle. Specific high-energy storms that create dramatic displays are unpredictable. The flow of energy from a typical solar storm takes approximately forty-eight hours to reach Earth.

As the electrically charged solar storm sweeps past the slowly rotating Earth with its magnetic field, a kind of electrical motor-generator action takes place. Energy from the solar storm is transferred to and further accelerates electrons and protons in Earth's outer atmosphere that travel along Earth's magnetic field lines toward Earth. The greater the storm the more it will force the oval belts to expand further from their respective magnetic poles, thereby creating auroras visible from more equatorial latitudes.

Auroras occur on all of the planets in our solar system that have both a magnetic field and an atmosphere. In a sense, aurora is the name we give to the solar-prompted, organized electrical discharge near the magnetic poles of these planets. Instruments aboard the *Voyagers I* and *II* spacecraft, and more recently on the Hubble Space Telescope, have been used to acquire electronic images of auroras in the atmospheres of Jupiter, Saturn, Uranus, and Neptune.

On an international scale scientists have recently coined the term "space weather" for their interest in how solar storms affect spacecraft within our solar system, and the impact of these auroras on Earth and on our technological societies. Specific problems include the failure of computer models to predict adequately the effects of different solar storms on individual spacecraft, or the changes induced in the circulation and chemistry of Earth's upper atmosphere by energetic auroras.

Bibliography

AKASOFU, S. I. "Aurora Borealis, the Amazing Northern Lights." *Alaska Geographic* 6 (1979).
———. "The Dynamic Aurora." *Scientific American* 260 (1989): 90–97.
EATHER, R. H. "Majestic Lights, the Aurora in Science, History, and the Arts." American Geophysical Union, Washington, DC, 1980.

NEAL B. BROWN

B

BAGNOLD, RALPH ALGER

Ralph Alger Bagnold was born at Stoke, Devonport, England, on 3 April 1896 to Ethel Alger Bagnold and Major (later Colonel) Arthur Henry Bagnold, Royal Engineers, and died on 28 May 1990 at age ninety-four. His sister Enid, six and a half years older, became a playwright and author, best known for her play *The Chalk Garden*, and for her children's classic, *National Velvet*.

Soldier, scientist, and engineer, Ralph Bagnold was another in a long tradition of British army officers who made significant contributions to areas of knowledge and to activities other than those directly involved with their military duties. He was an explorer—not only geographically but intellectually—who confessed to his close friends, early in his military career, that he would rather be a Fellow of the Royal Society than a major general. In his autobiography (Bagnold, 1990), completed shortly before his death, he stated: "My main urge, from boyhood onward, was curiosity. . . . Along with this, there has been a vague longing to discover something new."

Ralph Bagnold had an outstanding military career, serving in World War I with the Royal Engineers in the trenches of Flanders and France and in World War II, organizing and leading the Long Range Desert Group in North Africa in its behind-the-lines surveillance and destruction of Axis

troops and supplies. In scientific and engineering circles, he is renowned for his field and experimental studies of the transport of particulate material by wind and by water.

From an early age he showed great interest in learning to do things with his hands, and he received strong encouragement in such activities from his father, who as an experienced officer in the Royal Corps of Engineers understood the importance of the many skills and techniques involved in their broad spectrum of maintenance and construction activities. In 1914, upon graduation from Malvern College, Bagnold took the entrance examination for the army, ranking fourth among several hundred applicants. Following family tradition, he stated his preference for the Royal Engineers of the Regular Army.

England declared war on Germany on 4 August 1914, following which the course at the Royal Military Academy and at the Military Engineering School each was shortened from years to months. Upon completion of these in 1915, at age nineteen, Bagnold was posted to the British Expeditionary Force in France and began what eventually would be twenty-three years of active duty with the British army, first with the Royal Engineers, then with the Signals Branch of the engineers, and after 1923 with the Royal Signals Corps. For the first three years of his military service, with only occasional respite, Bagnold was involved in the deadly trench warfare of World War I.

Late in the war he was posted to the 19th Corps Signals, commanded by Lieutenant Colonel J. F. M. Stratton, who in civilian life was a professor of astronomy at Cambridge. When the war ended in November 1918, Stratton encouraged Bagnold to take advantage of the army's two-year educational leave program by attending Cambridge University. He graduated with an honor's degree in engineering in 1921, having enjoyed some of the relaxed, carefree life that he had missed while in the trenches in France.

Following graduation, Bagnold had a series of assignments in Ireland (1921–1923); Signals Training Center, Sussex (1923–1926); Egypt (1926–1928); India (1928–1931); School of Signals, Catterick (1931–1933); and Hong Kong (1933–1934). Throughout these years he took every opportunity to travel to unusual places. He especially enjoyed climbing, and on one occasion climbed both Monte Rosa and the Matterhorn in the European Alps with only one day's rest between. Ill health forced his return to England in 1934 and retirement from the army in 1935.

It was during his stay in Egypt that Bagnold and a few of his associates felt the lure of the desert. During a series of increasingly longer vacation and spare-time automobile excursions into the arid wastelands of the Middle East and North Africa— using first Model-T and later Model-A Fords—he developed techniques and equipment, such as the sun compass and the closed vehicle cooling system, for long-distance, self-sustained travel. It was this insight to desert travel that would serve the Allies well in World War II.

During his stay in Hong Kong, Bagnold wrote *Libyan Sands: Travel in a Dead World* (1935). It was a record of the desert expeditions that he and his associates had made to that time and documented the challenge and fascination of traveling in places so dry and barren that other humans may not have intruded for hundreds of years.

Having regained his health, Bagnold was recalled to active military service when another world war erupted in August 1939; he was posted to East Africa. A collision at sea resulted in his ship putting in to Port Said for repairs, affording Bagnold an opportunity to visit Cairo and friends stationed there. His presence came to the attention of General Sir Archibald Wavell who reassigned him to the signals section of an armored division. A year later, it was Wavell who responded to a proposal by Bagnold to field small, self-contained mobile units that could operate behind enemy lines in the desert to the west to report on movement of troops and supplies and to attack small, isolated outposts, forts, and airfields. Bagnold was given carte blanche to organize in six weeks what became known as the Long Range Desert Group (LRDG) that, using the element of surprise and hit-and-run tactics, played a major role in delaying the Italian offensive against Egypt, allowing Wavell valuable time to receive reinforcements of troops and supplies. In July 1941, Bagnold turned over field command of the LRDG to his successor and as a full colonel was assigned to headquarters in Cairo. Subsequently, he was reassigned and became deputy signals officer in chief and promoted to brigadier. He received the Order of the British Empire for his military service in World War II.

An assessment of the work of the LRDG was provided by Wavell in his dispatch of October 1941: "the group has penetrated into nearly every part of the desert Libya. . . . they have brought back much information, they have attacked enemy forts, captured personnel [and] transport and grounded aircraft as far as 800 miles inside hostile territory. They have protected Egypt and the Sudan from any possibility of raids, and have caused the enemy, in lively apprehension of their activities, to tie up considerable forces in the defense of distant outposts."

He has been described by those who worked with him during this period (Lloyd-Owen, 1981) as tough, wiry, reticent and undemonstrative, with frugal habits, inexhaustible stamina, and a natural modesty. He was a pefectionist and set for himself and others high standards of physical fitness, attention to detail, and sense of duty.

Just as Bagnold's extensive experience in desert travel resulted in the formation of the LRDG in time of international crisis, so did his familiarity with deserts lead to an abiding interest in the movement of sand by wind. Always keenly observant and extremely curious, he became fascinated by the repetitive pattern and the mobility of eolian features, from systems of small ripples to giant dunes. He also was aware that the process by which wind moved and shaped particulate material was not well understood, and he set about, in the years following his medical retirement from the army in 1935, to study the process in the laboratory and in the field, often using such devices as wind tunnels and manometers of his own design and construction. These studies culminated in 1941 in the publication of *The Physics of Blown Sand and Desert Dunes*, the first detailed treatise on the subject and still a classic in the field. Just before and after World War II, he also investigated the role of water in transporting particulate material.

In 1947, he was invited to become the director of research for Shell Refining and Marketing Company in the United Kingdom, but after two years he resigned to return to the independence and pace of his own research laboratory. In the 1950s, 1960s, and 1970s he was associated with the U.S. Geological Survey and his close friend, survey geologist Luna Leopold, in the study of water-transported solids. Several important papers resulted. He also traveled widely as his advice was sought on problems related to blown sand.

In 1946, Ralph Bagnold married Dorothy Plank, a long-time friend of his sister, Enid. They had two children, Stephen (born in 1947) and Jane (born in 1948) and five grandchildren. Mrs. Bagnold, "Plankie," died in 1986.

Numerous scientific and engineering awards and honors and honorary degrees were conferred on Bagnold: in addition to Fellowship in the Royal Society (1944), he was awarded the Gold Medal of the Royal Geographical Society (1934), the first G. K. Warren Prize of the U.S. National Academy of Sciences (1969), the Penrose Medal of the Geological Society of America (1970), the Wollaston Medal of the Geological Society of London (1971), the Sorby Medal of the International Society of Sedimentologists (1978), and the David Linton Award of the British Geomorphological Research Group (1981). Honorary degrees or fellowships were conferred by Imperial College, London (1971), the University of East Anglia (1972), and the Danish University of Aarhjus (1981).

Bibliography

BAGNOLD, R. A. *Libyan Sands: Travel in a Dead World.* London, 1935; republ. with epilogue, London, 1987.
————. *Sand, Wind and War: Memoirs of a Desert Explorer.* Tucson, AZ, 1990.
————. *The Physics of Blown Sand and Desert Dunes.* New York, 1941; Repr. New York, 1971.
KENN, M. J. "Ralph Alger Bagnold." *Biographical Memoirs of Fellows of the Royal Society* 37 (1991):55–68.
LLOYD-OWEN, D. L. *Providence Their Guide: A Personal Account of the Long Range Desert Group, 1940–1945.* Nashville, TN, 1981.

JAMES R. UNDERWOOD, JR.

BIG BANG

See Cosmology, Big Bang, and the Standard Model

BIOEROSION

The term "bioerosion" relates to biological destruction of hard substrates, and particularly to the boring, scraping, and etching activity of animals and plants. The resultant structures are trace fossils and are classified and named as such (*see* TRACE FOSSILS). As a group, however, trace fossils in hard substrates are sufficiently different from those in soft sediments (burrows) to deserve separate treatment.

Bioerosion is widespread in marine environments, and most of the following remarks refer to the seafloor. However, the process also occurs in brackish and fresh waters as well as in the terrestrial realm, but it is mainly restricted there to the work of microorganisms.

Substrates that are subjected to bioerosion fall into three categories: (1) carbonate rocks, skeletons (shells) of invertebrates, and red-algal crusts; (2) noncarbonate rock; and (3) plant material (wood and leaves). Correspondingly, the different properties of these materials determine the erosional means by which the organism attacks the substrate.

Two modes of bioerosion are recognized: chemical and mechanical. Chemical bioeroders are adapted to penetrate carbonate substrates. Few use acids; the majority secrete calcium-complexing chelating reagents. Many mechanical bioeroders employ tools that act as chisels, or grind and abrade; these animals are found in both carbonate and noncarbonate substrates. Others attack the substrate with sawing and biting actions, suitable for dealing with plant material.

The bioeroding organisms are very diverse. As it is costly in terms of energy to penetrate hard substances, most of the organisms that do so are highly specialized for the job. Also, the borings produced represent growth of static organisms as opposed to movement through the sediment as expressed by most soft-substrate trace fossils. Thus, in contrast to "soft" trace fossils, the morphologies of those in hard substrates to some degree allow the identification of their makers. In some cases this is possible at species level, but in the most simple borings it is doubtful even at phylum level. Individual types of "hard" trace fossils therefore receive trace fossil names, just as the "soft" ones do.

From the Precambrian to today, microbes have produced microscopic borings in carbonate grains and rocks. In the photic zone, cyanobacteria and green and red algae are ubiquitous. These primary producers are grazed upon by echinoids and mollusks, and from the Mesozoic to today they have left characteristic scratch patterns on the substrate surface (Figure 1). In deep water, beyond light penetration, bacteria and especially fungi are carbonate bioeroders.

Trypanites is the ichnogenus applied to simple single-apertured cylindrical borings that first ap-

Figure 2. Polished section of a stromatoporoid skeleton containing many cylindrical *Trypanites* borings. Cross sections (circular), oblique (elliptical) and longitudinal sections are seen to cut the laminar, dark host skeletal material. Devonian, Alberta. ×5.

peared in carbonate substrates in the Cambrian (Figure 2). Their tracemakers are not readily identified; several "worm" groups produce them today. Flask-shaped borings, Gastrochaenolites (Figure 3), on the other hand, are attributable to boring bivalves. They appear in the late Paleozoic and become increasingly important until today. Both mechanically- and chemically-boring species occur.

The name Entobia is used for complex borings consisting of ramifying passages, usually inflated as small, round chambers, open to the substrate surface by many pores (Figure 4). These are the work of boring sponges, making their appearance near the beginning of the Mesozoic. Entobia is restricted to carbonate substrates, and sponge bioerosion is an important ecologic factor, especially in coral reefs from Cretaceous to the present age.

Figure 1. Grazing traces produced on a shell surface by a regular echinoid. The five teeth of the echinoid incise five scratches arranged as a star at each bite. Repeated bites as the animal moves over the surface produce a characteristic pattern. Recent, Greece. ×10.

Figure 3. Gastrochaenolites: two borings of bivalves in dark limestone. The surface from which boring commenced is at left. The shells of the borer bivalves are preserved in the pale filling sediment. The net-like pattern of small borings at upper left, filled with the same pale sediment, is Entobia, bored by a sponge. Pliocene, Greece. ×3.

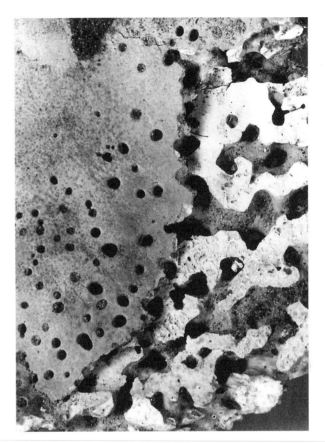

Figure 4. Entobia: the boring of a sponge in a bivalve shell. At left the shell surface is preserved, showing the numerous round apertures to the boring. Where the surface has broken away (right) the branching internal chamber system of the boring is exposed. Eocene, Greece. ×3.
All photos courtesy of Richard G. Bromley

Bioerosion became an important process during the "Mesozoic revolution" and played only a minor role in the Paleozoic. It is a useful tool in environmental interpretation. The presence of bioerosion in a shell or hardground surface demonstrates that the surface has been in contact with seawater for a period of time. Advanced bioerosion takes years to develop, and the process is stopped instantly by even slight burial under sediment (Bromley, 1992).

The photic zone is easily identified by the quality of the bioerosion. Algal borings are distinguishable from the work of nonphotic microbes, and the scratches left by animals grazing on the algae are characteristic. In contrast, deep-water bioerosion is characterized by low diversity, the absence of pho-tic elements, and the presence of certain diagnostic ichnospecies of Entobia.

Bibliography

BROMLEY, R. G. "Borings as Trace Fossils and *Entobia cretacea* Portlock, as an Example." In *Trace Fossils*, eds. T. P. Crimes and J. C. Harper. Liverpool, Eng., 1970.
———. "Bioerosion: Eating Rocks for Fun and Profit." In *Trace Fossils: Short Courses in Paleontology 5,* eds. C. G. Maples and R. R. West. Knoxville, TN, 1992.
WARME, J. "Borings as Trace Fossils, and the Processes of Marine Bioerosion." In *The Study of Trace Fossils,* ed. R. W. Frey. New York, 1975.

RICHARD G. BROMLEY

BIOSPHERE

The biosphere is a dynamic life-supporting envelope peculiar to Earth, comprising those portions of the lithosphere, hydrosphere, and atmosphere that are permeated by living organisms. It is the sum total of all of the ecosystems on Earth and is therefore the most encompassing unit in the ecological organization of life. This entry will examine the nature of the biosphere by considering its five most important properties: physical structure, biogeochemical interactions and feedback, hierarchical organization, diversity, and evolutionary history.

Understanding the biosphere is of increasing concern as human activities accelerate the rate of change in most, if not all, of its components. Changes in one portion of the biosphere may have unexpected but profound influences upon another portion. Within the life time of the individual, changes in the biosphere appear to be gradual. Sudden or catastrophic changes, such as floods, wildfires, or volcanic eruptions, have devastated portions of the biosphere on a local scale. In view of the scale of effects of human activities, it is increasingly important to monitor the biosphere globally and to use modeling procedures to construct and test models that are globally predictive.

Physical Structure of the Biosphere

One of the most basic properties of the biosphere is its layered physical structure. The layered architecture of the biosphere is a reflection of the layered structure of the geosphere and each of its three elements: hydrosphere, lithosphere, and atmosphere.

The hydrosphere includes the world ocean system, rivers, and lakes as its most obvious components (see EARTH'S FRESHWATERS and EARTH'S HYDROSPHERE). It also includes groundwater, snow, ice, and glaciers (see GROUNDWATER and EARTH'S GLACIERS AND FROZEN WATERS). The biosphere penetrates all of these components. It is most diverse and ubiquitous in oceans, where it extends throughout the water column, from the surface layers that cover approximately 71 percent of Earth, to more than 10,000 m in the deepest ocean trenches. It is distributed from the margins of ocean basins where they impinge upon the continents, to the vast reaches of ocean most distant from landmasses and any terrestrial source of nutrients, and from icy polar waters to warm tropical seas.

There are many physical and chemical factors that limit the distributions of organisms in the hydrosphere, and it is therefore especially interesting to observe the fringes of the biosphere and the kinds of organisms that are able to live under extreme conditions (see OCEANOGRAPHY, BIOLOGICAL). In regions of the deep sea where there is no light and therefore no primary productivity, there are species that are adapted to colonize and feed upon "food falls" as specific as whale carcasses or waterlogged wood. At other sites on the deep ocean floor, where water would normally be close to freezing, water as hot as 350°C rushes out of vents along with metal sulfides. These vents support rich communities of sulfur-oxidizing bacteria and invertebrates that have bacterial symbionts and metabolism based on chemical energy (chemosynthesis) rather than solar energy (photosynthesis).

The greatest productivity and biomass in the hydrosphere is concentrated in the water column by communities of organisms that passively float or actively swim through the medium (plankton and nekton). A second set of communities in the hydrosphere occupies the interface between the lithosphere and hydrosphere (benthos), supported by solid rock or sediment but carrying out their vital functions in water. A third set of communities occupies the interface between hydrosphere and atmosphere and includes organisms that are uniquely adapted to live in or on the surface tension at the water/air interface (neuston).

The atmosphere is a layered gaseous envelope extending from the surface of Earth to approximately 30,000 km, the point at which Earth's gravitational forces are no longer able to hold moving gas molecules in orbit (see EARTH'S ATMOSPHERE). It is only the innermost layer, or trophosphere, that is penetrated by large numbers of living organisms, and it is only this layer in which turbulent convection dominates and moisture is present in significant quantity. Energy is required to become airborne and to remain aloft; and organisms that spend substantial amounts of time in this envelope are either so small that they can be maintained passively in suspension by air movements (the aerial plankton) or have adaptations that actively keep them aloft. Aside from a few taxa that have evolved powered flight (insects, birds, bats), biological invasion of the atmosphere is primarily by aerial plankton or by reproductive structures such as pollen grains, seeds, and spores. The atmosphere is primarily a medium of dispersal rather than a preferred habitat, and for many minute organisms and propagules the problem of getting down may be more significant than that of getting up. No organism is able to complete its life cycle wholly within the atmosphere.

Most terrestrial organisms that function in the atmosphere are supported by the lithosphere and live at the interface between the two, where they play an important role in the exchange of matter and energy across the boundary. Most plants, for example, have structures that penetrate the lithosphere for water and nutrients as well as for support while playing a significant role in the exchange of gases.

Life does not penetrate very deeply into the lithosphere, and few organisms live totally encased in the rigid materials of Earth's crust. The boundary between lithosphere and hydrosphere is transitional in most places, a zone of porous unconsolidated sediment with water in the interstices between particles. Many organisms are able to burrow into this transitional medium, which supports a diversity of infaunal communities. Some organisms are able to bore into solid rock, although they maintain contact with the hydrosphere for carrying out life functions. On land, the interface between lithosphere and atmosphere is similarly transitional over much of Earth; it is typically weathered, particulate lithosphere or soil (*see* SOILS, FORMATION OF). This interface is not only host to unique communities of organisms, such as soil arthropods and nematodes, it is also a physical environment that is highly modified and structured by the activities of the organisms that inhabit it.

Biogeochemical Cycles and Feedback in the Biosphere

As a dynamic envelope, the biosphere is a zone of mass movements of materials through cycles that are critical to the construction and maintenance of organisms and ecosystems. The carbon cycle, nitrogen cycle, phosphorus cycle, and various mineral cycles all play major roles in the mass circulation of materials across boundaries between the atmosphere, lithosphere, and hydrosphere. These cycles draw upon reservoirs outside of the physical limits of the biosphere itself and divert materials into reservoirs beyond the physical limits of the biosphere. It is therefore appropriate to view Earth's biosphere as having a sphere of influence that is broader than its sphere of occupation.

For example, massive amounts of carbon once removed from the atmosphere by ancient plants are now stored deep within Earth's crust as hydrocarbons (fossil fuels such as coal and oil). Similarly, massive amounts of carbon removed from seawater to construct the skeletons of ancient organisms now reside deep within Earth's crust as limestone. And the ozone layer (O^3) within the stratosphere, at altitudes too high (10 to 50 km) to support communities of organisms, is a consequence of the interaction of sunlight with molecular oxygen (O^2) produced by photosynthetic organisms. Other Sun-Earth interactions, such as displays of the aurora extending upward into the ionosphere and thermosphere (500 km), are dependent upon oxygen generated in the biosphere.

Hierarchical Organization of the Biosphere

Understanding biosphere organization is a primary goal of the field of ecology. The ecological structure of life in the biosphere is hierarchical and characterized by interactions at different nested levels of biological organization. Individual organisms collectively form species populations that share a range of tolerances and functions (niche). Interacting populations form communities that are held together by an interconnected pattern of energy flow (food chains or webs). Communities are organized into provinces, which are geographic units bounded by barriers to distribution and characterized by patterns of diversity and productivity and are under climatic control.

Above the provincial level, it is difficult to recognize larger units with common ecological or functional properties, although terms such as realm, region, division, and domain are used in classifying and mapping biological diversity.

Diversity in the Biosphere

Beyond the fundamental question of how life arose (*see* LIFE, ORIGIN OF), there is no more intriguing question than that of why it diversified. Why are there so many kinds of plants and animals in the biosphere? It is clear that there are many more species than systematists have had time and resources to name and characterize. It also is clear that human impact on the biosphere is diminishing this diversity at an alarming rate, both through local effects and effects on global climate (*see* GLOBAL CLIMATIC CHANGES, HUMAN INTERVENTION). Currently the species is the unit that is used to measure diversity, and diversity is inventoried through lists or catalogs of species, with priority on identifying and managing those that are endangered.

Alternative measures of diversity may be more critical both to monitoring and conserving the richness of life in the biosphere. These alternative measures include estimates of genetic diversity, morphologic diversity, and functional diversity. It is not the sheer number of species or fundamental taxonomic units in the biosphere that is most interesting, but the sheer number of different kinds in terms of structural plan and functional range. Diversity is the least well understood property of the biosphere. The problem of diversity is imbedded in a larger problem, that of reconstructing the evolutionary history of the biosphere.

Evolutionary History of the Biosphere

The capacity for change is a fundamental property of the biosphere. The biosphere and physical Earth have coevolved, i.e., there has been feedback and reciprocity (*see* SOLID EARTH-HYDROSPHERE-ATMOSPHERE-BIOSPHERE INTERFACE). It is not only the biosphere that has changed, but also the atmosphere, lithosphere, and hydrosphere.

There are two parts to reconstructing the evolutionary history of the biosphere. The first is to establish the steps in the early assembly of the biosphere, which is closely tied to the origin and early history of life on Earth. The fossil record suggests that the hydrosphere was the initial site of biological diversification (*see* LIFE, ORIGIN OF). The oldest fossils of cyanobacteria-like microorganisms appear in the fossil record about 3.5 Ga, and microfossils from more than 400 Proterozoic localities indicate predominantly marine settings. Microbial life was present on land before the end of the Precambrian and had begun to play a significant role in weathering, erosion, sedimentation, and geochemical cycles. At the time of life's origin, the atmosphere was most probably predominantly carbon dioxide; and the change from a reducing to an oxidizing atmosphere is one of the most profound changes in the history of life and Earth. It opened

up a range of new opportunities in organismal complexity and organization that led first to the diversification of unicellular, photosynthetic eukaryotes and eventually to multicellular organisms (*see* HIGHER LIFE FORMS, EARLIEST EVIDENCE OF).

As with unicellular prokaryotic life, diversification and structuring of multicellular eukaryotic organisms and communities occurred in the hydrosphere before emerging onto land. The Phanerozoic fossil record of life in the biosphere is one of continual minor change punctuated by major large-scale reorganizations of life. The reorganizations follow mass extinction events, which are triggered by global changes in the atmosphere, hydrosphere, and lithosphere (*see* EXTINCTIONS). Some of these changes appear to have been terrestrial in nature and include tectonic events (major episodes of rifting or collision and suturing of tectonic plates), global climate change related to changes in atmospheric and/or oceanic circulation patterns, or continental glaciations. Others have been linked (although not without controversy) to extraterrestrial events, predominantly the impact of asteroids or other large bodies that have signatures in the form of craters, layers enriched in trace elements such as iridium that are rare in Earth's crust, microtectites, and shocked quartz.

Bibliography

ODUM, E. *Systems Ecology.* New York, 1983.
RICKLEFS, R. E. *Ecology,* 3rd ed. San Francisco, CA, 1990.
VALENTINE, J. W. *Evolutionary Paleoecology of the Marine Biosphere.* Englewood Cliffs, NJ, 1973.

CAROLE S. HICKMAN

BIRDS

Feathers distinguish birds from all other vertebrates and the presence of a full complement of feathers clearly identifies *Archaeopteryx lithographica* as a bird (Figure 1).

Bird Origins

This most-ancient bird fossil dates back to the Jurassic period about 145 million years ago (mya),

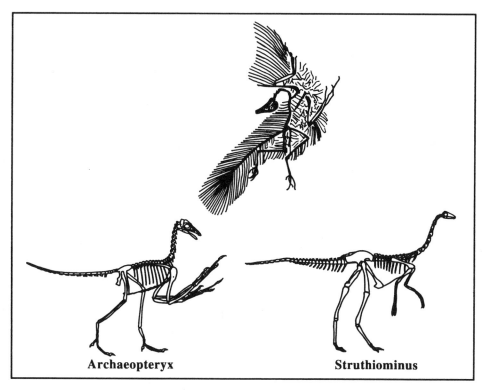

Figure 1. Above, drawing of the Berlin Museum specimen of *Archaeopteryx.* Below, skeletal reconstruction of *Archaeopteryx,* and the coelurosaurid dinosaur *Struthiomimus.*

when reptiles were the dominant terrestrial vertebrates. Several traits closely link *Archaeopteryx* and modern birds to extinct reptiles, particularly certain dinosaurs. Apart from the remarkable resemblance in the overall appearance of the skeleton of *Archaeopteryx* and some meat-eating dinosaurs (Figure 1), detailed similarities include: (1) a single sound-transmitting bone in the middle ear (mammals have three); (2) jaws comprised of five to six bones on each side (mammals have a single bone); (3) a distinct pattern of long-bone ossification and growth; (4) a specialized "intertarsal" ankle joint; and (5) hollow "pneumatized" bones (Ruben, 1991; Barreto et al., 1993).

Several soft tissues are also similar in modern birds and reptiles. These include scales on the hindlimbs and feet, a brain structure that resembles that of crocodiles, a pecten (a distinct, functionally obscure, internal eye structure), and nucleated red blood cells (mammalian red blood cells lack nuclei).

Physiological similarities between birds and reptiles include the production of uric acid as a waste product. Uric acid is relatively water insoluble and, among other functions, allows an embryo to dispose of waste within a shelled egg without danger of poisoning itself. This permitted reptiles and birds to lay eggs out of water, allowing access to terrestrial habitats unavailable to fish and amphibians. Reptile and bird eggs share a number of other structural features.

Further compelling evidence for the relationship between birds and reptiles is provided by protein and genetic analyses. These anlyses reveal far stronger affinities between birds and reptiles than between birds and any other group.

Following *Archaeopteryx,* the next well-known bird species to appear in the fossil record are *Iberomesornis* and *Concornis* from Spain, and *Sinornis* from China (Figure 2). These early Cretaceous era forms (about 135 mya) were probably capable of sparrowlike tree-perching and flapping flight. Among later Cretaceous birds were *Ichthyornis,* a toothed, ternlike seabird, and *Hesperornis* (Figure 2), a flightless, marine diving bird that was over 2 m long! These Cretaceous forms were extinct by

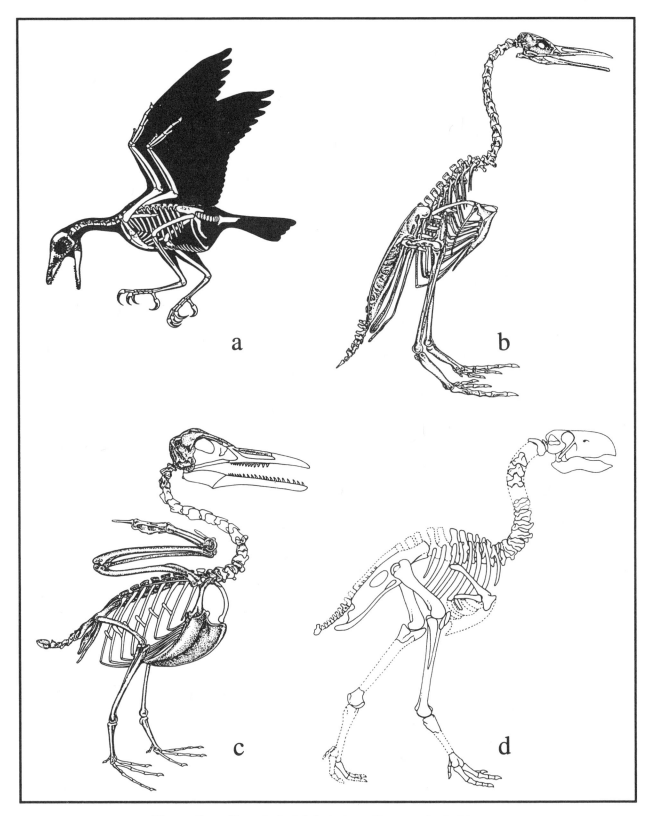

Figure 2. a. *Sinornis.* b. *Ichthyornis.* c. *Hesperornis.* d. *Diatryma.*

the end of the Mesozoic era. Subsequently, major adaptive radiations of birds occurred between 35 and 65 Ma, producing most of the modern avian orders. A spectacular species from these radiations that went extinct during the Tertiary period was *Diatryma*, a carnivore more than 2 m tall (Figure 2), with a huge, powerful bill (Gill, 1990).

Flight

The evolution of avian flight is controversial. One proposal (the "trees-down" hypothesis) suggests that "protobirds" were tree climbers that jumped between perches. They next evolved primitive flight feathers that allowed them to glide from perch to perch. Subsequently, powered, flapping flight developed. The main alternative proposal for the origin of bird flight is the "ground-up" hypothesis. In this scenario, protobirds were ground dwellers with weakly feathered forelimbs used for balance as they ran and leaped after prey. Eventually, many of the same forelimb motions generated for mid-leap balance were used to support short bursts of powered flight supported by elongated wing feathers (Caple et al., 1983).

Flight is energetically demanding. The body needs to supply high levels of oxygen to active muscles, and this requires substantial respiratory and circulatory capacity. A significant advance for birds over reptiles was a four-chambered heart. A four-chambered heart is more efficient in separating oxygenated and deoxygenated blood and hence in aerating tissues; this in turn allows for longer sustained activity and greater stamina. A second major advance was one-way air circulation through the lungs, rather than the "back and forth" air circulation of other terrestrial vertebrates. One-way air circulation maintains higher oxygen concentrations over the respiratory tissues than two-way circulation.

Endothermy

Additional metabolic efficiency was gained by endothermy, that is, the ability to maintain a relatively constant body temperature despite fluctuating environmental temperatures. Ecotherms such as reptiles rely on the external environment to obtain their body heat, although they may be relatively homeothermic if they have sufficiently large bodies to store that heat. In contrast, endotherms

such as birds control body temperature by high metabolism and high rates of internal heat production. Endothermy entails substantial metabolic costs, and birds expend much more energy than similar-sized reptiles in performing the same activities. Birds have greater stamina than reptiles, however, and this advantage seems to outweigh the substantial costs of endothermy. A lesser advantage of endothermy is that it permits animals to be active in colder environments than ectotherms. Surprisingly, there is some evidence that early birds may have been ectothermic (Ruben, 1991; Cinsamy et al., 1994). However, endothermy probably evolved in birds shortly after their origin.

Nesting

As endothermy evolved, it became more important for birds to keep their eggs' embryos at relatively stable temperatures. Stable, warm temperatures also speed up hatching. Primitive birds probably used ambient temperature to incubate eggs in the same way that reptiles do. Although brush turkeys in Australia and New Zealand bury their eggs in the manner of reptiles, they keep their eggs at relatively constant, warm temperatures by laying them over decaying vegetation or geothermal heat sources, and then scrupulously removing and adding material over the eggs.

Shallower soils may have permitted partial egg burial only, and parents may have had to shield eggs from sunlight or keep eggs warm at night. However, it is often easier to maintain constant egg temperature if the parent has skin contact with the egg on a more continuous basis, and this is what most extant birds do.

The first species that evolved continuous incubation likely laid their eggs on the ground, as do whippoorwills. The ground can be unsafe because of predators, and elevated sites may afford improved protection. Hence, fairy terns balance their single egg precariously on the branch of a tree. Enclosed hollows offer added defense against predators. Ashy storm petrels take advantage of hollows in rock to lay their eggs; numerous other species make use of natural hollows in trees. When such sites were unavailable, some species evolved behaviors to build their own hollows in sand (e.g., kingfishers, bank swallows) or trees (woodpeckers).

Many of the preceding species do not modify the substrate on which they lay their eggs, and most add no lining or insulation around their eggs.

In contrast, most birds build more elaborate nests to hatch and to hold their eggs. The simplest of these modifications are cups hollowed out of the sand, such as those of gulls and terns. Some ground nesters line these cups with vegetation. From these beginnings, more elaborate nests evolved. Some species, such as cave swiftlets, glue their nests to vertical surfaces (swiftlet nests made entirely of saliva give birds' nest soup its special flavor). Mud is sometimes used as a glue (e.g., cliff swallows). Mud also provides the structural strength in the cup nests of American robins and red-winged blackbirds. These latter species line the mud cup with strands of vegetation and feathers to provide a well-insulated, secure perch for their eggs. Many tropical species build domed nests to shield eggs from sun, rain, and the eyes of predators. Still greater sophistication is achieved in the pensile (hanging) nests of orioles, weaver birds, and other species. These nests are spectacular structures of woven vegetation, sometimes strengthened by spider webs or insect silk. The sociable weavers of Africa build their pensile nests abutting those of their neighbors, making huge nesting complexes reminiscent of apartment buildings.

Parental Investment

The survival of a bird's young depends on parental investment. Parental investment includes the energy that goes into nest-building, egg nutrients, incubation, guarding, and feeding of young. Species differ in how they invest in each type of care. For example, birds that lay large eggs and incubate their eggs for long periods tend to hatch young that can walk or swim shortly after hatching and do not need to be fed by a parent. In contrast, birds that lay small eggs and have short incubation periods tend to produce young that are unfeathered, blind, and helpless. The extent of parental feeding of hatchlings also varies substantially. Young brush turkeys are able to fly once they dig to the surface and may never see either parent. In contrast, toucans may not leave their nest cavity until they are fifty days old, and wandering albatrosses may not fly until 275 days old.

Bibliography

BARRETO, C; R. M. ALBRECHT; D. E. BJORLING; J. R. HORNER; AND N. J. WILSMAN. "Evidence of the Growth Plate and the Growth of Long Bones in Juvenile Dinosaurs." *Science* 262 (1993):2020–2023.

BENNETT, A. F., and J. RUBEN. "Endothermy and Activity in Vertebrates." *Science* 206 (1979):649–654.

CAPLE, G. R., R. P. BALDA, and W. R. WILLIS. "The Physics of Leaping Animals and the Evolution of Preflight." *American Naturalist* 121 (1983):455–476.

CHINSAMY, A., L. M. CHIAPPE, and P. DODSON. "Growth Rings in Mesozoic Birds." *Nature* 368 (1994):196–197.

RUBEN, J. "Reptilian Physiology and the Flight Capacity of *Archaeopteryx*." *Evolution* 45 (1991):1–17.

DAVE SHUTLER
JOHN RUBEN

BJERKNES, J. A. B. AND V. F. K.

Vilhelm Frimann Koren Bjerknes and Jacob Aall Bonnevie Bjerknes made up the Norwegian father and son "team" that created the foundation for modern meteorology by incorporating ideas of hydrodynamics and thermodynamics from physics into descriptions of atmospheric motion. These ideas, which constituted what became known as the Bergen school of meteorology, enabled scientists to describe atmospheric dynamics rigorously, thus providing the basis for long-term weather forecasting.

Vilhelm Bjerknes

Vilhelm Bjerknes was born on 14 March 1862 in Kristiania (now Oslo), Norway, to Carl Anton and Aletta Koren Bjerknes. The elder Bjerknes, who exerted a profound influence over his son's life, was a physical scientist who turned his research interests to questions of hydrodynamic forces on bodies in frictionless fluids. Vilhelm soon found himself collaborating with his father and working on similar problems during his early career in physics.

Vilhelm Bjerknes' formal education began at the University of Kristiania in mathematics and physics, where he completed his master's degree in 1888. He then went to Paris on a state fellowship, where he attended the lectures of mathematician Jules-Henri Poincaré on electrodynamics. It was there that he first became exposed to Heinrich Hertz's studies on electrical wave diffusion. In

1890 he moved to Bonn to work with Hertz, serving as Hertz's assistant and first scientific collaborator. This experience proved to be mutually invaluable to both scientists since Bjerknes' resonance curve experiments verified Hertz's theory and other experimental work performed by Bjerknes served as the basis for his own doctoral dissertation, for which he received his Ph.D. in Norway in 1892.

Bjerknes professional career began the next year, when he was appointed lecturer at Stockholm's School of Engineering. Two years later, he was named professor of applied mechanics and mathematical physics at the University of Stockholm, and he soon returned to his father's research problems in hydrodynamic forces. By the end of the century, he succeeded in integrating his own research with the work of William Thomson (Lord Kelvin) and Hermann Helmholtz concerning velocities of circulation and the conservation of a circular vortex. Bjerknes applied this integrated work to the largest fluid system, the atmosphere and the ocean. This served as the basis for "physical hydrodynamics," in which the laws of thermodynamics are incorporated into classical fluid mechanics to address atmospheric motion. The movement in the atmosphere is a result of the dynamic relationship between motion caused from the radiation of the sun's heat, which produces mechanical energy, and the production of heat caused by friction from atmospheric movements.

It was this work that created the beginnings of the Bergen school of meteorology. Assisted by hydrodynamic and thermodynamic theories, Bjerknes hoped to use knowledge of the present conditions of the atmosphere and the earth's ocean to calculate future weather conditions. For support, he traveled to the United States in 1905, where he presented his ideas in a series of lectures at the Massachusetts Institute of Technology (MIT) and the Carnegie Institute of Washington. Subsequently, the Carnegie Foundation awarded him an annual stipend to support his research, a stipend that continued until 1941.

In 1907, Bjerknes moved back to Norway, where he served as professor of applied mechanics and mathematical physics at the University of Kristiania. He soon developed the significant collaborative relationships that were to characterize his entire scientific career. First, by calling for more research in upper-air wind observations (he helped to popularize the use of balloons for meteorological research) and then by soliciting information from the nascent oceanographic community, Bjerknes began to create an international meteorological research community. His career and work achieved greater fame when he accepted the professorship of the newly organized Geophysical Institute at the University of Leipzig. However, with the outbreak of World War I, conditions in Leipzig became very difficult and Bjerknes returned to Norway, where he accepted the offer made by Fridtjof Nansen to start a new geophysical institute at the University of Bergen.

The years in Bergen were the most productive of Bjerknes' career. He founded the Bergen Geophysical Institute, which was the training ground for many of the world's leading meteorologists of the early twentieth century.

He also published his classic work, *On the Dynamics of the Circular Vortex with Applications to the Atmosphere and to Atmospheric Vortex and Wave Motion* (1921). It remains the clearest exposition of Bjerknes' basic meteorological ideas.

Bjerknes left the Geophysical Institute in 1926 to accept the professorship of mechanics and mathematical physics at the University of Oslo. He retired from active teaching in 1932, but continued to take an active role in departmental seminars and research discussions, especially as an advocate of modern meteorology, until his death on 9 April 1951.

Jacob Bjerknes

Vilhelm Bjerknes married Honoria Bonnevie in 1895 and on 2 November 1897 their son Jacob was born. The younger Bjerknes grew up in the atmosphere of his father's developing tradition in dynamic meteorology and weather forecasting, both of which came to serve as his own life's work. He attended the University of Kristiania until 1916 and then moved with his father to Leipzig, where he became part of the Bjerknes research team, which included his friend Halvor Solberg.

Jacob Bjerknes' research interest at this time involved questions concerning the physical characteristics of the "squall line," the area in which precipitation forms when shower clouds develop. Jacob Bjerknes concluded that the converging winds developed low-pressure centers, just as one observed in cyclonic weather systems. In fact, he argued that both the squall line and cyclone could be understood as part of one integrated weather system. This provided the basis for the cyclone model of airmass movement, a model that was modified by the Bergen school to create the three-dimensional view of the atmosphere that it used so successfully in analyzing changes in atmospheric motion.

The new cyclone model soon took on a rhetorical dimension as Jacob Bjerknes referred to the squall line first as a "battle line," then a "front," and finally a "polar front." All these terms took their basic form from World War I, but they soon became part of meteorological terminology with the publication of the paper by Bjerknes and Solberg, "The Life Cycle of Cyclones and the Polar Front Theory of Atmospheric Circulation." For this work, Bjerknes was awarded his Ph.D. in 1924 (after serving as chief of the Geophysical Institute since 1919!) by the University of Kristiania.

When his father left Bergen to return to Oslo, more of the responsibilities for continuing the work at the Geophysical Institute fell to Jacob Bjerknes. He became professor of meteorology in 1931, and continued his research on the three-dimensional structure of atmospheric motion. As part of his research activities, he accepted a one-year position at MIT, where CARL-GUSTAF ROSSBY had been since 1928. As World War II loomed, Bjerknes decided to accept an offer to visit the United States again. The decision was fortuitous, for shortly after he left, Germany occupied Norway. It was also timely because he was soon offered a position to build a meteorology program at UCLA as part of that institution's interest in aeronautical technology and worldwide weather forecasting.

Bjerknes recruited his countryman and his father's student, Jorgen Holmboe, to join him at UCLA, and together they soon developed a department in Los Angeles that was at the forefront of meteorological research in the United States. Holmboe and Bjerknes published "On the Theory of Cyclones," the ultimate theoretical paper on the cyclone theory in which the upper and lower atmosphere motions are integrated. This work, extended and elaborated by Bjerknes' first student, Jule Gregory Charney, served as the basic model for the first accurate computer-aided weather forecast (1950).

Jacob Bjerknes inherited not only his father's scientific interests, but also his collaborative style. Hence, the Bergen school was brought to the United States, where it flourished in southern California, both at UCLA and at Scripps Institution of Oceanography. In the 1950s, Bjerknes turned his research interests to the interrelationships between the atmosphere and the ocean, an interest he held in common with colleagues at Scripps (where geophysicist Harald Ulrik Sverdrup had served as director). In this work, he contributed to the understanding of how the variable heat supply of the equatorial oceans contributes to the annual variations in world climates. The late-twentieth-century work associated with "El Niño" phenomena is based on Bjerknes' pioneering work on the warming of the surface oceanic waters.

Bjerknes retired from UCLA in 1965 but he continued to live in the United States until his death in Los Angeles on 7 July 1975.

Bibliography

ELIASSEN, A. "Vilhelm Bjerknes and His Students." *Annual Review of Fluid Mechanics* 14 (1982):1–11.

FRIEDMAN, R. M. "Constituting the Polar Front, 1919–1920." *Isis* 73 (1982):343–362.

JEWELL, R. "The Meteorological Judgment of Vilhelm Bjerknes." *Social Research* 51 (1984):783–807.

MCPEAK, W. J. "Jacob Bjerknes (1897–1975), Norwegian Meteorologist." In *Notable Twentieth-century Scientists*, vol. 1. New York, 1995, pp. 179–181.

———. "Vilhelm Bjerknes (1862–1951), Norwegian Physicist and Geophysicist." In *Notable Twentieth-century Scientists*, vol. 1. New York, 1995, pp. 182–184.

PIHL, M. "Bjerknes, Vilhelm Frimann Koren." In *Dictionary of Scientific Biography*, vol. 1, ed. C. C. Gillispie. New York, 1972, pp. 167–169.

KEITH R. BENSON

BLACK HOLES

See Stars

BOMBARDMENT RECORD

See Impact Cratering and the Bombardment Record

BOWEN, NORMAN L.

Norman L. Bowen, considered the most outstanding petrologist of the twentieth century, was born on 21 June 1887 in Kingston, Ontario, Canada, the youngest son of William Alfred Bowen, an immigrant from London, and Eliza McCormick Bowen of Kingston. His father was a guard at the Kingston Penitentiary at the time of his birth. After serving as Sexton of St. George's Cathedral, his father then became proprietor of a bakery, which was later retained by Norman's older and only surviving brother, Charlie Lewis Bowen. His eldest brother was Alfred William Bowen.

As a boy, Bowen was attracted to the nearby quarries of both limestone and red granite. Abandoned quarries were often filled with water, and he used them for swimming in the summer and for ice skating in the winter. He attributed his interest in rocks to these boyhood haunts. Bowen attended Kingston public schools and the Collegiate Institute until 1900. Completing his junior matriculation in three years, he entered Queen's University in the fall of 1903. He completed the honors course in chemistry and mineralogy in 1907, and he was graduated with an M.A. degree, receiving the university medal in both subjects. Bowen's first geological work was done for the Ontario Bureau of Mines in the Larder Lake area under the supervision of Professor R. W. Brock, who presumably also served as director of the Geological Survey of Canada during the summer months. Essentially left on his own in the field at the age of twenty, Bowen demonstrated his ability to survive in the bush, to acquire mapping skills with no previous experience, and to develop considerable facility with a canoe. In the fall of 1907 after the ice formed on the lakes, he returned to Queen's University and spent two years in the School of Mining taking courses in mining engineering and geology. While completing the B.S. degree, which he received in 1909, he produced two papers based on three summers of fieldwork with the Ontario Bureau of Mines. The Canadian Mining Institute awarded one paper a first prize and its President's gold medal.

Even though the 1851 Exhibition Scholarship he won in 1910–1911 would have made it possible for him to do graduate work in Norway with J. H. L. Vogt and W. C. Brøgger, Bowen was discouraged by the language barrier. After a field season in the Gowganda Lake area, Ontario, he was accepted in the fall of 1909 at Massachusetts Institute of Technology (MIT), where he was attracted by fellow Canadian and inspiring teacher REGINALD A. DALY. In addition, C. H. Warren, H. W. Shimer, WALDEMAR LINDGREN, and T. A. Jaggar influenced his geological thinking, whereas A. A. Noyes, G. N. Lewis, and W. C. Bray provided instruction in chemistry (Bowen, 1942, p. 85). The philosophical basis for applying physical chemistry to mineralogical and petrological problems was probably imparted to Bowen by C. H. Warren. In the spring of 1910, Professor Jaggar, then departmental chairman, advised Bowen of an opening for a research student at the recently established Geophysical Laboratory of the Carnegie Institution of Washington, DC. Appointed as that laboratory's first pre-doctoral fellow, Bowen worked on the thermal phase relations in the nepheline-anorthite system, an important problem suggested by its director, Arthur L. Day. The work was accepted in partial fulfillment of the requirements for his Ph.D. from MIT, which was awarded on 4 June 1912.

Bowen served as an assistant to Daly in the summer of 1911, surveying the main line of the Canadian Pacific Railway and mapping the geology of the area covered by the Shuswap sheet. He led his own field party in the summer of 1912 in the Frazer River Valley from Lytton to Vancouver. In spite of the several opportunities to work as a volcanologist under Jaggar, as economic geologist under Lindgren, and as a field geologist on the Geological Survey of Canada, Bowen chose to satisfy his wish to work in both chemistry and geology as a staff member of the Geophysical Laboratory, and arrived there on 1 September 1912.

The next three years were not only important to Bowen, but resulted in the development of some of the most critical concepts in petrology. He used the new quenching technique of Shepherd, Rankin, and Wright (1909), whereby the phases grown in the laboratory at high temperatures are rapidly cooled to preserve the products; the phase equilibrium theory of H. W. Bakhius Roozeboom that constrains the relationship of crystals and liquid; and the preliminary results of Day, Allen, and Iddings (1905) on the plagioclase feldspars consisting of albite and anorthite. With this information, Bowen demonstrated their complete series of solid solutions, calculated the heat of melting of the end-members anorthite and albite, and explained their zoning (Bowen, 1913). Study of the system MgO-SiO_2 with Olaf Andersen (1914), the system diopside-forsterite-silica (1914), and diopside-albite-anorthite (1915) followed. Those systems were at the heart of the problems of the generation of the basic rocks, and were chosen with incredible perception and intuitive depth. On the basis of these studies his emerging ideas were collated in a classic memoir, "The Later Stages of the Evolution of the Igneous Rocks" (1915). The value of these studies was quickly recognized and he gained international respect at the age of twenty-eight.

As a result of his field studies in the Gowganda district, where he had observed accumulation of feldspar in a diabase, he turned to the monomineralic rocks, especially the anorthosites (1917). He attributed these to a process of crystal accumulation under gravity and cited examples from the Adirondacks and Quebec. With the onset of World War I, the Geophysical Laboratory turned its staff to the problems related to the manufacture of optical glass, supplies having been cut off from Germany. Bowen helped to put the secretive, cookbook method of making glass on a scientific base and dealt with the formation of "stones" and devitrification in glass in general. Bowen spent a brief period at the glass plants of the Bausch and Lomb Optical Company in Rochester, New York, and the Pittsburgh Plate Glass Company in Charleroi, Pennsylvania.

After the close of World War I, Bowen returned to teach at Queen's University as professor of mineralogy for seventeen months but found experimental work—with adequate facilities—more pleasurable and returned to the Geophysical Laboratory. In the next three years, Bowen published a series of outstanding papers on diffusion, the reaction principle (also known as Bowen's reaction series), and assimilation—all critical problems in petrogenesis. He demonstrated that diffusion did not play a significant role in the diversity of magmas, but was adequate to explain the reaction rims on inclusions. The reaction principle (1922), defining a continuous reaction series of crystals with liquid and a discontinuous reaction series, was the theoretical basis for the order of crystallization of minerals previously deduced in empirical grounds. He also showed that magmas could incorporate or assimilate large amounts of inclusions by reaction and precipitation but did not believe that process was essential to the diversity of rock types.

Collaborative studies with J. W. Greig (1924) led to the discovery of a principal phase in refractory ceramics, later called mullite. Work with G. W. Morey on the soda-lime glasses showed that a liquid formed near the composition $Na_2O \cdot 3CaO \cdot 6SiO_2$ (later called devitrite), having a low temperature and high viscosity, was important in the manufacture of glass. Fieldwork continued to attract him, and two trips to the island of Skye (Scotland) in 1923 and 1926 with Alfred Harker as a guide, gave him new insights on the intrusion mechanism of peridotite dikes and the limits in composition of olivine basalt liquids free of crystals. In 1927 he gave a series of lectures at Princeton University that were later published as *The Evolution of the Igneous Rocks*. It was acclaimed as the most influential book in petrology covering the principal processes in the generation of igneous rocks. Each chapter was analyzed on the book's fiftieth anniversary by experts on the various processes (Yoder, 1979). The analyses again illustrated his shrewd selection of pertinent problems and his wisdom in getting to the substantive issues in each problem integrating both field and laboratory data.

One of the common criticisms of experimental work at that time was the lack of iron in systems studied. To rectify this deficiency, Bowen joined with J. F. Schairer, a careful and very productive experimentalist, during the late 1920s and 1930s. Some motivation may have derived also from his running debate with Fenner (1929), who believed that magma differentiation led to an iron-rich residuum, following Teall (1897), and not necessarily one enriched in silica and alkalis as Bowen believed. With a new technique employing a pure iron crucible, the major systems $FeO-SiO_2$ (1932), $CaO-FeO-SiO_2$ (1933), $MgO-FeO-SiO_2$ (1935), and nepheline-silica-fayalite (1938) were studies that contributed not only to petrology but also to metallurgical slag technology. After a field season spent studying the alkaline lavas of the Rift Valleys of Africa and climbing the volcano Nyamaragira in Zaire (then called Namlagira, Belgian Congo) in the summer of 1929, Bowen and Schairer worked intermittently on the difficult systems involving alkali feldspar, silica, and the feldspathoids. These systems were pertinent to the rhyolites, trachytes, and phonolites of the East Africa volcanic fields and a preliminary diagram for $NaAlSiO_4-KAlSiO_4-SiO_2$ appeared in 1935, which was close to the one Bowen "deduced" in 1928. The importance of the diagram was revealed in 1937 under the label "petrogeny's residua system," which he described as a "sort of goal toward which all crystallization trends." No doubt he was satisfied that he had completely demonstrated his point of view, in opposition to Fenner, that "maximal fractional crystallization" led to compositions containing substantially alkali-alumina silicates.

In 1937, Bowen moved to the University of Chicago, presumably to develop experimental work in the universities, but also to avoid potential conflict with a prospective but unsuccessful candidate for the directorship of the Geophysical Laboratory. There he and his students focused on the petrologically important parts of the Na_2O-CaO-Al_2O_3-SiO_2 system. During his ten-year period at Chicago, of which he served two years as chairman of the Department of Geology and two years back at the Geophysical Laboratory to help in the war effort (1942–1944), he produced only five successful Ph.D. candidates: Two eventually became university professors and three went into industry. In addition, he produced in 1940 a classic paper titled "Progressive Metamorphism of Siliceous Limestone and Dolomite." The successive decarbonation reactions were worked out as a function of temperature and pressure, and he suggested that where curves crossed, a "petrogenetic grid" could be established. This benchmark paper evolved out of Bowen's lectures on metamorphic petrology—a new field of inquiry for an igneous petrologist applying the principles of physical chemistry!

The director of the Geophysical Laboratory, L. H. Adams, persuaded Bowen to return to experimental work. The enticement was presumably the new program after World War II that was to focus on volatile-containing systems. He was joined on the same date, 1 January 1947, by O. F. Tuttle, a new staff member, who had contributed to the war effort at the Laboratory from 1942–1945 and was particularly adept at designing hydrothermal equipment. Bowen apparently changed his opinion about the role of water because he had explicitly stated that volatiles were of "little importance in igneous differentiation" (1928, p. 302). The first paper, concerning MgO-SiO_2-H_2O, demonstrated that extrusion of a magma of serpentine composition was most unlikely. In order to investigate the "granite" system, a problem of continuing concern to Bowen (1948), it was necessary to first study the alkali feldspar join with water to facilitate crystallization, $NaAlSi_3O_8$-$KAlSi_3O_8$-H_2O (1950). Preliminary results on the alkali feldspars with silica, the "granite" system, were revealed in 1954, but the final results were written by Tuttle (1958) after Bowen's death by his own hand in Washington, DC, on 11 September 1956. They stated emphatically that granites were a residuum of fractional crystallization, but also noted that they might be the first liquid formed on selective fusion—a small concession to the transformist camp.

Few men have contributed so many new concepts to petrology and substantiated them with quantitative data. Bowen's knack for making the critical field observations and then testing them with simple laboratory experiments evolved into a methodology of lasting value. Although he disclaimed any experimental skills, he was indeed the catalyst in perfecting apparatus and techniques for phase equilibria studies. His greatness stems from the application of physical chemistry to complex geological field problems. For those contributions he was honored worldwide with medals, honorary memberships, honorary degrees, presidencies of societies, book dedications, and lectureships.

His marriage to Mary Lamont took place on 3 October 1911 after she had obtained her doctorate of medicine. Their only child, Catherine Lamont Bowen, was born on 18 December 1914 in Washington, DC.

Bowen was a quiet, gentle man with a droll sense of humor that was reflected in his writings. As a young man he was exceptionally strong and powerfully built. His students and colleagues held him in great awe, and few developed a close relationship with him. Modest and retiring, Bowen rarely answered questions immediately, but chose to think through the problem and then return with a very clear and logical response. He had sung in his church choir and was said to have had a fine tenor voice. He had a fondness for poetry and occasionally quoted passages in his speeches and writings. Even though Bowen disliked teaching and especially administration, he was most helpful to his students and all remembered his personal kindnesses and warm, sympathetic nature. In his later years he referred to his small physical stature, which his daughter Catherine attributed to a gradual collapse of his vertebrae of almost three inches. Nevertheless, he stood a head taller than the giants of science in the twentieth century.

Bibliography

BOWEN, N. L. "The Melting Phenomena of the Plagioclase Feldspars." *American Journal of Science*, 4th ser., 35 (1913):577–599.

———. "The Later Stages of the Evolution of Igneous Rocks." *Journal of Geology* (suppl.), 23 (1915):1–89.

———. *The Evolution of the Igneous Rocks.* Princeton, NJ, 1928. Repr., with intro by J. F. Schairer, New York, 1956.

———. "Progressive Metamorphism of Siliceous Limestone and Dolomite." *Journal of Geology* 48 (1940): 225–274.

BOWEN, N. L., and J. F. SCHAIRER. "The System FeO-SiO$_2$." *American Journal of Science*, 5th ser., 24 (1932): 177–213.

TUTTLE, O. F., and N. L. BOWEN. "Origin of Granite in the Light of Experimental Studies in the System NaAlSi$_3$O$_8$-KAlSi$_3$O$_8$-SiO$_2$-H$_2$O." *Geological Society of America Memoir* 74 (1958).

HATTEN S. YODER, JR.

BRACHIOPODS

Brachiopods are bivalved, marine invertebrates whose geologic record extends almost 600 million years (Ma), from the earliest Cambrian period to the Recent. Although superficially similar in morphology to clams (*see* MOLLUSKS) in that both are bivalved, the internal organs are completely different; they are phylogenetically unrelated and belong to separate phyla. Brachiopods are in fact most closely related to bryozoa, microscopic colonial invertebrates (*see* COLONIAL INVERTEBRATE FOSSILS). The key character linking these two outwardly dissimilar groups is their possession of a unique fleshy, tentaculated food-gathering structure called the lophophore.

Morphology and Ecology

All brachiopods are sessile (immobile) as adults. Some attach to the seafloor via a fleshy stalk, the pedicle, others lack a pedicle and rest, free-lying on the seafloor. Some brachiopods may have had a pseudo-planktic habit, floating attached to seaweed or other organisms. Brachiopods, in turn, provided appealing substrate for other encrusting organisms, commonly bryozoans or small corals. As sessile filter feeders, brachiopods were sensitive to their physical environment, especially substrate consistency and current energy. The shape and surficial features of the shell probably reflect adaptation to these environmental factors.

Brachiopods range in size from a few millimeters to several centimeters across. The bivalved shell may take a variety of shapes: biconvex, in which both valves are convex; plano-convex, in which one valve is convex and the other is more or less planar; and concavo-convex, in which one valve is convex outward and the other is concave. Surface features of the valves are also variable. Some brachiopods have very smooth shells in which the only ornament is very fine concentric growth lines; others have coarse ribbing or corrugation. Some brachiopods sported long spines from one or both valves. Brachiopod functional morphology, that is, relating the shape (morphology) of the organism to a particular use or function, is a field of active research, and there is still much to learn. Broad, flat shells (plano-convex) may have been an adaptation to life on a soupy, muddy substrate. This hypothesis is termed the "snowshoe effect." Spines may also have been an adaptation to muddy substrates, acting to increase surface area and thus support the brachiopod.

Geologic Record

The geologic record of brachiopods reveals that they were formerly much more abundant and diverse than they are today. In modern marine environments brachiopods tend to occupy cryptic or hidden niches, whereas in the Paleozoic they were obvious and numerically dominant members of the open seafloor community. The geologic record also documents wide morphologic adaptations (specializations), suggesting that in the past brachiopods occupied a wide variety of ecologic niches.

What caused the decline of brachiopods? Any one or a combination of factors may be responsible, including competition, predation, or habitat loss. Increased competition for food and nutrients from newly evolving, possibly more efficient groups of benthic filter feeders through geologic time could have led to the decline of the older, possibly archaic brachiopods. Modern predators on brachiopods include teleostean fish and certain crabs. The rise of these and possible earlier predators would also have contributed to a brachiopod decline. Habitat loss could take several forms: loss of shallow shelf habitat through sea-level fall, changes in climate resulting in unfavorable temperatures, or loss of suitable substrate. A change in the nature of shallow marine seafloor through time is another candidate. Early Paleozoic marine invertebrate faunas are dominated by epifaunal, sessile filter feeders, including brachiopods, clams, crinoids, sponges, corals, and bryozoans. During the Mesozoic era a significant faunal change occurred with the evolution and diversification of infaunal (living within the seabed) deposit feeders, especially among the mollusks (clams and snails). Occupation of the previously underutilized infaunal habitat led to an adapative radiation of these groups and profoundly affected the sessile, filter-feeding epifauna. Sessile, epifaunal organisms generally require a firm substrate for attachment and clear water for feeding. An abundant infauna may have destabilized the substrate by burrowing and muddying the waters, leading to the exclusion of the brachiopods and their kind. The modern seafloor is dominated by infauna, and the sessile epifaunal filter feeders, brachiopods included, are largely restricted to rocky or reef environments that afford firm attachment sites and clear waters for feeding.

Bibliography

Rudwick, M. J. S. *Living and Fossil Brachiopods.* London, 1970.

DANITA S. BRANDT

BRAGG, WILLIAM LAWRENCE AND WILLIAM HENRY

The physical properties of minerals are outward manifestations of their internal atomic architectures, and no individuals deserve more credit for discovering the blueprints for the mineral kingdom than the father-son team of W. H. and W. L. Bragg. William Henry Bragg was born on 2 July 1862 in Westward, England, and died 12 March 1942 in London; his eldest son William Lawrence was born on 31 March 1890 in Adelaide, Australia, and died 1 July 1971 in Ipswich, England. Together they established the technique of X-ray diffraction to unravel crystal structures at the atomic level, and they initiated the use of X-ray spectroscopy to determine the elemental compositions of solids. Working separately (though always in close communication), they employed these techniques to reveal the atomic structures of some of the most common and important minerals in the earth's crust and upper mantle.

William Henry Bragg was more a mathematician than a physicist by training. After attending secondary school at King William's College on the Isle of Man, he studied mathematics at Trinity College, Cambridge, from 1881 to 1884, graduating with first class honors. After an additional year at Cambridge, W. H. was selected to replace the retiring Horace Lamb as a professor of mathematics and physics at the University of Adelaide in Australia.

From 1886 until 1904, Bragg's duties as lecturer and administrator consumed his professional life. Because the university was very young (founded in 1875), Bragg devoted most of his time to the development of a scientific curriculum. Bragg taught himself most of the physics that he was required to know for his courses, and he built an X-ray tube for student demonstrations shortly after news of Wilhelm von Roentgen's discovery of X rays reached Australia in 1895. Bragg also was occupied with a growing family. After his marriage to Gwendoline Todd in 1889, two sons were born in quick succession: William Lawrence in 1890, and Robert Charles in 1891. A third child, Gwendolen Mary, was born in 1907.

Contrary to the adage that physicists perform their greatest work before the age of thirty, W. H. Bragg began his experimental research at the age of forty-one. Prompted by a desire to discuss radiation-induced ionization of gases in a presidential address to the Australasian Association for the Advancement of Science, Bragg initiated studies of the scattering and absorption behavior of α-, β-, γ-, and X-ray radiation. In the spring of 1904, he demonstrated that the initial velocities of α particles are characteristic of the radioactive source material, and this observation led to a rigorous method for the identification of radioactive substances. Bragg continued this line of research into 1907, and he soon thrust himself into the middle of one of the most vigorous and profound debates of the early twentieth century: Are γ rays and X rays corpuscular or wavelike? Results from W. H. Bragg's ionization experiments convinced him that these radiations are corpuscular, and his acceptance of their dual wavelike behavior came only through the experiments of his eldest son, William Lawrence.

W. L. Bragg was schooled at St. Peter's College in Adelaide and at Adelaide University. The Bragg family returned to England in 1909 upon the appointment of W. H. Bragg as Cavendish professor of physics at the University of Leeds. Like his father, W. L. Bragg entered Trinity College, Cambridge, and studied mathematics his first year, but he switched to physics during his second, graduating in 1911. The following year, William Lawrence learned of Max von Laue's success in observing diffraction effects from X rays scattered by a crystal of sphalerite (ZnS). Suspicious of these results, W. L. Bragg launched an X-ray investigation of the alkali halides and reinterpreted Laue's data to produce his own model for the diffraction process. In Bragg's conception of diffraction, planes of atoms within crystal structures serve as "X-ray mirrors" that reflect incoming rays at an angle equal to the angle of incidence. The reflected rays recombine in accord with Bragg's Law: $n\lambda = 2d \sin\theta$, where n is an integer, λ is the wavelength of the radiation, d is the distance between the atomic planes, and θ is the angle between the incoming (or outgoing) rays and the atomic plane (Figure 1).

In conjunction with the work of Laue, this seminal study struck down two outstanding problems at once. It established conclusively that X rays can be treated as wavelike entities, and it proved that crystals consist of arrays of atoms that are ordered with strict regularity in three dimensions. During the years 1913–1914, the Braggs feverishly pursued these two implications. With the aid of an ionization spectrometer constructed by William Henry, the two were able to obtain highly precise values for the X-ray wavelengths emitted by various metal targets upon electron impingement. Once these characteristic wavelengths were measured, the Bragg equation allowed for a host of interplanar d-spacings to be measured from X rays scattered by a single crystal. Through hard work and brilliant inference, these sets of interplanar spacings yielded the internal atomic arrangements of a number of minerals, including halite, diamond, fluorite, pyrite, calcite, and corundum. The methodology established by the Braggs remains fundamental to the science of crystal structure analysis today, and for this work W. H. and W. L. Bragg were jointly awarded the Nobel Prize in 1915.

These investigations were interrupted by the outbreak of World War I, and upon its close both Braggs resumed research encumbered with a heavy slate of administrative duties. In 1923 W. H.

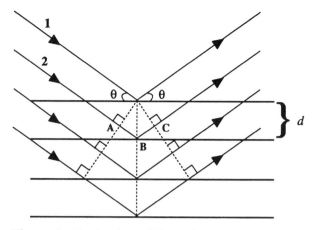

Figure 1. Derivation of Bragg's Law. Incident X rays are reflected by parallel planes within a crystal, and the reflected rays are of highest intensity when they interfere constructively. Total constructive interference occurs when the reflected rays are completely in phase, and the exiting rays are in phase when the difference in the path lengths traversed by the rays is equal to an integral number of wavelengths. In the diagram above, incident radiation of wavelength λ is reflected at an angle θ from a series of planes that are separated by distance d. The difference in path distances traveled by rays 1 and 2 is equal to AB + BC = 2AB = $2d \sin\theta$. Constructive interference occurs when $n\lambda = 2d \sin\theta$, where n is an integer.

Bragg transferred from the University College, London, where he had been Quain professor of physics since 1915, to serve as director of the Royal Institution in the Davy-Faraday Research Laboratory. From 1935 to his retirement in 1940, he served as president of the Royal Society. During these years he helped determine the structures of the silica (SiO_2) minerals, and he made contributions to organic crystallography. By the time of his death, he had been awarded numerous honors, including the Rumford Medal (1916), the Copley Medal (1930) of the Royal Society, and knighthood in 1920.

W. L. Bragg succeeded Ernst Rutherford as professor of physics at Manchester University in 1919, and for the next eighteen years he oversaw a group of young crystallographers who unraveled the atomic structures of the complicated silicate minerals. Though he claimed not to be a mineralogist, Bragg published a textbook entitled "Atomic Structure of Minerals" in 1937 with a revision in 1965. After a brief tenure as Director of the National Physical Laboratory from 1937 to 1938, W. L. Bragg again succeeded Rutherford as Cavendish Professor of Experimental Physics in 1939. His interests turned towards metal alloys and the structures of proteins, and it was under his directorship that James Watson and Francis Crick solved the structure of DNA. Bragg served as Fullerian Professor of Chemistry at the Royal Institution in London from 1953 until his retirement in 1966.

W. L. Bragg married Alice Hopkinson in 1921, and he was father to four children, Stephen Lawrence, David William, Margaret Alice, and Patience Mary. Like his father, by the time of his death he had accumulated a great many honors, including the Hughes (1931), Royal (1946), and Copley (1966) Medals of the Royal Society, knighthood in 1941, the Roebling Medal in 1948, and the Companion of Honor in 1967.

Bibliography

BRAGG, L., and G. F. CLARINGBULL. *Crystal Structures of Minerals.* London, 1965.

EWALD, P. P., ed. *Fifty Years of X-Ray Diffraction.* Utrecht, 1962.

LONSDALE, K. *Crystals and X-Rays.* London, 1948.

THOMAS, J. M., and D. PHILLIPS, eds. *Selections and Reflections: The Legacy of Sir Lawrence Bragg.* Northwood, Eng., 1990.

PETER J. HEANEY

BRONGNIART, ALEXANDRE

Alexandre Brongniart contributed significantly to the young science of geology during the first decades of the nineteenth century while also serving as director of the National Porcelain Factory at Sèvres (1800–1847). The son of Anne-Louise d'Egremont and the architect Alexandre-Théodore Brongniart, Alexandre was born in Paris, France, on 5 February 1770. He achieved success as a teacher of mineralogy, field geologist, ceramist, and administrator. Honors included election to the Académie des Sciences (1815), appointment to a professorship at the Muséum d'Histoire Naturelle (1822), and recognition as a commander in the Légion d'honneur (1845). He died in Paris on 7 October 1847.

Brongniart's ancestors had been ennobled in the early seventeenth century, and his parents operated in the upper echelons of Parisian society. The sculptor Jean-Antoine Houdon produced a bust of Alexandre as a child, and the artist Madame Vigée-Lebrun, a family friend, painted his portrait at various ages. Alexandre received an excellent education, with a scientific focus designed to help him enter the lucrative professions of medicine or pharmacy. While serving in the military during the tumultuous years of 1793 and 1794, Brongniart became interested in the geology of the Pyrenees Mountains. Despite his mother's concern about lack of financial rewards outside of the medical profession, Alexandre turned increasingly to zoology and geology. He published articles with a zoological focus and wrote geological and mineralogical pieces for the *Dictionnaire des sciences naturelles.* (See De Launay (1940), Rudwick (1970), and Ellenberger (1994) for details concerning Brongniart's eclectic output.)

Because of his background in chemistry and his field experience in geology, Brongniart received an appointment (1794) as a mining engineer for the government. In 1819 he became chief engineer of the Corps des Mines. Travels devoted to geology or ceramics took him all over Europe and England. After rigorous study of minerals and ores, Brongniart became a professor of mineralogy, initially at several Parisian lycées and colleges, then at the École des Mines and the Sorbonne, and finally held the prestigious chair of mineralogy at the Muséum d'Histoire Naturelle (1822).

In February of 1800, Brongniart married Cécile Coquebert de Montbret. Their son Adolphe-Théodore became a renowned paleobotanist. Daughter Herminie married the brilliant chemist and physiologist Jean-Baptiste Dumas, and Mathilde married Victor Audoin, a professor of entomology.

As the nineteenth century dawned, Brongniart established a close working relationship with Georges Cuvier (*see* CUVIER, GEORGES). Although less celebrated than his colleague, Brongniart played a critical role as one of the founders of biostratigraphy, a point stressed by G. Gohau (1990). Cuvier himself acknowledged that Brongniart made many of the observations and wrote much of the text incorporated in their *Essai sur la géographie minéralogique des environs de Paris* (1808). This essay was republished in 1810 and 1811, and greatly amplified in 1822 and 1835. The work powerfully demonstrated the merit of fossils as aids in interpreting Earth history. Not only did the authors state that organic remains in the rock never deceived them as indicators of relative age, they also used fossils to recognize environments of deposition, distinguishing freshwater from marine strata.

Even from a modern perspective, Brongniart and Cuvier's analysis of the Paris basin is impressive. The 1811 version included a geologic map of the region, and the 1822 edition contained 428 pages of text, plus plates displaying geological columns, fossils characteristic of particular strata, and colored maps portraying bedrock distribution. Brongniart, in solo-authored sections, demonstrated that rock strata in the Paris basin could be correlated with units in England and throughout Europe. Publication of the *Essai* postdated the unpublished commentaries of William Smith (1769–1839) on faunal succession, but the French work became well known in Europe before production of Smith's famous map of 1815.

Cuvier and Brongniart (1822) also called attention to breaks in the sedimentary record, what we now term disconformities, and they suggested that a great time interval might be involved (Conkin and Conkin, 1984). Brongniart had a reputation as a superb observer of nature who was cautious about applying theoretical constructs to what he saw. Nonetheless, he initially shared with Cuvier a belief in catastrophism, or sudden changes in the unfolding of Earth history, and they both accepted elements of the Neptunian theory of Abraham Gottlob Werner (*see* WERNER, ABRAHAM G.). Werner argued that all rock resulted from sedimentation from a universal ocean. He also felt that rock type could serve as a viable means of correlating strata from one place to another. Cuvier and Brongniart used the sharp boundaries produced by disconformities to strengthen their belief in a punctuated history of the earth, noting that geologic revolutions, possibly related to convulsive activity caused by the planet's internal heat, might be responsible for mountain genesis and organic extinction. Thus, they believed that the facts gathered in the field supported selected theoretical conclusions.

An innovative feature of the Cuvier and Brongniart analysis was their recognition of subtle variations in rock type within the same formation.

These differences, designated "facies" by Amanz Gressly (1814–1865) in 1838, were attributed to deposition under varying environmental conditions. The authors also used the evidence of alternating marine and freshwater deposits to undermine any notion of simple deposition from a universal marine sea. And they both felt that fossils served as better correlation tools than lithology or structural relationships of strata.

Brongniart's name is understandably associated with Cuvier because their Paris basin work represents a seminal statement in the evolution of the geosciences. But the linkage should not obscure the fact that Brongniart wrote noteworthy scientific articles on his own. His training in zoology led to a well-received paper on the classification of reptiles (1800) and allowed Brongniart to develop a detailed but unpublished study of the musculature of apes, comparing the muscle groups to human anatomy. Dedicated to education, he wrote a two-volume *Traité élémentaire de minéralogie* (1807) specifically designed for students in French lycées. He authored pioneering papers (1815, 1822) on trilobites, which he recognized as extinct. In a little-known but significant work (1823) on the geology of the Italian Alps, Brongniart argued that not all mountains were primitive. The book's plates illustrate Italian fossils that clearly demonstrate their affinity with Tertiary faunas of the Paris basin. Also noteworthy, but not as important to posterity as Brongniart had hoped, was his 1829 book *Tableau des terrains qui composent l'écorce du globe*. This "essay on the structure of the Earth" presented an outline of the epochs of Earth history, listing rock types and fossils characteristic of each stage in the planet's development. The Jovian period represented the modern Earth; older rocks dated from the Saturnian period and had class names such as Clysmian and Yzemian. This novel nomenclature appeared a few years before Adam Sedgwick (*see* SEDGWICK, ADAM), Roderick Murchison (*see* MURCHISON, RODERICK), and others established the geologic column, with its now-familiar names Cambrian, Ordovician, Silurian, and so forth. Some authors have made light of Brongniart's terminology, but we must see his effort in the context of its time and recognize that Brongniart, ever the classifier, was attempting to order all of Earth history.

Brongniart's ability to combine science and art contributed to his appointment and success as director of porcelain manufacture at Sèvres. At the outset, Brongniart fulfilled Napoleon's aspirations by revitalizing the manufacturing process, including the move from soft- to hard-paste porcelain. He added buildings, opened a museum, and instituted scientific procedures of testing and quality control. A neoclassicist in taste, he advocated painting on porcelain and appreciated the work of Jacques Louis David (1748–1825) and Jean Auguste Ingres (1780–1867). Brongniart's last publications were the *Traité des arts céramiques* (1844) and the *Description du musée céramique de Sèvres* (1845). The treatise forcefully demonstrates the depth of Brongniart's knowledge about all phases of ceramics production, from gathering clay in the field to production of specific glazes. The book abounds with chemical information, graphs, and detailed plates, similar to those in Denis Diderot's *Encyclopédie*, portraying each step in the manufacture of porcelain.

The personal portrait of Brongniart that emerges from accounts of family and colleagues is of an austere, observant, meticulous, quietly friendly, yet occasionally pessimistic individual. He enjoyed Sunday breakfasts with his students, often discussing specimens and current ideas. Constant Prévost (1787–1856), Omalius d'Halloy (1783–1875), and the American Lardner Vanuxem (1792–1848) profited from Brongniart's sympathetic teaching style. The Brongniarts entertained many celebrated figures of the era, including Michael Faraday, Alexander von Humboldt, André Ampère, Hans Christian Oersted, Jean-Baptiste Lamarck, Jean Auguste Ingres, and Eugène Delacroix. Work was a prime motivator, however, and for decades Brongniart compartmentalized his weeks by moving among residences at Sèvres, the museum, and his private home in Paris.

Bibliography

Conkin, B. M., and J. E. Conkin. *Stratigraphy: Foundations and Concepts.* New York, 1984.

Cuvier, G., and A. Brongniart. *Description géologique des environs de Paris.* Paris, 1822.

De Launay, L. *Les Brongniart: Une grande famille de savants.* Paris, 1940.

Ellenberger, F. *Histoire de la géologie, tome 2.* Paris, 1994.

Gohau, G. *A History of Geology.* New Brunswick, NJ, 1990.

Rudwick, M. J. S. "Alexandre Brongniart." In *Dictionary of Scientific Biography,* ed. C. C. Gillispie. New York, 1970.

KENNARD B. BORK

C

CAMBRIAN EXPLOSION

See Higher Life Forms, Earliest Evidence of

CAREERS IN THE EARTH SCIENCES: ATMOSPHERIC SCIENCES

When most people think of careers related to the atmospheric sciences or meteorology, they think of the weatherperson they see on television. While television weathercasting is certainly the most visible career associated with the field, there are many other opportunities available for men and women in the atmospheric science profession. Whether it is in preparing forecasts of daily weather and broadcasting them on television or radio, monitoring severe weather such as hurricanes and issuing warnings, assessing the impact of the air pollution from a power plant on the surrounding community, or studying the changes in Earth's climate, meteorologists and atmospheric scientists are directly involved in trying to make life better. It is this desire to do something worthwhile, as well as a fascination with the weather, that attracts most

meteorologists and atmospheric scientists to this profession.

Educational Requirements

Almost all career opportunities require at least a bachelor's degree in meteorology or atmospheric science. There are about sixty-five colleges and universities in the United States that offer bachelor degree programs in either discipline. To avoid confusion, it is worthwhile to say a little about the distinction that is made between the terms "meteorology" and "atmospheric science." Meteorology is the science of the atmosphere. It includes the study of the physics, chemistry, and dynamics of the atmosphere with the goal of trying to understand how the atmosphere works and to forecast its future state. The term "atmospheric science" is often used to describe a combination of meteorology and other branches of physical science, such as chemistry. Since the term "meteorologist" is often associated with individuals who work in forecasting the weather, a scientist who studies the chemical reactions taking place thirty miles above the surface of Earth may choose to be called an atmospheric scientist rather than a meteorologist. At the undergraduate level, there is virtually no difference between a program that offers a B.S. degree in meteorology and one that offers a B.S. in atmospheric science—colleges and universities have

63

simply chosen to call their programs by one name or the other.

As with any physical science, college students studying meteorology or atmospheric science take several courses in calculus, physics, and chemistry during their first two years before they begin taking courses in meteorology. High school students thinking about pursuing the atmospheric sciences should, therefore, take as much mathematics and science as possible before entering college in order to provide a background for their further studies.

Many career opportunities, especially those related to teaching at the college level or conducting research, require a master's or Ph.D. degree. Students who do very well in college are often encouraged to pursue a higher degree so that they will be able to seek these higher-level jobs. An important consideration for those thinking about obtaining a graduate degree is that most schools are able to provide financial support for students studying for a master's or Ph.D. degree. Thus, while most students must pay for their first four years of college (although there are some scholarships available, such as those offered by the American Meteorological Society), those students who are qualified for graduate study can usually obtain an advanced degree without paying any tuition and while earning a small living stipend.

Job Opportunities

A large percentage of all meteorologists work for the government in one way or another. Many of them are employed by the National Weather Service (NWS) or serve as meteorologists in the military. Forecasting the day-to-day weather and issuing watches and warnings for severe weather are the primary responsibilities of meteorologists in these positions. They work with some of the most sophisticated computers and equipment available to monitor the weather continuously and try to predict it accurately. All forecasters, including those we see on television, make use of the data collected by the military and NWS observation systems, and they all use the output from the NWS computer models.

The government does not represent the only source of jobs for individuals who want to forecast the weather. In addition to those meteorologists who work at television stations, there are many meteorologists who do forecasting for private companies. These companies make forecasts tailored to the needs of customers, such as power utility companies who want a specialized service not provided by the NWS forecasts. Private forecasting companies also provide forecasts for many radio stations across the country.

Research meteorologists and atmospheric scientists work primarily at government laboratories or university research centers. Their work includes the study of global issues, such as the way humans may be changing the climate as a result of atmospheric pollution, down to localized phenomena, such as trying to understand the airflow inside a tornado. Professors of atmospheric science at universities also undertake research in addition to their teaching duties.

A growing area of opportunities for meteorologists and atmospheric scientists is in private consulting firms. Consulting meteorologists carry out the environmental impact assessments required under the Clean Air Act and perform other activities in which their knowledge of the atmosphere is required.

Employment Outlook

The atmospheric sciences is a relatively small field in the United States with between 20,000 and 30,000 practicing meteorologists and atmospheric scientists, and about 1,000 new professionals enter the work force each year. Historically, there have been about as many positions available as there are new graduates looking for work so almost all are able to obtain a job in the field if they are diligent in their job search. Yet few are in a position to choose between several job offers and most have to be flexible concerning both the region of the country and the type of position they will consider. Changes occurring at the NWS and a growing political awareness of the environment are expected to improve this situation somewhat over the coming years.

Through the 1990s, the National Weather Service will undergo a massive program of modernization and restructuring so that it can take maximum advantage of the latest technology to better serve the public. The NWS anticipates hiring additional meteorologists with B.S. and M.S. degrees throughout this decade of modernization, so the job opportunities are better than they were throughout the 1970s and 1980s.

An ever-increasing concern about the environment by society, especially with regard to global topics such as the destruction of the ozone layer and the potential for climatic change, have resulted in an increase in funding of research related to these atmospheric issues. In addition, more stringent laws concerning air pollution control have led to an increase in the role played by air pollution meteorologists in industry so that environmental consulting companies are expected to continue being a good prospect for employment.

For more information on career opportunities in the atmospheric sciences, write to: American Meteorological Society, 45 Beacon Street, Boston, MA 02108–3693.

Bibliography

AMERICAN METEOROLOGICAL SOCIETY AND THE UNIVERSITY CORPORATION FOR ATMOSPHERIC RESEARCH. *1992 Curricula in the Atmospheric, Oceanic, Hydrologic, and Related Sciences.* Boston, MA, 1992.

AMERICAN METEOROLOGICAL SOCIETY. *Challenges of Our Changing Atmosphere.* Boston, MA, 1993.

KEITH L. SEITTER
STEPHANIE K. KEHOE

CAREERS IN THE EARTH SCIENCES: ENGINEERING GEOLOGY AND GEOLOGIC ENGINEERING

Toward the end of the twentieth century, the earth sciences reached a new high level of importance in nearly all activities involving engineered works and facilities designed to provide and protect human welfare and safety. Engineering geologists and geological engineers are foremost in collecting and assessing geological data for purposes of commercial and industrial development, environmental protection and remediation, and mineral and water resource recovery. The differences between these two geological disciplines are numerous, but in fact the titles themselves indicate the dominant differentiation.

Engineering geologists (EGs) are earth scientists who make use of existing classical geologic studies and who undertake special field exploration programs to collect geologic data that physically and chemically characterize sites for the purposes of human activity. Most human activity involved with the surface of the earth, with the exception of agriculture, requires collection and evaluation of geologic data. Geologic data are essential for environmental protection, resource management, and mitigation of geologic constraints that often adversely affect the health, welfare, and safety of people. Each year since the late 1960s, laws and regulatory requirements have combined with legally-based liabilities to further emphasize the need and obligation to collect and utilize geologic information. EGs serve well to identify and describe, both qualitatively and quantitatively, such essential geologic characteristics. As scientists, EGs do not provide a design-related function; their role is to characterize and to define the presence, three-dimensional location and bounds, and statistical makeup of the geologic elements of engineered design.

Engineering geologists are indispensable to all engineered works. EGs are found on the staffs of state and federal government scientific, regulatory, and construction agencies; major military commands; architectural engineering, environmental engineering, and geotechnical consulting firms; and at large construction companies and some large industries. Most EGs also serve as hydrogeologists. An EG must be, first and foremost, a competent field geologist with a broad knowledge of geologic fundamentals. Nearly all of their academic training is in this science; however, success in practice requires a good understanding of the design process of engineers, ability to read construction drawings, excellent communication skills, and an effort to learn the fundamentals of whatever industrial or mineral resource activity is being supported by their efforts.

Geological engineers (GEs) are, by nature, part engineer and part geologist, depending on individual emphasis. GEs are trained with the same lower-division science courses as EGs, but without the same upper-division specialty courses. GEs receive lower-division academic training equivalent to other engineers and then employ this training to describe quantitatively construction sites and to engage in design functions directly involving their special materials: rock, weak rock, soil, and water. Upper-division courses (last two years of the undergraduate program) concern the technical aspects of applied geology—in other words, the quantitatively based courses in geology applied to engineering. Academic training of GEs remains under the strict purview of the Accreditation Board for Engineers and Technologists (ABET) and graduates are eligible to take the Fundamentals of Engineering and Engineering Practice examinations, which lead to registration as Professional Engineers. In North America, GEs also are required to complete enough geology courses to qualify for registration as Professional Geologists and as Certified Engineering Geologists.

Relatively few university programs produce EGs (about fifteen) and GEs (twelve) in the United States. The situation essentially is equivalent for Canada.

Regulatory staffing for implementation of the post-1969 succession of federal and state environmental laws has created a burgeoning demand for EGs and GEs. Regulatory program manager positions filled by EGs and GEs are nearly unlimited, with the only real employment constraint being that of identifying and applying for individual positions. With the decline of the North American mineral industry, following passage of the federal NEPA (National Environmental Policy Act) in 1969 and major waste management laws, large numbers of displaced petroleum and minerals-industry geologists have retrained as regulatory specialists and as hydrogeologists. This trend peaked between 1983 and 1991, and those who were successful in the transition largely had been absorbed by 1993. Advancement in the consulting firms, the U.S. Army Corps of Engineers, U.S. Bureau of Reclamation, and state regulatory agencies traditionally is tied to the achievement of professional registration.

GEs have served in the petroleum and minerals industries, and the degree of Geological Engineer was created at the Colorado School of Mines in 1913 for that purpose. The state of Arizona recognized the application of geology in its first professional registration program, which was put in place in 1921 and included engineers; Arizona was followed by California in 1969, where registration of geologists also included specialty certification of engineering geologists.

Reflecting the ongoing demise of the North American minerals industry, employment opportunities for EGs and GEs in mine development and minerals recovery are nominal at best. Most of the work since about 1990 has been in the growing field of remediation of abandoned mines and mineral beneficiation plants and their refuse "piles"

and "dumps." Both specialties remain active in surface mining of coal and in industrial minerals recovery and mining of construction materials such as aggregates, silica sand, and limestone. In all cases, employment opportunities for EGs and GEs are relatively plentiful, yet special effort and attention is required by faculty members and graduating students to identify such positions, to invite potential employers to visit and lecture at campuses, and to determine the academic course content and relevant term-project assignments needed to supply properly trained professionals. The U.S. federal hiring process for EGs and GEs is complicated. Neither discipline is yet recognized by the federal Office of Personnel Management. EGs are lumped with physical scientists or geologists and GEs can qualify for these classifications as well as civil engineer or environmental engineer. In each case, the applicant must identify subject courses that he/she believes are relevant to those classifications (with the government's appropriate application form), to avoid a ten-point penalty on their overall rating score.

Nearly all (approximately 85 percent) of EGs and GEs currently work on assignments driven by some form of law or regulation. Prediction of site characteristics is an absolute requirement for all engineering design for activities in or on Earth's surface. Mitigation of geologic constraints (hazards) is not possible without the same prediction. Remediation of the environmental threats of uncontrolled waste disposal is patently impossible without the same work product by EGs and GEs. American environmental protection laws have evolved so that protection activities or remedial actions are now measured as successful in terms of their ability to control disposed waste masses and to maintain or improve groundwater and surface water quality. Design measures for this control must uniformly consider site and waste characterization, most of which is developed by EGs and GEs.

Both of these earth science specialties represent excellent and unusually broad avenues of potential employment and career opportunities. There is no reason, at the present time, for graduates of either specialty not to find suitable and challenging employment in almost any field of environmental regulation. Positions with industry and consulting engineering and geoscience firms will represent the most demanding of academic training and relevant employment experience, including professional registration as a geologist, engineering geologist, or engineer (geological or civil). A strong competence in oral and written communications will be required in all instances and generally will be perceived by employers as more important than computer literacy. Graduates of both disciplines should be competent in either hydrogeology (field-oriented/concepts emphasis) or geohydrology (numerical/computer analysis emphasis).

Bibliography

KIERSCH, G. A., ed. *Heritage of Engineering Geology—The First Hundred Years*. Geological Society of America, Centennial Special, vol. 3. Boulder, CO, 1991

WILLIAMS, J. W., and J. D. HIGGINS, eds. "Academic Preparation for Careers in Engineering Geology and Geological Engineering." Association of Engineering Geologists Special Publication, no. 2. Sudburg, MA, 1991, pp. 67–68.

ALLEN W. HATHEWAY

CAREERS IN THE EARTH SCIENCES: ENVIRONMENTAL SCIENCE

Earth science differs in many respects from other sciences such as chemistry and physics. Unlike chemists and physicists, who perform their own experiments, the earth scientist studies an environment already altered by natural processes. Earth scientists use a variety of tools and techniques to understand how a particular environmental setting was created. These Earth detectives comprise a wide range of subdisciplines that include geochemistry, geomorphology, geophysics, and at least fifteen other major types of specialization. While chemists and physicists usually conduct their work in a single building at one location, earth scientists travel to various areas on the globe to study a geological framework that differs in some respect from region to region. The subject matter of earth science can be very complex because of the many variables that are operating in a system that involves matter, dynamic Earth processes, and time.

Employment Opportunities

Earth scientists will always be needed because their discipline is so intertwined with society in obtaining necessary raw materials and overseeing geologic hazards and changes that threaten lives and property. A doubling of Earth's population in the next several decades will place increasing stress on the environment. Earth science will be in increased demand to discover new geologic resources to replace today's dwindling deposits and shrinking amounts of habitable land.

Because of the large number of subdisciplines within the earth sciences, specialists are employed in a number of different areas. Earth scientists are hired by government agencies, at all government levels from city and county to the fifty states and federal government. Many of the eighty thousand earth scientists in the United States are employed in the extractive industries of mineral and energy, such as petroleum and coal. Thousands are found in geoscience consulting firms, where some single firms employ several hundred scientists. Others are employed in the teaching profession, including high schools, colleges, and universities. Some jobs are strictly in offices and laboratories, whereas others are largely field-oriented.

Natural resources provide the backbone of civilization, and their location, mining, and extraction will always require large numbers of geoscientists. However, in the decades since the 1970s there has been a dramatic shift in the earth sciences toward involvement in environmental affairs. These positions comprise evaluation of geologic hazards (volcanoes, earthquakes, landslides, and floods), and of degradation by either natural or human causes. Hydrogeology (the study of water on and within earth materials) is the fastest growing specialization, as indicated by the number of want ads placed in trade journals, and by the number of scientists entering the field. Geoengineering is growing because society has become increasingly aware of the need to build safe structures and protect lives and property. Geotechnology (the use of industrial materials in a geologic context) is now recognized as a separate discipline. The National Research Council of the National Academy of Science has shown that this area has increased to such an extent that a geotechnology board has been formed to provide guidelines for further growth.

Training and Education

Geoscience enrollments in higher education during the 1991–1992 period were 18.8 percent higher than in 1990–1991, with a 22.8 percent increase in total number of degrees. This trend reversed an enrollment decline during the 1983–1990 period. B.A. and B.S. degrees with a major in the geosciences can form the basis for some entry-level positions, but advanced degrees and training are invariably required for more exacting jobs. The M.A./M.S. and Ph.D. degrees have generally become essential because of the increased complexity and sophistication of geological instruments and equipment and the skills necessary to understand, interpret, and report their results. However, as with many other types of academic endeavors, job performance is always enhanced by the experience of the scientist. Thus it is possible for earth scientists with B.A./B.S. degrees to rise to higher positions with greater authority after performing several years of meritorious work. Such advancement is hastened when communication skills, both verbal and written, can be favorably demonstrated.

The increased interest in earth science is a reflection of several practical considerations. The general public now has more education awareness of environmental problems and is demanding that greater attention be placed on solutions. Governmental laws and regulations necessitate various types of geoscience reports and monitoring systems. The infrastructure of such societal needs as water supply, waste disposal, pipelines, highways, bridges, and settlement patterns is in constant need of building, maintenance, and repair.

In the training of earth scientists it is important to stress the need for both up-to-date geology courses and ancillary courses. More attention needs to be placed on those courses that emphasize the physical and the historical aspects of geology. Mathematics and computer skills have become the language of science, and courses with this subject matter must become requirements.

Types of Environmental Problems and Global Issues

Earth scientists address the full complement of problems that beset society. The natural resources required for human endeavors must be located and extracted. Geoscientists become especially visi-

ble and necessary when consideration is given and decisions reached about the geologic hazards of volcanoes, earthquakes, landslides, and floods. Maps of hazard-prone areas must be prepared, and whenever possible suggestions made and protocols established to minimize losses to humans and property. Water resources comprise a large family of potential issues concerning location, storage, distribution, and contamination. Solid waste and its disposal in landfills require much input from the geologist. The loss of soil by human action in farming, deforestation, and construction projects creates problems that need resolution. In all these matters it is important that the earth scientist possess a sufficient database for possible prediction of deleterious impacts that may occur and to offer types of remediation.

The U.S. Global Change Research Program was enacted by Congress in 1989 to enhance the research activities of various federal agencies. It was funded initially at a level of $133.9 million, which increased to $1.1 billion in 1992 and to nearly $1.5 billion in 1994. Its objective is "to produce a predictive understanding of the Earth system, and to support national and international policy-making activities across a broad spectrum of global, national, and regional environmental issues." The Earth Summit Conference held in Rio de Janeiro, Brazil, in 1992, was organized to bring representatives from all nations to discuss environmental matters. A principal issue at the conference was an attempt to determine the point where economic goals are abandoned in favor of environmental considerations. The resulting eight-hundred-page document was touted as "the blueprint for action" in the twenty-first century. A common thread of sustainable development was woven into the fabric of environmental protection.

Pressing global issues that cross national borders, such as acid rain, desertification, ozone depletion, and ocean contamination, remain to be solved on the international level.

Bibliography

BARANOVIC, M. J. "Employment Opportunities in the Oil and Gas Industry." *Geotimes* 38 (1993):16–17.

CLAUDY, N. "Geoscience Careers—The Diversity Is Unparalleled." *Geotimes* 36 (1991):15–16.

COATES, D. R. *Geology and Society.* New York, 1985.

GERHARD, L. C. "Framing Policies in Resources and the Environment." *Geotimes* 3 (1994):20–22.

PRICE, J. G. "The Geosciences in Review: Outlook, Probability, Predictions, and the Profession." *Geotimes* 39 (1994):10.

SUITER, M. J. "Tomorrow's Geoscientists: Recruiting and Keeping Them." *Geotimes* 36 (1991):12–14.

DONALD R. COATES

CAREERS IN THE EARTH SCIENCES: EXPLORATION AND MINING

Discovery and development of mineral and fuel resources are heavily reliant on the application and integration of principles of basic science with fundamentals of earth science. The search for resources has traditionally required broadly educated and experienced earth scientists whose knowledge and education span the range from fundamental geological principles to the application of specialized aspects of chemistry (geochemistry) and physics (geophysics). Traditionally, the earth scientist has played an important role in the solution of many engineering problems attending development and mining of mineral resources. There are many regions on Earth where, for a variety of reasons dealing with politics, access, and potential of discovery, exploration has not been carried out or is incomplete. Conventional exploration involving the ground search of such regions by earth scientists was still taking place in the 1990s, and exploration is likely to continue well into the foreseeable future. The discovery, extraction, and use of minerals and fuels are baseline activities in the development of the economic well-being of nations and societies (*see* ECONOMIC GEOLOGY, HISTORY OF)

The future of exploration activities in the United States, as judged by trends of activity in the 1980s and mid-1990s, is less certain, however, as national debates concerning land use, ownership, preoccupation with environmental issues, and varied perspectives concerning the value of mineral resources weighed heavily upon decisions of land use. Whatever the outcome of these debates, the special knowledge and background of individuals whose expertise lies in mineral exploration lend themselves well to recognition and solution of a variety of mineral-resource related geological problems faced by the minerals industry and society. Employment opportunities in the western hemisphere, outside of the United States, were increasing in the mid-1990s.

Scope of Activity

Activities of the exploration or mining geologist have become increasingly diversified in the late decades of the twentieth century. As exploration in the United States has "matured" the search for resources has moved to less well explored regions. As attitudes and societal problems seem to be changing perspectives on resource extraction in the United States, the exploration geologist is faced with a future of changing relationships with society and expanding populations.

The earth scientist engaged in the traditional ways of exploration searches Earth's crust for concentrations of elements or minerals of value to society, or evaluates known concentrations or occurrences of mineral resources for useful chemical or physical properties. Others are engaged in the search for mineral fuels that include uranium and coal.

Exploration geologists (or geophysicists or geochemists) are involved in the quest for a broad range of naturally occurring commodities; these range from hidden bodies of concentrated base and precious metals with high intrinsic value, such as lead or copper and gold or platinum, to evaluation of exposed deposits of commodities of low intrinsic value, such as brick clay, limestone, or sand and gravel. The search takes place at two extreme ranges of scale and may involve the concern of groups with different levels of interest. At one extreme, the exploration geologist may work in the field for a corporate group in specific predetermined locations evaluating rocks and minerals in the context of their occurrence with a specific commodity. At the other extreme, the exploration geologist may work for some level of government or a corporation on broad geological problems of resource assessment.

A search for mineral resources may commence with an evaluation of the geology and geological history of a region, following which a more restricted region may be identified for on-site evaluation. Site evaluations involve application of many basic geological tools, including petrography, mineralogy, structural geology, and mapping, upon which interpretations of the existence of a mineral commodity may be based. The interpretations may be enhanced by consideration of information developed from chemical compositions of rocks, soils, and waters and by the data developed from the measurement of the physical properties of the crust and its varied rocks and structures.

The earth scientist involved in mining interfaces with engineers and engineering problems. At the fundamental level this scientist, called a mining geologist, develops and interprets data that allow engineers to extract rocks with the optimum economically recoverable levels of minerals or metals present. This activity, grade control, is coupled with continuing efforts to discover additional hitherto unknown resources within a mined volume of the crust. Of equal importance, the mining geologist applies engineering principles to the problems of rock extraction that enable safe and economic removal, and may contribute special knowledge to extractive metallurgy and waste control and containment.

Changing Trends in Activity

Search for mineral resources continues in many regions of Earth in the ways outlined above. In areas where a history of resource extraction exists, however, the unique perspective of the exploration and mining geologist has been adapted to the changing requirements involved in discovery, mine evaluation, and mining. The knowledge of surface and subsurface geohydrology is increasingly involved in evaluation of resources and their impact. The special knowledge of earth scientists of the way minerals react with water, together with principles of geohydrology will be increasingly brought to bear to solve special environmental problems in the 1990s that are associated with resource extraction (*see* POLLUTION OF GROUNDWATER AQUIFERS).

With a view to the future, the knowledge of the mineral industry geoscientist is focusing on new ways of mineral extraction, especially the in situ leaching of metals and minimizing environmental impact. In those regions of little exploration or of difficult access, the exploration geologist is turning increasingly to the new tools being developed in regional geophysics, including remote imagery and the application of the power of the computer to solve complex problems.

Education and Career Preparation

Traditional ways of exploring the planet coupled with the advent of new ways of looking at it require a fundamental background in mathematics and the basic sciences to stay abreast of evolving technology. The earth scientist in exploration must develop a level of intellectual comfort with the rocks, fabric, and history of the crust; such geologists work in all kinds and ages of geological materials and terranes. Their background is developed from both broad undergraduate and specialized graduate education and, just as important, from experience gained from observation on the ground.

The explorationist is becoming less isolated from society in pursuit of professional objectives. Heightened interpersonal skills, second-language fluency, and broad experience are important; whereas the exploration geologist once had to convince only his employers that a body of minerals occurred at a specific location, today the explorationist must work ever more closely with the public, regulators, and administrators in pursuit of those objectives. The field remains scientifically demanding with many opportunities for diverse kinds of employment in many regions. However, notwithstanding technological advances, the discovery of resources ultimately will always require experienced and educated persons to evaluate the rocks in the field.

Bibliography

DAVIS, P. A. "Geology in Crisis: The Geologic Profession in the Service Economy." *The Professional Geologist* 30 (1993):12–15, 18.
PRICE, J. G. "Some Perspectives on the Future of Geology in Mining." *The Professional Geologist* 31 (1994): 6–7.
WAIDLER, R. J. "Reflections of a Mining Geologist." *The Professional Geologist* 32 (1995):5–6.

SPENCER R. TITLEY

CAREERS IN THE EARTH SCIENCES: HYDROLOGY

Hydrology is the study of the occurrence, distribution, movement, and quality of water. Graduates in hydrology are in great demand because of increasing water needs for agriculture, industry, and cities. Hydrologists apply scientific knowledge to solve water problems that face society.

Hydrology is an interdisciplinary science that incorporates biology, chemistry, geology, mathematics, and physics. Hydrologists investigate the occurrence, availability, physical, biological, and chemical character of water and its relation with the environment. They investigate sites for waste disposal. Work at these sites involves geologic and hydrologic mapping, geophysical surveying, well drilling, taking samples and measurements in the field, determining hydrologic and geologic boundaries, designing computer models of water systems, interpreting results from models and from field evidence, and proposing action to protect groundwater.

Employment Opportunities

Hydrologists are employed in federal, state, and local governments and in the private sector. A 1988 survey by the American Geophysical Union of employment in hydrology gave the following profile: 27 percent, education; 7 percent, state and local government; 32 percent, federal government; 32 percent, industry, business, and self-employed. The employment breakdown was 53 percent in surface water and 47 percent in groundwater.

A 1990 survey of hydrogeologists by the National Water Well Association showed that 50 percent worked in private consulting firms, 20 percent in state government, 8 percent in federal government, 8 percent in industry, 5 percent in education, and 4 percent were self-employed. The mean annual earnings in 1990 were $38,257. About 26 percent had a bachelor's degree, 63 percent a master's degree, and 11 percent a doctoral degree.

Supply and Demand

There is a severe shortage of trained hydrologists to work on the environmental problems facing the nation. The Congressional Research Service estimates that 10,000 hydrologists are needed to deal with the protection and cleanup of contaminated groundwater and surface water in the United States.

The contamination of groundwater is one of the most significant environmental issues facing the

nation and is the focus of much public interest and concern. Contamination has caused the closing of more than 2,800 public and private wells. Federal and state agencies, as well as the private sector, will spend large sums of money to clean up contamination from abandoned waste sites.

The employment picture for hydrologists is very bright because of additional funding for water quality and water supply issues at all levels of government. The interest in hydrology is spurred by the U.S. Environmental Protection Agency's (EPA) regulatory programs, the Department of Defense (DOD), and Department of Energy (DOE) environmental restoration programs. Congress has recently passed legislation to protect the environment. The Drinking Water Act amendments provide funds for states to develop groundwater protection programs. The Superfund amendments authorize $8 billion for cleanup at abandoned hazardous waste sites. The Clean Water Act provides $18 million for projects to reduce water pollution. DOD recently expanded its Installation Restoration Program to clean up groundwater pollution at military installations. This program involves some 450 military bases and an estimated 800 hazardous waste sites. DOD plans to spend more than $500 million to mitigate groundwater pollution. The Department of Energy is carrying out similar programs to evaluate and mitigate hazardous waste contamination.

The following summary presents a breakdown of the kind of work carried out by various governmental agencies and in the private sector.

Corps of Engineers (Corps Eng.): Conducts feasibility studies, designs and constructs dams and navigation systems.

U.S. Environmental Protection Agency (USEPA): Manages extensive studies and issues regulations related to water pollution control involving groundwater and surface water. Conducts research on a variety of issues, including toxic and hazardous waste contamination.

U.S. Geological Survey (USGS): Conducts inventories, assessments, and research on the nation's surface water and groundwater. Monitors and appraises water resources in the United States.

Bureau of Reclamation (USBR): Constructs and maintains irrigation and hydroelectric systems.

Manages water supply, flood control, water quality, and pollution control projects.

Soil Conservation Service (SCS): Manages land-use activities to reduce the hazards of floods and sedimentation in rivers and streams. Develops conservation techniques to assure efficient use of water and land resources.

Department of Energy (DOE): Conducts research and development in areas of fossil energy, nuclear energy, and conservation and renewable energy. The siting of nuclear waste repositories requires geologic and hydrologic considerations.

Other Federal Agencies: The Nuclear Regulatory Commission (NRC), the Bureau of Land Management (BLM), the National Park Service (NPS), the Forest Service, the Tennessee Valley Authority (TVA) and the National Science Foundation (NSF) all have a need for hydrologists.

State Agencies: Have specific responsibilities in water resource management. Typically, state agencies allocate and regulate water use.

Private Sector: Consulting environmental or engineering firms offer many employment opportunities for hydrologists. The work includes assistance to governmental agencies, industrial site assessments, investigation at Superfund sites, and water supply investigations.

A number of technical societies are concerned with hydrology. These include the Geological Society of America, American Geophysical Union, American Society of Civil Engineers, National Ground Water Association, American Water Resources Association, International Association of Hydrogeologists, American Society of Testing Materials, American Water Works Association, American Institute of Hydrology, Association of Engineering Geologists, and American Association of Petroleum Geologists.

Training and Education

The basic education requirement for employment as a hydrologist is a bachelor's degree in physical or natural science or engineering. Courses should include five semester hours each of chemistry, physics, and differential and integral calculus, and twenty-five hours of the major specialty. It may be necessary to study for at least a year beyond the bachelor's degree. A master's degree in hydrology or a related field undoubtedly will enhance employment opportunities.

There are an increasing number of short courses, seminars, and training programs in hydrology offered by universities and scientific societies. Many newly developed academic degrees are available in environmental geology, environmental engineering, and environmental geochemistry.

Global Issues

Large quantities of usable water are needed to support modern, industrialized society. We do not have a national water shortage, but regional shortages occur as population increases in areas that have little water. Human activity contaminates water. Lack of good quality water can cause serious health problems and can affect food production.

With proper management, no one should be without usable water. Hydrologists are needed to provide this management. It is not possible to read a newspaper or look at a news magazine without seeing a story about an "environmental problem." We are constantly reminded about local and international environmental issues. An article in the United Nations environmental publication *Our Planet* presents a summary of the global environment in the next twenty years. The world population will grow by 1.7 billion, which will bring Earth's total inhabitants to 7 billion. These additional people will require food, clothing, shelter, and, of course, usable water.

Bibliography

BACK, W. "Opportunities in the Hydrologic Sciences." *EOS American Geophysical Union* 72, no. 45 (1991):491.

NATIONAL RESEARCH COUNCIL. "Opportunities in the Hydrologic Sciences." Washington, DC, 1991.

JOHN E. MOORE

CAREERS IN THE EARTH SCIENCES: OCEANOGRAPHY

Oceanography—the scientific study of the ocean—is a young field of science. It began with the British-supported *Challenger* Expedition (1872–1876), which systematically explored all the major ocean basins. Whalers, sailors, and other seafarers knew much about the ocean, but there was no repository for their knowledge, systematic collection of new data, or study of samples from the sea.

Most of the knowledge accumulated by seafarers was never recorded because of their lack of education or their fears that others might benefit from their knowledge. In other words, knowledge of the ocean was essentially a trade secret. For example, log books of early Portuguese and Spanish explorers were state secrets. This secrecy kept them from falling into the hands of others who might have profited from the information they contained.

As a field of employment oceanography dates from World War II (1939–1945), when ocean science became important in naval warfare, such as in the planning of amphibious landings. In the 1950s, oceanography emerged as a field of study when American universities began offering oceanography courses and degree programs. Initially, these programs trained researchers and scientists needed for the military. Both sides in the cold war spent heavily to develop new ways to detect and track the other's submarines. Later ocean sciences focused on studies of ocean processes. The emphasis shifted to protection and regulation of living resources (fisheries), and improving understanding and predictive capabilities of global climate change and its effects on Earth. With the collapse of the Soviet Union in the early 1990s national security activities diminished on both sides.

Jobs in oceanography normally require university training. The curricula are necessarily broad since the ocean is involved in almost all aspects of earth sciences—meteorology, geology, and physical processes controlling ocean currents are some examples. The boundaries between oceanography and its intellectual neighbors are diffuse and increasingly more difficult to define. Thus, students wishing to work in oceanography must take a broad course of study in the physical and life sciences.

Until the 1990s, most universities advised students to take a bachelor's degree in a traditional science field and to follow with a study of oceanography at the graduate level. Because of increasing opportunities for students with a B.S. degree to find employment, more American universities are offering undergraduate programs in oceanography.

Following are some major employment fields open to oceanographers.

Fisheries

Fisheries offered the first employment opportunities for people interested in the ocean. As early as the late 1800s, fish and whale stocks showed signs of being overexploited. Many countries established groups to study their fisheries and to devise ways to find new fish resources and to protect traditional supplies. Fisheries is still a major employment area and several universities have programs for fishery scientists.

Aquaculture

An emerging area of employment is aquaculture—growing aquatic organisms for commercial purposes. As wild fish stocks become depleted by overfishing or environmental changes, the increasing demand for fisheries products requires cultivation of both plants and animals. Although there are no domesticated marine organisms, growing marine plants and animals is a large business around the world, especially in Asia. China and Japan have well-established multibillion dollar industries. Aquaculture is also expanding to Europe and North America. In the United States, for instance, catfish, trout, and salmon aquaculture shows rapid growth. Demand for cultured products will grow as concerns increase about contamination of wild populations by diseases and by pollution. As aquaculture expands, it will need people trained in related fields of science.

Petroleum Industry

After 1945, when oil and gas fields on land had become highly developed, the petroleum industry began exploring shallow continental margins. In the 1990s, oil and gas production in many countries comes almost exclusively from offshore fields; Norway and Great Britain are prime examples. New technologies allow commercial oil and gas production from deeply submerged parts of continental margins. The most promising areas for finding new oil and gas lie on the continental margins. Thus, many marine scientists and engineers work in the oil and gas industry worldwide. This field is likely to continue as an important area of employment for people trained in oceanography.

Environmental Protection, Regulation, and Restoration

Many people trained in the marine sciences work to protect living resources and the environment. Due to their broad educational background, oceanographers find many opportunities in this field. Environmental regulatory agencies employ marine scientists at local, state/provincial, and national levels; oceanographers also work in research and in restoring damaged environments. For example, damaged wetlands can be restored, although this has proven to be difficult. Restoration projects range from planting and growing plants and animals that lived there before "reclamation" to research in understanding the complex systems involved.

Teaching and Research

Teaching and research, primarily at universities, has dominated ocean-science employment in the United States since the 1960s. A significant fraction of all oceanographers have doctoral-level training. With the end of the cold war and the accompanying reductions in defense spending on research, and slow growth in both universities and the commercial ocean research sector, this area has been growing very slowly. Many Ph.D.-level oceanographers have to take post-doctoral positions while they wait to obtain faculty or research appointments.

Emerging Employment Areas

Application of satellite-based remote sensing for ocean research and monitoring offers employment opportunities for oceanographers. In many countries, oceanographers prepare predictions of ocean conditions and other services. For instance, several countries prepare and distribute maps to their fishing industry showing where certain fish can be caught. Recreational fishermen and those com-

panies that supply their needs also use these products.

Computers have changed oceanography. Indeed, the availability of supercomputers has accelerated the development of new understanding and predictive capabilities. The enormous amounts of data generated from satellites and computer modeling require specialists in data management. Thus, oceanographers find new employment opportunities in using computers.

In short, oceanographers find work in many of the established sectors of the economy and also in such emerging areas as data handling and satellite-remote sensing.

M. GRANT GROSS

CAREERS IN THE EARTH SCIENCES: OIL AND GAS

Historically, most of the earth scientists in the United States have been employed to find, develop, and produce petroleum and natural gas. Today, there are fewer earth scientists in these careers than during the oil boom years of the early 1980s, and the percentage of total earth scientists is also smaller. The increasing use of technology to improve efficiency reduces overall needs for personnel. Still, employment in oil and gas exploration and development, while changing, is substantial and will remain so, both in the United States and globally.

Few professions combine scientific discipline and the excitement of discovery as does oil and gas exploration. If explorationists are wrong, the well is dry; if they are right, the rewards are immediate. Earth scientists, chiefly geologists and geophysicists, commonly use extensive data in deciding where to drill for oil and gas, but their imagination and ability to infer are paramount. These geologists are called "wildcatters." In undrilled or only sparsely drilled areas, inference and imagination are their principal tools.

Educational Requirements

In the early days of oil and gas exploration, most earth scientists working in industry held B.S. degrees, chiefly in geology. But since the late 1950s, graduate training has been generally required, so that most practicing oil and gas geologists now hold M.S. or Ph.D. degrees. All earth scientists involved in exploration and development must be grounded in scientific basics—mathematics, physics, and chemistry. As geologists and geophysicists also become involved more in oil and gas production and enhanced recovery, understanding of basic engineering principles and fluid behavior is important. Further, as oil and gas are market commodities, earth scientists who aspire to higher-level management positions in industry should be versed in economics, law, business, and public policy.

Oil and gas are finite, nonrenewable resources. As such, they are exhaustible (see HUBBERT, M. KING). Historically, estimates of the amount of oil and gas that can ultimately be found and developed have been low as subsequent discovery and production exceed earlier estimates. The principal reason for conservative estimates is that the role and pace of future technologies in discovering and recovering oil and gas are not always fully appreciated at the time of the estimate.

Natural gas in the United States during the 1970s and early 1980s, for example, was judged to be a rapidly depleting resource. National legislation was passed to preclude certain uses of natural gas. Estimates of remaining natural gas that could be found and produced in the United States were judged to be approximately 250 trillion cubic feet (Tcf), about a dozen-year supply at then-annual consumption rates (Hubbert, 1982). However, by the late 1980s and into the 1990s, gas supply had exceeded demand and estimates of gas that ultimately could be produced were five times greater in the mid-1990s than estimates a decade earlier. The National Petroleum Council (NPC) estimates 1,295 Tcf available at modest prices and with year 2010 technology (National Petroleum Council, 1992). Some other recent estimates exceed the NPC estimate. Perception of gas resource availability has changed with the accelerated pace of technology.

With an exhausted resource, there is little need to pursue efficiencies. But physical or even economic exhaustion of oil and gas lies in the distant future. Recent estimates of the amount of oil that could ultimately be found and produced in the United States likewise have been more optimistic than before. Again, the reason is the greater appreciation of technologies applied. In the early 1990s, the U.S. Department of Energy reported that, with existing technology and at prices of $20 per barrel, 99 billion barrels of oil (Bbo) remain to be accessed in the United States. With advanced technology at the same prices, the estimate was 142 Bbo (Fisher et al., 1992). At a price of $27 per barrel, the range was 130 to 204 Bbo, with existing and advanced technology, respectively. The upper estimate represents a volume about 25 percent greater than all the oil produced in the United States to date.

Although remaining oil and natural gas resources in the United States are substantial, even relative to the large amounts already produced, the remaining part of the resource base contrasts markedly with much of that already developed. Early reserve additions and production were characterized by giant fields providing large economies of scale. The remaining resource is convertible to producing reserves in relatively small increments. Future economies must come from efficiencies, achievable through technology and scientific know-how. This transition is well underway.

Employment Outlook

The United States can supply probable demand for natural gas and at least half of its oil needs from domestic production to the year 2050 or so, a time frame equivalent to three to four successive careers. The estimated global endowment of oil is about 6.9 trillion barrels. Of that amount, 675 Bbo has been produced and 1,000 Bbo exists as proven reserves. Of the remaining, nearly 1.4 trillion barrels is estimated to be produced ultimately, with almost equal amounts from new fields yet to be discovered and through improved recovery from existing fields. The amount of remaining oil now judged accessible globally at reasonable prices will support current levels of consumption for more than one hundred years.

Natural gas is a naturally clean fuel, and the products from crude oil, chiefly gasoline, are being made cleaner each year. The great versatility of oil and natural gas, the fact that they are or can be made environmentally sound, and the vastness of the remaining resource on a global scale mean that earth scientists will be significantly engaged in oil and gas exploration and development for at least the next one hundred years, perhaps much longer.

Earth scientists in the oil and gas industry of the future will not be the wildcatters of the past. As future U.S. and world economies will increasingly depend heavily on technologies and technological application, successful earth scientists in oil and gas will know, understand, and deploy technologies that are advancing at a rapid pace. With these technological advances comes a new opportunity for innovative and rewarding careers in the oil and gas industry.

Bibliography

FISHER, W. L., et al. *An Assessment of the Oil Resource Base of the U.S.* U.S. Department of Energy, DOE/BC-93/1/5P. Washington, DC, 1992.

HUBBERT, M. K. "Techniques of Prediction as Applied to the Production of Oil and Gas." In *Oil and Gas Supply Modelling,* ed. S. I. Class. National Bureau of Standards, Special Publication 631. Washington, DC, 1982.

NATIONAL PETROLEUM COUNCIL. *The Potential for Natural Gas in the U.S.* Washington, DC, 1992.

WILLIAM L. FISHER

CAREERS IN THE EARTH SCIENCES: PLANETARY SCIENCES

Planetary geosciences is a relatively new area of research. To understand the career potential in this field, it helps to have some knowledge of the origin of planetary geosciences, and the various roles played by the government, university and research institutions, and other contracting organizations. This entry provides a brief outline of the "Golden Age of Solar System Exploration" (as termed by Hinners, 1990), a review of the major types of organizations involved in planetary sciences, and information on the sort of work that is involved in a career in planetary studies.

Although geologic interpretations of the Moon had been around since the 1890s, the field did not really become a significant area of work for earth scientists until the United States made the commitment to land a man on the Moon. Prior to the 1960s, only a handful of scientists spent time studying meteorites and the Moon directly using telescopic observations. Early debates centered on whether the craters of the Moon were formed by impact or volcanism, the source of meteorites, and the origin of tektites and the Moon itself. The 1960s led to more refined scientific questions, and, with the establishment of a first stratigraphic system for the Moon, a way in which the near-surface materials could be assigned a relative age, and put into context for interpreting the geologic history of another planetary body.

Very few organizations took part in these early activities; in fact, one of the most influential and prescient planetary geoscientists, Ralph Baldwin, ran a machine shop and published books on the origin of lunar surface features in his spare time. The beginning of the space age saw also the emergence of a new area of geology in the U.S. Geological Survey, the Branch of Astrogeology. The Jet Propulsion Laboratory (JPL), run for NASA and other government agencies by the California Institute of Technology, concentrated on unmanned missions, while the Manned Spacecraft Center, later called the Lyndon B. Johnson Space Center of NASA, dealt with the Mercury through Apollo programs. Although some individual scientists at universities played key roles in advising the nascent space program, few universities were willing to establish faculty positions in such a new field.

When the first samples of the Moon were returned to Earth in 1969, earth scientists from around the world were invited to help analyze the rocks and soils. The early 1970s saw the success of unmanned missions to Mercury, Venus (by the Soviet Union), Mars, Jupiter, and Saturn. Detailed images of the outer planet satellites were returned by the *Voyager* spacecraft in the 1980s, and with *Voyager 2*'s successful encounters with Uranus and Neptune, the reconnaissance phase of planetary exploration in the solar system was completed (with the exception of Pluto).

The decline in the number of planetary missions in the 1980s was accompanied by a concurrent decrease in the number of available research positions in the planetary sciences. The number of government research centers that have full-time staff devoted to planetary studies has remained relatively constant, but even within those centers, the professional staff has sometimes been assigned other tasks. However, the importance of increased knowledge of our neighboring planets has been recognized in the academic community, and a few academic positions in the planetary geosciences have appeared in the 1990s. Recognizing the limited size of the field, most graduate advisors suggest that students have more "traditional" Earthbound thesis topics in the geosciences, but in certain universities, "tradition" is now expanding to include planetary topics.

Research careers in the planetary sciences range from those in government labs, academia, and research institutions to less abundant positions in the aerospace industry, the latter being reserved for those with engineering experience. NASA field centers, particularly Johnson Space Center and Goddard Spaceflight Center, primarily employ Ph.D.-level scientists who analyze data from past missions as well as propose new experiments for future missions. The Jet Propulsion Laboratory also maintains a group of planetary scientists involved in research, but is most active in engineering as it continues to be the lead organization for unmanned missions.

Most planetary geoscience research is done by those scientists who have completed their Ph.D., usually in geology and geophysics, but sometimes in related fields such as astronomy or physics. Very few positions are devoted solely to research; most have other duties such as teaching, administration, or tasks specific to the particular organization. Typically, those who complete their advanced degrees move on to postdoctoral positions, during which they may author one or more papers for professional journals. This allows students a chance to establish their areas of expertise and, at the same time, become known to the professional community. Not all students in the planetary sciences complete their Ph.D. Those who cease their formal eduction at the bachelor's or master's degree level have a wider opportunity for jobs dealing with science, but not always directly related to planetary studies. Careers for these individuals range from teaching at the secondary school level to editing and freelance writing. Exploration of the solar system is a popular topic for magazines and other media, and each planetary encounter seems to rejuvenate public interest.

With the exception of the few privately endowed research institutes and universities, the planetary sciences in the United States are tied to the political fortunes of NASA. Unlike the National Science Foundation, NASA is a mission-oriented agency, and the science supported by its grants and contracts is directed at specific areas of research. All the research proposals are reviewed by peers, but even highly ranked proposals may not be funded because of priorities set by the agency. The trend of the mid–1990s is toward smaller, less costly planetary missions that will accomplish a limited set of experiments, but even with such savings over the more complex spacecraft of prior decades, the amount of funding that goes into planetary exploration is limited by decisions internal to NASA, by congressional authorization, and by the overall state of the economy. Those considering a career in the planetary sciences would do well to keep Earth-bound work areas and alternative jobs as options to be considered.

Bibliography

HINNERS, N. W. "The Golden Age of Solar System Exploration." In *The New Solar System*, eds. J. K. Beatty and A. Chaikin. Cambridge, MA, 1990.

TED A. MAXWELL

CAREERS IN THE EARTH SCIENCES: TEACHING

Earth science, encompassing geology, oceanography, meteorology, space science, and astronomy, is the study of the composition, structure, and evolution of Earth, the solar system, and the universe. The enormous expanse of both space and geologic time are fundamental concepts not traditionally emphasized in other curricula.

Offering insight to the natural environment, earth science assists students in interpreting and evaluating their personal experiences in the world that surrounds them. It is particularly well suited to being taught by demonstration and experiment that involve students. Earth science provides students the opportunity to apply the scientific method in field and laboratory to the study of issues faced by society. Earth science, because it is so interdisciplinary, provides the broad background necessary for consideration of such issues as acid rain, greenhouse effect, ozone depletion, rain forest destruction, waste disposal, water and atmospheric pollution, and availability of natural resources. It also provides students with an awareness, understanding, and concepts for possible mitigation of such natural disasters as earthquakes, volcanoes, floods, tornadoes, hurricanes, and landslides.

Science and mathematics education in the United States is undergoing intense scrutiny to identify: (1) why more young Americans are not choosing to enter these challenging fields; and (2) why those students in the science and mathematics classrooms are not achieving at the level of their peers in other countries. Striving to distill wisdom from past experience, government and state and national education agencies—together with local schools, colleges and universities, and professional societies—are moving to raise standards, revise curricula, and identify teaching methods that are most effective. Such periods of change offer great challenge and opportunity for those in education.

Today's Teacher

Roles and methods of teachers are changing. Today's classes are characterized by interactive discussions and less formal lectures. Day-to-day activities of teachers are varied: planning and executing lessons, exercises, demonstrations, field trips; preparing and grading tests; assessing student performance; preparing report cards; supervising study halls and homerooms; supervising extracurricular activities; meeting with parents and school staff to discuss a student's performance.

Good teachers will always be in demand, and those who choose to study the teaching of earth science at the K-12 level not only learn to teach effectively but also encounter outstanding teachers, in both education courses and in science and mathematics, who often serve as role models. Effective teachers differ in their methods of presentation and organization and their style of interaction with students, but each will know the subject(s) well, each will be a skillful communicator, and each will demonstrate a strong interest in their student's progress and an earnest desire to help them.

Formal Education

Because the discipline of earth science encompasses geology, oceanography, meteorology, space science, and astronomy, basic courses should be taken in each area, with a bachelor's degree obtained in one of them. It is important for the pro-

spective teacher to be well versed not only in teaching technique and the psychology of learning, but also competent in, and comfortable with, the subject matter.

Certification requirements vary from state to state. In general, those institutions offering degrees in education will have programs that qualify its graduates for certification in the same state, including specific education courses and supervised student teaching. In addition, the National Science Teachers Association (NSTA) offers certification of science teachers who have successfully completed three years of teaching at the level for which they are requesting certification.

NSTA certification for earth and space science teachers requires: (1) completion of a minimum of thirty-two semester hours in earth and space science, with specific courses in each of the following: astronomy, geology, meteorology, oceanography, natural resources—environmental studies, and supporting courses in one or more of the above; (2) completion of a minimum of sixteen semester hours in supporting courses in biology, chemistry, and physics; (3) knowledge of mathematics through precalculus; (4) evidence of study that links geosciences to contemporary, historical, technological, and societal issues.

For graduates in education whose goal it is to teach, most states now have the additional requirement of a satisfactory grade on a comprehensive examination. A commonly used test is the National Teachers Examination administered by the Education Testing Service, Princeton, New Jersey.

Informal Education

Completion of the academic requirements, whether it be a bachelor's degree or a higher degree(s), marks the completion of just one phase of the educational experience. Continuing education, through short courses and workshops, or perhaps with additional formal course work from time to time, will constitute a longer, and arguably more important, phase of a teacher's education. Most K-12 school systems require their teachers to acquire additional academic credits throughout their professional careers, and these requirements are an integral part of the promotion and salary schedule. Professional societies play an important role in continuing education; two important ones in the earth sciences are the National Association of Geology Teachers (NAGT) and the National Earth Science Teachers Association (NESTA).

Supply and Demand: Short Term and Long Term

Because of the current need for science- and technology-literate citizens in the work force, and because earth science teacher preparation has been inadequate, great demand and opportunities exist for those teachers certified to teach K-12 earth science. This disparity will decrease somewhat over the long term as a result of the intensive effort underway in the 1990s to upgrade and enlarge earth science education programs. Nevertheless, the increase in both U.S. and world population, along with an increasingly complex world ever more dependent on science and technology, will ensure strong demand for K-12 earth science teachers. Their challenge will be to create curricula that satisfy the changing needs of their students, while developing the additional skills that effective teaching will require in the world of the future.

According to the *Occupational Outlook Handbook*, in 1992 approximately 3,255,000 teachers were employed, 90 percent of them in the public schools. Through the year 2005, employment of K-12 teachers will increase faster than average for all occupations. As the large number of teachers now in their forties and fifties retire toward the end of the decade, K-12 employment opportunities will increase substantially.

Curriculum Revision and Innovation

One of the most important out-of-classroom activities of teachers is serving on curriculum committees, and the National Science Teachers Association has provided guidelines for new and updated science curricula. According to the *NSTA Handbook* (1993), these curricula should:

1. Provide students the opportunity to study real-world problems and issues that touch their lives together with broader science and technological problems of society.
2. Provide, at all grade levels, field and laboratory study to enhance and extend classroom instruction.
3. Be integrated and correlated with respect to science, technology, mathematics, humanities, and the social sciences.

4. Develop awareness of careers in science and engineering.
5. Prepare at-risk, minority, and female students for careers in science and engineering.
6. Incorporate and integrate appropriate technology.
7. Utilize evaluation and assessment instruments and techniques that support and enhance the goals of science education.

Earnings

In 1992–1993, the average yearly salary of public-school teachers was $35,334; it is forecast to rise to $47,470 by 1997. Teachers in urban and suburban schools generally earn more than teachers in rural schools; public-school teachers usually earn more than those in private schools.

In many localities, public-school teachers belong to unions such as the American Federation of Teachers and the National Education Association and derive benefit from collective bargaining with school systems about salaries, work hours, and other conditions and terms of employment.

Impact of Economics and Politics

K-12 public education in the United States is largely tax supported and thus dependent on the mood of the electorate to provide necessary funding. Fortunately, most citizens always have held education in high regard, and even in time of economic stress, the education of the young traditionally has been given high priority.

Private K-12 education, employing about 10 percent of the teachers in the United States, also to some extent is tied to the economy of the country and to the areas in which the private schools are located. However, the middle to upper income families who support private schools are apt to be less affected by swings in the national, regional, and local economies.

At the college/university level, the term "earth science" includes the basic subdisciplines of geology: physical and historical geology, mineralogy and petrology, geomorphology, paleontology, structural geology, stratigraphy, geochemistry, and geophysics. Geologists have played a leading role in exploring for and extracting those natural resources needed in ever increasing quantity by an expanding world population. In recent years, increasing numbers of geologists have been involved in environmental geology, that aspect of geology concerned with such problems as solid and liquid waste disposal and with surface and subsurface contamination of soil and water.

Teaching at the College/University Level

Few persons enter college certain that they wish to prepare for a college/university teaching career in earth science. Most entering students have not encountered earth science at all, inasmuch as it is taught in very few middle schools and high schools. During their undergraduate years, especially if they become involved in undergraduate research and have faculty role models, students may begin to consider a career in academia. By the time undergraduates have received their baccalaureate degree, they will be aware that, ideally, they should obtain both the master's and doctoral degrees to be qualified for a successful career in academia.

What is involved in a career in academia in earth science? Academic activities commonly include teaching, research, and professional service to the academic department, college, and university and to the broader community. In smaller undergraduate schools and community colleges, emphasis will be on teaching with only a modest level of professional service, whereas in larger colleges/universities with graduate programs, teaching and research commonly will be emphasized equally with the expectation of a relatively high profile in professional service. One of the pleasures in teaching earth science at the college level is providing students with their first insight into the materials and processes that characterize their world. Most find the experience exciting and rewarding.

In addition to conducting classes and preparing laboratory exercises, experiments, and field trips, usually in a combination of undergraduate and graduate courses, the instructor spends time in preparing and grading examinations and assignments, evaluations of research papers, and in advising students. The instructor may also serve as thesis supervisor and major professor for one or more graduate students.

Faculty in graduate institutions, and in increasingly more undergraduate institutions, engage in research on some aspect of earth science of special interest. In more and more undergraduate programs, undergraduates are participating with faculty in their research programs and often assist in reporting, orally or in writing, the results of their research.

The mix of faculty responsibilities varies with the institution and with the individual. Research is a large component of faculty activity at a university, somewhat less at a four-year college, and a relatively small part of faculty responsibility at a community college. Usually, reduced involvement in research leads to an increased teaching load; in larger universities, some faculty appointments involve research only.

One of the principal attractions of a career in academia is the freedom that faculty enjoy to organize their activities. Many faculty routinely exercise the privilege of selecting when and at what hours their courses will be taught, and most choose their research focus or emphasis. Aside from meeting with their classes, maintaining regular office hours for consultation with students, and attending faculty and committee meetings, faculty have significant freedom in utilizing their time.

Career Preparation. Those who wish to seek careers in earth science education at the college/university level should consider the master's degree as the minimum qualification necessary. To qualify for faculty appointments at strong, four-year liberal arts colleges and large research universities, the Ph.D. degree is essential. Increasingly, two years of postdoctoral study and research are desired by potential employers.

In earth science, a master's degree typically takes approximately two years to complete, assuming that the student is employed half time by the academic department as a teaching assistant. Course work will dominate the first academic year of the two-year program, followed by a summer of field or laboratory work, or both, on thesis research. During the second year, course and laboratory work are completed and the thesis is written and defended. A master's program typically requires the completion of twenty-four semester hours of formal course work and six hours of thesis research.

Although the master's degree is a demonstration that the student can undertake a research project, organize it, see it through to completion, and report on it orally and in writing, the doctoral dissertation is expected to contribute new insight and knowledge to earth science. Ph.D. programs characteristically involve about 60 semester hours of additional course work and a dissertation on original research designed to test a hypothesis or model or address a fundamental issue in earth science. Ph.D. programs require from three to six years, or more, to complete.

Employment. In academia, employment traditionally is at the instructor, assistant professor, associate professor, or professor level. The rank of instructor usually signifies that the faculty member does not yet hold the degree in the field. In many instances, a person is hired as an instructor when the receipt of the doctorate is imminent and is promoted to assistant professor when the Ph.D. is received. A major challenge for faculty is the acquisition of tenure, the granting of which means they cannot be terminated without sufficient cause and without due process. Tenure-track appointments usually are made at the assistant professor level, following a probationary period of five to seven years, depending on the institution. The awarding of tenure at the end of the probationary period means that the faculty member's record in teaching, research, and contributions to the institution and to earth science have been outstanding. Tenure provides stability both to the institution and to the faculty, and most important, it assures faculty that their position will not be jeopardized by lecturing or undertaking research on unpopular or controversial subjects. In the United States, approximately 60 percent of full-time faculty were tenured as of 1994.

Because of increase in enrollments in colleges and universities, employment opportunities through the year 2005 in academia are expected to increase at about the same rate as the average for all occupations. In addition, the large number of current faculty that entered the academic profession in the 1950s and 1960s will retire during the 1960s, providing additional employment opportunities in academia until 2005.

Salaries. The American Geological Institute reported the following average salaries for 1995: professors, $63,763; associate professors, $45,306; and assistant professors, $38,713.

Impact of Economics, Politics, and Technology

In late 1982, the scarcity of crude oil on the world market, and thus its price, began to decline. That and the resulting downsizing and restructuring of U.S. oil companies resulted in a dramatic decrease in the employment opportunities for geologists in the petroleum industry. Many geology departments found themselves with too few students and too many faculty.

In the mid–1990s, departments were restructuring curricula to reflect the change in emphasis from providing industry with graduates who will play a major role in the extraction industries to those whose careers will focus on environmental geology, engineering geology, and hydrogeology (Roy, 1995).

Historically, geology graduates of traditional programs have had the breadth of background to take advantage of those career opportunities that seemed most attractive to them. A broad, solid background in science and mathematics will not change but will be augmented by computer and other technological skills, some of which cannot even be envisaged now, and applied to such pressing societal problems as waste disposal, preservation of the environment, acid rain, and rain forest devastation. Meanwhile, as world population increases, the need for additional natural resources also will increase. Inasmuch as geologists will be in the forefront of those locating and developing these resources, academic departments and their faculty will be challenged to provide their graduates with the theoretical insights and the practical skills necessary to achieve required levels of discovery and development of natural resources.

Bibliography

National Science Teachers Association. *NSTA Handbook 1993–1994.* Washington, DC, 1993.
———. *You Can Teach Science.* Washington, DC, 1995.
Occupational Outlook Handbook 1995–1996. Washington, DC, 1995.
Roy, E. C., Jr. "Earth Science Education and Employment." *Geotimes* 40, no. 4 (1995):15–16.
U.S. Department of Education. *Projections of Education Statistics to 2003.* Washington, DC, 1992.

JAMES R. UNDERWOOD, JR.

CAREERS IN THE EARTH SCIENCES: WRITING, PHOTOGRAPHY, AND FILMMAKING

With the explosion of new channels in cable television and recent expansion in new media such as CD-ROM and the Internet comes a greater demand for science programming. A solid market also still exists in traditional print media such as magazines, newsletters, and newspapers for science writers. For instance, the *New York Times* features a special weekly section devoted to the sciences that often includes articles on the earth sciences.

Skills and Course Study

Skills necessary for a successful career in writing and other media vary depending on the specific job sought. But there are several fundamental skills that serve anyone who is entering the field of communications.

Not all successful writer's and media producers have undergraduate or graduate degrees in earth sciences, but most have taken a fair number of science courses as part of their higher education. It seems that most science writers and producers have taken introductory courses in more than one science along with some upper level courses. Occasionally there are Ph.D.s who apply their knowledge to science programming for television.

Perhaps most important to a communications career is the ability to write and use other media such as television clearly and expressively. Television stations are more varied in their attention to the sciences, but reporters and producers must be able to understand the subject in order to create an accurate story about it.

Courses in creative and non-fiction writing, print and broadcast journalism, film and television production, and computer science are all helpful foundations for those choosing a career in any facet of science reporting or documentary. Majoring in one of these academic fields is sensible because there are many other skills involved in the technologically dependent fields of radio, television, and new media. Exposure to cameras, microphones, and editing equipment may be important during a student's academic career.

However, some students choose to intern in the field of their interest. They might spend a summer or a semester, sometimes for academic credit, working part-time for a communications company. These internships can lead to full-time positions after school.

Employment Opportunities

Students seeking an internship or future employment should approach a distribution outlet, such as one of the U.S. television networks (ABC, CBS, NBC, PBS, or FOX), various cable channels (Discovery, the Learning Channel, Turner, Disney), or even local access channels. Other opportunities may be found at local newspapers or television stations. Internships may be available with production companies (film, television, CD-ROM, online). Students may also look to their college or university's press or communications office for hands-on experience.

Most weekly and even monthly magazines have science sections. Others, such as *Science Week* and *Discover,* are devoted strictly to the sciences. Newspapers usually have a special weekly section addressing science issues. Trade and professional associations that produce newsletters are yet another source of employment.

Magazine-format shows (such as "National Geographic Explorer") and long-form documentaries (such as "NOVA") or much of the programming on the Discovery channel are directly devoted to the sciences.

Some universities, research institutes, corporations, and government agencies also hire science writers and producers for press releases and magazine articles. Video News Releases (VNRs) are a new form of communication that is expanding. Computer skills that include home-page development on the World Wide Web or CD-ROM production are now in high demand, and will continue to grow as new technology develops.

Children's programming in the sciences also shows continued strength. CD-ROM publishing is also growing and more primary schools are using this technology in science education.

In many ways a career in writing and other media can be exciting and fulfilling. Writers, reporters, and documentary producers often travel to exotic locations for their stories. Good physical conditioning is often necessary to work effectively. Some field experiences require the individual to be away from home for long periods, either on land expeditions or sea voyages.

Work in this career is not always glamorous, however. Much of the time devoted to work in these fields is spent at the library researching and reading, in front of the computer browsing the Web or specific databases, or on the phone interviewing scientists.

The National Association of Science Writers, the Association of Earth Science Editors, and the American Association for the Advancement of Science are good resources for anyone interested in pursuing careers in earth sciences communication.

Bibliography

BARNOUW, E. *Tube of Plenty: The Evolution of American Television*. New York, 1990.
FRANKLIN, J. *Writing for Story*. New York, 1994.
WALTERS, R. *Broadcast Writing*, 2nd ed. New York, 1994.

LARRY ENGEL

CARSON, RACHEL

Rachel Carson was a pioneer of the modern environmental movement in the United States. She was born in Springdale, Pennsylvania on 27 May 1907 and died in Silver Spring, Maryland on 14 April 1964. Carson made a career of her fascination with wildlife and concern for the environment, working for the U.S. Bureau of Fisheries and its successor, the U.S. Fish and Wildlife Service, from 1936 to 1952. Her best-known book, *Silent Spring* (1962), provided a catalyst that changed the way Americans think about their surroundings and particularly the impact of modern chemicals on the landscape. Rachel Carson played a significant role in the ideological enlightenment that led policy makers to focus on the serious study of environmental issues. The first Earth Day on 22 April 1970 was one outcome of the new environmental awareness, and a second was the creation of the Environmental Protection Agency (EPA) the same year.

Rachel Carson attended public schools in Springdale and nearby Parnassus, Pennsylvania.

Her mother taught her to enjoy the outdoors and fostered her daughter's interest in wildlife. Carson showed an early talent for writing, and on graduation from Parnassus High School enrolled in Pennsylvania College for Women in Pittsburgh to study English with the intention of becoming a writer. A course in biology rekindled her interest in science and led to a change to a science major. Carson was awarded her bachelor of arts in 1929, and went on to postgraduate studies at Johns Hopkins University. Commencing in 1930, she taught at Johns Hopkins summer schools for seven years. She joined the zoology staff of the University of Maryland in 1931, and obtained a master of arts from Johns Hopkins in 1932.

Carson developed a special interest in the life of the sea and undertook further postgraduate work at the Woods Hole Marine Biological Laboratory in Massachusetts. In 1936 she accepted a position as an aquatic biologist with the U.S. Bureau of Fisheries in Washington, D.C. She became editor in chief at the U.S. Fish and Wildlife Service, the successor to the Bureau, in 1947. During her years with the service Carson practiced her writing skills preparing many leaflets and informational brochures that publicized the central objective of the bureau: "to insure the conservation of the nation's wild birds, mammals, fishes and other forms of wildlife, with a view to preventing the destruction or depletion of these natural resources, and to promote the maximum present use and enjoyment of the wildlife resources that is compatible with their perpetuity."

Her first book *Under the Sea-Wind* appeared in 1941. The subtitle was "a naturalist's picture of ocean life" and the narrative told the life of the shore, the open sea, and the sea bottom. *Under the Sea-Wind* was well received both for the accuracy of its scientific content and its accessible style. Her second book, *The Sea Around Us* (1951), was delayed both by war work and a painstaking and prolonged period of research and writing. By her own admission, Carson was a slow writer who subjected her work to multiple revisions. As a general rule Carson declined offers from magazines to publish extracts from her books fearing that serialization would detract from the coherence of her arguments. However, before its publication in book form, *The Sea Around Us* was excerpted in the *New Yorker* in the summer of 1951.

When *The Sea Around Us* appeared in book form in July 1951, it was an immediate success. The work was greeted with praise for its literary style, approachability and informative content. In a New York *Herald Tribune Book Review*, Francesca La Monte described Carson's story of the sea as "one of the most beautiful books of our time." Writing in the New York *Times Book Review*, Jonathan Norton Leonard said of *The Sea Around Us:* "Its style and imagination make it a joy to read." *The Sea Around Us* provides a layman's geological guide through time and tide. It begins with an account of the contemporary understanding of the origins of the Earth and Moon, and then proceeds through the geological timetable, mapping the evolution of the planet, the formation of mountains and islands and oceans. Then follows the description of the sea, commencing with the surface and its inhabitants, and descending through lower depths to the sea bottom. Carson reveals the fascination of the hidden world of the oceans to nonscientists, exploring the mystery of the sea, its history and treasures. The book went into nine printings and was at the top of the nonfiction best-seller lists nationwide. It was selected by the Book of the Month Club, condensed for Reader's Digest, and translated into 33 languages. In the year of publication (1951), *The Sea Around Us* received the National Book Award for Non-Fiction.

The success of *The Sea Around Us* wrought major changes in Carson's life. In 1951 she accepted a Guggenheim fellowship that enabled her to take a year's leave of absence from her government job to start work on a third book. Her future secured by success, she resigned her position at the U.S. Fish and Wildlife Service in 1952 to devote herself full-time to her writing. The result was *The Edge of the Sea*, published in 1955. Conceived as a popular guide to the seashore, studying the ecological relationship of the seashore to animals on the Atlantic Coast of the United States, this work complemented her previous book and evidences Carson's growing interest in the interrelationship of Earth's systems and the holistic approach that would mark *Silent Spring*.

Silent Spring opens with a brief account of a fictional town in the American heartland "where all life seemed to live in harmony with its surroundings." Prosperous farms line roadsides alive with wildflowers and ferns. In winter, birds feed on colorful berries above the snowline. Then a blight spreads through the region. Sheep and cattle sicken and die. Chickens become ill. In the spring the hens brood but no chicks are hatched. The

birds disappear. Crops fail. The hedgerows wither and die.

The cause of this "creeping death" is the widespread use of pesticides. In *Silent Spring*, Carson propounds an eloquent argument for the careful and thorough consideration of both the short- and long-term effects of the use of chemicals for a range of applications. The book documents the negative effects that result from the use of pesticides, chemical fertilizers, and an array of chemical treatments designed to enhance production or simplify the production process.

The information contained in *Silent Spring* was not new. All the issues she covered had been discussed in the scientific journals. Carson's contribution was her presentation of "the overall picture" in a highly readable style. Carson highlighted the fact that while individual chemical products might be viewed as safe because they achieve what they were designed to do, the combined effects of an assortment of chemical products could be deadly. As an example, Carson cites the streams that became chemical soups carrying the outpourings of chemical treatment plants and the run-off from fields treated with pesticides and chemical fertilizers, killing algae, plant life, fish, and animals.

Carson began to awaken the public to the need to think beyond short-term quick chemical fixes and profit taking: "The central problem of our age has become the contamination of man's total environment with substances of incredible potential for harm—substances that accumulate in plants and animals and even penetrate the germ cells to shatter or alter the very material of heredity upon which the shape of the future depends." Carson used her considerable literary skills to bring these scientific concerns to the attention of the general public.

In detailed, well-documented accounts, *Silent Spring* revealed the vested interest of a chemical industry that marketed the effectiveness of a product designed to destroy pests without any reference to the irrevocable changes that would be wrought in the pest habitat. She emphasized that the ecology of the soil had essentially been ignored in the rush to apply chemical solutions to pest problems, as if the soil would remain unaffected by the poisons being poured onto it and channelled into it by affected insects. Carson also noted that it was well documented that some pests could and did develop resilience to the pesticides requiring ever more powerful pesticides resulting in an escalating toxic spiral. "The chemical war is never won, and all life is caught in its violent crossfire," she warned.

On publication, *Silent Spring* received much adverse criticism. The chemical industry united against Carson accusing her of ignorance, sensationalism, and distortion. There was even an attempt to convince her publisher that the book should not be published. More balanced reviews appeared in the scientific press including one in *Scientific American* which suggested that *Silent Spring* "may help us toward a much needed reappraisal of current policies and practices." Acceptance by the establishment was not long delayed. A 1963 report by the President's Science Advisory Committee was reviewed by the journal *Science* as "a fairly thorough-going vindication of Rachel Carson's *Silent Spring* thesis."

Bibliography

BONTA, M. *Women in the Field: America's Pioneering Naturalists.* College Station, TX, 1991.
BROOKS, P. *The House of Life: Rachel Carson at Work.* Boston, 1972.
CARSON, R. *Under the Sea-Wind.* New York, 1941.
————. *The Sea Around Us.* New York, 1951.
————. *The Edge of the Sea.* Boston, 1955.
————. *Silent Spring.* Boston, 1962.

PATRICIA DASCH

CATASTROPHIC IMPACT PROCESSES

See Impact Cratering; Impact Cratering and the Bombardment Record

CATASTROPHISM

See Famous Controversies in Geology

CAVES AND KARST TOPOGRAPHY

There is something about the pattern of hills and valleys and rivers that feels natural. Hills are uplands. Streams flow in the bottoms of the valleys that divide the hills. Small streams in the uplands flow down the valleys to join other tributaries forming still larger streams that in turn converge to become rivers. It is all very ordered.

Some landscapes do not follow this natural pattern. One such landscape is karst. Tributary karst streams may disappear into a cave entrance or into a pile of loose rocks leaving only a dry channel or no channel at all downstream. Such streams are called sinking streams and they often flow into blind valleys. Karst lands also contain bowl-like depressions called sinkholes. Some sinkholes are small, only a few feet deep and others are as large as valleys. Sinking streams and sinkholes collect water and drain it into underground conduits. The water returns to the surface at large springs.

Underground water flows beneath the karst lands through joints and fractures and in open conduits that behave much like natural storm drains. Some of these openings are completely water-filled and are not accessible. Some are air-filled but are inaccessible because there is no human-size access. Those fragments of active and abandoned underground drainage conduits that have human-size entrances are called caves. Caves give us at least a fragmentary view of the inside of the karst drainage system.

Most landscapes are carved by chemical and erosive attack by water. Soils are formed by the chemical breakdown of the minerals that make up the underlying bedrock (*see* WEATHERING). Valleys are formed when soil and weathered bedrock are carried away by moving water (*see* GEOLOGIC WORK BY STREAMS). The process of landscape sculpturing always contains a mixture of chemical attack and mechanical erosion. Karst landscapes form in rocks where chemical attack is the most important part of the process.

The rocks that most readily develop caves and karst landscapes are limestone, composed mainly of calcium carbonate; dolomite, composed mainly of calcium-magnesium carbonate; and gypsum, composed mainly of calcium sulfate. Limestone and dolomite are relatively insoluble in pure water and are resistant cliff-formers in arid terrains where there is little soil or vegetation. However, limestone and dolomite are moderately soluble in dilute acids so that some source of acid is required to create caves and karst. The most important acid is carbonic acid. Carbon dioxide from the atmosphere, from decaying organic matter in the soil, and from plant roots dissolves in infiltrating rainwater to make the water slightly acid. When the water reaches the limestone or dolomite beneath the soil there is a chemical reaction that causes the rock to dissolve.

$$CO_2 + H_2O \rightarrow H_2CO_3$$

$$CaCO_3 + H_2CO_3 \rightarrow Ca^{2+} + 2\ HCO^{3-}$$

The solid rock is carried away in solution as calcium ions (plus magnesium ions) and bicarbonate ions. Gypsum is about ten times more soluble in water than limestone and does not require acid to dissolve. As a result, gypsum rocks are usually dissolved away completely in humid climates, and we find gypsum karst only in arid regions such as west Texas, western Oklahoma, eastern New Mexico, and in many other countries.

When surface streams flow across limestone or dolomite, the water percolates downward into fractures and joints, gradually dissolving the rock and enlarging the joints into conduits. When the conduits become large enough to carry the entire flow of the stream, the surface channel is abandoned and a sinking stream in a blind valley results. Rainwater percolating through the soil must descend into the bedrock along joints and fractures because limestone and dolomite rock are impermeable. Again the rock is gradually dissolved away and a sinkhole is formed. After the joints and fractures have been enlarged by solution and after an underlying cave system has been formed, soil and insoluble residues from the bedrock can also be flushed down the sinkholes. As a result, karst landscapes often have areas of bare, exposed bedrock with no soil at all. The bedrock is sculptured by water into an irregular surface of pinnacles, small channels, and grooves known collectively as karren (Figure 1).

Underground, the conduit system is an integral part of the drainage system. The conduit system collects water from sinking streams and from sinkholes and delivers it to the karst springs. These conduits are constantly evolving. As time goes on, surface valleys deepen, stream gradients increase, new conduits form at lower levels and the former

Figure 1. Fluviokarst. The landscape contains a stream system, part of which occurs as normal surface streams. Some tributaries have been pirated underground leaving dry valleys, and some streams continue to downcut in their upstream reaches to form blind valleys. Closed depressions—sinkholes—form in the inter-stream areas. Extensive cave systems would likely occur beneath the land surface.

conduits are abandoned to become air-filled caves. Erosion of tributary valleys, sinkhole collapse, and other processes create entrances so that the caves become accessible to human exploration. At the same time, some cave passages become choked with clay and silt and others are blocked by rockfalls. The original integrated conduit system can rarely be explored completely. The accessible caves are only fragments of the original underground drainage system.

Some caves consist of single tunnels with few side passages. Some have a branching pattern similar to patterns of surface streams. Some take on maze patterns much like the pattern of city streets. The lengths, patterns, and passage shapes are recorded by cave mapping. Tens of thousands of cave maps have been prepared by cave explorers. There are many short caves and only a few long caves. The longest known cave in the world is the Mammoth Cave system in Kentucky, which had a reported length in early 1994 of 553 km.

Some caves are essentially horizontal because they formed at grade with surface streams. Some caves contain sequences of passages at different levels, connected internally by roof collapses or by pits and shafts. Shafts form by descending groundwater related to present-day topography and most are younger than the cave passages that they intersect. Vertical shafts are common in the caves of the eastern United States, where they have depths of more than 100 m. Caves in mountainous regions are often very deep with complicated sequences of steeply sloping passages, shafts, and underground waterfalls. The cave systems of the Huautla Plateau in southern Mexico reach depths greater than 1,200 m.

Water that has become very acidic while seeping through the soil dissolves a quantity of limestone when it comes into contact with the underlying bedrock. When these waters seep downward to drip from the cave ceiling, they release carbon dioxide into the cave atmosphere and redeposit their load of dissolved calcium carbonate in the form of stalactites, stalagmites, flowstone, and other forms collectively known as speleothems. Depending on the chemistry of the seeping waters, a variety of minerals may form. In all, more than 270 minerals have been found in caves, most of them very uncommon. Because the minerals form in a constant environment, they often appear as large, perfectly formed crystals of unusual shape.

The water that forms stalactites and stalagmites also deposits minute quantities of uranium. By analyzing the radioactive decay of the uranium, the age of the speleothems can be calculated (*see* GEOLOGIC TIME, MEASUREMENT OF). These methods have allowed the record of speleothem deposition to be traced back 350,000 years. Calcite speleothems form in northern climate caves during warm interglacial periods while deposition shuts down during periods of glacial advance.

Clay, silt, and sand are washed into caves through sinkholes and by sinking streams. The piles of layered cave sediments may remain undisturbed for thousands or millions of years. The sediment piles in the upper levels of multilevel caves may carry a record of conditions far back in the ice ages, although the ages of such sediments are difficult to determine. Paleomagnetic measurements sometimes provide a time marker. Mixed with the clay, silt, and sand, which are silicate minerals and nonmagnetic, may be a few grains of magnetic iron oxides. These grains act as tiny compass needles,

and as the sediment settles out of the water, these grains rotate to line up with the earth's magnetic field as it was at the time of deposition. Once the sediment piles are formed, the magnetic grains are locked in place. At irregular intervals, roughly a million years apart, the earth's magnetic poles reverse (*see* EARTH'S MAGNETIC FIELD). North becomes south and south becomes north. The last such reversal occurred 730,000 years ago. If it is found that the magnetic grains in a cave sediment pile point south instead of north, then the sediments must be at least 730,000 years old. If alternating normal and reversed sediments are found in a sequence of cave levels, then the age can be tracked backward to find the age of the oldest passages in the cave. By paleomagnetic dating it was found that the highest passages in Mammoth Cave were formed 2–3 million years (Ma) ago, at the beginning of the ice ages.

Life on Earth is very adaptable. Living organisms are found in just about every possible habitat and caves are no exception. Caves are a harsh and inhospitable environment that is offset to some extent by the constant conditions. Except for cave streams, which are subject to flooding, most cave habitats remain unchanged over long periods of time, giving organisms a chance to adapt to the special conditions. Most important among these conditions are the darkness and the very sparse food supply.

Some animals, birds, and small mammals simply use caves, particularly near the entrances, as shelter. Some animals live in caves for part of their life cycle but go outside to feed. The many species of bats are the most noticeable. Bats hibernate in caves, give birth in caves, and, as nocturnal animals, rest in caves during the daytime. They do not, however, depend on caves for food, although their droppings provide food for other cave-adapted organisms. Organisms that spend their entire life cycle deep inside caves include species of fish, crayfish, salamanders, and many smaller animals such as beetles, isopods, and amphipods. Because of the darkness, many of these species have lost eyes and pigmentation. Although populations are generally small, the number of cave organisms is surprisingly large. Because cave communities are much less complicated than surface communities, caves are useful as ecological and evolutionary laboratories where the interrelationships of organisms, environment, and food supply can be more easily studied.

Bibliography

COURBON, P., C. CHABERT, P. BOSTED, and K. LINDSLEY. *Atlas of the Great Caves of the World.* St. Louis, MO, 1989.

FORD, D. C., and P. W. WILLIAMS. *Karst Geomorphology and Hydrology.* London, 1989.

HILL, C. A., and P. FORTI. *Cave Minerals of the World.* Huntsville, AL, 1986.

JENNINGS, J. N. *Karst Geomorphology.* Oxford, Eng., 1985.

WHITE, W. B. *Geomorphology and Hydrology of Karst Terrains.* New York, 1988.

WILLIAM B. WHITE

CERAMIC MATERIALS

Ceramics are primarily produced by the treatment of nonmetallic minerals through various processes including heat, to produce articles of utilitarian or aesthetic properties. The word ceramics is derived from *keramos*, the Greek word for potter's clay or ware made from fired clay. The word ceramic, while retaining its original sense of a product made from clay, now also includes other materials besides clay, and other products made by the same general process of manufacture. A concise definition is, "Ceramics are products made from inorganic materials which are first shaped and subsequently hardened by heat" (Singer and Singer, 1963).

This entry addresses what ceramic materials are, what their history of use is, and how they are used today. In addition, the principal raw materials are described along with how these raw materials formed and where they are located.

Our present-day ceramic products have been developed as a result of an evolving use of raw materials and improved processing that has spanned many centuries. In primitive and early times the only ceramic raw materials were natural plastic clays, which could be shaped, dried, and fired. Ceramic artifacts are of major importance in archaeological studies. Interpretation of ceramic pieces can be particularly useful in making inferences about life in the past. The style and technology of ceramics are both important to the archaeologist. Style refers to a mode or manner of expression characteristic of an individual whereas

technology refers to the practice of a technical terminology by which a people provides itself with objects of material culture (Wright, 1985). In modern ceramics many raw materials in addition to clays play an important function, but the role of clay is still major. The principal raw materials used as ceramic ingredients include the following:

Clays	Magnesite
Silica (quartz)	Olivine
Feldspar	Dolomite
Nepheline Syenite	Bauxite
Talc	Graphite
Pyrophyllite	Zircon
Kyanite	Chromite
Wollastonite	

Ceramic Raw Materials

Clay is the most important ceramic material today as it was in past centuries. Pots made from clay can be traced back to the earliest historical periods. One of the major technological expansions of ceramics took place in China around the beginning of the Christian era (Norton, 1974). From a coarse, somewhat porous earthenware that was prevalent in early times, the Chinese developed a strong, watertight stoneware and a translucent porcelain of great artistic beauty. They did this by developing higher temperature kilns and by having the availability of a partially decomposed granite that contained kaolin, feldspar, and quartz, a natural body composition that when fired produced a strong white porcelain product (Figure 1). The area in central China where this weathered and decomposed granite was found is near the town of Jingdazhen. The town is still important for the manufacture of high-quality ceramic products. The term "china clay" was coined after Marco Polo returned to Europe with samples of this clay along with the beautiful pieces of porcelain produced from it. In England, a clay was discovered in the Cornwall area that is very similar in composition to the clay from China and hence was called "china clay," a term that is still used today. China clay and kaolin are terms used synonymously, particularly in Europe.

The term "clay" as used in the ceramic industry is generally applied to those materials which have the property of plasticity. There are several types of clay that are used as ceramic raw materials.

Figure 1. Large porcelain vase produced and decorated at Jingdazhen in China (photo courtesy of Haydn H. Murray).

These include kaolins, ball clays, fireclays, underclays, shales, bentonites, and various types of surficial clay deposits such as loess, glacial lake clays, and soils that can be used to make brick and drain tiles. Another clay used in ceramics is flint clay, which is non-plastic.

Kaolins are white firing clays that are comprised primarily of the mineral kaolinite. Kaolinite is a hydrated aluminum silicate $[Al_2Si_2O_5(OH)_4]$. Kaolinite forms as a result of the weathering of feldspars, a process in which potassium, sodium, and calcium are removed in solution and silica and alumina remain to form kaolinite (Murray, 1988). Kaolin deposits are classified as primary or secondary. A primary kaolin deposit is one that forms in place as a result of weathering or hydrothermal alteration of crystalline rocks. A secondary kaolin is one formed by erosion, transport, and deposition. The transported kaolin particles are usually deposited in quiet depositional environments such as lakes, lagoons, or oxbow river cutoffs.

Kaolins that are used for ceramics are found in many areas of the world. The most important kaolin deposits are in the states of Georgia and South Carolina in the United States, in the Cornwall area of southwestern England, in Bavaria in southern Germany, near Dresden in eastern Germany, in the Ukraine, in Spain near Guadalajara, in Japan near Nagoya, in China near Suzhou, in Indonesia on the island of Belitung, in New Zealand north of Auckland, in Australia north of Melbourne, in Brazil near São Paulo, and in Argentina in Patagonia.

Ball clays are sedimentary clays characterized by the presence of organic matter that gives them a gray or black color. They have high plasticity, high dry strength, a relatively long vitrification range, and are white or near white in color when fired. Kaolinite is the principal mineral constituent typically making up more than 70 percent of the minerals present. The term "ball clay" originated in England many decades ago from the mining practice of prying out a lump of clay, rolling it so that it formed a large ball, and loading it onto a horse drawn cart or wagon. The major ball clay deposits are found in western Kentucky and Tennessee in the United States, in the Dorset and Devon areas of England, and in the Westerwald area of Germany.

Flint clays are hard, refractory clays that break with a conchoidal fracture, and have little or no plasticity. They are comprised of almost pure kaolinite and are very refractory and so they are used mainly in high-duty refractory applications. They occur under coal seams, similar to fireclays and underclays, but are believed to have been precipitated in an acid swamp environment (Keller, 1964). Flint clays are produced in eastern Kentucky and Missouri in the United States, and in England, Germany, Australia, and China.

Fireclays are plastic clays that occur under coal or lignite. These clays generally are comprised of kaolinite, quartz, and a small amount of illite and/or chlorite. They have a fusion point above 1410°C (cone fifteen). The fusion point of a ceramic clay is where the clay begins to transform from the solid to liquid state and is a critical temperature to know before a ceramic body is fired. The method used to determine the fusion point is to compare the bending characteristics of the sample with those of a series of standard pyrometric cones, which fuse and then bend at known temperatures. Fireclay deposits are found in the United States in Pennsylvania, West Virginia, Ohio, Kentucky, Illinois, Missouri, and Colorado. The major fireclay deposits in Europe are in England and Germany. Fireclays are used for low, medium, and high-duty refractories.

Underclays or seat earths are plastic clays that have the same origin as fireclays but because they contain a higher percentage of illite and/or chlorite along with kaolinite and quartz, they are very plastic with a high green strength and have a fusion point lower than cone fifteen (1410°C). (Green strength is the strength of a clay after mixing it with water to make it plastic.) Some fire to an off-white color or a red color depending upon the iron content in the clay. They are used for making pottery, stoneware, and heavy clay products such as brick, sewer pipe, flue linings, conduit tile, and floor tile.

Shales are fine-grained sedimentary rocks that are comprised mainly of illite, chlorite, quartz, and occasionally some kaolinite. They generally fire to a red color because of their high iron content.

They are used primarily to make heavy structural clay products and floor tile. Shales are found in large quantities on every continent in the world.

Bentonite is a clay that contains a major proportion of the clay mineral smectite. Bentonite is usually an altered volcanic ash or altered tuff and the main smectite minerals are either sodium montmorillonite or calcium montmorillonite. Small amounts of smectite are used in ceramics primarily because of their low vitrification temperatures, and they are excellent plasticizers for those clays and shales that lack plasticity. Sodium montmorillonites are mined in Wyoming in the United States and calcium montmorillonites are mined in Texas and Mississippi and in England, Germany, France, and Italy in Europe.

Surficial clays such as loess (a windblown clay), glacial lake clays, and soil clays are used as ceramic materials primarily for making brick and drain tile. These clays are variable in composition, vitrify at a low temperature and have a short vitrification range. The reason these types of clays are used is because there are generally no higher-quality clays in a particular geographic region. The loess deposits are located on the eastern side of the major north-south rivers in the United States. The glacial lake clays are found in many locations across the northern tier of states in areas subjected to Pleistocene glaciation. Soil clays, of course, are ubiquitous and are found in most areas of the world.

Finely ground silica (quartz) in a major component (20–50 percent) of normal ceramic bodies for making earthenware, porcelain, and sanitary ware. Usually it is finely pulverized quartz sand, sandstone, or quartzite. Generally high-silica deposits must be very pure, containing 98 percent or more silica. The iron content must be very low. High-silica deposits are generally sedimentary sands and sandstones because they are easier to mine and process, but in some areas quartzites are used if high-silica sands and sandstones are not available. In the United States high-silica sands and sandstones are mined in New Jersey, South Carolina, Georgia, Tennessee, Pennsylvania, Ohio, Illinois, Missouri, Oklahoma, Texas, and California. Quartzite is mined in West Virginia. High-silica deposits are present on every continent.

Feldspar is the most important flux used in ceramic bodies and glazes. Feldspar is an alkali or alkaline earth, or potassium aluminum silicate mineral, and is igneous in origin. The most commonly used feldspars in the ceramic industry are potash and sodium feldspars. The main sources of feldspar for the ceramic industry are in North Carolina in the United States, and in the Scandinavian countries, Italy, France, and in the Czech Republic in Europe.

Nepheline syenite (an igneous rock that resembles granite but contains no quartz) is used in place of feldspar in many ceramic bodies and glazes. The chief mineral constituents are nepheline, potash feldspar, and albite. Nepheline syenite fluxes at lower temperatures than straight feldspar and is used in sanitary ware bodies, wall tiles, electrical porcelain, and whitewares. The major deposits are located in Ontario in Canada and in Norway.

Talc is a hydrated magnesium silicate that is used in electrical insulators and in other ceramic bodies such as automotive catalytic convertors, where low thermal expansion and high thermal shock resistance are needed, and in wall tiles in amounts up to 50 percent to reduce crazing (crazing is cracks in the glaze on the tile surface caused by differential shrinkage of the tile body and the glaze). In smaller quantities the fluxing action of the magnesia lowers the maturing temperature, reduces porosity, and increases the strength of semivitreous bodies. Talc is a secondary alteration product formed by the interaction of water and silica with magnesium in rocks such as dolomites and dunites. Important talc mining locations are in Vermont, New York, Texas, and Montana in the United States, and important foreign locations are in France, Finland, China, and Australia.

Pyrophyllite is a hydrous aluminum silicate that is used in wall tiles, in some whitewares, electrical insulators, and for special refractories. Like talc it is a secondary alteration product of alumina-rich rocks. Pyrophyllite deposits are mined in North Carolina and California in the United States, and in Brazil, China, Japan, and South Africa.

Wollastonite is an elongate calcium silicate mineral that is used to lower the maturing temperature of ceramic bodies. Wollastonite does not evolve any gas on heating so it can be used successfully in low temperature, single-fired products where the glaze melts at the firing temperature. Because there is no evolution of gas, the glaze remains intact and smooth. In wall tiles the firing shrinkage is reduced and the strength increased up to 50 percent because of the fibrous nature of the wollastonite crystals. Wollastonite is a contact metamorphic mineral formed when limestone and a silica-rich rock are heated and compressed. Wollastonite is mined primarily in New York State and in Mexico near San Luis Potosí.

Other ceramic materials listed above—kyanite, magnesite, olivine, dolomite, bauxite, graphite, and chromite—are discussed in REFRACTORY MATERIALS.

Ceramic Processing

After the ceramic raw materials are mixed in the proper proportions, the ceramic body is formed into the desired shape. There are several forming methods used (Norton, 1974), including slip-casting; plastic forming; extrusion forming; dry pressing; dust pressing; and isostatic pressing. After the body is formed it is dried; this is a very important step in the manufacture of ceramic articles. Although economics demand the fastest possible drying time, too fast a drying schedule causes differential shrinkage that in turn causes cracking. Thus the proper drying conditions are of critical importance. After drying, the next step is firing. Of all the steps in the production of ceramic articles, firing is the most critical. The ceramist and/or the ceramic engineer must know as much as possible about the reactions that take place at elevated temperatures. Of particular importance are energy states and the stability of the particular phases at various temperature levels, the heats of formation that can be either exothermic or endothermic, the vapor pressures of the various oxides, and the equilibrium conditions of the various phases. The effects of heat on ceramic bodies are complex. Below is a simplified table showing the relationship of temperature to some reactions that take place (Table 1).

Table 1. Reaction Temperatures in the Firing Process

Temperature °C	Reactions
up to 100	Loss of moisture
100–200	Removal of absorbed water
550 ± 50	Dehydroxylization of clays
500	Oxidation of organic material
573	Quartz inversion to the high temperatures form
850–1,250	New phases form, glass forms, and pores close

In general the body shrinks in size and the porosity decreases with increased temperatures up to the proper maturing temperature (the temperature is different for different body compositions). Thus the firing process is a science in itself. The setting of the ceramic pieces in the kiln, the type of kiln, the rate of heat increase, and the firing conditions (oxidation or reduction) are of critical importance. The proper selection of ceramic raw materials is most important in controlling the desired properties of the finished ceramic ware.

Bibliography

KELLER, W. D. "The Origin of High Alumina Clay Minerals—A Review." *Clays and Clay Minerals*, ed. W. F. Bradley. New York, 1964, pp. 129–151.

MURRAY, H. H. "Kaolin Minerals: Their Genesis and Occurrences." In *Reviews of Mineralogy*. Vol. 19, *Hydrous Phyllosilicates*, ed. S. W. Bailey. Washington, DC, 1988, pp. 67–89.

NORTON, F. H. *Elements of Ceramics*, 2nd ed. Redding, MA, 1974.

SINGER, F. and S. S. SINGER. *Industrial Ceramics*. New York, 1963.

WRIGHT, R. P. "Technology and Style in Ancient Ceramics." In vol. 1 of *Ceramics and Civilization*, ed. W. D. Kingery. Columbus, OH, 1985, pp. 5–26.

HAYDN H. MURRAY

CHARON

See Pluto and Charon

CHEMISTRY OF EARTH MATERIALS

Rocks are the end product of geological processes that occur at the surface of Earth and in its interior. Some rocks contain valuable elements in sufficient concentrations to be mined as ore deposits. Others have desirable mechanical properties that make them useful as building stones for construction of highways or as linings in blast furnaces. The job of geochemists is to determine the chemical compositions of rocks and other Earth materials and to study the chemical reactions and geological processes that cause these materials to form or to be destroyed. Earth materials also include soil, water, air, and fossil fuels (coal, petroleum, and natural gas). Humans and other life forms depend on the presence of these materials for their survival on Earth.

Synthesis of Atoms in Stars

The exploration of space has helped us to recognize that Earth is one of the nine planets in our solar system whose chemical composition reflects the conditions under which it formed. Earth formed 4.5 billion years ago (4.5 Ga) from a cloud of gas and dust called the solar nebula (*see* SOLAR SYSTEM). This cloud came into existence when one or several ancestral stars in our region of the Milky Way galaxy exploded as supernovas (*see* GALAXIES). These exploding stars scattered chemical elements into space that included not only hydrogen and helium, but also carbon, nitrogen, oxygen, and all of the other elements the stars had synthesized by nuclear reactions in their interiors. In this sense, Earth is composed of chemical elements that were made in stars.

The nuclear reactions that generate the energy stars radiate into space produce about 2,500 different kinds of atoms. These atoms are distinguished by the number of protons and neutrons in their nuclei. All atoms of a particular element have the same number of protons in their nuclei, but may contain different numbers of neutrons. Such atoms of a particular element are the isotopes of that element. The isotopes of an element have identical chemical properties, but they differ from each other by having different masses depending on the number of neutrons they contain in their nuclei. The masses of atoms are expressed in atomic mass units (amu), defined as one twelfth of the mass of the most abundant isotope of carbon. These atoms have six protons and six neutrons. This isotope is therefore known as carbon twelve.

Most of the atoms produced in stars are unstable and undergo spontaneous nuclear transformations that are accompanied by the emission of alpha particles, beta particles, and gamma rays from the nucleus. By the time Earth formed from the solar nebula, most of the unstable atoms had already decayed. Therefore, Earth is composed primarily of the stable atoms that make up the elements we know today. Nevertheless, a few elements still have unstable isotopes because these so-called radioactive atoms decay so slowly that they have not yet been transformed into stable atoms of other elements. The elements that have radioactive isotopes are uranium, thorium, radium, potassium, rubidium, and a few others (*see* ISOTOPE TRACERS, RADIOGENIC, and ISOTOPE TRACERS, STABLE).

The natural occurrence of radioactive atoms in the rocks of Earth is important for three reasons: (1) Radioactivity releases energy and is an important source of heat in the earth (*see* HEAT BUDGET OF THE EARTH); (2) the decay of radioactive atoms of the elements listed above can be used to measure the ages of rocks and of Earth itself (*see* GEOLOGIC TIME, MEASUREMENT OF); and (3) the gamma rays and nuclear particles emitted by naturally occurring radioactive atoms can damage cells and are therefore harmful to life. All organisms on Earth, including humans, are continually exposed to gamma rays and nuclear particles that are emitted by radioactive atoms in the rocks and soil of Earth, in the water we drink, and in the air we breathe (*see* PUBLIC HEALTH AND EARTH SCIENCE).

The chemical composition of the solar nebula, and hence of the solar system as a whole, is best preserved in the Sun (*see* SUN). The wavelength spectrum of sunlight has been analyzed very carefully to determine the chemical composition of the Sun. This information has been augmented by chemical analyses of stony and iron meteorites (*see* METEORITES) and of a wide variety of rocks from Earth. In addition, we can now determine the chemical composition of lunar rocks collected on the Moon by American astronauts during the Apollo program and by the Russian Lunakhod robot probes (*see* APOLLO ASTRONAUTS). About one dozen rock specimens from the Moon and a similar number from the planet Mars have also been found on the glacial ice sheet in Antarctica and elsewhere on Earth.

All of this information has been assembled to determine the so-called cosmic abundance of the elements. The results indicate that hydrogen and helium are the most abundant elements in the solar system even though these elements are not well represented on Earth, the Moon, and Mars. In general, the abundance of elements decreases as their atomic number (number of protons in the nucleus) increases. The cosmic abundance of the elements has been used to study the nuclear reactions that occurred during the life cycles of the ancestral stars whose violent explosion produced the solar nebula. In addition, knowledge of the cosmic abundance of the elements enables geochemists to recognize the effects of the chemical differentiation among the planets of the solar system and within the Earthlike planets Mercury, Venus, Mars, and the asteroids.

The Periodic Table of the Elements

The chemical properties of the elements depend primarily on the number of protons in the nuclei of their atoms because that number determines how many negatively charged electrons are required to neutralize the positive charges of the protons in the nucleus. It is the electrons that participate in chemical bonding, thereby defining the chemical properties of elements.

When Niels Bohr first explained in 1913 how the hydrogen atom works, electrons were thought to orbit the nucleus of an atom much like the planets of the solar system orbit the Sun. However, we now know that atoms must be described by means of quantum mechanics and that electrons cannot be treated as though they were small particles. The insights provided by quantum mechanics have enabled us to understand the chemical properties of the elements that are used to construct the periodic table (*see* frontmatter). It turns out that the regular and incremental changes in the chemical properties of the elements arise from the way electrons distribute themselves among the available orbitals of the atoms.

The most important characteristic of the periodic table is that elements having similar chemical properties are grouped together. For example, the so-called alkali metals (lithium, sodium, potassium, rubidium, and cesium) all form cations (positive ions) with a charge of +1 when they are dissolved in water. Similarly, the alkaline earths (beryllium, magnesium, calcium, strontium, barium, and radium) form cations with a charge of +2, whereas the halogens (fluorine, chlorine, bromine, and iodine) form anions (negative ions) with a charge of −1, and the noble gases (helium, neon, argon, krypton, xenon, and radon) do not form ions at all.

When these and other elements are acted upon by geological processes, the elements that form groups in the periodic table tend to remain together. For example, rocks that contain high concentrations of potassium commonly also contain the other alkali metals. Therefore elements from the same group in the periodic table exhibit geochemical coherence.

The Geochemical Classification of the Elements

Another form of association of chemical elements is well known to metallurgists who smelt ore to recover various metals. In this process three different immiscible liquids are produced: liquid iron, liquid sulfide (called matte), and liquid silicate (called slag). In addition, gases are released into the atmosphere. The geochemist VICTOR MORITZ GOLDSCHMIDT recognized that the elements can be classified by observing whether they are concentrated preferentially in the iron liquid, the sulfide liquid, the silicate liquid, or whether they associate with the gases. His work resulted in the so-called geochemical classification of the elements summarized in Table 1. The elements that preferentially dissolve in liquid iron are said to be siderophile (i.e., they "like" iron; *siderophile* is Greek for "iron loving"), those that are concentrated in the sulfide liquid are chalcophile, the elements that are most abundant in the slag are lithophile, and the elements that are released into the atmosphere are atmophile.

Table 1. The Geochemical Classification of the Elements

Siderophile	iron, cobalt, nickel, molybdenum, platinum-group metals, gold, carbon, phosphorus, and tin
Chalcophile	copper, zinc, silver, cadmium, mercury, lead, bismuth, gallium, indium, thallium, sulfur, selenium, tellurium, arsenic, and antimony
Lithophile	alkali metals, alkaline earths, halogens, boron, aluminum, oxygen, rare earth elements, scandium, titanium, vanadium, chromium, manganese, yttrium, zirconium, niobium, hafnium, tantalum, tungsten, thorium, and uranium
Atmophile	hydrogen, nitrogen, and the noble gases

The geochemical classification of the elements is useful because it helps us to understand how Earth formed. When Earth accumulated from the solar nebula, it was initially very hot, possibly completely molten. This allowed large masses of liquid iron to sink to the center of Earth to form the core. The silicate liquid surrounded the core and crystallized slowly to form the rocks of the mantle and the proto-crust of Earth. The sulfide liquid also sank but apparently did not form a continuous layer between the core and the mantle. The amophile elements and volatile compounds, such as water and carbon dioxide, escaped from the hot silicate liquid called the magma ocean and formed the first atmosphere of Earth.

The geochemical classification highlights some surprising facts:

1. Oxygen is a lithophile element and is concentrated in rocks rather than in the atmosphere;

2. carbon is a siderophile element, which means that a large amount of carbon probably now resides in the core of Earth; the existence of life on the surface of Earth therefore depends on the presence of carbon, which originally may have escaped into the atmosphere in the form of carbon dioxide or methane;

3. gold and the platinum group elements are also siderophile and were concentrated into liquid iron as it sank toward the center of Earth;

4. copper, zinc, lead, and silver associate with sulfur and occur primarily as sulfide minerals in ore deposits;

5. several elements can occur in two or more of the classes. For example, among the siderophile elements iron is also lithophile and chalcophile; cobalt, nickel, and mobydenum are chalcophile, whereas phosphorus, tin, gold, carbon, and the platinum-group metals are lithophile in addition to being siderophile.

The Task of Geochemists in Society

The tendency of the chemical elements to divide into the four classes of the geochemical classification helps to explain why the interior of Earth can be divided into the core, the mantle, and the crust. The core is composed of metallic iron and other siderophile elements. The mantle and crust are composed primarily of lithophile elements with lesser amounts of siderophile and chalcophile elements. The chemical differentiation of Earth implied by its internal structure has caused the rocks that form the continents to be enriched in only nine lithophile elements whose concentrations in weight percent are:

oxygen	45.5
silicon	26.8
aluminum	8.4
iron	7.1
calcium	5.3
magnesium	3.2
sodium	2.3
potassium	0.9
titanium	0.5
Total:	99.0

We see that all of the other elements together make up less than one percent of the crust of Earth. The list of scarce elements includes many important industrial metals, such as copper, zinc, lead, silver, gold, and uranium, to name a few.

The low abundance of these and many other valuable metals poses a challenge to earth scientists whose job it is to find deposits of natural resources, to plan their extraction from the earth, and to monitor the disposal of the waste products that result from the use of these resources.

In order to carry out these tasks, geochemists study the chemical reactions that occur at different stages of the rock cycle illustrated in Figure 1 (*see* ROCKS AND THEIR STUDY). The numbered arrows

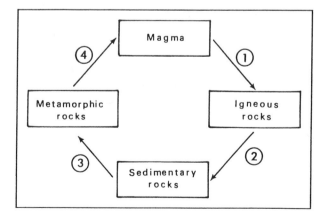

Figure 1. Chemical and physical processes by which Earth materials are transformed from magma to igneous rocks, to sedimentary and metamorphic rocks, and ultimately back to magma.

that relate magma and the three principal rock types to each other represent chemical and physical processes by which Earth materials are transformed from magma to igneous rocks, to sedimentary and metamorphic rocks, and ultimately back to magma, thus restarting the cycle.

Geochemists who are concerned with processes represented by arrow 1 in the diagram study the crystallization of minerals from cooling magmas to form igneous rocks, and on the alteration of these rocks by interaction with hot groundwater. The scientific subjects needed to understand these processes include physical chemistry of silicate melts, mineralogy, and water-rock interactions at elevated temperature. Arrow 2 in the diagram encompasses uplift and erosion, followed by chemical weathering, transport of weathering products by running water, deposition of sediment, and conversion of loose sediment to sedimentary rocks. The scientific disciplines needed to understand these processes include the chemistry of aqueous solutions, clay mineralogy, surface chemistry, hydrology, and the study of water-rock interaction at low temperatures. The conversion of sedimentary to metamorphic rocks (arrow 3) requires recrystallization of minerals under conditions of rising temperature and pressure during the formation of mountains. The minerals formed at this stage are, in some cases, identical to those that form igneous rocks (e.g., feldspar, mica, quartz, amphiboles, and pyroxenes). Under extreme conditions and in the presence of water and other fluxes, metamorphic rocks may begin to melt. The melt fraction segregates from the residual solids to form a new magma that is enriched in volatile compounds (e.g., water and carbon dioxide) and certain lithophile elements that form large ions. These are called the large-ion-lithophile (LIL) elements and include the higher atomic number alkali metals and alkaline earths and others. In this way, the igneous rocks that form from the new magma are enriched in the LIL elements relative to the metamorphic rocks from which the magma originated.

Each chemical element responds in its own way to this treatment by Earth depending on its properties. Some elements stay with the solid phases (e.g., oxygen, silicon, aluminum, etc.), while others preferentially dissolve in water (e.g., sodium, magnesium, calcium, chlorine, sulfur, etc.), or are released into the atmosphere (e.g., the noble gases). The gases, including helium and argon, are produced by decay of the radioactive isotopes of uranium and thorium (helium) and potassium (argon).

The operation of geological processes also gives rise to the so-called hydrologic cycle, which describes the movement of water by evaporation from the surface of the oceans to the atmosphere. The circulation of the atmosphere causes the water vapor to be transported to the continents where some of the water precipitates as rain, hail, or snow. The rainwater does geological work by transporting sediment and dissolved ions, and by facilitating chemical reactions with the minerals it comes in contact with. In addition, water sustains plants and animals and thereby makes life on Earth possible.

The circulation of water and the presence of life are recognized by geochemists as being complementary to the processes that affect rocks. The environments of Earth consequently have been divided into the lithosphere (rocks), the hydrosphere (water), the biosphere (life), and the atmosphere (air). To understand how Earth works, the properties of each of these separate entities and their mutual interactions must be understood. The four "spheres" together constitute the environment in which we live. A major task of geochemistry is to study the workings of the natural environment composed of rocks, water, life, and air in order to be able to predict their responses to particular anthropogenic or natural disturbances (see SOLID EARTH-HYDROSPHERE-ATMOSPHERE-BIOSPHERE INTERFACE).

The demand for natural resources increases with the size of the human population, which has been rising rapidly and may reach six billion by the end of the twentieth century. To meet this challenge, geochemists study how ore deposits form as a result of geological processes operating on the surface of Earth and in its interior. Success in this task is determined in part by application of the relevant principles of chemistry to the study of chemical reactions among solids, liquids, and gases over a wide range of temperature and pressure. These studies include experimental work in the laboratory, sophisticated analyses of rock and mineral samples, and careful observations of geological processes in the natural environment.

Geochemists are at the forefront of one of the most important challenges that confront humankind today. Increasingly, geochemists are being called upon to evaluate the environmental impacts of major construction projects or of the emission of gases and contaminated water by factories, cities, towns, and villages. Although the impact on the environment is localized in many cases, the effect is cumulative and global in scope. Consequently, geochemists now monitor the oceans, the land, and the atmosphere to detect contamination that may prove to be harmful to life. The discovery that the ozone layer in the atmosphere over the polar regions of Earth is deteriorating exemplifies this important aspect of geochemistry.

Perhaps even more serious than the partial loss of ozone is the increase in the concentration of carbon dioxide in the atmosphere, caused by the combustion of fossil fuels. The presence of carbon dioxide and other gases, such as methane and water vapor, increases the amount of heat absorbed by the atmosphere. The resulting warming of the global climate is expected to change weather patterns in the future and to cause a rise in sea level because ice sheets in Greenland and Antarctica may melt. As a result, coastal areas may be flooded, thereby removing fertile farmland from production and requiring inhabitants to relocate.

Bibliography

BROWN, G. C., C. J. HAWKESWORTH, and R. C. L. WILSON. *Understanding the Earth.* New York, 1992.

EMILIANI, C. *Planet Earth: Cosmology, Geology, and the Evolution of Life and Environment.* New York, 1992.

FAURE, G. *Principles and Applications of Inorganic Geochemistry.* New York, 1991.

LEWIS, J. S. *Physics and Chemistry of the Solar System.* San Diego, 1995.

SCHLESINGER, W. H. *Biogeochemistry: An Analysis of Global Change.* San Diego, 1991.

GUNTER FAURE

CHONDRITIC MODEL

See Earth, Composition of

CHROMIUM

Chromium (Cr) is a hard, corrosion-resistant, silvery-white metal that is an essential industrial commodity. When alloyed with iron (Fe), chromium yields stainless steel and this is the principal use of the metal. Depending on the specific use—household utensils, food containers, automobile parts, weapons, cutting tools—stainless steels contain between 12 and 36 percent chromium by weight. Other metals will also alloy with iron to produce tough stainless steels, but chromium is the least expensive and most abundant alloy metal. Other uses for chromium are in high-temperature furnace linings (as the mineral chromite), and various chromium salts used in chemical industries.

There is only one mineral from which chromium is commercially extracted, the mineral chromite, which is a member of the spinel mineral group having the general formula $A^{2+}B_2^{3+}O_4^{2-}$. The spinels have a wide compositional range and are characterized by extensive atomic replacement. In the case of chromite, A^{2+} can be any combination of Mg^{2+} and Fe^{2+}, while B^{3+} can be any combination of Cr^{3+}, Al^{3+}, Fe^{3+}. Chromites with high contents of Cr^{3+} and Fe^{3+} (Cr_2O_3 contents above 40 percent by weight) are desirable sources for alloys; those with Cr_2O_3 contents below 40 percent have high contents of Al^{3+} and are desirable for furnace linings.

The annual world production of chromite ore exceeds 13 million metric tons. In 1994 significant ore production came from twenty-two countries, with more than half of the total from South Africa and Kazakhstan. Other countries that produce important amounts of chromite are Turkey, the Philippines, Zimbabwe, Albania, India, and Finland.

There are two principal kinds of chromite deposits, stratiform deposits in layered igneous complexes, and podiform deposits.

Layered igneous complexes are bodies of mafic igneous rock that contain compositional layers (strata) as a result of crystallization. In some cases the layers are monomineralic. The most striking examples of monomineralic strata of chromite are in the Bushveld Igneous Complex of South Africa, and the Great Dike of Zimbabwe. The Bushveld Igneous Complex, which is the world's largest layered intrusive complex, is 9 km thick, has a surface area of 66,000 km^2, and contains fourteen chromite layers, six of which have been mined. The

Great Dike contains fewer chromite layers than the Bushveld Complex but, like the Bushveld, the layers are remarkably continuous and extensive. Other layered intrusive complexes known to contain potentially minable amounts of chromite are the Stillwater Complex in Montana and the Dufek Complex in Antarctica (although the Antarctic Treaty precludes mining).

Podiform chromite deposits are so named because they occur as pods of chromite enclosed in mafic and ultramafic igneous rocks. Such deposits are thought to form at the base of the oceanic crust or at the top of the underlying mantle. The deposits are located where fragments of oceanic crust and upper mantle have been thrust onto continental crust as a result of plate tectonics. The world's largest chromite mine, the Donskay mine in Kazakhstan, works podiform chromites in ultramafic rocks in the southern Ural Mountains. Other important podiform chromites are found in Turkey, Zimbabwe, Albania, Cuba, and the Philippines.

Bibliography

DIETRICH, R. V., and B. J. SKINNER. *Rocks and Rock Minerals.* New York, 1980.

BRIAN J. SKINNER

CLIMATOLOGY

Weather is the physical state of the atmosphere at any given moment in time. It is described by such phenomena as the movement of storms and clouds, the probability of precipitation, and changes in the humidity. Climate, on the other hand, is defined as the representative, or statistical, state of atmospheric conditions, the "expected weather conditions," over periods of years or longer. Climate is determined by the same variables as the weather: temperature, precipitation, winds, humidity; but the climate of a region is more than just the average of these variables. The climate includes the extremes in the weather as well as the range of expected weather. It also refers to the variability about the climatic average. Record high and record low temperatures, one hundred-year floods, century storms—these events are all present within the climate. While averages describe the normal, expected climate, the climatic extremes and climate variability encompass the full range of the climate for a region. A climatologist studies climate, climatic controls, and climatic change.

The climate system consists of five components: the atmosphere, the biosphere (including plants, animals, and humans), the lithosphere (Earth's surface), the hydrosphere (oceans and lakes), and the cryosphere (areas of snow and ice). Although most climate studies focus on the atmosphere as the primary component of the climate over timescales of weeks to years, each component affects the climate over different time and space scales. For example, most vegetation has an annual cycle of growth, while glaciers advance and retreat on timescales of thousands of years. For this reason, chemists, biologists, geographers, geologists, oceanographers, and atmospheric scientists all work in the area of climatology.

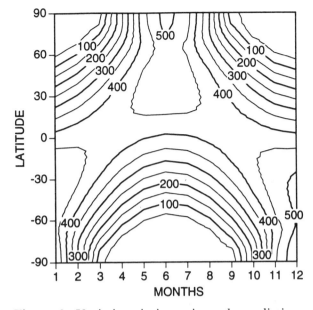

Figure 1. Variations in incoming solar radiation over latitude and time of year (beginning in January = 1). Values indicate a monthly average (in Wm^{-2}) for solar radiation at the top of the atmosphere.

Climate Controls

The distribution of climates on Earth is determined by several controls, or forcings. The most important control on the climate is the distribution and intensity of solar radiation at Earth's surface. Solar radiation is the energy received from the Sun in the form of shortwave electromagnetic waves or photons. The solar energy that reaches Earth provides energy to the climate system, driving the circulation of the atmosphere and oceans.

Incident solar energy is distributed unequally across Earth both in space and time (Figure 1) due to the geometric relationships between the Sun and Earth, including the rotation of Earth on its axis, the revolution of Earth around the Sun, and the tilt of Earth's axis of rotation relative to the plane of Earth–Sun orbit (the work "climate" originates from the Greek "climat," meaning "slope" or "tilt"). Solar radiation is further affected by atmospheric constituents, including aerosols, clouds, and gases.

Imbalances in energy across Earth's surface create the forces that drive the climate system, causing pressure imbalances that initiate atmospheric and oceanic motion. One can think of Earth as a large heat engine, balancing the unequal heating at the surface by redistributing that heat from the tropics

to the poles through atmospheric and oceanic circulation. Energy is redistributed primarily through the circulation around low-pressure weather systems. Low-pressure systems carrying storms across the midlatitudes are the means by which the atmosphere mixes warm and cold air. These systems, which produce much of the weather of the midlatitudes, swirl in a counter-clockwise direction around the center of the low pressure in the northern hemisphere (and in a clockwise direction in the southern hemisphere), pulling cool air into the low latitudes and pushing warm air into the high latitudes. The systems are pushed across the midlatitudes from west to east by the upper-air winds of the prevailing westerlies, thus mixing air across latitude and longitude.

Climate Change

Averaged over time and space, Earth maintains an energy balance and, excepting "normal" climate averages and extremes, should be globally neither warming nor cooling as a result of natural forcings, at least on a year-to-year basis. We know, however, through records of atmospheric temperature trends inferred from data recorded in deep ice cores, that Earth has gone through large changes in climate throughout its history (*see* PALEOCLIMATOLOGY). The changes in climate between glacial and interglacial periods occur over long time periods (hundreds of thousands of years) and, although they result in large changes in the climatic landscape of Earth, the average annual temperature of Earth changes relatively very little—on the order of only a few degrees. In the more distant past, Earth has varied over millions of years between relatively warmer and relatively cooler periods.

One of the more intriguing questions addressed by climatologists is the potential of future climate changes. Observations show the concentration of atmospheric carbon dioxide (CO_2) to be increasing since the beginning of the industrial revolution (Figure 2). Carbon dioxide, and other atmospheric gases such as water vapor, methane, and nitrous oxide, are called "greenhouse gases" because they trap much of Earth's energy within the atmosphere similar to glass trapping heat inside a greenhouse. Some concentration of atmospheric greenhouse gases is beneficial to maintain the present climate, but the concern is that a rapid increase in these gases may accelerate global warming beyond the present climatic equilibrium. Although there seems to be a connection between an increase in the atmospheric concentration of greenhouse gases such as carbon dioxide and ob-

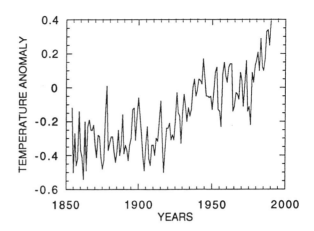

Figure 3. Trend in global surface air temperatures (shown as a difference from the 1960–1990 global average) taken from measurements.

served surface temperatures (Figure 3), observations of surface temperature are highly variable and any signal of climate change in the observations is difficult to detect. Therefore, some other means of investigating the problem must be used.

Climate Modeling

It is not possible to build laboratory models or experiments of the climate system. An alternative means of investigating problems of climate and climatic change is through the use of numerical models of the atmosphere. These models, called general circulation models, use mathematical equations to simulate the transports and exchanges of energy, momentum, and mass across the globe. Many of the physical processes (such as cloud formation, precipitation, surface hydrology, and vegetation) simulated in a climate model can be approached only by highly simplified representations of the true complex physical interactions between the components of the climate system. Therefore, the results of such models are approximations of the possible changes, limited by our knowledge of the physical relationships of the climate system and computer resources. However, results of such models can be used to gain insight into the potential consequences of increasing the atmospheric concentration of greenhouse gases such as CO_2 (see Plate 3).

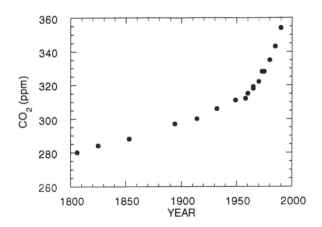

Figure 2. Time series of the trend in atmospheric carbon dioxide since 1800 as inferred from ice core analysis and by direct measurements.

Bibliography

GATES, D. M. *Climate Change and Its Biological Consequences.* Sunderland, MA, 1993.

HIDORE, J. J., and J. E. OLIVER. *Climatology: An Atmospheric Science.* New York, 1993.

SCHNEIDER, S. H., and R. LONDER. *The Coevolution of Climate and Life.* San Francisco, CA, 1984.

SUSAN MARSHALL

COAL

Coal is a combustible, organic, sedimentary rock formed from accumulations of plant material as peat in waterlogged areas (swamps, mires). This material becomes deeply buried under younger sediments and is transformed into coal during burial.

Minable coal beds originated from laterally extensive peat deposits that accumulated in thicknesses of tens of meters in swamps over many thousands of years. In its natural state, peat contains 70 to 90 percent water that can be partially squeezed out by hand. As additional layers of sediments (peat, mud, silt, and sand) accumulate, the peat is compacted, porosity is reduced, and water is squeezed out. If continued subsidence of an area permits acculation of hundreds to thousands of meters of overburden, peat is eventually transformed into rocklike coal by the overburden pressure and increasing temperature with depth.

Rank. While peat is near the surface, chemical agents, especially oxygen and enzymes, together with biological agents, primarily bacteria and fungi, attack the plant debris and decompose and transform it to varying degrees, especially the less stable components such as cellulose and proteins. The more stable plant components, such as resins, waxes, lignin, and tannins, tend to survive this initial transformation in the peat stage. As temperature rises with increasing depth of burial, the more stable components are converted into even more stable organic structures and peat becomes successively, lignite, subbituminous, and bituminous coal. If temperature during burial rises sufficiently, the organic material may eventually be converted to anthracite. This sequence from peat to anthracite is called the coalification series. The chemical and physical properties of coal change significantly during this maturation process (Table 1). The degree of coalification, commonly referred to as a coal's "rank," depends primarily on the maximum depth of burial and the attendant overburden pressure and temperature, and on how long the organic material was exposed to these conditions. Generally, millions to tens of millions of years are required for the coalification process, especially to attain a bituminous or anthracitic coal rank.

Type and Grade. Depending on the regional and local conditions of climate, water table fluctuations, and environments of deposition, different kinds of plants contribute to peat formation, and the plant material degrades to different degrees, leading to a wide range of types of peat and coal. Also, minerals may be washed into the swamp during flooding by streams or the sea, or blown in by wind, and thus diminish the purity of peat and coal, or its grade. The combustible portion conventionally must constitute at least 50 percent of a plant-derived rock, on a dry basis, for it to qualify as "coal"; otherwise it would be called a carbonaceous (coaly) rock.

Coal Lithotypes

To the naked eye, helped only by a hand lens if needed, several kinds of coal are easily distinguishable. A major distinction is between banded and nonbanded coals.

Banded coals are the most common. The banding is particularly well displayed in coals of bituminous rank. Shiny layers of vitrain alternate with semidull layers of clarain and dull layers of durain; vitrain, clarain, and durain are called lithotypes. Lithotype layers are usually only fractions of a centimeter to several centimeters thick. Vitrain is primarily derived from relatively large pieces of wood and bark of stems, branches, and roots. Clarain and durain together are sometimes referred to collectively as attrital coal, which is highly variable in appearance and composition. Attrital coal is commonly finely striated; it is composed primarily of fine fragments of coalified plant material. Dull attrital coal is commonly finely granular with a rough texture, and very hard. The duller coals tend to contain higher percentages of finely dispersed

Table 1. ASTM Classification of Coals by Rank (in box) and Other Related Properties

| ASTM Class | ASTM Group | Heating Value | | Agglom-erating | % VM (d, mmf) | R Oil max. (vitrinite) | % Moisture (moist, mmf) | % C (d, mmf) | % O (d, mmf) | % H (d, mmf) | FSI |
		Btu/lb (moist, mmf)	MJ/kg (moist, mmf)								
	Peat	1,000–6,000	2.3–4.0	No	62–72	0.2–0.4	50–95	50–65	30–42	5–7	
Lignite	Lignite B	Undefined–6,300	<14.6	No	40–65	0.2–0.4	40–60	55–73	23–35	5–7
	Lignite A	6,300–8,300	14.6–19.3				31–50	55–73	23–35	5–7
Subbituminous	Subbituminous C coal	8,300–9,500	19.3–22.1	No	35–55	0.3–0.7	25–38	60–80	15–28	4.5–6.0
	Subbituminous B coal	9,500–10,500	22.1–24.4				20–30	60–80	15–28	4.5–6.0
	Subbituminous A coal	10,500–11,500	24.4–26.7				18–25	60–80	15–28	4.5–6.0
Bituminous	High volatile C bit. coal	10,500–13,000	24.4–30.2		35–55	0.4–0.7	10–25	76–83	8–18	4.5–6.0	1–5
	High volatile B bit. coal	13,000–14,000	30.2–32.5		35–50	0.5–0.8	5–12	80–84	7–12	4.5–6.0	2–7
	High volatile A bit. coal	>14,000	>32.5	Yes	31–45	0.6–1.2	1–7	78–88	6–10	4.5–6.0	4–9
	Med. volatile bit. coal				22–31	1.0–1.7	<1.5	84–91	4–9	4.5–6.0	7–9
	Low volatile bit. coal				14–22	1.5–2.2	<1.5	87–92	3–5	4.5–6.0	1–9
Anthracite	Semianthracite	>14,000	>32.5	No	8–14	2.0–3.0	<1.5	89–93	3–5	3–5
	Anthracite				2–8	2.6–6.0	0.5–2	90–97	2–4	2–4
	Meta-anthracite				<2	>5.5	1–3	>94	1–2	1–2

VM = volatile matter; (d, mmf) = calculated on a dry, mineral-matter-free basis; R, Oil max.; (vitrinite) = measured reflectance of polished vitrinite under oil immersion; FSI = Free Swelling Index; Agglomerating = coals produce an agglomerate button in a crucible during the volatile matter determination that will support a 500-g weight without pulverizing, or a button showing swelling or cell structure, are considered "agglomerating"; any coal with an FSI of 1 or more is considered agglomerating.

Modified from Damberger et al., 1984.

mineral matter, particularly clay and quartz. Fusain is a charcoal-like, fibrous, crumbly material that usually occurs throughout a coal bed, both dispersed in other lithotypes and concentrated in highly lenticular layers that rarely exceed one or two centimeters in thickness. Fusain tends to be concentrated on selected bedding surfaces, along which blocks of coal tend to separate easily. The large pores of fusain are often filled with mineral matter. Fusain is the material in coal that dirties the hand.

Nonbanded coals lack layering and appear finely granular or pitchlike, and homogeneous, with a dull, greasy luster. This appearance reflects their composition of finely comminuted plant debris that was deposited in quiet water. Pollen and spore grains and algal remains are common components of nonbanded coals (e.g., cannel coal, formed primarily from spores, and boghead coal, formed primarily from algae). These coals can be positively identified only with the help of a microscope.

Impure coal has between 25 and 50 percent mineral matter by weight, on a dry basis. The mineral impurities range from fine-grained, detrital mineral matter (clay, silt, sand) that was washed or blown in during peat accumulation, to impregnations with metal sulfides (particularly pyrite) and carbonates (calcite, dolomite, siderite) that precipitated from circulating water subsequent to peat accumulation, filling open pores in the coal and any fractures that formed after the peat became coalified.

Macerals

The coal scientist needs a microscope to identify the plant-derived microscopic components of coal, called macerals (Table 2). The standard microscopic analysis for macerals is done on representative samples of finely crushed coal embedded in epoxy pellets; the surface of the pellet is ground and polished for observation under reflected light. Many macerals preserve original plant structures sufficiently well to permit their recognition as plant organs, at least in a general way, and their names have been chosen accordingly, for example: sporinite is derived from spores, cutinite from cuticles, alginite from algae, resinite from resin. Other macerals owe their origin to a specific alteration process: oxidation (e.g., during forest fires) for fu-

Table 2. Classification of Macerals*

Maceral Group	Maceral
Vitrinite	Collinite
	Telinite
Liptinite or exinite	Alginite
	Cutinite
	Resinite
	Sporinite
Inertinite	Fusinite
	Inertodetrinite
	Macrinite
	Micrinite
	Sclerotinite
	Semifusinite

* Based on ASTM Standard D 2796.

sinite, gelification (which obliterated much of the original cell structure) for vitrinite. Macerals by definition do not contain any visible mineral grains, but they may contain minerals of submicroscopical size. The macerals have been classified into three major groups based primarily on their technological behavior, particularly during the coking process: inertinites exhibit an inert behavior during the process, while vitrinites and liptinites are reactive components of quite different behavior. Note that all maceral names end with "inite" (while lithotypes have the suffix "ain").

Minerals

Noncombustible minerals, invariably present in coal, detract from its value. Some can be recognized with the naked eye—especially when they occur as distinct layers or "partings" in coal beds—or with the help of a hand lens. However, many mineral inclusions are so small and finely disseminated through the coal that they can only be identified under the microscope, or with the help of other special techniques, such as X-ray diffraction. A major advance in the recognition of minerals in coal was the development of the low-temperature ashing technique in the 1960s. The low-temperature asher oxidizes the organic matter of powdered coal samples in an oxygen-rich plasma at a temperature of 120 to 150°C, leaving the incombustible minerals in the residue. This temperature is low enough to retain most minerals unchanged. Minerals commonly found in coal are clays, pyrite, quartz, and calcite. Many other minerals are found in coal, but their frequency of occurrence generally is only moderate to rare.

Porosity and Internal Surface Area. The natural moisture content of coal roughly indicates its pore volume. Vitrinite, the dominant maceral in most coals, can be looked at as a hardened gel. Most pores of such materials have very small diameters, no larger than a few tens of Ångstroms ($1 \text{ Å} = 10^{-8}$ cm). Consequently, the internal surface area of coal is very large relative to its porosity, up to several hundred square meters per gram of coal. The small size of the pores makes it difficult to determine "true" values for the porosity, pore size distribution, and internal surface area of coals. The results depend strongly on the size of the molecule of gas or liquid adsorbate used and the temperature at which the measurement is made. One important consequence of this peculiarity of coal is that the methane invariably found in coal is primarily adsorbed as a thin layer on the very large internal surface from which it cannot be easily enticed to migrate toward a gas well. Coals of medium and low volatile bituminous rank tend to contain considerable amounts of methane that is released during mining, causing a potential hazard and increasing the cost of mining.

Technological Properties of Coal

Washability. All coal seams contain some incombustible mineral matter that occurs both as distinct "partings" of rock (e.g., shale) and finely disseminated mineral matter. Engineers subject representative samples of ground coal to float-sink tests in liquids of various specific gravities to determine how much of the sample floats and sinks at the selected gravities. From such tests, they can construct washability curves to project how much saleable coal of a desired ash content can be recovered from a coal deposit by operating a cleaning plant at various specific gravities. Raw mined coal may contain as much as 50 percent ash; cleaned coal generally will contain 5 to 15 percent ash, depending on the intended end use, and customer specifications.

Coking Properties. When certain coals are heated without access of air, they become plastic while releasing gases; a carbon-rich, porous coke is left behind. Medium volatile bituminous coals form the strongest coke and are prime coking coals for the production of metallurgical coke for blast furnaces. However, in practice medium volatile bituminous coals are blended with low and high volatile bituminous coals to form coke that has the desired properties.

Combustion Properties. Most coal is burned to generate steam in electric power and industrial plants. Therefore, a coal's heating value is of primary importance. Generally, the higher the rank the higher the heat content of a coal (see Table 1). The heating value of a coal is measured by burning a small sample in an adiabatic calorimeter bomb; the resulting temperature change indicates the heat content of the coal. The heating value depends on both the heat content of the organic portion of coal and the proportion of noncombustible matter, in particular moisture and mineral matter. Coal rank also influences combustion properties in general. The higher a coal's rank, the higher its ignition temperature and the lower its reactivity with air. The release of volatile matter is important during combustion; high rank coals, which have low volatile matter contents, burn with a short flame and more slowly than low rank coals. Another important property of concern during combustion is the mineral matter content and the properties of the ash formed during combustion. The ash fusion test is used to predict ash fusion behavior in the boiler. The test determines the temperatures at which the ash undergoes major phase changes, in particular the temperature range in which it becomes plastic. Coals with a high fusion temperature (greater than 1,540°C) are preferred over coals with low fusion temperatures (less than 1,100°C). Several empirical indexes that are calculated from chemical analyses of the ash are in use to predict the behavior of ash during combustion, particularly its propensity to form deposits within the boiler (slagging properties).

Coal Conversion. Conversion of coal to liquids or pipeline quality gas was thought to be the wave of the future in the wake of the "energy crisis" of the 1970s. Conversion technologies were developed and are available but are not competitive at recent prices of natural gas and oil.

Table 3. Coal Resources, Recoverable Reserves, and Production of Countries with Substantial Reserves and Production (in billion [10^9] metric tons)

	Resources in Ground	Recoverable Reserves	Production in 1991
Australia	785	90	0.218
Canada	67	9	0.071
China	2,737	114	1.090
Germany	332	80	0.351
India	115	63	0.238
Poland	199	41	0.210
South Africa	132	55	0.178
United Kingdom	186	4	0.093
United States	1,570	240	0.904
USSR (former)	5,502	241	0.684
Other	375	102	0.598
World	**12,000**	**1,039**	**4.635**

Coal Resources and Production

Coal is widely distributed throughout the world and many countries have minable resources. The first comprehensive compilation of the world's coal resources was completed for the International Geological Congress in Toronto in 1913. Since then the World Power and then World Energy Conferences or Council (WEC) have periodically updated the compilations. Table 3 is based on the latest published WEC survey. It should be kept in mind when reviewing a table like this that the standards for classifying coal resources in the ground and recoverable reserves vary between countries, as do the availability and quality of data. However, the table gives a sense of which countries are best endowed with coal resources.

The most recently available coal production data are included in the table for comparison. In most cases production could be maintained at current levels for several hundred years without exhausting known reserves.

Use of Coal

The greatest portion of coal is used for steam generation in electric power plants: 87 percent in the United States, 73 percent in the Organization for Economic and Cultural Development (OECD) countries, and 52 percent worldwide in 1992. Other important uses are in industrial plants (United States, 8%; OECD, 9%), and coke for iron and steel making (United States, 4%; OECD, 9%; world, 12%, combined with other industrial uses). Conversion of large quantities of coal into high-value petroleum-like liquids and pipeline quality gas is unlikely until its costs are more nearly comparable to the prices of oil and natural gas. However, the conversion of coal into medium Btu gas at power plants and its use in combined cycle plants (high-temperature gas turbine operated by combusted gas, followed by standard steam turbine) look attractive because it has the advantage of increased overall plant efficiency and permits relatively easy removal of sulfur during the gasification process.

Bibliography

DAMBERGER, H. H.; R. D. HARVEY; R. R. RUCH; and J. THOMAS, JR. "Coal Characterization." In *The Science and Technology of Coal and Coal Utilization*, eds. B. R. Cooper and W. A. Ellington. New York, 1984.

U.S. DEPARTMENT OF ENERGY, ENERGY INFORMATION ADMINISTRATION. *Annual Energy Review 1992*. Washington, DC, 1993.

VAN KREVELEN, D. W. *Coal, Typology, Chemistry, Physics, Constitution*. New York, 1961.

WARD, C. R. *Coal Geology and Coal Technology*. Boston, 1984.

WORLD ENERGY CONFERENCE. *1983 Survey of Energy Resources*. Twelfth Congress of the World Energy Conference, New Delhi, September 18–23, 1984. New Delhi, 1984.

HEINZ DAMBERGER

COAL MINING

Two basic methods are used to mine coal. In surface mining, a coal seam is uncovered by removing overlying layers of soil and rock. This overburden is placed back after coal extraction. In underground mining, coal seams buried deep in the ground are accessed through vertical shafts, horizontal tunnels (drifts), or inclined tunnels (slopes). Coal is mined and transported to the surface through the access openings.

Some coal properties, such as hardness and a tendency to break more easily along internal surfaces, are significant from the standpoint of mining. Others such as heat, ash, sulfur, and volatile matter contents affect coal's usage and its value, and thus the decision whether to develop a mine.

Surface Mining Techniques

Topography and stripping ratio (the amount of overburden in relation to the amount of coal) determine the surface mining method used. In hilly or mountainous terrain, for example, where the coal seam runs through a hill, coal is accessed by mountaintop removal (Figure 1a). The landscape changes permanently because hilltops are leveled and valleys filled. Mountaintop removal requires special permission from federal and state regulatory agencies if the original contour and vegetation are not to be restored, as required by law. This technique is mostly used in the Appalachian region of the United States, where mountaintop removal can create valuable flat surfaces for commercial, agricultural, or residential uses.

Contour surface mining is the technique used if mountaintop removal is not permissible or economically feasible (Figure 1b). Only the thinner portions of the mountaintop along the coal exposure are stripped away to mine coal. This is often followed by auger mining (Figure 1c) to maximize coal production. A large screw-like device "augers" coal from the hillside without removing the overburden.

The most widely used surface mining techniques are open pit and area mining, both developed to use large earth-moving equipment. The open pit technique is used to mine thick coal seams covered by relatively thin overburden. The pit remains open for long periods between cycles of overburden removal (Figure 2a). The area mining

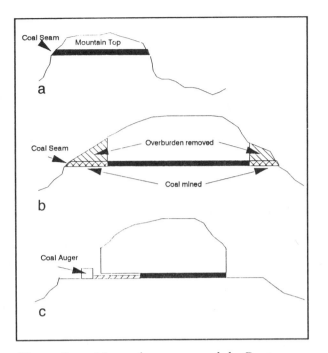

Figure 1. a. Mountaintop removal. b. Contour mining. c. Auger mining.

technique, used for thinner seams, is similar to the open pit, although the overburden must be removed more frequently to produce the same amount of coal that an open pit mine produces (Figure 2b). In area mines, the coal is exposed with an initial cut into the overburden, which is depos-

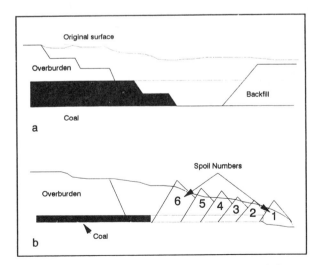

Figure 2. a. Open pit mining. b. Area mining.

ited to the side to create the initial spoil. While coal is still being mined from the first cut, an adjacent coal area is exposed and overburden from this second cut is deposited in the first mined-out area (second spoil). A series of cuts made by the advancing operation creates a series of spoils.

Where the overburden or coal is hard, mine operators may use explosives to loosen and break material before loading and hauling it. For other conditions, buckets of shovels or draglines are equipped with teeth strong enough to loosen the material while lifting it for loading. Bucket wheel excavators are suitable for the soft overburden conditions commonly encountered in mining lignite. Coal and overburden are transported in off-highway trucks at most surface mines, but large volumes of materials may also be transported by conveyor belts.

Special machines called scrapers remove the topsoil before excavating the overburden. Topsoil is stored separately. As mining operations advance, bulldozers level the first spoils and the topsoil is brought back. This step in land reclamation is legally required in the United States and must be followed by revegetation. The contour and use of the land must be comparable to its premining contour and use.

In the early 1970s, surface mining of coal in the United States began to expand rapidly, primarily in the thick, shallow deposits of subbituminous coal in Wyoming and Montana. Total U.S. coal production increased from 613 million tons in 1970 to more than one billion tons in 1990. Surface mining, which accounted for 6 percent of the total in 1970, rose to 33 percent in 1990. Nearly 80 percent of the increase in national coal production since 1970 came from surface mines in the western states.

Underground Mining Techniques

Coal is mined by underground techniques when surface mining is not economical because of the depth of the coal seam(s) or the nature of the overburden. Underground mines are accessed through a drift (Figure 3a) when the coal seam is nearly horizontal and accessible from the hillside. When the seam is inclined or deeper in the ground, a slope is driven from the surface (Figure 3b). Verti-

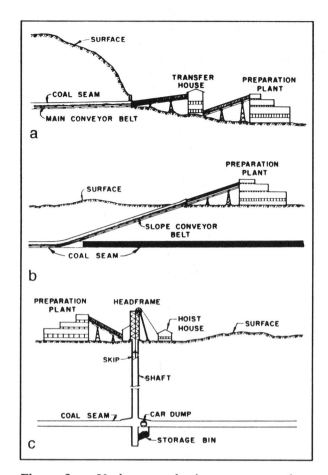

Figure 3. a. Underground mines are accessed through a drift. b. With a deep seam, a slope is driven from the surface. c. Vertical shafts allow access to coal seams located under deep overburden.

cal shafts are used when coal seams lie under very deep overburden (Figure 3c). Sometimes, a combination of slope and shaft is used for access.

Two basic designs, room and pillar (Figure 4) and longwall (Figure 5), are used in the development of modern underground mines. Machines called continuous miners cut access openings (entries and exits) into the coal and divide the seam into panels of convenient length and width. Each panel has several entries and exits that are used to convey workers, materials, and air into and out of the panel. Roads in the entry and exit systems are cross-connected by removing sections of coal separating them. Pillars of coal are left to support the roof over the entries.

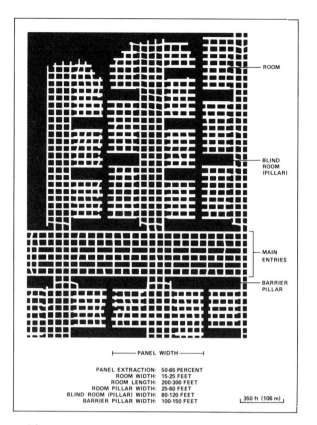

Figure 4. The room-and-pillar mine design.

Equipment for cutting rock and coal is fitted with water-spraying nozzles to suppress dust. Very fine rock and coal dust, if inhaled for a long time, can cause lung diseases such as pneumoconiosis ("black lung") among mine workers; however, modern coal mines are quite successful in dust suppression. As a safety measure against propagation of coal dust explosions the mine floor and walls are covered with limestone powder. Powdered limestone is also stored in strategic locations. In case of an explosion, the powder adsorbs heat and helps prevent it from spreading.

Mine ventilation generally keeps methane gas, associated with coal and released during mining, at harmless levels. Large fans installed at the mine entrance draw air from the mine. Fresh air is drawn in from a second opening, located optimally to serve for a long time as the mine expands with time. Often the air intake opening is located at the farther end of the mine. Legal standards have been established regarding the temperature, composition, and flow rate of air in mines.

Room-and-pillar coal extraction forms a checkerboard pattern: coal is left unmined (black pillars) to support overlying layers of rock; white areas represent mined-out coal (rooms). As Figure 4 shows, nearly one-half of the coal in the panel remains unmined. Reducing the size of pillars before mining moves out of the panel can reduce coal losses, but caution is required with this practice.

In longwall mining, coal is cut along an entire panel 120 to 180 m wide or more. A drum shearer is commonly used to cut coal in U.S. mines. A plow is used in some countries, especially where coal seams are thin. A chain of hydraulic props, organized as multiprop frames with strong shields, supports the roof and protects miners working at the coal face. A steel conveyor carries the coal away to the exit road system of the longwall panel from where it is carried by conveyor belts either to the shaft or directly to the surface. Shafts equipped with large bins called skips transport coal to the surface.

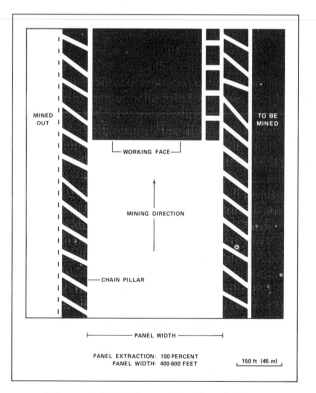

Figure 5. The longwall mine design.

No coal is left unmined between panel access roads, so the loss of coal in longwall mines is substantially lower than it is in room-and-pillar mining. As the mining machine advances along with the hydraulic roof support system, the roof collapses over the large panel area behind the system. As a result, subsidence on the earth's surface is inevitable, occurring soon after mining. This is a planned phenomenon.

The planned, relatively uniform subsidence of longwall mining contrasts with the incidental subsidence of room-and-pillar mining. Years after mining ceases, the floor below pillars may fail, old or weak pillars may crumble over a certain area, or the overburden cave into abandoned mine rooms. In shallow mines, up to 60 m deep, the collapse of overburden may propagate to the earth's surface and cause local pit subsidence. Buildings, roads, and other structures on the surface may be damaged, and some farmland lost. Whereas modern room-and-pillar designs attempt to prevent subsidence altogether, longwall mines are designed to produce subsidence under controlled conditions. When land settles quickly over a large area, damage to structures and crop land is less severe and can be dealt with immediately; structures can be protected while subsidence takes place.

Minimization of subsidence over longwalls is sometimes attempted by backfilling the voids with crushed rock blown in pneumatically or with water. Because of the high cost, backfilling is not practiced often.

Longwall mining requires costly equipment. It permits higher extraction rates (lower losses of resources), offers a safer working environment for miners, and often costs less per ton of coal mined than the room-and-pillar method. Longwall mining is the most widely used method of underground mining in Europe and Asia, and its use is increasing steadily in the United States.

Bibliography

CRICKMER, D. F., and D. A. ZEGEER, eds. *Elements of Practical Coal Mining*, 2nd ed. New York, 1981.
HUNT, S. R. "Surface Subsidence due to Coal Mining in Illinois." Ph.D. diss., University of Illinois at Urbana-Champaign, 1980.

SUBHASH B. BHAGWAT
HEINZ DAMBERGER

COASTAL PROCESSES

More and more people visit beaches each year, due in part to the growth in population living near the coast. It is estimated that about two-thirds of the world's population now live within an hour's drive of an ocean beach. In addition to their recreational value, beaches also act to protect developments built along the coast—parks, homes, hotels, and other structures. The beach does this by serving as a buffer between the developments and storm waves, causing the waves to break well offshore, and give up most of their energy before they actually reach the shoreline. In spite of this protection, erosion has become a significant problem in many coastal areas, especially during hurricanes and other severe storms. To preserve the beaches and reduce property erosion better, it is important that we understand the physical processes acting on the coast, which includes understanding the action of waves and currents, and how they affect beaches.

There is wide variety in the scenery of coasts and in the nature of beaches. Long stretches of the East Coast of the United States consist of barrier islands separated from the mainland by lagoons and bays. These barrier islands are low in elevation, with sand dunes as their highest points, and they may become completely submerged during hurricanes, which generate large waves and raise water levels because of the strong winds blowing water against the land. By contrast, the coasts of California, Oregon, and Washington are elevated, with sea cliffs backing the beaches in most areas. The problem on the West Coast is with sea-cliff erosion because homes and other developments are often built too close to the cliff edge, which may be slowly retreating as the cliff is episodically attacked by storm waves. The fundamental difference between the East and West Coasts is due to their contrasting tectonic settings. The West Coast is being uplifted by collisions between the oceanic and continental plates (*see* PLATE TECTONICS), while much of the East Coast is subsiding. The subsidence of the East Coast occurs in addition to the global rise in sea level due to the melting of glaciers, so that part of the coastal erosion results from the sea moving upward and over the low-lying barrier islands (*see* GLOBAL CLIMATIC CHANGES, HUMAN INTERVENTION; GLOBAL ENVIRONMENTAL CHANGES, NATURAL).

Beaches throughout the world are even more varied, showing marked differences in the grain sizes and mineral compositions of their sediments. Most beaches are composed of sand-sized particles of quartz and feldspar minerals, derived from the physical weathering of rocks, such as granite and schist, that are abundant on the continents (see WEATHERING AND EROSION; SEDIMENTS AND SEDIMENTARY ROCKS, TERRIGENOUS [CLASTIC]). The quartz and feldspar grains are light tan to white in color, but among them can be found a few dark grains of "heavy minerals" such as hornblende, augite, garnet, magnetite, and others. These heavy minerals are also derived from the weathering and erosion of rocks and, at times, can be traced back to specific rock sources. These rock sources can be within sea cliffs, the erosion of which supplies sand to the beach, while in other cases the minerals are traced back to a river that supplies sand to the beach, the river in turn having derived the sand from erosion of rocks hundreds of kilometers away within its watershed. An awareness of the sources of sediment to the beach can be important, because cutting off those sources by damming rivers or building seawalls along eroding cliffs may ultimately result in erosion of the beaches.

The dark heavy minerals are more susceptible to chemical and physical weathering than is the resistant quartz, and the loss of those dark minerals may result in a beach composed almost entirely of quartz sand grains. Such beaches, found in Florida and Alabama, as well as at beach resorts around the world, are intensely white in color. The only other sediment component common to those beaches are fragments of shells, which have been worn smooth by the waves. At the other extreme are the "black sand" beaches, composed almost entirely of heavy minerals. These beaches are normally rich in dark heavy minerals, but the minerals are further concentrated by the waves, which can easily pick up the low-density grains of quartz and feldspar and carry them offshore, leaving behind only the high-density heavy minerals. These black sand beach deposits sometimes contain valuable elements such as gold, platinum, magnetite (iron), and titanium. Another type of black sand beach can be found on volcanic islands such as the Hawaiian Islands. This type consists of particles of basalt and other volcanic rocks, often formed by the explosive disintegration of hot lavas as they flow into the cold water of the sea.

Although most of our familiar beaches are composed of sand, others consist of pebbles, cobbles, and even large boulders. In some parts of the world, these coarser-grained beaches are common. For example, beaches of flatish pebbles and cobbles called shingle are found along the coast of England. Indeed, the term "beach" is derived from an ancient Anglo-Saxon word that referred to shingle.

The mass of sand and pebbles of the beach is moved about by waves and currents and is shaped into a distinctive "morphology," the term used to describe the overall slope and contours of the beach. The dominant source of energy comes from waves reaching the coast, waves that may have been generated by storms thousands of kilometers away. The size of waves formed by the storm depends upon the speed of the wind, the fetch or area of the storm, and the duration of the storm (the time available to transfer energy from the winds to the ocean waves). The waves lose comparatively little energy as they cross the deep ocean, preserving it until they reach the coastal zone and break on the sloping beach. The wave energy is expended in the act of breaking and then as bores when the waves continue to break while crossing the surf zone and finally swash up the beach face. A wave first breaks when it reaches a water depth that is approximately equal to its height. The higher the wave the deeper the water in which it breaks, and generally the further offshore this occurs. This is one way in which the beach acts to protect the coast from storm waves, causing the higher waves to break earlier and to give up their energy further offshore. On steep beaches, the waves tend to break by plunging—the shoreward face of the wave becomes vertical and then curls over and plunges down. On lower sloping beaches the waves break by spilling, their crests cascading down as "white water" foam.

The slope and overall morphology of the beach depend on the grain sizes of its sediment and on the patterns of wave breaking. The general trend is that the coarser the beach sediment, the steeper the beach profile. Fine-sand beaches tend to have very low slopes, whereas pebble and cobble beaches may be so steep that they are difficult to walk on. With coarse-grained beaches, the forward surge of the breaking waves pushes the pebbles toward the land, tending to pile them up, but the return seaward flow is weakened because much of the water percolates downward into the large spaces between the pebbles. This leaves the coarse-grained beach with a steep slope. By contrast, there is very little water percolation down into a fine-sand beach, so the seaward return flow after a wave is nearly as strong as its landward surge, cutting the beach back down and maintaining the lower slope. The size of the waves is also important. Large waves act to cut back the beach to reduce its slope. Sediment eroded from near the shoreline or from the dry part of the beach is carried offshore to about the wave breaking position, where it is deposited to form an offshore bar. This can happen during an extreme storm like a hurricane, but on many coasts it occurs seasonally since the waves tend to be higher during the winter than the summer. Therefore, during the summertime of low waves, more sediment is found shoreward in the dry part of the beach, while during the winter of higher waves the dry beach is cut back and most of the sediment is moved offshore onto bars.

When waves reach the beach, their force generates different patterns of nearshore currents, depending upon whether the waves break with their crests nearly parallel to the shoreline or form large angles to the beach. When breaking parallel to the shore, seaward-flowing rip currents often form. These narrow currents can be very strong and are a hazard to swimmers because they may sweep them offshore to deep water beyond the zone of wave breaking. In the same way, the rip currents carry sand offshore, eroding the dry part of the beach. The shoreline will no longer be straight, but instead it will consist of a series of bays cut by the rip currents, with the distances between successive embayments being 30 to 500 m. Some embayments grow larger than others, due to the presence of stronger rip currents. The erosion of the rip embayment may threaten buildings or other structures backing the beach, especially when the embayment cuts entirely through the dry part of the beach and allows the waves to directly attack the developments.

When waves break at angles to the shoreline, rip currents are less likely to develop. Instead, the current is continuous along the length of beach. This type of current is important to the longshore transport of beach sediments. As each wave stirs up the sediment, the longshore current carries it along the length of the beach, and in this way the sand may drift for hundreds of kilometers along the coast. The longshore sediment movement is a nat-

ural process, and some beaches may acquire their sand in this way, the waves and currents having transported the sand from an eroding sea cliff or from the mouth of a river. Problems may develop when this longshore sand movement is interrupted, as would occur if jetties or some other harbor structures were built across the beach. These structures act much like dams to the longshore current and sand transport, so the sand accumulates where it is blocked and the beach there is observed to build outward into the sea. The problem occurs on the other side of the jetties, in a sense "downstream" from the dam. The sand blocked by the jetties can no longer reach the beach, while at the same time the waves remove sand and carry it further along the coast. The result can be extremely severe erosion near jetties or other harbor structures, which may lead to the complete disappearance of the beach and allow waves to reach homes and other developments.

To prevent the destruction of homes and perhaps even entire communities, several coastal protection measures are possible. Where the erosion is due to jetties blocking the longshore movement of beach sand, a bypass operation is often undertaken, ranging from simply trucking the sand past the jetties to the development of expensive systems of dredging the sand and then pumping it as a slurry onto the eroding beach. Whether or not the beach erosion is due to the construction of jetties, a similar response that is used with increasing frequency is beach nourishment, the practice of finding a new source of sand for the beach. This new sand commonly comes from offshore, from the seafloor of the continental shelf, where it is dredged and then pumped onto the beach. The widened beach will be improved for recreation, but it will also serve to protect properties better from storm waves. A variety of seawalls are sometimes built at the back of the beach to protect developments, especially where a beach nourishment project is not possible (usually for economic reasons). Unfortunately, seawalls can hinder access to the beach, and there is evidence that the walls can result in increased erosion of the beach during storms, even when they protect communities successfully.

Bibliography

BASCOM, W. *Waves and Beaches*, 2nd ed. Garden City, NY, 1980.

CARTER, W. G. *Coastal Environments.* London, 1988.

FOX, W. T. *At the Sea's Edge.* Englewood Cliffs, NJ, 1983.

KOMAR, P. D. *Beach Processes and Sedimentation.* Englewood Cliffs, NJ, 1976.

PAUL D. KOMAR

COLONIAL INVERTEBRATE FOSSILS

This entry describes three different groups of aquatic invertebrate organisms, phylogenetically unrelated but similar in that they exhibit colonial growth. Thus, even though unrelated by ancestry, members of all three groups respond to common constraints imposed by coloniality.

Sponges

Sponges belong to the phylum Porifera, a name descriptive of their major morphological feature, pores. Sponges are multicellular aquatic invertebrates that possess specialized cells but have no true tissues or organs. The sponge skeleton consists of thousands of spicules and/or a meshwork of organic fibers (spongin). Spicules may be calcareous, siliceous, or organic. The spicules may be arranged in an interlocking three-dimensional framework or dispersed throughout the spongin. Sponge classification is based almost entirely on spicule morphology (shape) and the arrangement of the spicules.

Sponges are most abundant in shallow marine environments, but some occur in deep water and fresh water as well. Adult sponges are exclusively sessile (stationary) filter feeders. Some sponges require a firm substrate; others are found living in muddy environments. Sponges are common inhabitants of modern reef environments and were among the earliest reef-building organisms.

Sponges are found in rocks as old as Cambrian (570 Ma) and range to the present. Because of their largely organic composition and delicate construction, intact sponges are relatively rare as fossils. The preserved record of sponges is dominated by their hard parts, the spicules. Sponges with well-defined three-dimensional spicule frameworks are sometimes preserved largely intact. The paucity of intact sponges in the fossil record is probably a taphonomic bias (*see* FOSSILIZATION AND THE FOSSIL RECORD) and does not reflect a true depletion of sponges in the geologic past.

Corals

Corals belong to a large and diverse phylum termed the "Coelenterata" or "Cnidaria" (pronounced "nye-daria"). The cnidarians are commonly viewed developmentally as the next step above sponges: multicellular, they possess tissues, but lack true organs. Members of the group characteristically possess tentacles, some with stinging cells (nematocysts). Some cnidarians are colonial, others are solitary. The three major classes within the phylum (Hydrozoa, represented by the polyp *Hydra*; Scyphozoa, which includes the jellyfish; and Anthozoa, the true corals) are distinguished primarily by their development, that is, the presence or absence of polyp (attached) and medusoid (free-floating) stages, and their symmetry. The predominantly soft-bodied Hydrozoa and Scyphozoa have low fossilization potential and are rare but important components of the geologic record. Possible trace fossils (body casts) of jellyfish have been described from the late Precambrian Ediacara fauna in Australia (*see* HIGHER LIFE FORMS, EARLIEST EVIDENCE OF). Thus, cnidarians are among the most ancient metazoan (multicellular) life forms.

The class Anthozoa is distinguished from other Cnidarians by their lack of a medusoid stage and their development of calcareous supports for the polyps. The calcareous skeleton, the corallum, is divided by vertical radial partitions, termed "mesenteries." In some Anthozoa the mesenteries are supported by calcareous septa. The class includes soft-bodied forms, such as the sea pens and sea fans, but the so-called stony corals, which possess a calcareous skeleton, constitute the bulk of the Cnidarian fossil record. The three major calcareous coral groups are the Tabulata, Rugosa, and Scleractinia.

Tabulate corals appeared in the Ordovician period and are last found in Jurassic strata. Exclusively colonial, the individual corallites, or compartments, that contain the soft-bodied organism are generally vertical tubes with shared walls. The common names of many taxa reflect the shape and arrangement of the corallites: *Favosites,* the "honey-comb" coral, *Syringopora,* the "organ-pipe" coral, and *Halysites,* the "chain-link fence" coral. Tabulate corals lack true septa, and their morphology is in many ways transitional between sponges and corals. Some researchers believe that tabulate corals are more closely related to sponges than corals. Several colonial fossil organisms once considered sponges are now classified with the tabulates.

Rugose corals, named for the wrinkled nature of the corallum, are commonly referred to as cup or horn corals for their slightly curved, conical form. Rugosans range from the Ordovician to the Permian and were both colonial and solitary in habit. Exclusively marine, these sessile organisms were sensitive to substrate consistency and current energy. Solitary, cup-shaped rugosans were probably oriented in life with their tapered end buried in soft substrate, or reclining on their convex side on firmer substrate. Rugosans were subject to being knocked over by currents, but they were resilient, modifying the growth of the corallum so that it eventually assumed an upright position. Rugosans are not present in the geologic record after Permian time. Some researchers believe that the rugosans did not die out, rather, they evolved into the modern scleractinian corals.

Scleractinian corals are the modern stony corals, originating in the Triassic period and dominant in today's shallow marine environments. Scleractinians may have arisen from rugose corals through an evolutionary mechanism termed "heterochrony." Heterochrony refers to evolutionary change in the sequence or timing in development. Rugosans passed through a juvenile stage of six-fold symmetry (as seen in the development of the septa) before they reached the four-fold symmetry of the adult. Scleractinians have six-fold symmetry throughout their development. Thus, it is possible that through evolutionary delay of the onset of four-fold symmetry, rugosans evolved into scleractinians.

Bryozoa

Bryozoans are sessile, colonial, aquatic invertebrates. The individual bryozoan animal, the zooid, is microscopic and lives in a skeletal compartment called the zooecium (plural, zooecia). Zooecium shape ranges from boxlike to tubular. The bryozoan colony or zoarium (plural, zoaria) is formed of aggregates of zooecia. The Bryozoa are classified largely by the shape and arrangement of the zoecia rather than by the overall morphology of the colony. Many bryozoa are polymorphic, that is, they possess differently developed zooids for different functions. For example, the zooid modified for feeding is called an autozooid. Despite their colonial habit and microscopic size of the individual, bryozoa are most closely related to the exclusively noncolonial, macroscopic brachiopods (*see* BRACHIOPODS). Both groups possess a unique structure, the lophophore, a fleshy, tentaculated organ used in feeding and respiration.

Modern bryozoans are sessile and aquatic. Most are marine but some live in freshwater lakes and streams. In the oceans they can be found at depths up to 5,400 m but they are most abundant in shallow waters. Most bryozoans appear to require a firm substrate for attachment, although a few mound-shaped fossil forms may have been free-lying on the seafloor.

As with other colonial invertebrates, colony shape probably reflects environmental parameters such as current energy, current direction, and substrate consistency. Major colony growth habits include massive branching and encrusting.

The phylum Bryozoa is the only invertebrate phylum that does not have a Cambrian record. The Bryozoa appear "full-blown," that is, their earliest fossil record comprises many different taxa, in the Ordovician period. The Permian extinction decimated bryozan taxa, but the group that survived re-radiated during the Jurassic period to give rise to the modern forms.

Bibliography

BERGUIST, P. R. *Sponges.* Berkeley, CA, 1978.
BOARDMAN, R. S.; A. H. CHEETHAM; and W. A. OLIVER. *Animal Colonies: Development and Function Through Time.* Stroudsburg, PA, 1973.
MOORE, R. C. *Treatise on Invertebrate Paleontology.* Part F, "Coelenterata." Lawrence, KS, 1981.
RYLAND, J. S. *Bryozoans.* London, 1970.

DANITA S. BRANDT

COMETS

A bright comet is one of the most spectacular astronomical events. Throughout history comets have left a strong impression on those who witnessed their apparitions. Their name comes from the Greek κομήρηs ("kometes"), meaning "the long haired one." Ancient Greeks thought comets to be atmospheric phenomena, part of the "imperfect" changeable Earth, not of the "perfect" immutable heavens. Many early civilizations considered comets omens of death and disease. Today we know that they are "icy conglomerates" as proposed in 1950 by Fred Whipple, that is, chunks of ice and dust left over from the formation of our solar system some 4.6 billion years (Ga) ago.

Comets are the most primitive bodies in our solar system. Because of their orbits and small sizes, comets have undergone relatively little processing, unlike larger bodies, such as the Moon and Earth, which have been modified considerably since they formed. This pristine nature of comets is evidenced in their high abundance of volatile compounds. The composition of comets contains a wealth of information on their origin and evolution as well as the origin and evolution of our solar system; hence, they are often referred to as cosmic fossils.

When a comet is far from the Sun, it is an inert icy body; as it approaches, heat will cause ices in the nucleus to sublimate, creating a cloud of gas and dust called the coma. Sunlight and the solar wind (a supersonic flow composed mainly of hydrogen ions streaming away from the Sun's outer layers) will push the coma gas and dust away from the Sun, creating two tails (Figure 1). The dust tail is generally curved and appears yellowish since the dust particles are scattering sunlight. The gas (or ion) tail is generally straight but can have considerable fine structure and appears blue because its light is dominated by emission from carbon monoxide ions (CO+). As a comet approaches the Sun, the tails will be behind it, but as it moves away they will lead. The appearance of comets in photographs can give the erroneous impression that they streak through the night sky like a meteor or a shooting star (Figure 2). In fact, comets move slowly from night to night in relation to the stars and can sometimes be visible for many weeks, like Comet Halley was during its 1985–1986 apparition.

Halley's is not the brightest comet; it is, however, the most famous, mainly because Comet Halley is the brightest of the predictable comets. Comets are classified into long- or short-period, depending on whether their orbital periods are greater or less than 200 years. Long-period comets

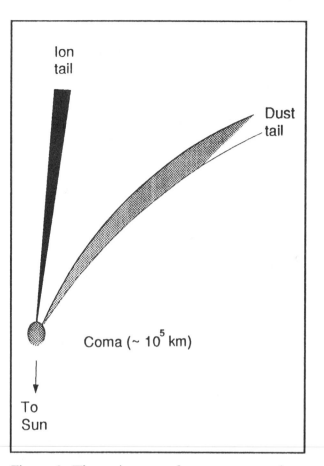

Coma (~ 10^5 km)

Figure 1. The main parts of a comet: coma, dust tail, and gas tail. The nucleus is approximately 1,000 times smaller than the coma.

are generally brighter and more active than short-period ones, but they are not predictable. Edmund Halley, an eighteenth-century British astronomer, was the first to calculate the orbits of comets by using Newton's recently developed theory of gravitation. Halley noted that the orbits of comets observed in 1531, 1607, and 1682 were quite similar and concluded that they were sightings of the same object during successive returns. He predicted the return of this comet in 1758 or 1759. Although Edmund Halley did not live to see it, the comet did return as predicted and was later named in his honor. Comet Halley's orbit has an average period of seventy-six years, perihelion between Venus and Mercury's orbits (0.59 astronomical units [AU]; 1 AU is the average distance between Earth and the Sun), and aphelion beyond Neptune's orbit (35 AU). Its orbit has an inclination of 162 degrees with respect to the ecliptic (the plane of Earth's

Figure 2. Photograph of Comet Kohoutek (1973f) showing two tails. The gas tail has a sharp edge, the dust tail is diffuse and extends to the right of the gas tail. (Photograph courtesy of S. Larson, University of Arizona.)

orbit); this means that Comet Halley orbits the Sun clockwise when seen from the north, while Earth orbits the Sun counterclockwise.

In 1986 an international flotilla of spacecraft were used to study Comet Halley. A detailed description of these spacecraft and their encounters with the comet is given by Reinhard (1990). In ad-

dition to space-based studies, an intensive campaign of ground-based observations was carried out. All these have yielded a wealth of new information about Comet Halley and about comets in general.

The Nucleus

One of the main results of the Comet Halley studies was the confirmation of Whipple's icy-conglomerate model. We now know that all of Comet Halley's activity originates in a discrete nucleus made up of ices and dust, and that water (H_2O) is the most abundant of the ices—about 80 percent. Comet Halley's nucleus turned out to be larger, darker, and less spherical than expected by most. It is peanut-shaped, approximately 18 km long and 8 km wide. The reflectivity or albedo is approximately 4 percent, which is as dark as coal. The direct observation of cometary nuclei using Earth-based telescopes is very difficult. When comets are near the Sun the nucleus is almost always masked by a bright gas and dust coma. On the other hand, when comets are at large, heliocentric distances, the nucleus is often too faint to be observed in detail. Nevertheless, the size, albedo, and approximate shape of several other cometary nuclei have been determined. Comet Halley's nucleus seems to be typical among short-period comets. All the nuclei studied in detail appear to have most of their surface covered by an inert mantle or crust. The active (exposed ice) fraction of their surface is small; in Halley this fraction is somewhere between 15 and 30 percent, the largest among the comets studied.

The development of a crust can suppress the activity in cometary nuclei and give them an asteroidal appearance. The best example to date is Comet Wilson-Harrington. This object was discovered in 1949 and was lost until it was rediscovered as minor planet 1979 VA, and later given the asteroid number 4015. In 1992, Brian Marsden identified 4015 and comet Wilson-Harrington as the same object. No clear cometary activity was detected from this object during its 1992–1993 apparition. The behavior of this comet has added credence to the long-held expectation that some Earth-crossing asteroids are extinct or dormant comets.

The internal structure of cometary nuclei is very uncertain. The current view is that comets are weakly bonded fractal assemblages of smaller "icy conglomerate" fragments, possibly welded into a single nucleus by gentle thermal processing. The compositional similarities between new and short-period comets suggest that cometary nuclei are not differentiated (i.e., the composition does not change with depth). However, small-scale differentiation is apparently present. For example, crust formation produces material rich in refractory elements and compounds, probably similar in composition to primitive meteorites. In addition, gas jets rich in certain compounds such as formaldehyde (H_2CO), reported in Comet Halley, and the temporal variability of several chemical species in Comet Encke are suggestive of some heterogeneity within the nuclei of these two comets.

Composition

The first composition of cometary nuclei is primarily inferred from studies of the coma components, namely neutral gas, plasma (ions), and dust. In 1985, water vapor was identified spectroscopically for the first time in Comet Halley using NASA's Kuiper Airborne Observatory and was found to be the most abundant volatile molecule, constituting about 80 percent of the coma molecules. It had been previously inferred from indirect evidence, such as the high hydroxyl (OH) abundance, that water ice was the most abundant ice in practically all comets. Other molecules which have been identified in Comet Halley and other comets with a reasonably high degree of confidence are listed in Table 1. The molecular abundances observed in Comet Halley and in most comets are in general similar to those observed in dense molecular cloud cores, which is the environment where star formation occurs. Thus, it appears that comets underwent little processing in the primordial solar nebula (the cloud of gas and dust out of which our solar system formed), and they preserve the best record of its original composition.

Ground-based studies of coma gases were, until 1986, the principal source of information about the composition of comets, and they remain the basis for comparing Comet Halley with other comets. Most of the gaseous species in comets accessible to ground-based telescopes are photo-dissociation (and to a lesser extent, chemical reaction) products of parent molecules that sublimate from nucleus ices. The abundances of several of these gaseous species including CN, C_2, C_3, OH, OH, NH, and NH_2 have been determined in the comae of more than one hundred comets. These studies indicate that comets are, to a significant extent, a chemically homogeneous set of objects. The "normal" abundances of the species are defined by roughly 50 to 70 percent of the objects studied, and Comet Halley falls in this group. A clear deviation from normal abundances was first noticed in Comet Giacobini-Zinner and has since been identified in several other comets. These deviations suggest that some comets may have formed in radically different environments, or were processed in a way that altered their chemical composition.

Ions provide clues about parent molecules and thus about the composition of the nucleus. For example, detection of species such as water ions (H_2O+) and carbon dioxide ions (CO_2+) were indicative of the presence of H_2O and CO_2 in a number of comets before these two neutral species were identified in Comet Halley.

Direct information on the composition of cometary dust particles was scarce prior to 1986. Studies of Halley's dust confirmed that some of the grains are silicates and discovered crystalline olivine ($(Mg, Fe)_2SiO_4$) as one of the silicate components. Also discovered in the spectrum of Comet Halley was an emission near 3.4 microns (μm). This spectral feature is believed to arise from C-H molecular stretch bands in organic compounds. However, the composition of the specific material is still unidentified. Some of this emission could be due to methanol (CH_3OH). The 3.4 μm feature has now been identified in several comets.

In situ measurements of solid particles in the coma of Comet Halley were carried out by three spacecraft (see Plate 2). These measurements indicate that the solid particles in Halley's coma have light element (hydrogen, oxygen, carbon, nitrogen, and sulfur) in abundances higher than the

Table 1. Parent Molecules Detected in Comets

Water (H_2O)
Carbon Monoxide (CO)
Carbon Dioxide (CO_2)
Diatomic Sulfur (S_2)
Hydrogen Sulfide (H_2S)
Methane (CH_4)
Hydrogen Cyanide (HCN)
Formaldehyde (H_2CO)
Methanol (CH_3OH)

most pristine meteorites (carbonaceous chondrites), thus confirming the primitive nature of cometary material. One of the unexpected results of the in situ measurement was the discovery that a large fraction of the dust is composed of organic material. (Here, the term "organic" refers to molecules containing mainly carbon and other light elements and is believed to have formed by abiotic processes. This term underscores the difference with minerals containing heavier elements.) This organic component was termed CHON because it contained almost exclusively the light elements carbon, hydrogen, oxygen, and nitrogen. There appear to be two end-member particle types, CHON and silicates. Most Halley grains are mixtures of these two types; in fact, silicate-core and CHON-mantle structure of the dust particles is suggested by the measurements. Such a core-mantle structure for cometary grains had been proposed by Greenberg (1980). The carbon that was "missing" from the gaseous fraction when compared with average solar system abundances was found in the CHON material (Delsemme 1991).

The abundances of different isotopes of the same element can also be diagnostic of the origin of the material. The deuterium to hydrogen (D/H) ratio in Halley is the same as terrestrial ocean water but significantly higher than that of Jupiter and Saturn. This result indicates that a significant fraction of Earth's water did not come directly from the solar nebula, as with Jupiter and Saturn, but from a different reservoir. Some of Earth's water may have been supplied by collisions with comets. If comets are indeed an important source of terrestrial volatiles, they may have also supplied a significant amount of organic material necessary for the origin and evolution of life on Earth. Although not proven, these ideas have sparked a great deal of interest among scientists and the general public alike.

Origin of Comets

Dutch astronomer Jan Oort (1950) noted that the source of new comets was a shell located between 20,000 and 100,000 AU. The existence of this "Oort" cloud is now widely accepted. However, this is not a likely place for comets to have formed. Many dynamicists have considered the origin and evolution of the Oort cloud (e.g., Weissman, 1990), and most agree that the region around Uranus and

Neptune is a likely formation site for comets. Furthermore, they argue that the formation of the Oort cloud is a natural by-product of the formation of these two giant planets, which would gravitationally scatter many of their own building blocks into the Oort cloud. Gravitational disturbances by passing stars would occasionally cause some of them to fall back into the inner solar system, where we see them as new comets. In addition to the Oort cloud, there may be another reservoir, one proposed in 1951 by GERARD KUIPER, who suggested there may be a ring of icy bodies beyond Pluto's orbit. This Kuiper Belt, if real, is considered by many dynamicists to be a more likely source for the short-period comets and might even replenish the Oort cloud. A structure like the Kuiper disk has been imaged around the star Beta Pictoris. If such a disk existed around our Sun, it would be too faint to have been detected from Earth. Two recently discovered minor planets, 1992 QBI and 1993 FW, are believed to have orbits beyond that of Pluto and could be the first detections of Kuiper Belt members.

Hazards and Resources

Many comets are in Earth-crossing orbits and collisions do occur. As mentioned earlier, comets may have provided a significant fraction of Earth's volatiles and organic compounds. It is now well established that an impact with an asteroid or comet created a large (100-km-radius) crater at the edge of the Yucatan peninsula 65 million years (Ma) ago. This impact, named the Chioxulub impact, might have caused the extinction of the dinosaurs. Considerable attention is being devoted to understanding hazards due to comets and asteroids (see IMPACT CRATERING and IMPACT CRATERING AND THE BOMBARDMENT RECORD). Preliminary estimates indicate that impacts large enough to cause global devastation occur on average every few million years. This global devastation would result from severe climate changes (multiyear winters) produced by dust, which would be ejected into Earth's upper atmosphere by the impact and would partially block sunlight. At this time it is impossible to predict when the next such impact might occur; plans are being made to study the population of potential hazards in enough detail to predict and prevent large impacts.

Comets are also being considered as near-Earth resources for human activity in space. Mining short-period comets and Earth-crossing asteroids is an attractive alternative to Earth-based or even lunar-based mining. For energy, it is less costly to extract raw materials from these low-gravity visitors than to bring these resources out of the gravitational wells of Earth and the Moon. Short-period comets could become the main source of water and other volatiles for near-Earth space activities.

The Impact of Comet Shoemaker-Levy 9 with Jupiter

The first collision between a comet and a planet ever witnessed by humans occurred in 1994. More than twenty observable fragments of Comet Shoemaker-Levy 9 struck Jupiter between July 16 and 22, producing a most interesting set of events.

The Comet. Comet Shoemaker-Levy 9, the fourth comet discovered in 1993, was identified by the team of Carolyn and Eugene Shoemaker, and David Levy. During a close approach to Jupiter in 1992, tidal forces split the comet into fragments, giving it a "string of pearls" appearance (Figure 3). This unusual morphology made it particularly interesting; however, it was not widely studied at first because it was sufficiently faint to make detailed observations difficult. Excitement swept the astronomical community in May 1993 when dynamicists determined that Comet Shoemaker-Levy 9 would strike Jupiter in July 1994. Predictions as well as

preparations began immediately. Even though the comet's speed with respect to Jupiter was accurately determined at 60 km per second, the diameter, mass, and density of the fragments were very uncertain, making detailed predictions difficult. Furthermore, the impact site was located on Jupiter's far side, just beyond its southwest limb, making Earth-based observations uncertain. However, NASA's *Galileo* spacecraft had a direct view of the impact sites.

The Impacts. More telescopes observed this event than any other in history, and the results were dramatic. The fragments and their corresponding impact sites were labeled alphabetically from A, the first to strike, through V. Fragment A, which, despite its relatively faint appearance, created one of the largest impact sites, is about twice as large as the diameter of Earth. All impacts occurred near 44°S latitude, and Jupiter's rapid rotation (once every ten hours) brought them from the far side into view within a few minutes. However, most impacts were unexpectedly dramatic and many observers did not have to wait for the planet to rotate. As many as three brightenings were observed for individual impacts. The first, called the "bolide," resulted from the friction between the incoming fragment and Jupiter's upper atmosphere, and was analogous to a meteor or "shooting star" on Earth. This seems to have occurred sufficiently high to clear Jupiter's limb and make it observable from Earth.

Figure 3. Comet Shoemaker-Levy 9's "string of pearls" appearance. The comet's twenty-one icy fragments stretched across 1.1 million km of space, or three times the distance between Earth and the Moon. When the picture was taken (17 May 1994), the comet was approximately 660 million km from Earth, on a mid-July collision course with the gas giant planet Jupiter.

Although analysis of most observations continues, much has been learned already. Many molecules not normally detected in Jupiter's atmosphere have been observed. The detection of diatomic sulfur (S_2) supports theoretical predictions of a layer of ammonium hydrosulfide (NH_4SH) clouds, which lie below the visible layer of frozen ammonia (NH_3) clouds. The iron, magnesium, and silicon detected are believed to have originated in the comet. There is some debate about the nature of the impacting body, however, and some argue that it could have been an asteroid instead of a comet. After several impacts occurred, water was detected by NASA's Kuiper Airborne Observatory, as expected from water-ice-rich cometary fragments. It was argued that the water detected could have been dredged up from a layer of water clouds below the ammonium hydrosulfide clouds, but recent data from the Galileo probe indicate that such a layer of water clouds does not exist. Astronomers hoped that seismic and acoustic waves produced by the impacts would help probe the planet's lower atmosphere and interior, but to date no such waves have been detected. On the other hand, the observable effects of the impacts on the upper atmosphere and their dissipation with time are providing much insight into the physical and chemical characteristics of Jupiter's upper atmosphere.

In the midst of the excitement we are reminded that human existence is possible due to a delicate balance among ominous natural processes. It is estimated by one of the discoverers of this comet, Eugene Shoemaker, that the largest fragment (G) was approximately 3 or 4 km across, and that its kinetic energy was equivalent to 6 trillion tons of TNT, i.e., hundreds of times greater that the world's entire nuclear arsenal. If an object such as Shoemaker-Levy 9 impacted Earth, it would produce effects similar to those of the Chicxulub impact 65 Ma ago.

Bibliography

DELSEMME, A. H. "Nature and History of the Organic Compounds in Comets: An Astrophysical View." In *Comets in the Post-Halley Era*, eds. R. L. Newburn, M. Neugebauer, and J. Rahe. Dordrecht, Netherlands, 1991.

GREENBERG, J. M. "What Are Comets Made of? A Model Based on Interstellar Dust." In *Comets*, ed. L. L. Wilkening. Tucson, AZ, 1982.

MARSDEN, B. *International Astronomical Union Circular* 5585 (1992).

OORT, J. H. "The Structure of the Cloud of Comets Surrounding the Solar System and a Hypothesis Concerning Its Origin." *Bulletin of the Astronomical Institute of the Netherlands* 11 (1950):91–110.

REINHARD, R. "The Halley Encounters." In *The New Solar System*, eds. J. K. Beatty and A. Chaikin. Cambridge, Eng., 1990.

SPENCER, J. R., and J. MITTON, eds. *The Great Comet Crash*. Cambridge, MA, 1995.

WEISSMAN, P. R. "The Oort Cloud." *Nature* 344 (1990): 825–830.

WHIPPLE, F. L. "A Comet Model I: The Acceleration of Comet Encke." *Astrophysical Journal* 111 (1950):375–394.

HUMBERTO CAMPINS

CONSTRUCTION MATERIALS

Construction materials are used in buildings; walkways, highways, runways, and canals (especially locks); culverts and bridges; and dams, jetties, and docks. Rocks and rock products have been among the materials utilized for such purposes since antiquity. Those that are predominantly nonmetallic and their uses in the construction industry are treated in this entry. Information about metals and petroleum products other than asphalt that are also used in construction are covered elsewhere in this encyclopedia.

Construction materials that consist of or are manufactured from nonmetallic minerals and rocks annually contribute more income to the world's economy than the combined total of all other mineral resources except the mineral fuels. Those that consist of nonmetallic minerals and rocks include building stones, dimension stone, crushed stone, sand and gravel, and asbestos. Those that are manufactured from minerals and rocks include cement, plaster, ceramic materials, glass, and lightweight aggregate.

Building stones, ranging from thumb-size cobbles to huge boulders, have served several purposes since early humans used large boulders for shelter more than 2 million years (Ma) ago. More recently, they have been used to build cobblestone roadways and courtyards, and today, boulders and large pebbles are used widely as facing stones for fireplaces, houses, and other buildings.

Dimension stones are specially shaped blocks and slabs of rock that are used for footings, foundations, roofs, veneers on both exterior and interior walls, flagging and flat "tiles," curbing, and monuments. Rocks are chosen to be used as dimension stone on the basis of their colors, textures, workability, and durability under diverse conditions of conjectured utilization. Almost all dimension stone is cut, shaped, and dressed rock that is quarried from open pits. Dressed surfaces comprise controlled fractures or are rough-sawn, ground, or polished.

Although prior to the nineteenth century the nearby availability of rock types controlled architectural styles characteristic of most regions, this is no longer true. Today, dimension stones with aesthetic appeal are shipped long distances. Three examples serve to emphasize this point: Larvikite ("blue granite") from southern Norway, marbles from Italy, and a flamboyant migmatite (the Morton gneiss) from Minnesota are used widely throughout the world. Several different rocks used as dimension stone are listed in Table 1.

Diverse granitic rocks, plus limestones, sandstones, marbles, and slates make up well over 90 percent of dimension stone production within the United States. Concrete has replaced dimension stone use in many buildings during the last several decades, but some experts predict that the trend may be reversed in the future because production of most dimension stone requires less energy in production than the manufacture of cement and other alternative construction materials such as steel and aluminum.

Crushed stone is the name applied to rock that has been crushed, ground, or otherwise broken into small, irregularly shaped fragments. The fragments are frequently size-sorted and in some cases compositionally controlled for particular uses. A large percentage of crushed stone is used as aggregate for asphalt and cement concretes. It is also used as fill, road base, railroad ballast, riprap (for jetties and for other shoreline and embankment protection), and as granules for roofing and terrazzo.

Limestone and dolostone (carbonate rock made of dolomite) are the most used rocks for crushed stone. They meet the strength specifications set up for most uses, yet limestone and dolostone are relatively soft rocks and consequently are not so hard on crushers as many other rocks. Other rocks used as crushed stone include granitic rocks, dolerite and basalt ("traprock"), sandstone, graywacke, quartzite, marble, slate, gneiss, amphibolite, volcanic cinders (scoria), and vein quartz. Also, concrete is sometimes crushed and recycled as

Table 1. Rocks Used as Dimension Stones

Rock Type	Rocks (examples only)
Igneous	"Granite,"[1] syenite (larvikite), diorite, gabbro, dolerite (=diabase = "traprock"), rhyolite, trachyte, andesite, basalt, diverse porphyries, ash tuff (e.g., peperino)
Sedimentary	Limestone (e.g., "oolitic" and recrystallized limestone[2]), travertine, sandstone (e.g., "brownstone"), puddingstone, conglomerate (e.g., "calico rock"), gypsum (alabaster), dolostone
Metamorphic	Marble,[3] slate, gneiss, amphibolite, serpentinite, soapstone, greenstone, migmatite[4]
Other	Laterite

[1] The term "granite" is often used in the marketplace to include several light colored igneous rocks that geologists identify as granodiorite, diorite, and syenite as well as granite per se.

[2] The so-called Tennessee marble is a widely used recrystallized limestone.

[3] Marble in the geological community is a metamorphic rock. In the marketplace, the term "marble" is often applied to limestones and dolostones that take a polish as well as to marble per se; see the preceding footnote.

[4] Strictly speaking, this rock consists of both metamorphic and igneous or igneoid constituents; an example is the widely used *Morton gneiss*.

Tabulation modified after Dietrich and Skinner, *Gems, Granites, and Gravels: Knowing and Using Rocks and Minerals* (1990) based in part on data compiled by Kempe (1983).

"crushed rock" for use in, for example, highway construction.

Nearly all rock used for the production of crushed stone is quarried from open pits. Some is mined from underground because of either local restrictions or economic considerations. Most of the restrictions are imposed by zoning codes. The economic considerations involve such things as costs of nearby mining versus haulage from remote sources, of extending surface quarries underground versus removing thick overburden, of selective recovery of certain strata versus disposing of undesirable associated rocks, and of the advantages of all-weather production versus weather-imposed, untimely shutdowns of surface operations. In any case, because of its generally low unit value, most crushed stone is used near where it is produced.

Sand and gravel are unconsolidated aggregates of sand grains and of sand grains plus pebbles, respectively. Most sand consists largely of quartz grains; most pebbles of most gravels are rocks that are relatively resistant to weathering and erosion.

Sand and gravel find their greatest use as road base for pavements and for fill. They also are used widely as aggregate for both asphalt and cement concrete. Approximately 97 percent of the sand and gravel produced in the United States was used for these purposes in 1989. Lesser uses in the construction industry are for roofing granules and exposed aggregate used for decorative concrete walkways, walls, and items such as waste-disposal receptacles.

Like crushed stone, most sands and gravels used for construction purposes are of low unit value and are used near their sources.

Asbestos is "a commercial term applied to a group of silicate minerals that readily separate into thin, strong fibers that are flexible, heat resistant, and chemically inert, and therefore suitable for uses where incombustible, nonconducting, or chemically resistant material is required" (Bates and Jackson, 1987). Serpentinite and banded ironstones are common host rocks.

The fibers are usually sorted and sold, on the basis of fiber length, to fabricators of asbestos-bearing, "fireproof" products. These products include roofing and siding shingles, floor tile, caulking, insulation, and textiles (e.g., asbestos paper for covering heat pipes).

From the 1970s certain uses of asbestos minerals have been banned, curtailed, or phased out in the United States and many other countries because of questions raised about possible health hazards connected with both their recovery and use. Research indicates that prolonged exposure to asbestos dust and asbestos products may lead to a type of pneumoconiosis, called asbestosis, or even to certain kinds of cancer, including lung cancer. Whether or not all asbestiform minerals produce the same health effects, there are, of course, those who consider health findings suspect or of minimal concern. In any case, some applications—certainly those that use encapsulated asbestos—are likely to continue (see PUBLIC HEALTH AND EARTH SCIENCE).

Asphalt, sometimes referred to as tar, is used as "asphalt cement" to produce asphalt concrete, the widely used paving material. Most asphalt concrete consists of asphalt produced by distillation of crude petroleum plus an aggregate of sand, gravel, or crushed rock; some of it, however, is natural asphalt recovered from a seep or a lake plus aggregate; and some of it is natural rock asphalt (e.g., bituminous sandstone), which has been quarried, crushed, graded, and—if necessary—blended.

Asphalt is also the base for other construction materials, such as "tar paper," granule-coated shingles, and pipe coatings.

Cement, when mixed with water and an aggregate (sand, gravel, or crushed stone) in appropriate proportions and allowed to cure, forms common concrete. For construction, wet concrete mixtures are poured and allowed to cure in place, or the mixtures are poured into precast forms, such as the pipes used as aqueducts and the concrete blocks used for walls. For many construction purposes, concrete is reinforced with metal bars.

Portland cement, which constitutes more than 90 percent of current hydraulic cement production, is produced from limestone and shale or appropriate proportions of substitutes such as calcareous shale, slate, marl, marble, chalk, aragonite sand, oyster shells, quartz sand or sandstone, shale, bauxite, clay, alkali wastes, and slag. The two rocks, or the appropriate substitutes, are combined—usually as crushed powders—according to a recipe giving 60–65 percent lime (CaO), 10–25 percent silica (SiO_2), and 5–10 percent alumina (Al_2O_3); the mixture is fired at approximately 1480°C to produce a clinker; the clinker along with 4–5 percent gypsum (which serves as a retarder) is crushed to give the powder marketed as cement. For some purposes, the composition of the powder is modified to produce some specialty cement; concretes made with these cements have special properties (e.g., high resistance to acids), or can be poured and cured under particular weather conditions (e.g., at low temperatures).

Ceramic construction materials, including brick and tile, have clay as their major raw material. Moist clay, plus additives, is molded into the desired shapes and then hardened by drying in the sun, for example, or by firing in a kiln. Most hard and durable, red-orange colored bricks are molded and fired mixtures of clay and sand that were fluxed and colored by their natural iron contents. Coloring agents are added to iron-free clays to produce other colors. Ceramic products with glazed surfaces have been sprayed with glaze and refired at about 1090°C.

Most brick- and tile-fabricating plants are located near sources of clay or clay-rich rock. Clay comprises a group of about two dozen mineral species, all of which are hydrous aluminum silicates. Most clays are plastic when wet and become rock-like when dried or fired. The clay deposits used are of several kinds, such as residuum (formed by weathering in situ); marine, lake, and estuarine deposits; and shale, slate, and other clay-rich rocks.

Most bricks and tile products are made from common clay or shale, most of which are made up largely of illite (one of the clay minerals) and/or chlorite (a claylike mineral) with or without noteworthy kaoline (another clay mineral). To be used, the raw material must be sufficiently plastic to be easily molded. For production, the raw clay or clay-rich rock is, if necessary, crushed and blended before being extruded and dried.

Although the main use of brick is to face buildings, some brick is used for walkways and patios. Tile is used for such things as floor, wall, and roof tiles and drainage pipes.

Plaster is a generic designation for materials, such as mud, quicklime, cement grout, and plaster of paris, that are used to coat and thus modify the outward appearances of diverse surfaces; today the term "plaster" almost always means plaster of paris.

The primary raw material for the production of plaster of paris is the mineral gypsum ($CaSO_4 \cdot 2H_2O$), most of which is mined from ancient evaporite deposits. Many geologists think that most gypsum has been formed from originally deposited anhydrite ($CaSO_4$) as a result of post-depositional hydration by meteoric waters. In any case, the gypsum is mined, crushed, and heated to about 190°C to drive off part of the water (i.e., to form $2CaSO_4 \cdot H_2O$), and the partially dehydrated product is pulverized to give the powder that is marketed as plaster of paris. Plaster per se is formed when plaster of paris is mixed with an appropriate amount of water to form a paste that hardens to fine-grained masses of gypsum.

In the past, most plaster was applied directly to interior lathed walls and ceilings; today, 95–98 percent of plaster is used as prefabricated wallboard, such as drywall. Drywall, which consists of a layer of plaster sandwiched between two sheets of heavy paper, is marketed as relatively large (e.g., 1 m × 2 m by 1.25 cm thick) sheets that can be handled rather easily and attached to walls and ceilings with nails, staples, or adhesives.

Glass has been used in ever-increasing amounts during the last two centuries, especially since World War II. Apparently first used for construction purposes by Romans who incorporated pieces left over from mosaics into the walls of their dwellings, glass now dominates the exterior facades of many, especially high-rise, buildings.

In addition, glass—as fiberglass—now has widespread use in the construction industry. Fiberglass is the major constituent of, for example, several insulation materials and plumbing ducts.

Lightweight aggregate consists of fragments of rock or rock products that have low bulk densities: pumice, pumicite (i.e., ash made up of pumice particles), scoria, and cinder—all products of volcanism—are used in their natural state; clays, clay-rich rocks, perlite (a glassy volcanic rock), and the mineral vermiculite are heated and their expanded or exfoliated products are used; some blast furnace slags having the desired properties are also used. Each of these materials is crushed and size-sorted, as necessary, before marketing.

Lightweight aggregate is used in both asphalt and portland cement concrete mixes, as filler in some plaster, and as road base. Its light weight makes it especially desirable for concrete used for bridges and high-rise structures because of the resulting reduction of strength requirements and thus the reduced cost of supporting framework. Additional attributes include the ease of handling products made with light aggregate, their low thermal conductivity, and their high sound absorption. The two latter properties have led to its widespread use in, for example, concrete-block walls and acoustical plaster and wallboard.

Bibliography

AMPIAN, S. G. "Clays." In *Minerals Yearbook*. Washington, DC, 1989.

BATES, R. L., and J. A. JACKSON, eds. *Glossary of Geology*, 3rd ed. Alexandria, VA, 1987.

BOLEN, W. P. "Industrial Sand and Gravel." In *Minerals Yearbook*. Washington, DC, 1989, pp. 889–903.

CLIFTON, R. A. "Asbestos." In *Mineral Facts and Problems*. Washington, DC, 1985.

DIETRICH, R. V., and B. J. SKINNER. *Gems, Granites, and Gravels: Knowing and Using Rocks and Minerals*. Cambridge, 1990.

KEMPE, D. R. C. "The Petrology of Building and Sculptural Stones." In *The Petrology of Archaeological Artifacts*, eds. D. R. C. Kempe and A. P. Harvey. Oxford, Eng., 1983, pp. 80–153.

MEISINGER, A. C. "Perlite." In *Mineral Facts and Problems*. Washington, DC, 1985.

———. "Pumice and Pumicite." In *Mineral Facts and Problems*. Washington, DC, 1985.

———. "Vermiculite." In *Mineral Facts and Problems*. Washington, DC, 1985.

MITCHELL, L. "Ceramic Raw Materials" In *Industrial Minerals and Rocks*, ed. S. J. Lefond, 5th ed. New York, 1983, pp. 33–39.

PRESSLER, J. W. "Gypsum." In *Mineral Facts and Problems*. Washington, DC, 1985.

TAYLOR, H. A., JR. "Dimension Stone." In *Mineral Facts and Problems*. Washington, DC, 1985.

TEPORDEI, V. V. "Crushed Stone." In *Mineral Facts and Problems*. Washington, DC, 1985.

———. "Construction Sand and Gravel." In *Minerals Yearbook*. Washington, DC, 1989.

RICHARD V. DIETRICH

CONTINENTAL CRUST, STRUCTURE OF

Although the earth's mantle comprises over 99.5 percent of the rocky (silicate) fraction of the earth, and extends to within a few kilometers of the surface, it can be seen at the surface in but a few limited slices. The crust serves as a geological skin that separates the mantle from the atmosphere and oceans. The rocks of the crust are less dense than the mantle and float atop the mantle in the same fashion that ice floats on water. The oceanic crust is composed of massive basaltic igneous rocks derived from the mantle, with an overlay of oceanic sediments (*see* OCEANIC CRUST, STRUCTURE OF). The continental crust consists of a variety of rocks with even lower density, which have been assembled into a complex collage by plate tectonic movements and erosional transportation. The base of the crust is known as the Mohorovičić discontinuity (the *M* discontinuity) and is determined by the analysis of seismic travel times from earthquakes and explosions.

The vertical movements of the crust, as it rides buoyantly atop the mantle, provide a general basis for understanding the evolution of crustal thickness and explain why most continental crust is very close to 35 km thick.

The height of the land surface and the depth to the *M* discontinuity are closely related. Because of the principle of isostatic balance, the crust assumes a level relative to the mantle datum that equalizes the total mass over vertical columns (*see* ISOSTASY). Typical continental rocks have a density of 2.70 g/cm^3, and the upper mantle density is about 3.35. The ratio of the submerged fraction of the crust to the total thickness is thus $(2.7/3.5) = 0.8$; about 20 percent of the crust is emergent above the mantle datum. As the continental crust is thickened by addition of new material, its surface elevation increases. For every 1 km of increase in surface elevation, the *M* discontinuity is deepened by 4 km.

Figure 1 illustrates the oceanic situation. The thin, basaltic, oceanic crust in effect consists of a thin volcanic carapace atop the mantle. The oceans cover this crust to a depth of 3–5 km over about 65 percent of the earth's surface. Sedimentation, volcanism, and tectonic processes have built structures on the oceanic crust, such as volcanic islands, oceanic plateaus, and deep-sea sedimentary fans, that reach toward the surface of the ocean and ride isostatically on this crust, which may be 10–15 km thick.

However, when these processes involve the accretion of much thicker assemblages of crustal material, as happens at major plate collisions, a crustal collage of 30–35 km or more is formed, and its buoyancy brings the surface above sea level (*see* MOUNTAINS AND MOUNTAIN BUILDING). During active mountain building, a thickness of up to 70 km can occur (e.g., Tibet), with elevations of 4-5 km. Rocks supporting such elevations are subject to

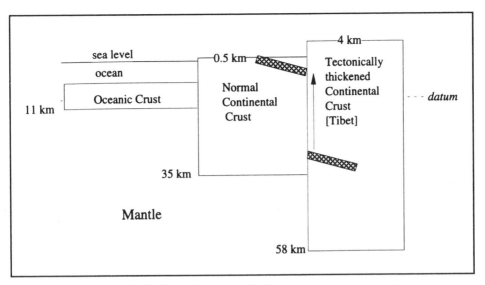

Figure 1. Evolution of crustal thickness—oceanic situation.

rapid erosion and to large-scale gravitational collapse (mega-landsliding) along shallow faults, and persist at best for a few tens of millions of years. As the excess elevation rocks are removed, the crust remains in isostatic balance, and the entire crustal section rises, as the *M* discontinuity shallows in a 4x mirror image of the removal of elevation. The end result is that the surface is eroded to near sea level, the *M* discontinuity rises to 35 km depth, and many formerly deep-seated rocks have risen to form the emergent surface of the crust (shaded block).

These rocks form the igneous and metamorphic "basement" of many continental areas, rocks that have been formed at high temperatures and pressures in the mid to lower crust during former mountain building. Heat generated by radioactive decay of trace quantities of radioactive elements in the mantle and crust leads to a normal temperature gradient of 20–30°C per kilometer increase in depth near the surface (*see* HEAT BUDGET OF THE EARTH). This temperature gradient tapers down at greater depths, but in any case, the temperature near the base of the normal crust can range from 500–700°C, or more. At these temperatures, the granitic rock assemblages typical of the crust approach partial melting. Even below the melting temperatures, the rocks become extremely weak, and respond to stress by ductile flow.

Consequently, the lower part of the ultra-thick crust tends to melt and to flow. Melts may readily migrate upward (due to their extra buoyancy) and leave behind rock assemblages that are less readily melted. The ductility permits the entire lower crust to flow laterally and in effect to erode away the deeper parts of the crust.

In summary, continental crust that is thickened beyond 35 km is unstable. Surface elevations are degraded and deep crustal roots are self-modified due to the high temperature conditions. The end result of this evolution is a 35 km thick crust, with slight elevation above sea level, and with common occurrence of formerly deep-seated igneous and metamorphic rocks at the surface.

When stable crust is further impacted by plate boundary processes, including both tectonic thickening and the intrusion of arc-type igneous rocks, the continued evolution of the crustal assemblage at near-melting temperatures leads to an even more stable end product. Interior Precambrian shields, or cratons, which compose about half the total area of the continents, are the result of multiple generations of tectonic and thermal evolution, and now form ultra-stable cores to the continents. These shields, in Africa, North and South America, Australia, India, northern Europe, and Antarctica, have undergone very little modification in the last 500 Ma and serve as platforms for the preservation in some cases of sedimentary rock sequences that are over 2 Ga old. Typical of the shields and stable cratons is the United States and Canada east of the Rocky Mountains, in which stablized ancient tectonic collages make up nearly all of the crustal section; atop these basement rocks, flat-lying or superficially folded younger sedimentary rocks constitute a thin cover to the continent—just a few kilometers thick. In contrast, much of the area of Asia today is composed of a young tectonic assemblage of fault-bounded terranes whose present geometry is the result of large-scale horizontal movements in the last 500 Ma.

How is this collage formed? When an oceanic plate collides with a continent, all of the crustal components—lower density material riding atop the mantle—gets compressed into imbricate slices along the edge of the continent, while the denser mantle simply descends beneath the continent (*see* PLATE TECTONICS). Figure 2a shows the thin sediment layer (S) of the ocean floor becoming compressed into a wedge (S1), which rides up on the preexisting continental crust (CC). Some of the sediments get plastered in a thin layer at the bottom of the crust along with the basaltic oceanic crust (S2 and B). This process can be continued and repeated until many million years of oceanic sediments have been carried into coastal mountain ranges. (The continental crust at the right of the figure should be considered stationary, as the oceanic crust moves from left to right.)

In Figure 2b, a volcanic edifice (V) lies atop the oceanic plate (note the isostatic depression of the *M* discontinuity). This is a layered stack of lavas and sediments that are too buoyant to sink into the mantle. When the collision occurs, this object is compressed and deformed into an imbricated lump (V1) attached to the edge of the preexisting continental crust. In both cases, the crustal material brought in by the collision has nowhere to go but to become piled up like the bank of snow in front of a snowplow. In becoming thickened, the surface rises up in mountains, and the compressed material descends to a depth of 25 km or more.

Volcanic island arcs constitute crustal units of nearly continental thickness that are formed upon oceanic crust, and acquire mass through the addition of volcanic lavas derived from the mantle. The collision of an oceanic island arc with a continent

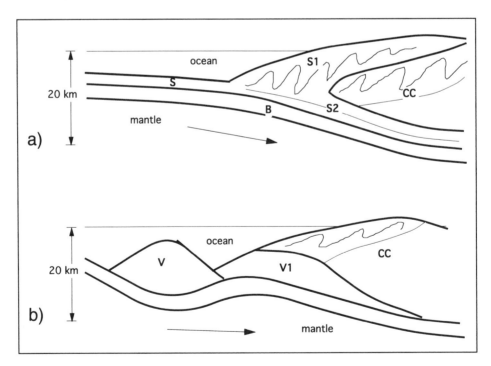

Figure 2. a. Compression of the thin sediment layer of the ocean floor into a wedge riding up on the preexisting continental crust. b. A volcanic edifice atop the oceanic plate.

carries the process in Figure 2 even further, in which the continent acquires a wide belt of new material and becomes extended tens or hundreds of kilometers.

In a similar process, different-sized blocks of continental material become involved in the same sort of collision. The island archipelagoes of southeast Asia consist of narrow blocks of continental thickness separated by oceanic basins. When a set of these structures collides with a continent, the continent may accrete hundreds of kilometers of new crust. Most of Alaska south of the Brooks Range consists of a set of tectonic slices that were formerly continental fragments (islands?) in the mid to South Pacific, and which were transported northward and collided with the existing continent (*see* TECTONIC BLOCKS).

Finally, a continental plate may collide with another continental plate. Since the 35 km thick crustal units are too buoyant to sink into the mantle, two things may occur: (1) their edges may collide, forcing up a linear mountain range, thickening the crust, and then stopping. The Pyrenees are an example of this kind of collision. The southern Appalachian Piedmont region now at near sea level is partly made up of a major sliver of Africa, which collided with North America about 250 Ma ago. (2) One of the continents is forced beneath the other, by underplating, as shown in Figure 3. The Tibetan plateau is the result of the actively continuing collision of India with Asia, which has thickened the crust to nearly 70 km and produced the characteristic 4–5 km surface elevation. Both types of collision tend to force continental material to

depths of 40–70 km, where high pressures and temperatures lead to significant metamorphism and melting. The extensive Grenville belt, over 500 km wide and 3,000 km long, running from Labrador through Quebec and Ontario, is a unit of Precambrian shield generally considered to have been produced by a Tibet-style continental collision 1 Ga.

Magmas formed in the mantle are significantly less buoyant than the mantle and rise readily up into the crust. Granitic magmas can penetrate to near the surface, where they produce massive intrusive igneous bodies often seen in mountain belts such as the Sierra Nevada. Basaltic magmas, being denser, become intruded near the base of the continental crust, where they add to the crustal volume by accretion from below. These magmas inject enough heat into the lower crust to melt (or remelt) crustal rocks, which then intrude into the shallow crust as granitic bodies.

The direct determination of the structure of the crust is done by the study of the travel times of seismic waves through the crust. Waves observed at distances of 80–200 km from the source (earthquake or explosion) travel obliquely to the base of the crust, become reflected from or refracted along the *M* discontinuity, and return to the surface. These make it possible to determine the gross structure of the crust in terms of seismic wave velocity and depth to the *M* discontinuity. The technique of reflection seismology, pioneered by the oil industry for remote detection of oil reservoirs, makes use of hundreds or thousands of small geophones to record energy reflected upward by artifi-

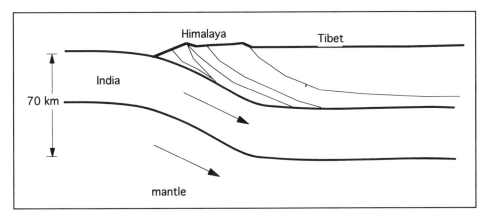

Figure 3. Collision of a continental plate with another continental plate, where one of the continents is forced beneath the other, by underplating.

cial sources (*see* GEOPHYSICAL TECHNIQUES). Reflection profiles on the continents have demonstrated two major structural principles: (1) In present or former zones of compressional tectonics, the crust appears to be made up of a collage of slices, dipping at angles generally less than 30°; this is in agreement with the structural style seen at the surface in outcrop. (2) In many areas, the lower third of the crust appears to be made up of a dense array of horizontal laminae. Since these laminae cannot be followed to the surface, they have remained somewhat of a mystery. They are often seen in areas of continental crust that have been subject to extension, such as the U.S. Basin and Range region and the Appalachian Piedmont. The laminae may be due to either: (1) gneisses, a granitic metamorphic assemblage that is often found in the cores of old mountain belts, which has been strongly sheared and deformed, resulting in a strongly foliated fabric, and on the larger scale, in horizontally oriented recumbent ductile folds; or (2) to the injection of basaltic magma from the mantle into horizontal cracks (sills) in the lower crust.

Future study of the structure of the continents is turning in the direction of analyzing the dynamic processes occurring during tectonism, for example, the study of fluids (water or CO_2) in the crust and the frictional behavior and physical properties of major faults. In addition, attention is turning to the question of the geologic structure and properties of the uppermost mantle, just below the *M* discontinuity, where it is certain that the plate tectonic history has left many characteristic geologic phenomena. For example, the U.S. Basin and Range, along with the Sierra Nevada, which constitute a striking and characteristic region of much of the western United States, are almost certainly being driven by processes originating in the subjacent upper mantle.

Bibliography

BALLY, A. W., and A. R. PALMER, eds. *The Geology of North America: An Overview.* Geological Society of America, DNAG series, vol. A. Boulder, CO, 1989.

BARAZANGI, M., and L. BROWN, eds. *Reflection Seismology: The Continental Crust.* American Geophysical Union, Geodynamics series, vol. 14. Washington, DC, 1986.

BROWN, G. C., and A. E. MUSSET. *The Inaccessible Earth: An Integrated View of Its Structure and Composition.* New York, 1993.

MOONEY, W. D., and L. W. BRAILE. "The Seismic Structure of the Continental Crust and Upper Mantle of North America." *The Geology of North America*, A (1989):39–52.

ROBERT A. PHINNEY

CONTINENTAL MARGINS

Continental margins are long narrow regions of continents that mark the transition of continental to ocean crust. The margins, although covered by ocean water, are geologically part of the continents. If the oceans were drained of their water, the continents, because they are composed primarily of low-density granitic bedrock that floats on the asthenosphere, would stand about 4.5 km above the high-density basaltic rock of the ocean floor. To an observer the margins of the continent would appear as a broad, gently sloping shallow platform, called the continental shelf, which extends from the former shoreline to its seaward edge, where a steeply dipping surface, the continental slope, descends to the ocean floor. Rising above the basalt-rich rock of the ocean floor and overlapping the base of the slope is a prism of continent-derived sediments, called the continental rise. Traditionally, these three components comprise the continental margin.

If the base of the continental slope is taken as the continent-ocean boundary (and not the area covered by ocean waters), continents would occupy about 40 percent, not 29 percent, of Earth's surface. About 25 percent of the modern continents are flooded by seawater, but that figure has varied considerably over geologic time. For example, the advance and retreat of glacial ice during the ice ages, over the past few million years, has caused sea level to rise and fall many times by 100 to 200 m. Each rise and fall of sea level was accompanied by a major marine transgression and regression of the shoreline, respectively.

The margins are an object of considerable interest because they are the repository of a wealth of information about Earth's history, including changes in climate, atmospheric composition, sea level, marine biota, water chemistry, and plate movement. The margins contain much of the world's petroleum resources and provide critical habitats for the world's most important fishing grounds. Approximately 70 percent of the world's population live on coastal borderlands, which many geologists consider part of the margins because they were flooded repeatedly in the geological past by transgressing seas. The margins are thus actively studied by large numbers of research geologists and geophysicists from academe, private industry, and from the U.S. Geological Survey (USGS) and the U.S. Office of Naval Research.

This entry focuses on the origin and evolution of continental margins.

Margin Types

Two major types of continental margins exist: passive margins, which occur in the stable interior of tectonic plates; and active margins, which form along active or recently active plate boundaries and where subduction is a major process. Each margin type has its own characteristic geologic features, as described below.

Passive Margins: Origin and Evolution. Passive margins form as continental lithospheric plates are rifted or broken apart and new oceanic magma, from the upper mantle, crystallizes between the two spreading continental plates (*see* OCEAN BASINS, EVOLUTION OF). The rifting process thus causes thinning and extension of the lithosphere. The Atlantic Ocean margins of North America, Africa, and Europe are examples of passive margins that started to form about 200 million years ago (Ma) when the interior of a large continent, called Pangaea, began to break up. It first split into two large plates, a northern one called Laurasia and a southern one called Gondwanaland. Today's Atlantic passive margins are considered to be tectonically quiet, lying thousands of kilometers from the Mid-Atlantic Ridge, where seafloor spreading, through active faulting and intrusion of oceanic magma, is still occurring along divergent plate boundaries. During their early formative stages in the Triassic period, about 240 Ma, however, the landmass that these margins now occupy stood 2.5 km above sea level and was the site of major plate boundary activity (Figure 1a). Basaltic magma, derived from the upper mantle, intruded and heated the crust which then expanded, rising to the elevation of a high plateau. The Colorado Plateau is an example, or analog, of what the east coast of North America looked like at that

time. The record of those ancient events, depicted in Figure 1, is now preserved in a discontinuous belt of rift strata that are exposed along the east coast from Nova Scotia to the Carolinas, or are buried beneath younger sedimentary rocks of the

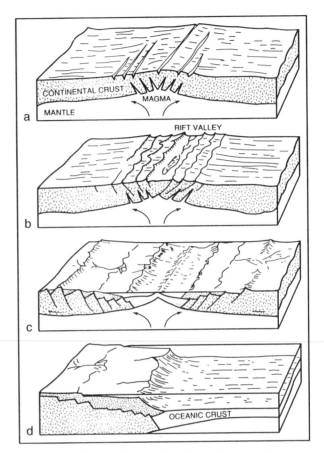

Figure 1. Stages in the breakup of a continent, the development of Atlantic-type passive margins, and the origin and the Atlantic type ocean floor.
a. Thermal uplift of a broad region with dike intrusion, e.g., Colorado Plateau.
b. Uplift of mantle, accompanied by normal faulting, dike intrusion, and formation of a rift valley, e.g., East African Rift Valley.
c. Formation of a new oceanic crust with continued faulting, uplift, and erosion of the rift shoulders and cooling and subsidence of the continental margins, e.g., the Red Sea.
d. Crustal thinning through faulting and cooling, accompanied by subsidence and sedimentation along the newly formed margin, e.g., the Atlantic passive margins and ocean. (Modified from Tarbuck and Lutgens, 1992.)

Atlantic coastal plain or the Atlantic passive margin, or both. The East African rift system is a modern example of continental rifting. Today in East Africa mantle plumes (columns of heated mantle rock), rising from the upper mantle, heat, elevate, and stretch Earth's crust. Continued convective drag at the base of the lithosphere thins and ultimately breaks the crust along normal faults. Displacement along these faults creates the familiar African rift topography with its highly elevated shoulders, rising about 2.5–3.0 km above sea level, and its downdropped valley floor, where thick lake and river sediment accumulates along with extensive lava flows that emanate from volcanoes and fissure eruptions within the rift valley. Faulting also provides channels for basaltic magma from the mantle to intrude and thus to extend the continental crust (Figure 1b).

As rifting and intrusion continue, a new oceanic floor is finally formed, ushering in seafloor spreading and the creation of the passive margin. The Red Sea Rift, with its newly formed oceanic crust and youthful passive margins, represents a modern analog of this stage (Figure 1c). As seafloor spreading develops and the thinned continental crust is displaced from the source of heat along the plate boundary, it cools and subsides, forming depositional centers for the accumulation of vast quantities of post-rift or passive margin strata. Derived primarily from the adjacent continents, much of this sediment consists of sand, silt, clay, and gravels that were transported to the coast by streams eroding adjacent continents. But where these landmasses stand low, as in the Persian Gulf and on the Bahamas Platform, the sediment mainly consists of the accumulation of marine shell animals, such as corals and mollusks, as well as evaporite minerals, such as salt and gypsum, which are precipitated directly from the overlying ocean waters. Sedimentation on the Atlantic passive margins began with the onset of seafloor spreading in the Middle Jurassic, about 185 Ma ago, and in places is in excess of 5.0–6.0 km thick. These strata constitute a window into Earth history, providing information about the early composition of the seawater, changes in shoreline and sea level, evolution of marine organisms, movement of continents and changing world climates, and so forth. This vast sedimentary wedge, comprising the continental shelf, continental slope, and, in part, the continental rise was built onto the submerged and fragmented continental crust (Figure 1d). These

marginal basins do not completely fill with sediment because, as shallow marine sediments are deposited, the underlying continental crust subsides isostatically into the mantle, providing additional room for sediment to accumulate.

Active Continental Margins. Active continental margins form primarily along convergent plate boundaries that are currently or have recently been tectonically active and where subduction of a lithospheric plate into the asthenosphere is, or has been, a dominant process. Active margins mark the continents bordering the Pacific Ocean. Subduction of the oceanic lithosphere along the rim of the Pacific has led to a wide range of plate responses and continental margin types, including: ocean to continental convergent margins (e.g., the Andean coast of South America); accreted terrane margins (e.g., the northwestern margin of North America from northern California to Alaska); and, to a lesser extent, the transform fault margins (e.g., the Gulf of California).

Subduction of the Nazca oceanic plate, for example, along the length of the Andean coast of South America, led to partial melting of the oceanic plate and the formation of andesitic magma that subsequently intruded the overlying continental crust to form extensive volcanoes and intrusives that comprise much of the Andes mountain range (*see* CONTINENTS, EVOLUTION OF). Subduction began about 140 Ma ago when the South American lithosphere broke away from Gondwanaland (see above) and began moving westward. As the subducting slab continued to descend into the mantle, a deep trench formed, which was later partially filled with low-density oceanic sediment that was scraped off the subducting plate. Figure 2 depicts the formation of the active continental margin and Andes mountain range toward the end of the Cretaceous period, about 100 Ma ago. Today, although igneous activity has shifted farther east, the active margin characteristically consists of a deep trench (Peruvian-Chilean trench), a subduction complex of metamorphosed sedimentary rock, a magmatic arc of igneous rocks, and a back-arc basin filled with sedimentary and volcanic rock (Figure 2).

Where subduction or transform faulting have tectonically transported microcontinents and island arcs that are too buoyant to descend into the mantle, these low density terranes are accreted to

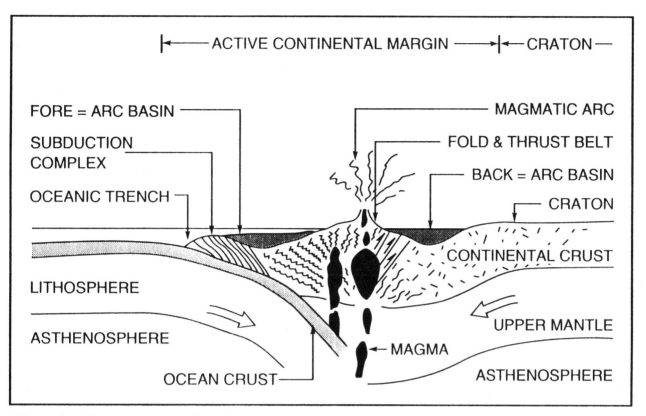

Figure 2. Schematic cross section of the active continental margin and Andes mountain range of South America, including: (a) near-surface features, such as the oceanic trench, subduction complex (accretionary wedge), fore-arc basin, magmatic arc, fold and thrust belt, back-arc basin, craton, oceanic and continental crust; and (b) deep-seated features, such as the asthenosphere, lithosphere, upper mantle, magma, and magma-generated batholith.

the margin. The continental margin of North America from northern California to Alaska is an example of an accreted margin. Island arcs form where two oceanic plates meet and one is subducted beneath the other; and microcontinents form where segments of continents are separated from the mainland by rifting or transform faulting. This process occurs today along the San Andreas fault, a large transform fault bordering the American and Pacific plates. Geologists postulate that, if the San Andreas fault continues to be active, the Los Angeles terrane will pass San Francisco in about 10 Ma and will be sutured to Alaska in about 60–80 Ma. In this context, the continental margin from Baja, California, to San Francisco is a transform margin, and the margin north of California to Alaska is accreted. Geologists infer that about 180 to 80 Ma ago numerous island arcs (e.g.,

the Japanese islands and the Aleutians) and microcontinents (e.g., New Zealand and Taiwan) dotted the eastern Pacific Ocean, much like they do today in the western Pacific. As the North American plate moved west during the breakup of Gondwanaland in the Southern Hemisphere (see above), the eastern half of the Pacific Ocean plate was subducted beneath the continent and the low-density islands of the Pacific were accreted to the margin. Today the western margin of North America, from California north to Alaska, consists of more than fifty disparate terranes or geological entities that were transported from nearby and distant landmasses and island arcs. The accreted margin of North America thus forms a mosaic of exotic terranes whose rock types and ages differ markedly from adjoining terranes and from which they are separated by faults.

Topography

As tectonic processes along passive and active margins differ significantly, so do their topographic profiles. Passive margins have a much broader continental shelf, a gentler continental slope, and, in the absence of active trenchs, a more marked continental rise. The width of continental shelves varies from only a few kilometers off California to about 500 km along the passive margin off Newfoundland. The average slope of the continental shelf is about 0.1°, where the average inclination of the continental slope is about 4–5°, but may exceed 25° along some active margins. Along both passive and active margins worldwide, the seaward edge of continental shelves and slopes is extensively eroded by turbidity currents, forming deep, V-shaped valleys, termed "submarine canyons." Turbidity currents develop when loose sediments on the continental shelf or slope become unstable, perhaps triggered by an earthquake or spring floods on land, and move downslope as a dense, chaotic, sediment-laden current at speeds in excess of 100 km per hour, and are thus capable of vigorous erosion of the outer margin. On active margins, where plate convergence is occurring, the topography is more diverse, including island arcs, deep-sea trenches, back-arc basins, and folded and thrust faulted borderlands.

Tectonic forces also shape the coast of continental margins. The coast may be uplifted or it may subside relative to sea level. Under converging plate activity, the coast typically is uplifted and unstable. Waves attack the coast, forming a topography of prominent cliffs, rocky headlands, irregular bays with small rocky beaches, and wave-cut terraces and wave-cut cliffs. If tectonic forces did not constantly act to uplift the coast, wave erosion ultimately would erode the coast to a flat, featureless plain. One of the most impressive pieces of data supporting the notion that the uplift has been long-standing and episodic is the presence of as many as thirteen cycles of wave-cut benches and cliffs along the southern coast of California. On the other hand, the east coast of North America and the Gulf of Mexico have been subsiding over the past 180 Ma. As discussed above, long-term subsidence is characteristic of the passive margin, and it occurs because of tectonic thinning through extension or thermal cooling. As the land subsides, sediments are deposited in the subsiding basin; consequently, the coastal region consists of broad submarine platforms with wide beaches, barrier islands, and well-washed, reworked sand.

Charting Continental Shelves

How are continental margins studied, as their seaward edge is overlain by 4.0–5.0 km of water and their early formative history, manifested in the rock record, may be covered by an additional 5.0–6.0 km?

Geologists and oceanographers obtain information directly by collecting samples of the seafloor and indirectly through remote sensing, which involves an array of geophysical instruments (*see* GEOPHYSICAL TECHNIQUES) that record data from the margin without being in direct contact with the seafloor. Samples of rock and sediment may be obtained by dredging the seafloor, coring the soft mud and sand of the ocean floor, and by drilling into the margin from a drilling platform or research vessel (*see* DRILLING FOR SCIENTIFIC RESEARCH). The basic tool for mapping the topography of the seafloor is the echo sounder, and the basic tool for determining the stratigraphy and structure of the seafloor is the seismic profiler. Both remote-sensing devices work on the same principle. Whereas the echo sounder emits a signal from a research vessel that bounces off the bottom and is then picked up by the vessel, the profiler uses a louder signal that penetrates into the substrate and is reflected from layers of rock within the margin. Magnetic, gravity, and seismic refraction surveys also are conducted at sea. Since the 1970s the seafloor, including the margin, has been studied by a fleet of manned submersibles that can withstand confining pressures at depths greater than 3 km. As the coastal borderlands mark the emergent part of the continental margin, geologists also gain insight into the history and processes that formed the margin by mapping both their surface and subsurface features.

Bibliography

DICKINSON, W. R. "Plate Tectonics and the Continental Margin of California." In *The Geotectonic Development of California*, ed. W. G. Ernst. Englewood Cliffs, NJ, 1981.

KLITGORD, K. D.; D. R. HUTCHINSON; and H. SCHOUTON. "U.S. Atlantic Continental Margin: Structural and

Tectonic Framework." In *The Atlantic Continental Margin*, eds. R. E. Sheridan and J. A. Grow. Boulder, CO, 1988.

TARBUCK, E. J., and F. K. LUTGENS, *The Earth*, 4th ed. New York, 1992.

WATKINS, J. S., and C. L. DRAKE. *Studies in Continental Margin Geology*. Memoir No. 34, the American Association of Petroleum Geologists. Tulsa, OK, 1982.

WARREN MANSPEIZER

CONTINENTS, EVOLUTION OF

Earth is unique in the solar system in that two-thirds of its surface is covered by ocean water. The water obscures a further uniqueness: the rocky crust beneath the oceans is very different from that of the continents. Oceanic crust is relatively young, ranging in age from zero at spreading ridge axes to a maximum of about 160 Ma (the age of the earth is 4,550 million years, or 4.55 Ga). Continental crust, however, is old, having mean and maximum ages of 1.7 and 4.0 Ga, respectively.

The meaning of the age difference is not clear-cut. Both types of crust are continuously created in some places and in other places destroyed. The processes of creation and destruction are approximately balanced. Nevertheless, small imbalances have probably changed the relative proportions of continental and oceanic crust over geologic time, especially during the early stages of Earth history. We must admit to being unsure of how much the continents have grown at the expense of ocean basins, or even if they have grown or shrunk overall. The assumption of net continental growth has generally been favored, but the evidence is inconclusive. The large age difference between continental and oceanic crust stems mainly from the fact that the rates of creation and destruction are much higher for oceanic crust than for continental crust. Oceanic crust is being "recycled" (created and destroyed) so rapidly that 200 Ma from now, virtually all the oceanic crust existing today will have been destroyed. In contrast, over 97 percent of existing continental crust will still be present 200 Ma from now.

Another difference between oceanic and continental crust is their thickness, and this is related to difference in their age. Oceanic crust is normally about 7 km thick; continental crust has an average thickness of about 38 km. As oceanic plates move away from spreading ridges, the subcrustal mantle rock cools and stiffens. Concurrently, its density increases due to thermal contraction. About 40 Ma after its formation, the overall density of the oceanic "plate" (crust plus cold stiff mantle) exceeds that of the underlying hotter and more plastic mantle, the asthenosphere. The oceanic plate is thereafter gravitationally unstable, and sinks end-on into the mantle at deep-sea trenches, a process referral to as subduction. Continental plates, on the other hand, resist subduction: their thick low-density crust renders them perpetually buoyant regardless of age. Although continental crust cannot be subducted intact, it can be recycled piecemeal. Subducting oceanic plates may carry down continentally derived sediments, for example, or continental rock scraped off at subduction zones. However, the strong resistance of continental crust to subduction is reflected in its high mean age.

The thickness of continental crust is ultimately dictated by sea level. Surfaces elevated above sea level are reduced by erosion. Surfaces depressed below sea level are filled in by sediments. Any overall change in continental crustal thickness requires a shift in the balance between ocean water volume and the holding capacity of ocean basins (i.e., total area multiplied by mean depth).

The formation of continental crust is a product of several processes, acting in various combinations. The four most important are trench accretion and arc, plume, and rift magmatism. Each process will be explained below.

Trench accretion occurs when material is scraped off the subducting oceanic plate and accreted to the front of the overriding plate. Much deformation—shredding, shearing, and crumpling—occurs in the transfer process. The accreted material is often sediment that accumulated in the trench, but it may have been carried to the trench as part of the subducting oceanic plate. If the accreted material consists entirely of continentally derived sediment, no net continental growth occurs. However, accreted volcanic plateaus, magmatic arcs, and plume-related seamounts represent mantle-derived crustal additions. At some trenches, incoming volcanic material is subducted but at others it is accreted: the presence of thick sediment seems to favor accretion. Deformation associated with accretion generally involves vertical thickening in response to horizontal compressive stress generated by plate convergence. The deformation is necessary for the accreted material to achieve continental crustal thickness. At some trenches, crustal material is actually removed from the overriding plate and carried downward into the mantle. This process is termed "ablative subduction," and its quantitative importance lies at the heart of the crustal growth issue.

Arc magmatism is caused by melting (magma generation) directly associated with subduction. Trenches are commonly arcuate in map view—the Aleutian Trench between Alaska and Kamchatka is a fine example. The arcuate form is related to the mechanics of bending a plate on the spherical surface of the globe. Lines of conical and typically highly explosive volcanoes occur parallel to trenches on the concave sides of the arcs. The explosive habit of arc volcanism is related to a high water content of the magma. The water is derived from the subducting oceanic crust, driven off by chemical reactions taking place around 100 km depth due to increased pressure and temperature. Earthquake locations reveal that subducting plates angle down beneath the concave side of the arc. Water released from the subducting plate infiltrates upward into the hot wedge of mantle material beneath the overriding plate. The presence of small amounts of water significantly lowers the mantle melting temperature, producing magma that ascends into the overlying crust, giving rise to surface volcanism. Provided that subduction replenishes the water supply, magmatism will continue in roughly the same arcuate zone. The crustal thickness of the arc increases progressively at the expense of the mantle melting column. The resulting evolutionary changes in magma composition are characteristic of magmatic arcs. The thickened crust also acts as a density filter, producing vertical gradients in crustal density and composition. Arcs developed entirely within ocean basins represent "juvenile" crust that may become added to continental margins by arc-continent collision, a form of trench accretion. Such collisions are occurring today on the northwest Australia and southeast China continental margins, involving the islands of Timor and Taiwan, respectively.

Plume magmatism is caused by narrow hot jets of ascending mantle material. They are often conceived as initiating at the base of the mantle, 2,900 km deep, as a result of heat transfer from the outer liquid-iron core. As plume material approaches the surface, it begins to melt due to pressure release. The resulting magmas collect within or upon the overriding crust, increasing its thickness. The Hawaiian islands and rocks at Yellowstone National Park exemplify plume magmatism in oceanic and continental settings, respectively. Plumes are most effective in crustal thickening where the plate is stationary with respect to the plume. Hawaii and Yellowstone are on moving plates but the African plate is at rest above a set of plumes. Consequently, Africa has many "intraplate" volcanoes and is unusually high-standing topographically. Given the great age of most continental areas, plume-induced melts must have been added to the crust repeatedly. Where a seafloor spreading ridge lies close to a plume, as in Iceland, the oceanic crust produced is 20 km or more in thickness, roughly three times the normal depth. Oceanic plateaus may originate in this way and may ultimately be incorporated into continents by trench accretion.

Rift magmatism is caused by pressure-release melting in mantle material ascending in response to lithospheric necking caused by horizontal tensional stress. The amount of melting depends strongly on the mantle temperature profile, as well as on the rate and degree of lithospheric necking. Where extreme necking associated with continental breakup occurs near a mantle plume, on the order of 25 km of crust may be added to the continental margin. Sources of the tensional stress on the lithosphere include the gravitational sinking of a plate at a trench and the gravitational collapse of a tectonically thickened continental plateau, such as Tibet or the Andean Altiplano. Magmatic arcs are highly susceptible to rifting, resulting in the formation of marginal seas (e.g., Japan Sea, Philippine Sea). Where arcs rift, pressure-release melting compounds the ongoing hydration (water-driven) melting, increasing the crust formation rate but shifting its thickness and composition in the direction of oceanic crust.

Arc, plume, and rift magmatism all involve primary mantle-derived melts having roughly the composition of basalt lava, similar to oceanic crust. However, the most abundant magmatic rocks in the upper continental crust are broadly granitic—richer than basalt in silica, alumina, and alkalis, and poorer in magnesium, calcium, and iron. Granites have strong concentrations of incompatible elements (e.g., potassium and rubidium, uranium and thorium), which have a strong affinity for melt. Most granites are products of secondary melting, involving preexisting deep crustal rocks heated by incoming mantle-derived melts. The generation of granitic melts in the deep crust and their subsequent ascent contribute to vertical crustal differentiation. The upper crust becomes increasingly granitic, while the lower crust is composed of basaltic material variably depleted by the removal of granitic melt. Ideally, the bulk crust should be basaltic in composition. It is difficult to estimate the composition of the lower crust because it cannot be directly observed. In only a few places on Earth are more-or-less complete sections of continental lower crust accessible for study. Most are in collision zones like the Alps and the Himalayas, and are exposed through the combined effects of crustal thickening and erosion. The sections are mostly basaltic in chemical composition, but there are also dense refractory rocks, consistent with the vertical gradients produced by the combined effects of crustal density filtering and secondary melting.

Whether the lower crust is quantitatively complementary to the upper crust cannot be established from the small number of lower crustal sections accessible for study. A more widespread, but fragmentary, sampling of the lower crustal is available in the form of xenoliths, rock fragments brought up to the surface by explosive volcanism at several hundred sites around the world. The tentative conclusion drawn from xenoliths, exposed crustal sections, and various geophysical observations is that the bulk composition of continental crust is not basaltic. There is insufficient dense residual material in the lower crust to balance the broadly granitic upper crust. The problem can be addressed in two ways. First, high-density lower crustal material may be selectively removed: it should sink gravitationally in basaltic magma, provided the magma is sufficiently free of residual mantle mush. Second, crust formation processes might have been different in the hot young Earth, leaving a lasting composition memory due to the slow recycling rate of continental crust. To test the second possibility requires investigation of the oldest crustal regions, the Archean shields.

Over geologic time, continents have broken apart and recombined repeatedly, in ever-changing combinations and configurations. Continental breakup and collision typically cause crustal reactivation over distances of hundreds to thousands of kilometers perpendicular to continental margins. Therefore, most continental areas have been structurally or magmatically reactivated several times. It might seem a hopeless task to attempt to isolate and infer the processes responsible for creating the oldest continental crust. Would the oldest rocks not be thoroughly distorted by later events? Paradoxically, crustal areas formed before about 2,000 Ma ago are relatively well preserved. They are protected by stiff refractory mantle roots, reflecting unusually high degrees of melt removal from the mantle when the crust was formed. Presumably, the thermal regime of the young Earth favored high degrees of melting. The ancient crust has been intensively studied: it hosts the lion's share of industrial and previous metals. The most rewarding study areas are in central Canada, southern Africa, and western Australia. Structurally and compositionally, the crust in these areas is easily interpreted in terms of processes active today. The deformed equivalents of oceanic plateaus, volcanic island arcs, rift volcanics, and trench sediments can be found within recognizable accretionary complexes. The accretionary complexes are typically invaded by sheets and plugs of broadly granitic material, intruded during the latest stages of structural consolidation. They might represent arc and rift magmatism at accreting continental margins. However, trace-element studies indicate that the granitic rocks were derived partly by melting of subducted oceanic crust. Remembering that the parental magmas in present-day arcs are mainly derived from mantle wedges, it follows that continental crust may have evolved over geologic time due to a proportional shift in the source of arc magmas. Silica-rich melts of crustal derivation were important early on and basaltic melts of mantle derivation became dominant later on.

It remains to be seen if the compositional paradox of continental crust indicates selective recycling processes, evolutionary changes in crust formation, or both. What are needed are more direct comparative studies of modern and ancient continental crust, and a more quantitative understanding of continental crust-forming and crust-destroying processes.

Bibliography

Burchfiel, B. C. "The Continental Crust." *Scientific American* (September 1983).

Windley, B. F. *The Evolving Continents*, 3d ed. New York, 1995.

PAUL F. HOFFMAN

CORALS

See Colonial Invertebrate Fossils

CORE, STRUCTURE OF

The structure and properties of the core, the dense metallic region at the center of Earth, are primarily determined from observations made in seismology and geomagnetism. In short, the core extends 3,483 ($\pm$5) km from the planetary center, nearly half the distance toward the surface, and consists of two parts: the 2,260-km-thick liquid outer core, surrounding the crystalline inner core (1,220 $\pm$10 km radius) at the earth's center (Table 1; *see also* EARTH, STRUCTURE OF). The entire core is thought to be an iron-rich alloy, and it is in the liquid outer core that Earth's main magnetic field is produced.

Seismological studies document the structure and density of the core in the greatest detail now available, with the travel times of body waves providing the best constraints on the radial dimensions of the inner and outer core. Travel times measured as a function of distance from earthquake epicenters can be used to determine the compressional- and shear-wave velocities (V_P and V_S) as functions of depth through the inner and outer core. Once these velocity distributions are known, the travel times of waves transmitted through or reflected off of the outer core and inner core yield direct measurements of the dimensions of these regions. In addition, the observation that horizontally polarized shear waves (SH) are not propagated into the core, upon entering from the overlying mantle, proves conclusively that the outer core is liquid; that is, the outer core responds like a fluid at seismic frequencies of approximately 10^{-3}–10^0 Hz.

The spectrum of Earth's free oscillations excited by large earthquakes provides key information complementing the body-wave measurements. Studies of the planet's oscillations confirm both the radial dimension and the fluid nature of the outer core, and it is the normal-mode frequencies which document that the inner core is crystalline—that it

has finite rigidity and is thus solid at seismic frequencies.

The free-oscillation frequencies also give the best constraints on the distribution of density throughout the interior, and the pressure at depth can be directly calculated from this density distribution. Pressure is thus known to increase from 1.36 million times atmospheric pressure (136 GPa) at the top of the core to 3.29 million times atmospheric pressure (329 GPa) at the inner-core boundary and 3.64 million times atmospheric pressure (364 GPa) at the center of the earth.

The core is almost a factor of two denser than the mantle, density jumping from 5.6 to 9.9 mg/m^3 at the core-mantle boundary. This density contrast, larger even than that between air and water or rock at the earth's surface, reflects the profound difference in the materials making up the mantle and core: rock (oxide minerals) and dense metal (iron alloy), respectively. In terms of contrasts in material properties, the boundary between the core and mantle can thus be considered the most significant of our planet.

In comparison, the small contrast in density between the outer core and inner core, 0.55 mg/m^3, suggests that these regions are basically composed of similar chemical constituents. That is, the inner core is viewed as the solidified (crystalline) equivalent of the outer-core liquid. Pressure generally increases the melting and solidification temperatures of materials, and the increase in pressure with depth apparently suffices to place the melting temperature of the core alloy, respectively, above and below the temperatures of the inner core and outer core. That the inner-core boundary is right at the melting point of the core material is qualitatively supported by the observation that seismic waves are strongly attenuated in the outermost several hundred kilometers of the inner core; though crystalline, this region is "mushy" at seismic frequencies, as though it were near the melting point.

Measurements of the geomagnetic field at and near the earth's surface demonstrate that more than 90 percent of the field originates within the planet. Both temperature and pressure act to suppress ferromagnetism; hence, the only known mechanism of producing the geomagnetic field is through a dynamo process involving the motion of an electrically conducting fluid. The presence of the earth's magnetic field is thus direct evidence that the outer core, by far the largest fluid region within the planet, is a metallic liquid.

Table 1. Seismological Structure of the Core

Region	Depth (km)	Radius (km)	V_P (km/s)	V_S (km/s)	Density (mg/m³)	Bulk Modulus (GPa)	Shear Modulus (GPa)	Pressure (GPa)
Lowermost Mantle	2891	3480	13.72	7.26	5.57	656	294	136
Core-Mantle Boundary								
	2891	3480	8.06	0	9.90	644	0	136
	2971	3400	8.20	0	10.03	674	0	144
	3071	3300	8.36	0	10.18	712	0	155
	3171	3200	8.51	0	10.33	748	0	165
	3271	3100	8.66	0	10.47	785	0	176
	3371	3000	8.80	0	10.60	820	0	186
	3471	2900	8.93	0	10.73	855	0	196
	3572	2800	9.05	0	10.85	889	0	206
	3671	2700	9.17	0	10.97	922	0	215
	3771	2600	9.28	0	11.08	954	0	225
Outer Core	3871	2500	9.38	0	11.19	986	0	234
	3971	2400	9.48	0	11.29	1016	0	243
Liquid	4071	2300	9.58	0	11.39	1045	0	252
	4171	2200	9.67	0	11.48	1074	0	261
Fraction of	4271	2100	9.75	0	11.57	1101	0	269
Earth =	4371	2000	9.84	0	11.65	1127	0	277
31% by Mass	4471	1900	9.91	0	11.73	1153	0	285
16% by Volume	4571	1800	9.99	0	11.81	1178	0	292
	4671	1700	10.06	0	11.88	1201	0	300
	4771	1600	10.12	0	11.95	1224	0	306
	4871	1500	10.19	0	12.01	1246	0	313
	4971	1400	10.25	0	12.07	1268	0	319
	5071	1300	10.31	0	12.13	1289	0	325
	5150	1222	10.36	0	12.17	1305	0	329
Inner-Core Boundary								
	5150	1222	11.03	3.50	12.76	1343	157	329
	5171	1200	11.04	3.51	12.78	1346	157	330
	5271	1100	11.07	3.54	12.83	1359	160	335
	5371	1000	11.11	3.56	12.87	1370	163	340
	5471	900	11.14	3.58	12.91	1381	165	345
Inner Core	5571	800	11.16	3.60	12.95	1390	168	349
	5671	700	11.19	3.61	12.98	1398	170	352
Crystalline	5771	600	11.21	3.63	13.01	1405	171	355
	5871	500	11.22	3.64	13.03	1411	173	358
Fraction of	5971	400	11.24	3.65	13.05	1416	174	360
Earth =	6071	300	11.25	3.66	13.07	1420	175	362
2% by Mass	6171	200	11.26	3.66	13.08	1423	176	363
1% by Volume	6271	100	11.26	3.67	13.09	1425	176	364
	6371	0	11.26	3.67	13.09	1425	176	364

Source: Preliminary Reference Earth Model (Dziewonski, A. M., and D. L. Anderson, *Physics of the Earth and Planetary Interiors*, vol. 25, 1991, pp. 297–356).

Iron is by orders of magnitude the most abundant element in the solar system that has both density and compressibility close to those observed seismologically for the core. The observed density, however, is about 10 percent lower than that of pure iron measured experimentally at the conditions of high pressure and temperature existing in the core. Therefore, the core is thought to consist primarily of iron, but alloyed with (contaminated by) other elements. Based on cosmochemical observations and models, including the observed compositions of meteorites, nickel (Ni), sulfur (S), and hydrogen (H) are considered the most plausible elements that might have been alloyed with iron at the time the core formed.

In addition, laboratory experiments show that liquid iron and iron alloys react chemically with crystalline oxides (including silicate minerals) at the conditions of the deep mantle and core. Oxygen is found to alloy readily with iron at these high pressures and temperatures, so it appears inevitable that the mantle has been reacting with—indeed, dissolving into—the outer core over geological time. Furthermore, scattering of seismic waves near the base of the mantle (D″ layer) indicates that the structure of this region is characterized by anomalously strong heterogeneity, heterogeneity that could well be produced by chemical reactions between the mantle and core. Hence oxygen, albeit added after core formation, is also thought likely to be an important alloying component of the core.

The melting temperatures of iron and iron alloys have recently been experimentally measured to pressures exceeding 50–100 GPa. Because the inner-core boundary is expected to be at (or near) the melting point of the core alloy, such measurements provide first-order constraints on the temperature at depth. Temperatures ranging between about 4,000 ($\pm$500) K near the top to 5,000–6,000 K at the center are thus inferred for the core. Ultimately, the largest source of uncertainty in these values comes from the lack of definitive constraints on the composition of the core. The composition of the alloy, in turn, is largely dependent on the geological history of the core; that is, when the alloying took place, before or after the accumulation of the metal at the center of the planet.

How the core became this hot is uncertain, but at least two sources of heat are thought to be involved. First, and potentially most significant, is the heat associated with the accumulation of the earth, including the emplacement of the core metal at the center of the planet. In particular, the giant impact thought to have splashed the Moon out of Earth toward the end of our planet's growth, some 4.5 billion years (Ga) ago, could easily have heated the interior sufficiently to explain the present temperature of the core (*see* EARTH-MOON SYSTEM, EARLIEST HISTORY OF). Second, the decay of naturally occurring radioactive isotopes, most likely present both in the core and the mantle, can account for the temperatures of the deep interior.

Cooling of the core over geological history is sufficient to drive convection of the outer core fluid; convection acts as a source of power for creating the geomagnetic field. Based on modeling the heat flow through Earth's mantle, the rate of cooling of the core is estimated to be a few hundred degrees celsius per billion years. In addition, crystallization of the outer core (with associated growth of the inner core) may also drive convection and help power the magnetic field. Whatever the mechanisms involved in producing the geomagnetic field, paleomagnetic observations on ancient rocks demonstrate that Earth's field has existed for at least 2.5 Ga, and perhaps longer than 3.0–3.5 Ga.

The most recent finding about the structure of the core is that the inner core is anisotropic: seismic waves travel faster, by a few percent, in the north–south direction than in directions within the plane of the equator. One possible explanation for this anisotropy is that it is caused by alignment of mineral grains within the inner core. Indeed, calculations suggest strongly that the inner core is convecting; though solid to seismic waves, it behaves like a fluid over geological time periods, just like the mantle. The resulting deformation within the inner core could produce a schistosity in the iron alloy that explains the observed seismic anisotropy. If correctly interpreted, the results suggest that tectonic motions take place from the surface right down to the center of our planet.

Bibliography

JEANLOZ, R. "The Nature of the Earth's Core." *Annual Review of Earth and Planetary Sciences*, 18 (1990):357–386.

MASTERS, T. G., and P. M. SHEARER. "Seismic Models of the Earth: Elastic and Anelastic." In *Global Earth Physics, a Handbook of Physical Constants*, vol. 1, ed. T. J. Ahrens. Washington, DC, 1995, pp. 88–103.

RAYMOND JEANLOZ

COSMIC DUST

See Interplanetary Medium, Cosmic Dust, and Micrometeorites

COSMOLOGY, THE BIG BANG, AND THE STANDARD MODEL

Cosmology is that branch of astrophysics that deals with the origin, structure, and evolution of the universe. The current best picture of the origin of the universe—the big bang theory—evolved from approximately 1910 to 1994 and involved the work of many people.

In astrophysics, observational data provide the constraining boundary conditions to theoretical speculation. While the tools of early twentieth-century observational astronomy were being honed to the tasks of measuring spectra, Doppler velocities (from positional shifts in spectral features that are proportional to velocity along the line of sight to a given object), and astronomical distances, the theoreticians of the day were making progress in their description of the structure of the universe. Chief among these cosmological speculators was Albert Einstein.

In 1905, Albert Einstein published three of the most profound papers of twentieth-century physics, including his "Special Theory of Relativity." His greatest intellectual triumph was presented to the world in 1916 in a paper titled "The Foundation of the General Theory of Relativity." The basic premise of Einstein's theory is that acceleration created by a gravitational field is indistinguishable from that produced by accelerated motion. The solutions to Einstein's field equations, cast in the mathematics of tensor calculus, specify the geometry of the universe based on the distribution of matter and energy, that is, the field equations are the fundamental expression of the general theory of relativity that allow us to explore the past and predict the future of the universe.

Since Einstein did not know the actual distribution of matter and energy in the universe, he began with a reasonable assumption called the Cosmological Principle, which Einstein stated thus:

"Aside from random fluctuations that may occur locally, the universe must appear the same for all observers." This is to say that Einstein assumed the universe was homogeneous (the same everywhere) and static (unchanging, on the large scale, in time).

From 1917 through 1927 alternative solutions were presented to the Einstein field equations by Willem de Sitter (1917), Aleksandr Friedmann (1922, 1924), and Georges Lemaitre (1927) that showed that not only static solutions were possible for the field equations, but nonstatic ones as well in which the universe could exhibit dynamic behavior such as expansion. These solutions did not receive much attention until 1931 when Sir Arthur S. Eddington, Lemaitre's mentor, advocated the reprinting of the 1927 paper, in English, in the *Monthly Notices of the Royal Astronomical Society*. By the early 1930s the observational evidence, obtained in the 1920s, strongly suggested that nonstatic universe solutions might be closer to reality than static solutions.

The first evidence for an expanding universe was presented in 1914 by Vesto M. Slipher, who announced the discovery of large Doppler velocities in fourteen spiral nebulae (*see* GALAXIES) at an American Astronomical Society meeting in Evanston, Illinois. Since this discovery preceded the advent of suitable methods of measuring distances to the galaxies using Cepheid variable stars (done by EDWIN HUBBLE in 1924), the full significance of the data could not be appreciated.

By 1929 Doppler velocities were available for a total of forty-six galaxies and, by then, Hubble had determined distances for eighteen of them. Although the data set was limited, Hubble constructed a graph of distance (x-axis) versus Doppler velocity (y-axis). A trend was evident—the greater the distance to a galaxy, the greater was its speed of recession. He calculated a linear fit to the data that has become known as Hubble's Law: $V = H \times d$, where H is known as the Hubble parameter, d = distance, and v = velocity. Hubble's original value for this quantity was 540 (km/s)/megaparsec (Mpc). Hubble, with his assistant Milton Humason, continued to refine such measurements into the 1930s.

The work of Hubble made a great impression on Einstein who, in February 1930, traveled to the Mount Wilson Observatory to confer with Hubble. Einstein had always favored a static model of the universe but was well aware of the expanding universe solutions of Lemaitre, de Sitter, and Friedmann. Following his consultation with Hubble, Einstein wrote: "New observations by Hubble and Humason concerning the redshift of light in distant nebulas [sic] make it appear likely that the general structure of the universe is not static" (Clark, 1971, p. 341).

The Hubble measurements thus had a natural interpretation in the context of the expanding universe solutions of Einstein's general theory of relativity. The classic example, used to assist students in understanding why an expanding universe would naturally give rise to a Hubble-type relationship, is the "raisin-bread" illustration. A little thought will show that if you place raisins in a lump of bread dough and then allow it to expand under the action of the yeast, that no matter which raisin you choose to use as the origin, all other raisins will appear to expand away from it as the dough rises. Further, the average speed of recession between any two raisins will be directly proportional to the distance separating them. In this case, the raisins are the "galaxies." This type of expansion is known as an homologous expansion.

In the context of the expanding universe solutions to Einstein's equations of general relativity, the clock may be run backward from the present state of the universe to an epoch when the universe was in a very dense state—a primordial form from which it expanded explosively. Thus, any theory of the origin of the universe of this type is called a "big bang" theory.

To many theoretical astrophysicists one of the most disturbing features of any of the big bang solutions is that they imply a beginning. A unique beginning is extremely difficult to deal with from the perspective of the philosophical cause and effect underpinnings of the scientific method itself, which does not adapt easily to handle the concept of a "first cause" very well. Indeed, a universe that has a beginning seems to be more the realm of theology than science as currently practiced.

Since the universe appears to be expanding, a natural question to ask is, Will the expansion ever stop? The answer to this question depends on the average mass-density of the universe that is calculated by taking all the mass in the universe, of all forms, and dividing by the total volume of the universe. In the "standard big bang" the value of the "critical density"—the density required to eventually bring the expansion to a halt—may be calculated from theory. The value of the critical density depends on the value of the Hubble parameter, H,

which, by the early 1990s was determined to be between 50 and 100 (km/s)/Mpc. If a mid-range value of 75 (km/s)/Mpc is adopted, the critical density is calculated to be 1.1×10^{-26} kg/m^3 (Weinberg, p. 476). If the average density of the universe is greater than this value, the expansion will eventually stop and the universe will collapse on itself—the "closed" universe case. If the average mass density is smaller than this value, the universe will expand forever and is said to be "open"; if it were exactly equal, the universe is described as "flat," that is, it will asymptotically approach zero expansion rate but never reverse and collapse.

Current observational evidence suggests the average mass density, due to visible matter, is only about 3.1×10^{-28} kg/m^3—only about 3 percent of the critical density. From this one must conclude that either the universe had a unique origin as a one-time event, or that there is mass present in the universe in a form not readily evident that may make up the deficit to achieve closure—the so-called missing mass. It has been speculated that such mass may exist in the form of black holes, massive neutrinos, or some other, as yet undetermined, form of mass. Perhaps the best observational evidence in favor of the existence of some form of dark matter is the rotation characteristics of many spiral galaxies that strongly suggest some form of hidden matter.

It would be incorrect to think that the entire astronomical community simply accepted the standard big bang without any thought to alternative explanations for the observations. In 1948, Hermann Bondi and Thomas Gold put forth an idea that would come to be known as the steady state theory for the creation of the universe. This theory neatly avoided the issue of a unique beginning to the universe by supposing that as the universe expands, new matter is spontaneously created at just the right rate to fill the void. The greatest, immediate objection to this theory, is that it violates the conservation of mass and energy, which states that the total energy and mass equivalent in the universe is a constant over all time.

Also in 1948 George Gamow predicted that a cosmological big bang should be evidenced by the presence of a low-temperature microwave background radiation—a feature not at all consistent with the steady state theory. Gamow estimated a modern-day background temperature for the universe of about 25 K; in 1964, revisions by a Princeton University research group, headed by Robert

H. Dicke, showed that the temperature should be less than 10 K.

In 1965 Arno Penzias and Robert Wilson were testing a new horn antenna (so termed because of its shape) that had been built at Bell Labs in New Jersey for the purpose of observing the ECHO 1 satellite, when they noticed an ever-present background signal that seemed to be coming from the whole sky. The radiation was consistent with the predictions of Dicke—the fossil radiation left over from the origin of the universe had been found. This discovery all but dealt a death blow to the steady state theory; Penzias and Wilson later received the Nobel Prize for their work.

The "inflationary universe" theory—a new twist on the big bang theory—has enjoyed a wide degree of attention among cosmologists during the 1990s. Scientists have suggested that the knotty problem of the origin of the universe, presented ad hoc in the standard big bang, is the result of a quantum mechanical fluctuation. In this scenario, the universe, 10^{-35} seconds after its appearance, would have existed in a quantum mechanical configuration termed a false vacuum—a state from which rapid expansion (inflation) is a natural consequence as well as the creation of all matter out of the energy of the event. This picture is not intuitive and requires the full mathematical perspective of quantum mechanics to appreciate.

In 1989 the Cosmic Background Explorer (COBE) satellite provided the best evidence to date for the general background radiation, at 2.73 K, and very slight anisotropies (variations from the average) on the order of a few ten millionths of a degree—sufficient to explain the formation of galaxies, clusters, and superclusters. The origin of such clumpiness, necessary to the formation of galaxies, clusters of galaxies, and so on, has not been completely addressed in either the standard big bang or the inflationary universe models.

In conclusion, one version or another of the big bang cosmology appeared to be nearest to the truth in the mid–1990s. Areas of greatest concern, still to be adequately dealt with, include (1) the so-called missing mass problem, and (2) the origin of large scale anisotropies in the early universe.

Bibliography

BERENDZEN, R.; R. HART; and R. SEELEG. *Man Discovers the Galaxies.* New York, 1976.

BERRY, M. *Principles of Gravitation and Cosmology.* Cambridge, Eng., 1976.

CLARK, R. W. *Einstein, the Life and Times.* Cleveland, 1971, p. 341.

EICHER, D. "Candles to Light the Night." *Astronomy* (September 1994):32–39.

GARDNER, M. *Relativity for the Million.* New York, 1962.

LEMAITRE, G. "Un univers homogène de masse constante et de rayon croissant." *Annales de la Société des Sciences de Bruxelles* 47 (1927). Trans. in *Monthly Notices of the Royal Astronomical Society* 91 (1931):483–490.

WEINBERG, S. *Gravitation and Cosmology (Principles and Applications of the General Theory of Relativity).* New York, 1972.

DAVID L. TALENT

CRYSTAL CHEMISTRY

See Mineral Structure and Crystal Chemistry

CUVIER, GEORGES

The French paleontologist Georges Cuvier was born at Montbéliard, Doubs, on 23 August 1769. Cuvier was the descendant of a Huguenot family that had been forced into exile in Switzerland. He was educated at Stuttgart, originally training for either a career in law or in the ministry. In 1788 he took a position as tutor to the family of the Comte d'Héricy, near Caen, Normandy, and about this time became interested in natural science. After being introduced to the naturalists Antoine Laurent de Jussieu and Étienne Geoffroy Saint-Hilaire in Paris, in 1795 Cuvier was appointed an assistant at the Jardin des Plantes, Muséum d'Histoire Naturelle, in Paris and was also made a member of the Académie des Sciences. With this start Cuvier quickly rose through the academic and governmental hierarchies. Among the many titles and positions that Cuvier held are the following: lecturer at the École Centrale du Panthéon (1796); professor of natural history at the Collège de France (1799); professor of comparative anatomy at the Jardin des Plantes (1802); permanent secretary of the Académie des Sciences (1803); appointed by Napoleon to the council of the Imperial University (1808); councilor of state (1814); elected to the Académie Française (1818); president of the Committee of the Interior (1819–1832); chancellor of the University of Paris (1821–1827); made a baron by King Louis Philippe (1831). Cuvier was named the minister of the interior just before he died in 1832.

During the first quarter of the nineteenth century Cuvier was a dominant force in the scientific community of France, and indeed throughout the western world. His specialty was comparative anatomy, especially of living and extinct vertebrates. Accordingly, Cuvier is sometimes considered the founder of modern comparative anatomy as well as the father of vertebrate paleontology. Cuvier paid particular attention to the relationships of different body parts in the same animal, emphasizing the integrated nature of the beast as a whole (his principle of the correlation of parts). Given a jumble of bones of different fossil species, Cuvier could sort the bones into the proper species, associating the correct toe bone, for instance, with the correct limb bone, the correct jaw, and so forth. It was sometimes suggested that from a single bone he could reconstruct an entire animal.

As an adjunct to his comparative anatomical studies, Cuvier was also interested in the classification of animals, including both extant and extinct forms. He became particularly interested in the study of fossil mammals, especially those found in the Tertiary rocks of the Paris Basin. He also studied reptiles, and in 1812 recognized a type of extinct "flying reptile" that he labeled a pterodactyl ("wing finger," since the wing membrane was stretched over an elongated finger).

With the French mineralogist and geologist Alexandre Brongniart (1770–1847), Cuvier documented the stratigraphic sequences and contained faunas of the Tertiary sediments in the Paris Basin (see BRONGNIART, ALEXANDRE).

Although Cuvier proposed his own classification of organisms, and acknowledged the close similarities among many species, he rejected the notion of evolution. According to Cuvier, any particular species was a well-integrated entity that was adapted to one particular way of life. Any significant change in a species would upset the harmonious relationships between the functioning parts of the organism, and between the organism and its environment; therefore, Cuvier argued that organisms could not significantly transform, transmutate, or evolve over time. Furthermore, Cuvier noted that French expeditions to Egypt had returned with mummified animals known to be thousands of years old, yet these animals were identical to living species. Thus the stability of species over time was established.

Cuvier's studies of extinct vertebrates, including mastodons, mammoths, Tertiary mammals, and Mesozoic marine reptiles, established that there were fossil forms that did not resemble any living organisms. Cuvier came to the conclusion that many of the fossil animals he reconstructed had become extinct (Lamarck tended to believe that they had not gone extinct per se but had evolved into the living forms on Earth today—see LAMARCK, JEAN-BAPTISTE). Furthermore, the older the rocks from which he collected, the less similar were the vertebrates to any extant animals. Thus Cuvier's work established that a sequence of successively different life forms had inhabited the surface of Earth. With modern hindsight we can interpret Cuvier's successive faunas in an evolutionary framework, but Cuvier saw each fauna as distinct and complete unto itself. There were not intermediate faunas or gradations between superposed faunas; indeed, Cuvier found abrupt breaks in the rocks separating superposed faunas. In particular, in the Tertiary strata of the Paris Basin, Cuvier documented a sequence of rocks that alternated between freshwater and saltwater deposits, indicating major and relatively abrupt changes in the position of land and sea.

To interpret his data, Cuvier postulated that a series of relatively quick changes had occurred in the Paris Basin (e.g., shifting from freshwater to ocean conditions), and during each "revolution" the old fauna had gone extinct and an entirely new fauna replaced it. Cuvier himself did not believe that the new faunas were miraculously created by God, but simply that they migrated into the local area from other regions. However, as more of the world was explored and more fossils discovered, the original Cuvierian migration theory became increasingly implausible (the areas where the migrants might have come from were never found). Cuvier's work lay the foundations for the concept of catastrophism (massive geological changes and upheavals, unlike anything seen in historical times, periodically occurring in the past) as a unifying theme in Earth history, and for miraculous creation as the origin of new species. Both theories were antithetical to the relatively slow, gradual transmutation or evolution of species as postulated by both Lamarck and Darwin (*see* DARWIN, CHARLES, and LIFE, EVOLUTION OF).

Cuvier was very productive and influential during his lifetime; he possessed the ability to persuade and influence those around him. Some of his major published works include: *Mémoires sur les espèces d'éléphants vivants et fossiles* (1796), *Tableau élémentaire de l'histoire naturelle des animaux* (1798), *Recherches sur les ossements fossiles de quadrupèdes* (1812), *Règne animal distribué d'après son organisation* (1817), *Mémoires pour servir à l'histoire des mollusques* (1817), *Discours sur les révolutions de la surface du globe* (1825), and *Discours sur la théorie de la terre* (1825). Civier died on 13 May 1832 in Paris, France.

Bibliography

BOURDIER, F. "Geoffroy Saint-Hilaire versus Cuvier: The Campaign for Paleontological Evolution (1825–1838). In *Toward a History of Geology,* ed. C. J. Schneer. Cambridge, MA, 1969.

BOWLER, P. *Evolution: The History of an Idea.* Berkeley, CA, 1984.

CUVIER, G. *An Essay on the Theory of the Earth.* Translated from the French by R. Kerr, with notes by R. Jameson. Edinburgh, 1813.

———. *The Animal Kingdom Arranged after Its Organization.* London, 1863.

LEE, S. B. *Memoirs of Baron Cuvier, with a List of His Writings.* New York, 1833.

ROBERT M. SCHOCH

D

DALY, REGINALD A.

In his principal contributions to geology, Reginald Aldworth Daly provided new understanding of the composition and structure of Earth's crust and upper mantle, the petrology of igneous rocks, and magma compositions, emplacement, and differentiation. He made additional contributions on the origin of coral reefs around atolls, glacial rebound from melting of ice caps, turbidity currents in deep submarine valleys, and the role of the asthenosphere in isostasy and convection and as a magma source. Daly was born on 19 May 1871 and died on 19 September 1957.

Daly was the youngest of four sons and five daughters; they all attended the public schools in Napanee, Ontario. His grandfather had emigrated from Ireland to Canada, and his father had attended Trinity College in Dublin before settling in Napanee. Daly recalled living "a simple life in a small town, under the direction of parents with high standards of conduct." He said he was "taught early to assume responsibility and take pleasure in hard work, and encouraged by uninterrupted schooling and opportunity for wide reading early in life."

Daly had no preference in his studies until after graduation with an A.B. degree in 1891 from Victoria College (later incorporated into the University of Toronto), where he won prizes in English

literature, astronomy, and science. In an extra year spent at Victoria to obtain an S.B. degree, he became very interested in geology, and he went on to Harvard University to do graduate studies in it. He got his Ph.D. at Harvard in 1896 and taught physiography there from 1898 to 1901. Under a Parker Traveling Fellowship from Harvard, he studied geology abroad for two years with Karl Harry Ferdinand Rosenbusch at Heidelberg, with Alfred Lacroix in Paris, and, on his own, inspecting geologically interesting localities in Europe. He also attended the Seventh International Geological Congress in Russia. In photographs of attendees of the Congress, young Daly appeared as a physically impressive, sartorially elegant presence among the somewhat worn-down gathering of geologists. In these and later trips abroad, he formed many personal connections that lasted throughout his life.

In 1901 Daly undertook an enormous reconnaissance field study for the Canadian Geological Survey of the geology along the western end of the American–Canadian boundary, the 49th parallel; the extent was 650 kilometers (km) almost entirely mountainous. Helped only by one field hand with no geological training, he mapped the geology of an 8–16 km wide swath during six field seasons, collecting 1,500 rock specimens and 1,300 photographs. After inspecting 960 thin sections, getting 60 chemical analyses, and performing a valiant analysis of his field and laboratory observations, he wrote twenty papers and a three-volume final report for the Geological Survey of Canada, covering the stratigraphy, structure, petrology, economic geology, and glacial geology of that huge area.

Daly returned, full of new ideas, to teach at Massachusetts Institute of Technology from 1907 to 1912, and then he taught at Harvard from 1912 to 1942. His lectures at both schools were based on his analysis of his Canadian fieldwork, and from this in 1914 he wrote his first textbook, *Igneous Rocks and Their Origin,* in which he described the chemistry, petrography, and physical properties of rocks, and the melting and differentiation of intrusive rock bodies. From his fieldwork, including that of Mount Ascutney in Vermont, he developed his concept of magmatic stoping to emplace igneous intrusions. He propounded the philosophy that "theoretical geology is the basis of practical geology," and that "the data for intelligent geological theory must be found chiefly in . . . areal mapping on the large scale." He was a superb lecturer, presenting his ideas clearly and concisely, and the students in his elementary course would often applaud at the end of lectures. He made a generous bequest to Harvard earmarked for meeting field expenses of graduate students.

Daly's scientific thinking was characterized by a bold imagination and by a steady effort toward synthesis. In his revised textbook of 1933, *Igneous Rocks and the Depths of the Earth,* he dealt with the origins, compositions, and structures of igneous bodies, including processes in the crust and upper mantle. The book *Strength and Structure of the Earth,* published in 1940, is devoted largely to summarizing and interpreting the geophysical evidence on isostasy, elasticity, the figure of Earth, and the nature of the interior of Earth.

After preparing his invited lectures, Daly wrote books intended for the educated general reader; they were copiously illustrated, packed with numerical data, and written with contagious enthusiasm, bearing the imprint of his imagination and intellectual force. The scope of subjects is revealed by the book titles: *Our Mobile Earth* (1926), from lectures at the Lowell Institute; *The Changing World of the Ice Age* (1934), from the Silliman lectures at Yale University; *The Architecture of the Earth* (1938), from the Harris lectures at Northwestern University; and *The Floor of the Ocean* (1942), from lectures at the University of Virginia.

There was one interlude in Daly's devotion to geology, when the United States entered World War I; he took leave from Harvard and went to France as the chief librarian of the Y.M.C.A. The *Comptes Rendus* of the Académie des Sciences remarked on his stint as librarian, "his interventions on behalf of our fellow countrymen were innumerable." In 1921–1922, he participated in the Shaler Memorial Expedition to the southern hemisphere to study the contrast between oceanic and continental rocks. After mapping for a month on Saint Helena island, he remarked, "What a pity Napoleon was not a geologist; he would have found St. Helena much more interesting." The expedition was the last of Daly's prodigious travels to do geologic fieldwork; he crossed America twenty-four times and the Atlantic Ocean fourteen times. His wife, Louise Porter Haskell, accompanied him on many of his travels; in addition she edited and typed most of his manuscripts and stimulated his interest in art and music. He was serious-minded but occasionally would offer a pun, and on social occasions he would relate stories taken from the many anecdotes and reminiscences acquired from his rich experience of places and people.

Daly received many honors. He received medals from the Philadelphia Academy of Sciences, the Geological Society of America, the American Geophysical Union, and the Geological Society of London. He received honorary D.Sc. degrees from Harvard University, University of Chicago, University of Toronto, and University of Heidelberg. He was a member of twenty-five American and foreign geological and other scientific societies. It is fitting that very recently the American Geophysical Union honored Daly by naming the annual invited lecture in its Volcanology, Geochemistry, and Petrology Section after him; the first lecture was given on 25 May 1993.

From his earliest papers, Daly showed an interest in the physical properties of rocks; he used his own measurements of density in explaining the origin, melting, and differentiation of igneous rocks. In time he realized that data from physical measurements of rocks and minerals under pressure and temperature conditions of the interior of the earth were required. In 1932, with help from his colleagues at Harvard, he initiated a program of research in geophysics and experimental geology; in forty years, this project produced many significant contributions to our knowledge of Earth's interior. A majority of these articles were written by his colleague and principal investigator, Francis Birch. In addition, Daly helped establish a modern seismological station near the town of Cambridge, Massachusetts. Daly was the principal advocate of the compilation "Handbook of Physical Constants," edited by Birch and published by the Geological Society of America in 1942. Daly directed only a few doctoral theses, and his principal influence was exerted through his writings, his lectures (at Harvard and invited), and his participation in scientific discussions and activities in different settings.

In a biographical memoir for the National Academy of Sciences (1960), Francis Birch described Daly's life and career in detail; many quotations from that memoir are used herein. Summing up the significance of the sixty-one years of Daly's scientific career, Birch said, "Perhaps the most enduring elements of his work will be the many contributions toward the quantification of the geological sciences, a transformation now conspicuously in process."

Bibliography

BIRCH, F. "Reginald Aldworth Daly." *Biographical Memoirs, National Academy of Sciences,* 34 (1960):30–64.

DALY, R. A. *Igneous Rocks and Their Origin.* New York, 1914.

———. *Our Mobile Earth.* New York, 1926.

———. *Igneous Rocks and the Depths of the Earth.* New York, 1933.

———. *The Changing World of the Ice Age.* New Haven, CT, 1934.

———. *The Architecture of the Earth.* New York, 1938.

———. *Strength and Structure of the Earth.* New York, 1940.

———. *The Floor of the Ocean.* Chapel Hill, NC, 1942.

EUGENE C. ROBERTSON

DANA, JAMES D.

Excellent training, background, travel, circumstances, and ambition enabled James Dwight Dana to become the leading American geologist and to rank among the most prominent American scientists of the nineteenth century. His most important contributions, scattered among his two hundred publications, included his lasting work on systematic mineralogy, paleontology, and geology from the Wilkes expedition; his influential textbook, *Manual of Geology*; construction and development of the geosyncline theory; and his many contributions to the developing framework of American science.

James Dwight Dana was born in Utica, New York, on 12 February 1813, the oldest of ten children born to James Dana and Harriet Dwight Dana. Dana was reared in a strict Christian environment. Starting at an early age, he collected and was fascinated by natural objects such as minerals, plants, and animals. His intensity of observation and collection set him apart, even before he was ten, and he was later influenced in school by an inspiring science teacher.

Owing to the reputation of the leading American science teacher of the day, Benjamin Silliman, Dana enrolled at Yale College in 1830 and studied a traditional curriculum of theology, philosophy, and science, among other topics. He was a serious student and spent his leisure time in the field and with Silliman's extensive mineral collection. The state of American geology was formative—the overriding conflict of the time between ABRAHAM GOTTLOB WERNER and JAMES HUTTON and their followers, concerned the origin of rocks, especially the volcanic rock basalt. Religion was a prominent issue in almost all aspects of geology and its acceptance; arguments over the creation of Earth and its inhabitants raged. Was the Genesis account correct? Could each new scientific development be reconciled with the Bible? Geology was a practical course of study that involved the development of natural resources and had little to do with theory or speculation.

Dana's first paper, "On the Condition of Vesuvius in July, 1834," resulted from the first of his extensive travels, as a mathematics teacher to midshipmen aboard the U.S.S. *Delaware* (1833–1834). The paper, like the majority of his scientific articles, was published in the foremost scientific journal of the day, the *American Journal of Science* (begun as Silliman's journal at Yale). Experiences from such travel laid the groundwork for some of his major studies, and provided the basis for his views of geology, especially of volcanoes and coral reefs, from a global and theoretical rather than a strictly descriptive and geographic perspective. Even more important than his Mediterranean cruise were his scientific duties with the Wilkes expedition (1838–1842). Dana had been recommended for the post by a friend, the well-known botanist Asa Gray. The Wilkes expedition—led by Lt. Charles Wilkes and authorized by a Senate bill to explore the Pacific—was a major event in American science; it circled the world, conducted natural history surveys in Polynesia, and confirmed the existence of the Antarctic continent. Similar in its consequences to CHARLES DARWIN's voyage on the *Beagle*, the Wilkes expedition served as the springboard for the development of Dana's career. Dana not only continued his systematic study of minerals—he published the first edition of *A System of Mineralogy* shortly before sailing with Wilkes—but also gathered the data and observations that resulted, over the next fourteen years, in his expedition reports *Zoophytes, Crustacea,* and *Geology*. These experiences, and his meticulous accounts of them, placed Dana as a prominent, though young, American scientist whose views were respected and sought. His increasing duties with the American Journal of Science, where he served as editor from 1846 until the 1880s, helped to expand his growing influence in American science.

Notable among Dana's contributions in this period were his writings on corals and coral reefs. Building on work begun by Darwin, Dana enlarged upon available observations and described three types of coral reefs—atolls, fringing reefs, and barrier reefs. He linked all three types to stages of evolution of coral growth on volcanic edifices, with island subsidence the dominant process causing the stages of growth, findings which largely, though not entirely, supported and enhanced Darwin's theory.

Along with his *Manual of Geology*, Dana's most lasting contribution to geology is his *System of Mineralogy*, a massive work that continued in many editions throughout his life, and was carried on by one of his sons, Edward Salisbury Dana. After his cruise on the *Delaware*, Dana returned to Yale in 1835 and became Silliman's assistant a year later. He wrote a piece on chemical nomenclature for minerals and sent it to Jöns Jakob Berzelius, a leading chemist in Europe. Berzelius suggested that Dana devise a new system, rather than modify the existing system. Dana did this, and the first edition of his *System* appeared in 1837. His system was a "natural" one, similar to that used in biology. Gradually, chemical compositions and physical characteristics, especially crystallography, became more important, characteristics fully evident in the 1850 edition. The *System* quickly became the standard reference and was supplemented by Dana's textbook for students, the *Textbook of Mineralogy*. Further enlargement and refinement of later editions led to a System of Mineralogy that today remains very much in use.

During his continued work on the *System*, the completion of the reports of the Wilkes expedition, the publishing of hundreds of papers on subjects as diverse as volcanos, coral reefs, mountain building, glaciers, and uniformatarianism, as well as teaching at Yale, Dana developed his theory of the Earth. He believed that thermal contraction of the earth was the primary driver for the development of Earth structures, including folded belts of mountains such as the Appalachians. Upon contraction of an originally molten Earth, subsidence ensued, and differential subsidence along adjacent basins caused fold mountains and their accompanying features such as folds and faults. Working independently of Dana as a geologist for the state of New York, James Hall made the seminal observation that the Appalachians were comprised of about 13,000 m of sediments but that the sediments were all of shallow-water origin (owing to contained fossil types and ripple marks, for example). Thus, subsidence must have taken place with deposition, with resulting features which were termed geosynclinals by Dana (later called geosynclines). Although the geologic process that changed the thick pile of sediment into folded linear mountains remained obscure for more than a century, the discovery, followed by Dana's own extensive advocacy of the hypothesis, resulted in what has been described as the first great American geological theory. The influential *Manual of Geology* described Dana's global views of geology and was critical in the evolution of geology from a descriptive to a more quantitative and theoretical science.

Dana made many lasting contributions to the developing institutions of American science. Though he was retiring by nature, and for the last thirty years of his working life suffered from unspecified ill health, his meticulousness, industry, and ambition drove him to influential positions in emerging American science. His most important roles included being a prolific contributor to and editor of the *American Journal of Science*; leader in the development of geology at Yale University, the oldest geology school in the United States; prominent in the development and leadership of the National Academy of Science; and prominent in the development of the U.S. Geological Survey (USGS) and its four major geographic and geologic surveys of the western United States, an area then poorly known (see KING, CLARENCE RIVERS, and POWELL, JOHN WESLEY). James Dwight Dana died on 14 April 1895, in New Haven, Connecticut.

Dana's scientific and religious lives were intimately interwoven. Like many of his colleagues, he expended considerable effort interpreting geologic concepts within his fundamental religious beliefs, a process that was evident in his "cephalization" theory, which attempted unsuccessfully to relate "higher" characteristics in animals to a greater "centralization of force" in the head and brain. Dana also corresponded with Charles Darwin on evolution and the ascent of humankind.

Bibliography

GILMAN, D. C. *The Life of James Dwight Dana.* New York, 1899.

PIRSSON, L. V. "James Dwight Dana." *National Academy of Science, Biographical Memoirs* 9 (1919):41–92.

PRENDERGAST, M. *James Dwight Dana: The Life and Thoughts of an American Scientist.* 2 vols. Los Angeles, 1979.

ROSSITER, M. W. "A Portrait of James Dwight Dana." In *Benjamin Silliman and His Circle: Studies on the Influence of Benjamin Silliman on Science in America,* ed. L. G. Wilson. Canton, MA, 1979.

E. JULIUS DASCH

DARWIN, CHARLES

Charles Robert Darwin is regarded by the world as the scientist who provided overwhelming data that evolution of organisms—change through time—occurs, but his contributions to geology have received less attention. Born on 12 February 1809 in Shrewsbury, England, Darwin entered the University of Edinburgh in 1825, where he took a course in geology under Robert Jameson, which he judged to be a dreadful subject. He was a member of both the geological Wernerian Society and the biological Plinian Society. He transferred to Cambridge in January 1828 and became increasingly interested in natural history. His formal field training in geology was a two-week trip with Reverend Adam Sedgwick to Wales in 1831; he may have attended some of Sedgwick's lectures at the university.

By good fortune Darwin became the traveling companion of Captain Robert FitzRoy on HMS *Beagle.* The ship left England 27 December 1831, spending most of the voyage in southern South America preparing Admiralty charts of both coasts and particularly of Tierra del Fuego. The ship continued westward and recrossed the Atlantic to check chronometers in eastern South America before docking in England 2 October 1836. Darwin was not the official naturalist of the voyage, though he soon assumed that role. His notebooks contain about four times as much material on geological observations as those on zoology. Darwin explored the Andes, noted fossil evidence of uplifted sedimentary strata, and experienced a dramatic earthquake in Chile where the coastline was raised more than a meter.

Shortly after Darwin's return he was elected to the Council of the Geological Society of London, and in 1838 he was elected one of the secretaries. That year he also presented a formal paper on the Concepción, Chile, earthquake that he had witnessed. Darwin was impressed with the concept of change in sea level through time and that year conducted fieldwork on the so-called parallel roads of Glen Roy, Scotland. He interpreted them as beaches formed during higher stands of the sea and presented a paper to the Geological Society in January 1839. With the general acceptance of continental glaciation and allied phenomena, these were reinterpreted in 1847 as former lake levels.

Throughout this period, Darwin was working on both his account of the trip and his views on the origin of coral reefs. During May 1839, only two and one-half years after his return from South America, he published "Narrative of the Surveying Voyages of His Majesty's Ships *Adventure* and *Beagle*. . . ." This pivotal work remains one of the finest travel books ever written, full of keen geological observations. Darwin married in 1839 and remained in London until September 1842.

In his posthumous autobiography Darwin writes that he thought of the concept of upward growth of corals with rise of sea level while he was still on the coast of South America and had not yet seen a reef. In May 1842, he wrote: "It is very pleasant easy work putting together the framework of a geological theory, but it is just as tough a job collecting & comparing the hard unbending facts." Despite arguments raised by the oceanographer Sir John Murray, Darwin's concept was generally accepted. He concluded that atolls are the result of growth of coral as a volcanic cone became gradually submerged. This theory has been vindicated by drilling through thousands of meters of coral into basalt.

Darwin published *The Structure and Distribution of Coral Reefs* in 1842, and then began writing *Geological Observations on the Volcanic Islands Visited During the Voyage of H. M. S. Beagle* (published in 1844). Throughout this time, Darwin was also supervising the distribution of natural history specimens and their description by specialists. Darwin's geological investigations were so highly regarded that in 1844 he was elected a vice-president of the Geological Society of London.

The third and final part of the geology of the Beagle voyage, *Geological Observations in South America,* was published in 1846, subsidized in part by Darwin himself. By that year, all of the biological specimens had been studied except for a single barnacle. Observation of this specimen would lead to the publication of a monograph of recent and fossil barnacles by the Palaeontographical Society (1881). One school of historians holds that because of this work in systematics, combined with his earlier publications in geology, scientists of the day were prepared to seriously consider Darwin's concept of evolution when it was first presented in 1859.

In *Origin of Species* (1859), Darwin speculated on the length of geologic time and suggested that erosion of the Cretaceous chalk Downs of southern England required about 300 million years. This figure upset the physicist William Thomson (Lord Kelvin) and led to Darwin's calculations on the age of Earth and a large body of geologic literature seeking to counter Kelvin's ever shortening concept of geologic time. Darwin removed his estimate from later editions of *Origin of Species* but did not participate in the violent arguments on the age of Earth.

In 1881, Darwin published a work on the formation of vegetable mold through the action of earth worms. This delightful little book may be viewed as a contribution to geomorphological processes and soil formation; it shows what a keen mind can do with seemingly prosaic observations.

Charles Darwin died on 19 April 1882, in Down, Kent, England.

Darwin's greatest legacy in geology is his work on coral reefs. Despite his observations on the Chilean earthquake, the gradual change in sea level he envisioned provided strong support for CHARLES LYELL's concept of uniformitarianism. More significantly, JAMES D. DANA accepted and amplified Darwin's views on reef formation. The emphasis on vertical movement of land and sea pervaded geologic thought until the more recent shift to the concept of seafloor spreading and lateral movement. Even more important, the general acceptance of evolution (and with it the concept of a great length of geologic time) has affected humanity's view of its own place within nature.

Bibliography

DESMOND, A. and J. MOORE. *Darwin.* New York, 1991.
HERBERT, S. "Charles Darwin as Geologist." *Scientific American* 254 (1986):116–123.

ELLIS L. YOCHELSON

DAVIS, WILLIAM MORRIS

William Morris Davis was born on 17 February 1850 in Philadelphia, Pennsylvania and died on 5 February 1934 in Pasadena, California. Best known for his comprehensive theory of the "geographical cycle," or cycle of erosion, and for the "genetic" description of landforms, Davis was also an indefatigable institution builder in American science at the turn of the century.

Davis was born to a Philadelphia family whose business interests were often overshadowed by the cause of the abolition of slavery at the behest of Davis' maternal grandmother, the abolitionist and feminist Lucretia Mott. Davis attended schools in Philadelphia and, for a brief period at the outbreak of the Civil War, in West Medford, Massachusetts. From 1866 to 1870 Davis attended the Lawrence Scientific School at Harvard University, where he earned a bachelor's degree in geology in

1869 and a master's degree in mining engineering the following year.

In later years Davis traced his scientific genealogy to the geologists Raphael Pumpelly, Josiah P. Whitney, and Nathaniel Southgate Shaler, each of whom gave Davis some instruction during his years at the Lawrence Scientific School. Davis also studied the rudiments of civil engineering with Henry Eustis. Eustis' curriculum emphasized descriptive geometry and landscape drawing.

From 1870 to 1873 Davis served as an assistant to the astronomer Benjamin A. Gould at the Argentine national observatory in Cordoba. Davis returned to Philadelphia in 1873 but left again in 1877 to take a position as Shaler's assistant at Harvard. Much of Davis' early career was in the shadow of Shaler, who was among the most popular professors in Harvard's history. Davis compensated for his less charismatic teaching style through research and publication in meteorology, geology, and physical geography. After an unpromising beginning he was appointed instructor in geology at Harvard in 1879, assistant professor of physical geography in 1885, and full professor in 1890. He was named Sturgis Hooper Professor of Geology in 1898 and retired from Harvard in 1912.

The bulk of Davis' earliest publications were in meterology, where most of his contributions clarified and synthesized the work of others. In 1894 he published *Elementary Meteorology* and ceased thereafter to teach meteorology at Harvard.

Beginning around 1888 and continuing until his death in 1934, Davis developed and promoted a system of landform analysis based on models of Darwinian evolution and pre-Darwinian systematics (*see* DARWIN, CHARLES). As was the case for his work in meterology, Davis' early work in physical geography, or geomorphology, was a clarification and synthesis of empirical and theoretical work by JOHN WESLEY POWELL, GROVE KARL GILBERT, and others. Davis rationalized what was then believed about the processes of subaerial erosion into a cycle of erosion—or, as he called it, a geographical cycle—wherein all features of relief were reduced, through time, to a peneplain. The cycle could be regenerated by uplift. The result was a system of causal, explanatory, or genetic description of landforms based on structure, process, and time or stage. Davis' system went well beyond the works of his predecessors.

After arranging early retirement from Harvard in 1912, Davis endeavored to rescue Charles Darwin's theory of coral reef formation from the state of doubt and scandal into which it had fallen, beginning in the 1860s. Building on earlier work by JAMES D. DANA, Davis attempted to demonstrate the usefulness of geomorphological evidence for deciding the problem, combining empirical studies of landforms with deductive inference. Although he studied several coral reefs and atolls firsthand, Davis based most of his work on existing literature and oceanographic charts. In 1928 he published *The Coral Reef Problem*, bringing together most of his previous work on the subject in the monograph.

Davis also enriched the visual language of the earth sciences through his use of block diagrams. Davis did not invent block diagrams, but he explored the limits of the genre by publishing examples that combined landscape and structure, showing change through a number of stages. As such, the block diagram was a compact means for expressing the geographical cycle. The full range of his technique is evident in *The Coral Reef Problem*.

Davis was involved in the founding of several associations of geologists and geographers, including the Geological Society of America (GSA) and the Association of American Geographers (AAG). He published in the earliest issues of these institutions' journals, as well as in those of the National Geographic Society. At meetings of the AAG, Davis developed a reputation for subjecting peers to immediate and often public criticism. To reach the growing population of high-school students, Davis published *Physical Geography* in 1898 and the shorter *Elementary Physical Geography* in 1902. He also published two sets of models, one of them with Shaler, to demonstrate landforms and geological structures. Davis also organized several field trips, including the Transatlantic Excursion of 1912. He used these occasions to argue for the geographical cycle and genetic description.

As much as any American scientist of his time, Davis was a player on a world stage. Although he wrote and published over six hundred papers and abstracts in English, he became eager around 1900 to have his work accepted by Europeans, and particularly by the Germans Albrecht Penck and Walter Penck. (Davis' conflicts with Walter Penck concerning the relative importance of climate for landscape evolution are well known to students of geology.) In 1909 Davis held a visiting professorship at the University of Berlin. In 1912 he published his most comprehensive exposition of the geographical cycle (in German only).

Davis remains an enigmatic figure in the history of geomorphology. As a system builder, he failed to convince thoroughly any but a handful of geographers and geologists that the pieces of his program cohered. A collection of his papers, edited by Douglas W. Johnson and published as *Geographical Essays* in 1909, appeared too early to include the more mature essays penned by Davis in the years 1911 to 1915. In the decades following his death in 1934, Davis' influence was eclipsed by the rise of quantitative geomorphology, though Davis' reputation was revised by geomorphologists beginning in the late 1960s.

Bibliography

CHORLEY, R. J., R. P. BECKINSALE, and A. J. DUNN. *The History of the Study of Landforms or the Development of Geomorphology*. Vol. 2, *The Life and Work of William Morris Davis*. London, 1973.
DAVIS, W. M. *Elementary Meteorology*. Boston, 1894.
——. *Physical Geography*. Boston, 1898.
——. *Elementary Physical Geography*. Boston, 1902.
——. *Geographical Essays*, ed. D. W. Johnson. New York, 1909. Repr. 1954.
——. *The Coral Reef Problem*. New York, 1928. Repr. 1969.

MARK L. HINELINE

DEBRIS IN EARTH'S ORBIT

Orbital debris, Weltraummüll, kosmichisky objecty: all names for an environmental problem with consequences for the continued safe, productive, and economical utilization of near-Earth space. While no legal definition of artificial orbital debris exists currently, one may consider it to include nonfunctioning spacecraft, rocket bodies, debris created by normal spacecraft operations (variously known as operational or launch debris, including launch debris), and fragmentation debris. Natural orbital debris consist of meteoroids and meteoritic and cometary debris.

Debris pose a hazard to spacecraft because the relative velocity between debris and their potential targets varies between less than a kilometer per second (km/s) to 16 km/s, with an average velocity in low Earth orbit of 10 km/s. Dust-to-centimeter-sized particles possess sufficiently large kinetic energies to cause surface damage erosion, similar to a "sandblasting" of the satellite surface, or even the catastrophic breakup of a satellite's structure. For example, an aluminum sphere of 1-cm diameter impacting a spacecraft at 10 km/s has the same kinetic energy as a 1,000-kg truck traveling at about 60 km per hour. An associated factor is the time debris spend intersecting the orbit of a prospective target. The longer the time, the more chances debris have to impact the target. This time varies between hours, in very low Earth orbit, to centuries in altitudes ranging from 500 and 2,000 km/s, the most heavily trafficked region of space. Lifetimes in geosynchronous Earth orbit, used by communication satellites, may be much longer. Natural cleansing mechanisms are provided by atmospheric drag; Earth, lunar, and solar gravity; and, for very small particles, the pressure of sunlight.

Historically, accidental or intentional explosions have accounted for the vast majority of fragmentation debris. By 1993 only one event out of 113—the fragmentation of the Soviet Union's *Cosmos 1275* in 1982—was generally accepted to have been caused by a random collision. Accidents arise primarily from design flaws. Residual fuels aboard rockets can corrode fuel tanks, leading to the mixing and explosive combustion of the fuels. Venting unused fuels on most U.S., European, and Japanese rockets is a simple way to prevent these explosions. Another source of accidental explosions has been the failure of the rocket motor during orbital maneuvers. Intentional explosions are usually associated with weapons testing or with a military spacecraft's failure that would prevent its recovery within certain geographic areas.

Operational debris are spawned during normal spacecraft deployment or operations and consist of shrouds, covers, lanyards, and other related items. Other debris originate in paint pigments liberated by solar and atmospheric erosion and the thermal expansion and contraction of the spacecraft's structure; manned spacecraft waste dumps; and particles from the firing of solid rocket motors.

Debris are detected by ground- and space-based sensors and the analysis of materials returned from space. Sensors include radars, visible and infrared telescopes, and impact detectors flown aboard spacecraft. Radar and telescopes excel at detecting large debris, while impact sensors sample the smaller objects. Impact sensors are unable to distinguish between artificial debris and natural micrometeoroids, though they are capable of measuring the mass and direction of the impactor. Telescopes and radar see micrometeoroids only when they re-enter the atmosphere and create the long ion trail of a "shooting star." Telescopes differentiate between these trails and re-entering debris by estimating the relative velocity: meteors may have velocities two to three times that of debris. Radars

offer the additional discriminators of range, velocity, and direction.

Data on impact rates, and hence the debris population, have been provided by materials exposed to the space environment, returned by astronauts, and analyzed in the laboratory. Scanning electron microscopy and chemical analysis have proven fundamental in identifying the texture and composition of sources for small debris, such as paint pigments or human waste, and for differentiating debris from natural micrometeoroid impacts on the basis of the chemical composition of residue within the impact crater.

Between the launch of *Sputnik 1* on 4 October 1957 and 1 January 1993, the U.S. Space Command's Space Surveillance Network (SSN) had cataloged 22,300 space objects, of which approximately 6,600 were in orbit as of the latter date. This population is composed of active payloads (6

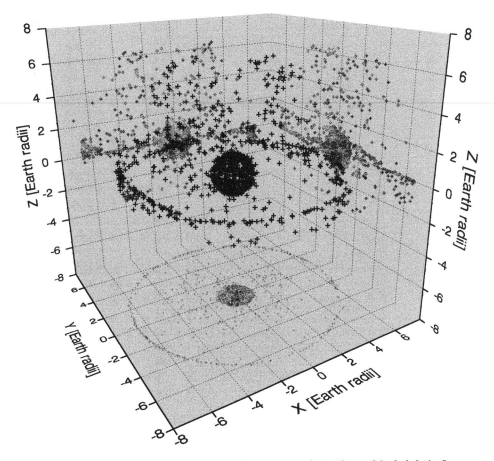

Figure 1. View of the cataloged environment of Earth's orbital debris from an altitude of 64,000 km.

percent), inactive payloads (23 percent), rocket bodies (16.5 percent), operational debris (12 percent), and fragmentation debris (42.5 percent). Figure 1 depicts the view of the cataloged environment as seen through the viewfinder of an imaginary 35-mm camera located over the United States at an altitude of 64,000 km. The relative congestion in low Earth orbit and the geosynchronous "rings" of communications satellites and other debris are readily apparent. Computer modeling indicates that while the cataloged objects account for about 99.93 percent of the 2 million kg of mass in orbit, they account for only 0.2 percent of the total number of objects, about 35 million. Debris between 1 mm and 1 cm in diameter are estimated to provide less than 0.03 percent of the mass but 99.67 percent of the number, with the remainder lying between 1 and 10 cm in diameter.

The primary source for debris tracking data is the SSN, which maintains orbital data on all space objects greater than approximately 10 cm in diameter through a worldwide network of radar and optical sensors. The National Aeronautics and Space Administration (NASA) has developed a suite of sensors to complement its analysis and modeling of the debris environment; these include a portable computer-driven Schmidt telescope, a 3-m-diameter liquid metal (mercury) mirror telescope, and the joint NASA/Air Force Haystack Auxiliary radar in Massachusetts. The former Soviet Union established a geographically limited counterpart to the SSN during the 1960s and tracks about two-thirds the number of objects as the SSN. Radars in Germany and Japan also observe debris.

Interest in the debris environment and the accompanying hazards presented to spacecraft has prompted research by NASA, the Department of Defense, academe, and the European, Japanese, and Russian space agencies. This research has concentrated on the characterization of the environment using sensor data, environmental forecasting using computer models, and the search for ways to prevent the creation of new debris or reduce existing debris.

Bibliography

KESSLER, D. J. "Earth Orbital Pollution." In *Beyond Spaceship Earth*, ed. E. C. Hargrove. San Francisco, 1986.

JOHNSON, N. L., and D. S. McKNIGHT. *Artificial Space Debris*. Malabar, FL, 1987.

PHILLIP ANZ-MEADOR

DESALINATION

As the earth's population has grown and people have achieved a higher standard of living, the amount of freshwater needed worldwide has increased dramatically. In many countries, needs have strained available freshwater supplies, and have restricted economic growth and development. Historically, seawater and brackish water (called saline water) were not considered suitable for human, agricultural, or industrial use due to high salt levels. However, over the last half century the process of desalination has been established as a reliable technique for removing salts from seawater and brackish water, thus making these water resources available. Although the concept of desalination has been around for almost two thousand years, only in the last forty years has man developed the technology for desalination water treatment facilities. Three major desalination technologies have been perfected for commercial use: distillation, reverse osmosis, and electrodialysis. Ion exchange and freezing processes can also be used for desalination, but neither has shown much potential for commercial application. In general, distillation works best for desalting seawater, whereas electrodialysis and reverse osmosis are more suitable for brackish water. A major consideration in the use of desalination is its high cost, often from two to fifty times more expensive than conventional water treatment. Because desalination is inherently energy-intensive, the increases in energy costs that occurred during the 1970s have partially offset most improvements made in equipment and operating efficiencies. However, despite the substantial costs involved, the availability of desalted water can be an economic boon in many water-short areas.

Saline Water

Saline water contains significant quantities of total dissolved solids (TDS), as well as small quantities of organic material and dissolved gases. Most of the TDS are inorganic salts, which by definition are compounds made up of positive metallic ions (calcium, sodium, ammonia) bonded to negative ions (chloride, sulfate, carbonate). These salts are quite soluble in water, and are derived from the erosion of soils on or beneath the earth's surface, and from volcanic activity, which releases hydrogen chloride, hydrogen sulfide, carbon dioxide, and so on from deep within the earth. After being dissolved in water, these materials are transported through rivers and streams to the ocean. In the oceans, the average level of TDS has been measured at 34,400 milligrams per liter (mg/l), a concentration close to 4 percent by weight. The oceans and seas make up 97 percent of the world's water supply, which is constantly recycled by evaporation and condensation to produce rain and snow.

Classification of Saline Water

For the design of desalination systems, water is divided into four classes according to salt concentration: freshwater, brackish water, seawater, and brine. Freshwater is defined as having a TDS below 1,000 mg/l, brackish water from 1,000 to 35,000 mg/l, and seawater 35,000 mg/l and above. Brine is the concentrated salt solution remaining after pure water has been removed by desalination.

The oceans and seas of the world are interconnected, and the composition of seawater is relatively consistent. However, in specific areas the composition of seawater can vary considerably, although the relative proportion of various chemical constituents to the chloride ion is fairly constant. Land masses, heating and cooling, ocean currents, winds, and tides can all affect the concentration of a saltwater body at any given location and time. Since desalting seawater is more expensive than desalting brackwish water, seawater is used for desalination only when sufficient brackish water is not available.

Brackish water can be found in lakes, rivers, and streams, and below the earth's surface in groundwater supplies. Brackish water usually occurs naturally due to several basic processes, including (1) freshwater flowing through or over mineral deposits and dissolving them; and (2) freshwater and seawater intermixing where they interface in bays, estuaries, and/or groundwater formations. Deposits of rock salt, gypsum, and limestone develop over time as brackish water evaporates, but then are gradually redissolved by rain to form new brackish water supplies. Human activities that produce brackish water include various domestic, industrial, and agricultural uses that generate wastewaters high in TDS. As successive use and reuse of water occurs, salinity continues to build up eventually making the water unusable.

Unlike naturally occurring saline water, domestic wastewater is not uniform in composition, and contains whatever people dispose of in sewers, from human wastes to highly potent cleaning agents and industrial chemicals. Treated domestic wastewater has been used as feedwater for desalination plants in California, most of which use the reverse osmosis process. Drainage from irrigated land is also a potential source for desalination. Agricultural drainage generally requires pretreatment to remove pesticides, herbicides, and chemicals that could adversely affect the desalination process.

Desalination Processes

The major desalination techniques in use today can be classified as either distillation or membrane processes, including reverse osmosis and electrodialysis. The choice of a process and its related costs are determined by the chemistry and physical condition of the feedwater, the production rate required, and the energy source available to power the desalination plant.

Distillation. Distillation is a simple process in which saline water is boiled, and the steam produced is collected and condensed to produce salt-free water. Salts are not vaporized and remain in the brine. For an efficient distillation process, some of the steam heat is transferred back to the salt water to improve thermal efficiency. The most widely used distillation process, multistage flash (MSF) distillation, is based on the principle that water will boil at lower temperatures when subjected to lower pressures. In the MSF process, heated seawater flows into a chamber where the pressure is just low enough to boil some of the water and flash off hot vapor. The vapor is then condensed on tubes that carry fresh, cool seawater into the system. The seawater not evaporated then moves into a second chamber, at a lower pressure, and the process repeats. A large MSF plant may have fifty or more flash chambers. Multiple effect distillation (MED) is a modification of the MSF process in which the largest portion of pure water is produced by boiling, and it uses the steam generated in one vessel as the heat source for the next. Vapor compression (VC) systems use heat from the compression of vapor, rather than the direct exchange of heat from boiler-produced steam. Solar distillation uses the Sun's energy to evaporate saline water in large surface-area basins called solar stills. The vapors condense on the inner surface of basin covers and the runoff is collected in troughs. Inexpensive land and intense sunshine are required to make this process economically feasible.

Reverse Osmosis. Reverse osmosis (RO) uses pressure to force pure water through a special membrane, leaving the dissolved salts behind on the feedwater side of the membrane. Energy is required primarily for pressurization, and no heating or liquid-to-vapor change is necessary. Operating pressures and product water quality vary with feedwater salt concentration and the type of membrane employed. Four types of membranes are generally used: plate and frame, tubular, hollow fine fiber, and spiral. Spiral and hollow fine-fiber membranes have completely dominated the industry to date, and large installations, over 1 million gallons per day (mgd), or 3,800 cubic meters per day (m^3/d), have successfully been used to treat brackish water for many years. Other types of membranes that operate at higher pressures have been developed for seawater desalination. In some applications, seawater RO systems may be cost competitive with distillation, but for very large facilities, over 5 mgd (19,000 m^3/d), distillation is more economical.

Electrodialysis. Electrodialysis (ED) is based on the fact that most dissolved minerals will dissociate into electrically charged ions. By means of two special membranes that allow the passage of only positive (cation) or only negative (anion) charged ions, and a direct current electric field, salts can be removed from saline feedwater. The two membranes are arranged in alternate layers with water passages between them, and electrodes at the top and

bottom to introduce the required electric current. After salt separation, alternate water layers contain either purified water or brine. Because of the electrical energy required, ED is normally used to purify only brackish, low salinity water (1,000–4,000 mg/l TDS). A process improvement called electrodialysis reversal (EDR) was introduced in 1972.

Besides solar stills, other renewable energy systems such as solar voltaic, wind, wave power, and geothermal have not been applied commercially for desalination. However, interest in these techniques has increased sharply in recent years, and several experimental projects are now underway.

Disposal of By-Products. Desalination processes produce two resultant streams. One is the product stream, which has a low level of TDS, and the other is the brine (or reject) stream, which has a higher level of dissolved salts than either the feedwater or the product water. The brine must be disposed of in an environmentally acceptable manner. For facilities located close to the ocean, disposal does not generally pose a problem, but it can be a serious problem for inland locations. The most common disposal methods include evaporation by solar (or conventional) thermal means, injection into saline groundwater zones, and pumping into a saline water source.

History of Desalination

Progress has been rapid in the field of desalination. In 1967, the worldwide installed capacity was only about 250 mgd (946,000 m^3/d), and distillation was the most widely used process for the desalination of both brackish water and seawater. ED had enjoyed some commercial success in brackish water desalination with several installations in North America and the Middle East, but RO was still in the developmental stage. By 1972, the worldwide installed capacity had doubled to about 500 mgd (1,890,000 m^3/d). Distillation still dominated both brackish water and seawater desalination, but ED and RO were beginning to make significant advances into brackish water applications. By 1980, the worldwide installed capacity had again doubled to over 1 billion gpd (3,780,000 m^3/d), and although distillation was still more prevalent, economic constraints were beginning to restrict its use almost exclusively to seawater. ED and RO were being used more for desalination of brackish waters due to lower costs. In the early 1990s, over 7,500 desalination units with a total capacity of 3.7 billion gpd (14 million m³/d) had been installed worldwide or were under contract. The main market for desalination used specifically to produce drinking water has been the water-short Middle East. In the United States and Europe, desalination has been selected primarily to treat industrial water sources. However, population expansion in Florida and drought conditions in California are creating interest in desalination to enhance potable water supplies there. The cost of desalinated water has largely eliminated its use for normal crop irrigation.

The U.S. Army Corps of Engineers developed considerable expertise in design and construction of desalinization plants during the 1970s when the United States provided construction support to Saudi Arabia. Corps desalinization projects included large treatment systems suitable for cities and towns, as well as small units required by military troops in the field. The large plants employ the MSF distillation process and have performed well over the years, even, for example, when threatened by the oil spill that occurred during Operation Desert Storm (1990). The smaller units were used by all three branches of the armed services during Desert Storm, and are based on the RO desalinization process. The larger Corps desalination plants were constructed in Saudi Arabia and include a 1.67 mgd MSF distillation unit at Jedda Naval Base, and a 2.0 mgd MSF distillation unit at Jabal Naval Base, both installed in the 1970s. Smaller systems include multiple 10,000 gpd RO units deployed at various locations in Saudi Arabia during Desert Storm (1990–1991), and several 2,000 gpd RO units installed at An Shas Air Force Base, Egypt, in the 1970s.

Bibliography

Buros, O. K. *U.S.A.I.D Desalination Manual.* Washington, DC, 1980.

———. *The Desalting ABCs.* Topsfield, MA, 1990.

Eubank, F. C. *Corps Accomplishments in Desalination.* Washington, DC, 1993.

Howe, E. D. *Fundamentals of Desalination.* New York, 1974.

Porteous, A., ed. *Desalination Technology.* New York, 1983.

Spiegler, K. S., ed. *Principles of Desalination,* 2nd ed. San Diego, CA, 1980.

FREDERICK C. EUBANK

DESERTIFICATION

See Land Degradation and Desertification

DIAGENESIS

Diagenesis is the sum of physical, inorganic chemical and biochemical changes in a sedimentary deposit after its initial accumulation, excluding metamorphism (*see* METAMORPHIC PROCESSES). The boundary between diagenesis and metamorphism is not well defined. A key diagenetic change is the conversion of sediments into sedimentary rocks (*see* SEDIMENTS AND SEDIMENTARY ROCKS, CHEMICAL AND ORGANIC; SEDIMENTS AND SEDIMENTARY ROCKS, TERRIGENOUS). It therefore is the link between sand and sandstone, mud and shale, carbonate sediment and limestone. Diagenesis involves, among other things: (1) compaction; (2) addition of new material; (3) removal of material, and transformation of material by (4) change of mineral phase; or (5) replacement of one mineral phase by another (*see* MINERALS, NONSILICATE, and MINERALS, SILICATES).

The addition of new material raises questions as to where the material came from and how it spread through the sediment. The removal of material creates new pore spaces and may provide the substance of cements for some other sites of precipitation, millimeters or kilometers away. Thus the addition and removal of material are closely allied to the contrasting processes of dissolution and precipitation. Removed material may be the source of the cement. For example, the degree to which calcite has cemented carbonate sediments commonly is related directly to the extent to which aragonite particles of this sediment have been dissolved. Common examples of replacement involve calcite and quartz. Quartz may replace calcite or calcite may replace quartz.

Diagenesis and fluids are intimately related. The fluids influence diagenesis through dissolution and precipitation, and diagenesis in turn influences the spaces that can be occupied by fluids.

Compaction

Sediments usually undergo at least some compaction. Reduction in pore space accompanies the reduction in volume of the sediment and the expulsion of its interstitial waters. Compaction affects all sediments, but changes are most pronounced in fine-grained sediments, such as very fine-grained (less than 2 μm) platelets that before compaction maintain a loose packing governed by surface electrostatic charges. The initial porosity of the open-packed, fine-grained, clay-rich sediments can be as much as 70 percent. Compaction of the fine-grained sediments depends on the rate at which water can be expelled. Finally, the mechanical readjustment of the platelets is such that they become preferentially oriented parallel to one another. However, loss of porosity with increasing depth, even in shales, appears to be in large part a chemical process involving precipitation of cements rather than the result of mechanical compression.

Carbonate sediments respond to burial more by a solution-precipitation mechanism than to physical readjustment or crushing of particles. Studies of ancient limestones show that most particles, including those in lime mud, have not been deformed. Because of its fine-grain size, we might expect that lime mud would show large-scale compaction effects similar to those of clays. However, lime muds do not behave like sediments composed of clay minerals. In the depositional environment, many lime muds become pelletized by the feeding action of organisms. Ancient lime muds only rarely show evidence that the pellets have been squashed. This observation is explained by early cementation, which transforms the lime muds into solid limestone before compaction.

The factors that influence compaction in sands are mostly shapes and sorting of particles and depth of burial. During compaction, sand particles respond by shifting into more dense packing arrangements, hence porosity decreases.

Pressure solution is a process in which a solid dissolves at its contact with another solid and with pore water because increased pressure has increased its solubility. This kind of dissolution may yield tightly sutured contacts between particles and thus weld the rock and reduce pore space. This process may occur during load or deformation compaction. Stylolites, consisting of the insoluble residue of material dissolved in limestones, may provide evidence for pressure solution (Figure 1); some limestones have been subjected to large-scale pressure solution. Pressure solution is also common in sandstones.

Carbonate Diagenesis

In modern carbonate sediments, the average porosity is in the range of 60 to 70 percent. By contrast, porosities of ancient limestones usually are less than 2 percent. In carbonate sediments, therefore, the volume of space may exceed that of sedimentary particles, whereas in limestones the porosity is vanishingly small. Although lime muds, surprisingly, are nearly as compatible as clay muds, compaction accounts for relatively little, in places almost no, reduction in pore volume. In the transformation of carbonate sediment into limestone,

Figure 1. Stylolites parallel to bedding of limestone seen in polished slab cut nearly perpendicular to bedding. Location not known.

the most important process is the introduction from the outside of calcium-carbonate cement into existing pores. This cement is generally derived by dissolution of calcium carbonate elsewhere and travels in solution to the site of precipitation.

Percolating freshwaters are usually slightly acid and hence take aragonite into solution with facility. The aragonite of a fossil may be dissolved completely, and an empty space retains the shape of the original fossil, forming a mold. The new pore space, whose shape is that of a previous aragonite fossil, is an example of moldic porosity. Moldic porosity in limestone commonly, but not always, is an indicator of the former existence of aragonite fossils and of their dissolution by freshwater. By contrast, fossils composed of calcite are less prone to dissolution. Hence in the fossil record, calcitic fossils are more likely to be preserved than those composed of aragonite.

The carbonate minerals composing limestones are very susceptible to being recrystallized. After these minerals have recrystallized it may be difficult or impossible to decide what kind of original carbonate sediment was ancestral to the limestone being studied. Because the former particles take on new shapes as crystals, recrystallization is commonly referred to by the name neomorphism (neo = new; morph = shape). We can define neomorphism as a transformation between one mineral and itself or a polymorph, which results in the growth of new crystals that are larger or simply different in shape from the previous particles or crystals. Generally, in neomorphism crystals of calcite replace original particles that were composed either of aragonite or of calcite.

Limestones may be replaced by dolomite, $CaMg(CO_3)_2$. Despite a voluminous literature, the origin of dolomite is a complex topic. Dolomite may form under a variety of conditions. It forms in modern regoliths, both in saline soils and in the zone of fluctuations of freshwater. In places dolomite composes 90 percent of the authigenic carbonates in soil, and it may be widespread in freshwater lacustrine deposits. Dolomite may extend as caliche across carbonate bedrock. Dolomite may result from biological activity; this activity involves bacterial processes. Dolomite may form even under deep-sea conditions or by saline fluids derived from the mantle. Yet despite the observation that dolomite may form under many kinds of depositional conditions, most dolomites in the rock record formed under conditions of hypersalinity.

Sandstone and Shale Diagenesis

Silica cement consists of several mineral phases, the most common of which is quartz. Typically, the quartz cement tends to grow outward from each particle as if particle and cement were a single crystal. In other words, the crystallographic arrangement of the quartz particle being cemented governs the crystallographic orientation of the quartz forming the cement. Such quartz cement added in crystallographic continuity with a quartz particle is called an authigenic overgrowth (Figure 2). Because of the crystallographic continuity between particle and quartz overgrowth, it may be difficult to find the boundary between them. In thin sections viewed under the petrographic microscope the boundary between particle and overgrowth can be recognized if solid or liquid inclusions (vacuoles), iron oxides, or clay minerals rim the particle as a dusty border.

The stage of early burial is commonly, but not necessarily, characterized by precipitation of an incomplete cement. Much void space remains among the particles. As burial progresses and pressure

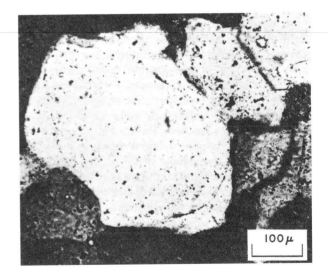

Figure 2. Photomicrograph of petrographic thin section cut from quartz-cemented sandstone. Large round quartz particle (outline marked by "dusty" border) contains authigenic overgrowth which grew in crystallographic continuity with the particle. Concavo-convex contact between two particles of quartz is at lower left. Simpson Sandstone (Lower or Middle Ordovician), Anadarko Basin, Oklahoma.

dissolution becomes effective, quartz particles begin to interpenetrate and form an interlocked granular framework whose particles display concavo-convex and sutured contacts (Figure 3). With further burial these contacts become deeply interpenetrated, even microstylolitic. Ultimately, particles become completely intergrown and welded. They form a fabric that is a precursor to the fabric of metamorphic quartzites. Such complete interpenetration results from the association of elevated temperatures and pressures, and the presence of interstitial waters of high ionic concentration that are found at considerable depth of burial or where structural deformation is in progress. Although such a sequence of progressive interpenetration leading to metamorphic quartzites is common, the quartz of some deeply buried sandstones and even quartzites does not display the diagnostic features resulting from pressure dissolution. These rocks may have been cemented early in their diagenetic history.

An inverse relationship exists between the amount of quartz cement and the abundance of detrital clay matrix. Evidently, the presence of clay minerals retards the growth of quartz cement. One likely explanation for this relationship is the low porosity and permeability of sandstones rich in clay matrix. In addition, the presence of clay-mineral crystals on the surfaces of quartz particles reduces the surface area of "clean" quartz on which quartz overgrowths can nucleate.

The configuration of the contacts between particles is related to (1) original packing, and (2) the amount of pressure dissolution. The number of contacts per particle and kind of contact—whether tangential, long or straight, concavo-convex or sutured—depends on pressure increase resulting from weight of overburden or from structural deformation (Figure 3); the effects of structural deformation are more pronounced than those of weight of overburden. Tangential contacts are the result of original packing. Long or straight contacts result from the interplay of three factors: (1) original packing; (2) pressures; and (3) precipitated cement. Concavo-convex and sutured contacts are generally the result of pressure. Strained quartz particles show many more concavo-convex and sutured contacts than do non-strained quartz particles. This behavior probably results from the greater solubility of strained crystals.

The major source of silica is thought to be that dissolved from dissolution of particles that interpenetrate, especially where particles have been "welded" to each other. In some sandstones, the amount of silica cement precipitated balances the amount dissolved during particle interpenetration. In such sandstones effective pressures and point contacts of quartz particles increase the solubility of the quartz. Thus, at these contacts, especially where clay films are present between the particles, silica dissolves. As silica is liberated, the pore waters become supersaturated with respect to silica, and quartz is reprecipitated as overgrowths (Figure 2). Increase of pH above 9.5 and increase of temperature raise the solubility of silica. In some freshwater aquifers, waters having pH above 9.5 have been measured. Such high pH values, as with increased temperatures, may lead to a condition of supersaturation with respect to silica in pore waters and thus to the ultimate precipitation of quartz. This happens if subsurface waters are able to transport solutions rich in silica from a site where silica is being dissolved to another site where it is precipitated.

An additional source for silica is mud and shale.

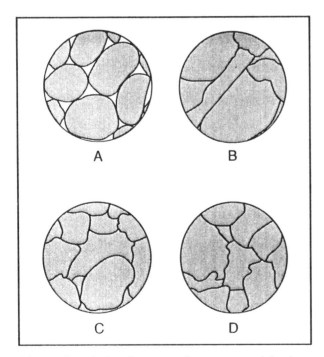

Figure 3. Kinds of contacts between particles in sandstones. A. tangential. B. long or straight. C. concavo-convex. D. sutured. (Adapted from J. M. Taylor, "Pore-Space Reduction in Sandstones," Plates 1 and 2.)

When these fine sediments are compacted they release silica-supersaturated waters. In sandstones the amount of silica cement has been found to increase toward a contact with a shale. Likewise, mineral transformations in which clay minerals, feldspars, and other silicates react with meteoric waters that trickle into the subsurface may provide dissolved silica that can be reprecipitated as quartz cement in sandstones. Another source is dissolution of volcanic glass.

The low level of silica concentration in the sea is maintained mostly because silica-secreting organisms, such as diatoms, sponges, and radiolaria, remove silica. Trapped marine waters are thus undersaturated in silica or at best in equilibrium with quartz. Yet the organisms that secrete silica tests provide a source of silica for cement. These organisms synthesize their tests from opal, an amorphous polymorph of silica with 8 to 10 percent water. Their tests become entrapped in the silica-undersaturated seawater. The silica tests are dissolved. Ultimately, the interstitial waters become supersaturated with respect to quartz, and precipitation of authigenic overgrowths may commence. Molds of tests, such as those of radiolaria, may testify to the former presence of these organisms.

Argillaceous muds, with an initial porosity of 70 to 90 percent, are compacted during shallow-burial diagenesis (depth down to 500 m) to mudstone (or shale if fissile) with a porosity of about 30 percent. The decreases in porosity and water content are rapid down to about 250–300 m burial depth and slow below that depth. Diagenesis passes into low-grade metamorphism when recrystallization of sheet silicates occurs at about 200°C and about 10,000 m burial depth. During burial the electrolyte concentration of pore fluid decreases and its composition changes. With increasing depth, pore water exhibits a strong depletion of Mg^{2+} and K^+ and a corresponding enrichment in Ca^{2+}. Dissolution of plagioclase and formation of Mg-rich smectites appear to be responsible for these changes. During deep-burial diagenesis smectite converts into illite, or into a mixed-layer illite-smectite, having a high proportion of illite layers with increasing burial depth. This process is accompanied by other minor changes, such as decomposition of coarser-grained mica and K-feldspar, decomposition of kaolinite, and formation of chlorite or chlorite interlayers. Kaolinite is eliminated during deep-burial diagenesis, whereas chlorite and/or corrensite is formed. The deepest stage of diagenesis is characterized by a uniform clay-mineral association "illite-chlorite."

Bibliography

FOLK, R. L. Some Aspects of Recrystallization in Ancient Limestones. In *Dolomitization and Limestone Diagenesis*, eds. L. C. Pray and R. C. Murray. Society of Economic Paleontologists and Mineralogists Special Publication 13. Tulsa, OK, 1965, pp. 14–48.

FRIEDMAN, G. M., J. E. SANDERS, and D. C. KOPASKA-MERKEL. *Principles of Sedimentary Deposits: Stratigraphy and Sedimentology*. New York, 1992.

FÜCHTBAUER, H. *Sedimente und Sedimentgesteine*. Stuttgart, 1988.

LARSEN, G., and G. V. CHILINGAR. *Diagenesis in Sediments and Sedimentary Rocks 2*. Amsterdam, 1983.

TAYLOR, J. M. "Pore-Space Reduction in Sandstones." *American Association of Petroleum Geologists Bulletin*, 34 (1950):701–716.

GERALD M. FRIEDMAN

DINOSAURS

Dinosaurs are a category of four-legged land vertebrates, the dominant large animals on land from their origin in the Late Triassic (Carnian), circa 235 Ga, until their extinction at the end of the Mesozoic, 65 Ga; they are generally classified as reptiles. There were hundreds of different forms. Some were the largest land animals of all time, probably exceeding 100 tons in weight, while others were no bigger than a chicken. They varied also in other ways and were often of bizarre appearance. Some were bipedal in stance and/or gait, some were quadrupedal; some were adorned with bony armor, knobs, crests, spikes, and horns, while others were simply covered in scaly skin; and they included not only carnivores (ferocious hunters, fish-eaters, scavengers, and probably cannibals) but also great numbers of wholly vegetarian forms (grazers, browsers, and fruit-eaters). Yet others were egg-stealers and omnivores. While it is likely that the carnivores were solitary or hunted in small packs, most of the herbivores probably stayed together in herds.

As far as is known, however, no dinosaurs lived in the sea (although some may have been semiaquatic in freshwater). None glided or flew (except that, if every group includes all its own descendants, as is required by cladistic classification theory, then birds, which originated from theropod dinosaurs, must themselves be regarded as both dinosaurs and theropods too). None lived a burrowing life.

Geological Distribution

Dinosaur remains, therefore, are generally found in continental rocks of Mesozoic age, usually in lowland sediments—especially those of swamp, lake bed, and riverine origin. They occur less often in upland and subaerial sediments, and even more rarely—when a carcass has floated out to sea—in marine deposits.

The composition of the dinosaur fauna varied with time, so that each stratigraphic stage, from the earliest dinosaur deposits in the Carnian (Upper Triassic) to the end of the Mesozoic, is characterized by a unique assemblage of species. This enables an expert to determine the age of a dinosaur bone deposit by identifying its component groups.

Geographical Distribution

The geographical distribution seems to have been cosmopolitan, throughout both supercontinents, Laurasia and Gondwanaland, until about the middle of the Cretaceous period. It seems that differing faunas subsequently evolved in different continents (as then existing), doubtless because of the continuing disintegration at that time of the global Mesozoic landmass.

Popular Misconceptions

Three popular misconceptions are: (1) that all dinosaurs were very large; (2) that "the dinosaur" is one particular species of animal (generally a sauropod); and (3) that any large extinct animal may be called a dinosaur, favorite candidates in this connection being the giant marine reptiles (plesiosaurs, ichthyosaurs, and mosasaurs), the flying reptiles (pterosaurs, including pterodactyls), and large mammals (especially mammoths). None of those is a dinosaur, although pterosaurs are fairly closely related. A fourth popular misconception is that dinosaurs were slow, clumsy, stupid, and gen-

erally inefficient, and hence were so uncompetitive that they soon became extinct (indeed, "dinosaur" is often used in this sense as a pejorative synonym for anything obsolescent). Yet the dinosaurs as a group survived for 170 Ma, were highly successful, and gave rise to the equally successful birds.

First Discoveries

The first record of the discovery of what we now know to be a dinosaur bone is from Oxfordshire, England (Plot, 1676), although there must have been many unrecorded discoveries before then. In 1822 Dr. Gideon Mantell and his wife found large teeth (and subsequently bones) in Sussex, England; in 1825 Mantell named the animal *Iguanodon* and was the first to recognize the hitherto unsuspected existence of "gigantic herbivorous reptiles" in Mesozoic rocks. However, another English dinosaur had been found earlier, before 1818, again in Oxfordshire, and this was also the first to be named (*Megalosaurus* Buckland, 1824); it later turned out to have been a carnivore. The word "Dinosauria" (Greek, "terrible lizards") was not coined until 1841, when Professor Richard Owen read a paper at the Plymouth meeting of the British Association for the Advancement of Science (published 1842); he specified *Megalosaurus, Iguanodon,* and *Hylaeosaurus* as constituting the group. Later nineteenth-century discoveries came predominantly from North America, eventually in great numbers from the region to the east of the Rocky Mountains (many publications by Edward Cope and by O. C. Marsh), and some were relatively small. Thus, Mantell's "gigantic herbivorous reptiles" were not all gigantic, not all herbivorous, and some modern authorities would not even classify them as reptiles.

General Position in the Classification

The dinosaurs belong to the Archosauria (the so-called ruling reptiles), one of the major divisions of the class. The only archosaurs living today are the crocodilians, although the birds, too, are archosaur descendants. The archosaur skull is of the diapsid type (i.e., it has two large "windows" behind the eye socket) and is typified by the presence of an additional "window," the antorbital fenestra, just in front of the eye socket; in many later forms, however, including many dinosaurs, the antorbital fenestra is secondarily closed. Dinosaurs and crocodilians alike evolved in Late Triassic times from an older group of reptiles called the thecodontians, the first and most primitive members of the Archosauria; a typical thecodontian was not unlike a modern crocodile in size, appearance, and skeletal structure, but it was thoroughly terrestrial in habit.

Diagnosis: "Fully Improved" Stance and Gait

Earlier reptiles (like more primitive reptiles today) rested with the belly on the ground and with the limbs protruding from the sides of the body ("sprawling" position); when walking or running they raised the trunk just clear of the ground and moved forward by throwing the whole body into horizontal serpentine waves and simultaneously "rowing" themselves along with their limbs. The dinosaurs, by contrast, had brought the limbs under the body, rotating the elbow backward and the knee forward so that both fore and hind limbs lay entirely in a vertical plane and supported the trunk well clear of the substrate. When walking or running, the entire limb moved backward and forward in that vertical plane, parallel to the direction of movement.

This "fully improved" limb position is the most striking characteristic of the dinosaurs, both when the animal was stationary ("stance") and when it moved ("gait"). Indeed, this has been arbitrarily chosen as the primary diagnostic of the Dinosauria, there being no other clear character that would serve to diagnose that category.

Major Groupings

Almost all dinosaurs fall neatly into one of three major groupings: (1) the Theropoda (bipedal carnivores); (2) the Sauropodomorpha (mainly gigantic quadrupedal herbivores, the Sauropoda, but including also the earlier, partly bipedal Prosauropoda); and (3) the Ornithischia (a wide variety of herbivorous forms, large and small, quadrupedal and bipedal, but firmly linked together by a number of unique anatomical characteristics).

The earliest dinosaur beds in the Carnian already contain both theropods and sauropodomorphs, but no reliable ornithischian material is known before the beginning of the Jurassic. All three major groupings continue to the very end of the Cretaceous.

Theropoda. The Theropoda, the bipedal carnivores, are conventionally classified into the small, lightly built Coelurosauria and the large, heavily built Carnosauria. The artificial nature of this distinction is now generally recognized, yet recent attempts to classify the theropods cladistically (such as Gauthier, 1986) are no more satisfactory. Subgroups that possess a distinct identity include the Ornithomimidae ("ostrich dinosaurs"), the fish-eating Baryonychidae, the egg-stealing Oviraptoridae, and, just before the end-Cretaceous extinction, the gigantic Tyrannosauridae with their fearful jaws, six-inch teeth, tiny fore limbs and huge predatory hind feet. It was the Theropoda that gave rise to the birds.

Sauropodomorpha. The main component of the Sauropodomorpha is the Sauropoda of the Jurassic and Cretaceous, the familiar "dinosaurs" of popular understanding—a vast elephant-like body supported on four columnar limbs, tapering forward into a long neck and relatively tiny head and backward into a long whip-like tail. Entirely herbivorous, the largest of them probably weighed over 100 tons. It used to be thought that their great size necessitated a semiaquatic existence in lakes, marshes, and swamps, where the water would help support their weight; but it is now realized that the evidence favors a wholly terrestrial existence, indeed life in the water would create more problems than it solved. The sauropods were preceded chronologically by the Prosauropoda of the Late Triassic and Early Jurassic, generally somewhat smaller and some partly bipedal, which used to be thought of—probably wrongly—as the sauropods' ancestors.

Ornithischia. The Ornithischia form a very coherent clade, characterized by several unique anatomical features—in particular a retroverted pubis which subsequently developed an anterior ramus to produce a bird-like, tetraradiate pelvis. The entire group was herbivorous.

There were no undoubted ornithischians before the Early Jurassic. One of the three main components is the wholly quadrupedal Thyreophora, which include the mainly Jurassic Stegosauria (plated dinosaurs) and the mainly Cretaceous ankylosaurs (armored dinosaurs). The second is the Ornithopoda; these include the partly bipedal ornithischians, large and small, such as the Iguanodontidae and the Heterodonosauridae, respectively, and also the great Hadrosauridae, or duck-billed dinosaurs, so abundant in the Late Cretaceous. (Some of the duckbills possessed bony crests on top of their skulls; the function of these crests remains uncertain although several suggestions have been made.) The third component is the so-called Marginocephalia of the Cretaceous, only tentatively associated with each other; it comprises the mainly bipedal Psittacosauridae ("parrot dinosaurs"), the wholly bipedal Pachycephalosauridae (dome-headed dinosaurs), and the quadrupedal Ceratopia (horned dinosaurs, often wrongly called Ceratopsia). The ceratopians resembled huge reptilian rhinoceroses, roaming the Late Cretaceous plains in vast herds just before the dinosaurs died out altogether. Surprisingly, it is not these "bird-hipped" dinosaurs that are believed to be ancestral to the birds.

Soft Parts and Trace Fossils

While most dinosaur fossils consist of petrified hard parts (bones and teeth), several other categories do exist. A very few specimens have mummified soft tissues; in others some of the internal structures have rotted entirely, and the consequent cavities have filled with mineral material closely replicating such organs as the brain, nerves, blood vessels, and inner ear. "Negative" impressions of the animal's exterior on the surrounding mud (its scaly skin, footprints, dragging tail) are sometimes found, as are the fossilized products of its activities such as feces (known as coprolites), eggs, and nests; unfortunately, it is often impossible to assign such extraneous fossils to a particular type of dinosaur. Also known are fossilized stomach contents such as partly digested bones, teeth, scales, twigs, leaves, and cones, together with the rounded stones (gastroliths) that some dinosaurs swallowed and used to help grind up the food in their stomachs.

Footprints and Trackways

Dinosaur footprints, sometimes forming continuous trackways, have been studied intensively. They, too, are often difficult to assign to a particular dinosaur, all the more so as bones and footprints are rarely preserved in the same deposits; consequently, footprint specialists have found it necessary to develop a parallel nomenclature for

them. The study of footprints (ichnology), especially en masse, has provided much information on such subjects as stance and gait, speed and length of stride, herding, and parental care.

Temperature Regulation

Until about 1970 dinosaurs were thought to have been ectothermic (cold-blooded) rather than endothermic (warm-blooded) simply because they were classified as reptiles and because all modern reptiles are ectotherms. John Ostrom (1969) and later Robert Bakker (1972) then developed the idea that dinosaurs might well have been endotherms; Bakker even went so far as to suggest that dinosaurs ought to be classified as a class of vertebrates on their own, separate from Reptilia. The alleged evidence for the supposed endothermy included the success of the earliest dinosaurs in competition with other groups of large terrestrial animals, especially the advanced mammal-like reptiles (Therapsida); their bird-like and mammal-like stance and gait, and likewise their bone histology; the low predators to prey ratio in the population, suggesting a high meat requirement by the former; and their having given rise to the undoubtedly endothermic birds (see below). Most of this "evidence," however, can be refuted, or the arguments based on it are illogical (Charig, 1976). The view most widely held at present is that some of the smaller, extremely active theropods might have been endothermic to some extent, but no others; the largest ones, especially the sauropods, were "inertial homoiotherms," able to maintain a fairly constant body temperature by virtue of their vast bulk and low surface to volume ratio.

If dinosaurs were indeed ectothermic, their presence in a deposit would give a good indication of a tolerably warm climate.

Reproduction: Eggs, Nests, and Parental Care

Sexual dimorphism appears to have been fairly common among the dinosaurs. Some dinosaurs may have been viviparous, for adult skeletons are known in which baby skeletons lie within the ribcage; this phenomenon, on the other hand, may be the result of cannibalism. Most dinosaurs, in any case, are presumed to have laid eggs. The eggs have been found in various parts of the world (notably Provence, the Gobi Desert, and Montana) and were generally laid in clutches with a regular arrangement; they are often in nests, at least partly buried, and some contain the bones of embryos. In Montana several different types of dinosaur eggs have been found; they vary considerably—in size, in the surface texture and ornamentation of the shell, in the number of eggs in the clutch, in their arrangement (varying from straight lines to spirals to concentric circles), in the development of the nest (varying from nonexistent to a shallow scraped-out hollow to a basin built of mud and lined with vegetation), in the probable method of incubation, in the degree of development of the young at hatching (varying from helpless to almost independent), and, as a corollary of the last, in the degree of parental care. Some of the hatchlings must have been fed by their parents for several weeks. The hadrosaur *Maiasaura* nested in regularly spaced colonies, returning to them year after year.

Relationship to Birds

The many resemblances between dinosaurs and birds were noted long ago (Huxley, 1868), but the suggestion that birds evolved from dinosaurs was soon scotched on the grounds that birds possess clavicles (conjoined as the furcula or "wishbone") while dinosaurs do not. However, the suggestion was resuscitated by Ostrom (1973), with a wealth of convincing evidence and with the clavicle difficulty overcome; the theory is now accepted by most scientists, though not all. Avian origins are believed to lie among small, lightly built, bipedal theropods; whether these creatures developed their flying abilities from the trees down or from the ground up is still a matter of controversy.

Dinosaur Extinction

It is generally believed that, at the end of the Cretaceous, the dinosaurs died out suddenly and simultaneously worldwide (birds, of course, excepted). Why they did so is the most often asked question in the whole of paleontology and has led to endless speculation; Charig (1993) lists some eighty-nine suggested causes of the end-Cretaceous extinctions as published. The explanation most widely held today by non-paleontologists is the meteoritic impact disaster theory (Alvarez et al., 1980) (*see* EXTINCTIONS). However, in view of the fact that many diverse groups of organisms died out at that time while many other similar groups seem to have been wholly unaffected, no suggested cause is sufficiently selective in its alleged effects upon living organisms to be considered the prime or only cause; disaster theories are especially untenable because, while there may well have been catastrophes of either terrestrial or extraterrestrial origin (or both) at the end of the Cretaceous, there is no real evidence to support the view that dinosaur extinction occurred everywhere, suddenly, at the Cretaceous-Tertiary boundary (Charig, 1989). Vertebrate paleontologists prefer to believe that the end-Cretaceous extinctions, including dinosaur extinctions, did not differ in kind from the other extinctions that have been taking place throughout the history of Life; they were not even unique in their magnitude. There is no special cause to be sought.

Bibliography

ALVAREZ, L. W.; W. ALVAREZ; F. ASARO; and H. V. MICHEL. "Extraterrestrial Cause for the Cretaceous-Tertiary Extinction." *Science* 208 (1980):1095–1108.

BAKKER, R. T. "Anatomical and Ecological Evidence of Endothermy in Dinosaurs." *Nature* 238 (1972):81–85.

BUCKLAND, W. "Notice on the *Megalosaurus* or Great Fossil Lizard of Stonefield." *Transactions of the Geological Society of London* (2), no. 1 (1824):390–396.

CHARIG, A. J. "'Dinosaur Monophyly and a New Class of Vertebrates': A Critical Review." In *Morphology and Biology of Reptiles*, eds. A. d'A. Bellairs and C. B. Cox. London, 1976.

———. *A New Look at the Dinosaurs*. London and New York, 1979.

———. "The Cretaceous-Tertiary Boundary and the Last of the Dinosaurs." In *Evolution and Extinction*, eds. W. G. Chaloner and A. Hallam. London, 1989.

———. "Disaster Theories of Dinosaur Extinction." *Modern Geology* 18 (1993):299–318.

COLBERT, E. H. *Dinosaurs, Their Discovery and Their World*. New York, 1961.

GAUTHIER, J. A. "Saurischian Monophyly and the Origin of Birds." In *The Origin of Birds and the Evolution of Flight*, ed. K. Padian. San Francisco, 1986.

GILLETTE, D. D., and M. G. LOCKLEY, eds. *Dinosaur Tracks and Traces*. Cambridge, Eng., and New York, 1989.

HARLAND, W. B.; R. L. ARMSTRONG; A. B. COX; L. E. CRAIG; A. G. SMITH; and D. G. SMITH. *A Geologic Time Scale 1989*. Cambridge, Eng., 1990.

HUXLEY, T. H. "On the Animals Which Are Most Nearly Intermediate Between Birds and Reptiles." *Annals and Magazine of Natural History* 4 (1868):66–75.

NORMAN, D. B. *The Illustrated Encyclopedia of Dinosaurs.* London, 1985.

OSTROM, J. H. "Terrestrial Vertebrates as Indicators of Mesozoic Climates." *Proceedings of the North American Palaeontological Convention* (D) (1969):347–376.

———. "The Ancestry of Birds." *Nature* 242 (1973):136.

PLOT, R. *The Natural History of Oxfordshire, Being an Essay Towards the Natural History of England.* Oxford, 1676.

THOMAS, R. D. K., and E. C. OLSON, eds. *A Cold Look at the Warm-blooded Dinosaurs.* Boulder, CO, 1980.

THULBORN, R. A. *Dinosaur Tracks.* London, New York, 1990.

WEISHAMPEL, D. B.; P. DODSON; and H. OSMOLSKA, eds. *The Dinosauria.* Berkeley, Los Angeles, and Oxford, Eng., 1990.

ALAN J. CHARIG

DRILLING FOR SCIENTIFIC RESEARCH

Throughout history, drilling devices have probed Earth's crust to answer questions regarding unknown environments beneath the surface. Principal targets of most drilling ventures are water supplies, mineral deposits, and energy resources essential for human survival. Since the majority of drilling projects are funded by private capital seeking economic returns, use of a drill as a scientific research tool is not common practice. In order to obtain subsurface rock, fluid, and gas samples necessary to understand the physical and chemical composition of Earth, however, scientific drill holes are required.

Questions regarding Earth's interior have been raised since recorded history. Evidence of human curiosity about what lies beneath Earth's surface is found in biblical references to earthquakes and other crustal phenomena; Greek scholars' descriptions of Earth's underground regions; Babylonian writings envisioning a flat Earth with celestial bodies hanging above; and ancient Hindu, Muslim, and Buddhist writings referring to the center, or "navel of the earth," with awe. Theories about the composition of Earth's interior have abounded.

In 1692, the astronomer Edmund Halley concluded that Earth was composed of hollow spheres filled with luminous gas, and leakage of this gas produced aurora borealis/northern light displays. Captain John Cleves Symmes expanded the hollow-Earth concept by proposing that openings to the center of Earth existed at the north and south poles. Symmes's 1882 request for funds from the U.S. Congress to verify his theory lacked sufficient support.

Earth's subsurface has inspired fictional classics, including Jules Verne's *Journey to the Center of the Earth* (1864), Sir Arthur Conan Doyle's *The Lost World* (1911), and Edgar Rice Burrough's *Tarzan at the Earth's Core* (1929), as well as movies such as *Crack in the World* and *Earthquake.*

Growing interest in Earth's composition has been stimulated by scientists confronted with enigmas that can only be resolved by physical probes of Earth's crust. In 1881, CHARLES DARWIN expressed need for a drill hole to produce "cores for examination from a depth of 500 to 600 feet" from the Pacific and Indian oceans. It was not until 1897 that Darwin's request was acted upon when a 348-m-deep hole was drilled in the Ellice Islands' atolls. Results of the drilling were inconclusive but prompted the drillings of additional holes in the region. Scientific drilling was endorsed in 1902 by GROVE KARL GILBERT, director of the Carnegie Institution, who proposed "a deep boring into plutonic rock" to investigate subsurface temperatures. Gilbert recommended that a 1,830-m hole be drilled at a site in Georgia underlain by granitic rocks, but the cost was considered excessive and questions were also raised about the technical feasibility of the project, as the deepest hold drilled in the world at that time was 854 m.

In 1922, Dr. Thomas A. Jaggar, founder of the Hawaiian Volcano Observatory, directed the drilling of four holes into the floor of the Kilauea volcano on the island of Hawaii. Sensitive instruments lowered into the holes recorded rock temperatures at depths of 7 to 24 m. Jaggar's pioneer efforts in scientific drilling initiated a series of subsurface penetrations of active volcanos in the Hawaiian Islands that provided critical data regarding formation of the islands and the nature of ocean crusts. Jaggar also advocated the establishment of "ocean observatories" composed of groups of large ships linked together from which drilling into ocean bottoms would be conducted, but the concept was given little attention. It was not until the early

1960s that a long-term scientific ocean drilling program was finally implemented.

Ocean Drilling Program

Scientific research drilling is classified by two general types: ocean and continental. Ocean drilling consists of drilling through ocean waters from ships and platforms and drilling on island bases.

During the period 1934–1936, Japanese scientists drilled to a depth of 432 m on an island south of Japan in an attempt to further define the relation of coral growths to underlying rock bases. U.S. drilling for similar scientific goals at the Bikini Island atoll in 1947 failed to achieve the objectives within the 780 m penetrated by the deepest hole. Sporadic but successful ocean drilling efforts in the 1950s stimulated U.S. geoscientists to design the world's first major ocean drilling project. Since the crust is estimated to be 32 to 48 km thick under most of the continents, and may be as thin as 4 km in some ocean bottoms, it was concluded that a drill hole into the ocean floor was the most practical method of sampling mantle rocks. The National Science Foundation (NSF) agreed to finance a feasibility study to lower a drill pipe through several thousand feet of water to drill into the ocean floor at a site north of the Hawaiian Islands. The ultimate target was the crust/mantle boundary known as the Mohorovicic discontinuity, or Moho, and the proposed drilling program was referred to as Project Mohole.

For three years, discussions and debates were held regarding the value of the technically complicated and expensive Project Mohole. In 1960, funds were provided by NSF. Some technical challenges were resolved by drawing upon the expertise of oil and gas industry engineers. For example, stabilizing the drill ship on the ocean surface by use of a dynamic positioning system, designing a string of drill rods that would flex with vertical movement of the drill ship, and reentering a drill hole on the ocean bottom after replacing a worn drill bit, were among the complications addressed. The ocean floor was successfully penetrated by a scientific drill hole through water depths of over 3,355 m in early 1961.

Proponents of Project Mohole proceeded with plans for construction of a larger vessel capable of more extensive drilling to the Moho disconformity. As technical details were refined, however, estimates of project costs increased five-fold. With growing opposition to Project Mohole from both within the scientific community and by members of Congress, proponents backed off from proposing a full-scale drilling effort to penetrate the Moho. A compromise proposal under which more modest drilling would be undertaken over a three-year period was submitted to the NSF for support. As a result, a smaller vessel capable of drilling shallow holes under a much-reduced budget was approved. While the scaled-down program was being negotiated, a consortium of U.S. universities formed the Joint Oceanographic Institutions for Deep Earth Sampling (JOIDES) and successfully drilled fourteen holes into Atlantic Ocean floor sediments in early 1965. When a consolidated Deep Sea Drilling Project (DSDP) was funded by the NSF in 1966, JOIDES was selected to provide scientific oversight for the program.

Experience gained from Project Mohole and JOIDES drilling was drawn upon in the design of a new drilling vessel, the *Glomar Challenger,* launched in 1968. This customized drill ship was used by JOIDES associates for ocean drilling until it was replaced by the *JOIDES Resolution* in 1984 (Figure 1). During its sixteen years of service, the *Glomar Challenger* essentially confirmed the theory of sea-floor spreading, and produced over 96 km of drill core. The Federal Republic of Germany, France, Japan, the United Kingdom, and the USSR became participants in the ocean drilling program between 1974 and 1976.

Since its formal inception in 1964, the former DSDP, renamed the Ocean Drilling Program (ODP), expanded to include twenty countries that provided funding totaling over $35 million in 1993. The ODP has produced important data related to:

- Character and composition of deep portions of the oceanic crust
- Nature of normal-faulted continental margins versus those characterized by excessive volcanism
- Structure and volcanic history of island arcs
- Ocean circulation, paleoclimate, and hydrocarbon characteristics associated with black shale deposits
- Patterns of ocean circulation response to changing ocean boundaries
- Evolutionary change in marine organisms

The first eight years of *JOIDES Resolution* operation supported forty-eight internationally staffed

Figure 1. *JOIDES Resolution*—Ocean Drilling Program 10-story drill ship/oceanographic laboratory. (Ocean Drilling Program—Texas A & M University, 1990.)

expeditions in the Pacific, Atlantic, and Indian oceans, during which 696 holes were drilled at 281 sites. The drilling produced over 77 km of drill cores from which over 4,000 samples had been distributed to scientists throughout the world by August 1993. The core samples and scientific instrument measurements within the drill hole provided databases for many of the approximately 4,500 scientific/technical papers that have been published on topics related to ocean drilling and its implications. In November 1991, the deepest hole in the ocean crust (2,000 m) was drilled from the ten-story *JOIDES Resolution* floating laboratory.

Ocean Drilling Program participants from the academic, government, and private sectors throughout the world continue to resolve remaining scientific drilling technical challenges, such as:

- Drilling and logging technologies capable of functioning at high temperatures and in highly corrosive environments
- Vessels equipped to drill in polar latitudes
- Ability to drill in shallow-water atolls
- Fluid samplers, packers, and other tools allowing measurement of rock and fluid properties in situ within drill holes

Continental Scientific Drilling

The concept of continental drift, from which plate tectonics theories evolved, was not given serious credence by the geoscientific community until the 1960s. The plate tectonics hypothesis states that the extreme outer shell of Earth's upper crust is composed of a series of large plates constantly moving and interacting at their boundaries. This concept of continent movement has now been generally accepted and scientific drilling confirmed its validity.

Ocean floors offer access to 5 percent of Earth's crust that represents the most recent 200 million years of geologic history. Since rock formations representing the older 95 percent of Earth history can only be probed from continent-based drill sites, geoscientists recognized the need for continental scientific drill holes. An upper mantle project involving deep drilling on the continents was proposed at the International Union of Geodesy and Geophysics in 1960. Successful drilling of exploratory and production wells to depths of over 9,455 m by the oil and gas industry encouraged the earth science community to proceed with preparation of drilling proposals. Workshops held by the National Academy of Sciences in the 1970s identified subsurface features requiring drill hole access, including:

• Fault and earthquake mechanisms
• Hydrothermal systems and active magma chambers
• Heat flow and thermal structures within the crust
• Ambient stress within the North American plate
• Structure and evolution of the crystalline continental crust

As a dedicated continental scientific drilling program in the United States became a reality, specific drill sites to examine such environments were proposed by geoscientific consortia.

In the late 1970s, the Department of Energy (DOE) initiated a drilling program to examine thermal regimes, focusing on regions containing major crustal spreading centers and volcanic calderas. These included the Salton Sea Trough of California, Valles Caldera in New Mexico, and the Long Valley region of California. A DOE hole drilled in the Salton Sea Trough in 1985–1986 attained a depth of 3,222 m in a brine-rich geothermal environment. During the drilling, downhole measurements determined differential fluid flow rates and compositions, gravity and temperature variations, vertical seismic geophysical variants, and the geothermal potential of the site area. DOE drilling in the Valles Caldera volcanic structure involved downhole hydraulic fracturing to produce cracks between drill holes penetrating hot rock environments. Fracturing allowed water to be circulated through drill holes to extract heat from the host rock and be pumped to the surface.

The Continental Scientific Drilling Committee organized in 1980 by the National Academy of Sciences prepared a series of reports prioritizing continental scientific drilling objectives. The Committee's advisory panels recommended "that a Continental Scientific Drilling Program be initiated to achieve expanded knowledge and understanding of the uppermost part of the crust of the earth." A dedicated drilling program was formally initiated in the United States in 1985, when significant funding was provided by the NSF. A consortium of universities organized a private company to manage the government-financed effort, and proposals for drilling sites were solicited from the scientific community. The most productive of the resulting drilling projects was located at Cajon Pass in southern California, where the deepest scientific hole drilled on the North American continent attained a depth of 3,512 m. This drilling project, begun in 1986 and stopped in 1988, was located within two miles of the San Andreas fault zone, which represents the contact of the massive Pacific and North American continental plates. Earthquakes of various magnitudes have occurred along the San Andreas fault zone throughout modern history. Major earthquakes strike this region on an average of about once every 140 years, and less destructive movement occurs along the fault zone more frequently and erratically. The Cajon Pass scientific drilling project acquired subsurface data regarding the nature of the fault, including pressure and temperature relationships, physical components of the host rock environment, and other critical information that could contribute to more reliable earthquake predictions. The project, which used a large oil-field rotary drill adapted to acquire core samples under relatively sterile conditions, was a scientific first in the study of the San Andreas fault zone at variable depths. Drill hole data permitted the integration of downhole geophysical and geochemical measurements with analyses of retrieved core and cutting samples. Interpretations of geophysical seismic reflectors in continental basement rocks were made by direct sampling and downhole logging. The Cajon Pass drilling resulted in a better understanding of the geologic and geophysical setting of this earthquake-prone region of southern California.

Proposed drill sites in the United States include locations in the midcontinent region, Valley of 10,000 Smokes in Alaska, southern Colorado, Texas Gulf coastal region, Hawaiian Islands, and the Appalachian Mountain region.

The most mature continental scientific drilling program has been in the former Soviet Union, where drilling of a superdeep hole on the Kola Peninsula in northwest Russia began in May 1970. By January 1993, this hole had attained a world's record depth of over 12,261 m. Besides important scientific data obtained from the Kola hole, innovative drilling and downhole measurement techniques developed to penetrate over 12 km of Earth's crust have been applied at other former USSR drilling sites. These technologies include the use of heat-resistant, high-speed turbodrills to eliminate the need to rotate the entire drill rod string; bit changes without pulling the string of rods; special alloy lightweight/high-strength drill pipe; automated pipe handling capable of pulling eight miles of drill rods, changing a bit, and returning to the bottom of the hole in sixteen hours; and downhole instruments capable of withstanding high temperatures and extreme pressures. Each of the initial twelve USSR scientific drilling projects had both basic and applied scientific objectives because the government investment in the program was expected to reap economic benefits as well as contributions to the geosciences. Besides its super-deep drilling efforts, Russia has implemented other scientific research drilling projects, including drilling in the bottom and on the shores of Lake Baikal in southeastern Siberia. This multinational project is designed to acquire paleoclimatic data from the largest, deepest, and one of the oldest lakes in the world.

Another ambitious scientific drilling project was initiated in September 1987 in northeast Bavaria (Germany) (Figure 2). Cores recovered from approximately 90 percent of the first hole drilled at this site were studied by geoscientists throughout the world. The drill hole provided new information regarding the subsurface geologic/geophysical setting of this portion of Central Europe, including discovery of steep rock unit foliations, porous zones in basement rocks containing brine solutions, and higher thermal gradients than anticipated by project scientists. The initial drill hole was stopped at a depth of 4,003 m in April 1989, and a second hole was collared nearby in October 1990. High downhole temperatures required modifications in drilling fluids and downhole instruments for use in the second hole as the drill attempted to achieve a minimal depth of 10,000 m.

Rock instability, steeply dipping formations, enlargement of the drill hole ("breakouts"), and discing of the drill core that complicated drilling of the first hole were considered in design of the deeper hole. Over 300 geoscientists and engineers from eleven countries participate in Germany's scientific drilling project. The German government announced intentions to make the deep drilling rig available for scientific research drilling elsewhere in the world after completion of its second hole.

Besides the long-term commitments and major investments represented by the Russian, German, and U.S. continental scientific drilling programs, other countries have been involved in drilling projects with scientific goals, including the following:

- Belgium has conducted scientific drilling to acquire subsurface data for its national geologic mapping program, to inventory mineral resources, and to identify natural gas storage sites.
- In 1963, the Canadian government core drilled the Muskox intrusion in the Northwest Territories as a scientific drilling project. Extensive seismic reflection geophysical surveys have been conducted across Canada, and scientific drill holes are proposed to depths of up to 4,500 m along the geophysical lines to provide ground-truth for interpretive data.
- France began a scientific drilling program in 1982 and has drilled a series of progressively deeper holes with both scientific and economic goals. Downhole measurements are conducted in each drill hole, and core samples are shared with geoscientists from other nations. Some of the holes drilled in France have been cored to depths of over 3,500 m.
- Scientific drilling in Great Britain, begun in 1948, has focused on obtaining data for the selection of solid low- to intermediate-level radioactive waste disposal sites, the national geological mapping program, inventorying mineral resource reserves, and identifying alternative energy sources such as hot dry rock environments.

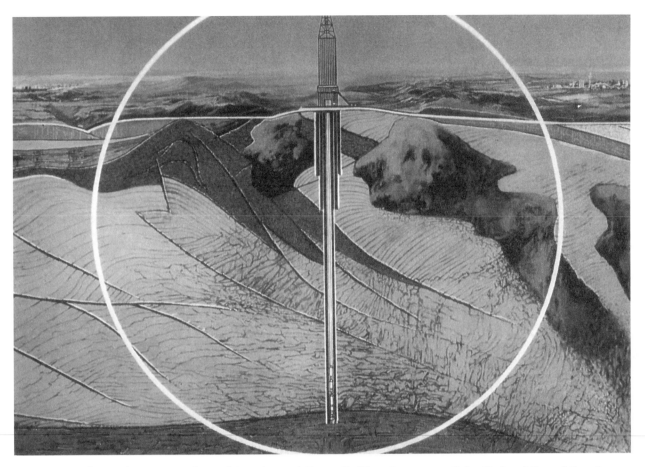

Figure 2. Schematic cross section of Continental Deep Drilling Program of the Federal Republic of Germany ("KTB"). Bavaria, September 1988. (German Continental Deep Drilling Program.)

- Iceland has drilled over one thousand wells to determine the extent of its geothermal energy supply. Prior to the Iceland Research Drilling Project of 1977–1978, 120 wells deeper than 1,000 m were drilled for geothermal research purposes. These drill holes provided important downhole temperature and fluid flow data.
- A series of sites has been mapped and detailed scientific drilling plans prepared to examine Japan's island arc system. The Japanese Deep Earth Gas Project is expected also to provide valuable scientific data from its proposed drill holes.
- The New Zealand government has sponsored shallow scientific drilling while evaluating its geothermal energy resources. A proposal has been made to drill holes up to depths of 600 m in the margins and crater of an active volcano to investigate magmas, brines, and vapors.

- In 1980, the Swedish government decided to abandon its twelve nuclear power reactors by the year 2010 as a demonstration of its opposition to the use of nuclear fuels. As part of its subsequent search for other energy sources, the government and private citizens financed the drilling of a hole seeking abiogenic (nonfossil) gas. Methane concentrations derived from Earth's mantle were expected to have migrated upward into the fractured crust. The resulting hole, which achieved a vertical depth of 6,770 m, was drilled in the Siljan impact structure. Although the economic goal of the project was not achieved, scientific data from the hole aided in interpreting the structure and composition of the meteorite crater.
- Switzerland has drilled holes up to 2,500 m to investigate potential sites for radioactive waste disposal.

Summary

Drilling for scientific research purposes may be complicated, time-consuming, and costly, and may not achieve the specific scientific objectives sought by project participants. Serendipity can play a role in a project, however, and unexpected results have sometimes proven to be more valuable than the original scientific goals.

Besides important scientific data produced by research drilling, a number of technical advancements have been made in the course of drilling that have industry applications:

- Sophisticated downhole geophysical logging tools and improved interpretation of measurements, particularly in igneous and metamorphic rocks
- Improved core and rotary drill bits to penetrate hard and fractured rock formations
- Motor-driven wireline-retrievable core barrels using fluids to rotate the barrel
- High-temperature fluid samplers and drilling muds
- Instruments to measure downhole high-pressure stresses
- On-site computer processing of downhole data on shipboard and at surface drill sites

The significance of scientific research drilling results extends beyond the scientific and technical communities and into our daily lives, providing benefits such as:

- Better understanding the causes and potential sites of natural catastrophes such as earthquakes
- Identification of suitable nuclear and toxic waste disposal sites
- Data on the formation of energy resources including petroleum, natural gas, coal, and uranium
- Alternative energy sources, such as geothermal and hot dry rocks
- Paleoclimatic data for predicting climate patterns
- Access to mineral genesis environments to interpret how and where mineral deposits form

Drilling for scientific research has been carried out intermittently and to a limited extent compared with other major scientific programs throughout the world. However, as questions continue to be raised regarding the nature of the planet we live on and environments beneath its surface, scientific drilling programs will become more ambitious. With this expansion will come challenges to scientists and to technicians to accomplish project objectives.

Bibliography

AMSOC COMMITTEE. *Drilling Thru the Earth's Crust.* National Academy of Sciences—National Research Council Publication 717, 1959.

ANDREWS, R. S., and T. E. PYLE. "A Survey of Recent Technology Developments in International Continental Scientific Drilling Programs." *Scientific Drilling* 1 (1990):310–323.

CONTINENTAL SCIENTIFIC DRILLING COMMITTEE. *Priorities for a National Program of Continental Drilling for Scientific Purposes.* Board of Earth Sciences, National Research Council. Washington, DC, 1984.

CROMIE, W. J. *Why the Mohole—Adventures in Inner Space.* New York, 1964.

FUCHS, K.; Y. A. KOZLOVSKY; A. I. KRIVTSOV; and M. D. ZOBACK, eds. *Super-Deep Continental Drilling and Deep Geophysical Sounding.* New York, 1990.

GEOLOGICAL SURVEY OF CANADA. Paper 66–13. *Drilling for Scientific Purposes.* Ottawa, Canada, 1966.

KOZLOVSKY, Y. A., ed. *The Superdeep Well of the Kola Peninsula.* New York, 1984.

MATTHEWS, S. W. "Scientists Drill Sea to Pierce Earth's Crust." *National Geographic* (November 1961):686–697.

G. ARTHUR BARBER

DU TOIT, ALEXANDER

The distinguished American geologist REGINALD A. DALY of Harvard University once called Alexander Logie du Toit "the world's greatest field geologist." In the nation of South Africa, du Toit continues to be revered for the extraordinary extent and quality of his geological mapping. Du Toit's work forms the basis of much of the geological understanding of the southern third of Africa. In the world beyond Africa du Toit is best known for his eloquent advocacy of the continental drift hypothesis proposed by the German scientist ALFRED WEGENER a hypothesis eventually proved correct by plate tectonics but for many years rejected by the geological establishment of the northern hemisphere, especially in the United States.

The name du Toit is an old and distinguished one in South Africa. The family originally lived near Lille in France, but fled to Holland because of religious persecution in the seventeenth century. From Holland two du Toit brothers went to South Africa in 1687 as Huguenot settlers. Only one of the brothers, François, had male descendants and from him all South African du Toits are descended.

Alexander Logie du Toit was born on 14 March 1878, at Rondebosch, near Cape Town. He was educated in Cape Town, graduating with a bachelor's degree from the University of the Cape of Good Hope (today the University of Cape Town). From South Africa he went to Glasgow, Scotland, where he studied mining engineering, and then, in 1900 and 1901, to London, where he studied geology at the Royal College of Science. After a brief period as a lecturer in geology at the University of Glasgow, du Toit returned to Cape Town in 1903 and was appointed Assistant Geologist to the Geological Commission of the Cape of Good Hope. He was employed by this commission and by its successor, the Union Geological Survey, which was formed in 1912 following the formation of the Union of South Africa, and served until 1921, when he was made Chief Geologist to the Department of Irrigation. During the years 1903 to 1921 du Toit carried out the field mapping for which he has become famous. The results of his work have been analyzed in detail by T. W. Gevers (1950) in his presentation of the first Alexander du Toit Memorial Lecture. Du Toit was particularly intrigued by the rocks of the Karroo basin, an assemblage of sedimentary and igneous rocks that are widely distributed throughout southern Africa. The Karroo System, or its equivalent, is also known in India and the other continents of the Southern Hemisphere. Du Toit eventually became Africa's, and then the world's, greatest authority on the Karroo System. This work finally led him to embrace the continental drift hypothesis, which brought him world fame.

Du Toit first studied the Karroo System in the Cape Province, where the base of the system is the Dwyka Tillite, a Carboniferous-aged sediment of glacial origin. Beneath the tillite are glacially polished pavements. Du Toit saw this remarkable evidence for ancient continental glaciation during his first field season in 1903. For the rest of his career du Toit continued to gather and synthesize the evidence for Carboniferous glaciation in southern Africa, India, South America, and Australia. In 1921, in a paper in the *South African Journal of Science*, du Toit first acknowledged his belief in the theory of continental drift. From that time on he spared no effort gathering and extending evidence to support the concept that a larger continent located near the South Pole, called Gondwanaland, had broken up near the end of Karroo deposition and that the pieces had drifted to their present positions. In 1927 du Toit published an influential volume on the geological similarities between Africa and South America, and in 1937 he published his best-known work, *Our Wandering Continents*. Du Toit dedicated *Our Wandering Continents* to the memory of Alfred Wegener and in it he marshaled the evidence in favor of continental drift. Together with Wegener's own work on continental drift, *Our Wandering Continents* is the most eloquent and closely argued work in support of continental drift prior to the revolution in geology created by the plate tectonics paradigm.

Alexander du Toit was a quiet, low-key man of slight build. He was modest, abstemious, and exceptionally energetic. Legends of his extraordinary zeal and dedication for work abound. From 1927 onward, when du Toit resigned from government service, he consulted for mining companies, particularly the De Beers Company, and he traveled widely. In this same period du Toit wrote and published a popular textbook, *The Geology of South Africa* (1926). Alexander du Toit died on 25 February 1948 in Cape Town.

Bibliography

DU TOIT, A. L. "Land Connections Between the Other Continents and South Africa in the Past." *South African Journal of Science* 18 (1921):120–140.

———. *A Geological Comparison of South America and South Africa.* Carnegie Institution Publication no. 381. Washington, DC, 1927.

———. *The Geology of South Africa.* Edinburgh, 1926; 2nd ed., 1939.

GEVENS, T. W. "The Life and Work of Dr. Alexander L. du Toit." *Proceedings of the Geological Society of South Africa* 52 (1949):annexure 1–109.

HAUGHTON, S. H. "Memorial to Alexander Logie du Toit." *Proceedings of the Geological Society of America*, (1950):141–149.

BRIAN J. SKINNER

E

EARTH, COMPOSITION OF

The bulk composition of Earth is an interesting datum to attempt to determine. From the study of chondritic metorites, we know that chemical fractionations occurred during formation of the solid grains from the solar nebula. Did these same fractionations affect the matter that went into building the planets? If not, why not? Knowledge of the bulk composition of Earth (and the other planets) will help us understand the earliest history of the solar system. Because the earth has differentiated, we cannot find a sample that is representative of the bulk Earth. The large-scale structure of Earth (core, mantle, crust, hydrosphere, and atmosphere) show that differentiation caused extreme fractionations of the elements within Earth. The purpose of this entry is to discuss how estimates of the bulk composition of Earth are made and what the gross chemical structure of Earth is.

The Chemical Structure of Earth

Only the very surface of Earth, including the oceans and atmosphere, are readily available for study. Together, these represent an insignificant mass fraction of Earth (less than 0.5 percent), almost all of which is the crust. To get some idea of

the bulk composition of Earth then, we need to find some way to probe its interior. Fortunately, Earth itself provides several means to do this. One clue comes from the study of seismic waves resulting from earthquakes (*see* EARTH, STRUCTURE OF). The velocities of seismic waves increase as the density of the material they travel through increases. By studying the travel times (or velocities) of a large number of seismic waves, geophysicists have been able to construct a density structure for Earth. These models show that the core of Earth is made largely of metallic iron (*see* CORE, STRUCTURE OF). Additional evidence in support of this comes from the moment of inertia of Earth and the existence of Earth's magnetic field. The moment of inertia of a homogeneous sphere can easily be calculated from basic physics. The moment of inertia of Earth departs from this value in a way consistent with Earth having a high-density, metallic iron core. Models for the generation of Earth's magnetic field require that the outer core consists of an electrically conductive fluid, with liquid metallic iron being the most plausible candidate (*see* METEORITES).

Geophysical data, then, indicate that Earth's core is made largely of iron, and that the core constitutes almost 32 percent by mass of Earth. However, the seismic data also indicate that a portion of the core is composed of an element lighter than iron. Several candidates are commonly debated, with hydrogen (H), carbon (C), oxygen (O), and sulfur (S) being most prominent. Iron (Fe) and troilite (FeS) form a single melt phase at high temperatures, so it is likely that when the core formed on Earth, much of the S went with the iron into the core. It is also probable that much of Earth's chalcophile element inventory followed S into the core. Recently, laboratory experiments have shown that FeO is soluble in molten iron at the high pressures in the core. Therefore, oxygen, too, is probably a light element in the core. We know that certain elements tend to be concentrated in the metallic iron phase of meteorites. These are the siderophile elements, including many of the row 4, 5, and 6 transition elements (*see* CHEMISTRY OF EARTH MATERIALS). These elements are highly depleted in the mantle and crust compared to chondritic meteorites, yet it is implausible that Earth never accreted them. It is very likely that the siderophile elements are concentrated in the core of Earth. Generally, these siderophile elements are in too low an abundance to make their presence in the core known through analysis of geophysical data. Nickel (Ni) and cobalt (Co) are in higher abundance (probably about 6 percent and 0.3 percent, respectively), but their densities are similar to iron so they will not be detected by geophysical means.

Estimating the composition of the mantle, which makes up almost 68 percent by mass of Earth, is an easier chore. This is because we have actual samples of the uppermost mantle available for study, and because the pressure and temperature conditions throughout the mantle are within the realm of laboratory experimentation. Hence, posited causes for the density structure of the mantle are readily subjected to experimental testing. The samples we have for study include slivers of the top few kilometers of the mantle that have occasionally been tectonically emplaced onto the crust (called ophiolites), and xenoliths, rocks brought up from depths of up to approximately 200 km in kimberlites and basalts. Mantle samples are ultramafic rocks, composed mostly of the mafic minerals olivine and orthopyroxene. High temperature and pressure experiments show that these same compositions can explain the density structure of the mantle down to the depth of the core (*see* EARTH, STRUCTURE OF). Hence, we can use the compositions measured on the available upper mantle rocks to estimate the bulk composition of the mantle. The major constituents of the mantle are O, Si (silicon), Mg (magnesium), and Fe with minor amounts of Na (sodium), Al (aluminum), Ca (calcium), and Cr (chromium).

Several researchers have made estimates of the bulk silicate fraction of Earth (mantle plus crust). E. Jagoutz and colleagues at the Max Planck Institute for Chemistry showed that among the many ultramafic xenoliths analyzed, several have major element compositions that appear to fall along extensions of fractionation sequences defined by chondritic meteorites. They therefore suggested that these few ultramafic rocks represented unmodified, primitive Earth rocky material. Analyses of these xenoliths constituted their estimated bulk Earth silicate composition. (The xenoliths are depleted in siderophile elements compared to chondrites, so they can only represent the bulk silicate fraction of Earth, at best.) A. E. Ringwood took another tack. He made the assumption that the typical ultramafic rock was essentially the bulk Earth silicate minus a basaltic component. Using the compositions of ultramafic rocks and basalts and a knowledge of petrology, he estimated the amount of basalt the average ultramafic rock lost, and added this back in to come up with a bulk primitive mantle composition he called "pyrolite."

The bulk composition of the crust, which constitutes about 0.4 percent by mass of Earth, is still easier to estimate, because we can obtain many samples of essentially all levels of the continental and oceanic crusts. All that is needed is to estimate the abundances of the various rock types in the crust. This can be done using a combination of areal estimates of rock types from geologic maps and from geophysical data on the lithologic structure of the crust (*see* CONTINENTAL CRUST, STRUCTURE OF; OCEANIC CRUST, STRUCTURE OF). The oceanic crust is formed at mid-ocean ridges through the intrusion and extrusion of basalts, and a realistic estimate of its composition is that it is basaltic. Numerous researchers have made estimates of the bulk composition of the continental crust. Although there are differences between the various estimates, all are grossly similar to intermediate volcanic rock, andesites, in composition. This means that compared to oceanic crust, the continental crust is richer in Si, Al, Na, and K (potassium). The continental crust contains the highest concentrations of the incompatible trace elements, that is, those that do not easily fit into the structures of minerals in the mantle (*see* TRACE ELE-

Table 1. Comparison of Cl chondrite abundances (the most primitive meteorite composition) and bulk Earth abundances of the elements. The column labeled Earth/Cl is the bulk Earth abundance divided by the cosmic abundance, both normalized to Si. The percentages appearing in the last column are those of elements that are greater than 0.1 percent of the bulk Earth.

Atomic Number	Symbol	Cl Abundance ($\mu g/g$)	Earth Abundance ($\mu g/g$)	Earth/Cl (Si normalized)	Percentage of Earth
1	H	20,200	78	0.004	
2	He	*	0.0013	0.132	
3	Li	1.5	2.7	1.336	
4	Be	0.0249	0.056	1.669	
5	B	0.87	0.47	0.401	
6	C	34,500	350	0.010	
7	N	3,180	9.1	0.003	
8	O	464,000	285,000	0.614	28.3
9	F	60.7	53	0.648	
10	Ne	*	0.000025	0.123	
11	Na	5,000	1,580	0.234	0.2
12	Mg	98,900	132,100	0.991	13.1
13	Al	8,680	17,700	1.513	1.8
14	Si	106,400	143,400	1.000	14.2
15	P	1,220	2,150	1.308	0.2
16	S	62,500	18,400	0.218	1.8
17	Cl	704	25	0.026	
18	Ar	*	0.0704	0.177	
19	K	558	170	0.226	
20	Ca	9,280	19,300	1.543	1.9
21	Sc	5.82	12.1	1.543	
22	Ti	436	1,030	1.753	0.1
23	V	56.5	103	1.353	
24	Cr	2,660	4,780	1.333	0.5
25	Mn	1,990	590	0.220	0.1
26	Fe	190,400	358,700	1.398	35.6
27	Co	502	940	1.389	0.1
28	Ni	11,000	20,400	1.376	2.0
29	Cu	126	57	0.336	
30	Zn	312	93	0.221	
31	Ga	10	5.5	0.408	
32	Ge	32.7	13.8	0.313	
33	As	1.86	3.6	1.436	
34	Se	18.6	6.1	0.243	
35	Br	3.57	0.134	0.028	
36	Kr	*	0.0000043	0.076	
37	Rb	2.3	0.58	0.187	
38	Sr	7.8	18.2	1.731	
39	Y	1.56	3.29	1.565	
40	Zr	3.94	7.42	1.397	
41	Nb	0.246	0.501	1.511	
42	Mo	0.928	2.96	2.367	
44	Ru	0.712	1.48	1.542	
45	Rh	0.134	0.32	1.772	

Table 1. Continued

Atomic Number	Symbol	Cl Abundance (μg/g)	Earth Abundance (μg/g)	Earth/Cl (Si normalized)	Percentage of Earth
46	Pd	0.56	1	1.325	
47	Ag	0.199	0.08	0.298	
48	Cd	0.686	0.021	0.023	
49	In	0.08	0.0027	0.025	
50	Sn	1.72	0.71	0.306	
51	Sb	0.142	0.064	0.334	
52	Te	2.32	0.94	0.301	
53	I	0.433	0.017	0.029	
54	Xe	*	0.000010	0.056	
55	Cs	0.187	0.059	0.234	
56	Ba	2.34	5.1	1.617	
57	La	0.2347	0.48	1.517	
58	Ce	0.6032	1.28	1.574	
59	Pr	0.0891	0.162	1.349	
60	Nd	0.4524	0.87	1.427	
62	Sm	0.1471	0.26	1.311	
63	Eu	0.056	0.1	1.325	
64	Gd	0.1966	0.37	1.396	
65	Tb	0.0363	0.067	1.369	
66	Dy	0.2427	0.45	1.376	
67	Ho	0.0556	0.101	1.348	
68	Er	0.1589	0.29	1.354	
69	Tm	0.0242	0.044	1.349	
70	Yb	0.1625	0.29	1.324	
71	Lu	0.0243	0.049	1.496	
72	Hf	0.104	0.29	2.069	
73	Ta	0.0142	0.029	1.515	
74	W	0.0926	0.25	2.003	
75	Re	0.0365	0.076	1.545	
76	Os	0.486	1.1	1.679	
77	Ir	0.481	1.06	1.635	
78	Pt	0.99	2.1	1.574	
79	Au	0.14	0.29	1.537	
80	Hg	0.258	0.0099	0.028	
81	Tl	0.142	0.0049	0.026	
82	Pb	2.47	0.143	0.043	
83	Bi	0.114	0.0037	0.024	
90	Th	0.0294	0.065	1.640	
92	U	0.0081	0.018	1.649	

*For the rare gases (He, Ne, Ar, Kr and Xe) the Cl abundances are altered by exposure to the solar wind and cosmic rays. The Earth/Cl abundance ratios of these elements are based on the isotopes least affected by exposure to the space environment.

MENTS). All the larger lithophile elements fit into this category, as do the light elements Li, Be, B, and C.

The easiest reservoirs for which to estimate a bulk composition are the oceans and atmosphere because these fluids are fairly well mixed, easily sampled and analyzed (*see* EARTH'S ATMOSPHERE, CHEMICAL COMPOSITION OF; SEAWATER, PHYSICAL AND CHEMICAL CHARACTERISTICS OF).

Estimating the Bulk Composition of Earth

There are several ways to calculate an estimate of the bulk composition of Earth. The one I will use was developed by E. Anders, then at the University of Chicago. Anders started from a detailed knowledge of the compositional variations among chondritic meteorites to define the chemical fractionations that occurred during formation of primitive nebular materials. Compositional variations among chondrite groups include fractionation of refractory lithophile elements from volatile lithophile elements, fractionation of siderophile elements from lithophile elements, and others. Anders posited that these fractionations were caused when chondritic parent bodies accumulated differing quantities of seven basic components of nebular matter: (1) early condensate; (2) remelted silicate; (3) unremelted silicate; (4) remelted metal; (5) unremelted metal; (6) troilite; and (7) volatile-rich component. If this is indeed the cause for elemental fractionations among the various chondrite groups, then perhaps the same model could be used to estimate the bulk composition of Earth. In order to do this, one needs to be able to estimate how much of each of these hypothetical components Earth acquired. This can be done using geophysical data, knowledge of the compositions of Earth rocks, plus geochemical reasoning.

We know from the study of meteorites that refractory lithophile elements are almost always in cosmic ratios, that is, all chondrites have ratios of La/Th (lanthanum/thorium), Sm/Hf (samarium/hafnium), Ba/Ta (barium/tantalum), for example, that are equal to our best estimate for the bulk composition of the solar system. Two of the refractory elements, U (uranium) and Th, are radioactive and make a considerable contribution to the heat budget in Earth. Therefore, measurements of heat flow out of Earth can be used to estimate the amount of U and Th in the bulk Earth. One isotope of K is also radioactive, and K is thus a significant heat source. Potassium is not a refractory lithophile, so we cannot assume Earth has a cosmic K/U ratio. However, both K and U are highly incompatible elements. This means that they do not easily fit into most mineral structures, so that during melting to form basalts, they are quantitatively partitioned into the magma. Therefore, for all intents and purposes, measurement of the K/U ratio in basalts gives a good estimate of Earth's value. Using geochemical and geophysical arguments such as these, Anders arrived at his best estimate of the abundances of the seven components in Earth, and therefore the bulk composition of Earth (Table 1).

There are other ways to estimate the bulk composition of Earth, depending on the different assumptions one believes are reasonable. Many estimates are generally similar to that proposed by Anders; Earth is slightly rich in refractory lithophile and siderophile elements, and is depleted in the moderately volatile and volatile elements compared to Cl chondrites.

Bibliography

BOTT, M. H. P. *The Interior of the Earth: Its Structure, Constitution, and Evolution,* 2nd ed. Amsterdam, 1982.

BROWN, G. C., and A. E. MUSSET. *The Inaccessible Earth: An Integrated View of Its Structure and Composition,* 2nd ed. London, 1993.

RICHARDSON, S. M., and H. Y. SWEEN, JR. *Geochemistry: Pathways and Processes.* Englewood Cliffs, NJ, 1989.

TAYLOR, S. R., and S. M. McLENNAN. *The Continental Crust: Its Composition and Evolution.* New York, 1985.

WEDEPOHL, K. H. "The Composition of the Continental Crust." *Geochemica et Cosmochimica Acta* 59 (1995): 1217–1232.

DAVID W. MITTLEFEHLDT

EARTH, HARMONIC MOTIONS OF

In a harmonic motion the entire Earth vibrates at one frequency. Such a vibration is described as "forced" or "free," depending on whether it is driven by a sustained periodically oscillating effect, or is started by an impulse (i.e., suddenly, and then left alone). A tidal deformation of the solid Earth, forced by the gravitational attraction of the Moon or Sun, can be represented as a wave of vibration traveling around the rotating Earth (*see* EARTH, MOTIONS OF). An asteroid's impact on Earth's surface, or an earthquake, may be regarded as an impulse (*see* EARTHQUAKES). A very large earthquake, or a very large asteroid impact, sets the whole planet ringing like a huge bell, by exciting free oscillations, which are also waves of vibration. Such impulsively generated harmonic motions are also called normal modes of Earth. Every normal mode has a characteristic frequency (or eigenfrequency) and deforms Earth in a characteristic geometric pattern, making the amplitude of vibration a characteristic function (or eigenfunction) of latitude, longitude, and depth below Earth's surface. If the effects of Earth rotation are ignored, vibrations in a normal mode become standing waves, in phase everywhere, and can be regarded as three-dimensional analogs of the patterns exhibited by a freely vibrating violin string.

To understand the details of a harmonic vibration, focus on a small sample of the vibrating Earth. During the vibration, according to Newton's law of motion, the mass acceleration of that small segment of Earth is balanced by restoring forces, which tend to return the mass to its undisturbed location. The nature of the restoring forces, together with the distribution of density in Earth, determines both the spectrum of frequencies of the normal modes, and the geometric pattern of the eigenfunctions. The restoring forces in seismic vibrations are provided primarily by Earth's elasticity (i.e., its incompressibility and shear rigidity, measures of its resistance to compression/expansion and twisting, respectively).

Seismic body waves generated by an earthquake radiate out and, much like light waves, are refracted by, and reflected from, surfaces at which Earth's material properties change sufficiently sharply. Reflection at the outer surface of Earth gives rise to the formation of surface waves of two distinct types, named after the English scientists Love and Rayleigh (*see* EARTH, COMPOSITION OF). Seismic waves of the same type and frequency, traveling in opposite directions around Earth, can interfere constructively to yield a standing wave pattern. Such standing waves are equivalent to a subset of Earth's seismic free oscillations.

Any motion of Earth started impulsively can be represented as a combination of normal modes, and by mathematical techniques (Fourier analysis) even a complex motion can be resolved into its normal mode constituents. Different kinds of global vibrations can be triggered by large earthquakes, by the redistribution of mass in the atmosphere, and perhaps by more subtle events such as changes in the topography of the core-mantle boundary (CMB). The instrument used to record such a harmonic motion depends on how that vibration disturbs Earth. The vibratory motion itself may be picked up by a seismic strain meter, or may cause a small fluctuation in the strength of gravity itself and be recorded by a gravity meter, or may bring about a small variation in the orientation of Earth relative to its rotation axis, registered by very-long-baseline radio interferometry (VLBI). By using such records from a widespread array of instruments, the relevant material properties of Earth's interior, consistent with the data, can be inferred. This procedure (greatly facilitated by high-speed computing) is termed inversion of the data to obtain an Earth model. Current models assume Earth's undisturbed state to be one of hydrostatic equilibrium (i.e., static pressure forces are balanced by the weight of overlying material).

Analysis of the instrumental records of the vibrations following a great earthquake permits deduction of a spectrum of normal modes (whose relative amplitudes are characteristic of the particular earthquake and its location relative to the recording instrument). For most of the normal modes very accurate calculations of the eigenfrequency and eigenfunction can be made by ignoring Earth's rotation and modeling it as spherical in shape, with every material property a function only of radius. The vibration of elastic spheres was studied mathematically as long ago as the 1880s, but not until the great Chile earthquake of 1960 were instruments sensitive enough, in the frequency range characteristic of most of Earth's free oscillations, to register the latter unambiguously. Even after a very large earthquake (i.e., surface wave magnitude 8), a typical normal mode amplitude does not exceed a few millimeters. Since 1960 over one thousand distinct normal modes have been identified in the vibrations generated by large earthquakes. Because elasticity provides the major restoring force for these free oscillations, they are sometimes referred to as the "acoustic" or "seismic" modes. Just as there are two kinds of surface waves, there are two kinds of seismic free oscillations. They are classified by the kind of deformation they give to Earth's surface, as recorded by a globally distributed network of strain meters.

The normal modes, which include Love waves, are called toroidal (or torsional), involve a purely twisting motion (i.e., no compression/expansion, hence no changes in density), and have no component in the radial direction. In one of the simplest toroidal modes, two hemispheres twist in opposite directions, that is, disturbed small mass samples move on parallels of latitude (Figure 1). This ring-like pattern gives rise to the name "toroidal." As long as Earth is modeled as spherical and nonrotating, shear rigidity provides the only available restoring force for these modes, and toroidal vibrations can exist only in the solid parts of Earth (not in its liquid core). When Earth rotation is taken into account, the Coriolis force (that also causes circulating flows to develop around regions of low or high pressure in the atmosphere) provides the restoring force needed for toroidal modes in the liquid core.

Spheroidal modes (including standing Rayleigh waves) involve compression/expansion as well as twisting, and so are affected by incompressibility as well as shear rigidity. Because these modes are accompanied by fluctuations in density, there is a

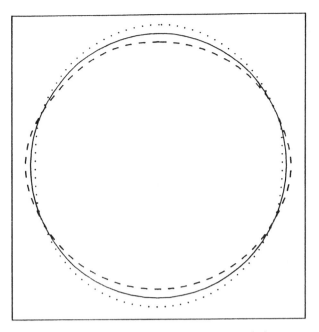

Figure 2. Deformation of spherical Earth in "football" spheroidal mode. During one cycle of vibration Earth's shape oscillates from oblate (- - -) to prolate (···) and back again.

gravitational contribution to the restoring force. Hence they can also be detected by measuring the associated small variations in gravity (amounting to at most a few parts per billion for a large earthquake). The simplest spheroidal modes involve purely radial motion (with no twisting), as if Earth were to breathe in and out like a balloon being alternately slightly inflated and deflated. Also easy to visualize is the "football" mode (Figure 2), in which Earth's (normally spherical) shape oscillates between an oblate spheroid (like a Gouda cheese) and a prolate one (like a football). Its period, about 54 minutes, is the longest of all the seismic normal modes.

The seismic reverberations of Earth die away in a few days, as their energy is dissipated by friction. The effect of friction (i.e., anelasticity) is measured by the number Q, where $2\pi/Q$ is the fraction of the energy of a free oscillation dissipated during one vibration period. Roughly speaking, a decaying oscillation loses 96 percent of its amplitude in Q cycles. Frictional damping depends on location in Earth, so each normal mode has a Q characteristic of its amplitude pattern. Anelasticity can be included among the material properties (density, incompressibility, and shear rigidity) obtained as

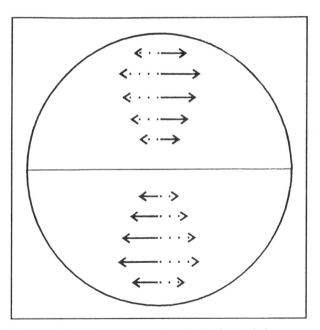

Figure 1. Deformation of spherical Earth in a simple toroidal mode. During one cycle of vibration the hemispheres twist in opposite directions, from (— —) to (···) and back again.

functions of depth throughout Earth's interior by inversion of seismic free oscillation spectra.

Lateral heterogeneity (i.e., material properties that depend on latitude and longitude as well as depth) and Earth's rotation are effects which complicate the foregoing description of seismic free oscillations. Both these effects cause each mode predicted for the nonrotating spherically stratified Earth model to be split into several distinct modes whose frequencies and eigenfunctions are closely grouped around the values calculated for the simpler model. The normal mode spectrum of the latter Earth model is said to be a degenerate version of the spectrum for the more realistic model. For free oscillations with periods much longer than those of the seismic subset, the Coriolis force can no longer be treated as a minor perturbation but becomes a significant, and for many modes the dominant, restoring force. Long-period normal modes are neither purely toroidal nor purely spheroidal, but instead their motions combine both types, linked by the Coriolis effect.

The first free oscillation observed (though not described as such at the time) was a normal mode of exceptionally long period, the Chandler wobble (*see* EARTH, MOTIONS OF). Crucial to this mode's existence are the real Earth's rotation and its nonspherical shape, which closely resembles a slightly oblate ellipsoid (i.e., its equatorial diameter is greater than that from pole to pole). This equatorial bulge stabilizes the rotation axis against large rapid departures from its present geographical alignment, and makes Earth respond like a gigantic gyroscope to any deformation tending to change the internal distribution of its mass. In the Chandler wobble Earth's equatorial bulge swings to and fro (with amplitude up to 0.15 arcseconds and a period of 437 days) about the plane at right angles to Earth's axis of rotation (Figure 3). This mode was first detected in 1891, having manifested itself through small opposite changes of latitude at places nearly 180° apart in longitude. How the Chandler wobble is excited is still uncertain, but it is most likely sustained by small changes in the distribution of mass in the solid Earth (due to large earthquakes) and/or in the atmosphere (associated with nonseasonal, geographically varying changes in atmospheric pressure). Uncertainty as to the excitation mechanism leaves the damping time of this mode poorly determined, but it is probably at least a few decades.

This mode was mathematically predicted by

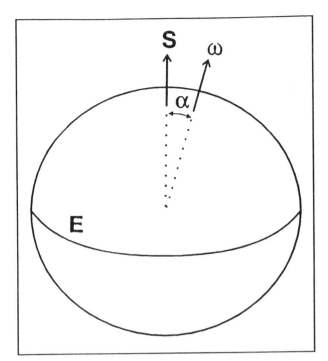

Figure 3. In the Chandler wobble, Earth's symmetry axis (S) traces out a cone (with apex angle 2α) about the rotation axis (ω), causing the plane of the equatorial bulge (E) to oscillate (with amplitude α) about the plane perpendicular to ω.

Leonhard Euler in 1755, with a period of 305 days for a perfectly rigid Earth. The difference between the presently determined period of 437 days (to which the oceans and the liquid core make nearly cancelling contributions) and Euler's value is almost entirely due to the imperfect (i.e., finite) rigidity of the mantle. This property permits the equatorial bulge to move slightly (in response to the small departure of the symmetry axis from the rotation axis) toward the plane at right angles to the rotation axis, thus reducing the gyroscopic restoring torque acting on Earth. The principal restoring force provided by gyroscopic stability being thus modified by elasticity, the Chandler wobble is a combination of toroidal and spheroidal vibrations coupled together by Earth's rotation.

Forced nutations are driven by the torques exerted on the equatorial bulge by the same lunar and solar gravitational forces as produce the tidal distortions of Earth. A typical gyroscopic response of the rotating Earth, they are small-amplitude "nodding" motions of Earth relative to the stars, like those of a spinning top (hence the name). The largest has amplitude 9.2 arcseconds and a 18.6-year period (*see* EARTH, MOTIONS OF).

Other free harmonic motions are predicted by the theory of Earth's dynamics but (as of 1994) have not been directly and decisively observed, because they are largely confined to the liquid core, with very small amplitudes in the much more rigid mantle. They include:

1. The free core nutation, which is possible because the presence of the liquid core makes Earth more like two tops, one spinning inside the other, coupled by fluid pressure torques acting across the CMB (*see* EARTH, MOTIONS OF). Its predicted period of about 460 days is sensitive to the CMB oblateness, and indirect evidence for this resonance (from gravimetry, and VLBI measurements of oscillations in the space orientation of Earth's rotation axis) suggests a slightly greater oblateness than accords with hydrostatic equilibrium

2. The Slichter mode, in which the solid inner core sloshes back and forth in the surrounding liquid core, in translational motion relative to Earth's center, with a predicted period of several hours. The dominant restoring force for this mode is gravitational, due to the jump in density across the inner core boundary

3. Inertial modes of the liquid core (with periods all longer than twelve hours), for which the Coriolis effect of rotation provides the necessary restoring force

4. Gravity modes of the liquid core, that can exist only if the liquid core is negatively buoyant (i.e., has a slightly overstable gradient of density versus depth) over some range of depths

Direct evidence of 2–4 is being sought from several years of data recorded by a (still far from global) array of superconducting gravity meters (operated at cryogenic temperatures to provide magnetic levitation of the sensor against gravity, and to reduce thermal noise).

Bibliography

GUBBINS, D. *Seismology and Plate Tectonics.* Cambridge, Eng., 1990.

ROCHESTER, M. G., and D. CROSSLEY. "Earth's Third Ocean, the Liquid Core." *Eos, Transactions of the American Geophysical Union* 68 (1987):481–492.

MICHAEL G. ROCHESTER

EARTH, MOTIONS OF

A person on the surface of Earth undergoes a complex motion in space. This motion is composed of the orbital motion of Earth around the Sun and the ever-changing rotation of Earth. Through an understanding of these motions we are able to learn about the evolution of Earth's orbit, the internal structure of Earth, and the nature and variability of the forces applied to the solid part of Earth.

Orbital Motion

Earth moves around the Sun in a near-circular orbit, taking a little over 365 days to complete one full circuit. The Moon takes about twenty-nine days to complete one orbit around the earth. The equator of Earth is inclined at about 23.5 degrees to the plane of Earth's orbit. It remains approximately fixed in direction as Earth orbits, thus causing the seasons on Earth. The Moon's orbit plane is inclined at some 5 degrees to Earth's orbit, and the changing positions of the Sun and the Moon cause time-dependent tides in the oceans of Earth and in Earth itself. It is almost certain that the orbits and periods of rotation observed today were not always this way. Determining the evolution of Earth's orbit is an important part of understanding past climate changes on Earth and provides part of the solution for predicting future changes. The details of predicting the changes in the lunar orbit and the rotation of Earth depend critically on the mechanisms for energy loss in the Earth-Moon system. Most of the energy lost today is due to friction between the tides in shallow seas and the seafloor, but over the last 3 billion years (Ga) the dominance of this mechanism is not clear (Carton, 1989; Comins, 1991; Folger, 1993; Lambeck, 1980). The predictions from these types of models reveal that the length of year 3 Ga ago may have been as long as 400 days, and that periodic variation in the orbit may have been responsible for substantial climate variations in the past (Zimmer, 1992). Yet another problem addressed by models of the evolution of the lunar orbit is the origin of the Moon. Was the Moon once a part of Earth that was thrust into orbit by a giant meteor impact, the leading theory, or was it a passing planetoid that was captured in Earth orbit because it came too close? Despite many years of study and several space flights to the

Moon, no one knows the answer to this puzzle (Hartmann, 1989; Naeye, 1994).

Rotational Motion

While the rotation of Earth appears to be constant and uniform, it is variable on all time scales. These variations are of three types: changes in the rate of rotation; changes in the direction of the rotation axis in space; and changes in the position of where the rotation axis pierces the surface of Earth (Wahr, 1986). The largest of these variations, called precession, is caused by the gravitational attraction of the Moon and Sun on the equatorial bulge of Earth. This attraction causes the rotation axis to trace out a cone in space, much like a tilted spinning top does. It takes 26,000 years for the rotation axis to complete one circuit of the cone. Two other natural oscillations of the direction of rotation axis occur with much smaller amplitudes: the Chandler wobble, named for the person who in 1891 first observed the motion; and the "free core nutation" (FCN), which results from the relative rotation of the fluid core and mantle of Earth. While the existence of the FCN mode was theoretically predicted near the end of the nineteenth century, its effect on the rotation of Earth was not definitively observed until the mid-1980s. By this time, observations of the rotation of Earth, one thousand times more accurate than the optical astronomical observations, were being made with powerful radio telescopes using a technique called very long baseline interferometry. These measurements were not only able to detect the presence of the fluid core but were also able to determine the difference between polar and equatorial radii of the core with an accuracy of about 50 m. They showed that the core was more flattened (by about 500 m) than was expected, indicating that forces inside Earth played an important role in determining the shape of Earth (Weisburd, 1987). By 1990, changes in Earth's rotation due to tidal currents in the oceans were measured. These measurements can now be used to test the quality of global ocean tide models.

Besides oscillations of position of Earth's rotation axis, there is steady drift of its position. Between 1900 (when the position of the rotation axis could first be accurately defined) and 1990, the rotation axis drifted about 8 m to the west. Most of this change can be explained by the viscous re-

bound of Earth after the removal of the ice sheets of the last iceage 10,000 years ago. Part of the variation may also be due to the sea level changes during the last one hundred years, and changes of the position of the rotation axis provide one means for monitoring global sea level changes (Peltier, 1988).

In the 1940s it was realized that the rotation rate of Earth was not constant apart from long-period changes that were known about before this time. At this time very accurate quartz crystal clocks became available, and relative to these clocks it became apparent that there were seasonal changes in the earth's rotation rate, that is, Earth rotated faster during the Northern Hemisphere's winter than during its summer. It would take another twenty years before it was established definitively that these variations in the rotation rate were due to the exchange of angular momentum between the solid Earth and the atmosphere. By 1990, with the advent of measurement techniques that could measure the length of day (the time, measured by an atomic clock, for Earth to complete one rotation) with an accuracy of 10 microseconds (μs), and the development of weather forecasting models that could be used to estimate the angular momentum of the atmosphere, it was realized that nearly all rotation rate changes with periods of less than about two years could be explained by the interaction of Earth and its atmosphere (Hide and Dickey, 1991). To determine changes in the rotation rate before 1940, a variety of techniques are employed. To calculate changes from the 1700s forward (after the invention of the telescope) observations of the Sun, inner planets, and the Moon are used. For calculations of the rotation rate of earlier eras, ancient eclipse data are used. All of these measurements indicate that the rotation rate of Earth is decreasing, with the most likely explanation of the slowing being dissipation of energy through friction between ocean tides and the seafloor. One consequence of this energy loss is that the Moon is receding from Earth at a rate of about 4 cm per year. This increasing distance to the Moon has been directly measured by accurate laser range measurements to corner-cube reflectors placed on the Moon by the Apollo astronauts in the early 1970s. Another consequence of Earth's rotation rate decreasing is that the second, defined in the 1940s, no longer represents 1/86,400th of the rotational period of Earth (there should be 86,400 seconds in a day), and therefore about once per year now we need to add a second to one day (a so-called leap second) to keep atomic time synchronized with the rotation of Earth (Marshall, 1987; Cleere, 1994).

In addition to the very rapid variations due to the atmosphere and the long-period variations due to energy loss in the oceans, there are fluctuations in the rotation rate on decade time scales that are too large to be caused by the atmosphere and too variable to be due to the changes in the rate of energy loss. These variations are most likely due to exchanges of angular momentum between the fluid core and the solid Earth. Estimates of the fluid flow at the top of the core derived from measurements in the changes on Earth's magnetic field combined with the shape of the core mantle boundary from seismic observations yields variations in the rotation rate in close agreement with the observed variations going back to the turn of the twentieth century (Jault and Le Mouël, 1989).

Rotational variations on Earth occur on all time scales and are measures of changes in the planet itself and the forces acting on it. Understanding today's variations can lead to models of past variations and will be useful in assessing the impact of humans on the planet because of variations in the forcing from the atmosphere and changes in sea level.

Bibliography

CARTON, W. "Oppolzer's Great Canon of Eclipses." *Sky and Telescope* 78 (1989):475–478.

CLEERE, G. "Making Time." *Natural History* 103 (1994):86.

COMINS, N. "The Earth Without the Moon." *Astronomy* 19 (1991):48–53.

FOLGER, T. "The Fast Young Earth." *Discover* 14 (1993):32.

HARTMANN, W. K. "Birth of the Moon." *Natural History* (1989):68–77.

HIDE, R., and J. O. DICKEY. "Earth's Variable Rotation." *Science* 253 (1991):629–650.

JAULT, D., and J.-L. LE MOUËL. "Core-mantle Boundary Shape: Constraints Inferred from the Pressure Torque Acting Between the Core and Mantle." *Geophysical Journal International* 101 (1990):233–241.

LAMBECK, K. *The Earth's Variable Rotation: Geophysical Causes and Consequences.* Cambridge, Eng., 1980.

MARSHALL, E. "A Matter of Time." *Science* 238 (1987):1641–1643.

NAEYE, R. "Moon of Our Delight." *Discover* 15 (1994): 72.

PELTIER, R. "Global Sea Level and Earth Rotation." *Science* 240 (1988):895–901.

WAHR, J. "The Earth's Inconstant Rotation." *Sky and Telescope* 71 (1986):545–549.

WEISBURD, S. "Opening Doors to the Core, and More." *Science News* 131 (1987):9.

ZIMMER, C. "Peeling the Big Blue Banana." *Discover* 13 (1992):46–47.

THOMAS A. HERRING

EARTH, ORIGIN OF

Solar Nebula Accretion

The origin of Earth began with the formation of the solar system. The solar system and stars in general form by collapse of the cores of dense molecular clouds in the galaxy. Molecular clouds consist mainly of gas with a small amount of dust. The collapsed disk of gas and dust that resulted in our solar system is called the solar nebula (*see* SOLAR SYSTEM; SUN). The formation of the solar system is best dated at 4.559 ± 0.004 billion years ago, or 4.559 Ga (giga annum), determined by lead isotope age dating of calcium-aluminum-rich inclusions from the Allende CV carbonaceous chondrite meteorite.

Materials similar to the chondrite meteorites were the building blocks of the inner planets, Mercury, Venus, Earth, and Mars (*see* METEORITES). These meteorites indicate that during the early stages of solar nebula evolution, solid material (interstellar dust) was concentrated at the midplane of the nebula. Some of the metal and silicate dust components clumped together to form millimeter-sized objects, or dust balls. Processing of the solid material during the formation and evolution of the nebula included melting, evaporation, condensation, and aerodynamic sorting. Melting of the clumps of material by an unknown process produced the rounded objects called chondrules, a major component of most chondritic meteorites. The heating process involved rapid "flash" heating and cooling within hours. Numerous heating mechanisms have been suggested, including lightning, chemical energy, frictional heat, and reconnecting magnetic field events similar to processes that form solar flares.

The next stage in the evolution of the solar system was the accretion of the chondritic material into asteroid-sized objects. This "planetesimal hypothesis" involved the gradual coalescence of many tiny bodies into planetary embryos. Heating of some of these planetesimals caused metamorphism or melting of the chondritic material. The subsequent accretion of these objects into larger and larger planetesimals ultimately led to the formation of the planets themselves. Nebular gas and ice beyond the asteroid belt were also building blocks for the outer planets. During this process, Jupiter and Saturn were large enough to accrete some of the gas from the solar nebula and are therefore called the gas giants. At the heliocentric distance (distance from the Sun) of Jupiter and beyond, temperatures were low enough for ices to be stable. This critical distance is sometimes called the "snowline" in the solar nebula. The outer planets and satellites, therefore, are thought to contain a large amount of ice. Uranus and Neptune, which did not accrete much nebular gas, are often called the ice giants. The rocky inner planets—Mercury, Venus, Earth, and Mars—did not accrete large amounts of either nebular gas or ice and are called the terrestrial planets.

Most meteorites come from parent asteroids that probably formed in the asteroid belt between Mars and Jupiter (*see* ASTEROIDS; METEORITES). The asteroids in this outer region experienced a range of heating from essentially none to temperatures that caused complete melting. In those asteroids that melted, the metallic iron (alloyed with 5 percent nickel on average) in the original chondritic material sank to the center to form metallic cores. Broken-up pieces of these cores that hit Earth are known as iron meteorites. If the heat source that melted asteroids was electromagnetic induction due to an early intense solar wind, the planetesimals closer to the Sun may have been more strongly heated. If the heat source was radioactive aluminum (Al^{26}), a wave of planetesimal growth propagating outward over a time interval corresponding to several half-lives of Al^{26} ($t_{1/2}$ 730,000 years) may have also led to a gradient in heating with heliocentric distance. Thus, during the formation of Earth the accreting planetesimals may have already differentiated to form mantles and cores.

The next stage, leading to the formation of the planets themselves, was dominated by the accretion of very large planetesimals. The best evidence for this episode is the wide range of axial tilts of the planets relative to the perpendicular to the plane of their orbits around the Sun. Tilts ranging from 0° to 97° (Earth's axial tilt is 24°) were caused by the random momentum of the last few large planetesimals to accrete to each planet. These large tilts imply that the late accreting material was contained in only a very few large objects because the combined effects of the accretion of large numbers of small objects would cancel out. Theories for the origin of the Moon, and the high density of Mercury, also involve the impacts of planet-sized planetesimals.

Origins of the Earth and Moon

The Earth–Moon system has several unique features compared to the other planets and planet-satellite systems. The Moon's bulk density is low compared to Earth, suggesting a low total iron (FeO in the mantle plus Fe-metal in the core) content. The chemical composition of the Moon, revealed by the samples from the U.S. Apollo and Soviet Luna missions, is also unusual. The Moon contains no water, is strongly depleted in volatile elements such as potassium (K), lead (Pb), and bismuth (Bi), and is probably enriched in refractory elements such as calcium (Ca), aluminum (Al), titanium (Ti), and uranium (U). The FeO (iron oxide) content in the Moon is approximately 50 percent greater than that of the upper mantle of Earth. However, the isotopic ratios among the oxygen isotopes, and the abundance of moderately siderophile elements (elements with an affinity for Fe-metal), such as tungsten (W), cobalt (Co), and nickel (Ni), are similar in Earth's mantle and the Moon. This combination of properties has led to significant difficulties for theories of the origin of the Moon.

Theories for the origin of the Moon include capture, coaccretion, fission, and the single-impact theory (*see* MOON). Capture is unlikely because if the Moon formed in an independent orbit around the Sun it would have difficulty losing enough energy to be captured in orbit around Earth. The Moon's low density is also inconsistent with an independently formed solar system object. Formation of the Moon by coaccretion from a disk of material around Earth that contains normal chondritic metal contents cannot explain the low density of the Moon. In addition, this theory cannot easily provide the required angular momentum needed to form a circum-terrestrial disk. Material in solar orbit would hit the disk around Earth on both sides almost equally; thus half of the material would have head-on collisions, causing material in the disk to fall onto Earth. The fission hypothesis involves the derivation of the Moon from the terrestrial mantle due to rapid early rotation of Earth. The rapid rotation could result from spin-up of Earth caused by core formation, much like a spinning ice skater pulling in his or her arms. This theory has the advantage of explaining the low density of the Moon, but it cannot explain some of the compositional differences, such as the different FeO contents. In addition, the overall angular momentum of the Earth-Moon system is not as large as that predicted by the fission theory.

The single impact of a planet-sized object with the proto-Earth to form the Moon has been widely adopted as the most likely explanation (Newsom and Taylor, 1989). The single-impact theory can explain the differences and similarities between Earth and the Moon, although this theory has so many adjustable parameters that agreement between theory and observation is almost guaranteed. By mixing material from the proto-Earth with the impactor, differences in the FeO contents can be explained. The low density of the Moon is explained by the incorporation into the Moon of metal-free material from Earth's mantle. The material incorporated into the Moon from the mantle is free of metal because core formation in the proto-Earth removed the metallic iron. The material incorporated into the Moon from the impactor is also free of metal because the impactor's core is incorporated into Earth and not into the Moon. The low volatile content of the Moon is explained by degassing of the material left in orbit around Earth after the impact, which later accretes to form the Moon. Additional chemical changes to the outer portion of the Moon are caused by the formation of the lunar magma ocean (a melting event that affected the outer portion of the Moon) and the formation of a very small metallic lunar core of iron and nickel, less than 5 percent by weight of the bulk Moon, from metal that did not end up in Earth.

The Earliest Earth

The events of the first half billion years of Earth's history are known only from indirect evidence due to the lack of rocks dating back to that early period (*see* OLDEST ROCKS IN THE SOLAR SYSTEM). The accretion of Earth lasted from 50 to 200 million years (Ma) after the initial formation of the solar system. The impact of large planetesimals into the proto-Earth was certainly occurring during that time period. Some of these impactors may have approached the size of Mars, based on the tilt of Earth's rotational axis and the possible impact origin of the Moon. These large impacts by themselves caused extensive melting of early Earth, and the accepted existence of an early magma ocean on the Moon argues for a similar terrestrial magma ocean. Unlike the Moon, the terrestrial magma ocean did not leave a clear chemical record of itself in the upper mantle, although the structure of the outer portion of Earth above a depth of 670 km (known as the transition zone and upper mantle) has been interpreted as the remnants of a solidified magma ocean. The lack of a chemical record may indicate either that the solidification of the terrestrial magma ocean did not result in the normal chemical changes, or fractionations, as seen in the solidification of smaller igneous bodies during later Earth history, or that the solidified material was later thoroughly mixed back into the mantle.

A key event during the early evolution of Earth was the segregation of iron-nickel metal toward the center of Earth to form Earth's core. Because metal cores are thought to have formed in the small asteroid parent bodies of the iron meteorites, core formation in the proto-Earth probably began at a very early stage of accretion. The segregation of metal from the silicate (rocky) portion of Earth depleted the bulk silicate Earth (mantle + crust) in the chemical elements that have an affinity for iron metal and iron sulfide, such as Co, Ni, W, molybdenum (Mo), copper (Cu), gold (Au) and the platinum metals. The original abundances of these metallic elements in Earth are related to those measured in chondritic meteorites. The abundances of the metallic elements in the bulk silicate Earth are determined by chemical analysis of upper mantle inclusions found in continental volcanics and by geochemical modeling of trace element abundances in mantle-derived volcanic rocks and the continental crust. The proposed models to explain the depletion of metallic elements for Earth can be easily divided into two classes: equilibrium models, in which the observed depletions of metallic elements in the mantle are established by equilibrium between metal and mantle silicates, and disequilibrium models (Newsom and Sims, 1991). The disequilibrium models include multiple-stage additive models in which the first stage is the formation of a mantle that is highly depleted in metallic elements due to extensive core formation under reducing conditions during high temperature accretion in the presence of a thick atmosphere (Ahrens, 1994). The abundances of metallic elements are controlled by the addition of late veneers to the highly depleted mantle. The veneers may consist of late accreting material from the solar nebula, or even the mantles or metal cores of large planetesimals, such as the one needed in the impact origin of the Moon theory. Another type of disequilibrium model is the dilution model, which involves the mixing into the mantle of silicates that have been stripped of metal and metallic elements in a localized part of the mantle, thereby lowering or diluting the remaining metallic elements in the mantle as a whole. The theory does not explicitly locate the region where the stripping occurred, but a connection with magma ocean processes seems a likely possibility. Unfortunately, current evidence does not allow us to discard any of the above models. Both of these classes of models involve processes at higher pressures and temperatures for which experimental data are still very scarce. Seismic observations and experimental investigations of processes at the present core-mantle boundary may provide additional clues to the chemical state of Earth's core (Jeanloz and Lay, 1993).

The current composition of Earth's atmosphere is very depleted in the noble gases helium (He), neon (Ne), argon (Ar), krypton (Kr), and xenon (Xe) relative to solar system abundances. This depletion implies that either the gases in the solar nebula had been dispersed by a strong solar wind before the proto-Earth had grown to a size sufficient to capture a primitive atmosphere, or that the early atmosphere of Earth was blown off by the large impacts discussed above (Ahrens, 1994). The isotopic composition of the noble gases He, Ar, and Xe in Earth's present atmosphere also indicates that the atmosphere was created by degassing from the mantle primarily during the first half billion years after accretion (4.5–4 Ga ago). This early atmosphere did not contain free oxygen, however, until 2.9–2.5 Ga ago, when oxygen-producing bacteria became abundant.

The existence of Earth's oceans has great implications for the early history of Earth. Liquid water cannot be in equilibrium with iron-metal, arguing that Earth's oceans formed late, after core formation had occurred. The question not yet resolved is whether the water was accreted in the form of a late veneer of ice-rich comets from beyond Mars or from inner solar system planetesimals that retained some volatile elements (Taylor, 1992).

The earliest history of the continental crust is not known for certain. The oldest continental rocks date back to about 4 Ga, half a billion years after the formation of Earth, and the bulk of the present continental crust was formed 2–2.5 Ga ago. The existence of some continental crust during the first half billion years is indicated by neodymium (Nd) isotopic data, but the presence of a large volume of continental crust similar in composition to today's crust is very unlikely. Some scientists have proposed that an early basaltic crust was present, which was much later mixed back into the present mantle.

Bibliography

AHRENS, T. J. "The Origin of the Earth." *Physics Today* 47 (1994):38–45.

JEANLOZ, R., and T. LAY. "The Core-Mantle Boundary." *Scientific American* 268 (1993):8–55.

NEWSOM, H. E., and K. W. W. SIMS. "Core Formation During Early Accretion of the Earth." *Science* 252 (1991):926–933.

NEWSOM, H. E., and S. R. TAYLOR. "The Single Impact Origin of the Moon." *Nature* 338 (1989):29–34.

TAYLOR, S. R. *Solar System Evolution.* Cambridge, Eng., 1992.

HORTON E. NEWSOM

EARTH, SHAPE OF

The study of Earth's shape and gravity is a central part of the discipline of geophysics and is also referred to as geodesy. A knowledge of the shape and gravity field provides information on the density distribution and structure of the planet's interior. It also provides the information necessary for predicting the trajectories of artificial satellites around Earth.

The two issues of shape and gravity field of the planet are closely related. When geophysicists describe the shape of Earth they do not usually refer to the geometric or topographic form of Earth's surface but to surfaces of constant gravitational potential, which is intimately linked to the gravity field.

Geodesy is one of the oldest branches of geophysics, but interest in the study of the gravity field and shape of planets has been renewed in the last few decades because of the development of satellite methods for measuring these quantities, and because these measurements relate directly to the study of Earth's deep interior, providing, for example, constraints on mantle convection and tectonics. One of the exciting recent developments has been the ability to measure time-dependent changes in the gravity field and shape of Earth.

Earth's Gravity Field

Newton's law of gravitation forms the basis for discussions of Earth's shape and gravity field, and complexities introduced by Einstein's general theory of relativity are generally unimportant here. If we have two point masses m_1, m_2 separated by a distance r, then Newton's law for the force δF of gravitational attraction is

$$\delta F = G \frac{m_1 m_2}{r^2}$$

where G is the gravitational constant, found experimentally to be equal to about 6.670×10^{-11} N m^2/kg^2 (m^3 kg^{-1} s^{-2}).

Earth can be represented by a very large number I of point masses m_i so that a small mass m outside Earth will experience a total force of

$$F = Gm \sum_{i=1}^{I} \frac{m_i}{r_i^2}$$
$$= Gm \int_m \frac{dm}{r^2}$$

where r_i is the distance of m_i from the mass and the integral is over the mass of Earth. The mass m at a position P will experience an acceleration according to Newton's first law of motion $F = ma$ and hence will experience the acceleration of gravity, usually denoted by g, of

$$g = G \int \frac{dm}{r^2}$$

Gravity is a vector quantity acting in a direction toward the center of Earth.

If the acceleration is defined in a coordinate frame that is fixed to and rotating with Earth (e.g., the usual latitude, longitude, and height coordinates) the total acceleration at P is the sum of the gravitational attraction and the centrifugal force, or

$$g = G \int \frac{dm}{r^2} - \frac{1}{2} \Omega^2 d$$

where Ω is the angular velocity of Earth around its spin axis and d is the distance of P from the rotation axis. It acts in a direction orthogonal to the rotation axis and increases with increasing d. The average value of g is about 9.8 ms^{-2}, of which the centrifugal force represents about 0.3 percent. Despite this latter term being small it has played an important part in shaping Earth. At the equator the centrifugal force is opposite to the gravitational attraction and tends to reduce the total gravitational acceleration whereas at the pole this term vanishes. If the planet were a rotating fluid this difference would be enough to cause the body to flatten so that the equatorial radius would be greater, by about 21 km, than the polar radius. Gravity at the equator would be about 9.780 ms^{-2} and at the poles it would be about 9.832 ms^{-2}. That the actual shape of Earth closely resembles such an ellipsoid of revolution attests to the power of the small centrifugal force and demonstrates that, over long timescales, Earth acts much like a rotating fluid body.

The ellipsoidal shape would represent well the Earth if the density distribution within it were radially symmetric. Because of tectonic processes and mantle convection, however, lateral variations in density do occur (see MANTLE CONVECTION AND PLUMES; PLATE TECTONICS). Relatively cold downwelling convection currents in the mantle, for example, will be of higher density than warm upwelling currents and gravity can be expected to be higher than normal in the former case and lower than normal in the latter case.

The gravity field of a planet can therefore be expected to exhibit considerable spatial variability and this is indeed observed with variations reaching amplitudes typically of 10^{-3} ms^{-2}. The field can be measured with special gravity meters that measure, for example, the acceleration of a mass that is allowed to fall freely within an evacuated volume. More traditional is the measurement of the change in length or torque of a mass that is suspended from a spring. As the gravitational attraction is increased so is the length of the spring increased or its "twist" decreased. These latter instruments measure only changes in gravity, not absolute gravity.

Results for the gravity field of the Australian continent illustrated in Figure 1, in the form of an illuminated relief map, are typical of results for other parts of the world. Some of the principal features here are linear anomalies in the center of the continent and the anomalies along the margins. But many of the smaller anomalies are equally important in that they delineate geological structures in the crust and they contain much information on the tectonic history of the continent as well as providing a valuable contributing data source for mineral and hydrocarbon exploration.

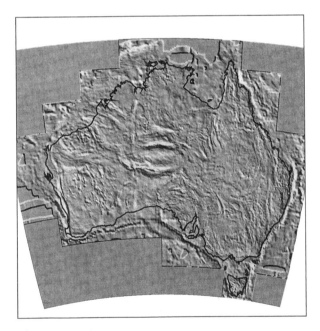

Figure 1. The gravity field over the Australian continent and adjacent continental shelves. The field shows the departures in gravity from the field generated by an ellipsoid of revolution. Maximum anomalies are about 0.2 ms^{-2}. Features that appear "raised" are local gravity highs, while "troughs" are gravity lows. (Based on the original gravity measurements compiled by the Australian Geological Survey Organisation.)

Gravity measurements are laborious and costly and it has not been possible to cover the entire globe. But over the past few decades an important and powerful new method has become available for measuring the gravity field. Consider a satellite orbiting a planet. The main force operating on a satellite will be gravity, and as it passes over mass concentrations or deficits the gravitational forces tug and push the satellite, which responds by accelerating or decelerating along its orbit. By measuring with great precision the successive positions of the satellite in its orbit, it becomes possible to map the accelerations of the spacecraft and the acceleration of gravity. Now it has become possible to map the global gravity field of Earth and other planets with very high precision. Such results are often illustrated by means of "geoid" maps.

The Shape of Earth: The Geoid

It is often convenient to consider not the gravity field but the gravitational potential U of Earth. The potential scalar is defined as

$$g = -grad\ U$$

where *grad* defines the mathematical operator known as the gradient operator. The reasons for introducing the concept of potential are largely mathematical but it allows for a relatively simple definition of the shape of Earth. Surfaces of constant potential, or equipotential surfaces, are surfaces to which the gravity vectors are everywhere perpendicular. Hence there is no component of gravity along the surface. A fluid surface, in the absence of nongravitational forces, will adjust itself so that it is everywhere normal to the gravity vector and it represents an equipotential or level surface. If not, the fluid would run "downhill" until a level surface is reached. Thus, in the absence of winds or currents, the oceans of the world represent an equipotential surface. This is usually called the "geoid," and it forms a natural definition of the shape of Earth. That is, in a geophysical sense, the shape is defined as an equipotential surface that corresponds to the (mean) ocean surface or the surface that the oceans would occupy if the latter everywhere covered the earth. A simple intuitive picture would be the shape that the ocean surface would take if the water volume were increased so as to cover all land that is now emerged.

Like the gravity field, the geoid exhibits much spatial variability. Consider, for example, a higher-than-average density in some part of the mantle. Gravity on Earth's surface over this feature will be increased and a constant potential surface over this anomaly will lie further away from the center of Earth than elsewhere.

Earth's shape in a first approximation resembles an ellipsoid of revolution and this forms a good reference for describing the geoid as the departure of the equipotential surface from a mean ellipsoidal surface. The geoid is therefore represented as a map of highs and lows relative to the best-fitting ellipsoid. This geoid will reflect largely the density structure of Earth. Highs in the geoid point to areas of excess mass and lows point to areas of mass deficits compared to the globally averaged radial mass distribution of the planet. Typically these "geoid anomalies" attain ±100 m. Figure 2 illustrates such a global geoid map. Its features include a large "low" south of India and another in the Caribbean Sea. Highs occur over Papua New Guinea and the North Atlantic. An important feature of these satellite geoid maps is that they reflect

well the broad-scale structure of Earth, whereas the gravity maps based on surface measurements reflect more detailed structure. Thus the two results are very complementary and the latter, despite the success of the satellite results, remain important. These and other geoid anomalies provide powerful constraints on the structure of Earth's interior.

Inferences on Earth's Interior

The past years have seen much improvement in the measurement of gravity using a variety of satellite methods, with the results that the gravity field or geoid is now one of the best-known physical quantities of Earth. The interpretation of this field is more of a problem. The existence of the geoid anomalies indicates that density anomalies occur but it is a feature of potential theory that different mass distributions can give rise to the same geoid. Thus the interpretation of the anomalies is not unique and a complete discussion of their significance requires other geophysical data, as from seismology, magnetism, or heat flow.

Figure 2. The global geoid derived from the analysis of satellite orbit perturbations. The anomalies are with respect to a best-fitting reference ellipsoid. Contours are at 10 m intervals. (Based on original data computed at the Goddard Space Flight Center, NASA.)

Broadly, the existence of the geoid anomalies points to substantial density anomalies in the mantle and to the occurrence of mantle convection. If the mantle did not convect then the density anomalies would adjust, the mantle material would flow so as to minimize the stresses in the mantle, and the shape of the planet would approach that of a rotating fluid, the ellipsoid of revolution. Thus the existence of the geoid anomalies generally points to the occurrence of mantle convection and much of the current emphasis on convection modeling is to produce mantle models that are consistent with the geoid or gravity data.

The smaller details seen in Figure 1, for example, contain much information on the near-surface structure of Earth, on the crust and uppermost mantle. Many of the features correlate with the boundaries of geological or tectonic provinces, some of which date back several billion years. This gravity map therefore contains a record, albeit a difficult one to interpret, of the geological history of Earth. Processes that control this history may also provide suitable environments for mineral deposits or for oil and gas fields to form, and the gravity measurements are an important element in the search for these resources.

Bibliography

LAMBECK, K. *Geophysical Geodesy: The Slow Deformations of the Earth.* Oxford, 1988.

PHILLIPS, R. J., and K. LAMBECK. "Gravity Fields of the Terrestrial Planets: Long-wavelength Anomalies and Tectonics." *Reviews of Geophysics and Space Physics* 18 (1980):27–76.

TORGE, W. *Geodesy.* Berlin, 1991.

KURT LAMBECK

EARTH, STRUCTURE OF

Earth is made up of five main layers with distinct compositions and, hence, different densities and other properties. The innermost region, the core, is a sphere of radius 3,480 km, made up of an iron-rich metallic alloy with an average density of 11.0 Mg/m^3. (One Mg/m^3, or million grams per cubic meter, is equivalent to 1,000 kilograms per cubic

meter or, in more familiar units, to 1 gram per cubic centimeter.) The core is surrounded by the mantle, a shell of rock with an average thickness of 2,867 km and an average density of 4.5 Mg/m^3. The mantle is overlain by a relatively thin veneer of less dense rock, the crust, with an average thickness of 16 km and an average density of 2.8 Mg/m^3. The hydrosphere, made up of water and ice, with an average density of 1.0 Mg/m^3 and average thickness of 3 km, rests on top of the crust. The outermost layer is the atmosphere, which has a density of 0.001 Mg/m^3 at the earth's surface, and becomes increasingly less dense with elevation, its density falling by a factor of two for each 5 km increase in elevation.

During its accretion from the solar nebula, the energy released from bombardment by meteorites heated our planet to beyond its melting point, creating, in effect, a giant blast furnace. The dense metal that makes up the core sank to Earth's center, leaving a slag of molten rock floating on top. This rearrangement of the planet released additional gravitational energy that led to a catastrophic runaway of this differentiation process. Throughout Earth's subsequent evolution, during and after the slag of molten rock solidified into the mantle and crust, variations in density resulting from variations in temperature, as well as composition, have continued to drive convective motions in the core, mantle, hydrosphere, and atmosphere. For example, cold, hence dense tectonic plates (lithosphere) sink into the hot asthenosphere at island arcs in a process called subduction, resulting in motion of the tectonic plates. And at volcanoes, molten magmas rise through the mantle and into the crust, increasing the volume of the crust and releasing water into the oceans and gases into the atmosphere. Some of this material is incorporated into the tectonic plates and recirculated into the mantle. The dynamics of the system depend on its density structure, which generates stresses, and its mechanical structure—its strength.

The density of planetary material is influenced by a number of factors. Most important is its chemical composition. The metallic alloy making up the core is at Earth's center because it is more dense than the overlying mantle rocks, which are composed of chemical compounds containing substantial oxygen, an element that is less dense than most metals. The compounds constituting the mantle are enriched in the denser elements magnesium (Mg) and iron (Fe) relative to the overlying crust, which is more enriched in silicon (Si), aluminum (Al), and other lighter elements. The hydrosphere overlies the crust because it is made of water, which is made up of hydrogen (H) bonded to oxygen (O). (Table 1 summarizes the most important constituents and other properties of the major regions making up our planet.)

In addition to the kind of atoms that comprise a material—its chemical composition—the way in which these atoms are arranged—its petrology, or the phases it contains—is important in determining a material's density. A familiar example is water, which has a density of 1.0 Mg/m^3 in its liquid state, a density of 0.9 Mg/m^3 in its solid state (ice),

Table 1. Earth Structure

Region	Main Chemical Components	Main Phases	Radius (km)	Density (Mg/m^3)	Pressure (GPa)	Temperature (K)	Strength (MPa)	
Inner	Iron	Solid	0	13.09	363	5600	?	
Core	Iron Oxide	Metal	400	13.05	360	5535	?	
	Sulfur		800	12.95	349	5469	?	
	Nickel		1222	12.76	329	5400	?	
Outer	Iron	Molten	1222	12.17	329	5400	0	
Core	Iron Oxide	Metal	1600	11.95	307	5232	0	
	Sulfur	Alloy	2000	11.65	278	5055	0	
	Nickel		2400	11.29	244	4878	0	Fluid
			2800	10.85	207	4701	0	
			3200	10.33	166	4524	0	
			3480	9.90	137	4400	0	
Lower	$(Mg, Fe)SiO_3$	Perovskite	3480	5.57	137	4400	1	
Mantle	(Mg, FeO)	Periclase	3600	5.51	130	3689	3	
(D″)	Iron?	Metal	3700	5.46	124	3096	10	
Lower	$(Mg, Fe)SiO_3$	Perovskite	3800	5.41	118	2750	20	
Mantle	$(Mg, Fe)O$	Periclase	4000	5.31	107	2666	20	Mesosphere
			4400	5.11	86	2499	20	
			4800	4.90	66	2331	20	
			5200	4.68	47	2164	20	
			5600	4.44	28	1996	20	
			5711	4.38	24	2000	20	
Transition	$(Mg, Fe)_2SiO_3$	Spinel	5711	3.99	24	2000	2	
Zone	$Mg_3(MgSi)Si_3O_{12}$	Garnet	5800	3.94	20	1933	2	
	$(Mg, Fe)SiO_3$	Pyroxene	5900	3.81	16	1811	2	
			5971	3.72	13	1700	2	Asthenosphere
Upper	$(Mg, Fe)2SiO_4$	Olivine	5971	3.54	13	1700	0.1	
Mantle	$(Mg, Fe)SiO_3$	Pyroxene	6100	3.46	9	1608	0.1	
	$Mg_3(MgSi)Si_3O_{12}$	Garnet	6200	3.40	5	1575	0.1	
			6275	3.35	3	1550	0.1	
			6300	3.35	2	1525	10	Lithosphere
			6325	3.41	1	1067	100	
			6340	3.44	1	793	250	
			6347	3.46	1	672	180	
Crust	$Ca(Al_2Si_2)O_8$	Plagioclase	6347	2.90	1	672	180	
	$K(Al, Si_3)O_8$	Orthoclase	6356	2.90	0	500	100	
	SiO_2	Quartz	6356	2.60	0	500	100	
			6368	2.60	0	280	9	
Hydro-	H_2O	Water	6368	1.02	0	280	0	
sphere	(Na, Ca)	Salt	6371	1.02	0	280	0	
Atmo-	N_2	Air	6371	0.001	0	280	0	
sphere	O_2		6500	0	0		0	

and a density of 0.0007 Mg/m^3 in its vapor state—less than the average density of the atmosphere. The hydrosphere is more dense than the atmosphere not because it is made up of intrinsically denser compounds, but because these compounds are arranged in an intrinsically denser phase (liquid) than are the atoms in the gaseous atmosphere.

The pressure a material is subjected to affects its density in two ways. First, increasing pressure increases the density of material by squeezing the atoms closer together. In addition, changes in pressure can cause changes in phase, where the atoms in a compound are rearranged into a more compact structure. For example, the materials making up the lower part of the mantle have their atoms arranged in more closely packed geometric structures than those in the upper mantle, where the minerals are characterized by a more loosely packed arrangement of atoms. The density of the mantle increases by about a factor of 5/3 from the uppermost to the lowermost mantle. Of this increase, about a factor of 4/3 is the result of the ordinary effects of compression and about 5/4 is the result of rearrangement of atoms into different phases.

The effects of temperature on density are comparable to the effects of pressure but in an opposite sense. Increasing temperature leads to thermal expansion, resulting in a decrease in density. In addition, changes in temperature can lead to changes in phase, with higher temperatures generally favoring less densely packed atomic arrangements. Because both pressure and temperature increase with depth in the earth, their effects offset each other to some extent.

The material making up the earth, like the children's toy Silly Putty, deforms over time when subjected to the stresses that result from gravity acting on variations in density. The way in which a material deforms in response to applied stress is called its rheology. The rheology, or strength, of Earth material depends on composition, phase, pressure, and temperature. Convection in the earth's mantle drives plate motions. The details of the interaction of the stresses generated by density differences and the resistance to deformation determine the style of tectonic activity. For example, although Venus and Earth have comparable temperature and density variations, their type of tectonics is quite different because of the greatly differing rheological properties of their outer regions.

The remainder of this entry is a description of estimates of Earth structure, presented for the major regions, or subdivisions, of these five layers. These regions are defined according to their density, chemical composition, most important mineral phases (petrology), pressure, temperature, and strength (rheology). Variations in these properties as a function of radius are given in Table 1, plotted in Figure 1, and discussed in the following section.

To a first approximation, Earth has a structure like an onion, with successive, nearly spherical layers. While this article discusses in detail only the average structure as a function of radius, there are many important lateral variations in structure as well. One of the largest departures from spherical symmetry is the ellipticity that results because Earth is spinning; rotation distorts the planet into an elliptical shape, so that the average equatorial radius is 21 km greater than the average polar radius. Variations in ocean depth or surface elevation are about half this magnitude. The crust varies in thickness by over 75 km—almost five times its average thickness. There are important variations in temperature and strength that are associated with mantle convection and plate motions. More details on some of these topics are given elsewhere in this encyclopedia.

Core

The core is made up of a metallic alloy, primarily iron. The temperature at Earth's center is 5,600 $\pm$ 1,000 degrees Kelvin (K), dropping to 4,400 $\pm$ 600 K at the top of the core, the core-mantle boundary (CMB). The pressure at Earth's center is 363 GPa (3.63 million atmospheres), dropping to 1.35 GPa at the CMB. Despite the high temperatures, at a radius less than 1,222 km, the pressure is sufficiently high (greater than 3.29 GPa) that the metallic alloy freezes, forming the solid inner core. The outer core, which lies above the inner core boundary (ICB; radius = 1,222 km) and below the CMB (radius = 3,480 km), is made of molten metal.

The temperature of the entire planet, including the core, is decreasing through time, both because radioactive heat production is decaying and because the heat released by planetary formation is being transported to the surface. As a result of this cooling, the inner core continues to grow. The material that freezes out of the outer core to form the inner core is more nearly pure iron in chemical composition, leaving behind molten metal enriched in the alloying elements sulfur (S) and oxygen. This material is less dense than the overlying fluid, so it rises through it, a process known as compositionally driven convection. The resulting motion of the metallic, electrically conducting outer core causes the geodynamo, which generates Earth's magnetic field.

The density at Earth's center is 13.1 Mg/m^3, decreasing to 12.8 Mg/m^3 at the top of the inner core. The density at the bottom of the outer core is 12.2 Mg/m^3, so there is a density contrast of 0.6 Mg/m^3 across the ICB. The density at the top of the fluid core is 9.9 Mg/m^3. Virtually all of the variation in density through the core is the result of pressure variations, or self-compression, except for the density contrast at the ICB, which results from the change in phase and composition between the solid iron inner core and the molten alloy outer core. Because the core is a fluid with a viscosity comparable to water, it can support no significant lateral variations in temperature or density.

Mantle

The mantle, composed of magnesium- and iron-enriched silicate rocks, floats on the fluid core. The region above the CMB (radius = 3,480 km), and below a depth of 660 km (radius = 5,711 km), is the lower mantle. It is composed primarily of minerals consisting of $MgSiO_3$ in the perovskite structure and MgO. The average temperature at the top of the lower mantle is 2,000 ± 400 K and the pressure is 24 GPa (240,000 atmospheres). The average density just above the CMB is 5.57 Mg/m^3, decreasing to 4.38 Mg/m^3 at the top of the lower mantle. Most of the variation of density and temperature with radius through the lower mantle is quite smooth, resulting primarily from the extreme pressure variations—squeezing a substance increases both its density and temperature. There are lateral variations in temperature in the mantle of hundreds of degrees Kelvin.

In a region just above the CMB, with a thickness of several hundred kilometers, there are large variations in properties that cannot be explained just by self-compression. This region, termed the D'' region by the seismologists who first noticed its effects on the propagation of seismic waves, shows large variations in structure, laterally as well as radially. There may be temperature variations of as much as 1,600 K across this region, which forms the thermal boundary layer between the hotter core and the cooler mantle. Such a thermal boundary layer is a likely spot for mantle plumes to originate. There may also be chemical reactions occurring between the metallic core and the silicate mantle that result in the D'' layer having a composition intermediate between that of the mantle and core.

The region above the lower mantle, between a radius of 5,711 km and 5,971 km, is the transition zone. There is an exceptionally large change in density with radius in this region resulting primarily from pressure-induced phase transitions. The density changes abruptly at a depth of 660 km as the result of minerals changing structure from the perovskite structure characteristic of the lower mantle (density of 4.38 Mg/m^3) to the spinel and garnet structures (density of 4.08 Mg/m^3). The density at the top of the transition zone decreases relatively rapidly with increasing radius as the result of additional distributed phase changes, with minerals in the garnet structure changing to the pyroxene structure, reaching a density of 3.77 Mg/m^3 at the top of the transition zone. At a depth of 400 km (radius of 5,971), there is another important phase change, with the material in the upper mantle above 400 km depth predominantly in the olivine structure. Some garnet remains, with pyroxenes the other important minerals.

The temperature decreases with radius from 2,000 ± 200 K at the base of the transition zone to 1,550 ± 100 K at a depth of about 100 km as the result of the decrease in pressure over this interval. There is a rapid decrease in temperature with depth through the outer 100 km of the planet through the thermal boundary layer that is associated with the tectonic plates. At a depth of about 100 km the pressure is sufficiently low and the temperature sufficiently high that partial melting of the mantle can occur beneath mid-oceanic ridges and island arcs. Lateral variations in temperature approach 1,000 K in the upper 100 km.

The density in the upper mantle just above 400 km depth is 3.54 Mg/m^3. Density decreases with radius, reaching a value of 3.38 Mg/m^3 at a depth of about 100 km before increasing, due to a decrease in temperature of 900 K to a value of 3.42 Mg/m^3 at the top of the upper mantle.

Crust

The crust, which is the most heterogeneous layer of the planet, lies above the mantle. The average density in the lower crust is 2.9 Mg/m^3, decreasing to 2.6 Mg/m^3 in the upper crust. The crustal thickness, like Earth's topography, has a bimodal distribution, with a distinct contrast between oceanic and continental regions. Oceanic crust has an average thickness of 7 km; the lower 6 km formed as the result of igneous activity at mid-oceanic spreading centers and the uppermost kilometer formed by sediments raining down as the crust moves from spreading center to subduction zone. Ocean crust older than 150 Ma is very rare. In contrast, continental crust has an average thickness of about 35 km. Its greater thickness and low density keep continental crust from being subducted into the mantle, although it is reworked during the process of mountain building when continents collide. The oldest rocks preserved in the continental crust are 4 Ga old. Continental crust reaches a maximum thickness of 80 km beneath the Tibetan plateau.

Hydrosphere

The hydrosphere, which covers 70 percent of Earth's surface, is made up mainly of the oceans, ice sheets, and groundwater reservoirs. Most of the mass of the hydrosphere is in the oceans, which if spread uniformly over the planet would have a depth of 3 km. River water flowing from the continents carries salts and other minerals, which are concentrated by evaporation in the oceans. The ocean structure is stratified, both in temperature and in salinity. For example, cold freshwater from melting of ice in polar latitudes sinks to the bottom of the ocean, where it underlies water that is warmer, but more saline. Compositionally driven convection results in eventual mixing between these water masses.

Atmosphere

The atmosphere is a gas composed primarily of nitrogen and oxygen. Because it is highly compressible, its density decreases rapidly with elevation. The mass within the lowermost 5 km of the atmosphere is comparable to the mass of the remainder of the atmosphere, which extends hundreds of kilometers into space. This mass is also comparable to the mass in the upper 5 m of the hydrosphere.

The atmosphere is divided into two main layers. The lower, denser layer is the troposphere, with an average thickness of about 10 km. Temperature falls off relatively quickly with elevation within this layer, reaching a minimum at the tropopause. Our weather is the result of motions driven by tiny density variations within the troposphere. The upper layer, the stratosphere, has even smaller temperature variations.

Strength of the Solid Earth—"Dynamic" Structure

The strength of the solid Earth varies dramatically as a function of position. The rheology of the rocks that make up the crust and mantle depends on temperature, pressure, composition, and the level of the stress applied. Near the surface, where temperature and pressure are both relatively low, when the stresses are sufficiently high, rocks fail by brittle failure. This is the process that generates earthquakes. Near the surface, at temperatures below about 750 K, the strength of the rocks is determined mainly by friction resisting motion on faults. In the absence of fluids, the strength is approximately a third of the weight of the overlying rocks. But many faults fail at lower stresses, perhaps because they are lubricated by high pore fluid pressures. There also seems to be a relationship between the distance that a fault has slipped over geologic time and its strength, with major plate boundary faults, which have accommodated substantial displacements, weaker than other faults.

At higher temperatures, rocks deform by ductile flow, like Silly Putty, and can be characterized by an effective viscosity. For ductile materials, the "strength" depends on the rate at which they deform. The strengths tabulated and plotted here are for a rate of deformation of 1 percent per million years—a deformation rate characteristic for the convecting mantle. Rocks become increasingly ductile with increasing temperature, with their strength decreasing by about a factor of 10 for each increase in temperature by 100 K. Because temperature increases rapidly with depth near the surface of the earth, there is an accompanying rapid decrease in strength with depth. The cold, strong rocks near the surface make up the lithosphere, the layer that forms the tectonic plates.

The weaker, more ductile layer beneath is the asthenosphere. Because of the great contrast in strength, the lithospheric plates move over the asthenosphere without deforming, except at weak plate boundaries. The thickness of the lithosphere is on average about 75 km, but its thickness varies greatly, from only about 10 km beneath mid-oceanic ridges to hundreds of kilometers beneath old continents. In most places the entire crust and part of the upper mantle are included within this layer,

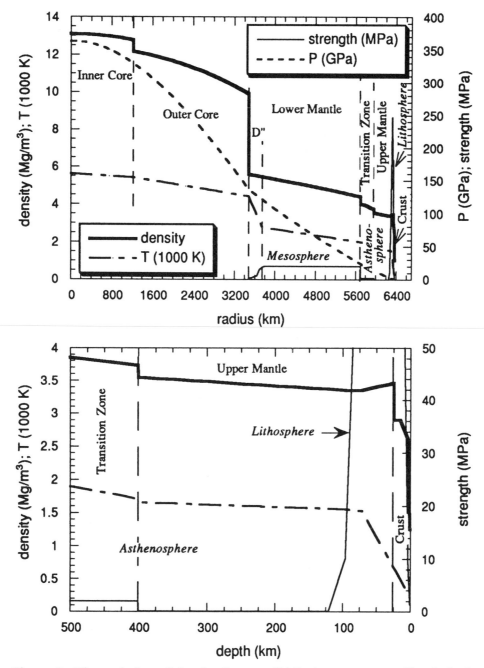

Figure 1. The variation of density (heavy solid line), temperature (dot-dashed line), pressure (dotted line), and strength (light solid line) as a function of radius (top panel) and depth (bottom panel). The petrologic regions are indicated in regular type; the "dynamic" regions are indicated in italics.

within this layer, which is defined according to its strength, not its composition. In places where the upper mantle is unusually hot, the lithosphere may include only the upper part of the crust.

Beneath about 100 km depth, the temperature increases more slowly with depth, and the effect of increasing pressure on rock strength becomes important. The phase changes that occur across the transition zone also affect the strength. The net result is that the strength of mantle rock increases with depth, as shown in Figure 1. The stronger region beneath the asthenosphere is called the mesosphere. The D'' region is probably weaker than the overlying mantle because of the rapid increase in temperature with depth across this layer.

Most boundaries in the "dynamic" structure are determined primarily by the thermal structure, rather than the petrologic structure. The mechanically strong lithosphere includes both crust and upper mantle. The D'' layer is weak primarily because it is a hot thermal boundary layer. But the boundary between the asthenosphere and the mesosphere is likely a result of the difference in strength of minerals making up the transition zone and the rest of the mantle. Although the qualitative variation of strength with depth is reasonably well known, the quantitative values of the strength distribution are uncertain by a factor of three or more.

Bibliography

ANDERSON, D. L. *Theory of the Earth*. Boston, 1989.

JEANLOZ, R., and T. LAY. "The Core-Mantle Boundary." *Scientific American* (May 1983):26–33.

JEANLOZ, R., and S. MORRIS. "Temperature Distribution in the Crust and Mantle." *Annual Reviews of Earth and Planetary Sciences* 14 (1986):377–415.

PRESS, F., and R. SIEVER. *Understanding Earth*. New York, 1994.

STACEY, F. D. *Physics of the Earth*, 3d ed. Brisbane, Australia, 1992.

BRADFORD H. HAGER

EARTH AS A DYNAMIC SYSTEM

Although the study of planet Earth has most often been pursued as though the Earth system could be understood as comprising a number of essentially non-interacting component parts, this view is invalid at best and seriously misleading at worst. These component parts, which consist respectively of the solid Earth, the oceans, the atmosphere, the biosphere, the cryosphere, and the near space environment (magnetosphere), interact strongly across a wide range of spatial and temporal scales. Their interactions and the resulting phenomena to which the interactions give rise encompass physical, chemical, and biological processes. Models of individual phenomena require input from each of the basic sciences, a fact that has led to the recognition of earth science as a primary multidisipline. The increasing concern of humankind with the problems of environmental degradation and climate change has sharply increased the importance of deepening our understanding of planetary form and function. This concern has found expression in a wide range of large-scale international programs, such as the World Climate Research Program (WCRP) and the International Geosphere-Biosphere Program (IGBP), as well as national initiatives, such as the Earth Observing System (EOS) of the U.S. National Aeronautics and Space Agency (NASA). These vast undertakings are all focused upon the problem of developing an understanding of the way Earth works as a dynamic system. This entry focuses on a series of specific phenomena that have been selected to illustrate the interconnections between Earth system components that govern global change on a wide range of spatial and temporal scales.

No single process better illustrates this interconnection than global warming (*see* GLOBAL CLIMATIC CHANGES, HUMAN INTERVENTION). Examination of surface temperature variations over the past century reveals an increase in the mean surface temperature of Earth of approximately 0.5°C. Although debate continues as to the cause of this temperature rise, the observation itself is seldom seriously contested. The results of atmospheric general circulation model experiments have been employed to demonstrate that significant temperature increase is expected as a consequence of increasing concentrations of atmospheric "greenhouse" gases, primarily carbon dioxide (CO_2) and methane (CH_4). The increase in these gases has resulted from the burning of fossil fuels, an influence on the global climate system that has been active only since the beginning of the industrial revolution in the late eighteenth century. Over the past several decades the concentration of CO_2 in the atmosphere has been monitored carefully at a number of stations across a wide range of latitudes from the South Pole to the North Pole. When the increasing levels of such radiatively active trace gases are included in model calculations of the evolving temperature field, it becomes possible to approximately reproduce the 0.5°C surface warming that has occurred over the last century (Houghton et al., 1996). This verification of an increase in the "greenhouse effect," however, has required a careful accounting for the dynamical influence of the oceans and for the influence of atmospheric (sulphate) aerosols. Both effects serve to decrease the warming that would have been expected if the planetary atmosphere were not strongly coupled to the oceans and if high concentrations of sulphate aerosols were not lofted into the atmosphere in heavily industrialized regions and in regions subject to high seasonal levels of biomass burning. The buffering influence of the oceans on surface warming is associated with the warming of the oceans themselves, whereas the buffering by atmospheric aerosols is due to the efficiency with which these particulates scatter incident solar radiation back to space. The fraction of the incident energy that is available for warming the low-level atmosphere is thus reduced. Before the incorporation of these influences, models were unable to explain the observed warming that has occurred over the last century. Although active debate continues on the issue of global warming, especially concerning the role of clouds, the phenomenon itself is clearly an excellent example of interactive Earth system evolution.

Related to the issue of global warming and providing an excellent second example of Earth system interconnections in global change is the issue of global rise in sea level. As the mean surface temperature of the planet has increased over the past century and warming has been significantly buffered by the oceans, it is not unreasonable to expect a rise of mean sea level as a consequence of warming. The rise in sea level could be the result of either one (or both) of two mechanisms—the "steric" effect of the thermal expansion of the oceans or the "eustatic" effect associated with the melting of polar ice sheets or of smaller ice sheets and glaciers. The former influence raises sea level of a fixed mass of ocean water by increasing its volume through thermal expansion, the latter by increasing the water mass itself. In fact, recent analyses of modern tide gauge data, based on individual records exceeding fifty years in length, have strongly suggested that global sea level is rising at a rate near 1.8 ± 0.6 mm yr^{-1}. However, it has been necessary to fully account for one extremely prominent solid Earth process in the course of analyzing the data. Glacial isostatic adjustment produces a variation of relative sea level that is distinct from that which could be associated with global warming. This source of contamination of modern tide gauge recordings is caused by the last deglaciation event of the current ice age, an event which began approximately 20,000 years ago and ended approximately 6,000 years ago. At the last glacial maximum, global sea level was reduced by approximately 100 m, a reduction effected by the removal of the water from the ocean basins that was required to build the vast continental ice sheets that existed over northwestern Europe and the northern part of the North American continent at that time as well as the increased ice cover over Antarctica. Although these additional ice sheets had essentially disappeared by 6,000 years ago, their disappearance continues to contribute a significant influence on tide gauge recordings of secular sea level change, a consequence of the fact that the earth's shape continues to change as a result of isostatic rebound (see ISOSTASY). The rate at which the planet's shape evolves is controlled by the viscosity of the planetary mantle, a physical property that can be directly measured by analysis of geological data that constitute memories of the response to the deglaciation event. The most useful information in this regard consists of relative sea level data, which record the time-dependent changes of the elevation of land with respect to the surface of the sea (the geoid). If the tide gauge data are not properly filtered so as to remove this effect of postglacial isostatic adjustment, the tide gauge data are so geographically variable that their climatological significance is questionable. When the data are filtered so as to remove the effect of the postglacial isostatic adjustment process, however, the signal is found to be less geographically variable. This is an excellent example of the fact that the dynamical influence of the solid Earth may be highly significant, even on processes that evolve on timescales of decades to centuries.

Similarly connected to the problem of global warming is the problem of the global carbon cycle. Climate research is concerned with the expected impact of increasing atmospheric levels of CO_2. Even in the absence of an anthropogenic effect on

CO_2 variability, the concentration of this trace gas would not be unchanging. Atmospheric CO_2 levels have been decreasing for the last several tens of millions of years, primarily due to the increased sequestration in $CaCO_3$. It is believed that the global cooling of the lower atmosphere that would be caused by such diminution of CO_2 was a principal cause of the onset of late Quaternary glaciation and for the intensity of the quasi-periodic 100,000 year cycle of glaciation and deglaciation that has dominated the last 1 million years (Ma) of this period. In the last century of the Holocene epoch, in which the anthropogenic effect has been operating, the main issue connected with the carbon cycle concerns the issue of the sink or sinks of CO_2. This factor must be invoked to understand the reason why the increase of CO_2 in the atmosphere has been significantly less than would be expected on the basis of the strength of the known sources (fossil fuel combustion, biomass burning, etc.). Tracer-calibrated models of the rate of oceanic uptake of CO_2 indicate that the oceans could not be responsible for the entirety of the required sink. It has therefore been suggested that the continental biosphere (forests, etc.) must be expanding as a consequence of enhanced growth due to the influence of CO_2 fertilization. An understanding of the CO_2 cycle is thus dependent on an understanding of the dynamic interactions between the atmosphere, oceans, and biosphere, where the biosphere includes not only surface vegetation but also humankind. The carbon cycle is an example of a biogeochemical cycle through which interlinkages between physics, chemistry, and biology operate so as to control chemical-biological reservoirs and the fluxes through which the reservoirs are connected.

The carbon cycle is also deeply involved in a further Earth system phenomenon for which a complete mechanistic understanding is still lacking, namely the late Quaternary ice-age cycle itself. The recent increase in knowledge of this large scale climate oscillation represents one of the most important advances in the earth sciences since the development of the ideas of plate tectonics and continental drift.

It is now well understood, on the basis of oxygen isotope data from both deep-sea sedimentary cores and deep ice cores from Antarctica and Greenland, that the last 1 Ma of Earth history have been dominated by an intense cycle of ice-age recurrence in which continental ice sheets have ad-

vanced on a timescale of approximately 90,000 years and collapsed on a timescale near 10,000 years, thus completing a cycle in a period of 100,000 years. Detailed analysis of variations in oxygen isotopic composition of deep-sea cores provides an indirect measure of changes of continental ice volume with time. The oxygen isotopic composition of the shells of microscopic oceanic organisms (mainly foraminifers) becomes enriched in the heavier oxygen isotope (^{18}O) during times of cooler climate. This is caused by changes in the evaporation rate of surface waters (see ISOTOPE TRACERS, STABLE). Because water containing the lighter isotope, ^{16}O, is more easily evaporated, the water left behind by evaporation becomes heavier (enriched in ^{18}O). During times of cooler climate—when isotopically light water becomes sequestered in continental ice sheets—the oceans are enriched in ^{18}O. This change is reflected in the oxygen that goes into making the shells of the foraminifera, and is preserved when their shells finally end up in the deep-sea sediments after they die. Variations in the oxygen isotopic composition of the foraminifera in deep-sea core samples therefore measure changes in climate, and indirectly the presence of continental ice sheets. Variations in oxygen isotopic composition of foraminifera occur on timescales that can be related to changes in the solar insolation due to changes in the geometry of Earth's orbit around the Sun. Variations in oxygen isotopic composition are seen on a timescale of 41,000 years, the timescale for the variation in orbital obliquity, and on the timescales of 23,000 years, and 19,000 years due to the modulation of the influence of orbital precession by the variation of orbital eccentricity (see EARTH, MOTIONS OF).

Continental ice volume variability on these timescales is entirely expected based upon the Milankovitch theory of ice ages, a theory which holds that these epochal events are entirely controlled by solar insolation change due to changing orbital geometry. A fundamental difficulty with this theory concerns the problem of explaining why the dominant periodicity of the ice age occurs at the period of 100,000 years. The dominance of the 100,000-year period in continental ice sheet volume would appear to require climate system nonlinearity in order to be explained. Recent research on this important issue has led to the recognition of a number of dynamic interconnections between elements of the Earth system that may be involved in generating the 100,000-year cycle within the cryosphere as a response to orbital insolation forcing. These mechanisms involve feedbacks both within the atmosphere and within the oceans that appear to be strongly implicated in the ice-age cycle. The first of these feedbacks involves the greenhouse gases CO_2 and CH_4, both of which are observed to be significantly reduced during glacial conditions according to measurements in air bubbles in ice cores from both Greenland and Antarctica. Because atmospheric CO_2 is reduced during glaciation, it enhances cooling, and because it is increased during deglaciation, it also enhances warming; the feedback loop is therefore positive. The second feedback implicated in the glacial cycle is that connected to the deep, cold, saline water circulation in the Atlantic Ocean. Based upon data from deep-sea cores, it is clear that the rate of North Atlantic Deep Water (NADW) formation was significantly reduced during full glacial conditions. Since the process of NADW formation leads to significant heating of the overlying atmosphere (averaging 100 W/m^2 in the modern climate system), the fact that NADW production was low during full glacial conditions leads to further Northern Hemisphere cooling, thus constituting a further positive feedback loop on the glaciation cycle. A third positive feedback loop to which attention has recently been drawn is one involving atmospheric aerosols, specifically dust from the continents. On the basis of both ice cores and deep-sea sedimentary cores it is well known that deglaciation events are preceded by episodes of extremely high atmospheric dust concentration. If the dust load, indicative of extremely cold, dry climate conditions, were to decrease the albedo of the ice sheets in their ablation zones and thus enhance ablation rates, a significantly increased rate of deglaciation could result. Developing an understanding of the 100,000-year ice-age cycle may be the greatest of all challenges to our understanding of Earth as a dynamic system.

On timescales of the same order as the age of Earth itself (that is, on the order of 10^9 years), the dominant interactions within the Earth system involve the solid Earth to a critical degree. Here the most important issues are connected with the processes that govern the rate of planetary cooling (*see* HEAT BUDGET OF THE EARTH) and the processes that occur as a consequence of this cooling. Governing the behavior of the system on these time-scales is the process of mantle convection through which the rate of interior heat loss is controlled. Fundamental to the understanding of this process is the fact that the convective circulation within Earth is one that intimately involves the motion of material at the planet's surface and which is evident in the processes of continental drift and sea-floor spreading (*see* PLATE TECTONICS). That the mantle of the planet is able to deform as a viscous fluid is, of course, fundamental to the thermal convection process. Vital to the understanding of this process is knowledge of the effective fluid viscosity of the planetary mantle. This can be measured using the changes in relative sea level caused by the glacial isostatic adjustment process discussed above. The value so determined (near 10^{21} Pa · s) is precisely that which is required in the successful construction of a convection model of the drift process. Modern work on the mantle convection process has led to an appreciation of the extent to which the endothermic (heat absorbing) phase transition from spinel to a mixture of perovskite and magnesiowustite that occurs at 660 km depth (*see* MANTLE, STRUCTURE OF) could lead to the occurrence of a strongly time-dependent mode of radial mixing with the flow strongly layered in the sense that distinct convective circulations exist above and below this depth. This layering could be episodically disrupted by intense "avalanche" events during which cold material from the transition zone abruptly descends into the lower mantle. Irrespective of the importance of this process for the cooling history, it is clear that secular cooling of the planet as a whole is required to understand the formation of the magnetic field, a feature of the planet that has existed since shortly after planetary formation approximately 4.5 billion years or Ga (*see* EARTH'S MAGNETIC FIELD). Although it is currently believed that the dynamo process in Earth's outer core that maintains the planet's magnetic field is forced to a significant degree by composition buoyancy associated with the growth of the solid inner core, this mechanism would not come into play at all if it were not for the fact that Earth as a whole is cooling. The fluid outer core and the overlying "solid" mantle are therefore strongly interconnected dynamically.

Earth is a dynamic system of strongly interacting components that exhibits a wide range of phenomenology in which individual elements participate to varying degree. To understand the form and function of Earth requires appreciation of inter-system linkages as well as the operation of its individual components. Typical issues in global change involve the linkages, rather than simply the systems' component parts.

Bibliography

HOUGHTON, J. T., ed. *Intergovernmental Panel on Climate Change: IPCC Second Scientific Assessment of Climate Change*. Cambridge, Eng., 1966.

HOUGHTON, J. T., G. K. JENKINS, and J. J. EPHRAUMS, eds. *Intergovernmental Panel on Climate Change: The IPCC Scientific Assessment*. Cambridge, Eng., 1990.

OJIMA, D., ed. *Modelling the Earth System*. Boulder, CO, 1992.

PELTIER, W. R., ed. *Mantle Convection*. New York, 1989.

W. RICHARD PELTIER

EARTH-MOON SYSTEM, EARLIEST HISTORY OF

During most of this century, geologists had virtually no understanding of the first third of Earth's history. Historical geology texts devoted about 80 or 90 percent of their content to the last 15 percent of Earth's history, the post-Cambrian period in which the classic fossil record had permitted a stratigraphy to be deduced by the founders of modern geology in the 1800s.

As for the first third of the geologic record, it has been mostly destroyed by Earth's active geologic processes of erosion and plate tectonics, including mountain building and subduction.

Since the 1960s, however, the blank early chapters of Earth's early geology have been filled in from two extraterrestrial sources. First, the American Apollo and Soviet Luna missions brought back lunar samples from nine lunar sites. These revealed events in the earth-moon system during the first third of its existence, from 4,400 to 3,000 million years (Ma) ago. (This is 4.4 to 3.0 billion years or Ga in the American system, where "billion" = 10^9.) Because of the time spans involved in this article, it will be convenient to work in units of millions of years. Second, increasingly sophisti-

cated studies of meteorites filled in the earliest details by revealing planetary formative events from about 4,550 to 4,000 Ma ago. Also, increasingly old, but rare, deposits have been located on Earth itself, such as 3,960-Ma-old rocks in the Northwest Territories of Canada, found in 1989, and even older minerals (zircons) in Australia (*see* OLDEST ROCKS IN THE SOLAR SYSTEM).

In other words, the exploration of the Moon and the study of meteorites are beautifully complementary to the study of Earth's geology. As a result of this progress, we are now in a much better position to sketch the first third of Earth's history than were scientists only a generation ago.

Six subjects of interest are (1) the original accretion of Earth; (2) the origin of the Moon; (3) the cratering history during the early period; (4) the evolution of the early atmosphere, which was different from the modern atmosphere; (5) the origin of continental masses; and (6) the origins of life.

Earth's Accretion

The accretion of the planetary material occurred within a few tens of millions of years, about 4,550 Ma. Various isotopic studies of meteorites, lunar samples, and terrestrial samples give the 4,550 Ma figure as the date when the Sun and planets formed; and meteoritic studies show that the solid, parent bodies of the meteorites accreted from cosmic dust grains to asteroid-sized bodies in a period of roughly 20 Ma. Astronomical studies confirm that such dust exists around all newly formed stars. The accretion itself was accomplished by collision of the grains, which lose energy as they collide and aggregate as a result of their gravitational attraction. The aggregated bodies, growing from a centimeter to 100-km scale, are generally called planetesimals. Notable studies of the accretion process are by Greenberg et al. (1978), and Wetherill (1990). Isotopic studies of terrestrial material suggest that Earth reached most of its present size within as little as 50 Ma, by 4,500 Ma. The accretion of Earth involved impacts by both small and large planetesimals.

Lunar Origin

According to modern theories, the origin of the Moon was the most traumatic event suffered by Earth throughout its entire geologic history. This modern view is that the primordial Earth was hit by one of the largest nearby planetesimals, perhaps as big as Mars, and that this blasted material from the rocky mantles of both bodies into orbit around Earth. The Moon accreted quickly from that material. This collision probably happened toward the end of Earth's accretion, when there had been enough time for other planetesimals to grow, but after Earth's iron core formed. A plausible date is 4,520 to 4,500 Ma. (Some lunar rock chips have been dated as old as 4,440 +/− 20 Ma, so that the Moon must be older than this; see Carlson and Lugmair, 1988.)

This theory was first suggested by Hartmann and Davis on the basis of accretion calculations and lunar rock chemistry in 1975, and independently by Cameron and Ward in 1976, but it was not widely adopted until an international conference on lunar origin in 1984 (see Conference volume edited by Hartmann, Phillips, and Taylor, 1986). Pre-Apollo theories, such as co-accretion of the Moon alongside Earth, capture of the Moon after its formation elsewhere, or rotational fission of the Moon off Earth's equator, failed, in the face of lunar rock evidence, for various chemical and dynamical reasons. Evidence in favor of the giant impact theory includes: (1) relative proportions of the different oxygen isotopes are exactly the same in the Moon as in Earth, but different in other solar system bodies, implying that the Moon formed from terrestrial, or terrestrial-zone material; (2) the bulk composition of the Moon is similar to the bulk composition of Earth's mantle; and (3) the Moon lacks volatile constituents, which is explained by the tremendous heating of the material blown out by the impact.

According to computer models by researchers such as J. Melosh, W. Benz, A. Cameron, and their colleagues, the giant impact should have heated much of Earth's mantle material to temperatures much higher than would have occurred otherwise, had there been no impact. Current testing of the hypothesis centers around geochemical studies to see whether Earth's mantle shows evidence of this traumatic heating.

Intense Early Bombardment

As can be seen from the accretion models and the giant impact hypothesis for lunar origin, bombardment and cratering of the Earth by planetesimals were important in Earth's early history, as was also true for all other planets.

We have direct proof, by counting craters at different dated lunar landing sites, that the rate of impact cratering on the Earth averaged about 500 to 1,000 times higher 3,900 or 4,000 Ma than it does today, and that it declined thereafter.

The behavior of cratering before this time is the subject of controversy between two models. The more straightforward model assumes a relatively steady decline of impact rates from the period of planetary accretion, 4,550 to 4,000 Ma, followed by a slower decline until today. In support of this, Hartmann (1975, 1980) pointed out that the impact mass flux required to accrete the whole Earth in about 50 Ma is of the order 10^9 times higher than the measured mass flux today; if one assumes decline of the impact rate due to the sweep-up of the planetesimals as they hit planets between 4,500 and 4,000 Ma, one obtains a mean half-life of the planetesimals of about 20 to 25 Ma. Wetherill (1975, 1977) made purely dynamical studies of the attrition rate as the planetesimals were swept up by planets and obtained exactly the same figure. Thus, a logical and self-consistent model is that the impact rate simply declined from the beginning, as the planetesimals were swept up by the planets. This picture explains the extreme abundance of observed craters on the oldest surfaces in the solar system.

A remarkably different view was pioneered by Tera, Papanastassiou, and Wasserburg (1974), a group at Cal Tech who dated the first lunar rock samples. Much to the surprise of lunar geochemists, who had instructed astronauts to look for "genesis rocks" that would betray the origin of the Moon, virtually no lunar rocks were found older than 4,000 Ma. Therefore, these investigators hypothesized that a brief, intense episode of cratering—a so-called cataclysm—occurred about 4,000 to 3,900 Ma and destroyed all earlier rocks while creating most of the lunar craters we see today.

As a counter to this, Hartung (1974), Hartmann (1975), and others argued that the first model could also explain the observations. Due to the intense cratering, near-surface rocks formed before a certain critical age (about 4,000 Ma) would have very low probability of surviving until the present; but the cratering rate was dropping so fast between 4,500 and 3,900 Ma that rocks formed only 100 or 200 Ma after the critical date would have survived.

More recently, Ryder (1990) revived the cataclysm model on the grounds that there are very few fragments of impact-melted material older than 4,000 Ma. He argues that there was almost no cratering between 4,500 and 4,000 Ma, and that this quiescent period was followed by a cataclysm from 4,000 to 3,900 Ma, which produced most of the craters we see.

Physics aside, the controversy between these models has led to confused semantic usage. The term "early intense bombardment" was coined even before the Apollo flights to refer to cratering during the accretion and post-accretion era; it implies that the earlier the date, the more intense the bombardment. The terms "late bombardment" or "terminal cataclysm" were introduced by those writing about the cataclysmic models. Today, these contradictory terms are used almost interchangeably by speakers who fail to distinguish which model they are assuming.

The difference between these two cratering models of early Earth is not academic. First, it makes a big difference in Earth's environment 4,000 to 3,900 Ma—a crucial period in which organic chemical processes may have been leading toward living material. The extremely high cratering rates in this period, required by the cataclysm model, would have been destructive not only to the crust and oceans, but also to life. Second, the cataclysm model offers a different picture of the history of the planetesimals, and raises dynamical issues of how planetesimals were "stored" somewhere and then appeared suddenly in the inner solar system. Future critical dating of rock units among ancient surfaces on the Moon and other bodies may resolve this controversy about early cratering.

Early Atmosphere

Following the work of geochemist Harold Urey and others in the pre-Apollo era, most researchers believed that Earth's early atmosphere was a hydrogen-rich, reducing gravitational concentration of the solar gases from the surrounding solar nebula. This belief influenced, for example, the choice of ambient conditions for early test-tube chemical experiments on the origins of life.

A more current view is that, whether a local concentration of hydrogen-rich gases ever existed at the beginning, the original accretional bombardment strongly heated the surface, aided the sepa-

ration of iron into a core, and led to strong outgassing of volatiles, probably by 4,500 Ma. This atmosphere would have somewhat resembled the gases emitted today by deep-seated, mantle-tapping volcanoes. Today, these volcanoes are observed to emit mostly H_2O and CO_2. Modern models, for example by Matsui and Abe (1986), thus picture an early atmosphere of hot steam and carbon dioxide.

According to these models, as the accretional bombardment lessened and Earth cooled, the water condensed into early oceans, leaving a thick CO_2 atmosphere. (Evidence for a "native" CO_2-dominated, oxygen-poor atmosphere on early Earth includes the presence of CO_2 atmospheres on the neighboring planets Venus and Mars.)

Probably by 4,000 to 3,000 Ma, much of the CO_2 dissolved in ocean water, creating a weak carbonic acid solution. This acidic ocean water attacked rocks through reactions discussed by Urey, leading to deposition and "storage" of much of Earth's original atmospheric CO_2 in the form of deep-sea carbonate rock sediments (*see* UREY, HAROLD). Inventories of the total CO_2 in this form show that Earth has about the same total mass of CO_2 as Venus, except that Earth's has gone into its carbonate sediments and oceans, while Venus's remains in its atmosphere.

A still more extraordinary change in Earth's atmosphere came during Earth's middle age, as plants evolved and consumed more of the CO_2, releasing oxygen. Geochemists believe that the oxygen content rose dramatically around 2,500 to 2,000 Ma, from near zero to its present value of 21 percent by volume (Budyko, Ronov, and Yanshin, 1985). Evidence for this is the lack of highly oxidized "redbed" sedimentary deposits before this period.

Most people do not realize that Earth's atmosphere has evolved so dramatically during Earth's history, and that the barren, CO_2-dominated Earth in the first third of its history would have seemed quite an alien planet, compared to the fertile, oxygen-rich planet which today teems with plants and animals.

Origin of Continents

The evolution of continental crust on Earth is poorly understood. Melting and partial differentiation of the mantle presumably concentrate lower-density rock units in the crust, but did this produce homogeneous or heterogeneous crust? Lunar petrology showed that the Moon started with a "magma ocean" surface, probably melted by the heat input of the intense bombardment, yielding a relatively uniform feldspar-rich crust. If the Moon accreted in Earth orbit, however, it accreted faster than Earth. Uncertainty about Earth's continents was dramatized at a mid-1980s conference on continental evolution, where the program cover was a graph of continental area versus time during Earth's history; the curves, drawn from various papers in the literature, were scattered over the entire diagram. Some researchers assert that the crust started without continents and that they grew either rapidly or slowly; others have proposed something more like an initial global crust that broke and folded into continental blocks.

Probably it is erroneous to start with the question of whether endogenic processes could have made a uniform crust or discrete continents, because the intense early cratering assures that the early crust was pocked by craters and giant impact basins up to 1,000 km across, as on the Moon. Frey (1977) and others have suggested that such giant impact basins thinned the crust in some places, piled up thicker continental deposits in others, and may have even influenced the course of mantle convection.

The issue is dramatized by evidence from similar-sized Venus, which appears to have much smaller raised areas (granitic continents?) and no well-developed plate tectonics or plate-margin mountain belts.

Origin of Life

Following early work by A. I. Oparin, J. B. S. Haldane, H. Urey, and others, it is universally recognized that natural chemical processes must have produced amino acids and other important organic molecules—building blocks of life—on the early Earth (see also LIFE, EVOLUTION OF). In 1955 Stanley Miller showed that a flask containing water, methane, and other "primitive atmospheric gases," with sparks representing lightning on the early Earth, synthesized amino acids. The same result has been produced under other conditions, even with impacts instead of sparks as the energy source. In support of all these conclusions, extraterrestrial amino acids have been found in carbo-naceous meteorites, and comet nuclei have been found to contain a rich array of organic compounds, especially in so-called CHON particles, rich in carbon, hydrogen, oxygen, and nitrogen, in Halley's comet.

Recognition of intense cratering has influenced discussions of the problem of life's origins. Certain microfossil materials, believed to represent the earliest known life-forms, date from 3,500 Ma, but life was often assumed to have started earlier, shortly after Earth's formation. However, Maher and Stevenson (1988) and others have proposed "impact frustration of the origin of life," a delay of several hundred million years due to the unstable environment. NASA researcher Kevin Zahnle proposes that life may have started, been extinguished, and re-started in several episodes due to giant basin-forming impacts that could have sporadically vaporized oceans until 4,000 or 3,500 Ma. Growing evidence for major extinctions associated with lesser (still large-scale) impacts in the dwindling stages of cratering, even as recently as the 65-Ma-old KT boundary impact, support the emerging view that the random impacts of interplanetary debris may have played a role in life's evolution.

Bibliography

BUDYKO, M. I., A. RONOV, and A. YANSHIN. *History of the Earth's Atmosphere.* New York, 1987.

CAMERON, A. G. W., and W. WARD. "The Origin of the Moon." Houston, TX, 1976.

CARLSON, R. W., and G. LUGMAIR. "The Age of Ferroan Anorthosite 60025: Oldest Crust on a Young Moon." *Earth Planetary Science Letters* 90 (1988):119.

FREY, H. "Origin of the Earth's Ocean Basins." *Icarus* 32 (1977):235.

GREENBERG, R., J. WACKER, W. HARTMANN, and C. CHAPMAN. "Planetesimals to Planets: Numerical Simulation of Collisional Evolution." *Icarus* 35 (1978):1.

HARTMANN, W. K. "Lunar Cataclysm: A Misconception?" *Icarus* 13 (1975):299.

———. "Dropping Stones in Magma Oceans: Effects of Early Lunar Cratering." In *Proceedings of the Conference on Lunar Highlands Crust*, eds. J. Papike and R. Merrill. New York, 1980.

HARTMANN, W. K., and D. R. DAVIS. "Satellite-sized Planetesimals and Lunar Origin." *Icarus* 24 (1975): 504.

HARTMANN, W. K., R. PHILLIPS, and G. TAYLOR. *Origin of the Moon.* Houston, TX, 1986.

HARTUNG, J. "Can Random Impacts Cause the Observed Ar 39/40 Age Distribution for Lunar Highland Rocks?" *Meteoritics* 9 (1974):349.

MAHER, K. A., and D. J. STEVENSON. "Impact Frustration of the Origin of Life." *Nature* 331 (1988):612.

MILLER, S. L. "Production of Some Organic Compounds under Possible Primitive Earth Conditions." *Journal of the American Chemists Society* 77 (1955):2351.

RYDER, G. "Lunar Samples, Lunar Accretion and the Early Bombardment of the Moon." *Eos* 71 (1990): 313.

TERA, F., D. PAPANASTASSIOU, and G. WASSERBURG. "Isotopic Evidence for a Terminal Lunar Cataclysm." *Earth Planetary Science Letters* 22 (1974):1.

WETHERILL, G. W. "Late Heavy Bombardment of the Moon and Terrestrial Planets." *Proceedings of the Lunar Science Conference* 6 (1975):1539.

———. "Evolution of the Earth's Planetesimal Swarm Subsequent to the Formation of the Earth and Moon." *Proceedings of the Lunar Science Conference* 8 (1977):1.

———. "Formation of the Earth." *Annual Review of Earth and Planetary Science* 18 (1990):205.

WILLIAM K. HARTMANN

EARTH OBSERVING SATELLITES

During the 1980s, ambitious plans were laid for the beginning of a new era in earth studies. In the 1990s, a series of U.S. Presidential Initiatives created the U.S. Global Change Research Program (USGCRP), which includes the Mission to Planet Earth (MTPE) program as its largest component. Led by the National Aeronautics and Space Administration (NASA), the MTPE program seeks to improve our understanding of the earth as a system, and our ability to assess and predict the environmental, social, and economic impacts of natural and human-influenced processes—the overall goal to establish the scientific basis for national and international policy-making in response to changes in the Earth system.

Observations of Earth from space—whether from satellite instruments or photographs taken by astronauts—have changed the way we view our home planet. Images from space can provide a truly global perspective, showing the limits of a world dominated by water and shielded from space by a thin atmosphere. From space, Earth can be better seen as a unified whole; the land, oceans, and atmosphere in tandem provide our life-sup-porting environment. Over the last several decades, an increasing pace of scientific advancement in earth studies has coincided with enhanced public awareness of the global and regional aspects of environmental issues. Changes in Earth's climate over time had been documented and continue to be informed by sources such as tree rings, gases trapped in the polar ice caps, glacial landforms, and stratification in ocean sediments and in rocks. Today, scientists track processes related to global change, and possibly climate change, virtually as they occur.

Some of these global change processes have natural causes, but many may be related to the rapid increase and pervasiveness of human activities over the last few centuries. The industrial revolution and concurrent population explosion of the eighteenth century extended human influences over the entire planet. Environmental impacts that were once local effects have been compounded into regional and global phenomena. We cannot yet fully comprehend all the ways in which global-scale human activities stress the earth. Researchers lack the long-term, consistent measurements of key biological, chemical, and physical variables to help define changes in the earth system. These kinds of data can distinguish human impacts from natural causes. Other practical benefits include improvements in our ability to forecast weather, manage natural resources, protect public health, and plan for crops, forests, fisheries, wildlife, and human needs.

Mission to Planet Earth

NASA's Mission to Planet Earth focuses primarily on obtaining global observations from spaceborne instrumentation and combining these derived data with in situ measurements to develop models of Earth as a system. This integrated, comprehensive, and sustained program to document the planet supports exploratory studies of the physical, chemical, biological, and social processes that influence Earth system. Improvements over current capabilities in the range, detail, and frequency of observations are also sought to develop and test integrated conceptual and predictive models of Earth.

Three major tasks comprise Mission to Planet Earth—improving observations, data, and science. The program has been designed to exploit integrated observational capabilities (i.e., spaceborne, airborne, and ground- and sea-based measurement systems); to build a comprehensive data and information system, thereby making data more useful and readily available; and to train the next generation of earth scientists and engineers. A new generation of interdisciplinary earth scientists needs to be cultivated to analyze the data collected, build models of earth system, and provide the unique perspective required to improve understanding of Earth. Scientists are learning to cross traditional discipline boundaries, and in so doing create more confidence in our ability to predict the Earth's future state given various scenarios. Phase I of the program is well underway, with these three tasks integrated within a series of flight missions (Table 1). Phase II began in 1998, with the coordination of these tasks with the Earth Observing System (EOS) program (Table 2). In addition, the second phase of Mission to Planet Earth will continue support for smaller, unique projects and focused

Table 1. MTPE Phase I Missions: NASA Contributions

NASA Mission	Launch Status	Mission Objectives
ERBS Earth Radiation Budget Satellite	Operating	Earth radiation budget, aerosols, and ozone concentration
TOMS/Meteor-3 Total Ozone Mapping Spectrometer	Operating	Mapping and monitoring ozone (jointly with Russia)
UARS Upper Atmosphere Research Satellite	Operating	Chemistry and dynamics of stratosphere and mesosphere
ATLAS/Space Shuttle Atmospheric Laboratory for Applications and Science	Operated 3/92 and 4/93; October 1994	Chemistry and dynamics of atmosphere, and solar radiation
TOPEX/Poseidon Ocean Topography Experiment	Operating	Ocean surface circulation (jointly with France)
LAGEOS-II Laser Geodynamics Satellite II	Operating	Monitoring crustal motion and Earth rotation (jointly with Italy)
SIR-C/X-SAR/Space Shuttle Spaceborne Imaging Radar-C/X-Band SAR	Two missions in 1994	Mapping of land, vegetation, snow, ice, and oceans, and land topography (jointly with Germany and Italy)
LITE/Space Shuttle Lidar In-space Technology Experiment	September 1994	Atmospheric aerosols and their contribution to climate change
TOMS/Earth Probe Total Ozone Mapping Spectrometer	Launch in 1995	Mapping and monitoring ozone
SeaWiFS Sea-Viewing Wide Field Sensor	Launch in 1995	Role of phytoplanktons in ocean biology
NSCAT/ADEOS NASA Scatterometer	Launch in 1996	Role of surface winds in ocean circulation (joint with Japan)
TOMS/ADEOS Total Ozone Mapping Spectrometer	Launch in 1996	Mapping with monitoring ozone (joint with Japan)
TRMM Tropical Rainfall Measuring Mission	Launch in 1997	Precipitation, clouds, and radiation at low latitudes (joint with Japan)
Landsat-7 Land Remote-Sensing Satellite	Launch in 1998	Remote sensing of land surfaces

Table 2. Components of the Earth Observing System

Major Components	Purpose
EOS-AM Series (first launch 1998) Morning Crossing	Clouds, radiation, and aerosols; surface temperatures of land and ocean; vegetation and ocean phytoplankton; global biological productivity; and chemistry of the troposphere
EOS-COLOR (tentative launch 1998)	Ocean phytoplankton and biological production
EOS-ALT Radar Series (first launch 1999)	Ocean circulation
EOS-PM Series (first launch 2000) Afternoon Crossing	Clouds, radiation, and aerosols; precipitation and humidity; snow cover and sea ice; surface temperatures of land and ocean; vegetation and ocean phytoplankton; and global biological productivity
EOS-CHEM Series (first launch 2002)	Chemistry of the troposphere and stratosphere
EOS-ALT Laser Series (first launch 2003)	Glaciers, ice sheets, and land surface topography
EOS Data and Information System (first release in October 1994)	Data product generation, data archive and distribution, and information management from nine major data centers and numerous associated computing facilities; spacecraft and instrument operations
Interdisciplinary Science Investigations (selected in 1990)	Twenty-nine teams to further understanding of key Earth system processes; to develop predictive capabilities through numerical modeling; to increase the utility of satellite data; and to prepare for use of new types of data expected from new technologies
Instrument Science Teams (selected in 1990)	Twenty teams to obtain observations of key Earth system processes; to define scientific requirements for each EOS instrument for flight series; to generate algorithms for extracting information content of observations; and to develop new data products and establish their scientific utility
Global Change Graduate Student Fellowships (established in 1990)	Ensures a pool of highly qualified earth scientists and engineers to disseminate data generated by EOS

scientific investigations that require specific observations or timeliness that cannot be achieved by EOS or by other national and international programs.

Many small satellite programs already provide valuable data. A series of Total Ozone Mapping Spectrometer (TOMS) instruments has been monitoring global ozone levels since 1978; they are improved and replaced periodically to continue the time series. Another small project—the joint U.S./

Italy Laser Geodynamics Satellite II (LAGEOS-II)—was launched by NASA in 1992, and aids in measuring Earth's surface (i.e., continental drift/tectonics) and gravity field. The Sea-Viewing Wide Field Sensor (SeaWiFS) reinitiates the ocean color observations first made by the Coastal Zone Color Scanner (CZCS) in the 1978–1986 time frame.

Phase I observatories also include larger, multi-instrument satellites. Since 1991, the Upper Atmosphere Research Satellite (UARS) has measured atmospheric chemistry and radiation, including unique observations of the chemistry and dynamics of polar stratospheric ozone holes. The joint U.S./France Ocean Topography Experiment (TOPEX)/Poseidon was launched in 1992 to study ocean currents, and has provided the first global view of annual sea level changes and other atmosphere–ocean interactions.

Prior to the EOS era, two major collaborations between NASA and the space agencies of Japan will take place. The Advanced Earth Observing System (ADEOS) will be launched by Japan in 1996, carrying two NASA instruments—a TOMS instrument for ozone and the NASA Scatterometer (NSCAT), which will measure wind speed and direction over the oceans. The rest of the payload (one French and five Japanese instruments) will observe atmospheric chemistry, radiation, and ocean color and temperature. In 1997, the joint U.S./Japan Tropical Rainfall Measuring Mission (TRMM) will be launched to help understand the role of tropical regions on Earth's climate by quantifying the distribution of two-thirds of the global precipitation estimated to fall in the tropics and sub-tropics each year. TRMM will enhance understanding of the El Niño-Southern Oscillation phenomenon, which is responsible for interannual climate variations that produce drought, storms, and floods on a global scale. The next in the Land-Remote Sensing Satellite (Landsat) series will also be a part of Mission to Planet Earth. Landsat-7, scheduled to launch in 1998, will extend the time series of high-resolution land surface measurements begun in 1972.

The space shuttle provides short demonstration flights for new instruments. The utility of multifrequency Synthetic Aperture Radar (SAR) systems has been proven in viewing land and oceans under all weather conditions. A space shuttle flight carried the Lidar In-space Technology Experiment (LITE) and demonstrated the use of space-based light detection and radar (lidar) technology for ob-servations of aerosols, clouds, and the earth's surface. The shuttle has also carried well-calibrated research instruments for detailed studies from space. The Atmospheric Laboratory for Applications and Science (ATLAS) has flown since 1992, in both a research and calibration/validation capacity to verify the Upper Atmospheric Research Satellite (UARS) and the operational weather satellites atmospheric ozone and radiation measurements.

Aircraft missions serve a similar role by providing data and test beds to validate the capabilities of instruments that may eventually be flown in space. MTPE field experiments have taken advantage of the flexibility of aircraft to look at processes at various spatial and temporal scales. The most detailed field studies combine precise sampling of the environment, remote-sensing "snapshots" at intermediate scales from aircraft, and longer term monitoring at larger scales from satellites. Field programs have included studies of tropical rain forest burning, cloud systems, ocean carbon cycling, polar stratospheric ozone chemistry, volcanoes, grasslands, boreal forests, tropical ocean–atmosphere interactions, and glaciers.

In addition, the MTPE program supports research using other sources of existing satellite data from other U.S. agencies and other nations. Major global data sets of interest to earth scientists have been collected by the U.S. National Oceanic and Atmospheric Administration (NOAA) and the U.S. Department of Defense (DoD), and by space agencies in Europe, Japan, and France. These data include observations from weather satellites and high-resolution land remote-sensing systems. The availability and utility of existing satellite data sets has been improved by an interagency Pathfinder Program. For example, the Landsat Pathfinder has already yielded a significant scientific breakthrough. NASA acquired over 1,000 high spatial resolution Landsat scenes of the world's tropical forests collected between the early 1970s and the present. The extent and rate of tropical deforestation in the Brazilian Amazon region was quantified, and the study resulted in considerably lower estimates than those that had previously been made based on statistical extrapolations or lower resolution remote sensing. An analysis of all the world's tropical rain forest should be complete by the mid-1990s. Anticipated future work of the Landsat Pathfinder involves the study of land-use changes in North America over the last two decades. Other ongoing Pathfinder projects include work on high-priority data sets for vegetation, snow cover, sea surface temperatures and ice, radiation, clouds, atmospheric temperature and humidity, precipitation, and wind speeds.

The Earth Observing System

Mission to Planet Earth Phase II begins in 1998, with the first launch of an EOS satellite. As Table 2 reveals, the EOS flight series will make observations covering a broad range of variables critical to understanding global climate change. EOS satellites will carry payloads of one to six instruments pulled from a pool of twenty-three different EOS-funded instruments. Instrument complements have been developed based on simultaneity and synergy requirements, and associated accommodation constraints. Each series will be replenished with a new satellite every three to six years to provide a minimum of fifteen years of continuous global observations. EOS has been designed to focus on the most pressing issues in global climate change research, including:

- Water and Energy Cycles—Cloud formation, dissipation, and radiative properties, which influence response of the atmosphere to greenhouse forcing; and large-scale hydrology and moisture processes, including precipitation and evaporation
- Oceans—Exchange of energy, water, and chemicals between the ocean and atmosphere, and between the upper layers of the ocean and deep ocean, including sea ice and formation of bottom water
- Chemistry of Troposphere and Lower Stratosphere—Links to the hydrologic cycle and ecosystems, transformations of greenhouse gases in the atmosphere, and interactions inducing climate change
- Land Surface Hydrology and Ecosystem Processes—Improved estimates of runoff over the land surface and into the oceans; sources and sinks of greenhouse gases; exchange of moisture and energy between the land surface and atmosphere; and changes in land cover
- Glaciers and Polar Ice Sheets—Understanding their role in sea level change and global water balance
- Chemistry of the Middle and Upper Stratosphere—Chemical reactions, solar-atmosphere relations, sources and sinks of radiatively impor-

tant gases and their impact on global ozone concentration/distribution
- Solid Earth—Volcanoes and their role in climatic change

Scientists and program managers planning EOS based these priorities on the scientific uncertainty that now exists, particularly as it affects the accuracy of climate predictions. The EOS program sponsors studies of the extent, causes, and regional consequences of global climate change by providing not only the means for observing Earth, but also the improvements in data and information management, research, and assessment.

The EOS program can be considered the most ambitious single project ever dedicated to Earth system science and global climate change research. The development of EOS started in 1989, with the selection of instrument science teams, interdisciplinary science teams, and the onset of planning for the space component and a state-of-the-art data and information system. Building on existing and near-term missions, the EOS program supports scientific studies and enhanced access to existing data even before the launch of its first spacecraft. By developing computing and communications systems to receive, process, store, and distribute the large quantities of data generated by space-based observing systems during the EOS era, Mission to Planet Earth benefits nearer term objectives as well. The EOS Data and Information System (EOSDIS) strives to make data access easy, to preserve the knowledge base of data and research results, and to foster collaboration among earth scientists. To make earth science data understandable to as large a user base as possible, EOSDIS makes its operation transparent to the end user, removing the need for each user to have a detailed knowledge of remote-sensing instruments and methods. Along with scientific research, the user community for information and services could include those interested in education, commercial applications, resource inventories and planning, policy decision, and disaster assessment.

Researchers supported by Mission to Planet Earth study many aspects of the earth system, including radiation budgets, climate dynamics, clouds, hydrology, ecosystem dynamics and biogeochemical cycles, atmospheric chemistry, solid earth processes, oceanography, and polar ice dynamics—the foundation of which is a vigorous research and analysis program that promotes areas needing the most immediate attention and builds long-term research capabilities. A significant fraction of the national talent and infrastructure for conducting research in the earth sciences receives support directly or for limited times via the various NASA research centers. Science teams for each mission and/or instrument work on instrument design, development, calibration, and data applications. In addition, the EOS program supports interdisciplinary science investigation teams to focus on understanding key earth system processes and the development of predictive numerical models, and to increase the use and utility of existing satellite data in preparation for the new types of data expected from next-generation instrumentation.

When considering the system of satellites, instruments, information systems, and scientists assembled for Mission to Planet Earth, it is important to remember the vision that inspires the program. It is the vision of humanity taking greater responsibility for Earth by making informed decisions based on sound scientific understanding to enact environmental policy.

Bibliography

ASRAR, G.; S. G. TILFORD; and D. M. BUTLER. "Mission to Planet Earth: Earth Observing System." *Global and Planetary Change* (1992):3–8.

COMMITTEE ON ENVIRONMENT AND NATURAL RESOURCES RESEARCH. *Our Changing Planet: The FY 1995 U.S. Global Change Research Program.* Washington, DC, 1994.

DOZIER, J. "Planned EOS Observations of the Land, Ocean, and Atmosphere." *Atmospheric Research* 31(1994):329–357.

GURNEY, R. J.; J. L. FOSTER; and C. L. PARKINSON, eds. *Atlas of Satellite Observations Related to Global Change.* Cambridge, Eng., 1993.

UNNINAYER, S., and K. BERGMAN. *Modeling the Earth System in the Mission to Planet Earth Era.* Washington, DC, 1993.

GHASSEM R. ASRAR

EARTHQUAKES

An earthquake is a vibration of the ground caused by the passage of elastic waves through Earth. These "seismic" waves can be detected and recorded by instruments known as seismographs, even if they are too weak to be felt. The prefix "seis-," from the Greek *seismos* meaning "shock," is used in many words having to do with earthquakes. Thus, seismology is the scientific study of earthquakes, while seismicity refers to the frequency, strength, and geographical distribution of earthquakes.

Causes of Earthquakes

Although seismic waves may have a variety of natural or artificial origins, including the movement of magma within volcanoes, chemical and nuclear explosions, and rock bursts in mines, by far the most prevalent cause of earthquakes is tectonic stress in the lithosphere, the uppermost 70 km or so of the solid Earth. Geologists use the term "tectonic" to refer to forces and processes responsible for mountain-building and deformation of rock. The origin of tectonic forces is poorly understood but seems to lie in the interaction between the lithosphere and the underlying, more fluidlike asthenosphere. According to the widely accepted plate tectonics theory (*see* PLATE TECTONICS), the lithosphere is divided into various segments (plates) that are moving over, and in some places plunging into, the asthenosphere. The latter process is called subduction. Resistance to plate motion at the boundaries of plates causes stress to build up in the lithosphere.

Investigation of the 1906 San Francisco earthquake by the California State Earthquake Commission headed by A. C. Lawson resulted in the first definite attribution of a tectonic earthquake to slippage of rock along a geologic fault (the famous San Andreas fault). The geologist GROVE KARL GILBERT had suggested an association between earthquakes and faults as early as 1884 (Bolt, 1985). In 1910, Harry Fielding Reid, Professor of Applied Mechanics at the Johns Hopkins University, proposed the elastic rebound theory, which holds that the energy derived from tectonic stress is stored in the form of elastic stress until it is suddenly released by rupture of a fault. This process may be compared to stretching a rubber band until it breaks.

The mechanics of fault rupture during an earthquake is a complex process only partially understood. To a first approximation, the source of the seismic waves may be thought of as a planar surface of finite area, on either side of which rock slips in opposite directions during an earthquake.

The slip is essentially parallel to the fault plane, which may be vertical or inclined at some angle. The sense of slip may be horizontal (strike-slip), up and down the dip of the fault plane (dip-slip), or some combination (oblique-slip). The amount of slip, which may be as much as several meters in great earthquakes, varies over the active area of the fault. Seismologists have found the concept of seismic moment—a quantity directly proportional to the product of the average amount of slip and the active area of the fault—to be a useful measure of the size of an earthquake, as discussed more fully below in connection with the concept of earthquake magnitude.

Viewed on a global scale, the source of the seismic waves can be considered to be a single point within Earth, known as the hypocenter, or focus, of the earthquake. The point on Earth's surface directly above the hypocenter is called the epicenter. Hypocenters usually are located only in the uppermost 70 km of Earth, where rock is relatively brittle. Subduction zones, however, are the sites of deep-focus earthquakes, with hypocenters as much as 700 km below Earth's surface.

Intensity and Magnitude

Early attempts to classify earthquakes by size were based on the effects of the earthquakes, and several such earthquake intensity scales remain in use today. The one generally used in the United States is a modification of the intensity scale developed in Italy by Guiseppe Mercalli and hence is known as the Modified Mercalli (MM) scale. The MM scale classifies earthquake intensity on a scale of one to twelve, usually represented by Roman numerals, based on what people report that they felt or heard, how much damage was done, and various other effects such as eruptions of sand or mud from the ground. A complete description of the Modified Mercalli Scale can be found in the books listed in this entry's bibliography.

Although maximum MM intensity values are useful for comparing one earthquake to another, and especially for evaluating how local geology and topography influence earthquake effects, intensity is not the most reliable means of classifying earthquakes by size. The earthquake magnitude scale, developed in the 1930s by Professor Charles Richter at the California Institute of Technology and known popularly as the Richter scale, is more ob-jective and reliable than intensity scales and can be applied even to earthquakes with epicenters in un-populated regions.

Magnitude is calculated from measurements of the amplitude of seismic waves recorded by a seismograph. Richter assigned a magnitude value of three to an earthquake that produced a maximum amplitude of 1 mm on a certain type of seismograph located 100 km from an epicenter. He then defined each ten-fold increase or decrease in maximum amplitude on the same type of seismograph at the same distance as one whole step on the magnitude scale. The key to the success of the Richter scale is the fact that maximum recorded amplitudes decrease in a consistent manner with increasing distance from the epicenter, so a magnitude value can be calculated from measurements made at distances other than 100 km. The Richter scale is not a ten-point scale, as seems to be commonly believed, but is open-ended, like the Celsius or Fahrenheit scales used to measure temperature. Small earthquakes may have negative magnitude values, while the largest earthquakes have magnitudes exceeding nine. This is not the upper limit of the scale but merely reflects an apparent limit to how much elastic energy the earth can store and release at one time. Various studies of the relationship of magnitude to energy suggest that each whole-number increase on the Richter scale represents slightly more than a thirty-fold increase in seismic energy, but the precise relationship between energy and magnitude is complex and not well understood.

Richter developed the magnitude scale originally for use only in southern California to compare the sizes of relatively local earthquakes. Later, the concept of magnitude was applied in other regions and to earthquakes recorded at great distances from their epicenters. These extensions of the Richter scale necessitated certain changes in the methods used to arrive at a magnitude number, with the result that seismologists now recognize several different kinds of magnitude. Ideally, all methods should give the same numerical result for a given earthquake, but in practice this is not always the case. A serious problem with most methods is that the scale saturates at high magnitudes. That is, all great earthquakes appear to have magnitudes of about 8.5, even though some may actually be larger than others in terms of energy released. To get around this problem, seismologists have begun using the concept of moment magnitude, especially for large earthquakes. The moment of the earthquake—a number proportional to the area and slip of the fault—can be found from analysis of seismograms and then used to calculate a magnitude number. Moment magnitude is consistent with magnitudes arrived at by more traditional methods up to values of about 7.5 but continues to rise without saturation for the largest earthquakes.

Seismicity of Earth

Earthquakes are not rare phenomena. Dozens of earthquakes are recorded worldwide every day, but most cause no damage to buildings or injury to people because they are of small magnitude or their epicenters are in remote regions. The actual number of earthquakes is, no doubt, considerably greater than the number recorded because of the difficulty in detecting earthquakes with magnitudes less than about four. Richter and his colleague Beno Gutenberg discovered that the number of earthquakes occurring each year down to magnitude four increases logarithmically with decreasing magnitude. They estimated the annual number of earthquakes with magnitude values of at least four to be over 7,000 (Gutenberg and Richter, 1954). During the twentieth century (the era of seismographic recording), the number of earthquakes of magnitude seven or greater has averaged twenty per year, ranging from a high of forty-one in 1943 to a low of six in 1985.

Although historical experience suggests that earthquakes occur more commonly in some places than in others, it was not until the installation of the World Wide Standardized Seismograph Network (WWSSN) in the late 1950s that seismologists realized just how well-defined the world's earthquake belts are. The impetus behind the WWSSN was the desire of the United States to have a means of detecting nuclear weapon tests, but the stations, operated by scientists of the host country with financial support from the United States, also greatly improved seismologists' ability to detect and locate earthquakes.

The great majority of earthquakes occur at active plate boundaries. Earthquakes do occur in the interior regions of plates, however, especially in East Africa, Central Asia, Australia, and eastern North America.

The Earthquake Hazard

No natural phenomenon holds more potential for sudden death and destruction than a great earthquake. The earthquake that struck Shaanxi Province, China, in 1556, killing an estimated 830,000 persons, probably was the greatest natural disaster of all time, not counting plagues and famines. The official death toll from the 1978 earthquake that totally destroyed the Chinese city of Tangshan was "only" 243,000, but many experts believe that the actual number may have been more than three times larger. Some other notable earthquake disasters, with the year and number of persons estimated to have lost their lives, are:

Aleppo, Syria (1138; 100,000)
Naples, Italy (1626; 70,000)
Lisbon, Portugal (1755; 60,000)
Messina, Italy (1908; 58,000)
Ningxia Province, China (1920; 200,000)
Tokyo, Japan (1923; 99,300)
Yungay, Peru (1970; 67,000)
Armenia (1988; 24,900)

The United States has been relatively fortunate with respect to earthquake disasters, with only the 1906 San Francisco earthquake causing more than about 100 deaths.

People may be killed or injured in an earthquake in various ways. Many people are killed by the collapse of buildings, caused either by the direct effects of the seismic waves or by the phenomenon known as liquefaction, in which water-saturated soil loses its bearing strength when shaken. Earthquake waves may trigger landslides that can bury entire towns; this is what happened to Yungay, Peru, in 1972. Other causes of injury and destruction in earthquakes are seismic sea waves (tsunamis) that sometimes are generated by earthquakes with offshore epicenters, and fire that may follow an earthquake.

The intensity of ground shaking in an earthquake depends not only on the size of the earthquake and distance from the epicenter, but also—and sometimes critically—on the nature of the surficial geologic materials on which buildings are founded. As a general rule, damage is much more severe in areas underlain by unconsolidated sediment or thick soil than it is in areas where bedrock is present at or near Earth's surface. In the Mexico

City earthquake of 1985, only those parts of the city built on filled ground over beds of lake sediments were seriously damaged.

Earthquake Hazard Mitigation

Earthquake hazard reduction begins with efforts to identify and characterize so-called seismic source zones. These may be known faults or simply regions where earthquakes have occurred in the past, even if they cannot be attributed to specific faults. Regional networks of seismograph stations have proven to be extremely valuable in this work. The first regional seismographic network, operated by the University of California at Berkeley, consisted of fifteen stations in central California linked together by telephone lines. Later, the U.S. Geological Survey (USGS) began installation of a network south of San Francisco that eventually included about 200 stations (Bolt, 1985). Many additional networks now exist in California and other regions.

Probability theory commonly is applied to the characterization of seismic source zones. For example, in 1987 the USGS published estimates of the probabilities of earthquakes of various magnitudes occurring along various segments of the San Andreas fault before the year 2018. Probabilities also have been estimated for the acceleration of the ground exceeding certain values in various regions within a stated period of time.

An important aspect of earthquake hazard reduction is the design and construction of earthquake-resistant structures. A fundamental principle is that buildings must be designed to resist horizontal loads as well as vertical ones. Much progress is being made in earthquake engineering, but a detailed discussion of this field is beyond the scope of the present article.

Another facet of earthquake hazard mitigation is preparation for the earthquakes that inevitably will strike. This work involves the development of emergency response plans, the training of fire, police, and medical personnel, and education of the populace about what to expect in an earthquake and what precautions can be taken in the home, school, and workplace.

Efforts at developing means of predicting the specific time and place of an earthquake have met with little success so far. Among the phenomena that have given some promise of foretelling an earthquake are:

- Changes in the elastic properties of rock in the vicinity of a potential hypocenter as revealed by changes in the velocities of seismic waves passing through the rock
- Tilting of the ground surface
- Changes in the strengths of local magnetic or electrical fields
- Changes in the yields of wells, the flow of springs, or the amount of radon in ground water
- An increase in the occurrence of small earthquakes
- Unusual behavior of wild and domesticated animals

None of these phenomena, however, has proven to be reliable as an earthquake predictor.

Bibliography

Bolt, B. A. *Earthquakes.* New York, 1993.
———. *Earthquakes and Geological Discovery.* New York, 1993.
———. "The Development of Earthquake Seismology in the Western United States." In *Geologists and Ideas: A History of North American Geology,* eds. E. T. Drake and W. M. Jordan. Boulder, CO, 1985.
Gere, J. M. and H. C. Shah. *Terra Non Firma: Understanding and Preparing for Earthquakes.* New York, 1984.
Gutenberg, B., and C. F. Richter. *Seismicity of the Earth and Associated Phenomena,* 2nd ed. Princeton, NJ, 1954.
Howell, B. F., Jr. *An Introduction to Seismological Research.* Cambridge, Eng., 1990.
Thomas, G., and M. M. Witts. *The San Francisco Earthquake.* Briarcliff Manor, NY, 1971.

CHARLES K. SCHARNBERGER

EARTHQUAKES AND SEISMICITY

An earthquake is a sudden violent movement of one rock mass past another. Typically this movement takes place on a planar surface known as a fault and is a shearing slip of one side of the fault with respect to the other in response to elastic forces stored in the rock mass (*see also* EARTHQUAKES). The total amount of slip and the area of the fault that ruptures determine the size of the earthquake. The number and sizes of earthquakes define the seismicity of a region.

Most earthquakes are believed to occur as part of a cycle of elastic strain (or stress) accumulation that culminates with sudden release by slip on a fault. This is called elastic rebound theory. The jerky, "stick-slip" motion of a block being pulled by a spring is a useful analogy; in this case the block slips forward only when sufficient force has built up in the spring to overcome the frictional resistance of the block to slide. When the accumulated force is released, the block comes to rest; if you continue to extend the spring, the cycle begins all over again.

When an earthquake occurs, part of the energy released takes the form of "seismic" elastic waves (sound waves are another example of elastic waves), which travel away from the earthquake source or hypocenter. Seismology is the study of the passage of these waves through the earth. From measurement of these waves we can determine the location and size (magnitude) of the earthquake, and also the type of fault and fault movement that caused the earthquake.

There are two main types of seismic waves: body waves and surface waves. Body waves travel through the entire body of the earth, whereas surface waves are guided near the surface of the earth (upper few kilometers to a few hundred kilometers depth). Because seismic waves from large earthquakes travel completely through and around the earth, earthquake seismology is our most powerful technique for studying the structure of the earth's interior. Seismology defines the major internal divisions of the earth—the crust, mantle, and core—and shows that a liquid outer core surrounds the earth's solid inner core.

Detection and timing of seismic waves by seismometers, instruments that very accurately measure and record small ground motions (motions in the range of only 0.1 mm to 10+ cm) caused by the passage of seismic waves, allow for accurate determination of the location of the earthquake as well as its magnitude. Body waves travel most rapidly in the earth, and it is their arrival that marks the local perception of an earthquake; for earthquakes at distances less than 1,000 km the body waves travel at speeds of between 4 and 8 km/s through the earth's crust.

Earthquake Locations and Plate Tectonics

An integrated global network of seismometers, which provides uniform global detection coverage for larger earthquakes, was established in the 1960s. This global seismometer network confirmed earlier work, indicating that earthquakes did not occur randomly around the globe but rather were largely restricted to discrete bands of activity (see Plate 4). Data from this network also confirmed that earthquakes can occur as deep as about 700 km, which is about one-ninth of the earth's radius of 6,300 km.

Two main categories of earthquakes have been defined: shallow to moderate focus (depth) events and deep focus events. The shallow to moderate focus events occur at depths generally less than 70 km and are by far the most destructive and the most common (78 percent of all located earthquakes since 1964). The deep focus events occur in the depth range 70–700 km. The deepest earthquakes occur in inclined zones of seismicity extending from near the earth's surface along the margins of oceans to a depth of more than 600 km (first noted by Professor Kiyoo Wadati in Japan in 1935). The global pattern of discrete bands of seismicity, the occurrence of inclined zones of deep earthquakes, as well as the type of earthquake faulting occurring in different regions were among the fundamental observations that firmly established the great unifying theory of modern earth science, PLATE TECTONICS.

Plate tectonics defines the motion of the outermost layer of the earth, the earth's crust and the uppermost mantle (lithosphere), which is broken into tectonic "plates" typically 100–200 km thick. These tectonic plates move relative to one another with somewhat regular average rates over geologic time (on the order of 1–20 cm/yr) and cause repeated earthquakes along plate boundary faults.

Three main types of plate boundaries have been identified: divergent, convergent, and transform. Each type of plate boundary is associated with distinct types of faulting, so that fault movements in California, for example, are different than those occurring deep beneath Japan. Divergent plate boundaries (generally coinciding with mid-ocean ridges) are the site of upwelling material and the formation of new crust; these boundaries have only shallow earthquakes (generally less than 5–10 km). In contrast, convergent plate boundaries result in shortening, which is accommodated either by formation of collisional mountain belts, such as the Himalayas, or by subduction zones where cold oceanic plates are subducted under an adjacent plate. These subduction zones are delineated by the inclined zones of deep earthquakes discovered by Wadati. Most of the world's active volcanoes occur above these inclined seismic zones and are a result of melting of the cold oceanic slab as it descends into the warmer mantle. At transform boundaries, the tectonic plates slide horizontally past one another, such as along the San Andreas fault zone in California where the Pacific Plate moves northwest past the North American Plate.

Earthquake Magnitudes

The concept of earthquake magnitude was developed by CHARLES RICHTER, a seismologist at CalTech, in the early 1930s. Prior to that time, quantification of the size of an earthquake focused on how an earthquake was felt or the degree of damage it produced. Several numerical scales were developed to grade earthquakes; perhaps the best and still most widely used is the modified Mercalli intensity scale. Mercalli intensities are given in Roman numerals (I to XII) and are based on levels of local perception (e.g., felt only by those at rest, or difficult to stand during, etc.) or the level of damage (e.g., damage to masonry, or complete collapse of buildings, etc.). A major drawback of this method of quantifying earthquake size is that intensity varies with distance from the earthquake source, and reports of damage depend on population density, local site conditions, and construction quality; thus, a whole range of intensity values characterize a single earthquake.

To avoid these drawbacks and to quantitatively compare the size of separate events, Richter borrowed the concept of a magnitude scale from astronomers who used a similar scale to classify the brightness of stars. He decided to rely on the direct measurement of the maximum amplitude of ground motion at a given distance, which depends on the energy released in the earthquake. Knowing that this amplitude varied as a function of distance from the earthquake source, a simple correction could be made (depending on the distance of the seismometer from the earthquake) so that all earthquakes could be described by a single number related to the energy released. This led to the still popular Richter magnitude scale.

The seismometers Richter used have been largely replaced by much more sophisticated and capable instruments. These newer instruments can record the entire seismic wave train. This is important because Richter magnitude determinations are not meaningful for events much above about magnitude 7, because the earth does an increasingly poor job of transmitting the body waves for these very large events that the Richter magnitude determination requires. A number of magnitude scales are in use by seismologists relying on measurements of different parts of the wave train. The preferred magnitude scale for large events, the so-called moment magnitude, is proportional to the total amount of energy released by fault slip in the earthquake; however, seismologists have adjusted all other magnitude scales to be in good agreement to the Richter magnitude over the range for which that scale is most applicable.

All earthquake magnitude scales are logarithmic, which means that one unit difference in magnitude corresponds to a difference of a factor of 10 in local ground motion. However, the amount of energy released in earthquakes varies by a factor of 32 for each unit of magnitude. The 1995 earthquake at Northridge, California, had a magnitude of 6.7, whereas the largest instrumentally recorded earthquake was the 1960 Chilean event with a magnitude of 9.6. This 2.9 difference in magnitude means that the Chilean earthquake produced ground motions about 800 times larger than the Northridge earthquake and were caused by the release of 23,000 times the energy.

The global frequency of occurrence of earthquakes by magnitudes is given in Table 1. The minimum magnitude for a felt earthquake is generally about magnitude 2.5–3.0, provided the earthquake occurs close by (fewer than 40 km distant). Although roughly 1,000 magnitude 2–3 events are estimated to occur each day, most of these go unfelt because they occur far from where people are living.

Intraplate Earthquakes

While most of the world's earthquakes occur along plate boundaries, significant seismicity occurs in broad (up to 1,000 km wide) zones directly adjacent to plate boundaries (e.g., Mediterranean, western United States, Iran, China). However, the mid-plate regions of many plates are also seismically active. The mid-plate seismicity is generally

Table 1. Frequency of Occurrence of Earthquakes Based on Observations Since 1900

Descriptor	Magnitude	Average Annually
Great	8 and higher	1
Major	7–7.9	18
Strong	6–6.9	120
Moderate	5–5.9	800
Light	4–4.9	6,200 (estimated)
Minor	3–3.9	49,000 (estimated)
Very Minor	<3.0	2–3 about 1,000 per day (estimated)
		1–2 about 8,000 per day (estimated)

From National Earthquake Information Center QED tables.

diffuse, and far less common in the interior "cores" of the continents which contain the oldest rocks (cratons). The faults that cause mid-plate earthquakes are rarely exposed at the surface or identified at depth. The historic record indicates that these earthquakes occur infrequently. The cause of mid-plate earthquakes is not well understood; however, most events have been shown to occur in response to the same broad scale forces that drive plate motions.

Mid-plate earthquakes are often deadly because they often hit areas poorly prepared for earthquake shaking. A recent example is the 1993 M = 6.4 Latur earthquake in central India, which killed more than 11,000 people, most in poorly constructed homes. The Latur earthquake struck a region with no historical record of earthquakes. Furthermore, three of the most powerful earthquakes in U.S. history occurred along the Mississippi Valley in the central United States in 1811–1812, in a mid-plate zone that still exhibits a high level of seismicity. The effects of these three magnitude 8+ earthquakes were felt over nearly the entire eastern half of the United States. One of the earthquakes caused church bells to ring in Boston, 1,000 miles away.

Effects of Earthquakes

A number of factors determine how strongly an earthquake is felt and the resulting level of damage. Earthquake magnitude is the most obvious factor. The larger an earthquake, the more energy released, thus, the larger the ground motion and the longer periods of a high level of shaking. Distance from the earthquake is the next most obvious factor, as the amount of ground motion from an earthquake falls off rapidly with distance. However, for similar magnitude earthquakes at a similar distance, a number of additional factors influence the level of shaking at a given site. The most important source effect is depth; in general, the deeper an earthquake, the less its effects are felt at the surface, although it is usually felt over a broader area.

As both the 1985 Mexican earthquake and the 1989 Loma Prieta, California, earthquake vividly demonstrated, local site conditions factor heavily in determining the level of shaking and damage. Some of the most severe damage occurred in Mexico City at a distance of roughly 400 km from that earthquake. The dramatic building destruction and freeway collapse in San Francisco and Oakland occurred more than 100 km distant from the Loma Prieta hypocenter. The high levels of distant shaking and damage were due to unstable, water-saturated mud in the Loma Prieta case, and a thick basin of soft sedimentary rock in the Mexico City case.

The deadliest earthquake on record occurred in 1557 in central China, striking a region where most people lived in caves carved from soft rock. These dwellings collapsed during the earthquake, killing an estimated 830,000 people. Ten historic earthquakes have had death tolls exceeding 100,000 people. Due to the growth of population, five of these deadly events have occurred this century (three in China: 1976, 1927, and 1920; one in Japan in 1923; and one in Italy in 1908); the earlier, deadlier earthquakes occurred in China (in 1290 and 1556), Iran (856, 893), and Syria (1138).

Earthquake Prediction

The earthquake topic that probably most intrigues the public imagination is earthquake prediction. Seismologists have suggested that a variety of physical phenomena occur prior to an earthquake, including: changes in the speed of body waves of small earthquakes; unusual electric or magnetic signals; chemical changes in soil gas and water; changes in the level of water in wells; and anomalous animal behavior. In addition, some studies have attempted to correlate earthquake occurrence with regular external forces acting on the earth, such as tidal fluctuations. Unfortunately, none of these methods or techniques have been shown to be repeatedly reliable in controlled scientific tests. For example, reports of anomalous animal behavior prior to earthquakes are generally anecdotal and reported after the fact. Would such "unusual" behavior have even been noted had an earthquake not occurred? (*See also* EARTHQUAKES.)

On 4 February 1975 the Chinese successfully predicted a M = 7.5 earthquake near Haicheng in the Manchurian province. Nearly three million people were evacuated from their poorly constructed homes and when the earthquake struck, only 300 people were killed, rather than thousands. The prediction by Chinese seismologists was based on a variety of phenomena, including an ongoing and increasing level of moderate seismicity; uplift and tilting of the ground surface; changes in the earth's magnetic field; and unusual animal behavior that included snakes being found dead at the surface, having left their winter hibernation in underground holes. Other suggested premonitory phenomena, such as changes in the seismic velocity of crustal rocks, were apparently not observed. And the anomalous animal behavior, particularly the death of snakes and rodents, may be explained by the rising groundwater levels as measured in the region's wells.

As successful as this prediction was, other Chinese scientists carefully monitoring many similar types of phenomena failed to predict a M8 earthquake in Tangshan, China, roughly a year later on 28 July 1976. This earthquake struck in a densely populated region 160 km east of Beijing, killing an estimated 250,000 people, making it the deadliest earthquake this century and perhaps the second deadliest earthquake on record.

Earthquake prediction remains a goal of seismologists; however, it is now clear that the earth is a very complex machine and no one simple observation will clearly warn us of an impending earthquake. Research into the physical processes acting on and in earthquake faults will help to unravel the workings of this complex machine.

Bibliography

FOWLER, C. M. R. *The Solid Earth.* Cambridge, Eng., 1990.

FROHLICH, C. "Deep Earthquakes." *Scientific American* (August 1989):48–55.

RICHTER, C. F. *Elementary Seismology.* San Francisco, 1958.

WALKER, B. *Earthquake.* Alexandria, VA, 1982.

MARY LOU ZOBACK

EARTH'S ATMOSPHERE

We live in Earth's atmosphere. It is the air that we breathe and into which we dump unwanted gases, the products of industrial as well as metabolic activity. The atmosphere brings us wind and weather, and it protects us from the fierce radiations of space. How much atmosphere is there? What is it made of? How do its properties vary from place to place? Does it have a history or are its properties fixed for all time? How do we learn about Earth's atmosphere and its history?

Earth's atmosphere is a layer of gas held on the planet by the force of gravity. Atmospheric pressure, which we can measure with a barometer, corresponds to the weight of the overlying air. Pressure is force per unit of area. Consider the balance of forces at the bottom of a mercury barometer. Inside the barometer stands a column of mercury that on average at the ground is 74 centimeters (cm) high. Because the density of mercury is 13.6 grams per cubic centimeter (g/cm^3), the mass of mercury in the column is $13.6 \times 74 = 1,006$ grams per square centimeter (g/cm^2). Weight is equal to mass times gravitational acceleration. The downward force of the mercury is the mass times Earth's gravitational acceleration of 980 cm per second per second. This is the force that Earth's gravity exerts on the mercury, and it is equal to the pressure at the bottom of the column. But the pressure of the mercury at the bottom of the barometer equals the pressure of the atmosphere. Without this balance, the height of the column of mercury would change. So the mass of atmosphere per unit area outside the barometer is equal to the mass of mercury per unit area inside the barometer. To calculate the mass of the whole atmosphere it just remains to multiply the mass per unit area, deduced from average pressure at the surface, by the surface area of Earth, 5.1×10^{18} cm². The answer is 5.1×10^{21} g. This is very much less than the mass of the ocean, 1.4×10^{24} g, or the mass of the solid part of Earth, 6.0×10^{27} g. The large difference in masses is important. Because the atmosphere contains relatively little material, its properties can be altered by the release of gas from ocean or solid Earth. A release that has relatively little effect on the average composition of ocean or solid Earth can have a large effect on the atmosphere.

Even though Earth's atmosphere is a very thin layer spread out over a very large sphere, its properties vary more rapidly with height than with position on the globe. This rapid variation with height is imposed by Earth's gravity. Atmospheric pressure is equal to the weight of the air above. As altitude increases there is less air above and therefore less weight. Pressure therefore decreases with increasing altitude. But air responds to a decrease in pressure by expanding. Air at low pressure is less dense than air at high pressure, other properties being equal. Air density, therefore, decreases with increasing altitude as a direct consequence of the decrease in pressure. The two changes are connected, however, because the weight of the air above changes less rapidly with altitude at high altitudes where density is small than at low altitudes where density is relatively large. So pressure and density decrease with altitude rapidly at the bottom of the atmosphere and then more and more slowly; the atmosphere tails off gradually into the near vacuum of space rather than exhibiting a clear top like the top of the ocean. The ocean is different because water does not expand significantly when pressure is reduced. Although pressure increases with depth in the ocean, the density of seawater remains almost constant.

On average, pressure and density decrease by 50 percent for each increase in height by 5.8 km. Thus, at a height of 5.8 km, air pressure and density are only one half of their values at the ground. At a height of 11.6 km they are only one quarter. Although Earth's atmosphere extends thousands of kilometers above the surface, most of the air is close to the ground. If the atmosphere, like the ocean, were a layer of constant density with density equal to the value at the ground, the layer would be just 8.4 km thick.

Air temperature also varies rapidly with altitude. In the lowest 10 to 20 km of the atmosphere,

where we live, the air is stirred by winds driven by horizontal variation in pressure and by vertical motions driven by differential heating and cooling. When air moves up in the atmosphere, it is subjected to lower pressure, it expands, and as it expands, its temperature falls. In the same way, air that moves down to lower altitudes and regions of higher pressure is compressed. Its temperature rises. The result is that temperature decreases with altitude in the lowest 10 to 20 km of the atmosphere at an average rate of 6.5°C per kilometer of altitude. Average surface temperature is about 15°C. Average temperature at a height of 10 km is about −50°C. This decrease of temperature is reversed at higher levels as a result of the absorption of solar radiant energy in the ultraviolet region of the spectrum. For example, atmospheric ozone absorbs near ultraviolet radiation from the Sun, protecting living creatures from this harmful radiation. The absorbed energy causes atmospheric temperature to increase from about −50°C at a height of 20 km to about 0°C at a height of 50 km.

We can find out what the atmosphere is made of by direct measurements on atmospheric gases using a variety of methods. Such measurements have shown that air is a mixture principally of nitrogen, oxygen, and argon, in proportions that do not vary from place to place or from day to day. The proportions—by numbers of molecules, not by weight—are 78 percent nitrogen, 21 percent oxygen, and 1 percent argon. In addition, there are a host of gases that are generally present in smaller and variable concentrations (see EARTH'S ATMOSPHERE, CHEMICAL COMPOSITION OF). Of these, the most important and abundant are water vapor and carbon dioxide, both of which affect the climate. Other gases that are present in small concentrations in unpolluted air include methane, hydrogen, carbon monoxide, ozone, nitrous oxide, nitric oxide, nitrogen dioxide, ammonia, sulfur dioxide, and hydrogen sulfide. Polluted air contains enhanced concentrations of many of these gases as well as more than a thousand chemical compounds that are products of industry and other human activities.

Our understanding of the history of Earth's atmosphere is based largely on inference and deductions from general scientific principles. Direct evidence of atmospheric evolution is meager. It is clear that Earth, like the other inner planets, is not made out of the same material as the Sun, which is mainly hydrogen. Careful comparison of the average composition of Earth, including the solid part, the ocean, and the atmosphere, with the average composition of the Sun shows that Earth is relatively deficient in gases, particularly the inert gases helium, neon, argon, krypton, and xenon, which do not form solid or liquid chemical compounds. We infer that Earth was formed by the accumulation, under the action of gravity, of a collection of little planets without atmospheres at a time when most of the gas in the solar system had already been gathered into the Sun or into the outer planets. The gases and liquids that now constitute Earth's atmosphere and ocean were chemically bound in solid form in the little planets.

The bombardment of the growing Earth by little planets converted gravitational energy into heat. High temperatures caused chemical reactions that drove the gases and liquids to the surface, out of the solid part of the planet. During the period of bombardment, about 4.5 billion years (Ga) ago, Earth's surface was probably hot enough to prevent the condensation of the water that now constitutes the ocean. Earth's atmosphere, therefore, began hot, massive, and mostly steam. Chemical reactions at these high temperatures would have driven into the atmosphere all of the carbon that is now combined in rocks in the form of carbonate minerals and organic debris, so carbon dioxide would have been abundant also.

Bombardment largely ended when Earth had collected all of the material in the local region of the solar system, a process believed to have lasted only a few tens of millions of years. Without the heating provided by bombardment, surface temperatures fell, the water vapor condensed to form the oceans, and Earth was left with an atmosphere composed mainly of carbon dioxide. This carbon dioxide reacted over a period of hundreds of millions of years with seawater and with the materials of Earth's surface to form solid carbonate minerals, the constituents of limestone. Removal of carbon dioxide into solid compounds left nitrogen as the most abundant atmospheric gas. All of these changes took place before the formation of the oldest rock yet found on Earth. There is no direct evidence for this history. It represents a scientific best guess of how Earth's atmosphere ought to have originated according to our knowledge of the properties of Earth materials and the early history of the solar system. The subsequent history of Earth's atmosphere has been profoundly affected by life and concerns largely the evolution of chemical composition.

Bibliography

BROECKER, W. S. *How to Build a Habitable Planet.* Palisades, NY, 1985.
GOODY, R. M., and J. C. G. WALKER. *Atmospheres.* Englewood Cliffs, NJ, 1972.

JAMES C. G. WALKER

EARTH'S ATMOSPHERE, CHEMICAL COMPOSITION OF

A chemical compound contains different elements in fixed ratios to one another. Carbon dioxide, for example, has one atom of carbon for every two atoms of oxygen. Water has two atoms of hydrogen for every one atom of oxygen. Air is not a chemical compound. Rather, it is a mixture of compounds. The proportions of the different compounds in the mixture can vary from place to place or with time. In fact, the concentrations of the most abundant constituents of Earth's atmosphere—nitrogen, oxygen, and argon—do not vary perceptibly with location and are believed, on theoretical grounds, to vary only slowly with time. Most of the other gases in Earth's atmosphere do vary significantly in response to processes that add gases to the atmosphere or remove gases from the atmosphere. Average atmospheric composition is summarized in Table 1. Percent by volume is equivalent to the percentage of molecules in a mixture. Just under 21 percent of all air molecules, for example, are oxygen. The concentration of water vapor varies from nearly zero to a few percent, depending on air temperature and the balance of evaporation, cloud formation, and precipitation.

Table 1 includes only the most abundant gaseous constituents of Earth's atmosphere. Even unpolluted air contains a large number of other compounds, gases, liquid droplets, and solid particles that are usually present in smaller concentrations than those shown. Polluted air contains enhanced concentrations of nitrogen oxides, hydrocarbons, sulfur compounds, particles, ozone, and more than a thousand other compounds of industrial origin.

Table 1. Average Atmospheric Composition

Constituent	Chemical Formula	Percent by Volume in Dry Air
Nitrogen	N_2	78.08
Oxygen	O_2	20.95
Argon	Ar	0.93
Carbon dioxide	CO_2	0.0355
Neon	Ne	0.00182
Helium	He	0.000524
Methane	CH_4	0.00017
Nitrous oxide	N_2O	0.00003

Most atmospheric gases are colorless, odorless, and tasteless. Our senses do not inform us about any changes in their concentrations. Water vapor is an exception. We are made aware of high humidity by the failure of perspiration to evaporate. We can also sense some of the constituents of air pollution. They cause noticeable irritation of the eyes and the respiratory tract, and they affect visibility. But for the most part, we are unaware of changes in atmospheric composition, but our lack of awareness does not mean that composition is not changing.

Direct measurements of the concentrations of atmospheric gases over extended periods of time have revealed systematic changes in several important constituents. The best known of these is carbon dioxide. In 1958 Charles D. Keeling began regularly repeated measurements of atmospheric carbon dioxide at Mauna Loa on Hawaii, a site chosen for its pristine Pacific air, free of pollution. The first few years of measurement revealed a pronounced seasonal change in concentration from a low of 313 parts per million (or ppm) in October to a high of 318 parts per million in May. The units, 318 parts per million, mean that of every one million molecules of air, 318 were carbon dioxide molecules. This seasonal change is a consequence of photosynthesis mainly by land plants. During the summer, plants grow by using the energy of sunlight to convert carbon dioxide and water into plant matter and oxygen. The concentration of carbon dioxide in the atmosphere decreases. During the winter, photosynthesis stops, but plants and animals continue to gain energy through respiration, which converts organic matter and oxygen to carbon dioxide and water. The concentration of carbon dioxide in the atmosphere increases.

Is there a corresponding change in the concentration of oxygen in the atmosphere, increasing during the summer and decreasing during the winter? As far as we know there is. For every molecule of carbon dioxide removed from the atmosphere by photosynthesis a molecule of oxygen is added. But it has not been possible to measure the seasonal changes in oxygen because they are too small compared with the large amount of oxygen in the atmosphere. As the table shows, there are 700 oxygen molecules in the atmosphere for every carbon dioxide molecule. The seasonal change in carbon dioxide is 5 parts per million out of 315, or not quite 2 percent. The seasonal change in oxygen must also be 5 parts per million, but out of 210,000 parts per million this is a percentage change that has been too small to measure.

It did not take Keeling long to discover that there was a steady increase in the carbon dioxide concentration superimposed on the seasonal ups and downs. Each month's average values were almost 1 part per million higher than the average for the same month one year earlier. The annual average concentration has increased steadily from 315 parts per million in 1958 to 357 parts per million in 1993. This increase is a consequence of human activities, partly the clearance of forests but mostly the burning of fossil fuels, coal, oil, and natural gas. Forest clearance and fossil fuel combustion both transfer carbon from storage in either the forest or the ground to the atmosphere as carbon dioxide. It has now become possible to analyze the composition of ancient air preserved in bubbles in polar glaciers. Measurements of this kind have shown that the concentration of carbon dioxide in Earth's atmosphere was only 280 parts per million prior to 1750. The concentration began to climb in the eighteenth century as a result, at first, of widespread deforestation in Europe, North America, and Asia and later as a result of the accelerating use of fossil fuel to power the industrial revolution. By 1990, carbon dioxide had increased by 25 percent from its preindustrial value and the annual increase was getting larger each year. A continuation of these trends is expected to double the carbon dioxide concentration by the middle of the twenty-first century and to cause concentrations as much as five times the preindustrial value within a few centuries. All of this is a matter of concern because carbon dioxide in the atmosphere upsets Earth's global energy balance. The average temperature of the planet depends on a balance between the absorption of sunlight and the emission of infrared radiation by the planet to space. Carbon dioxide absorbs some of the radiation emitted by the ground, returning heat energy to the ground. This perturbation of the energy balance must cause climate change, although we do not know for sure whether the planet will warm or some other feature of climate will change.

Our concern over the greenhouse effect, as this phenomenon is called, is enhanced by the observation that atmospheric concentrations of methane, nitrous oxide, and the whole family of chlorofluorocarbon gases are also increasing at rates that are readily measured (see GLOBAL CLIMATIC CHANGES, HUMAN INTERVENTION, and GLOBAL ENVIRONMENTAL CHANGES, NATURAL). The increase in chlorofluorocarbon concentrations is clearly a result of human activities because these gases are produced only by industry. There are no natural sources. The increases in methane and nitrous oxide are also believed to be caused by humans because the rising concentrations run roughly parallel to increases in population and in economic activity. All of these gases interfere with Earth's global energy balance in the same way as carbon dioxide, by absorbing infrared radiation and returning extra heat to the ground.

The composition of ancient air trapped in bubbles of Antarctic ice has been measured back in time to 170,000 years ago, through two ice age glacial advances. The measurements reveal significant variation of the concentrations of carbon dioxide and methane associated with the advance and retreat of ice age glaciers. Both gases exhibit concentrations that decrease as the ice advances and increase during the warm periods between glaciations. At the height of the last ice age, some 18,000 years ago, the concentration of carbon dioxide was only 210 parts per million. It increased rapidly at the end of the ice age to 280 parts per million. The causes of these observed changes in atmospheric composition are not yet well understood, although it is likely that the carbon dioxide change was caused by changes in the composition of seawater and that the methane change was caused by changes in the distribution of vegetation and swamps. Humans are not implicated in the ice age changes in atmospheric composition.

There are no reliable records of atmospheric composition from earlier periods of Earth history. For the history of the atmosphere we must rely on the chemical and mineralogical composition of sedimentary rocks of various ages interpreted in the light of our understanding of how these compositions depend on the properties of the atmosphere and our understanding of the processes that might cause atmospheric composition to change with time. There is general agreement that the concentration of carbon dioxide must have varied a good deal during the course of Earth history but little agreement on the magnitude and time of these variations.

For oxygen the situation is in some ways more clear. Essentially all of the oxygen in the atmosphere is produced by green plants in the process of photosynthesis. There could have been little oxygen in the atmosphere before the origin of life and the origin of photosynthesis. Indeed, we have not yet discovered uncombined oxygen anywhere in the universe other than on Earth. So the concentration of oxygen in the atmosphere has increased as a result of life's activities. The time of this increase is indicated fairly clearly at about two billion years ago by a variety of changes in sedimentary rocks that can be attributed to reactions between minerals and oxygen. This increase in oxygen concentration as a result of biological processes is, as far as we know, the most significant event in the geological history of the atmosphere.

Bibliography

BRIMBLECOMBE, P. *Air Composition and Chemistry*. Cambridge, Eng., 1986.

WALKER, J. C. G. *Evolution of the Atmosphere*. New York, 1977.

———. *Earth History: The Several Ages of the Earth*. Boston, MA, 1986.

JAMES C. G. WALKER

EARTH'S ATMOSPHERE, STRUCTURE OF

Earth is surrounded by a gaseous envelope called the atmosphere. The vertical structure of Earth's atmosphere is described by its composition, and by the distribution of mass (pressure) and temperature.

Composition of Atmosphere

The atmosphere is a mechanical mixture of gases called air. The dry air component of Earth's present-day atmosphere is mostly composed of nitrogen (N; 78.08 percent by volume), oxygen (O; 20.95 percent), and argon (Ar; 0.93 percent). Although on geologic timescales (millions of years) the ratio of these gases changes, on the human timescale they may be considered constant. The remaining 0.04 percent of day air consists of so-called trace gases that are measured in parts per million by volume (ppmv). Several of these trace gases influence the radiative balance of the planet (greenhouse effect). Of primary importance are carbon dioxide, ozone, and methane. In the lowest 80 km, a zone called the homosphere, the mixture of these gases remains in the same proportion as at the surface. The atmosphere above 80 km, where the composition is heterogeneous, is known as the heterosphere. The layer from 80 to 220 km is largely composed of nitrogen, and above this are layers of oxygen, helium (He), and hydrogen (H). In addition to dry air, the lower atmosphere contains highly variable amounts (in both space and time) of water vapor. The amount of water vapor in the atmosphere is temperature-dependent; warm air can hold more water vapor than cold air. Water vapor comprises up to 4 percent of air by volume near the surface, but only a few ppmv above 10 to 12 km. Water vapor is the most important contributor to the greenhouse effect and thus plays a major role in determining the temperature of the lower atmosphere (see also EARTH'S ATMOSPHERE, CHEMICAL COMPOSITION OF; GLOBAL ENVIRONMENTAL CHANGES, NATURAL; GLOBAL CLIMATIC CHANGES, HUMAN INTERVENTION).

Distribution of Mass

The density (mass per unit volume) of a gas at constant temperature is proportional to pressure (Boyle's Law) and at constant pressure is inversely proportional to temperature (Charles' Law). Combining these laws leads to the relationship between pressure (p), density (ρ), and temperature (T) known as the equation of state, $p = \rho RT$, where $R = 287 \text{ J kg}^{-1} \text{ K}^{-1}$ is the gas constant for dry air. Due to the compressibility of air, lower layers of the atmosphere are much more dense than those above. Figure 1 (left panel) shows the density of air as a function of height from the surface to 100 km. Fifty percent of the total mass of the atmosphere lies in the lower 5 km, 80 percent below 10 km. The average density of air is 1.2 kg m^{-3} at the surface, 0.7 kg m^{-3} at 5 km, and 0.4 kg m^{-3} at 10 km. The average pressure (the force per unit area exerted by the weight of the atmosphere) at sea level is equal to the mass of the atmosphere (5.14×10^{18} kg) multiplied by the acceleration of gravity divided by the surface area of the earth, or slightly more than 1,000 millibars (mb), where 1 mb = 100 N m^{-2}. The global mean value for sea level pressure is 1,013.25 mb. Density stratification (decrease in density with altitude) is maintained by a balance between the pressure gradient force (which is directed upward) and the pull of gravity (which is directed downward). This force balance is expressed by the hydrostatic equation, $dp/dz = -\rho g$, where dp/dz is the rate of change of pressure with height, g is the acceleration of gravity, and ρ is density. From the hydrostatic equation and the equation of state, it follows that pressure in Earth's atmosphere decreases nearly exponentially with height. This relationship between pressure and height is shown in Figure 1 (right panel).

Distribution of Temperature

The vertical temperature distribution of Earth's atmosphere (Figure 1, center) is divided into four layers: troposphere, stratosphere, mesosphere, and thermosphere. The boundaries at the top of the lower three layers are called the tropopause, stratopause, and mesopause, respectively. The height of these boundaries varies in both time and space (primarily with latitude and season). The heights shown in Figure 1 are for the so-called standard atmosphere that is representative of typical (average) mid-latitude conditions.

The global average surface temperature for Earth is approximately 15°C due to the greenhouse effect. Incoming solar radiation entering Earth's atmosphere is predominantly short wave.

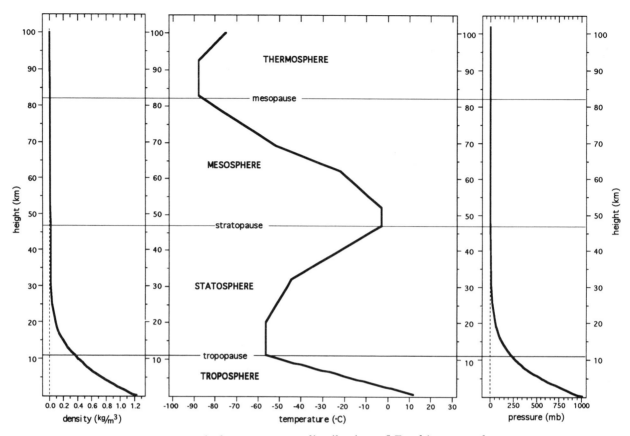

Figure 1. Vertical temperature distribution of Earth's atmosphere.

Radiation leaving Earth is long wave, or infrared. Earth's atmosphere absorbs infrared radiation due to the molecular properties of water vapor, carbon dioxide, methane, and other trace gases. While incoming and outgoing radiation tend to remain nearly in balance, the net effect of Earth's opacity to infrared radiation is to warm the surface of the planet. If not for the presence of an atmosphere, the surface temperature on Earth would be distinctly below the freezing point of water (less than −20°C).

In the troposphere, temperature decreases with altitude from the surface up to the tropopause (approximately 8 km in polar regions, 16 km in the tropics). When gas expands, its temperature decreases; when gas is compressed, its temperature increases. This change in temperature of a gas due to expansion or compression is called an adiabatic temperature change. Thus when air in the atmosphere rises, it expands and cools adiabatically;

when it descends, it is compressed and heats adiabatically. The decrease in temperature with increasing altitude is called the lapse rate. The dry adiabatic lapse rate on Earth is approximately 10°C/km. When air temperature reaches the dew point (i.e., the temperature at which air becomes saturated with water vapor), continued cooling results in condensation of water (cloud formation) and a release of latent heat to the atmosphere. The heat released slows the cooling with altitude to around 6°C/km (the saturated adiabatic lapse rate). The temperature at the tropopause is approximately −55°C in mid to high latitudes, and −75°C in the tropics.

In the lower stratosphere, temperature remains nearly constant (isothermal) from the tropopause up to around 20 km. In the upper stratosphere, extending to 50 km, temperature increases by up to 4°C/km rising to near 0°C at the stratopause. This stratospheric temperature increase is caused by the presence of ozone. In the layer between 20–60 km, ozone is generated by photochemical reactions. The long-term flux of ozone is downward from the source region in the stratosphere to sink at the surface. The maximum concentration of ozone is between 15 and 30 km. Ozone absorbs ultraviolet radiation from the Sun and consequently the stratosphere is heated.

Above the stratopause, in the mesosphere, temperature again decreases with height up to the mesopause at 80 km where the temperature approaches −90°C. In the thermosphere, a region that has no well-defined upper limit, temperatures again increase with height due to the absorption of very short wave solar radiation.

Bibliography

AHRENS, C. D. *Meteorology Today: An Introduction to Weather, Climate, and the Environment*, 5th ed. St. Paul, MN, 1994.

HOLTON, J. R. *An Introduction to Dynamic Meterology*, 3rd ed. San Diego, CA, 1992.

LUTGENS, F. K., and E. J. TARBUCK. *The Atmosphere*, 6th ed. Englewood Cliffs, NJ, 1995.

PEIXOTO, J. P., and A. H. OORT. *Physics of Climate*. New York, 1992.

WALLACE, J. M., and P. V. HOBBS. *Atmospheric Science: An Introductory Survey*. New York, 1977.

KIRK A. MAASCH

EARTH SCIENCE INFORMATION

Earth science information is an umbrella term for the resources that earth scientists and others turn to for data and information. It includes what are now thought of as traditional information sources such as books, journals, government reports, theses, dissertations, and literature indexes in various formats. It also includes a variety of nontraditional information sources such as maps, field trip guidebooks, open-file reports, well logs, seismic records, and drill core logs. Earth science information also includes sets of geochemical and geophysical data, earthquake information, and satellite and space mission data. These materials exist as paper copy, microfilm or microfiche, or may be digitized and stored on magnetic tape, floppy disks, and compact disc read-only memory (CD-ROM). The user of earth science information must be multi-platform literate, possessing the skills to access information regardless of whether it is found in a book, a journal article, an online World Wide Web (WWW) site, an anonymous File Transfer Protocol (FTP) site, or on CD-ROM.

Preservation

Unlike information in some scientific fields, earth science information tends to remain valuable even when it is several hundred years old. Nevertheless, use of older earth science material, as well as contemporary information in unusual formats (such as oversized colored maps), are difficult to preserve. Brittle paper documents, especially those printed on paper produced from wood pulp, are rapidly disintegrating. The usual solution of making microform copies for archival preservation is not yet suitable for oversized maps and for the color representation necessary to earth scientists, although the problem is being studied. Newly developed techniques, such as CD-ROM, have not yet proven to meet archival standards. Digitized data pose an additional problem: computer hardware and software develop so quickly that already some valuable data sets stored on magnetic tape have become unusable because the hardware to access them is no longer available or the software itself is obsolete.

Research and technological development is underway that may result in archival-quality stable color film and the ability to handle large-format materials. Even when the preservation technology becomes available, however, the cost of implementing preservation solutions may be excessive. Meanwhile, the older paper resources are in danger of disintegrating before a solution is at hand.

Collections

Earth science information is usually found in libraries, where indexes and catalogs (in card, print, or online format) are available so that users can find what they need. Online catalogs are easier to update, usually easier to use, and are considered the standard today. Library databases linked electronically to the catalog records and contents of other libraries help those institutions share resources through interlibrary cooperation and lending. Telefacsimile machines enable quick online access to the increasingly available full-text articles and standard-sized materials not held by the user's library. Online catalogs and the Internet allow researchers to browse the catalogs of libraries around the world, as well as to discover collections of data, computer software, and full-text sources using GOPHER (a user-friendly non-graphical way of getting access to the Internet), WWW indexes, and other electronic/Internet applications, right from their own offices. Librarians and information specialists are trained to help patrons with these indexes and catalogs, with using the various hardware necessary to access them, and with finding and using the primary information sources once they are identified.

Finding Earth Science Information

Research often begins with basic information sources such as dictionaries, encyclopedias, handbooks, and directories. The earth sciences have many specific resources of this type. Examples include Bates and Jackson's *Glossary of Geology,* P. W. Thrush's *Dictionary of Mining, Mineral, and Related Terms.* N. J. Hyne's *Dictionary of Petroleum Exploration Drilling & Production,* R. W. Fairbridge's *Encyclopedia of Earth Sciences Series,* Dana's *System of Mineralogy,* and the American Geological Institute's *Directory of Geoscience Departments.*

Increasingly, researchers also begin at a Web site, frequently Yahoo's Geology and Geophysics index, found on the Internet at:

> http://www.yahoo.com/Science/
> Geology_and_Geophysics/.

From here, the user can jump to various catalogs, Web pages devoted to specific topics such as volcanoes, earthquakes, or geophysical fluid dynamics, or to home pages of agencies that produce and archive data.

Bibliographic Databases

Researchers find valuable information by talking with colleagues and by perusing the literature, but more comprehensive information can be found by using the numerous indexes to earth science information. In addition to online catalogs, there are specialized bibliographic indexes to earth science journal and report literature and to data sets. Just like reports and maps, these indexes appear in a variety of formats: printed (e.g., *The Bibliography and Index of Geology*); as online services available over the Internet, or through dial-up services (e.g., the source database known as *GeoRef*), or on CD-ROM (e.g., *GeoRef*, from Silver Platter Information Inc.).

GeoRef, produced by the American Geological Institute, is a bibliographic index to the broad range of international earth science literature found in journals, government reports, maps, field-trip guidebooks, conference proceedings, and dissertations. In recognition of the importance of older materials, *GeoRef* indexes earth science literature from 1785 to the present, although its international coverage really only begins with the 1933 literature. Other earth science indexes include GEOBASE (the online equivalent to *Geological Abstracts*); GEOARCHIVE (*Geotitles Weekly* and *Geoscience Documentation*); SWRA (*Selected Water Resources Abstracts*); GEODE; *IMM Abstracts*, from the Institution of Mining and Metallurgy, and TULSA (*Petroleum Abstracts*).

Electronic Data Storage

Electronic media are used, more and more, to store bibliographic data, full text, numerical data, and images. Information is accessible over phone lines, through the Internet, on magnetic tapes mounted on mainframe computers, or on CD-ROMS, laser disks, floppy disks, and hard disks for microcomputers, workstations, and personal computers (PCs). Electronic storage allows information—especially numeric data sets—to be used differently than that of printed sources. Retrieval is always faster and generally more comprehensive. Queries can easily be tailored to combine concepts difficult to combine in printed sources, such as a quest for information on Chilean copper reserves written in English.

Such numerical data can be fed into a geographic information system (GIS) or other software and used in combination with other information to aid research and interpretative efforts; for example, seismological data, fault locations, strain data, and geology (all collected by different researchers and stored in separate data sets) can be overlaid on a single map to aid current interpretations of earthquake potentials. An excellent example of such cross-disciplinary work is a series of maps of areas of Los Angeles County, California, showing buildings damaged in the January 1994 Northridge earthquake, and the languages spoken by their inhabitants. The language data were collected in the 1990 census; the street grid data were provided by a private company; the damage information was collected and inputted by both the city of Los Angeles and Los Angeles County. The maps were developed by the California Office of Emergency Services so that the Federal Emergency Management Agency (FEMA) could determine the language proficiency needs of their field advisors.

Consideration must necessarily be given to standardization and the translation of data formats, data elements, and scales of magnitude to allow the integration of data from a variety of sources. This is developing quite rapidly, as pressure to conform to standards increases.

Of major concern in this area is the question of data integrity. As we move into a more digitized world, there is not yet any definite way to verify that data copied from one application and pasted into another is legitimate. What if numbers in a data set have been changed? Where is the verifiable original to check? This is especially important in mapping and geographic information: data collected and stored at a small scale such as 1:250,000 can be used to produce a large-scale map at 1:24,000, but the impression of detail will be false, because the data only reflect the information at the larger, less detailed scale.

Earth Science Data Sets

Examples of earth science data sets include information available in the Mineral Data System, National Water Information System II, Geotherm, National Coal Resources Data System, SIG-QUAKE, HYDRODATA, and the various World Data Centers. Many organizations have been established to collect and manage earth science data sets, and there are a myriad of data sets that reside with individuals, compiled during their research. Some efforts have been taken to have organizations archive the data generated by their researchers, one example being the Southern California Earthquake Center's SCEC Data Archive. Image data from NASA's Magellan Venus project are available on the Planetary Data System. The phe-

nomenal amount of quality data available in these data sets, and the increasing ability to display and analyze them, undoubtedly will continue to revolutionize the understanding of the earth sciences.

Bibliography

GUINNESS, E. V., S. SLAVNEY, and T. C. STEIN. "The Magellan Data Project Access System." In *Changing Gateways: The Impact of Technology on Geoscience Information Exchange*, eds. B. E. Haner and J. O'Donnell. Geoscience Information Society Proceedings 25. Alexandria, VA, 1995.

LEE, S. W. "The Planetary Data System." *Reviews of Geophysics* 29 (1991):338–340.

O'DONNELL, J., and C. R. M. DERKSEN. "Overlooked Sources of Information: Electronic Media." In *Geoscience Information—Essential Foundation, Critical Skill*, ed. J. V. Lerud. 1991.

DENA HANSON
JIM O'DONNELL

EARTH'S CRUST, HISTORY OF

The temperature of Earth's interior has decreased over geologic time. This decrease has occurred because ever-smaller amounts of heat are produced by the radioactive decay of isotopes of uranium, thorium, and potassium, and because heat is continuously lost from Earth's surface to space (*see* EARTH, COMPOSITION OF). The Earth's crust has evolved through time in response to this cooling. The rate of production of oceanic crust at mid-ocean ridges has decreased, as has the thickness of that crust. The rate of accretion of new continental crust has decreased because this material is ultimately derived from thick oceanic crust. The depth of volatile loss at subduction zones has increased as the subducting oceanic crust has become colder and less buoyant, leading to changes in subduction zone magmatism (*see* PLATE TECTONICS). Continental magmatism has changed in composition and in volume. The composition of sediments has evolved in response to changes in continental surface area and mid-ocean ridge spreading rates.

Soon after Earth accreted from the solar nebulae 4.56 billion years ago (4.56 Ga), it differentiated into two layers—a dense, molten iron-rich core and a more buoyant, solid, silicon-magnesium (Si-Mg) rich mantle (the latter with an ultramafic composition of 40 to 50% MgO by weight). As solid state convection began to cool the mantle and to transport material from deeper to shallower levels, partial melting of the mantle in areas of upwelling began to occur, creating magmas that rose to the surface to form oceanic crust. This crust was probably mafic in composition (i.e., greater than 7% MgO) and was created and consumed by plate tectonic processes broadly similar to those operating today (i.e., mid-ocean ridges and subduction zones).

Continental crust begins as especially thick oceanic crust, formed at volcanic island arcs associated with subduction zones (the Aleutian arc is a modern example), at oceanic hotspots (the Hawaiian island chain is a modern example), and at oceanic plateaus (Iceland and the Ontong Java plateau in the western Pacific are modern examples). This thick crust remains on Earth's surface more-or-less permanently because it is too buoyant to be recycled into the mantle at subduction zones. Many such pieces of thick crust have been accreted together by plate tectonic motions to become the continents. Subsequent volcanism, which causes remelting of crustal rocks, has also caused major changes in their chemistry, enriching them in Si and aluminum (Al); hence, continental rocks are referred to as sialic. However, the way that this enrichment occurs is poorly understood. Upper crustal enrichment by partial melting of the lower crust and overall crustal enrichment by addition of granitic magmas are both important, but the relative contribution of the two processes is unknown. Very old grains of minerals from continental rocks have been found in ancient sediments, indicating that continents began to form very early in Earth's history (4.3 Ga) (*see* CONTINENTS, EVOLUTION OF, and OLDEST ROCKS IN THE SOLAR SYSTEM).

The rigid units that make up the "plates" of plate tectonics are not composed solely of Earth's crust but also include a layer of the shallowest, coolest mantle (with the combination being referred to as the lithosphere). The major factor controlling the buoyancy of the lithosphere is the density of the lithospheric mantle, not the density of the crust itself. This effect is due to the depletion of the mantle of dense minerals during the partial melting process. Continents are thus in some ways

analogous to icebergs, with the ocean corresponding to the deep mantle, the ice to the lithospheric mantle, and the snow on the top of an iceberg to the crust. An iceberg floats because its root of ice is buoyant; similarly a continent is elevated because its lithospheric mantle root is buoyant. Although the snow and the crust are the lightest parts of iceberg and continent, respectively, they do not make them float.

Evolution of Magmatism: Basalts and Komatiites

Earth's upper mantle is made up of three major minerals: olivine, garnet, and pyroxene, each of which melts over different ranges of temperature. Mg is concentrated in the olivine, and Al and potassium (K) are concentrated in garnet and pyroxene. At higher temperatures, more of the olivine melts, and more Mg-rich rocks are formed. At lower temperatures and in more water-rich environments, proportionately more garnet and pyroxene melt, producing Al- and K-rich rocks.

Komatiites are ultramafic volcanic rocks that are almost exclusively confined to the period prior to 1.8 Ga (Figure 1), and are thought to represent

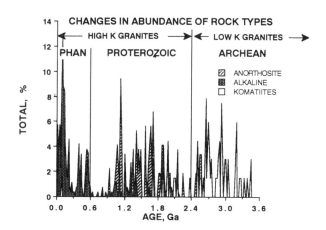

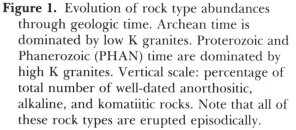

Figure 1. Evolution of rock type abundances through geologic time. Archean time is dominated by low K granites. Proterozoic and Phanerozoic (PHAN) time are dominated by high K granites. Vertical scale: percentage of total number of well-dated anorthositic, alkaline, and komatiitic rocks. Note that all of these rock types are erupted episodically.

partial melting of unusually hot patches of the mantle. The oldest komatiites are the most Mg-rich (28% MgO) and the most Al-poor, implying that they formed by partial melting of the mantle at very high temperatures. The youngest komatiites (0.06 Ga) have only 18 percent MgO, indicating that they formed by partial melting at somewhat lower temperatures. The change in komatiite composition is a direct consequence of the cooling of Earth's mantle (estimated to be about 100°C since 3.0 Ga).

The large amount of partial melting that occurs at especially hot patches of the mantle also leads to the production of especially thick oceanic crust, approximately 60 km for the 28 percent MgO komatiite and approximately 31 km for the 18 percent MgO komatiite. This thick crust provides the raw material from which the continents are constructed. The decrease in mantle temperature over geologic time has produced a corresponding decrease in the rate of production of new continental material and a decrease in the crustal thicknesses of newly accreted material. The thickness of the preserved continental crust averages around 41 km for pre-2.4-Ga crust and decreases to 38 km for post-1.8-Ga continental crust.

Although komatiites have received a great deal of attention, the most common mafic rock in pre-2.4-Ga continental crust is low aluminum, thoeleiitic basalt. This type of basalt forms from melting of mantle that has only small amounts of water. These rocks often occur in long linear belts called greenstone belts. Some of these greenstone belts may represent fragments of ancient oceanic plateaus or hotspot island chains that were accreted to the continents. Others may represent flood-basalt magmatism (the Columbia River flood basalts in the northwestern United States and the Deccan Traps of India are modern examples).

Evolution of Plate Tectonics and Its Effects on Magmatism

The main way that Earth loses its heat is through the production of new, hot oceanic plates at mid-ocean ridges, the cooling of those plates at Earth's surface, and the consumption of old, cold plates at subduction zones. Because heat production was higher before 2.4 Ga, oceanic plates were produced and consumed at least three times faster than at present. As a result, the average oceanic plate reached a subduction zone at an age of about 20 million years (Ma), compared to a present age of about 60 Ma. Because oceanic plates were younger and hotter, they lost their volatiles (e.g., water and carbon dioxide) at shallower levels. Consequently, the compositions of subduction zone magmas have evolved over geologic time (Figures 1 and 2).

Shallow water loss at subduction zones increases the probability that the mafic crust of the subducting oceanic plate, or the continental crust of the overriding plate, will melt. Melting of a mafic crust produces a granite that is very low in K (i.e., containing only sodium-calcium [Na-Ca] feldspar, rather than K feldspar). Low K granites are very common in areas of pre-2.4-Ga continental crust, and decrease in abundance as the age of the crust decreases. Modern granites have nearly equal proportions of K and Na-Ca feldspar. This change in granite composition is a direct result of the evolution of plate tectonics over time.

As the depth of release of volatiles has increased over geologic time, the effects that those volatiles have on magmatism have changed too. In the Precambrian, volatiles were released at shallow depths into the lithospheric mantle of the overriding plate. Because this part of the mantle had already had a partial melt extracted (i.e., when it was created), little new melting occurred. However, as volatiles were released at greater and greater depths, they began to affect the deeper, convecting mantle that lies immediately beneath the subducting plate. The convecting mantle is continually recirculated and has more garnet and pyroxene than the overlying lithospheric mantle. It melts more readily and produces more magma. Therefore, the decline in mantle temperature over time leads to an increase in the volume of subduction-zone related island arc volcanism.

The magmas that result from deeper melting of a volatile-rich mantle are enriched in Al and K, compared to normal mid-ocean ridge magmas. The mafic rocks that result are called high-Al basalts. The granites and silicic volcanics have higher contents of K. High-K silicic rocks and high-Al basaltic rocks have become more abundant with geologic time and are the most abundant type of magmas at modern continental arcs.

Anorthosites are Al-rich rocks that are composed almost entirely of Na-Ca feldspar. Anorthosites are very rare before 2.4 Ga, are common in

227

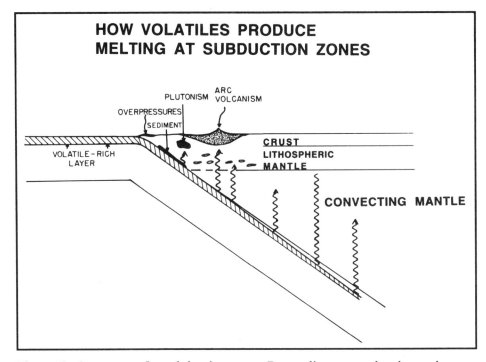

Figure 2. Structure of a subduction zone. Depending upon the thermal environment, volatiles (water and carbon dioxide are the most common) will enter different parts of the overlying plate. If the subducting plate and/or the underlying mantle are very hot, volatiles primarily go into the upper plate and produce melting there. If the subducting plate and mantle are cold, most volatiles go directly into the convecting mantle. The regions that receive the most volatiles are the most likely to melt.

the middle Precambrian, and then become rare again after 0.7 Ga (Figure 1). The oldest anorthosite occurrences are very small, whereas the middle Precambrian anorthosite bodies are extremely large, sometimes hundreds of kilometers across. These middle Precambrian anorthosites appear to have a continental origin and may have formed through remelting of a high Al mafic crust. Hence the increase in the supply of Al-rich magmas at continental margins with geologic time may have contributed to the "anorthosite-event" in the middle Precambrian. However, we do not know why anorthosites subsequently became rare again during more recent geologic time. Hence, anorthosites are still somewhat of a geological enigma.

Mantle convection was more vigorous during the early Precambrian than it is today, owing to the higher mantle temperatures, and it led to faster creation of oceanic plates at mid-ocean ridges and

faster consumption of these plates at subduction zones. Subduction zones, including that at the edges of continents, were more abundant so that the probability of collision between two continental blocks was higher than today. Subduction-related mountain ranges (the Andes are a modern example) were more common, too, as were coarse-grained sediments produced by the rapid erosion of those mountains. Early Precambrian sediments are typically coarse-grained ones: arkoses, graywackes, and muddy sandstones.

As the Earth aged, plate production and consumption rates decreased. As a result, the edges of continents became tectonically quieter, with fewer subduction zones and fewer continental collisions. In addition, the continents increased in surface area as small, isolated continents were accreted together into larger ones. The exact rate of continental growth is still controversial. However, it is agreed that over 50 percent of present-day continental surface area was present by 2.4 Ga. As a result, post-2.4-Ga continents had larger surface areas and less topographic relief than early Precambrian continents. The lower average relief of the post-2.4-Ga continents led to fine-grained sediments, such as clean sandstones and shales, and to biologically derived sediments, such as limestones.

Hotspots and Plumes: Their Effects on the Geological Record

Small regions of hot upwelling mantle (50 to 100 km across) are termed hotspots. Hotspots beneath oceanic plates produce island chains. Hotspots beneath continents produce alkaline rocks, which range in composition from alkali basalts (only slightly more alkaline than normal oceanic basalts) to kimberlites (extremely alkaline). However, not all hotspots produce magmatism in the overlying plate. If the overlying plate is either too thick or moves too quickly relative to the underlying mantle, the hotspot will be unable to melt through and there will be no magmatic activity (see VOLCANISM).

A second type of hot upwelling region is termed a mantle plume. These are much larger than hotspots (500 to 2000 km across). When a mantle plume upwells beneath an oceanic plate, it produces an oceanic plateau. When a mantle plume rises beneath a continent, it produces rifting of the continent and flood basalt volcanism.

The evidence for early Precambrian mantle plumes is relatively strong. Three pre-3.0-Ga oceanic plateau terranes have been identified on the continents, with the oldest occuring at 3.5 Ga (although these identifications are somewhat controversial). Each of these oceanic plateau terranes has minor komatiites and voluminous low-Al thoeleiitic basalts of the types that now occur within oceanic crust.

In the last 100 Ma, we see many episodes of flood basalt volcanism. Ancient flood basalts are more difficult to identify. Often the lava flows themselves are eroded, and only the feeder dikes remain. These feeder dikes are referred to as massive dike swarms. The oldest, well-dated dike swarms occur at 2.6 Ga, but there are claims that some flood basalts are as old as 3.0 Ga. Because only a small percentage of the area of continents is over 3.0 Ga in age, this claim (if true) implies that

there have been continental flood basalts as long as there have been continents.

The large mantle plumes that produce plateaus and flood basalts are active for geologically brief intervals (3 to 14 Ma). After that time, plume activity becomes much weaker, and the radius of the upwelling decreases. The plume then becomes a longer lived hotspot that persists for as long as several hundred Ma. In most cases, the magmas produced by hotspots are alkaline.

There have been some claims that alkaline rocks, and by implication hotspots, have become more abundant with geologic time. This claim is tempered by concerns about the likelihood of loss of ancient alkaline rocks by erosion. Most exposures of alkaline rocks, in particular those that form on stable continents, have very small diameters (less than 1 km). The intrusions are often inverted cone-shaped, and their surface area decreases during erosion. As a result, alkaline rocks are not easy to find in deeply eroded crust.

Indirect evidence for early Precambrian alkaline rocks is found in the occurrence of diamonds in 2.6 Ga sedimentary rocks in South Africa. These diamonds were presumably brought to Earth's surface in a kimberlite, and later concentrated in placer deposits in rivers and streams as that kimberlite eroded away. Hence the kimberlite formed prior to 2.6 Ga. Other early Precambrian sedimentary deposits of diamonds have been identified in Northern Africa, South America, Europe, and Asia. These diamond placers provide indirect evidence for widespread occurrences of early Precambrian alkaline magmatism.

As the surface area of continents increased, continents became a larger proportion of the surface area of any given plate. Plates containing large continents generally move more slowly across the underlying mantle than those that are wholly oceanic, owing to the drag associated with the continental root. Paleomagnetic data indicate that the very largest continents, such as the middle Precambrian "supercontinent," moved slowest of all. Hotspots could more easily melt through the continental portion of these plates. Hotspot volcanism remelted the crust and produced a type of granite that is termed anorogenic because it is not associated with either a subduction zone or with a continental collision. The middle Precambrian supercontinent produced the most widespread anorogenic granites in the entire geological record.

However, a few anorogenic granites have formed more recently on the nearly stationary African continent.

Episodic Versus Steady Evolution

Although the general trends in sedimentary and igneous rock compositions record a secular trend that is a response to the decreasing internal temperature of Earth, there is also evidence for fairly rapid fluctuations over much shorter time intervals. The most recent such episode is the late Cretaceous (124 to 84 Ma) "super-plume" event, which was a period of unusually vigorous volcanism: Seafloor spreading rates increased dramatically (the wider mid-ocean ridges displaced so much seawater that they caused widespread continental flooding); many large oceanic plateaus formed (and were accreted to continents, leading to a surge in continental growth); and alkaline magmatism was more common.

The exact cause of the super-plume event is unknown. However, the late Cretaceous was also a period of an unusually steady magnetic field, which has been taken as evidence that some unusual event was occurring in Earth's core (from whence the magnetic field originates) that possibly increased the heat transport from core to mantle.

The Phanerozoic geological record contains evidence for three to four other super-plume events. There is not enough data to confirm the presence of super-plumes in Precambrian time. However, alkaline magmatism shows an apparent episodicity as far back in time as there is adequate sampling, about the middle Precambrian (Figure 1). Furthermore, the isotopic data on continental growth provide strong evidence that the continents have grown episodically throughout the preserved geological record. Only two hundred oceanic plateaus with the surface area of the Ontong Java plateau, which formed during the Cretaceous super-plume event, would be required to make the total surface area of the continents. Therefore, short-term fluctuations as well as long-term evolutionary trends are likely to be important contributors to the rock record.

Bibliography

LARSON, R. L. "Geological Consequences of Super-plumes." *Geology* 19 (1991):963–966.

NISBET, E. G. *The Young Earth: An Introduction to Archean Geology.* Boston, 1987.

TAYLOR, S. R., and S. M. MCLENNAN. *The Continental Crust: Its Composition and Evolution.* Oxford, Eng., 1985.

WINDLEY, B. F. *The Evolving Continents.* Chichester, Eng., 1994.

DALLAS HELEN ABBOTT

EARTH'S FRESHWATERS

The total quantity of water on the globe is approximately 1.35×10^9 km^3, and this quantity is fixed. Almost all (97.2%) of the world supply is saltwater in the oceans (Table 1). Thus only the small remaining fraction is freshwater. Icecaps and glaciers today account for the largest portion of freshwater (about 70%), and groundwater constitutes most of the remainder. Freshwater in rivers and lakes represents a tiny fraction of the whole, about 0.00007 percent, or 0.006 percent of the freshwater. These numbers are important because human activities are not only dependent on water but, for the most part, use water from limited freshwater sources—streams, lakes, and groundwater.

Glaciers have waxed and waned, and because the total quantity of water remains constant, the amount in the oceans has fluctuated during geo-

Table 1. Quantities of Water on the Globe

	Volume $\times 10^3$ km^3	Freshwater %
Reservoirs:		
Oceans	1,350,000	
Atmosphere	13	0.04
Land:		
Rivers	1.7	0.006
Freshwater lakes	100	0.3
Inland seas, saline	105	
Soil moisture	70	
Groundwater	8,200	30
Icecaps/glaciers	27,500	69
Biota	1.1	

logic time. During periods of glaciation, the water in icecaps and glaciers stored on the landscape increases, while that in the oceans decreases (*see* EARTH'S GLACIERS AND FROZEN WATERS). There have been many fluctuations in these quantities during the recent glacial period, the Quaternary, beginning roughly 2 million years ago (2 Ma). During the time of maximum glaciation, sea level dropped 130 m and the volume of water stored in icecaps and glaciers increased to 70 million km³, roughly 40 million km³ more than at present (Table 1).

The Hydrologic Cycle

The quantities of water stored in various locations on the globe provide only a static picture of the freshwater. The hydrologic cycle (Figure 1) describes the movement of water between these reservoirs over the globe. Water is evaporated from the sea and from the ground surface to the atmosphere, where it is transported in the atmosphere and precipitated on land and sea, then recycled to the atmosphere. The average yearly cycle shows that most of the large quantity of water evaporated from the sea is precipitated back to the sea, while only one-tenth is precipitated on the land. Precipi-

tation is evaporated from the land and transpired (evaporated from the leaves of plants) to the atmosphere. Evapotranspiration is the combination of evaporation from water surfaces and transpiration by vegetation. Roughly 60 percent of the precipitation that falls on land is returned to the atmosphere through evapotranspiration, and the remaining 40 percent is runoff, water that runs over the surface of the ground or beneath the ground to rivers. Eventually nearly all of the runoff reaches the sea. A relatively small amount flows to closed basins on continents such as the great Salt Lake in the United States. Runoff is renewable water.

Two continents, Asia and South America, account for 50 percent of the total runoff to the oceans, amounts controlled by the size of the land area and high rainfall. Low runoff in Africa is dictated primarily by limited precipitation (Table 2), and Antarctica, whose landmass is significantly less, is the driest continent.

For a given region, the hydrologic budget describing the hydrologic cycle can be expressed as follows:

$$\text{Precipitation} = \text{Runoff} + \text{Evapotranspiration} \pm \text{Change in Storage}$$

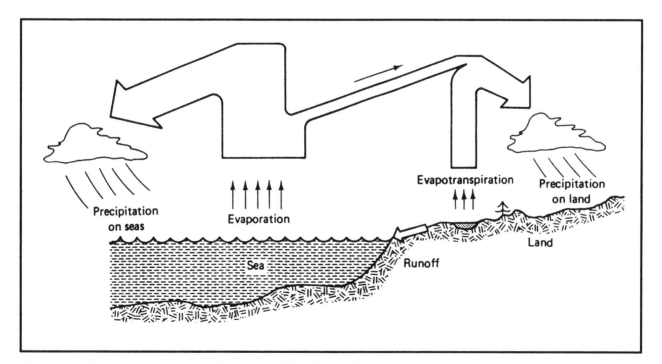

Figure 1. The hydrologic cycle. Units are 10^{12} m³/year (from G. M. Masters, *Introduction to Environmental Engineering and Science*, 1991).

Table 2. Magnitude of Runoff from the Continents

Territory	Annual Stream Flow (mm)	Annual Stream Flow (km³)	Percentage of Total Runoff	Area (km² × 10³)	Specific Discharge (1 s¹ km²)
Europe	306	3,210	7	10,500	9.7
Asia	332	14,410	31	43,475	10.5
Africa	151	4,570	10	30,120	4.8
North and Central America	339	8,200	17	24,200	10.7
South America	661	11,760	25	17,800	20.9
Australia and Tasmania	45	348	1	7,683	1.4
Oceania	1,610	2,040	4	1,267	51.1
Antarctica	160	2,230	5	13,977	5.1
Total land area	314	46,768		149,022	10.0

Source: *Nature and Resources* 26, no. 3 (1990):35.

Over a period of years, the storage term is zero where there is no trend toward an increase or decrease in the quantity of water being stored in the soil or ground. Runoff is simply the difference between precipitation and evapotranspiration. Thus, in a dry warm climate where evapotranspiration is high, there is no runoff, on average, over the period of a year. Stream beds are usually dry. Stream flow occurs only over short periods of time during heavy or intense rainfalls in which either the rate of precipitation exceeds the infiltration capacity of the soil, or the total quantity of precipitation exceeds the ability of the soil to absorb the rainfall.

Climate controls the distribution of precipitation and, to a large extent, the kind of vegetation that grows in a given region and hence the evapotranspiration. In the United States, for example, on the basis of a yearly average, evapotranspiration equals precipitation in the western great plains (western Kansas and eastern Colorado). Toward the Southwest, the potential for evaporation exceeds the precipitation and many streams are ephemeral, while to the east precipitation greatly exceeds potential evapotranspiration and streamflow becomes perennial.

Variations in the timing and spatial distribution of runoff on the surface of the globe are a function of climate, as well as of local topography. Thus rainfall varies from nearly zero in deserts, such as parts of the Sahara and northern Chile, to as much as 2,250 cm per year in the region of northern India subject to heavy monsoonal rains. The flow of rivers, representing the residual between precipitation and evapotranspiration, is even more varied. Both precipitation and river flow are subject to significant temporal variations over short time periods and from year to year. Such variation is readily seen in the records of annual mean river flow of the Brazos River in a dry area of Texas and the Pascagula in Mississippi (Figure 2). While stream flow is variable in each, the largest variation occurs in the semiarid region of the Brazos. The variability is much less in the wetter Pascagula area. These significant variations in space and time demand adaptations in the use of freshwater by human beings on many parts of the globe. From very low flow to large floods, water quantity varies a hundred- to thousand-fold or more at a given river location. Reservoirs are constructed to mitigate the impact of such variability. For the world as a whole, the total volume of reservoir storage capacity is 5,500 km³, equivalent to about 15 percent of the mean annual runoff.

Chemical Composition of Runoff

Contrary to common wisdom, precipitation or rain does not have the composition of "pure" or distilled water. Rather, because of the existence of gases, primarily carbon dioxide in the atmosphere, precipitation is somewhat acid, having a pH of about 5.7 rather than the neutral pH of 7. At the

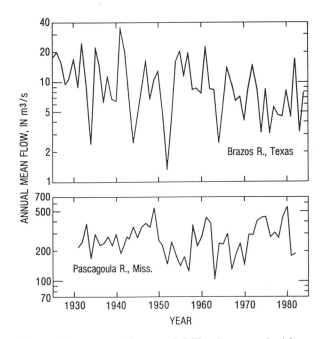

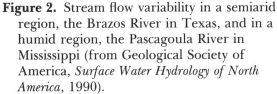

Figure 2. Stream flow variability in a semiarid region, the Brazos River in Texas, and in a humid region, the Pascagoula River in Mississippi (from Geological Society of America, *Surface Water Hydrology of North America*, 1990).

same time, because there are particles and salts in the atmosphere derived from both the ocean and the land surface, precipitation also may contain a variety of salts. The closer one lives to the ocean, the more likely that precipitation there will contain sodium chloride (NaCl) and other salts picked up by the wind blowing over the ocean surface. While these natural variations in the composition of precipitation are important, human activities introduce additional constituents, such as sulfur dioxide from the burning of fossil fuels, hydrocarbons and nitrous oxides from automobiles, and ammonia from feedlots (*see* POLLUTION OF ATMOSPHERE). These contributions may be large. Thus, the pH of rainwater in some parts of the United States, Europe, and regions of China and India on occasion may be as low as 2.5 with mean annual values of 3 and 4.

While variations in the chemical composition of rainfall are important, the chemical content of surface runoff of rivers and streams is controlled largely by the underlying geology and soils of a region. As precipitation falling on the land moves over the surface, or through the ground, provid-

ing runoff to streams and to the ocean, the water picks up salts from the surrounding rocks. The rocks themselves are weathered by chemical, physical, and biological processes occurring under different climatic and physiographic conditions. Water is the major influence on these processes. Thus the composition of surface waters reflects the rock composition of the landscape and the climatic conditions of the region.

This is broadly shown by a comparison of common ions in average river waters and a range of rock types (Table 3). In simple terms, the most mobile or easily leached elements in weathering of many rocks—calcium, sodium, magnesium, and potassium (roughly in that order)—are enriched in river water, whereas silicon and iron and aluminum are more concentrated in the residual soil or rock. The dissolution of rocks is influenced by the available water, composition of the rocks, biota, organic matter, topography, and the structure of the rock and soil that controls the permeability of the soil.

Because the composition of rocks is highly variable, and climatic conditions quite diverse, the composition of different river waters is also diverse. For example, Table 4 indicates that the Colorado River has very high concentrations of sodium, calcium, and magnesium compared to the average for the world's rivers, or to the Amazon. This disparity is due to the fact that much of the drainage basin of the Colorado is underlain by sedimentary rocks containing limestone ($CaCO_3$), gypsum ($MgSO_4$), and halite (NaCl) and other salts. In contrast, the Amazon drains a large area underlain by igneous rocks containing low concentrations of

Table 3. Major Elements in Rocks Compared with the Average Composition of Rivers and Lakes

Range of Major Elements in Rocks World Average of Four Rock Types (Weight Percent)		*Average Rivers and Lakes*
Si	19–32	12
Fe & Al	4–19	3
Mg	0.6–26	3
Ca	0.7–7	20
Na	0.6–3	6
K	0.03–3	2

Table 4. Range in Composition of River Waters (ppm)

Ion	World Average Rivers	Maximum and Minimum Values from Specific Rivers	
		Colorado	Amazon
Na^+	6.9	124	3.1
K^+	2.1	11.8 wn	0 c
Ca^{2+}	15.0	94	6.5
Mg^{2+}	3.9	30	1.0
Cl^-	8.1	113	3.9
SO_4^{2-}	10.6	289	0.44
HCO_3^-	55.9	183	22.5
SiO_2	13.1	25.3 wn	4.9 h

wn = White Nile; c = Columbia; h = Hudson.

these elements and is subject to the intense chemical weathering of the tropics. And except for the Andes Mountain region, the area is one of low relief. Table 4 also shows that potassium is absent in samples from the Columbia River, which drains a region of basalt rock which is without potassium-bearing minerals.

Where precipitation is high, weathering of the land surface is more intensive and more material can be dissolved over a period of time. Although the concentration of solutes in rivers in regions of high precipitation is low, the total quantity carried in solution is large because the river flow is large. Thus the quantity of solutes increases with the quantity of runoff. Vegetation is also, in part, determined by the amount of precipitation, as is the quantity of organic matter. By regulating biota, climate also indirectly influences the solubility and transport of solutes through precipitation, infiltration into the soil, and ultimately by way of runoff. The landscape or topography also influences the kind of constituents carried in rivers and streams. For example, where the landscape is rugged and the relief great, a given rainfall is likely to run off more easily, carrying with it erosion products. Thus, materials that have been broken up by frost action and weathering may be readily transported to streams. To the extent that the transport of these materials from the surface is rapid, they do not remain in place to allow chemical dissolution to proceed for long periods of time. Under these circumstances, the dissolved constituents may be less than the particulate or sedimentary materials car-

ried in streams. This difference too will be reflected in the composition of the surface waters.

In addition to geology, climate, and topography, the composition of rivers and lakes can be significantly altered by products of human activities, ranging from atmospheric deposition and industrial effluents to surface runoff from farms and cities (*see also* POLLUTION OF GROUNDWATER AQUIFERS; POLLUTION OF LAKES AND STREAMS). This is shown strikingly in the Rhine River, flowing in one of the most heavily industrialized regions in the world. While concentrations of silica (SiO_2) and bicarbonate (HCO_3^-) reflect the natural background (Figure 3), ions such as chloride, sodium, and nitrate are derived principally from anthropogenic sources.

Table 5. Chemical Composition of Comparison of River Water and Seawater

Ion	World Average Rivers (*ppm*)	Seawater (*ppm*)
Na^+	6.9	10,800
K^+	2.1	407
Ca^{2+}	15.0	413
Mg^{2+}	3.9	1,296
Cl^-	8.1	19,010
SO_4^{2-}	10.6	2,717
HCO_3^-	55.9	137
SiO_2	13.1	0.5–10

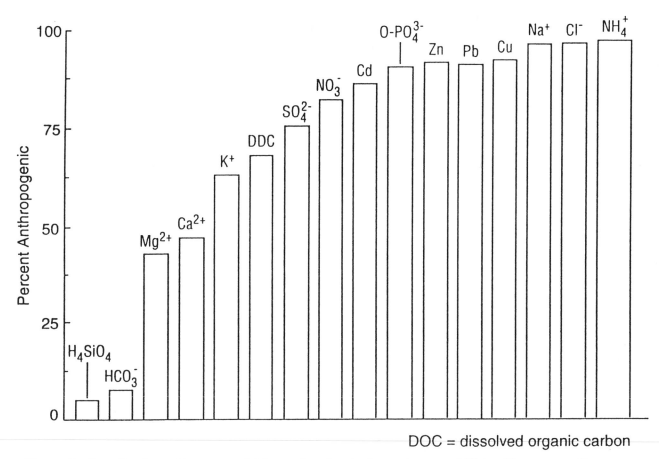

DOC = dissolved organic carbon

Figure 3. Contribution of human activities to the chemical constituents of Rhine River water (from R. Siever and W. Stumm, 1982, *Quality of Water—Surface and Subsurface, Scientific Basis of Water-Resource Management*, 1982).

Chemical Composition of the Ocean

Two puzzles, one geophysical and the other geochemical, inspired speculation since ancient times. If rivers flow to the ocean, why does the ocean not rise continuously? And if rivers carry salts to the sea, why is the sea not ever more salty? As shown earlier, evaporation removes water from the ocean. The constancy of the concentration of major ionic constituents in the ocean is more difficult to explain and not yet fully understood. The basic composition of seawater, compared to river water, is shown in Table 5 (*see* SEAWATER, PHYSICAL AND CHEMICAL CHARACTERISTICS OF). Calcium and silica are removed from the sea in the shells of organisms. Sulfate is reduced within sediments and incorporated into insoluble minerals in bottom muds.

Only recently has it been recognized that the composition of the ocean is controlled in large measure by the reactions of seawater with hot basaltic magmas that rise from hydrothermal vents on the ocean bottom, supplying the basalt rock that floors the oceans. Concentrations of sodium, potassium, magnesium, and chloride found in the vicinity of the vents, combined with laboratory experiments simulating the reactions of magma with seawater that produce ionic concentrations similar to those found in the ocean, confirm the important control exercised by the magma. Like freshwaters, seawater also contains trace quantities of all natural elements found in the composition of the earth, often in very small concentrations. These trace chemicals provide important clues to oceanic processes.

Water Use

The variation in space and in time of quantities of freshwater over the surface of the globe affects the quantities of water available for human use. In addition, because the level of activities of human societies and the numbers of people also vary across the surface of the globe, the quantities of freshwater available per capita in different parts of the globe are even more variable. Djibouti is a small country with a modest population, located in the arid Middle East. Canada, in contrast, is a vast country with a small population, in a temperate, well-watered region. These two countries illustrate that the range of available water for a population within a given nation may be very large.

For the world as a whole, freshwater runoff is approximately 40,000 km^3 per year. Of this, human beings withdraw approximately 3,500 km^3 for use, about 1/10th of the available annual recycling surface water. Still less, approximately 2,000 km^3, or 5 percent of the total, is actually consumed or lost in human activities. These figures suggest that a great deal of renewable freshwater is available for human use. But the global comparison is misleading because it masks very great spatial varia-

Table 6. Typical Drinking Water Quality Parameters and Their Associated Health Effects

Parameter	Health Effects	Typical Conc. Lake or Stream	Drinking Water Standard
Microbiological:			
Total coliforms, no./100 ml	Indicator organisms, not necessarily disease causing	<100	1
Turbidity, NTU	Interferes with disinfection	1–20	1–5
Inorganic:			
Arsenic, mg/l	Nervous system and skin problems	<0.01	0.05
Barium, mg/l	Circulatory problems	<0.01	1
Cadmium, mg/l	Kidney trouble	<0.01	0.01
Chromium, mg/l	Liver/kidney problems	<0.01	0.05
Lead, mg/l	Nervous system and kidney problems (highly toxic to infants and pregnant women)	<0.01	0.05
Mercury, μg/l	Nervous system and kidney problems	<0.01	2
Nitrate, mg/l	Methemoglobinemia	<1.0	10
Selenium, μg/l	Gastrointestinal effects	<1	10
Silver, μg/l	Skin discoloration	<1	0.05
Fluoride, mg/l	Skeletal damage	<1	4
Organic:			
Endrin, μg/l	Nervous system/kidney problems	<1	0.2
Lindane, μg/l	Nervous system/kidney problems	<1	4
Total trihalo-methanes, μg/l	Cancer risk	<50	100
Benzene, μg/l	Cancer	<1	5
Other:			
pH	Corrosivity (not health)	6–8	6.5–8.5

Source: B. T. Ray, *Environmental Engineering* (1995), p. 241, Table 8.

tions noted earlier. The distinction between withdrawal and consumptive use is important. For example, about 99 percent of the water withdrawn for use in the production of hydroelectric power passes through the turbines and returns directly to the stream system. In contrast, in the process of irrigating crops, as much as 60 to 70 percent of the water withdrawn for irrigation may be evaporated or transpired by the crops, and only 30 to 40 percent of the amount withdrawn is returned. In the case of irrigation 60 to 70 percent is "consumed" in the sense that while it returns to the atmosphere, it is no longer available for use in the region.

The amounts of water withdrawn for various uses depend not only on the available supply, and to some extent on the quality of the supply, but also on the technology and efficiency with which water is used. Thus there are enormous variations in the quantities of water withdrawn, for example, for specific industrial uses, and variations often are determined by the value of the water in various uses in different parts of the world. In some rich oil regions of the Middle East, for example, water from the sea is desalinated at considerable cost because the value of the water is high and the economy can support this process. Elsewhere, such high costs could not be borne by a poorer economy or one engaged in the production of commodities deemed less valuable by society.

Drinking Water Quality

Humans use water for a myriad of purposes, from small amounts for drinking and household use to withdrawal of enormous quantities for the production of crops or for electric power generation. The chemical and biological quality demanded for such uses differs markedly. Industrial boilers may require water of very low salt concentrations to prevent the formation of scale by deposition of salts. Similarly a limited range of water quality is appropriate for human consumption (see WATER QUALITY).

Based on the concentration of total dissolved solids, a rough classification defines freshwater (less than 1,500 mg/l), brackish water (1,500 to 5,000 mg/l), and saline waters greater than 5,000 mg/l. Total dissolved solids of 500 mg/l is considered a limiting standard for human consumption. Potability, the term used to characterize water quality for human consumption, is defined by a

Table 7. Secondary Standards for Drinking Water

Contaminant	Level
Chloride	250 mg/l
Color	15 color units
Copper	1 mg/l
Corrosivity	Noncorrosive
Foaming agents	0.5 mg/l
Iron	0.3 mg/l
Manganese	0.05 mg/l
Odor	3 threshold odor number
pH	6.5–8.5
Sulfate	250 mg/l
Total dissolved solids	500 mg/l
Zinc	5 mg/l

variety of measures based primarily on distinctive parameters associated with human health (Table 6). In addition to standards for chemical quality, as Table 7 indicates, drinking water standards also include microbial thresholds designed to minimize the presence of pathogens as well as criteria for a wide variety of organic compounds. Additional standards reflect a variety of other qualities (Table 7). Some of these criteria mirror the earliest historical standards: Does the water look acceptable, does it smell, and does it taste bad? While these measures have been quantified, matters of aesthetics, smell (odor), and taste continue in importance.

Where raw waters are treated to achieve drinking water quality standards, a variety of techniques are used depending upon the composition and concentration of the constituents in the raw water. Increased understanding of the physiological affects of various constituents, including their interaction with chemicals used in treatment, has led to increasing attention to the composition of raw and treated waters and to tightening the standards for contaminants that may affect human health.

Bibliography

DINGMAN, S. L. Physical Hydrology. New York, 1994.
GLEICK, P. H., ed. Water in Crisis: A Guide to the World's Fresh Water Resources. New York, 1993.
KRAUSKOPF, K. B., and D. K. BIRD. Introduction to Geochemistry. New York, 1995.

M. GORDON WOLMAN

EARTH'S GLACIERS AND FROZEN WATERS

Ice covers approximately 10 percent of Earth's surface, yet ice sheets and glaciers contain over 75 percent of the planet's freshwater. Far from being static, ice moves slowly, constantly responding to and creating changes in the environment. The huge ice sheets presently covering Antarctica dramatically influence global climate and oceanography. Glaciologists study the glacier ice, its properties, and movements. Glacial geologists examine the products of ice movement of the present and past to understand the ever-changing global climate and environment.

The majority of the world's ice lies on the Antarctic continent, which consists of 86 percent of the total ice cover. Most of the remainder (11%) covers Greenland (Hambrey, 1994). If the Antarctic ice were to melt, the global sea level would rise 66 m (Denton et al., 1991). The remaining 3 percent of the ice is distributed as ice caps or mountain glaciers in places such as Alaska, the Canadian Arctic, Patagonia, New Zealand, the Himalayas, and the European Alps.

What Is a Glacier?

A glacier or ice sheet is an accumulation of ice, air, water, and rock debris (sediment). It is of sufficient mass to flow due to its own internal deformation. Glaciers flow slowly, covering tens of meters to thousands of meters per year. Ice can cover a continent, such as Antarctica, or it can fill a small valley between two mountains.

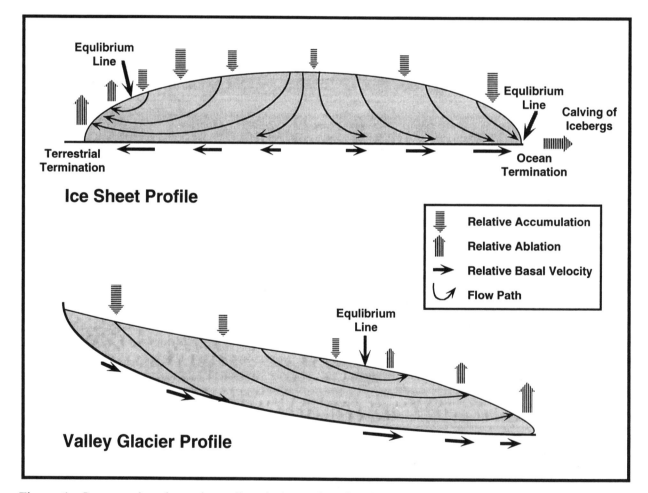

Figure 1. Cross section through a valley glacier and an ice sheet illustrating the locations of the accumulation and ablation zones and the equilibrium line (modified from Sugden and John, 1976).

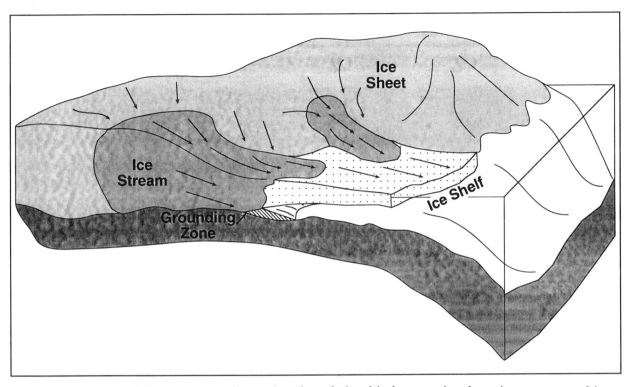

Figure 2. Diagram of the ice system illustrating the relationship between ice sheet, ice stream, and ice shelf (modified from Alley, 1990).

All glacial systems have accumulation and ablation zones (Figure 1). These are separated by the equilibrium line. In the accumulation zone, mass is added to the glacier or ice sheet; no mass is lost. Usually, this area is near the top of the glacier or ice sheet. In the ablation zone more mass is lost than gained, through melting, evaporation, or calving (formation of icebergs through separation or breaking). No mass is lost or gained at the equilibrium line. Temperature and precipitation are key elements in establishment and maintenance of a glacier. For a glacier to grow (advance), snow accumulation must exceed ablation. Glaciers shrink (retreat) when ablation outweighs accumulation.

Ice flows or moves by two primary mechanisms, internal deformation and basal sliding. It is important to realize that ice always moves in the same direction, whether it is advancing or retreating. Ice behaves like a plastic; it deforms under its own weight because of gravity. Movement of a glacier by internal deformation is slow, on the order of tens of meters per year. In basal sliding, the base of the ice sheet is near the pressure melting point,

and some water is present to reduce the basal friction and allow the ice to slide. Movement by basal sliding is ten times faster than movement by internal deformation. Basal sliding influences how much a glacier erodes the landscape and the types of depositional features that result.

Water may flow toward the terminus of the glacier in a series of tunnels in the ice, or may exist as a thin sheet under the ice. Water is not the only material that will enhance sliding; sediment also facilitates movement at the glacier base. If the glacier has an underlying bed of water-rich sediment, it can move as effectively as if it were riding on a water layer.

Types of Ice

Ice Sheet. Ice sheets, or ice caps, are the thickest and largest accumulations of ice. They virtually cover the underlying land topography (Figure 2). Ice cap usually refers to the smaller of the two. Ice sheets and caps are comprised of ice domes, slow moving accumulation regions, and ice streams or outlet glaciers, rapidly flowing conduits of ice. Today, ice sheets cover Greenland and Antarctica. The Antarctic Ice Sheet is divided into two parts, the East Antarctic Ice Sheet and the West Antarctic Ice Sheet, separated by the Transantarctic Mountains. The East Antarctic Ice Sheet reaches over 4 km in thickness.

Ice sheets can be either land-based or marine-based. A land-based ice sheet is one in which the base lies primarily above sea level. The East Antarctic Ice Sheet and the Greenland Ice Sheet are examples. In contrast, the bottom of the West Antarctic Ice Sheet, a marine-based ice sheet, essentially lies *below* sea level; in some locations, this ice sheet lies as much as 2,000 m below sea level!

Ice Shelf. An ice shelf is a floating ice mass attached to the coast along at least one edge (Figure 2). Other types of glaciers are grounded; their bases are in contact with the substrate. The grounding line is the point where the ice sheet ceases to be attached to the substrate and begins to float. The Ross Ice Shelf in Antarctica is the largest ice shelf today. It is approximately the size of Texas.

Valley Glaciers. Ice that flows through a valley forms a valley glacier. Valley glaciers commonly are steep, and usually are tens of kilometers long. Valley glaciers join progressively larger valley glaciers, forming a network similar to a river tributary system.

Ice Tongues. Outlet glaciers, ice streams, and valley glaciers may extend, floating, into the sea. These seaward extensions are called ice tongues. Eventually, the length of the tongue becomes long and unstable. The tongue, or a portion, will break off (calve) and float to sea as an irregularly shaped iceberg.

Icebergs. In Antarctica, ablation, or loss of ice mass, occurs almost exclusively through the formation of icebergs. Icebergs also occur in Arctic waters. The form and size of the icebergs vary tremendously. Iceberg B-9, calved from Ross Ice Shelf, Antarctica, in 1987, was approximately twice the size of Rhode Island (Keys, 1990; *Houston Post,* 1987).

Sea Ice. Sea ice, or pack ice, is a general term for the thin ice cover that forms over large parts of the polar seas. Sea ice includes grease ice, frazil ice, pancake ice, and pack ice. Formation of pack ice begins in late summer as the sea surface temperature cools. The water surface thickens as unorganized ice crystals, called frazil ice, form. Thin, delicate, transparent plates of grease ice develop from the frazil ice. Grease ice evolves into pancake ice. Plates of pancake ice resemble large lily pads; upturned edges form where the plates meet. The plates grow in size, reaching a meter or two across. The pancakes suture, forming a thin sheet of white across the water surface. The ice thickens as more frazil ice attaches to the base of the pack. Wind and waves constantly break the pack into jagged, closely associated pieces.

Antarctic and Arctic sea ice differ. Arctic sea ice develops within the landlocked Arctic basin. This protects the ice and permits the growth of thicker, multiyear sea ice. In contrast, approximately 85 percent of the Antarctic sea ice melts annually (Sugden, 1982). Most of the pack ice that forms the following season is thin, first-year ice. Pack ice thicknesses in Antarctica range from 1 to 3 m. Each winter, the sea ice around Antarctica grows until the size of the continent more than doubles.

Ice-Related Landforms and Features

As ice moves, it sculpts the land, removing material from one location and carrying it to another. In this way, through erosion and deposition, a glacier leaves a definitive mark on the landscape to record its presence. These features are used by glacial geologists to reconstruct past "ice ages" (*see* AGASSIZ, JEAN LOUIS RODOLPHE). Glacial geologists categorize the features by determining if the feature is erosional or depositional, and where it formed with respect to the ice mass. Features can be subglacial (formed below the ice), transitional (formed where the ice stagnates), or proglacial (formed in front of the ice).

Glacial Erosion. The glaciers wear down, or erode, the landscape because they move. The internal deformation of glaciers does not leave an imprint on the landscape, but the basal movement of ice, accompanied by water, is a powerful erosive agent. Two forms of erosion, plucking and abrasion, contribute to the removal of Earth's materials. Plucking involves direct removal of material from the rock basement. The particles can be quite large; boulders the size of houses occur in debris fields. Abrasion involves erosion of the substrate by material contained in the ice sheet base. As the ice moves over the rock basement, it freezes particles into its base, resulting in a gritty texture (similar to sand paper). The particles scrape, or abrade, the underlying substrate further. As ice passes over bedrock, the action of abrasion leaves a polished surface. The slow scraping and grinding produce rock flour.

Erosional features are created by the glacier removing rock material and transporting it away; these features always are subglacial and provide a clue about the direction of ice flow, the nature of the base of the ice, and in some cases, the thickness of the ice. Erosional features can be large-scale, such as the Great Lakes of North America, or they can be small, such as the striations engraved in pebbles.

When a river cuts through the landscape, it cuts a "V" shape. An ice stream or outlet glacier carves a "U" shaped trough (Figure 3). Glacial troughs or valleys tend to have rounded bottoms, scoured by glacial abrasion and plucking.

At the head of valley glaciers there commonly exists a circular depression, a cirque, that represents the accumulation region of snow that fed the glacier. As snow accumulates in the depression, it gradually turns to ice and flows down-valley, widening and deepening the depression (cirque glacier). Where two cirques meet, the thin, knife-like edge between them is called an arete. Where three or more cirques join, the peaked juncture between them is called a horn.

As the ice passes over basement obstacles, it may slowly grind the obstacles into streamlined features, aligned in the direction of glacial movement. Such features include rock drumlins, whalebacks, and roches moutonnees.

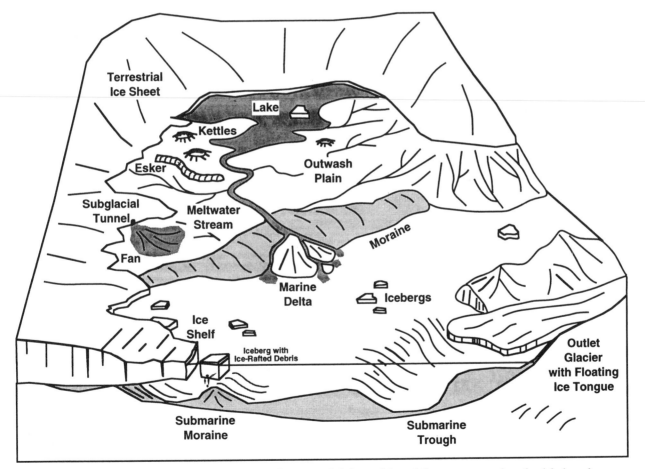

Figure 3. Illustration of selected marine and terrestrial depositional features associated with ice sheets (modified from Edwards, 1986).

Striations are the smallest features formed by subglacial processes. They are scratches in the rock surface carved by materials held in the glacier base as the glacial ice passed over the rock. Hence, striations indicate direction of glacial movement. Because the striation is essentially a line, there are actually two possible directions in which the ice could have been moving. Larger-scale striations, on the order of a few meters deep and wide, and hundreds of meters long, are called grooves.

Glacial Deposition. Depositional features are created by release of sediment from the glacier. They vary in scale from thin morainal ground cover to huge terminal moraines such as the terrain of Cape Cod, Massachusetts. Depositional features include landforms directly created by ice, but also encompass the by-products of galciation, such as outwash plains.

Till is a term used for the poorly sorted sediment deposited in association with a glacier. It contains materials collected from the path along which the glacier flowed. Pebbles in the till often are striated and have broken edges (faceting). Till is deposited under the glacier, and also at the glacier front when the ice pauses or retreats.

Moraines are accumulations of glacial material (till) that have surface expression. Several different types of moraines exist. Till sometimes takes a shape that reflects the base of the glacier ice; it may have long furrows or grooves parallel to the ice flow direction. This is called fluted ground moraine. Rogen moraines are subglacial moraines formed perpendicular to ice flow.

Drumlins are another example of ice sculpting. These are large mounds of glacial debris streamlined parallel to ice flow. In many places, they have a rough teardrop shape with the tail pointing in the direction of ice movement.

Eskers are long linear depositional features that record water flow in channels under, within, or on top of the glacier (Figure 3). Meltwater in the glacier commonly forms a network of tunnels that eventually reach the terminus. In general, the conduits parallel the direction of ice movement. While the tunnel is active, water, carrying large quantities of sediment, flows through it. The sediment becomes lodged in the tunnel, eventually filling it. As the ice retreats, these ridges of sediment are exposed.

Transitional depositional features refer to those features that form in a setting of melting ice. They originate in front of the actively retreating ice. When glaciers retreat, a region commonly exists where the ice breaks apart into discreet blocks that slowly melt in place. The landscape in this setting is disorganized and characterized by high water flow and an abundance of standing meltwater. Often, there is material on the surface of the melting ice block. As the ice melts, this material slowly comes into contact with the underlying substrate. The debris forms piles and sheets. Some of the material is shed from the ice into standing water. Water serves to sort the material, forming layers of different-sized material (stratification). Kames are mounds of stratified material. In this same landscape, some of the ice blocks are covered by thick piles of glacial debris. As the ice melts, a hole, or kettle, forms in the debris (Figure 3).

The proglacial environment is complex, especially in the regions of warmer glaciers that produce more meltwater. It encompasses terrestrial environments, streams, lakes, and the ocean. The deposits include moraines, deltas, fans, and thick packages of sediment deposited in the marine realm (Figure 3). Other than where streams rework and erode previously deposited material, this is a predominantly depositional environment.

It is important to note that the proglacial environment moves with the ice edge. Therefore, as the ice advances, the zone of proglacial deposition moves forward as well. As the ice retreats, the proglacial setting follows the retreating margin "backward." This means that when the glacier retreats, the proglacial deposits are laid on top of the subglacial deposits.

End moraines are accumulations of glacial debris that mark the edge of the ice; they are long ridges of till perpendicular to ice flow (Figure 3). End moraines can be quite large. They mark the maximum extent of ice advance and pauses in the retreat. The longer the glacier pauses, the larger the moraine. End moraines mark the transition from the subglacial environment to the proglacial environment.

Medial moraines and lateral moraines form adjacent to the glacier, and are in contact with the ice. Lateral moraines are mounds of till that form at the margins of outlet or valley glaciers. They parallel ice flow and form by the scraping and tumbling-down of debris to the ice edge. Where two valley glaciers or outlet glaciers flow together, their lateral moraines flow together also, forming medial moraines, and occupying a position in the center of the newly merged, larger valley glacier. Medial moraines also parallel ice flow.

Outwash is a general term for all of the debris carried by meltwater from the glacier. An outwash plain is the terrestrial area in front of a glacier characterized by outwash deposits (Figure 3). Outwash also can be deposited in lakes or the ocean. Commonly, there is a large volume of standing water in the proglacial environment. An end moraine may form a dam and trap meltwater between itself and the glacier. This forms a lake. Several lakes of varying sizes may characterize the proglacial setting.

Subglacial, englacial, and supraglacial tunnels carry meltwater through the glacier. When meltwater streams reach the glacier edge, they no longer are confined by the tunnel walls. The water flows out and spreads across the landscape, or into the marine or lake environment, subdividing into smaller streams (distributaries). As the streams spread, they lose velocity and cannot carry particles as large as when they were flowing quickly. Thus, the streams deposit the larger grains.

The deposition of material from the streams results in the creation of a fan or delta. These are large, triangular-shaped accumulations of meltwater sediment deposited in front of the glacier (Figure 3). The largest clasts (boulders) occur closest to the glacier edge. Farther from the glacier, the materials become finer (sand and clay). Fans may form in terrestrial, marine, or lacustrine environments below the water surface; deltas form in standing bodies of water and their top stays at the water surface.

The finest-grained product of the glacial environment, rock flour, travels the greatest distance. Even the slowest moving meltwater streams transport this very fine-grained mud. Rock flour is probably the sediment size produced in the greatest quantity by glacial erosion; it accumulates in large thicknesses in the proglacial environment. It is commonly picked up by wind, and later deposited to form the distinctive sedimentary units called loess.

Ice-rafted debris is material carried in the ice and dropped into a lake or marine setting as the ice melts. Ice rafting occurs over a wide distance. If the glacier terminates in a body of water, the material at its nose will slowly melt out and drop through the water to the substrate below, close to the ice margin. Alternatively, icebergs, derived from glaciers reaching the water's edge, may carry glacial debris great distances from the original source (Figure 3).

Bibliography

ALLEY, R. "West Antarctic Collapse—How Likely?" *Episodes* 13 (1990):231–238.

DENTON, G. H., M. L. PRENTICE, and L. H. BURCKLE. "Cainozoic History of the Antarctic Ice Sheet." In *The Geology of Antarctica*, ed. R. Tingey. Oxford Monographs on Geology and Geophysics, No. 17. Oxford, Eng., 1991, pp. 365–433.

EDWARDS, M. "Glacial Environments." In *Sedimentary Environments and Facies*, ed. H. G. Reading. Oxford, Eng., 1986.

"1,400-square mile Antarctic Iceberg Could Threaten Shipping." *Houston Post*, December 11, 1987.

HAMBREY, M. *Glacial Environments*. London, 1994.

KEYS, J. R. "Ice." In *Antarctic Sector of the Pacific*, ed. G. P. Glasby. Elsevier Oceanography Series, vol. 51. New York, 1990, pp. 95–122.

SUGDEN, D. *Arctic and Antarctic*. Totowa, NJ, 1982, pp. 128–162.

SUGDEN, D., and B. S. JOHN. *Glaciers and Landscape*. London, 1976.

STEPHANIE S. SHIPP

EARTH'S HYDROSPHERE

The Earth's hydrosphere is a discontinuous shell of water that includes the oceans, rivers and lakes, groundwater, ice caps, and atmospheric water vapor. The hydrosphere is essential to life on Earth, and unique among the planets in our solar system. Marine sediments at least 3.8 billion years old (Ga) and algal fossils (stromatolites) as old as 3.4 to 3.5 Ga are found in the geological record. This indicates that water has been present for much of Earth's history. Has it been present since the origin of Earth? When and how did it arise? These questions are the subject of this entry.

The compound water (H_2O) is so familiar to us in all its phases (solid, liquid, and gas) that we are likely to be aware of its importance to life on Earth. We are aware of concerns regarding the health of this natural resource (such as the detrimental effects of pollution). Most people are probably even familiar with the various compartments of the hydrologic cycle (atmospheric water vapor, surface water, groundwater, oceans), which probably may not be said about the subdivisions of the atmosphere. What may not be so familiar is the geological importance and history of the water within the hydrosphere (Table 1). Water is essential to many geological processes, such as landscape evolution, weathering, transport of sediments, diagenesis, and lithification of sediments (*see* DIAGENESIS; SEDIMENTS AND SEDIMENTARY ROCKS, TERRIGENOUS; and WEATHERING AND EROSION). Water is the primary agent of mass transport within and above the crust of Earth (*see* EARTH'S GLACIERS AND FROZEN WATERS; GEOLOGIC WORK BY STREAMS). It is important that we have some idea of how and when the hydrosphere originated, and how it evolved through geological time.

The Earth's hydrosphere contains approximately 1.4 billion km^3 of water. As mentioned above, the geological record indicates that water has been actively eroding rocks and depositing sediments for at least 3.8 Ga. However, the absence of preserved sedimentary rocks older than that may not necessarily indicate that the hydrosphere was not present in some form at an earlier time. Scientific curiosity regarding the origin of this shell of water has a very long history. Geologists have also been interested in water for a long time. As a result, several hypotheses have been suggested and examined for the origin of Earth's hydrosphere: (1) the "primordial" hydrosphere hypothesis; (2) the weathering hypothesis; (3) the outgassing hypothesis; and (4) the extraterrestrial accumulation hypothesis.

As the materials that eventually formed Earth solidified around 4.5 Ga (*see* EARTH-MOON SYSTEM, EARLIEST HISTORY OF; SOLAR SYSTEM), it is probable that some water was initially present, possibly resulting in the formation (after condensation) of a primordial atmosphere and hydrosphere. However, if our present hydrosphere evolved from this primordial atmosphere, then its present composition should be similar to that of the primordial atmosphere and oceans. The primordial atmosphere and oceans most likely contained much greater quantities of rare gases (neon, argon, krypton, and xenon) than are found today. This appears to invalidate the primordial atmospheric origin for the hydrosphere, since any loss of volatiles over the eons would have preferentially removed the lighter constituents (water has an atomic weight of 18 versus 20 for neon). In addition, the

Table 1. Estimated Distribution of Water in the Hydrosphere

	Volume (km³)	*Percent of Total*	*Percent of Freshwater*
Water in Land Areas:	47,971,710	3.5	
Lakes:			
Fresh	91,000	0.007	0.26
Saline	85,400	0.006	
Rivers	2,120	0.0002	0.006
Marshes	11,470	0.0008	0.03
Soil moisture	16,500	0.0012	0.05
Groundwater:			
Fresh	10,530,000	0.76	30.1
Saline	12,870,000	0.93	
Biological water	1,120	0.0001	0.003
Icecaps and glaciers	24,364,100	1.76	69.6
Atmosphere	12,900	0.001	0.04
Oceans	1,338,000,000	96.5	
Total	1,385,984,610	100	

Source: Maidment, D. R. "Hydrology." In *Handbook of Hydrology*, ed. D. R. Maidment. New York, 1993.

large collisions of matter that occurred during the later stages of accretion would have removed any primitive atmosphere. It is generally believed, therefore, that the present atmosphere and hydrosphere of Earth are entirely secondary in nature, formed sometime after the initial solidification of Earth, due to outgassing or accretion.

Weathering of crystalline rocks could have provided some of the components, including water, of the hydrosphere. Once early Earth had cooled, and rocks of various types had solidified at the surface, decomposition of these rocks may have produced free water that was originally chemically bound. The problem of the origin of the hydrosphere so intrigued a geologist working for the U.S. Geological Survey in the late 1940s that he spent considerable time looking at geological evidence and performing chemical mass balance calculations to examine the possibility of a weathering origin for the hydrosphere. WILLIAM W. RUBEY was studying the origin of marine phosphate rocks, and was curious about whether the salinity of the oceans had varied much in the past. Rubey compared the compositions of different crustal rock types. He found that sedimentary rocks are much richer in water and carbon dioxide than the other types. He then compared the total amounts of several substances—including water, carbon dioxide, chlorine, nitrogen, and sulfur—in the atmosphere, oceans, and sedimentary rocks with estimates of the amounts of these substances that could have resulted from weathering of igneous rocks, and found a large discrepancy (Rubey, 1951). There was far too much water, carbon dioxide, and chlorine to be accounted for by simple weathering of igneous rocks. In other words, there had to be some other source for the water and other "excess volatiles" that comprise the hydrosphere. Rubey postulated that outgassing of Earth from volcanoes and mid-ocean ridges could be responsible.

Rubey's idea has become the favored hypothesis for the origin of Earth's hydrosphere, but several questions remain. Did this outgassing occur all at once, early in the history of Earth, or gradually through time? Rubey himself reasoned that the process could be gradual and continuous. The oceans appear to have had a fairly constant salinity over time, as indicated by the continuous presence of fossils of marine organisms that are sensitive to changes in salinity. Furthermore, at the current rate of supply of water vapor from volcanoes, only about 1 percent of the water would need to be retained over 4 Ga to account for the present hydrosphere. What happened to the rest? This is a difficult question to answer, if we accept the "slow soak" form of the hypothesis (that accumulation was gradual). Most geologists recognize the contribution of gradual outgassing, but favor a rapid initial outgassing phase ("fast soak" hypothesis). Evidence comes from the timing of astronomical and geological events in the early history of Earth. During this time, the initiation of plate tectonics, the close proximity of Earth and the Moon, and differentiation of the core and mantle could all have produced a rather rapid initial outgassing.

If volcanic activity is responsible for the origin of Earth's hydrosphere, then Earth's mantle must be the source of the water (*see* EARTH, COMPOSITION OF). Is this a valid suggestion? Examination of stony meteorites (*see* METEORITES), thought to resemble the mantle, and other geological inferences indicate that the mantle contains 0.1 to 0.5 percent water. Based on the relative amounts of mantle rocks and hydrosphere water, the mantle must have lost approximately 0.031 percent of its mass as water to produce the observed amounts of water in the hydrosphere, which is quite reasonable. The question then is, where did the mantle get its water? This question presents an interesting problem. Very little water was present initially in the region around the Sun from which Earth and other terrestrial planets formed since water was never condensed (the early planets were extremely hot) and was probably lost along with other volatiles. Some water may have remained in hydrated minerals, but much may have come from meteors or comets from beyond Mars (Taylor, 1992), perhaps late in the early accretional episode in Earth's history.

This last idea, that Earth's water has an extraterrestrial source, has lately been revived in a new and interesting manner, which provides the final hypothesis for the origin of the hydrosphere. Louis Frank, a space scientist at the University of Iowa, spent much of the early 1980s examining images of data obtained from an ultraviolet photometer aboard the satellite *Dynamic Explorer I*. These images show Earth's "dayglow," produced by the adsorption and reradiation of solar energy by atomic oxygen in the outer portions of Earth's atmosphere. In addition, these images appeared to show short-lived dark spots or holes. Frank suggested that these holes are created by the entry into Earth's atmosphere of house-sized comets, balls of ice, and other matter. Examination of the images led Frank to conclude that these small comets are entering Earth's atmosphere at a rate of about 20 per minute, or about 10,000,000 per year (Frank et al., 1986). At this rate, Earth would receive about 2.5 cm of additional water every 1,000 years, more than enough to fill the oceans in 4 Ma. Although this theory is extremely controversial (Dessler, 1991, and Frank and Sigwarth, 1993), it is interesting to note that the debate about something so fundamental, water on Earth, continues to engage and intrigue scientists.

Bibliography

DESSLER, A. J. "The Small-Comet Hypothesis." *Reviews of Geophysics* 29 (1991):355–382.

FRANK, L. A., and J. B. SIGWARTH. "Atmospheric Holes and Small Comets." *Reviews of Geophysics* 31 (1993): 1–28.

FRANK, L. A.; J. B. SIGWARTH; and J. D. CRAVEN. "On the Influx of Small Comets into the Earth's Upper Atmosphere I. Observations. II. Interpretation." *Geophysical Research Letters* 13 (1986):303–310.

RUBEY, W. W. "The Geologic History of Sea Water—An Attempt to State the Problem." *Bulletin of the Geological Society of America* 62 (1951):1111–1148.

TAYLOR, S. R. "The Origin of the Earth." In *Understanding the Earth*, eds. G. C. Brown, C. J. Hawkesworth, and R. C. L. Wilson. Cambridge, Eng., 1992.

JEFF P. RAFFENSPERGER

EARTH'S MAGNETIC FIELD

Geomagnetism or the study of Earth's magnetic field has a long history. The magnetic field was the first property attributed to Earth as a whole, aside from its roundness. The findings of William Gilbert, physician to Queen Elizabeth I, were published in 1600, predating Newton's gravitational *Principia* by about 87 years. The magnetic compass had been in use, beginning with the Chinese, since about the second century B.C.E., but temporal variations in the magnetic field were not well documented until the seventeenth century. A formal separation of the geomagnetic field into parts of internal and external origin was first achieved by the German mathematician Karl Friedrich Gauss in the nineteenth century. Gauss deduced that by far the largest contributions to the magnetic field measured at Earth's surface are generated by internal rather than external magnetic sources; only sources internal to Earth are discussed here.

The magnetic field is a vector quantity, possessing both magnitude and direction; at any point on Earth a compass needle will point along the local direction of the field. Although we conventionally think of compass needles as pointing north, it is the horizontal component of the magnetic field that is directed approximately in the direction of the north geographic pole. The difference in azimuth between magnetic north and true or geographic north is known as declination (positive eastward) and may be as much as several tens of degrees. The field also has a vertical contribution; the angle between the horizontal and the magnetic field direction is known as the inclination and is by convention positive downward. At Earth's surface the field is approximately that of a dipole located at the center of Earth, with its axis tilted about 11° relative to the geographic axis. The magnitude of the field, the magnetic flux density passing through Earth's surface, is measured in microteslas, abbreviated as μT, with $1 \ \mu T = 10^{-6}$ T, and is about twice as great at the poles (about 60 μT) as at the equator (about 30 μT). Inclination ranges from $+90°$ at the north magnetic pole to $-90°$ at the south magnetic pole. Contours of locally orthogonal components of the geomagnetic field for 1990 are shown in Plates 5, 6, and 7. The magnetic field plays an important role in protecting us from cosmic ray particle radiation, because the incoming ionized particles can get trapped along magnetic field lines, preventing them from reaching Earth. One consequence of this is that rates of production of radiogenic nuclides (*see* ISOTOPIC TRACERS, RADIOGENIC) such as ^{14}C and ^{10}Be are inversely correlated with fluctuations in geomagnetic field intensity.

The internal field can be divided into contributions from the crust and those originating in the fluid outer part of Earth's core. At Earth's surface the crustal part is orders of magnitude weaker than that from the core, but remanent magnetization carried by crustal rocks has proved very important in establishing seafloor spreading and plate tectonics, as well as a global magnetostratigraphic timescale. The much larger part generated in Earth's core exhibits secular or temporal variations (see Plates 5 through 7), observable on timescales greater than a year; shorter period variations are attenuated by their passage through Earth's moderately electrically conducting mantle. On very long timescales (about 10^6 years) the field in the core reverses direction, so that a compass needle points south instead of north, and inclination reverses sign relative to today's field. The present orientation of the field is known as normal, the opposite polarity is reversed.

The present and historical magnetic field is measured at observatories, by surveys on land and at sea, and from aircraft. Since the late 1950s a number of satellites, each carrying a magnetometer in orbit around Earth for months at a time, have provided more uniform coverage than previously possible. Early satellites only measured the magnitude of the field: however, it was shown in the late 1960s that measurements of the field's direction are also required to specify the field accurately. Prehistoric magnetic field records can also be obtained through paleomagnetic studies of fossil magnetism recorded in rocks and archeological materials.

The Crustal Field

How Rocks Become Magnetized. The crustal magnetic field arises from magnetization carried by a number of magnetic minerals (e.g., members of the ternary system magnetite, rutile, and hematite; *see* MINERALS, NONSILICATES) that occur naturally in the rocks that make up the crust. Although some minerals are magnetically viscous (i.e., their magnetization changes rapidly to follow the direction of any ambient field), many rocks carry an imprint of the ambient magnetic field at the time of their formation, which remains stable over geological timescales. Two important mechanisms for acquiring this fossil magnetization or magnetic remanence are temperature changes and depositional processes. When a rock is heated, it gradually loses its magnetization. The Curie temperature of a mineral, the temperature above which all magnetic order is lost, varies widely with chemical composition and structure. Thermoremanent magnetization is acquired when magnetic material cools from above its Curie point in a magnetic field and records its direction at the time of cooling. Thermoremanent magnetization is commonly found in basalts and lava flows. Remanent magnetization can also be acquired by sediments, as small magnetic particles are incorporated into the sediment; on average these will align with the ambient field during deposition and become locked into position during subsequent dewatering and compaction of the sediment. Chemical, mineralogical, and metamorphic processes after rock formation can also influence the magnetization, sometimes causing the signal to be partially or completely overprinted by a later magnetic field.

Paleomagnetism, Plate Tectonics, and Magnetostratigraphy. Paleomagnetism, the study of the fossil magnetic record, has important applications in the earth sciences. Taking oriented samples of crustal rocks and measuring their directions of magnetization in a magnetometer allow the determination of the local paleofield direction when the rock acquired its magnetization. During the late 1950s and 1960s it became increasingly apparent that these observations did not support the existence of a predominantly dipolar field configuration like that existing today, unless the continents had moved around on Earth's surface since the time the rocks were formed. These observations provided important evidence supporting the theory of plate tectonics. Even more compelling evidence was found in the patterns of seafloor magnetic anomalies in the crustal field. Ocean floor basalts extruded at mid-ocean ridges (*see* PLATE TECTONICS) rapidly cool and acquire a thermoremanent magnetization. The material is carried away from the ridge by plate motion so that age increases with distance from the ridge axis. Successive normal and reverse polarity epochs of the magnetic field will have normally and reversely magnetized basalt associated with them. The associated magnetic field can be measured with a magnetometer towed behind a ship; anomalies in the magnitude of the magnetic field are observed with distinctive lineated or striped anomaly patterns approximately parallel to ridge axes. The pattern of polarity changes in the geomagnetic field as a function of distance from the ridge axis can be correlated with land-based sections dated by radiometric methods; distance is transformed to age, providing the basis for the global magnetostratigraphic timescale that spans the approximate time interval 0–106 Ma, the age of the oldest ocean basins.

Together with the assumption that averaged over sufficient time the geomagnetic field directions observed approximate those of a dipole aligned along Earth's rotation axis, paleomagnetic observations and seafloor magnetic anomalies provide constraints used in continental reconstructions over geologic time. Paleomagnetism is also useful for studying regional tectonic problems.

The Core Field

The great bulk of the geomagnetic field arises from sources in the liquid outer core of Earth. Although the field at any location is often approximated by that expected from a magnetic dipole situated at the center of Earth, there are additional significant contributions to the field; these are called the non-dipole parts of the field. Historical and contemporaneous measurements indicate how both the dipole and non-dipole contributions to the field have changed over the past few hundred years. There has been a gradual decrease in the magnitude of the dipole moment of about 8 percent in the last century, and Plates 5 through 7 show the secular variation is concentrated in the Atlantic hemisphere; some features in the magnetic field appear to drift slowly (fractions of a degree per year) westward, while other features stay stationary but grow or decay in magnitude with time. These short-period secular variations are correlated with decadal changes in the length of day and are believed to arise from transfer of angular momentum between Earth's mantle and core, though the origin of the coupling (chemical, thermal, topographic, or electromagnetic) remains uncertain.

Paleosecular Variation and Reversals. The full spectrum of behavior exhibited by the geomagnetic field can only be determined from paleomagnetic and associated geochronological studies. Typically only observations of field direction are possible, but when the magnetic remanence acquisition process can be simulated in the laboratory, the magnitude may also be estimated. During times of stable magnetic polarity, secular variation causes the orientation of Earth's dipole axis to change by some 10–15° on timescales of a few hundreds to thousands of years; corresponding local fluctuations in field intensity may be as much as factor of two or three, but the global dipole moment is less variable. A large departure from the approximately geocentric axial dipolar state for the field is known as a geomagnetic excursion and is usually accompanied by strong fluctuations in

intensity of the field. Because they are of short duration (a few thousand years or less), it is difficult to find geological records of an excursion that confirm it is a globally synchronous event; the most recent global geomagnetic excursion, known as the Laschamp event, appears to have taken place about forty thousand years ago. Excursions are often thought of as aborted reversals, and it is likely that a continuum of behavior exists ranging from normal secular variation through excursions to full polarity reversals. During geomagnetic reversals local field intensity typically drops to 10–20 percent of its stable polarity value; the whole process including intensity decrease and subsequent recovery may take several thousand years. The latest transition, from the Matuyama reversed polarity epoch to the present normal polarity epoch (the Brunhes epoch), took place at about .78 Ma.

The time between successive reversals is extremely irregular, but over long periods there are systematic changes in geomagnetic reversal rates. Figure 1 shows a fairly steady rise in average reversal rate since about 84 Ma. Prior to that there was a long period during which no known reversals occurred, the Cretaceous Long Normal Superchron. Another such reversal-free interval occurs from about 320–250 Ma when the field was consistently reversed in polarity; other intervals may also exist. The transition to such states may reflect changing physical conditions for the geodynamo, related to Earth's cooling history and growth of the inner core. The existence of the geomagnetic field since about 3 Ga is well documented, and such information as we have about its intensity during Archean and Early Proterozoic times suggests that it was roughly the same order of magnitude as it is today.

Models of the Geodynamo. Paleomagnetic observations have led to acceptance of the idea that the geomagnetic field must be generated by a self-sustaining dynamo process. The core temperature (about 5,000 K) is too high for the field to be generated by magnetic remanence. The outer part of the core is known to be liquid from seismic observations, and consists predominantly of iron (*see* PHYSICS OF THE EARTH), with a high electrical conductivity of around 3×10^5 siemens/m. The flow by convection of electrically conducting material through Earth's magnetic field in the liquid outer core provides a mechanism for generating a changing geomagnetic field. The longevity, variability, and reversible nature of the geomagnetic

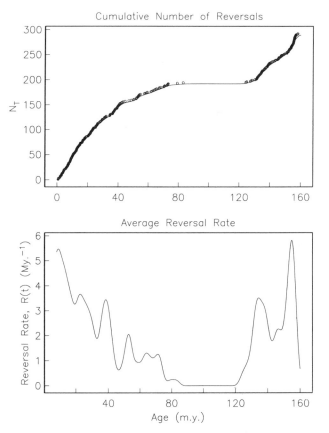

Cumulative Number of Reversals

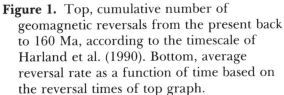

Average Reversal Rate

Figure 1. Top, cumulative number of geomagnetic reversals from the present back to 160 Ma, according to the timescale of Harland et al. (1990). Bottom, average reversal rate as a function of time based on the reversal times of top graph.

scales from those seen in the geomagnetic field and although precession of Earth can produce fluid instability, a suitable mechanism is needed to transfer momentum from the mantle to the core.

Bibliography

Cox, A., and R. B. Hart. *Plate Tectonics: How It Works.* Cambridge, MA, 1986.

Glen, W. *The Road to Jaramillo*, Stanford, CA, 1982.

Hale, C. J. "Paleomagnetic Data Suggest Link Between the Archean-Proterozoic Boundary and Inner-core Nucleation." *Nature* 329 (1987):233–237.

Harland, W. B., R. L. Armstrong, A. V. Cox, J. E. Craig, A. G. Smith, and D. G. Smith. *A Geologic Time Scale, 1989.* Cambridge, Eng., 1990.

Langel, R. A. "International Geomagnetic Reference Field, 1991 Revision." *Physics of the Earth and Planetary Interiors* 70 (1992):1–6.

CATHERINE CONSTABLE

field can all be generated by such a mechanism provided there is an adequate energy source (estimates range from 10^8–10^{11} W) to overcome the ohmic diffusion in the core and maintain fluid motion. Fluid motion could be powered by either thermal buoyancy from heat sources in Earth's core (decay of radiogenic isotopes or cooling of the core, resulting in freezing of the outer core to form the inner core) or by a gravitationally powered dynamo (from Earth's own gravitational energy) in which inner core freezing is accompanied by separation into a solid heavy part, leaving behind a lighter buoyant fluid that drives convection in the same way as heat drives thermal convection. A possible alternative is a rotationally powered dynamo from tidal friction or precession of Earth; however, tidal friction occurs at quite different time-

ECHINODERMS

The phylum Echinodermata comprises such morphologically disparate organisms as starfish and brittle stars (asteroids and ophiuroids), sea cucumbers (holothurians), sea lilies and feather stars (crinoids), sea urchins, sand dollars and sea biscuits (echinoids), and a variety of bizarre extinct forms whose life habits and close relations are poorly known. The phylum covers such a range of morphological diversity that there is no single characteristic shared by all members. One non-morphological trait shared by all modern echinoderms, and presumably by their fossil relations as well, is that they are exclusively marine. The term "echinoderm" literally means "spiney-skinned" and refers to the spine-bearing nature of many members of this phylum. Generally, echinoderms possess a distinctive radial pentagonal (five-fold) symmetry and an endoskeleton formed of many plates, or ossicles. All modern echinoderms possess an internal hydrostatic system used for locomotion and food gathering called the water vascular system.

There are about fifteen classes of extinct echinoderms and only five classes of extant (living) echinoderms. The phylum is traditionally informally divided into two major groups based on the life habit of the adult forms: eleutherozoan (free-living) and pelmatozoan (stalked, sessile). The different classes of echinoderms are distinguished on the basis of such characteristics as body symmetry, body shape, position of the mouth and anus, and the number and arrangement of plates.

Geologic Record

Echinoderms originated in the Cambrian period, and most groups had appeared by Ordovician time. Complete echinoderm specimens are rare as fossils because of the many-plated nature of the skeleton and the life habit (living exposed on the seafloor) of many members of this group. The paucity of certain fossil echinoderms (e.g., asteroids) probably does not reflect true fossil diversity but is largely a taphonomic artifact (*see* TAPHONOMY).

Crinoids

The most abundant and diverse group of the stalked echinoderms, crinoids are much-sought-after fossils for their delicate beauty and rarity as complete specimens. The hundreds of individual plates are held together in life with living tissue. Upon death and decay of the organic material, the skeleton rapidly disarticulates into individual ossicles. Crinoids originated in the Cambrian and persist into modern oceans, but not nearly in the numbers of their ancestors. The Late Paleozoic era (Mississippian and Pennsylvanian periods) was an especially productive time for stalked echinoderms. So abundant were their remains that thick limestone beds composed predominantly of crinoid skeletal material formed in the shallow seas of the midcontinental United States. Fabulous slabs of these fossils are found in many museums of natural history. These limestones are known by various names, but all can be described by the geographic area from which this material has been quarried as "Indiana Limestone." This rock was early recognized for its superior qualities as a building stone, and Indiana Limestone can be found in the foundations of some of the world's most famous buildings (for example, New York's Empire State Building). More than one icon of modern civilization rests on the remains of these delicate invertebrates from a long-ago sea.

The abundance of late Paleozoic stalked echinoderms stands in marked contrast to the paucity of modern stalked echinoderms. In fact, it was long thought that stalked echinoderms were extinct. They certainly did not inhabit accessible, shallow, tropic environments that were their Paleozoic bailiwick. During the late 1800s modern stalked crinoids turned up in deep-sea dredges. It was not until the 1960s that deep-sea photographs and submersible expeditions found stalked crinoids growing in place, in the dark, cold, abyssal depths of the ocean. Modern shallow-water crinoids (comatulid crinoids) are known, but they lack stems; they are basically an oral ring with arms. Comatulids are found in protected nooks and crannies in reefs, living a cryptic lifestyle—coming out at night to feed and hiding during the day. Some comatulids made up for lack of a stem (which would place the organism higher in the water column and expose it to more food) by climbing up tall sea fans and perching on the edge, arms spread in an efficient "filtration fan." What happened to cause the shift in morphology and habit from the shallow-water, stalked crinoids of the Paleozoic to the cryptic, stemless comatulids of the modern ocean?

The answer may be ecologic displacement; that is, the shallow-water Paleozoic stalked echinoderms may have been displaced through time to deep-water environments. What agent caused the displacement? Biological factors, specifically predation, may be responsible. Comatulid remains have been found in fish gut contents. The evolution of modern bony fish may have brought with it increased predation pressure on sessile, stalked crinoids, eliminating them from the shallow fish-filled environment. Vulnerable stalked crinoids survived in the predator-poor environment of the deep sea, where we find them today. Comatulid crinoids, by virtue of being stemless and cryptic, were able to minimize predation pressure and thus inhabit the more treacherous shallow-water environment.

Extinct Echinoderms

More major groups of echinoderms are extinct than are living today. Many extinct echinoderms have morphologies that are so different from anything with which we are familiar that the only appropriate term to describe them is "bizarre." Some of these extinct groups were fairly long-lived, geologically speaking, others were here and gone in a geologic instant. When echinoderm diversity is considered through all of geologic time, interesting patterns emerge. Early echinoderm diversity or disparity (number of different major types) is high. All echinoderm groups seem to have appeared by the end of the Cambrian period. This richness in morphologic variation falls off through the Paleozoic, and only ancestors of the few modern groups survived the terminal Permian extinction. Why did these groups persist and the others perish? The pattern of high disparity early in the evolutionary history of a lineage, progressively dropping off through time, appears to be typical of many groups of organisms and indeed may be a characteristic of how evolution works. The early history of a phylum may be characterized by many "experimental" designs, thus explaining their great disparity. With time, some experiments prove successful, others may prove less so and gradually or quickly slip into extinction. We cannot yet say with certainty what specific factors favored or disfavored a particular echinoderm type, and there is no reason to think that any one set of factors was responsible for the demise of all "unsuccessful" echinoderm designs. The wide variety of echinoderm form probably reflects an equally wide range of life habits, and by their very existence, all echinoderm groups were successful for at least some interval of geological time.

Bibliography

GOULD, S. J. *Wonderful Life.* New York, 1989.
MACURDA, D. B., JR., and D. L. MEYER. "Sea Lilies and Feather Stars." *American Scientist* 71 (1983):354–365.
MOORE, R. C., ed. *Treatise on Invertebrate Paleontology.* Parts U, S, and T. Lawrence, KS, 1966, 1967, 1978.
NICHOLS, D. *Echinoderms,* 4th ed. London, 1972.
SMITH, A. *Echinoid Palaeobiology.* London, 1984.

DANITA S. BRANDT

ECLIPSES AND OCCULTATIONS

An occultation occurs when—from the perspective of an observer—one body (such as the Sun) is blocked from view by another (such as the Moon). By longstanding popular usage, however, an occultation of the Sun by the Moon has been termed a "solar eclipse." A true eclipse occurs when a body disappears (or at least dims substantially) because light has been blocked by another body from reaching it. A lunar eclipse is a true eclipse, since the Moon dims because Earth has blocked the sunlight before it reaches the Moon. If you were on the Moon during a lunar eclipse, you could see an occultation of the Sun by Earth. This illustrates how eclipse and occultation phenomena depend on the observer's perspective (Figure 1). An eclipse will be visible to any observer who can see the body (because an eclipsed body is not illuminated), while an occultation will be visible only to observers that lie within the shadow of the occulting body. Hence lunar eclipses are visible from anywhere on Earth that the Moon is above the horizon, while solar eclipses (occultations) are visible only over a narrow path on Earth (or to an astronaut in space).

Solar eclipses are visible from Earth on average about every one and a half years. Since the Moon blocks out the bright solar disk (photosphere), these events give astronomers a chance to study the Sun's faint, hot outer atmosphere, called the corona. If one observes a solar eclipse, special care should be used to filter harmful ultraviolet rays from the Sun, since permanent damage to the retina can occur from direct observation of the Sun (not just during an eclipse, but at any time). Since the orbits of Earth and the Moon are slightly elliptical, the angles subtended by the Sun and Moon are different for different eclipses. When the angle subtended by the Moon is larger than that subtended by the Sun, the light from the photosphere is totally blocked for an observer in line between the centers of the two bodies. This phenomenon is called a total eclipse. If the observer is not aligned with their centers, however, only a partial eclipse will be seen, and the solar corona will not be visible because it is much fainter than even a small amount of remaining photosphere. A special type of partial eclipse occurs when the angle subtended by the Moon is less than that subtended by the Sun. Even if the observer is perfectly aligned with the centers of the Sun and Moon, all the sunlight will not be blocked because the Sun looks larger. This is called an annular eclipse because the visible photosphere forms an annulus around the Moon. The lunar limb is lumpy, due to peaks and valleys, so when the Sun is barely occulted, portions of the bright photosphere shine through the valleys. This phenomenon is called "Baily's beads." When only a single bead is visible, it appears particularly bright,

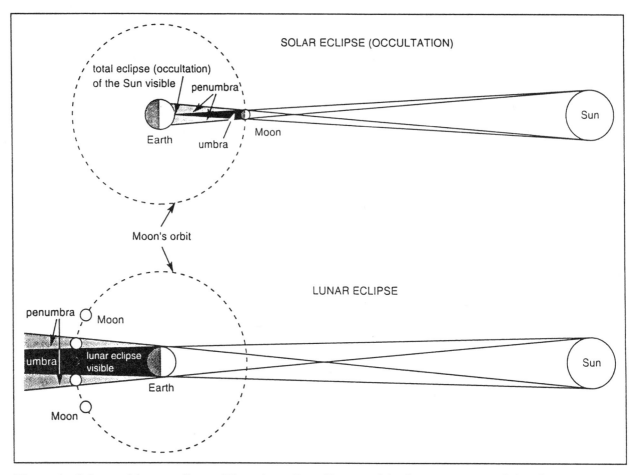

Figure 1. Solar and lunar eclipses. The solar eclipse illustrated in the upper part of the diagram is visible over only a narrow path on Earth because the Moon and Sun subtend nearly the same angle as seen from Earth. In the lower diagram, the Moon is seen full before it enters the penumbra of Earth's shadow, where it is partially eclipsed. The eclipse is total in the umbra. The size of the Sun, lunar orbit, and distance to the Sun have been necessarily exaggerated in this diagram.

and this has been termed the "diamond ring." Solar eclipses depend on the Moon's orbit around Earth, and they repeat in cycles of about 18 years (called the Saros), because of the precession of the lunar orbit.

Lunar eclipses last much longer than solar eclipses because Earth's diameter is nearly four times that of the Moon. Although lunar eclipses are of less scientific value than solar eclipses, they often present interesting sights to an observer. The Moon does not disappear completely during a lunar eclipse because it remains faintly illuminated by sunlight scattered through Earth's atmosphere. Depending on the amount of dust in Earth's atmo-

sphere at the time, the Moon can appear in various shades of orange.

Occultations and eclipses need not involve the Sun or the Moon. When Earth passes through the equatorial plane of a planet, one can observe occultation and eclipse phenomena involving the planet and its moons. By carefully recording the light intensity and timing these events, one can learn about the atmospheres, surfaces, and orbits of the bodies involved.

Another type of occultation occurs when a planet or other solar system body passes in front of a star, blocking the light from the observer. These stellar occultations are of special value to planetary astronomers since they can be used for study of the occulting body with thousands of times more spatial resolution than is possible with an ordinary image taken with an Earth-based telescope, such as the Palomar 5 m telescope or even the Hubble Space Telescope. Stellar occultation observations were used to discover the rings of Uranus in 1977 and to learn the temperature of the atmospheres of most planets. Neptune's rings and Pluto's atmosphere were discovered by stellar occultation observations as well. Stellar occultations of stars bright enough for visual observation of the phenomenon occur rarely, although stellar occultations by the Moon occur regularly. Since the Moon has no atmosphere, however, the starlight does not dim slowly, but is cut off abruptly.

Bibliography

ELLIOT, J., and R. KERR. *Rings: Discoveries from Galileo to Voyager.* Cambridge, MA, 1987.

PASACHOFF, J. M. *Contemporary Astronomy*, 2nd ed. Philadelphia, 1981.

JAMES L. ELLIOT

ECONOMIC GEOLOGY, HISTORY OF

The *Glossary of Geology* (Bates and Jackson, 1987) provides two definitions of economic geology. The first is broadly stated as "the study and analysis of geologic bodies and materials that can be utilized profitably by man, including fuels, metals, nonmetallic minerals, and water." Brian J. Skinner (1995), as president of the Society of Economic Geologists, has suggested that engineering and environmental aspects of geology might appropriately be added to this list. Moreover, he concludes that "economic geology is not the study of a special class of facts, it is the study of all the facts of geology from a special point of view, and specifically from the point of view of utility and management." Skinner's second definition, in contrast, is more narrowly defined as the "application of geologic knowledge and theory to the search for and the understanding of mineral deposits." This entry will be restricted to mineral deposits as described by the more specific and commonly accepted second definition.

The universal need for mineral commodities amply justifies the terms "search" and "understanding" of the second definition, and thus provides employment for economic geologists in several areas. Career opportunities are most abundant in the industrial sector where activities are largely directed toward mineral exploration, property evaluation, mine development, and mine production. The majority of economic geologists are associated with mining companies, but others are affiliated with petroleum, geothermal, chemical, agricultural, railroad, and timber companies; with smaller exploration syndicates and geotechnical firms; or work as private consultants. Many state geological surveys and mine bureaus employ economic geologists, and larger numbers are found in federal agencies such as the U.S. Geological Survey, Bureau of Mines, Forest Service, and Bureau of Land Management, among others. Their work with such government organizations ranges from basic geologic mapping, through regional resource assessments and detailed property examinations, to fundamental and highly specialized research. Economic geologists employed at colleges and universities contribute to instructional programs, research endeavors, and employment opportunities for students.

Mineral Deposits Versus Ore Deposits

Technically, any body of rock is a mineral deposit, but few contain sufficient quantities of valuable minerals (asbestos, chromite, diamond, garnet, quartz, sapphire, etc.) or of minerals containing sufficient concentrations of valuable metallic elements (aluminum, copper, gold, iron, lead, zinc, silver, etc.) and/or nonmetallic elements (fluorine, phosphorus, sulfur, etc.) to render them of economic importance, or potentially profitable. Economic geologists engaged in the search for mineral deposits direct their efforts to finding ore bodies. An ore body is a unique type of mineral deposit: one that is well defined in size (three dimensions) and grade (concentration of the valuable components), and from which one or more minerals, metals, or nonmetals may be profitably extracted. Factors that may influence profitability, in addition to size and grade, include many variables such as location, climate, topography, depth below the surface, geometry of the ore body, method of mining, mineralogy of the ore, metallurgical technology, commodity prices, availability of power and water, environmental constraints, royalties, taxation, and so on. Preliminary economic analyses of new mine developments commonly involve great uncertainties because a number of these factors may vary other time and be unpredictable. Moreover, exploration is expensive in relation to the rate of success. Probably fewer than one in a thousand mineral deposits examined in the United States survives the lengthy and costly progression from a raw mineral prospect to a proved and developed ore body ready for mine production. Because of such uncertainties, unfavorable statistics, and the highly mechanized and automated nature of the industries, it is not surprising that *Fortune* magazine consistently ranks petroleum and mining as the most capital-intensive of all industries: in 1993 they required financial investments of approximately $670,000 and $480,000 per employee, respectively. More detailed information on the role and duties of economic geologists in the minerals industry may be gleaned from the textbook *Exploration and Mining Geology* by W. C. Peters (1987).

Search Guided by Understanding and Research

The effective exploration for new ore deposits is neither random nor haphazard. It is based on the careful geologic study of known ore deposits, the prevailing knowledge of various ore-forming processes, and the new insights gained from applied and theoretical research investigations. An excellent reference for such information is the textbook *The Geology of Ore Deposits* by J. M. Guilbert and C. F. Park, Jr. (1986). Economic geologists for more than a century have attempted to develop a genetic classification of mineral deposits. The rationale is that if one understands the geologic environment and processes by which a particular ore deposit has formed, then it is likely that additional deposits of this type may be found in geologically similar terrain nearby or elsewhere. Of the many classifications devised, and with each characterized by differing detail and features of emphasis (host rock, minerals, textures, metals, etc.), most contain major subdivisions related to magmatic, sedimentary, hydrothermal, and weathering processes of ore formation.

Although a worthy scientific endeavor and one of great practical value, no single classification has proved to be universally satisfactory. The reasons

for this dilemma include divergent interpretations of basic geological and analytical data, the multiplicity of environments in which some mineral may form, and the ever-increasing need for additional research utilizing state-of-the-art analytical techniques. Because of these and other shortcomings, many economic geologists prefer a pragmatic classification that avoids explicit genetic connotations. Examples might include names based on the major economic element or mineral, the rock type, the geographic locale, and many subtypes of these and other deposits. For geologists having the appropriate knowledge through experience or the literature, such terms may convey qualitatively useful information pertaining to the host rock, age, mineralogy, metals, geometry, and possible genesis for each type of these deposits.

Regardless of potential shortcomings of the classifications, the understanding of and search for mineral deposits are based on research that is best performed on deposits that have been fully developed (exposed) for mine production. Ideally, the effort begins in the field or mine with the construction of detailed geologic maps of the deposit. These normally portray in three dimensions the relationships of the ore to various host rocks and structural features of the area, and the maps serve as the basic framework upon which subsequent interpretations of practical and/or theoretical value are made. Outstanding examples of mine maps are recorded in the literature for mineral deposits at Bingham, Utah; Buchans, Newfoundland; Butte, Montana; Climax and Henderson, Colorado; Highland Valley, British Columbia; El Salvador, Chile; Sudbury, Ontario; Twin Buttes, Arizona; and elsewhere. Now uncommonly, the geologic mapping by industry personnel is followed by numerous and varied topical research investigations conducted by academic and government geologists.

Research undertaken in the study of mineral deposits is nearly as varied as that performed throughout all disciplines of the earth sciences. This includes applied, experimental, and theoretical approaches to the many branches of geochemistry, geophysics, mineralogy, petrology, and other areas. The reason for this breadth is because ores are unique types of rock that differ from their commonplace brethren only by virtue of rarity, unusual mineralogy, and intrinsic value. Elegant compendia edited by Barnes (1967, 1979) and Skinner (1981) give broad overviews of research directions, technical methodology, and applications to economic geology (*see* MINERAL DEPOSITS, EXPLORATION FOR; MINERAL DEPOSITS, FORMATION THROUGH GEOLOGIC TIME; MINERAL DEPOSITS, IGNEOUS; MINERAL DEPOSITS, METAMORPHIC).

Geochemistry is pervasive throughout nearly all facets of the minerals profession. Most prevalent are routine analyses of rock for metals at all levels of concentration that are used to delimit the ore zones within mineral deposits and of rock, soil, stream sediment, and vegatation to guide both regional and local exploration programs. More sophisticated techniques that utilize electron microprobe, mass spectrometric, neutron activation, plasma emission spectroscopy, and other instrumentation applied to ores, rocks, and discrete mineral crystals serve to precisely define compositions and to determine provenance (source), temperature-pressure conditions of formation, and ultimately the genesis (origin) of such materials. For example, isotopic ratios of many elements are determined by mass spectrometry. Those of the radiogenic elements such as the parent-daughter pairs represented by potassium-argon (K-Ar), rhenium-osmium (Re-Os), rubidium-strontium (Rb-Sr), samarium-neodymium (Sm-Nd), thorium-lead (Th-Pb), and uranium-lead (U-Pb) are used to deduce the absolute ages of host rocks and ores (*see* GEOLOGIC TIME; GEOLOGIC TIME, MEASUREMENT OF), and as isotopic tracers to identify the likely source regions (mantle versus deep or shallow crust) from which the magmas and metals were derived (*see* ISOTOPIC TRACERS, RADIOGENIC). In contrast, the isotopic ratios of stable elements, particularly those of carbon (C), hydrogen (H), oxygen (O), and sulfur (S), vary not because of radioactive decay, but as a consequence of temperature-dependent fractionation with various chemical and physical processes (*see* ISOTOPIC TRACERS, STABLE). Differences in the isotopic ratios between two or more oxygen- or sulfur-bearing minerals may serve as geothermometers. Isotopic ratios of carbon and sulfur may discriminate between magmatic or biogenic sources of these elements, and the geologic environment in which host rock or ore mineral formed. Ratios of hydrogen and oxygen comprise a useful "isotopic duo" for identifying the provenance of aqueous fluids (waters of magmatic, metamorphic, meteoric, or oceanic origin) in active geothermal systems or trapped as a hydrous component of, or as small inclusions in, the minerals of rocks and ores. Examination of the fluid inclusions using a heating-freezing stage mounted on a petrographic microscope provides temperatures of mineral formation, identification of daughter minerals, and compositional data pertaining to the ore-forming fluids. Our current understanding of ore-forming processes in magmatic, hydrothermal, and sedimentary environments is not based exclusively on these and other empirical studies. Important insights and constraints are increasingly derived from experimental laboratory investigations of various rock, mineral, and fluid systems under which conditions of temperature, pressure, and composition are controlled or purposely varied. Moreover, the application of thermodynamics has contributed enormously to our knowledge of the stability and compatibility of minerals, the stability of aqueous chemical species in solution, and the transport and deposition of metals and minerals throughout a spectrum of geologic environments.

Geophysics and other forms of remote sensing, such as aerial photography conducted from aircraft and satellites, have contributed significantly to the search for mineral deposits. Geophysical surveys are normally undertaken to find buried or hidden mineral deposits, and are conducted near the final stage of an exploration project when the area of interest is relatively small. Electrical, electromagnetic, and magnetic methods are most widely applied, whereas gamma-ray spectrometry, gravity, and seismic methods are used more selectively. Choice of the method or methods employed is dictated largely by the presumed type, geometry, and mineralogy of the deposit being sought. It should be emphasized, however, that chemical, not physical, properties of minerals and rocks serve as the best guide to ore. With few exceptions (magnetite, one of several ore minerals of iron, is magnetic, and uranium-bearing minerals are radioactive), the ore minerals do not exhibit unique geophysical signatures where buried. Nonetheless, geophysics offers the only glimpse of the subsurface, albeit indirectly, in the absence of more costly drill holes or underground development. Moreover, satellite-borne imaging spectrometry using sensors such as the Landsat thematic mapper and Landsat multispectral scanner has successfully identified several minerals in surface rocks and soils that are commonly associated with ore. With further development, these remote-sensing techniques may ultimately become the favored exploration method for inaccessible regions. The details of geophysical theory and its application to exploration are given in the textbook *Applied Geophysics* by W. M. Telford and others (1990).

The Future

Geologic mapping will continue to identify areas of potential for the discovery of new mineral deposits. However, because most deposits having obvious surface indications have largely been discovered, except for those residing in geographically remote regions (including the Moon, asteroids, and other planets), the search for new ore bodies has become and will continue to be increasingly difficult. Those deposits that remain undiscovered are mostly deep and concealed. Thus, a major future challenge will be the replacement of resources now being mined and rapidly consumed by the industrial nations. To be successful, economic geologists will have to apply the results and concepts gained from research in mineralogy, petrology, geochemistry, geophysics, and geostatistics to new high-risk exploration strategies based on sound geologic insight and inference. In North America there have been impressive discoveries over the past two decades: diamonds in the Northwest Territories; gold and silver at many locations in Nevada, California, and elsewhere; copper-silver, in western Montana; molybdenum (Mo) at Henderson, Colorado; nickel-copper-platinum (Ni-Cu-Pt) at Voisey Bay, Labrador, and Subdury, Ontario; zinc-copper-lead-silver (Zn-Cu-Pb-Ag) at Crandon, Wisconsin; zinc-lead-silver (Zn-Pb-Ag) at Red Dog, Alaska; and numerous other finds. Barring unforeseen environmental and political constraints, the future replenishment of these and other resources through basic research and successful exploration endeavors seems assured.

Bibliography

BARNES, H. L., ed. *Geochemistry of Hydrothermal Ore Deposits*. New York, 1967; 2d ed., 1979.

BATES, R. L., and J. A. JACKSON, eds. "Economic Geology." In *Glossary of Geology*, 3d ed. Alexandria, VA, 1987.

GUILBERT, J. M., and C. F. PARK, JR. *The Geology of Ore Deposits*. New York, 1986.

PETERS, W. C. *Exploration and Mining Geology*, 2d ed. New York, 1987.

SKINNER, B. J., ed. *Economic Geology Seventy-Fifth Anniversary Volume, 1905–1980*. Lancaster, PA, 1981.

SKINNER, B. J. "Presidential Perspective—What's in a Name." *SEG Newsletter*, 21 (April 1995).

TELFORD, W. M., L. P. GELDART, and R. E. SHERIFF. *Applied Geophysics*, 2d ed. Cambridge, Eng., 1990.

CYRUS W. FIELD

EKMAN, VAGN WALFRID

Vagn Walfrid Ekman was born on 3 May 1874 in Stockholm, Sweden, the youngest son of F. L. Ekman, a pioneering Swedish scientist in the fields of physical and chemical oceanography. The senior Ekman died in 1890 when Walfrid was 15 years of age. Walfrid entered Uppsala University and received a B.A. degree in mathematics and physics in 1896. After graduation Ekman moved to Stockholm, where he because a devoted auditor of the lectures on hydrodynamics given by Vilhelm Bjerknes (*see* BJERKNES, J. A. B. and V. F. K.). In 1899, in his first publication, Ekman took up a problem that had attracted the attention of his father: the salty-water bottom current that travels upstream in rivers as they enter the sea. In this first paper, Ekman used the idea of turbulent (eddy) viscosity, a concept introduced in geophysics a few years earlier and very likely imparted to him by Bjerknes.

Contemporary students of physical oceanography know that an "Ekman layer" is generated by the stress of the wind acting on the surface of the ocean on a rotating Earth. The discovery of the Ekman layer had its root in FRIDTJOF NANSEN's observations aboard the *Fram* during his voyage on the Arctic Ocean (1893–1896). In an attempt to understand his many observations, Nansen contacted Bjerknes about two particular problems. The first, known as "dead water," concerned the effect that sometimes hindered the headway of ships, including that of *Fram*. Bjerknes realized that internal waves must be involved and suggested that Ekman investigate the problem. Ekman's work verified Bjerknes assumption and led to the publication of two papers in 1904.

The second problem that Nansen sought to resolve is best described in Ekman's own words in his 1905 paper (see below): "On studying the observations of wind and ice-drift taken during the drift of *Fram*, Nangen found that the drift produced by a given wind did not, according to the general opinion, follow the wind's direction but deviated 20°–40° to the right. He explained this deviation as an obvious consequence of Earth's rotation; and he concluded that the water-layer immediately below the surface must have a somewhat greater deviation than the surface layer, and so on. . . ." According to Bjerknes, Ekman arrived at a mathematical solution that same evening. The solution was published in 1902 as part of both Nansen's article in English and Ekman's doctoral thesis in Swedish. Ekman elegantly combined his knowledge of Earth's rotation with the new eddy-viscosity concept, resulting in the recognition of a rather shallow wind-influenced layer, the Ekman layer, with its spiral staircase of smaller and smaller step vectors. The total integrated wind drift, known as the Ekman transport, is directed 90° to the right of the wind direction in the Northern Hemisphere, and 90° to the left in the Southern Hemisphere.

Some oceanographers were pleased with Ekman's demonstration that wind-driven currents are shallow, because they preferred to explain deep currents by thermal processes. However, in 1905 Ekman published a much enlarged and now famous version of the theory of wind-driven currents, demonstrating that the presence of coasts makes it possible for the wind to influence deep water indirectly. If, for instance, a wind in the Northern Hemisphere blows from the south with the coast to its right, then the Ekman transport causes a rise in sea level toward the coast. The ensuing pressure gradient drives a deep current that runs along the coast, that is, in the same direction as the wind. Near the seafloor, due to bottom friction, there is a deviation to the left so that in a steady state as much water is transported outward in an Ekman bottom layer as inward in the Ekman (surface) layer. The necessary vertical transports were later called Ekman pumping. In 1906 Ekman noted that in the atmosphere there are similar components of a frictional ground wind layer, equivalent to the ocean Ekman bottom layer, but only referred to as the Ekman layer by present-day meteorologists. Above it an upper wind is flowing along the isobars.

From 1902 to 1908 Ekman worked on the staff of the International Council for the Exploration of the Sea (ICES), headquartered in Copenhagen. Working at ICES Central Laboratory in Oslo, he constructed new instruments either in collaboration with its director, Nansen, or alone. In 1908, after long and careful experimental work, Ekman published a formula giving the compressibility of seawater as a function of pressure, temperature, and density. When the work was repeated by others in the 1970s, Ekman's original belief that some old values he had used may have been less accurate was confirmed. This completed Ekman's otherwise perfect work.

In 1910 Ekman became professor of mechanics and mathematical physics at the University of Lund, where he remained until 1939. His attentions were now directed to topics other than oceanography, but in 1922 Ekman initiated a second period of oceanographic activity. He began to focus his attention on horizontal variations, including *vertikale Wirbelstärke*, the vorticity of planetary and topographic components. Despite difficulties in clarifying the problem, and an analysis that is somewhat hard to follow, his intense efforts inspired a younger generation. CARL-GUSTAF ROSSBY referred to Ekman when he published his potential vorticity theorem, and at the same time he introduced a simplified latitude-dependence of Earth's rotation, the beta-effect. Later, HENRY STOMMEL made use of Ekman's bottom friction studies and the beta-effect to construct a solution of a wind-driven circulation in an idealized ocean, a goal Ekman had aimed at and strongly contributed to, but had not attained. Vagn Walfrid Ekman died on 9 March 1954 in Gostad, Stockaryd, Sweden.

Bibliography

WELANDER, P. "Ekman, Vagn Walfrid." In *Dictionary of Scientific Biography*, vol. 4, ed. C. Gillispie. New York, 1971, pp. 344–345.

ARTUR SVANSSON

ELSASSER, WALTER M.

The scientific career of Walter M. Elsasser spanned six decades in the middle of the twentieth century, from the time of the quantum mechanics revolution in physics through the plate tectonic revolution in earth science and beyond. Elsasser made important contributions to each of these revolutions and, in addition, laid important foundations for several new fields in geophysics.

Born in Mannheim, Germany, on 20 March 1904, Walter Elsasser began his career in science as an undergraduate at the University of Munich, enrolling in the cycle of theoretical physics courses taught by Arnold Sommerfeld in the early 1920s. In 1924 Elsasser went on to Göttingen, then the international center for the emerging subject of quantum mechanics, where he began his graduate studies with Max Born and James Franck. In 1925 he published a short article in *Die Naturwissenschaften*, where he interpreted the famous experiments on electron scattering by Clinton Davisson and Charles Kunsman as indicating the wave-like nature of matter. This paper preceded by two years the publication of the decisive study by Davisson and Lester Germer, which earned those authors a Nobel prize in physics. This incident typified Elsasser's role in science, wherein he would introduce a seminal idea into a new field, and then leave the development of that idea (and most of the subsequent credit) to others.

Elsasser emigrated from Germany in 1933 and began working in nuclear physics at the laboratory of Frédéric Joliot in Paris. While in Paris, he acquired an interest in the theoretical foundations of biology through discussions with the philosopher Théophile Kahn and through his own exploration of Jungian psychology. During the 1930s he began to develop the concept of irreducible complexity in biological organisms, from which he argued that it is intrinsically impossible to construct an ordering of living organisms in strictly logical or mathematical terms. This line of reasoning led him away from a purely reductionist view of life-forms and on a life-long quest to find a more suitable theoretical framework for biological systems, a task which, along with geophysics, took him away from his origins in theoretical physics.

In 1936 Elsasser came to the United States and was introduced to the field of geophysics while employed in Theodore Von Karman's laboratory at Caltech, which later became the Jet Propulsion Laboratory. In the ensuing years Elsasser moved frequently from one position to another, beginning at the RCA laboratories during World War II. Following the war he held professorships at numerous American universities, including the University of Pennsylvania, the University of Utah, the University of California, Princeton University, the University of Maryland, and finally at the Johns

Hopkins University. Elsasser died in Baltimore on 14 October, 1991 at the age of 87.

Elsasser made fundamental contributions in many areas of geophysics, including atmospheric radiation, formation of Earth, plate tectonics, and terrestrial and stellar magnetism. He approached geophysics from the perspective of a natural philosopher, always seeking explanations for apparently complex phenomena in terms of simple physical processes. His hallmark was an uncanny ability to reduce a problem to fundamental terms by stripping away the unnecessary complications and exposing the essential underlying principles.

His approach is best exemplified by his pioneering work on the dynamo theory, which explains the origin of Earth's magnetic field. By the 1940s, observations of nearly ubiquitous stellar and planetary magnetic fields seemed to indicate that magnetism was a fundamental property of many massive objects in the cosmos. Indeed, Albert Einstein ranked the origin of planetary and stellar magnetism as one of the most significant unsolved problems in physical science. The British physicist P. M. S. Blackett had advanced a unified field theory that associated planetary magnetism with rotation and gravity. In effect, Blackett was advocating an addition to Maxwell's equations, which for fifty years were thought to constitute a complete description of all electromagnetic phenomena.

Competing with Blackett's explanation was the so-called self-excited dynamo theory, postulated in general terms by J. J. Larmor two decades earlier. The dynamo theory held that an electrically conducting fluid in motion could sustain a magnetic field indefinitely through a positive feedback in which secondary magnetic fields induced by the fluid motion continuously augment the primary field, in apparent violation of Lenz's rule but otherwise explicable by Maxwell's laws. Initially the concept of a self-excited fluid dynamo did not have a large following, in part because critical elements of the theory were either vague or missing, and also because several "anti-dynamo" theorems proved by the mathematician T. G. Cowling indicated that large classes of geometrically simple fluid motions could not act as dynamos.

In a sequence of papers beginning in the late 1940s, Elsasser put the dynamo theory on firm ground, first by developing the mathematical tools necessary for construction of quantitative dynamo models and, second, by proposing explanations for the known behavior of the geomagnetic field in terms of the predictions of the theory. He introduced the concept that a fluid dynamo becomes self-sustaining through a process in which magnetic energy is continually exchanged between the poloidal part of the magnetic field (that is, the part of the field emerging from the core and extending through the mantle to the surface and beyond into space) and the hidden toroidal part of the magnetic field which lies within Earth's electrically conducting, iron-rich core (*see* CORE, STRUCTURE OF). Elsasser recognized that a positive feedback between these two components of the magnetic field requires induction by motion of the liquid outer core. He further proposed that the liquid outer core motion is the result of thermal convection. He postulated that planetary and stellar dynamos require three basic ingredients: (1) an electrically conducting fluid with large linear dimensions; (2) convection to drive the fluid motion; and (3) rotation to organize the fluid motion and facilitate the exchange of poloidal and torodial magnetic energy. Since these three ingredients are common in stars and planets, he was able to explain why most of these objects behave as self-sustaining dynamos. In addition, he isolated the critical dimensionless parameter governing the internal dynamics of the geodynamo, the ratio of the Lorentz and Coriolis forces affecting the fluid motion. Later Paul H. Roberts named this ratio the Elsasser number and it has come to be recognized as a fundamental parameter in modern dynamo theories. Subsequent discoveries of complex behavior in the paleomagnetic field, such as polarity reversals and long-term secular variation, have enhanced the significance of the dynamo theory, and direct numerical simulations of convection and magnetic field generation in the core have led to an essentially complete acceptance of Elsasser's dynamo theory for Earth's magnetic field.

In addition to his ground-breaking contributions in geophysics, Elsasser wrote several books on biology, outlining a holistic theory of the organism that went beyond the traditional and accepted reductionist interpretation of life-forms. In these he introduced such novel concepts as irreducible heterogeneity, in an effort to explain the diversity of behavior found even within the same species. These books were published during the heyday of molecular biology and generally received cold reception from influential biologists. However, this attitude may be changing. Due to the infusion of new concepts from chaos theory and nonlinear dynamical systems, ideas such as irreducible heterogeneity are finding a wider application in the life sciences today.

Bibliography

ELSASSER, W. M. *The Chief Abstractions of Biology*. Amsterdam, 1975.
―――. *Memoirs of a Physicist in the Atomic Age*. New York, 1978.

PETER OLSON

ENERGY, GEOTHERMAL

Geothermal energy is the extractable and commercially usable heat of Earth. Typically this heat is extracted through the medium of water, either water produced directly from within Earth, or water that has been injected into the ground to scavenge heat from rock and soil. The source of this heat is related to the origin and composition of Earth. Earth may be seen as a great heat engine, deriving its energy from the decay of radioactive elements present within rocks of Earth's crust and mantle. Principal among these radioactive elements are an isotope of potassium (^{70}K), present in trace amounts in most crustal rocks and believed to be present in certain mantle rocks, plus isotopes of uranium (^{234}U, ^{235}U, ^{238}U), thorium (^{232}Th), and numerous others in trace amounts. In addition to this internally generated heat, there may be residual heat remaining from the formation of the planet, and the upper several meters of the ground surface is heated by incoming solar radiation.

Earth's heat is expressed tangibly at the surface as volcanic eruptions, hot springs, geysers, and fumaroles (steam vents). In addition, rock temperature increases everywhere with depth below the upper several meters affected by solar radiation and past climatic cycles. The rate of increase in temperature with depth (the geothermal gradient) varies widely according to the regional and local geology. Areas within the stable continental interiors, far from tectonic plate boundaries and intraplate rifts, exhibit a gradient as low as 20°C per kilometer, and locally possibly even less. At the boundaries of crustal plates, and at mantle plumes (hot spots) and along intraplate rifts, the gradients may reach several hundred degrees per kilometer of depth. Most attempts to extract Earth's heat have occurred along or near plate boundaries, hot spots, and intraplate rifts, although there is a significant and growing use of the normal geothermal gradient of the continental interiors, 20–40°C per kilometer. Most recently, there has been the successful commercial utilization of thermal energy stored in the upper few tens of meters of the ground (ground-source heat pumps).

Geothermal energy commonly is used both directly, as thermal baths and spas, and as water pumped through pipes to heat buildings and for use in industrial and agricultural processes; and indirectly, in the generation of electricity. The use of naturally occurring thermal springs as baths surely goes back into prehistory. The use of geothermal spring water to heat buildings can be traced with confidence to Roman times and to ancient China. Throughout the middle ages and modern periods, there has been the episodic extraction of various chemicals from geothermal springs, fumaroles, and shallow well waters, most notably boric acid, sulfur, alum, and common salt (NaCl), in Italy, Japan, and elsewhere. In recent decades geothermal waters have been pumped from drilled wells and used to heat homes, offices and factories, greenhouses, and fish hatcheries, as well as swimming pools, and as process heat in such diverse operations as laundries, dryers and dehydrators of wood pulp, seaweed and agricultural products, and in chemical extraction activities. The range of water temperature used in these activities commonly is between 20 and 80°C, occasionally extending significantly above 100°C.

The principal countries making use of geothermal heat in such operations are, in decreasing order, Japan, Iceland, Hungary, France, New Zealand, Russia, the United States, China, and Italy. Most of these countries are characterized by the presence of crustal plate boundaries, intraplate rifts, or hot spots within or closely adjacent to their territory. However, Hungary and France are not; in these countries, waters reflecting the normal gradient of stable continental interiors (50–100°C at 1.5 to 2.5 km) are produced from wells and used.

In the United States, Sweden, Denmark, and elsewhere in Europe, emphasis now is being given to using the temperatures found at merely a few meters in depth for home heating and cooling.

Temperatures at these depths typically are between 10 and 30°C. The process involves the drilling of a shallow well or the emplacement of coils in a shallow trench, the connection of these to piping within the building, and to a heat pump, and the circulation of tap water through the entire system. During the winter, heat is extracted from the ground and supplied to the building, and during the summer heat is removed from the building and returned into the ground (cooling).

Heating districts have been formed in many countries to provide either this ground-source temperature control or heating via the more conventional pumping of geothermal well waters. Over 80 percent of the population of Iceland currently obtains conventional geothermal home heating. By contrast, Japan, the principal nonelectric user of geothermal energy, instead uses it as the source of heat for spas and swimming pools. In China, the supply of heated water for fish farms is the principal use of geothermal energy.

The use of geothermal energy, more specifically steam produced from drilled wells, to generate electricity began at Larderello, Italy, on an experimental basis in 1904 and on a commercial basis (0.25 megawatts [MW] of electric power) in 1913. This has continued up to the present day, with over 500 MW of power generation stations now installed in the Larderello region of the Tuscany province of Italy. Experimental generation of geothermal electricity was attempted in the 1920s in Japan, the United States, Indonesia (then the Dutch East Indies, under colonial administration from Holland), and elsewhere. These attempts were unsuccessful, principally because of inadequate technology for handling the high temperature and often mineralized and corrosive fluids, and the high unit cost of producing and transmitting electricity to often-distant markets.

It was after World War II that serious attention again turned to the development and use of geothermal electricity. In 1994, the world's total installed geothermal generating capacity was approximately 6,500 MW. Not all of this capacity was in regular use, some plants being experimental facilities, some being operated only when needed, and some not being adequately supplied with geothermal fluid for continuous operation. Therefore, the daily output from these 100-odd power plants located in about 75 fields in 22 countries is about 5,000 MW per hour, or just under 80 percent of the maximum power capacity of these plants. Over 40 percent of all geothermal generating capacity is located in the United States, the world's principal producer of geothermal electricity. Over half of this capacity is installed at The Geysers, the world's largest developed geothermal field (about 1,800 MW), located about 60 km north of San Francisco, California. Following the United States, the largest producers of geothermal energy in 1994 were: the Philippines (998 MW), Mexico (753 MW), Italy (637 MW), Indonesia, New Zealand, and Japan, in turn followed by El Salvador, Nicaragua, Iceland, Costa Rica, Kenya, and Iceland, and then by China, Turkey, and Russia. There have also been small (less than 4 MW) geothermal plants installed in Guadeloupe (by France), Greece, Argentina, the Azores (by Portugal), and very small (less than 1 MW) geothermal plants in Argentina, Thailand, and Zambia.

In addition to The Geysers, fields with over 200 MW of installed generating capacity include Cerro Prieto, Mexico (600 MW); Salton Sea, California; Larderello, Italy; and Tiwi, Bulalo, and Tongonan, the Philippines. Numerous discovered fields remain to be developed in several of these countries. Geothermal fields also have been discovered by drilling, but are not yet developed, in Chile, Djibouti, Guatemala, Ethiopia, Honduras, and Bolivia. Countries with the most extensive plans for geothermal power development over the next decade are the Philippines, Indonesia, and Mexico, followed by El Salvador, Costa Rica, Kenya, Japan, Italy, and the United States. The rate of geothermal power generation has been about 7 percent per year since the early 1960s. This represents a doubling of installed generation capacity every ten years. This rate is expected to continue for at least the next decade.

Individual power plants range in size from less than 1 MW to about 110 MW, and to 140 MW in a single case in the United States. Power generation uses three methods. In the simplest, dry steam is produced from wells (vapor-dominated geothermal systems) and sent via pipes directly to the turbine generator, where electricity is generated. Examples are: The Geysers; Larderello; Matsukawa, Japan; Olkaria, Kenya; and Kamojang, Indonesia. In the most common method, hot water is flowed from wells (water-dominated systems), and a fraction of the water is allowed to flash to steam at a suitable wellhead pressure (typically between 7 and 35 atmospheres). This steam is sent to the turbine generator. In the third method, developed commercially in the mid-1980s, geothermal water is pumped from wells under high pressure and not allowed to flash to steam. Heat exchange occurs in special tubing between the geothermal water and an organic working fluid possessing a low-temperature boiling point. The vapor of that working fluid is put through the turbine-generator set, and then condensed and cooled for reuse. Because two fluid cycles are involved, this process is known as binary-cycle generation. It has become widely accepted in the United States (East Mesa, California; Soda Lake, Nevada) and is being adopted in other countries.

Binary-cycle generation is best suited for medium-enthalpy (heat content) geothermal fluids with temperatures ranging between a minimum of about 120°C and a maximum of about 180°C (the current maximum operating temperature for downhole pumps). The principal advantage of binary-cycle method is that it allows the generation of electricity from what otherwise would not be a commercial resource. Flash-steam generation is used for higher-enthalpy systems having temperatures in excess of 180°–200°C. The maximum temperature of hot-water geothermal systems currently in use is 360°–380°C. Vapor-dominated systems all are at about 240°–250°C, as a consequence of fluid thermodynamics and constrained reservoir permeability.

In all three methods, the used geothermal fluid is reinjected into the subsurface reservoir from which it was produced, for reasons of environmental protection and to supply pressure support to the geothermal reservoir. Despite this, several fields have experienced significant pressure declines, reflecting depletion. In others, injection has resulted in local cooling of the reservoir.

Geologically, high-enthalpy geothermal systems are found in regions of youthful or active volcanism and igneous intrusive activity, associated with tectonic spreading centers, subduction zones, mantle plumes, and intraplate rifts. The conductive cooling of fluid outflow from high-enthalpy centers often results in the creation of medium- or low-enthalpy reservoirs, usually characterized by surface thermal springs. Thermal springs also may be found at great distances from active plate boundaries, the product of deep circulation of meteoric water through permeable strata or along fractures. Medium- and low-enthalpy geothermal reservoirs are often present where deep circulation occurs in areas of crustal extension within continental interiors (examples: Basin-and Range Province of western United States; Rhine Graben of Europe; East African Rift; basin-and-range and rifted terranes of central Asia).

Most geothermal reservoirs are characterized by fracture permeability, locally enhanced by chemical solution porosity, or by initial volcanic porosity (lava tubes or rubble zones). Primary sedimentary permeability is rare, principally because rock-water reactions at high temperatures result in mineral precipitation in pores and fractures, or in metamorphic recrystallization of the host rock. The resulting geothermal fluid typically is mineralized (SiO_2, Na, K, Ca, Cl, SO_4, HCO_3), and capable of depositing mineral scale within fractures, wellbores, and surface piping. Geothermal fluids under some circumstances are corrosive and have measurable to significant (0.5 to over 10 percent) concentrations of noncondensible gases (principally CO_2, with H_2S, N_2, CH_4, etc.). Remedial technology has been developed to suppress scaling, remove scale, reduce corrosive activity, and remove H_2S as elemental sulfur.

Exploited reservoirs range in depth from a few hundred meters to perhaps 1.5 km for low-enthalpy systems, and to just over 3 km for high-enthalpy systems. Few geothermal wells have been drilled to appreciably greater depths. Exploration of geothermal energy largely has focused on areas of recent volcanism (especially silicic volcanism), and around thermal springs and fumaroles. Accidental discoveries have been made during the drilling of oil and water wells or during mineral exploration. Exploration methodology is based on geologic mapping (structures, igneous chronology, fluid flow paths), geochemistry of thermal waters (reservoir temperature, fluid phase, and composition), and a variety of geophysical techniques (gravimetry, electrical resistivity, spontaneous potential, magnetotellurics, microseismic monitoring—*see* GEOPHYSICAL TECHNIQUES), to identify possible reservoir target zones. Drilling of sets of slim holes to several hundred meters is common for temperature measurement for use in three-dimensional conceptual modeling. Numerical simulation of initial-state and proposed development conditions, based on test data from deep drillholes, is the present basis for the estimation of reserves.

Research is ongoing into the extraction of geothermal energy from systems of very low permeability (hot dry rock), liquid magma, and very deep (greater than 5 km) wells in permeable basins.

None of these shows immediate signs of commercial utility. A hot dry rock test facility has been established in New Mexico, wherein cool water is injected into one deep hole under sufficient pressure to create fractures that allow the fluid to be produced from a second nearby deep hole. En route, the fluid has scavenged heat from the fractured rock. Similar experiments have been tried in France, Germany, England, and Japan.

Bibliography

DUFFIELD, W. A.; J. H. SASS; and M. L. SOREY. *Tapping the Earth's Natural Heat.* U.S. Geological Survey Circular 1125. Washington, DC, 1994.

RYBACH, L., and L. J. P. MUFFLER, eds. *Geothermal Systems: Principles* and *Care Histories.* New York, 1981.

JAMES B. KOENIG

ENERGY FROM STREAMS AND OCEANS

Water flowing in streams and rivers is a major source of the world's energy. Hydroelectric power stations, which capture the energy contained in water as it falls from a higher elevation to a lower one, produce about 7 percent of the world's energy, slightly more than nuclear power plants generate.

Virtually all hydroelectric power stations require the construction of dams, which store water in reservoirs upstream. Dams enable engineers to regulate the flow of water at a desired rate. Water from the reservoir is directed through conduits, called penstocks. The rushing water then spins a turbine to generate electricity. The greater the water flow and the farther it falls, the more energy it can generate. In a few places, however, the drop in the water level and the rate at which the water flows are naturally sufficient to generate power without building a dam.

Most of the world's hydroelectric stations use dams that are from tens of meters high to several hundred meters high, and scores of them, sometimes called "megadams," rise more than 150 m. The world's tallest dam, the Nurek Dam in Tajikistan, is over 300 m high. Water stored behind large dams may create huge artificial lakes that previously did not exist and which change the earth's natural landscape. The proposed Three Gorges Dam on the Yangzte River in China, for example, which when complete would be the world's largest producer of hydroelectric power would create a 560-km lake upstream.

Hydroelectric power has many advantages over other methods of generating electricity. Water is a renewable resource that does not depend on depletable fuels. Large hydroelectric generating stations also produce huge amounts of electricity without creating hazardous emissions or waste products. The Guri Dam on the Carana River in Venezuela, currently the world's largest power station of any type, can generate as much electricity (10.6 million kilowatts or kW) as ten large, coal-fired or nuclear power plants (Figure 1).

Hydroelectric stations, however, pose other types of problems for the environment. Filling the reservoir behind a dam may flood large areas of land, which may require the relocation of many people and threaten archaeological artifacts. Dams may prevent fish from reaching their natural spawning areas, and they may cause soil erosion downstream when engineers release large amounts of water during periods of heavy demand for electricity. Dams also trap silt in rivers that would otherwise flow downstream. This not only deprives alluvial plains and deltas of rich new soil, but the build-up of silt at the dam site may shorten the dam's effective life. In some instances, large dams have also collapsed, causing widespread death and property loss. For these reasons, many people oppose large dams in favor of small ones that produce relatively small amounts of electricity for local use.

Dams are also very capital intensive. Building a large dam, therefore, may pose heavy financial burdens, especially on developing nations that often must construct extensive electricity grids to ac-

Figure 1. Fort Gibson Dam located near Wagner, OK, on the Grand River. It has a generating capacity of 191 million kWh (courtesy of U.S. Department of Energy).

company them. Still another disadvantage of hydroelectric stations is the unreliability of their electrical output during droughts. To ensure reliable energy, electric utility companies in the United States typically use hydroelectricity in conjunction with other types of generating stations.

Canada, the United States, Brazil, Russia, and China together produce about half of the world's hydroelectric energy. Canada, which generated 312 billion kilowatt hours (kWh) of hydroelectricity in 1992, is the world's largest producer. Brazil, however, depends on hydroelectric dams for a larger percentage of its electricity (96 percent). Several countries depend on water power for virtually all of their electricity, including Norway (99 percent) and Zaire (97 percent).

Another novel method of obtaining energy from water in motion is to harness the energy of tides. Since 1966, a tidal power plant with a capacity of 240,000 kW has been operating on the Rance River, an estuary of the English Channel in northwestern France. The difference between the high tides and low tides at this site is sufficiently large (8.5 m) to support a power plant. The incoming tide of the estuary flows through a dam, driving turbines to generate electricity. The water is trapped behind the dam, then released when the tide ebbs, spinning the turbines once again and generating more electricity.

History

Energy from water was one of humankind's earliest and most important energy sources. The ancient Greeks and Romans used simple wooden waterwheels to grind grain. During the Middle Ages, large waterwheels, some capable of generating 50 hp, were used to run machinery. Waterwheels also helped fuel the industrial revolution. Many towns and cities grew up near streams and rivers, where waterwheels could supply energy to run mills and machine shops.

During the 1800s, steam engines, which could provide energy more reliably (e.g., during droughts or when rivers froze in winter) grew more popular than waterwheels. But waterpower soon took off again with the development of the electric generator and a growing public demand for electricity. A hydroelectric dam built in Appleton, Wisconsin, in 1882 reestablished water power as an important source of energy.

Over the next several decades, water power became the most important source of electricity in the United States as many electric companies built small, hydroelectric stations near major population centers. By the late 1920s, however, energy from water faced growing competition again from fossil fuels, as steam-electric power plants grew in size and efficiency. To enable hydroelectric stations to compete, engineers began to increase the size of dams in order to take advantage of economies of scale. Large dams, however, required government assistance to raise the necessary capital. In 1930, Congress authorized construction of the nearly 220-m-high Boulder Dam (later renamed Hoover Dam) on the Colorado River along the border between Arizona and Nevada. Completed in 1936, Hoover Dam is the tallest concrete dam in the United States. Its erection marked the beginning of what is sometimes referred to as the "age of dams."

Over the next four decades, generation of electricity at hydroelectric stations rose steadily as engineers built many large dams in the United States. The Tennessee Valley Authority (TVA), a federal corporation created by Congress in 1933, built a network of large dams to generate power, control flooding, and provide water for irrigation. In 1942, the U.S. Bureau of Reclamation completed construction of the Grand Coulee Dam on the Columbia River in Washington. With an electrical generating capacity of 6.5 million kW, the Grand Coulee Dam is the largest single source of water power in the United States.

Generation of electricity at U.S. hydroelectric stations reached a peak of 335 billion net kWh in 1983, before declining as a result of a long drought in western states. By then, most of the suitable sites for large dams in the United States were also already in use. Utility companies, including the TVA, which was created to exploit hydroelectric potential, were already turning almost entirely to other methods of generating electricity. A 1978 federal law known as the Public Utilities Regulatory Policies Act, however, has encouraged the construction of small-scale hydroelectric stations by requiring that large electric utilities purchase their electrical output and distribute it through established transmission networks. Today, as reliance on other methods of generating electricity continues to grow, hydroelectric dams contribute less than 10 percent of U.S. electricity production.

The amount of electricity generated at hydroelectric stations in the United States is expected to remain stagnant over the next several decades, although annual output may vary as a result of changes in precipitation. Worldwide generation of hydroelectricity, however, continues to grow steadily, especially in developing nations.

Bibliography

"Dam Building Fervor Wanes in U.S. But Picks Up Steam in Third World." *The Los Angeles Times*, Feb. 27, 1994.

U.S. ENERGY INFORMATION ADMINISTRATION. *Annual Energy Outlook 1994, With Projections to 2010*. Washington, DC, 1994.

———. *International Energy Annual 1992*. Washington, DC, 1994.

ROBERT LIVINGSTON

ENERGY FROM THE ATOM

Energy from the atom, also called nuclear energy, is the most powerful source of energy used today. When a nuclear weapon is detonated, the resultant uncontrolled release of man-made nuclear energy is an earthly version of the immense, violent forces that allow the Sun to heat and light the solar system. If nuclear energy is carefully controlled, however, it can produce useful energy. Most nuclear energy is used to generate electricity. This is done by using nuclear fuel to produce heat that boils water into steam to drive a turbine. The method differs from the way a conventional power plant generates electricity primarily because the heat is derived from nuclear processes rather than from the burning of fossil fuels such as coal, oil, or natural gas, or biofuels such as wood.

At the end of 1993, 430 commercial nuclear power plants, also called nuclear reactors, were operable in thirty countries (Table 1). Many modern nuclear plants generate a million kilowatts of electricity or more—enough to meet the needs of a small city. They produce 17 percent of the world's electricity, or about 6 percent of the global primary energy supply. Nuclear energy remains highly controversial, however, and its growth around the world has largely stagnated. Safety concerns and

rising costs have dampened interest in it in many countries. A few countries have rejected it as a source of energy (*see* HAZARDOUS WASTE DISPOSAL).

In the United States, 109 commercial nuclear plants produced 639 billion kilowatt hours (kWh) of electricity in 1994, which was about one-fifth of all the electricity generated. No further nuclear plants are expected to be built, however, until the myriad problems facing the nuclear energy industry can be resolved. Nuclear energy's share of U.S. electricity production, therefore, is likely to decline slowly for the rest of the century and beyond as other energy sources meet incremental demand for power. Even so, nuclear energy will remain a significant source of electricity in many localities.

Nuclear energy also remains attractive to some other industrialized countries (e.g., France, Japan) that have little or no indigenous fossil fuel resources, as well as to some developing countries (e.g., China, the Republic of Korea) where demand for electricity has been growing rapidly.

Whether interest in nuclear energy returns in the United States or continues to decline will depend on many factors. These include the availability of safer nuclear plant designs and an industry infrastructure needed to build them; the rate of growth in electricity demand; the cost and availability of alternative forms of energy; public opinion; and whether solutions can be found to unresolved problems, especially the safe storage of radioactive waste. Some energy experts doubt that nuclear energy provides enough benefits to justify the economic and safety risks. Others believe the problems facing nuclear energy, which may appear formidable today, will gradually be resolved. They believe it could be a useful source of energy for future generations as the world's population grows and as the need increases for energy sources that do not burn carbon-based fuels and, therefore, produce no air pollution or greenhouse gases.

Fission

All of the world's commercial nuclear plants employ nuclear fission to generate energy. Fission occurs when a neutron pierces the nucleus of a heavy atom, such as a uranium (U) atom, splitting it into new, smaller atoms, plus some free neutrons. The combined weight of the new atoms, called fission fragments, and free neutrons is a tiny bit less than the weight of the parent atom. The missing mate-

Table 1. Nuclear Energy Around the World—Nuclear Power Plants by Nation (1993)

Country	Operating Reactors	Output (Billion kWh)	Percentage of Electricity Generation
Argentina	2	7.2	14.2
Belgium	7	39.5	58.9
Brazil	1	0.4	0.2
Bulgaria	6	14.0	36.9
Canada	22	88.6	17.3
China	2	2.5	0.3
Czech Republic	4	12.6	29.2
Finland	4	18.8	32.4
France	57	350.2	77.7
Germany	21	145.0	29.7
Hungary	4	13.0	43.3
India	9	5.4	1.9
Japan	48	246.3	30.9
Kazakhstan	1	0.4	0.5
Korea	9	55.4	40.3
Lithuania	2	12.3	87.2
Mexico	1	3.7	3.0
Netherlands	2	3.7	5.1
Pakistan	1	0.4	0.9
Russia	29	119.2	12.5
Slovakia	4	11.0	53.6
Slovenia	1	3.8	43.3
South Africa	2	7.2	5.5
Spain	9	53.6	36.0
Sweden	12	58.9	42.0
Switzerland	5	22.0	37.9
Taiwan	6	33.0	33.5
Ukraine	15	75.2	32.9
United Kingdom	35	79.8	26.3
United States	109	610.3	21.2
Total	**430**	**2093.4**	

Source: International Atomic Energy Agency.

rial is converted into vast amounts of energy, as physicist Albert Einstein predicted in his famous equation: $E = mc^2$, in which energy (E) is equal to the mass (m) multiplied by the square of the speed of light (c). While the mass of an atom is not great, the speed of light is a very large number. Hence, the fissioning of a single atom of uranium releases about 200 million electron volts—enough energy to make a grain of sand jump visibly. It is also about 70 million times more energy than an atom of carbon in coal or oil releases when it burns chemically. When a series of fission reactions, called a chain reaction, occurs, millions of atoms split apart and a very large amount of energy is produced. After one atom fissions, some of the freed-up neutrons strike the nuclei of other atoms. As these other atoms split, more neutrons are freed, and they go on to strike still more nuclei, and so on. Eventually, the chain reaction becomes self-sustaining.

A variety of nuclear plant designs are in use around the world today to convert fission energy into electricity. They differ primarily in the methods they use to encourage the fission process, the type of fuel they use, and the safety mechanisms they employ. In most nuclear plants, including all of those in the United States, fuel pellets are stacked inside fuel rods made of zirconium or stainless steel. Groups of fuel rods that have been fastened together, called fuel bundles, are loaded inside a steel tank, called a reactor vessel. This is where fission takes place. The most common nuclear fuel is uranium, but plutonium (Pu) and thorium (Th) can be used. The atoms of these heavy elements are well suited to undergo fission, because they have many protons and neutrons. To further encourage a chain reaction, the nuclear fuel is surrounded by a moderator, typically water or graphite, which slows down free neutrons so they are more likely to hit the nuclei of other at-

oms. When nuclear plant operators want to halt the chain reaction, they insert control rods made of neutron-absorbing materials such as boron (B) into the reactor vessel. By absorbing neutrons, the control rods allow fewer fissions to occur.

Most of the energy released when atoms fission is in the form of heat. So much heat is produced that it would quickly damage the nuclear fuel, and possibly cause it to melt. To prevent this, a coolant—either water or a gas, such as helium (He)—is piped through the core. This keeps the nuclear fuel from overheating. The coolant also carries off heat so it can be used to boil water into steam (Figure 1).

The rest of the energy released during the fission process is in the form of radiation. This is why

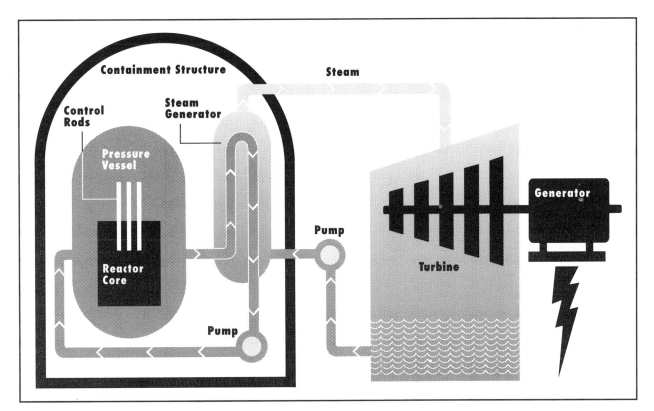

Figure 1. Light water reactors. The most common type of nuclear plant around the world is the light water reactor. Fission takes place inside the steel reactor vessel. Ordinary water (rather than heavy water) circulates through the vessel to prevent the nuclear fuel from overheating, and also to "moderate" the nuclear reaction (i.e., slow down free neutrons in order to make fissions more likely to occur). After absorbing heat from the nuclear core, the coolant water, now approximately 600°F, is piped to a steam generator. There, its heat is transferred to a secondary supply of water, which boils to make steam to drive a turbine.

Source: U.S. Council for Energy Awareness.

fresh nuclear fuel, which is mildly radioactive when it is loaded into a nuclear plant, becomes highly radioactive during use. This radioactivity is a potential health hazard. To prevent it from escaping a nuclear plant, engineers take many safety precautions. In most types of nuclear plants, a containment building, typically a 0.9-m-thick, domed structure, surrounds the steel reactor vessels. The containment building serves two purposes: to protect the reactor from airplane crashes and other external threats, and to reduce and delay the release of radioactivity to the environment in case a pipe carrying radioactive water breaks or the reactor vessel leaks during a serious accident. After nuclear fuel has been used, it remains highly radioactive and must be disposed of in a manner that will keep it isolated from people and the environment for many thousands of years. When the used fuel, called spent fuel, is moved from the reactor to a storage site, it is carried inside heavy, lead-shielded shipping containers.

Uranium, a naturally occurring element found in abundance in the earth's crust, is mined much like ores of coal or iron, or obtained as a by-product of phosphate mining. A few nuclear plants, such as those used in Canada, operate with natural uranium. In most nuclear plant designs in use around the world, however, the natural uranium is processed in several stages to make it easier to fission. Nearly all of the atoms in natural uranium are the isotope 238U, which is difficult to split, while only 0.7 percent are the lighter isotope, 235U, which fissions more readily. To make fuel for nuclear plants, the natural uranium is converted into a gas, then sent to an enrichment plant where membranes or centrifuges filter out some of the abundant 238U atoms. The enrichment process is analogous to making cream by skimming off milk. What is left is uranium with 2.5 percent to 3 percent of the rarer 235U atoms. Finally, the lightly enriched gas is reconverted into a solid and fabricated into nuclear fuel.

Some nuclear plants in Europe and Japan operate with fuel containing a uranium/plutonium mix. While scientists have found tiny amounts of plutonium in nature, all plutonium used as fuel for nuclear plants is man-made. It is produced inside nuclear reactors as a by-product of the fission process. To recover it, technicians use remote controls to immerse the highly radioactive spent fuel in chemicals that dissolve it. The plutonium is then separated and mixed with fresh uranium into new nuclear fuel. The U.S. government, however, opposes the recycling of plutonium in U.S. commercial nuclear plants and discourages its use internationally. Unlike lightly enriched nuclear fuel used in commercial nuclear plants, separated plutonium could be used by an unfriendly country to build a nuclear weapon. Another concern is that plutonium could be stolen by a terrorist group.

Nuclear fission is also used to lesser degrees for purposes other than to generate electricity. Some nuclear plants in northern Europe also provide steam directly to district heating systems. Small nuclear reactors provide power for submarines, cruisers, and aircraft carriers for the navies of several countries. Small nuclear reactors are also used for research or to produce radioactive isotopes that diagnose disease, treat cancer, and perform a variety of other tasks. Nuclear energy also provides small amounts of electricity to power instruments on board spacecraft exploring the outer solar system, where light is too dim for solar power. This is done with radioisotope thermoelectric generators, which convert heat from decaying ^{238}Pu into electricity.

Nuclear Safety

If nuclear fission reactors are not operated safely, serious accidents can occur capable of destroying the power station itself and endangering workers and the public.

The world's worst nuclear accident occurred in Ukraine in April 1986 at Unit 4 of the Chernobyl nuclear station, which employed a reactor design developed in the former Soviet Union. The reactor raced out of control, resulting in a nonnuclear explosion and fire that destroyed the plant. A vast amount of radiation also escaped from the damaged building. More than thirty workers were killed by the explosion or as a result of receiving fatal doses of radiation. Millions of other people were exposed to radiation in amounts above those present in nature, and large areas of land in Ukraine and adjacent countries were severely contaminated, necessitating an enormous cleanup effort and the resettlement of many people. Doctors monitoring the health of 150,000 people from surrounding areas who were exposed to the most radiation have observed higher fatality rates from radiation-induced illnesses, such as leukemia and other cancers, which can take many years to develop. The catastrophic costs of the Chernobyl accident are apparent in data showing that in 1993, seven years after it occurred, the government of Ukraine was spending 12 percent of its income mitigating the effects. The remains of the Chernobyl Unit 4 reactor and its heavily damaged nuclear fuel are enclosed in a concrete-and-steel sarcophagus to prevent further leakage of radiation.

Other serious nuclear accidents have occurred as well. In 1957, highly radioactive liquid waste stored in a tank at the Chelyabinsk-40 nuclear facility near the town of Kyshtym, Russia, began to boil. Eventually, the liquid waste, a by-product of plutonium produced for military purposes, exploded, releasing radiation that contaminated large areas of the southern Ural Mountains.

The 1979 accident at Unit 2 of the Three Mile Island nuclear station in Pennsylvania resulted in partial melting of uranium fuel. The accident destroyed the plant and caused a financial disaster for the utility company owners, but it did not result in a large-scale release of radiation or cause any measurable increase in deaths or illnesses. Most of the radiation was trapped inside the plant's containment building.

The threat to public health posed by the large-scale release of radioactive materials has focused renewed international attention on improving the safety of nuclear plants, particularly older designs like those used at the Chernobyl station, and on finding remote sites where radioactive wastes can be safely sequestered. In the United States, most spent fuel remains temporarily stored at nuclear station sites in shielded, water-cooled tanks located below ground, or in air-cooled canisters that are placed above ground on concrete pads. The U.S. government is studying Yucca Mountain in Nevada as the possible site for a permanent, underground nuclear waste repository, but even if the site is found acceptable it is not scheduled to open until 2010 at the earliest.

Fusion

Nuclear fusion—the joining together of lightweight atoms to form a heavier atom—may one day be used to produce energy, but it is not yet practical. The energy of the Sun comes from the nuclear fusion of two hydrogen isotopes, deuterium and tritium, which combine to form helium. The weight of the new, heavier element is less than the combined weight of the two original elements. The lost matter is converted into energy during the process.

If fusion can be controlled on Earth, it could provide a virtually limitless source of electricity, but accomplishing this poses a huge challenge. Temperatures of at least 100 million degrees are needed to join deuterium and tritium nuclei, which is why the fusion process is referred to as a thermonuclear reaction. Such temperatures occur on Earth only in a plasma, a special form of matter consisting of a gas made up of free nuclei and free electrons. Current fusion research focuses on a technique, known as magnetic confinement, in which the extremely hot plasma is thermally insulated from the walls of a containing vessel through powerful magnetic fields created with electrical currents. The plasma must be held in place long enough for atoms to fuse and then maintain a self-sustaining state, known as ignition.

Experimental fusion reactors in the United States and Europe have produced small amounts of energy. In May 1994, the Tokamak Fusion Test Reactor at Princeton University in New Jersey briefly generated nine million watts of thermal energy for about one-quarter of a second (Figure 2). However, this was only about 40 percent as much power as required to create the fusion. Much larger reactors are needed to reach breakeven, the point at which the amount of energy produced is equal to the amount of energy required to sustain the fusion reaction.

The first fusion reactors capable of producing electricity at reasonable cost will probably not be available until at least 2040.

History

Scientists in Europe began exploring the nature of matter and energy in the late nineteenth and early twentieth centuries. In December 1938, two German chemists, Otto Hahn and Fritz Strassmann, who were conducting a tabletop experiment, produced the first known artificially created fission reaction when they bombarded uranium with neutrons. Since they could not see atoms, they did not realize they had achieved fission, and they could not explain why their experiment produced barium (Ba), an element much lighter than uranium. Austrian physicist Lise Meitner and her nephew Otto Frisch theorized that a uranium atom had

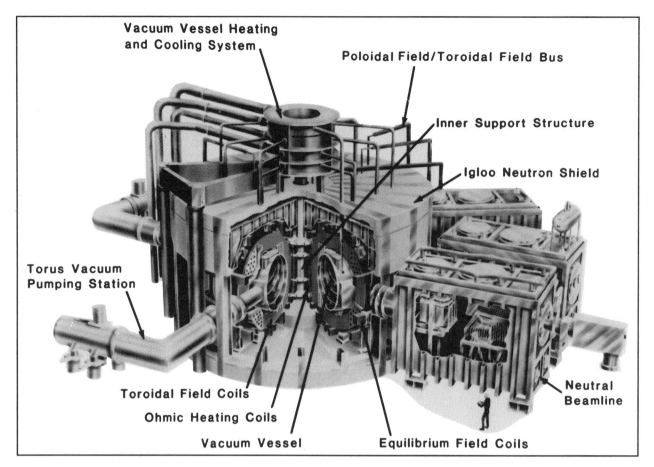

Figure 2. Cutaway drawing of the Tokamak fusion test reactor (Princeton University, plasma physics laboratory).

Source: U.S. Department of Energy.

split during the Hahn/Strassmann experiment into two smaller fragments: one consisting of barium, the other of krypton (Kr). Their theory led to great excitement among scientists around the world. With World War II about to break out, however, the initial excitement turned to profound concern that Nazi Germany would exploit nuclear fission to build an atomic bomb. This led the United States and its wartime allies to launch the Manhattan Project, a huge government program aimed at building an atomic bomb before German scientists did.

A team of scientists under the direction of Italian physicist Enrico Fermi produced the first controlled nuclear chain reaction in a room beneath a stadium at the University of Chicago on 2 December 1942. Two and a half years later, the United

States exploded the first experimental atomic bomb, which was made with plutonium, in the desert near Alamagordo, New Mexico. On 6 August 1945, an atomic bomb made with highly enriched (weapons grade) uranium destroyed the Japanese city of Hiroshima. Three days later another plutonium bomb destroyed Nagasaki. Japan surrendered, and World War II was over. These were the only two atomic bombs ever used in warfare.

After the war, the United States tried to keep atomic energy a secret, but by 1949 the Soviet Union exploded its first atomic bomb, prompting an arms race. In 1952, during the Korean War, the United States exploded an experimental thermonuclear weapon. Less than a year later, the Soviet Union exploded its first thermonuclear device. As both countries began to build huge arsenals of nuclear weapons and test them in the atmosphere, and as other nations began to explore the secrets of the atom, sentiment grew around the world to use energy from the atom for the benefit of mankind instead of for destruction. This idea crystallized into the Atoms for Peace plan proposed by U.S. President Dwight D. Eisenhower to the United Nations in 1953. Soon experimental nuclear reactors using various designs were under construction in the United States, the United Kingdom, and Russia, and a different type of race was under way—a race to use atomic energy to generate electricity. In 1955, a test reactor at the Idaho National Engineering Laboratory generated electricity used by the nearby town of Arco, Idaho. The era was a time of great promise for nuclear energy. Some optimists predicted energy from the atom would transform society, making deserts bloom and providing cheap electricity.

In 1957, the United States started up the small, 60,000-kW Shippingport station near Pittsburgh, Pennsylvania. Its success led to the construction of more nuclear plants in the United States, each capable of generating several hundred thousand kilowatts of electricity. By the late 1960s and early 1970s, engineers began to increase the size of nuclear plants—up to a million kilowatts or more—to take advantage of economies of scale. A few such large nuclear plants were already in operation when the Arab oil embargo and oil price shocks of the early 1970s prompted electric companies in the United States and other countries to order hundreds more.

Soon, however, the U.S. nuclear energy program began to encounter serious problems because of rising construction costs caused by inflation and stricter safety regulations. Orders for new plants ground to a halt. After the 1979 Three Mile Island accident led to even more expensive safety requirements, a majority of the proposed U.S. nuclear power projects, including many in advanced stages of construction, were canceled. Nuclear energy's contribution to U.S. electricity supply nevertheless grew throughout the 1980s, as some electric companies that had invested large sums of money in nuclear power projects pushed them to completion.

In addition to high costs and safety concerns, other unresolved problems continue to hamper the further use of energy from the atom in the United States. In some states, including California, Oregon, Maine, and New York, laws bar new nu-

clear plant construction until a method is available to dispose of spent fuel permanently.

Following the Persian Gulf War, experts from the International Atomic Energy Agency in Vienna discovered that Iraq was close to producing a clandestine nuclear weapon, using sophisticated equipment purchased from other countries under the guise of a peaceful nuclear energy program. This underscored the need for a tighter system of international agreements, which most of the world's nations have adhered to for decades, to halt the spread of nuclear weapons.

Bibliography

CLARK, R. J. *The Greatest Power on Earth: The International Race for Nuclear Supremacy from Earliest Theory to Three Mile Island.* New York, 1980.

PRICE, T. *Political Electricity: What Future for Nuclear Energy?* Oxford, Eng., 1990.

RHODES, R. *The Making of the Atomic Bomb.* New York, 1986.

TELLER, E. *Energy from Heaven and Earth.* San Francisco, 1980.

U.S. ENERGY INFORMATION ADMINISTRATION. *Annual Energy Outlook 1994, With Projections to 2010.* Washington, DC, 1994.

———. *International Energy Annual 1992.* Washington, DC, 1994.

ROBERT LIVINGSTON

ENERGY FROM THE BIOMASS

Biomass, living and nonliving plant matter, is the most abundant and accessible renewable fuel source available worldwide. The leaves of green plants capture solar energy and convert water and atmospheric carbon dioxide to stored chemical energy through a process called photosynthesis. These energy-rich complex carbon compounds are used in the cell wall to build biomass in the form of the stems, branches, and roots. The energy contained in the annual biomass growth is equivalent to about ten times the total amount of energy used annually worldwide. Actual use of biomass fuels for energy varies in different countries. Biomass use, as a percent of total energy used, ranges from 3 percent in the United States to 38 percent in some developing countries. The worldwide average is 15 percent. It was only during the mid–1800s that fossil fuels displaced biomass fuels as a primary energy source. Due to global pollution concerns and questions about long-term supplies of fossil fuels, biomass energy is being rediscovered and reengineered.

The most common biomass resources currently used for energy include fuelwood harvested from native forests and agricultural residues. Other examples of biomass energy resources include: bark, sawdust, food processing wastes, newspapers, urban tree trimmings, and animal dung. Recycling of wastes and residues to produce biomass energy is desirable but can provide only a limited portion of the world's total energy demand. Additionally, excessive removal of natural plant growth has undesirable environmental consequences. Therefore, crops must be grown for energy to provide for large-scale production of electricity and liquid fuels from biomass. Research into genetics and bioengineering of trees and grasses is tremendously increasing the potential for environmentally sound production of large amounts of biomass.

Each biomass resource has different fuel characteristics that must be considered in designing bioenergy conversion systems. The ash content or mineral matter varies considerably, with the lowest being in clean wood and the highest in straws. Clean wood chips, wet sludge, and dry straw require very different handling systems to feed the materials into the conversion process. Differences in proportions of cellulose, hemicellulose, and lignin (cell wall materials) lead to preferences for different types of feedstocks for thermal and biological conversion technologies. Conversion technologies that can handle the widest variety of biomass feedstocks are likely to be the most successful.

The most frequent use of biomass is in direct combustion for the production of heat and steam. The steam is used directly for district heating and process head applications or alternatively to drive a steam turbine and a generator to produce electricity. Combustion devices range from simple cookstoves to advanced technology boilers, which produce steam at a rate of 150 tons/hour under high pressure and superheated temperatures. The most common combustors simply burn wood chips that are fed to the boiler on a fixed or moving grate. There has been a recent trend to use bubbling and circulating fluidized bed combustors, which can operate well with a wide range of fuel types and mixes. Net thermal plant efficiencies of most biomass steam-electric systems are only 20–30 percent compared with coal systems at 35–45 percent. Higher net efficiencies are possible with dry biomass and high temperature and pressure boiler operating conditions.

The production of liquid transportation fuels from biomass has been under development since the oil price fluctuations and supply disruptions that occurred in the late 1970s. Brazil quickly commercialized production of ethanol from the sugars in sugar cane, and by the mid–1980s, 50 percent of the cars and trucks in Brazil were fueled soley with ethanol. At the same time, the United States commercialized production of ethanol from the starch in corn grain. Ethanol entered the U.S. transportation fuel market in the 1980s primarily as a blend with gasoline. In both countries, the subsidized commercial production of ethanol has provided farmers with a good alternative market for their crops. To reduce subsidies and avoid the use of food crops, research has focused on ways to produce ethanol from the cellulose and hemicellulose (cell wall polymers) found in the stems and leaves of crops, wastes, and residues. The costs of transforming this type of biomass into sugars and ethanol have been significantly reduced with breakthroughs in pretreatment and fermentation technologies, and are forecast to be competitive with fossil fuels in the near future.

Thermal gasification offers a way of using a diverse range of biomass resources in the production of both electricity and liquid fuels with high environmental performance. Thermal gasification uses controlled levels of heat to convert biomass feedstocks into carbon- and hydrogen-rich fuel gases that can be utilized directly in chemical synthesis, gas turbines, and fuel cells. Research and development are improving the gas cleanup and filtration necessary to use the gas in high-performance gas turbines. Methanol, an alternative liquid fuel that is produced today from natural gas, can be produced by catalytically combining carbon monoxide and hydrogen gases from the thermal gasification of biomass. A derivative of methanol (in the form of methyltertiarybutylether, or MTBE) is being added to gasoline as an oxygenate to meet air emission requirements.

Numerous other bioenergy conversion technologists exist. In Europe, biofuel oil products derived from oil seed crops such as soybeans and rape seed (canola) are used as a diesel substitute.

Anaerobic digestion of animal wastes and industrial waste waters is widely used to produce methane, fertilizer, and compost as well as for water treatment and pollution prevention. The biological conversion of organic wastes to methane and carbon dioxide by anaerobic fermentation also takes place in municipal solid waste landfills. In the United States there is over 500 MW of electricity generation installed on landfills, which are being mined for their methane. The process is also used on intensive livestock operations and at industrial processes such as breweries.

Biomass energy is available to any country with biomass wastes and land suitable for growing biomass crops. Greatly improved biomass conversion technologies allow highly efficient conversion of biomass resources to electricity and liquid fuels. New innovations are setting the stage for biomass energy to meet a significant portion of the ever-increasing energy demand of people all over the world.

Bibliography

BROWER, M. *Cool Energy: Renewable Solutions to Environmental Problems.* Cambridge, MA, 1992.

JOHANSSON, T. B., H. KELLY, A. K. REDDY, and R. H. WILLIAMS, eds. *Renewable Energy: Sources for Fuels and Electricity.* Washington, DC, 1993.

KITANI, O., and C. W. HALL. *Biomass Handbook.* Newark, NJ, 1989.

LYNN WRIGHT
RALPH P. OVEREND

ENERGY FROM THE SUN

The Sun is the source of most of the energy used on Earth, but perhaps not in the way most people are accustomed to thinking about solar energy. This energy begins with nuclear fusion reactions deep within the Sun's interior. It flows outward to the surface of the Sun, then into space in all directions in the form of electromagnetic radiation. The temperature at the surface of the Sun is 5,500°C, but by the time a tiny portion of this energy reaches Earth, approximately 150 million kilometers away, it arrives at a relatively diffuse rate of about 1.37×10^6 ergs per square centimeter per second—a figure known as the solar constant. While this is more than 10,000 times as much energy as humans produce, it is not very intense since it is spread over a large area. About a third of the radiation is deflected by clouds or scattered by the atmosphere back into space. The rest is collected by the atmosphere, oceans, land, and plant life.

People use the Sun's energy indirectly in myriad ways. Green plants collect sunlight and employ it to convert carbon dioxide and water into energy-storing compounds, a process known as photosynthesis. People eat the plants, or animals that ate the plants, then use the food energy to perform work. The remains of plants that decayed and were buried beneath the ground millions of years ago are the source of fossil fuels, which provide most of mankind's energy today. The Sun even plays a role in producing wind energy, since the uneven solar heating of the atmosphere, ocean, and land creates wind (*see* ENERGY FROM THE WIND). The same can be said of hydroelectric dams, since the Sun's heat evaporates water in the oceans, which returns to Earth as precipitation. Some of this precipitation falls over land and drains into rivers and streams (*see* ENERGY FROM STREAMS AND OCEANS). Virtually the only forms of energy used on Earth that did not originate with the Sun are tidal power, geothermal power, and nuclear energy (*see* ENERGY FROM THE ATOM).

People throughout history have also used energy from the Sun directly to perform simple tasks. The ancient Romans used solar energy to heat public baths at Pompeii. They lined channels carrying water to the baths with black slate tiles that collected the Sun's warming rays. The Mesa Verde cliff dwellers in Colorado built rock overlays that shaded their habitats from the high summer Sun, but allowed the rays of the lower winter Sun to enter. Until recently, however, mankind mostly overlooked the Sun as a source of energy, as the need for large amounts of energy to supply growing populations and support higher standards of living spurred the introduction of increasingly concentrated forms of energy—first wood, then peat, coal, oil and natural gas, and uranium. Then in the 1970s and 1980s, various problems with the more concentrated energy sources led to a renewed interest in solar energy in the United States and some other industrialized nations. Many scientists believe carbon dioxide gas released by the burning of fossil fuels contributes to a greenhouse effect, which may cause global warming (*see* POLLU-TION OF THE ATMOSPHERE). Like the roof of a greenhouse, the atmosphere allows sunlight in. As carbon dioxide and other gases accumulate in the atmosphere, however, they prevent infrared light, which generates heat, from escaping back into space.

This article discusses the ways in which scientists seek to exploit solar energy—energy that does not contribute to global warming or produce other noxious emissions. Since sunlight is diffuse by the time it reaches Earth, however, some of the methods devised for capturing it in large amounts require significant quantities of land and equipment.

Direct Uses of Solar Energy

The simplest methods for capturing solar energy directly are passive (i.e., they require no artificial collection devices). Modern builders orient buildings to face the Sun, locate windows favorably, and use appropriate thermal insulation. They also employ window glazings and films to reduce lighting and air conditioning requirements. Active devices that capture solar energy are grouped into several categories.

Low-temperature collectors provide heat up to 43°C. They are used mostly to warm water in swimming pools. Typically, they consist of a black plastic or rubberlike sheet with tubing through which water is circulated. The heat of the Sun is transferred directly from the black absorbing material to the water circulating through the tubing to supply heat to the pool.

Medium-temperature collectors provide greater heat, usually 60 to 82°C. One type, called a flat plate collector, is a relatively simple, rooftop-mounted device that captures sunlight to provide domestic hot water, space heating, or air conditioning (which requires a source of heat to operate). A typical flat-plate collector consists of a black metal plate covered with sheets of glass. The black plates absorb heat from the Sun. The glass sheets help prevent the heat from escaping. When the plates get hot, they heat a fluid, which serves as a heat-transfer medium.

High-temperature collectors collect sunlight from a wide area, then concentrate it. An early, experimental solar furnace built in the 1960s near Odeillo, France, in the Pyrenees Mountains, uses many mirrors to heat water to temperatures near 3,300°C. It is used primarily for materials research.

A variety of other experimental high-temperature collectors have also been built around the world, primarily to test their ability to generate electricity for utility companies. A series of solar trough collectors built on four hundred hectares of land in the Mojave Desert in California in the 1980s, for example, use parabolic reflectors to concentrate sunlight onto oil-filled black pipes to produce fluid temperatures of about 204–288°C. The heat raises steam to power a turbine to generate electricity. When operating at peak capacity, it can generate more than 360,000 kilowatts (kW). At rates prevailing for fossil fuels in 1993, however, it cost about twice as much as electricity generated with natural gas.

An experimental power tower near Barstow, California, generates electricity using a large number of mirrors on the ground to reflect sunlight to a boiler mounted at the top of a high tower. At peak capacity, this raises enough steam to generate 10,000 kW of electricity. The station employs more than 1,800 heliostats that automatically rotate the mirrors throughout the day so they face in a direction to collect maximum sunlight (Figure 1). Future plans call for using molten salt, which retains heat, instead of water or oil as the "working fluid" (the fluid that is heated by the Sun and then used to produce steam). This is done so the plant could operate continuously even if sunlight is temporarily interrupted by clouds, as well as for up to four hours after sunset. Larger power towers capable of generating 100,000 to 200,000 kW are needed, however, to make costs competitive with conventional sources of electricity. Even then, the number of power towers may eventually be limited by the large amounts of land required.

A Dish/Stirling system, another type of experimental solar thermal collector, may be able to generate economical electricity in quantities small enough to be useful for remote water pumping and village power needs. The Dish/Stirling system uses parabolic reflectors in the shape of a dish to focus the Sun's rays onto a receiver mounted above the dish at its focal point. The solar energy ultimately heats a fluid to power a small Stirling engine, which converts thermal energy directly to electricity. Operating at about 815°C, a single dish module can generate up to 50 kW of electric power.

Photovoltaic (PV) cells, also called photoelectric cells, convert sunlight directly to an electrical current, rather than by indirectly employing the Sun's

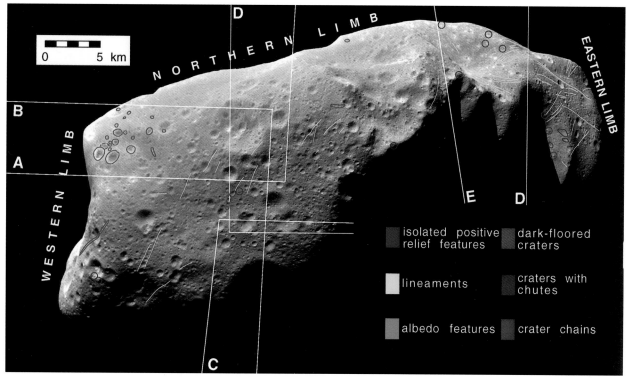

Plate 1. Preliminary geologic map of Ida.

Plate 2. Hubble color image of multiple comet impacts (NASA).

MODEL TEMPERATURE CHANGE 2×CO₂ MINUS 1×CO₂

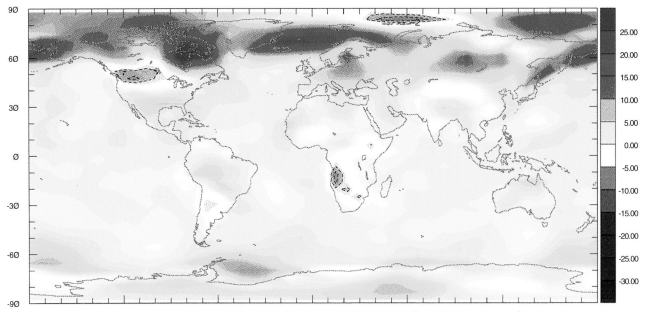

Plate 3. General circulation model results showing the change in January surface air temperature between a present-day concentration ($1 \times CO_2$) and a doubled concentration of atmospheric CO_2 ($2 \times CO_2$).

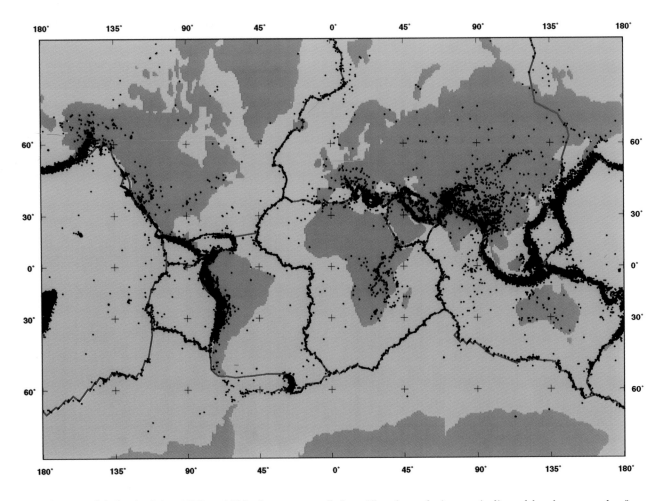

Plate 4. Global seismicity, 1970 to 1990, shown as small dots. Plate boundaries are indicated by the network of medium gray lines surrounding the globe.

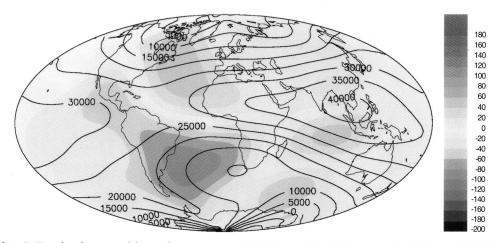

Plate 5. X or local geographic north component of the International Geomagnetic Reference Field for 1990 (contours) and its rate of change (colors). The magnetic flux intensity (in units of nanoTesla, $=10^{-9}$ T) for the X-component is shown by the black contour lines: it is characteristically low near the poles and highest near the equator, and mostly positive. The contour interval is 5,000 nT. Colors indicate the local rate of change of X, measured in nanoTesla/year; the scale is indicated by the color bar on the right. The field model is described in Langel (1992).

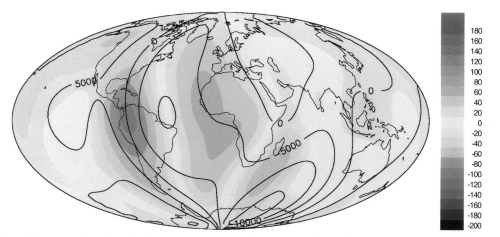

Plate 6. Same as for Y or local east component. Note the much smaller magnitude, and different spatial variability of Y compared to X.

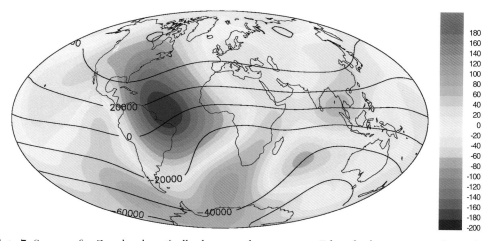

Plate 7. Same as for Z or local vertically downward component. Z has the largest range, from about -60,000 nT near the South Pole to 60,000 nT near the North Pole, reflecting the predominantly dipolar structure of the field. The contour interval is 20,000 nT. Z also shows larger rates of change in the flux intensity than either X or Y.

Plate 8. A 5-ct faceted and flawless diamond surrounded by natural diamond crystals. Collection: Jack Greenspan. © Erica and Harold van Pelt photographers.

Plate 9. An irregularly shaped ruby crystal with a weight of 39.5 g and a height of 46 mm. Inset: Two faceted rubies of 3- and 2-ct size, respectively. Collection: Los Angeles County Museum of Natural History. © Erica and Harold van Pelt photographers.

Plate 10. An irregularly shaped crystal accompanied by three differently colored, faceted sapphires. All are from Sri Lanka. The sizes of the cut stones are: yellow, 10.45 ct; blue, 4.5 ct; pink, 3.9 ct. Collection: Pala International. © Erica and Harold van Pelt photographers.

Plate 11. Natural gem-quality turquoise nodule and cabochon cut stone of 32-ct size. Collection: Pala International. © Erica and Harold van Pelt photographers.

Plate 12. Peridot crystal of 4-cm height and a faceted peridot of 8.5 ct. Both are from Myanmar. Collection: Pala International. © Erica and Harold van Pelt photographers.

Plate 13. Faceted garnets, ranging in size from 3 to 48 ct, showing the wide variety of colors of this silicate group. © Erica and Harold van Pelt photographers.

Plate 14. "Imperial topaz" crystal and two cut stones, from Ouro Preto, Minas Gerais, Brazil. The sizes of the cut stones are 21 and 28 ct, respectively. © Erica and Harold van Pelt photographers.

Plate 15. Natural blue topaz crystal, 17 cm in height, and a faceted stone of 182 ct, both from Brazil. Collection: Pala International. © Erica and Harold van Pelt photographers.

Plate 16. Emerald crystal, 2.2 cm high, and an emerald cut stone of 1.66 ct. Both from Muzo, Colombia. Collection: Jack Greenspan. © Erica and Harold van Pelt photographers.

Plate 17. Light blue-green colored crystals and cut stone of aquamarine, and two greenish-yellow crystals and cut stone of "golden beryl" also known as "heliodor." The cut stones are 38 and 43 ct, respectively. Collection: Pala International. © Erica and Harold van Pelt photographers.

Plate 18. Cut tourmalines ranging in size from 2 to 56 ct showing the very wide array of colors exhibited by this gem mineral. Most of the stones show a single color but several show a pronounced zonation in color from green to pink, and blue green to red. These are known as "bicolor" or "particolored" stones. © Erica and Harold van Pelt photographers.

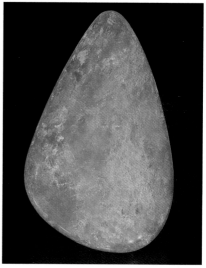

Plate 19. Cabochon cut opal, 43.9 ct, from Anamooka, Australia. Collection: Michael Scott. © Erica and Harold van Pelt photographers.

Plate 20. Woody texture of a block of subbituminous coal from the Powder River Basin, Wyoming. (Scale bar equals 5 cm.)

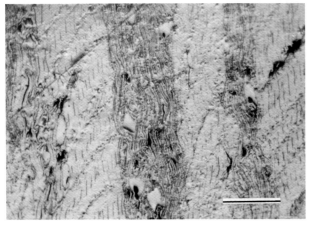

Plate 21. Photomicrograph of a thin section of bituminous coal in transmitted light from the Illinois Basin. Red components are vitrinite macerals, which are derived from the compressed cell walls and humic gels from degraded plant tissues; yellow components are liptinite macerals; serrated layers are leaf cuticles; black components are oxidized macerals or opaque minerals. (Scale bar equals 0.05 mm.)

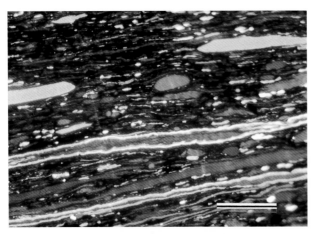

Plate 22. Photomicrograph of a polished section of subbituminous coal in reflected light from the Appalachian Basin, Wyoming, showing layers of compressed plant cells that have resulted in the formation of the maceral vitrinite. (Scale bar equals 0.05 mm.)

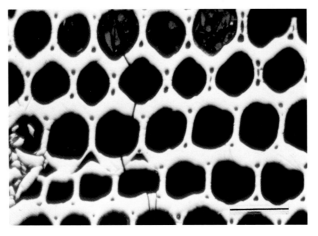

Plate 23. Photomicrograph of a polished section of subbituminous coal in reflected light from the Appalachian Basin, showing a cross section of plant cells that have been carbonized, resulting in the maceral inertinite. (Scale bar equals 0.05 mm.)

Plate 24. *Voyager 2* mosaic of Enceladus showing several craters near the upper left of the picture (NASA).

Plate 25. Photograph of Saturn and its rings taken by the Hubble Space Telescope. (Photo courtesy Lunar and Planetary Photographic Sciences.)

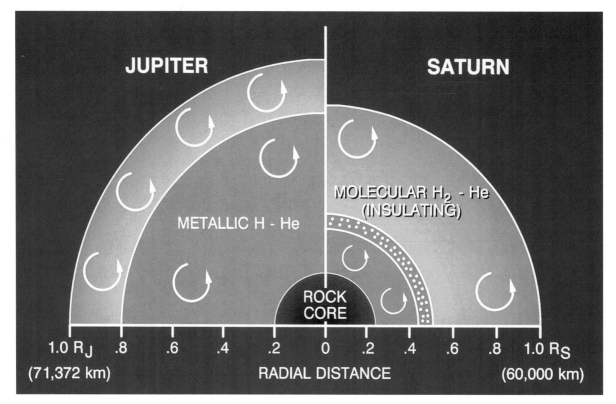

Plate 26. Cross sections of the giant planets. (Courtesy of J. Connerney, NASA/Goddard Space Flight Center.)

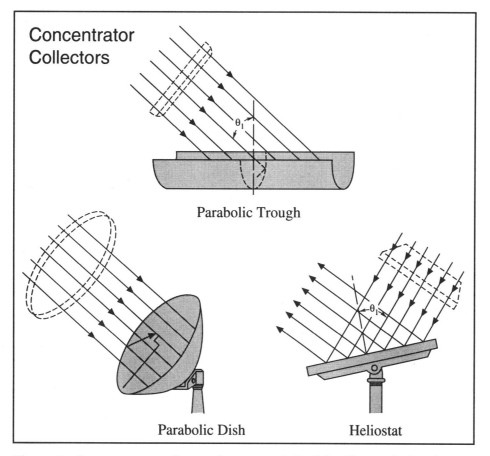

Figure 1. Concentrator collectors (courtesy of the Solar Energy Industries Association).

heat. PV cells provide small amounts of electricity to run calculators or watches. When larger amounts of power are needed, single cells are grouped into modules, which cost less to manufacture. PV cells are made from crystalline silicon or other semiconductor materials. Some materials convert up to 34 percent of the sunlight that strikes them into electricity. To lower costs, however, PV cell manufacturers seek semiconductor materials that offer the best combination of efficiency and ease of manufacturing.

PV cells were employed as early as the 1950s as a reliable, lightweight method of generating electricity for Earth-orbiting satellites. Today, they continue to provide most of the power used on satellites, including those that explore the inner solar system where sunlight is sufficient. In the future, PV cells may also be used to provide power for manned space stations on the Moon or Mars.

Electricity generated with PV cells costs more today than conventional, grid-connected electricity, but PV cells are being used increasingly in niche applications. Several electric companies in the western United States, for example, install PV systems to provide power for homes in remote locations far from existing power lines (Figure 2). Thousands of PV systems also provide people in developing countries with small amounts of electricity—enough, typically, to power a television set and reading light. Other niche markets for PV cells include remote communications equipment, airport runway lights, and billboards.

In mid-1995, plans were also under way to build the first, large-scale powerplant that would generate electricity from PV cells. If built, it could begin operating in the southern Nevada desert in late 1996 and eventually produce 100,000 kW of power, enough to supply the needs of a small city.

Figure 2. Photovoltaic module used for water pumping (photo courtesy of Idaho Power Company).

Further advances in technology and manufacturing techniques, as well as larger markets to support larger factories that can take advantage of economies of scale, are needed, however, before PV cells will be cheap enough to provide a significant portion of the world's energy needs.

Some visionaries have suggested covering large, orbiting satellites with PV cells. They would collect sunlight, which would be beamed in the form of microwaves to receiving stations on Earth.

Biomass and Biofuels

When trees and other plants are used to produce energy, they are called biomass. A few small biomass power plants in the United States capable of generating about 16 million kW were in use in the United States in 1993. Other "waste" products like wastepaper and lawn scraps may also be used as power-plant fuel.

When biomass is converted into fuels, like ethanol or methanol, it is called biofuel. Manufacturers in Europe and Japan use plants like rapeseed, which produce large quantities of oils called lipids, to make biodiesel fuel. The United States uses several billion gallons of biofuel (produced primarily from corn) to produce additives that are blended with petroleum-based gasoline in order to reduce pollutant emissions and improve engine performance.

By the year 2000 the first energy crops may be harvested from large energy plantations that will yield biofuels at a low enough cost to be used by vehicles in the United States, or as fuel for electric generating stations. One approach envisions the use of fast-growing hybrid poplar trees. They would be dried in an air-supported fiberglass dome, then burned in sections. Switchgrass, a fast-growing grass, is also under study, as are drought-resistant strains that would reduce irrigation costs. Eventually, aquatic plants (microalgae) may also be used. Locating energy plantations near electric generating stations would reduce the prohibitive cost of transporting large volumes of biomass long distances.

Bibliography

HUBBARD, H. M. *Progress in Solar Energy Technologies and Applications.* American Solar Energy Society. Boulder, CO, 1994.

U.S. DEPARTMENT OF ENERGY. *Biofuels at the Crossroads: Strategic Plan for the Biofuels Systems Program.* Washington, DC, 1994.

————. *Biofuels for Transportation: The Road from Research to the Marketplace.* Washington, DC, 1994.

U.S. GENERATING ACCOUNTING OFFICE. *Efforts Under Way to Develop Solar and Wind Energy.* Washington, DC, 1993.

ROBERT LIVINGSTON

ENERGY FROM THE WIND

Energy from the wind is a small but rapidly growing resource for generating electricity throughout the world. The wind produces electricity by spinning the blades of a wind turbine that turns a generator. Wind turbines may be deployed individually, when small amounts of electricity are needed, or sited together by the thousands on wind farms located in windy areas away from large obstructions. Since the mid–1970s, the U.S. government has supported a vigorous research and development effort to make wind turbines more efficient, cost effective, and reliable. This has yielded improved designs that can generate electricity for five cents or less per kilowatt hour, a price that is competitive with electricity produced by some conventional, grid-connected sources. Lower costs, in turn, are broadening the potential applications of wind energy, which before had been largely restricted to a few places (California's mountain passes, Hawaii's coastlines) or to niche markets (e.g., remote water pumping).

As of early 1995, a number of electric companies in southern and northern California, the Pacific Northwest, Texas, the Midwest, and the Northeast had contracted, or were negotiating contracts, to purchase thousands of wind turbines, or to purchase wind-generated electricity from independent power producers. By the end of the decade, these and other wind farms planned by electric utilities are expected to double the amount of grid-connected wind-generated electricity in the United States. In 1994, U.S. grid-connected wind-generating capacity was 1.7 million kilowatts (kW), which was approximately equal to the output of several large, central-station power plants.

Commercial interest in energy from the wind is also growing in other countries, especially Europe, which by the end of 1993 had wind-generating capacity of about 1 million kW, much of it in Denmark, Germany, the United Kingdom, Netherlands, Spain, and Italy (Table 1). Even the most enthusiastic electric companies, however, are likely to limit their reliance on wind energy to a portion of their total power generation. Since electricity cannot be stored easily, utility companies must produce it at the time it is needed, which may not always coincide with when the winds are favorable.

Wind turbines are also owned by thousands of individuals who use them to provide electricity for homes, farms, and small businesses. Typically, they generate a small amount of electricity (5–20 kW), but some larger systems generate a few hundred kilowatts. They reduce the amount of electricity the owners must purchase from a utility. When the winds are favorable, they may even generate a small surplus of electricity, which the owner can sell back to the utility.

Table 1. World Grid-Connected Wind Energy Capacity (1994)

Country	Estimated Capacity (MW)
United States	1717
Germany	643
Denmark	540
India	182
Netherlands	153
United Kingdom	147
Spain	72
Sweden	38
China	30
Greece	28
Italy	22
Portugal	9
Ireland	7
Ukraine	4
Brazil	2
Mexico	2
Other Countries	55

Source: American Wind Energy Association.

Small wind turbines are also used for many remote applications where grid-connected electricity is unavailable. They provide power, for example, for telecommunications stations, rural homes, and water-pumping stations. In many developing countries, they provide electricity to rural villages. Some remote wind turbines are colocated with a photovoltaic (PV) module (see ENERGY FROM THE SUN), since periods of low wind may coincide with optimal conditions for a PV system. Remote wind turbines are also frequently colocated with diesel generators, which provide back-up power when there is too little wind.

Wind Turbine Design

There are two basic types of wind turbines. The most common design has two or three propeller-like blades that rotate on a horizontal axis mounted at the top of a tower (Figure 1). The blades, also called rotors, capture the wind's energy and use it to turn a shaft connected to a drive train, which consists of a gearbox and a generator. Other equipment includes controls, electrical cables, and ground support equipment. Some wind turbine designs, however, employ blades that rotate

Figure 1. Horizontal axis wind turbines (photo courtesy of American Wind Energy Association).

around a vertical axis. One vertical-axis wind turbine uses curved blades that make it look like a giant, upside-down eggbeater (Figure 2).

Improved blade designs have helped increase the efficiency of wind turbines. So have advanced power electronics and "smart" controls, which allow the speed of the blades and generator to vary with the speed of the wind. Wind turbines are also growing in size and capacity. Most new models today have towers reaching 50 m. The average capacity of new wind turbines has grown to about 350–500 kW, which is considered an optimal size for developing economies of scale in many locations.

Sites for large wind farms may require average wind speeds of at least 21 kilometers per hour (km/h) in order to generate electricity efficiently. Many areas of the world have wind resources equal to the task. Small wind turbines may work well where wind speeds average 17 km/h.

History

Since the dawn of civilization, people have harnessed the wind. Early Egyptian paintings from 3000 B.C.E. depict ships with masts and sails to assist oarsmen. Wind energy remained the primary method of propelling commercial ships until the late nineteenth century, when steam propulsion gained ascendance. The first simple windmills were probably used in ancient Persia to grind grain or pump water for irrigation, but they were very inefficient. They employed sails fastened to a horizontal wheel that revolved around a vertical shaft. They captured energy only when the sails faced in a favorable direction (i.e., about half the revolution of the wheel).

When windmills spread to Europe in the 1100s, builders boosted efficiency by standing them upright (i.e., by fastening the sails, or blades, to a horizontal axis connected to a vertical shaft). The

Figure 2. Vertical axis wind turbines (photo courtesy of American Wind Energy Association).

revived interest in energy from the wind. Since there is no fuel to burn, wind turbines produce no noxious emissions or waste products. They can, however, pose other environmental problems that require resolution. In a few areas, such as national parks, wind farms have been rejected as visually obtrusive. Birds can be killed if they fly into the blades of a wind turbine.

The goal of the U.S. Department of Energy is to encourage the development of wind turbines that employ lightweight materials and advanced power electronics and rotor designs that can generate electricity for four cents per kilowatt hour in areas with 20 km/h winds. The use of energy from the wind would then likely grow at an even faster pace.

Bibliography

EGGLESTON, D. M., and F. S. STODDARD. *Wind Turbine Engineering Design.* Florence, KY, 1987.
NAAR, J. *The Modern Windpower.* New York, 1989.
SHEPPARD, D. B. *Historical Development of the Windmill.* Washington, DC, 1990.
U.S. GENERATING ACCOUNTING OFFICE. *Efforts Under Way to Develop Solar and Wind Energy.* Washington, DC, 1993.

ROBERT LIVINGSTON

sails could then be brought into the wind by rotating the vertical shaft. By the 1400s, many windmills operated in Holland to drain water from marshes and lakes. Many are still in use today.

The first windmills to generate electricity were built in Denmark in the late nineteenth century. Many windmills were also employed at the time by farmers and ranchers in the western United States and played a role in "winning the West." Their numbers began to decline when electrification programs in the 1930s brought inexpensive electricity generated at hydroelectric and fossil-fueled central power stations to rural areas.

The energy shortages of the 1970s and the heightened interest in the 1980s for ways to generate electricity with fewer environmental impacts

ENERGY USE AROUND THE WORLD

Energy is one of the most vital and valuable commodities. Energy is used to perform work, to provide heat and light, and to transport people and goods.

Vast industries operate to supply energy commercially on a global scale. Crude oil may be pumped from wells in the Saudi Arabian desert, piped to a Persian Gulf harbor, then transported in giant oceangoing tankers thousands of miles to the United States. There it may be refined into gasoline, aviation fuel, diesel oil, and other fuels. Without this energy, factories would close; farmers could not plant crops, harvest them, or transport food to cities; and people could not get to and from their jobs. Oil and other forms of energy are so important to modern nations that devising strategies to assure their ample supply is a major preoccupation of most governments.

The world consumed 343 quadrillion British Thermal Units (Btu) of commercial energy in 1992. A Btu is the quantity of heat needed to raise the temperature of one pound of water by 1°F at 39.2°F (4°C). Fossil fuels (petroleum, coal, natural gas) supplied 87 percent of this energy. Large hydroelectric dams and nuclear power plants provided most of the rest. Other sources such as solar, wind, geothermal energy, and small hydroelectric installations provided a tiny fraction of the total, but their contribution has begun to grow rapidly. Over the next several decades, they may provide a few percent of the world's commercial energy.

Industrialized nations use a disproportionately large portion of the world's commercial energy. The United States and Canada, for example, have only 5 percent of the world's population, but they consume 27 percent of all energy. They are also responsible, however, for one-fourth of global economic activity, and they have the world's highest standards of living—both of which require large amounts of energy. About one-half of the world's population has no access to commercial energy. In many parts of the developing world, people collect firewood for such simple human needs as cooking and hot water.

Availability of Energy

Over the next several decades, demand for energy is likely to grow significantly as the world's population increases and as developing nations seek to raise the living standards of their citizens. The world's resources appear sufficient to meet this growing demand, but the global energy supply is subject to political, economic, and environmental factors that affect its availability or usefulness regionally.

One problem is that the world's energy resources are not evenly distributed. One-half to two-thirds of all the oil discovered around the world, for example, lies in the sparsely populated Middle East. The United States has only a little more than 2 percent of the world's crude oil reserves. Some other major industrialized countries that use large amounts of oil, such as France and Japan, have none. The United States imports more

than half of its oil; France and Japan, all of theirs. They and other oil-importing nations worry that oil-rich nations may decide some day to limit petroleum exports for political reasons or to reserve it for their own future use. They also worry that wars or other upheavals could disrupt world oil trade.

The use of fossil fuels also poses environmental problems. This is particularly true of coal, and, to a lesser extent, of oil. Many local laws limit their use or require costly pollution-control measures, particularly in developed countries. Burning even the cleanest fossil fuel, natural gas, also emits gases that could cause global warming as they accumulate in the atmosphere (see POLLUTION OF THE ATMOSPHERE). Like glass in a greenhouse, such gases allow sunlight to warm the earth but prevent heat from escaping back into space. No measures have been devised for preventing greenhouse gas emissions; hence, the local use of fossil fuels also has worldwide environmental implications.

Ultimately, fossil fuels are also finite resources. They were formed millions of years ago when dead plants decayed and were buried beneath the earth by natural processes, then subjected to immense pressure and heat. While such natural processes are ongoing, obviously they cannot replenish the world's fossil fuels at the rate they are being used.

The world's major energy sources are discussed below in order of their importance to global energy supply.

Petroleum. Oil provides about 38 percent of the world's energy, more than any other source. People use oil so extensively because it is relatively easy to extract from the ground and to transport, and because it is so versatile. Besides energy, oil is a source of valuable products such as plastics and synthetic rubber. Oil also causes less air pollution than coal, although its use poses other potential environmental problems such as oil spills from pipelines or tankers. Approximately 1 trillion barrels of oil (1 barrel = 158.98 liters) have been discovered around the world and were considered recoverable under economic conditions prevailing in 1992. This is sufficient to last more than forty years at 1992 rates of consumption. Advanced technologies, such as the use of steam or chemicals to push more oil out of existing wells, may expand the world's recoverable oil resources in the future. Usable petroleum can also be extracted by heating oil shale and tar sands, which contain about five hundred times as much petroleum as all the world's known sources of oil. While processing oil shale has yet to become practicable, modern production methods have made it commercially attractive to refine tar sands in northern Alberta into crude oil. Oil squeezed from tar sands accounts for one-fourth of Canada's crude oil production.

Coal. Coal is the most plentiful of all the fossil fuels. More than 1.15 billion tons of coal are known to exist around the world—enough to last for more than two hundred years. This coal has five times as much energy content as all the world's crude oil, but coal provides only 26 percent of the world's commercial energy. Coal is bulky, which makes it more difficult to mine and transport than oil or gas. Hence, most coal is burned at power plants that typically are located near railroad lines or ports. To a lesser extent, coal is also used to provide steam for industrial processes. Coal is also dirty. Burning it releases sulfur and other pollutants to the atmosphere unless expensive pollution controls are employed. Current methods for burning coal also result in relatively high greenhouse gas emissions. The United States and the former Soviet Union each have about 23 percent of the world's known coal reserves. China, however, depends on coal for more of its energy, and produces more coal, than any other nation. Many scientists worry that rapid growth of coal consumption to meet China's rising demand for energy will lead to a significant increase in emissions of greenhouse gases. China's extensive use of coal already gives it one of the world's highest per capita rates of carbon dioxide emissions.

Natural Gas. Natural gas is a premium fuel around the world today, providing about 24 percent of the world's commercial energy. Because it contains little sulfur, natural gas burns more cleanly than other fossil fuels, although its use still results in the formation of greenhouse gases. Like oil, natural gas is relatively cheap to produce and easy to transport. Natural gas also provides a wide range of other valuable products, such as fertilizers. Usually, it is transported in gaseous form in pipelines that pose little threat of spillage. However, some natural gas is liquefied at very low temperatures, then transported in specially constructed ships. About 4.9 quadrillion cubic feet of commercially extractable natural gas are known to exist around the world—enough to last for more than sixty-five years. Eastern Europe and the former USSR have the world's largest reserves.

Hydroelectric Energy. Hydroelectric dams provide about 7 percent of the world's energy, all of it in the form of electricity. Water power requires no fuel and, therefore, poses no air pollution or energy security concerns. However, dams can operate only where water flows from a higher elevation to a lower one. Many of the best sites around the world for large dams have been used up (*see* ENERGY FROM STREAMS AND OCEANS). Construction of large dams can also flood large areas of land and, therefore, pose social problems. The proposed Three Gorges hydroelectric project in China, for example, would displace many people from their homes and flood large areas of land.

Nuclear Energy. Nuclear power plants operate in thirty countries and produce about 6 percent of global energy supply, all of it in the form of electricity. Since nuclear plants use the heat released when atoms split, rather than the heat produced by burning fuels, they emit no air pollution or greenhouse gases. Nuclear energy remains highly controversial, however, and its growth around the world has largely stagnated because of safety concerns and rising costs. A few countries, such as France and Japan, which have little, if any, indigenous fossil fuel resources, continue to plan new nuclear plants. Nuclear fusion, which uses the heat released when isotopes of hydrogen fuse, may some day provide the world with a virtually inexhaustible source of energy from ordinary seawater, but commercial energy from fusion is not likely to be available until at least 2040 (*see* ENERGY FROM THE ATOM).

Renewable Energy. Solar, wind, and geothermal energy are referred to as renewable sources because, unlike fossil fuels, they do not use depletable fuels. The most promising of these is wind energy. Recent engineering advances enable wind turbines to generate electricity more efficiently than in the past (*see* ENERGY FROM THE WIND). Solar energy is collected by a variety of natural and artificial devices. The most promising solar technologies involve the growing of crops, called biomass, to produce solid, liquid, or gaseous fuels for power-plants or vehicles, and the use of photovoltaic cells, which convert sunlight directly into electricity (*see* ENERGY FROM THE SUN). Geothermal power plants use the energy contained in hot water or hot, dry rock deep within the earth to generate electricity. Such power plants, however, are limited to sites where geothermal heat is commercially exploitable.

New Sources of Energy

To assure an ample supply of usable energy to meet the world's needs, scientists seek ways to find and produce more oil, to burn coal with less pollution, and to develop new energy technologies. For example, utility companies are testing fuel cells, battery-like devices in which gas or liquid fuels combine chemically to generate electricity. Fuel cells would use fossil fuels more efficiently and cleanly. Since nothing burns in fuel cells, they cause little pollution. They also generate little waste heat, which allows the same amount of electricity to be produced with less fuel. Scientists are also exploring ways to extract hydrogen from ocean water by passing an electric current through it, a process known as electrolysis. Since the only by-product from burning hydrogen is water, hydrogen could provide the world with a limitless, pollution-free fuel. Current methods of producing hydrogen, however, are too expensive to allow its commercial use.

Conservation and energy efficiency help to slow the rate at which the world's energy resources are being used. Millions of people practice energy conservation every day when they install insulation in their homes, turn off lights that are not in use, and keep the tires of their automobiles inflated to proper pressures. Energy efficiency generally refers to the use of engines that use fuel more stingily, lighting and variable-speed motors that use less electricity, and industrial machinery that recovers and reuses waste heat. As a result of public demand for greater energy efficiency, for example, automobiles today go much farther on a gallon of gasoline than they did two decades ago. But while such improvements have slowed the growth rate of U.S. energy consumption, they have not halted it.

Altering traditional patterns of energy use can also allow plentiful and usable sources of energy to replace those in short supply. Automotive engineers, for example, are seeking to develop cars and trucks that run on electricity, natural gas, or alcohol fuels. This would reduce both air pollution and the need to import oil from other nations.

History

Since prehistoric times, people have lightened their workloads and improved the quality of their lives by using energy. Energy was probably used for the first time around 500,000 B.C.E. when our ancestors learned how to control fire to cook food and provide warmth. As humans evolved from hunters and gatherers into farmers around 10,000 B.C.E., the need for energy began to increase. People put animals (elephants, water buffalo) to work pulling plows and carts carrying grain. Wood, however, was the major source of energy, and it would remain the world's major source for most of history. As civilizations flourished and population grew, people in parts of the Middle East, India, and China burned so much wood that large areas were stripped of forests, resulting in soil erosion and flooding. This was an early indication of the environmental impacts associated with the use of energy.

To a lesser extent, the ancient world also employed energy from wind and water. As early as 3,000 B.C.E., ships with masts and sails captured energy from the wind to assist oarsmen. The ancient Persians built simple windmills to grind grain or pump water for irrigation. To a limited extent, the ancient world also used fossil fuels. The Babylonians employed bitumen, a semisolid, oozy substance that seeped to the surface of the earth, to provide light and heating for cooking. The ancient Chinese used coal for domestic heating and cooking and natural gas to provide heat and light for the palaces of emperors. The first water wheels appeared in the latter stages of the Roman Empire, where they were used to grind grain.

By the Middle Ages, water wheels were used widely in medieval Europe to provide energy for a variety of manufacturing processes. The Dutch built many windmills to pump water as part of land reclamation projects. In the twelfth century, Europeans mined coal, but its use was limited. In fourteenth century England, the burning of coal in home fireplaces caused so much pollution that laws prohibited its use. One man may even have been executed for violating the law.

By the eighteenth century humankind's need for energy overweighed environmental concerns. One factor was that England began to run out of firewood. A second factor was the industrial revolution, which began with the invention of the steam engine. Soon, energy use in Europe and the United States grew rapidly. In the nineteenth century, coal became the main fuel of industrialized society.

The modern petroleum era began when the first commercial well began producing oil in Pennsylvania in 1859. Initially, petroleum replaced whale oil as fuel for lamps. The invention of the internal combustion engine, however, opened up vast new markets for oil in the twentieth century, and it soon overtook coal as the world's major source of energy. Oil became so vital to the economies of many nations that several wars have been fought over it. Japan attacked Pearl Harbor in December 1941 in part because it had no indigenous sources of oil and wanted to protect its flank as it moved to conquer the oil-rich Dutch East Indies. The 1991 Gulf War was fought in large part to prevent Iraq's President Saddam Hussein from gaining control of the oil-rich Persian Gulf.

Bibliography

DEUDNEY, D., and C. FLARIN. *Renewable Energy.* New York, 1983.

PRICE, T. *Political Electricity: What Future for Nuclear Energy?* Oxford, Eng., 1990.

TELLER, E. *Energy from Heaven and Earth.* San Francisco, 1980.

U.S. ENERGY INFORMATION ADMINISTRATION. *Annual Energy Outlook 1994, With Projections to 2010.* Washington, DC, 1994.

——. *International Energy Annual 1992.* Washington, DC, 1994.

WORLD ENERGY COUNCIL. *Energy for Tomorrow's World.* Washington, DC, 1994.

YERGIN, D. *The Prize: The Epic Quest for Oil, Money, and Power.* New York, 1991.

ROBERT LIVINGSTON

ENGINEERING ASPECTS OF EARTH SCIENCES

When we think of the interactions between applied scientists (engineers) and scientists, we imagine a one-way transfer of information: from scientist to engineer. The scientist explores for new information and the engineer uses this information. A scientist views humanity's volume of knowledge from the perspective of what is unknown. The engineer, however, views humanity's knowledge from the perspective of solving a problem. Whenever the engineer determines that the information needed to solve a certain problem is not available, then he or she becomes the explorer for information. Thus both the engineer and earth scientist are explorers for new information, but their search is driven by different objectives—to increase knowledge or to solve a problem (*see also* ENGINEERING GEOLOGY, HISTORY OF).

The history of the development of the earth sciences is laced with close encounters with engineers and the engineering techniques. As technology advanced and civilizations grew from primitive to modern, the demand for new materials, safe building sites, water, waste disposal, and many other resources increased. Underground mining for copper, which began at least 15,000 years ago in the Sinai Peninsula (Kiersch, 1991), established the need to understand how the deposits were formed so that the miners could find new ones (*see* MINING TECHNIQUES, PAST AND PRESENT). This demand fueled the study of mineralogy. Tunneling, which began about 5,000 years ago, generated a need to predict what was ahead, thus the demand for geologic mapping and subsurface exploration. The quarrying of rock for use as building and monument stone is an ancient practice. Early builders and artists recognized that different types of rock would respond to the quarrying operation or artist's hammer in different ways. The location of most of the towns and cities in the world can be traced to the geological conditions at the site. A dependable water supply, good transportation routes, and a defendable position were the primary factors. Many of today's cities were located for these attributes.

Leonardo da Vinci (1452–1519) is well recognized as a famous painter; however, few know of him as a hydraulic engineer concerned with canals and water projects (Clements, 1981). While viewing the excavations for these projects, he noticed shells in the rocks of the Tuscany hills and concluded that these hills were once below the sea. His studies of rivers and floods led him to conclude that valleys are cut by the rivers that flow through them, sediment transported by the rivers becomes smaller and more rounded as it moves toward the sea, and that these sediments are eventually deposited in the sea where they become the sedimentary rocks that are uplifted again to form mountains. Leonardo da Vinci's solutions to flood hazards and water supply problems led him to the study of how river systems work. His original engineering studies formed the basis of the scientific understanding of river systems.

During the late 1700s and early 1800s, the construction of canals was hampered by numerous geologic problems, mostly landslides and groundwater flooding. While working as a surveyor on canal projects in England in the 1790s, William Smith recognized that the various rocks encountered along the canal routes contained different fossils (Kiersch, 1991). Smith realized that he could relate the fossils to the rock units and use the fossils to determine the sequence of deposition of each rock unit. With this knowledge, Smith was able to map the geology of parts of England and to predict the engineering conditions that would be encountered at any site. Because he could understand where the rock units were located in the subsurface by investigating the rock units exposed at the surface, Smith was also able to predict where water existed in the subsurface. Thus Smith, in an attempt to better survey and define the ground conditions for engineering projects, actually developed the basic scientific foundations for the study of stratigraphy and of sedimentary rocks (*see* STRATIGRAPHY, HISTORY OF). Smith's engineering surveys and field observation and measurement techniques form the foundation of field work carried out by modern earth scientists.

By the mid–1880s major cities had grown up around the world, but their water supply and wastewater disposal practices were primitive. A worldwide cholera epidemic caused by drinking polluted water demanded that engineers develop methods to clean up the drinking water. A French hydraulic engineer, Henry Darcy, was experimenting with sand filters that could be used to rapidly and effectively filter the cholera bacteria from the water supply in Dijon, France. Darcy discovered

that the rate of water movement through the filter was related to the difference in height of the water at both ends of the filter and the length of the filter. Darcy's experiments led to the development of an equation that could be used to design filters for water treatment plants. This equation, known as Darcy's law, is also one of the basic equations used by groundwater geologists in the study of aquifers (*see* GROUNDWATER).

The concentration of people into cities required improved agricultural production. Agriculture had evolved from the practice of each family planting and gathering the food they needed to a situation where a few farmers produced the majority of a community's food. One critical requirement for improved agricultural production is a dependable supply of irrigation water. The practice of building dams and canals along rivers to provide irrigation water was well developed 4,500 years ago. The Native Americans (Hohokam) who lived in what is today southwestern Arizona and northwestern Mexico were expert irrigation farmers who changed the desert into an agricultural center more than 2,000 years ago (Maxwell, 1991). The expansions of Spanish missions in the 1500s reintroduced these Old World designs to the Americas. The missions along the San Antonio River in Texas, for example, still use the dams and canals built during the 1700s. As the English empire expanded around the world, English engineers were deeply involved in the design, construction, and operation of large-scale irrigation projects. A critical engineering problem facing these engineers was obtaining the desired volume of water from a canal without causing erosion and destruction of the canal. Engineers used large laboratory models to develop mathematical relationships between channel size, slope, and material. The most successful of these relationships is Manning's equation. Earth scientists rely on this equation in determining the flow conditions in natural streams and rivers (*see* SURFACE WATER, STORAGE AND DISTRIBUTION).

The study of how the landscape developed, particularly how hillsides evolve, has been a fundamental question in the earth sciences since the time of Leonardo da Vinci. Studies carried out in the late 1800s and early 1900s concentrated on the description and classification of landslides. The study of the mechanics of how these slope processes worked, however, had to wait for the development of soil and rock mechanics by engineers. Karl Terzaghi, a European highway materials engineer, started studying the behavior of unconsolidated materials (soils) in 1908 in response to a need for better roads and railroads. With the publication of his book on soil mechanics in 1925, Terzaghi provided the first tools that earth scientists needed to analyze the mechanics of slope movements. Terzaghi continued to apply engineering theories and principles to the solution of landslide mechanics, and his papers published in the 1950s and 1960s provided an understanding of how landslides start and how they move down a slope. These studies were truly interdisciplinary because Terzaghi needed the geologist's description and interpretation of conditions before he could analyze landslide mechanics. Because Terzaghi recognized the importance of a team effort of geologists and engineers, he became a role model to many of today's practicing engineers and earth scientists (Fleming and Varnes, 1991). A similar history occurred in the development of our ability to analyze rock slides. Early rock slide descriptions and classifications had to wait for the development of rock mechanics by engineers before rock slope processes could be evaluated (see GEOMORPHOLOGY, HISTORY OF).

Earthquakes have caused major losses of life and property throughout human history. The relationship between earthquakes and faults, however, was not fully recognized until 1884. The mechanics of earthquake processes were not identified until 1910 (Bonilla, 1991). Catastrophic failures or near failures of earth fill dams, after earthquakes in California in 1917 and 1971, led engineers to investigate the dynamic properties of granular materials. These engineering analyses determined that a water-saturated sand deposit could become a liquid (liquefy) when subjected to the dynamic loading caused by an earthquake. This discovery has allowed earth scientists to study the earthquake history of many locations around the world by analyzing the characteristics of sedimentary deposits and looking for evidence of flows.

Other examples of how engineering studies have impacted the earth sciences are evident in the history of construction aggregate development. Engineers noticed that certain kinds of aggregates would react with cement in a concrete mix and turn a solid gravel particle into a jelly-like mass, thus destroying the concrete. The solution to this problem aided the advancement of geochemistry because earth scientists became aware of reactive

minerals. Thermodynamics, the study of heat and heat transfer, is an important consideration in many engineering applications, such as gas and diesel engines, heating and cooling buildings, power plants, and electronics. These thermodynamic technologies were applied to protect the permafrost during the design and operation of the Trans-Alaska pipeline, which carries heated oil from the North Slope of Alaska to the seaport at Valdez (see THERMODYNAMICS AND KINETICS).

History has continued to show that the results of engineering studies carried out to solve one problem often solve many others. Since the 1950s the rate of development of new technology has in some cases outrun the rate of development of the earth sciences. New computer capacity continues to expand as electrical and electronic engineers, with the assistance of mathematicians, develop ever faster microprocessor chips that fit into smaller spaces. However, an incomplete understanding of earth processes limits the use to which we may put these new tools. Electronic and chemical engineers have developed highly sensitive detectors that can measure concentrations as small as 1 part per billion or even 1 part per trillion (1 part in a trillion is equivalent to a 15.24 cm step along a path from Earth to the Sun). The earth is a complex natural system that is highly variable, and therefore a measurement to 1 part per trillion may have no meaning—or it may be highly significant. Our understanding of the geochemistry of the earth is not sufficiently advanced to determine the answer.

The need to solve earth science problems in order to design, construct, and safely operate engineering projects has driven many earth science research projects. This earth science research, however, has also greatly improved the earth scientist's understanding of the history of earth systems. The barriers that once existed between the applied scientist (engineer) and the scientist are rapidly disappearing as interdisciplinary research and design teams work together to solve the many problems affecting modern society. Earth scientists have learned to use the results of the engineer's investigations and equations to solve their scientific questions. The engineers, in turn, have learned to use the results of the earth scientist's field and laboratory studies to better define the physical conditions affecting their project sites. The engineer's need for the best available answer to a specific design problem, such as canal erosion, has led to a better understanding of how rivers work by the earth scientist. The complex interactions between civilizations and the earth systems will be solved with interdisciplinary teams of engineers and scientists.

Bibliography

BONILLA, M. G. "Faulting and Seismic Activity." In *The Heritage of Engineering Geology: The First Hundred Years*, ed. G. A. Kiersch. Boulder, CO, 1991.

CLEMENTS, T. "Leonardo da Vinci as a Geologist." In *Language of the Earth*, eds. F. H. T. Rhodes and R. O. Stone. New York, 1981.

FLEMING, R. W., and D. J. VARNES. "Slope Movements." In *The Heritage of Engineering Geology: The First Hundred Years*, ed. G. A. Kiersch. Boulder, CO, 1991.

KIERSCH, G. A. "The Heritage of Engineering Geology: Changes Through Time." In *The Heritage of Engineering Geology: The First Hundred Years*, ed. G. A. Kiersch. Boulder, CO, 1991.

MAXWELL, J. A., ed. *America's Fascinating Indian Heritage*. Pleasantville, NY, 1991.

CHRISTOPHER C. MATHEWSON

ENGINEERING GEOLOGY, HISTORY OF

Engineering geology represents the concepts of geology applied to the welfare and works of people, excluding petroleum and mineral resource exploration and production but including exploration for and production of industrial minerals and materials of construction. Engineering geology is the characterization of sites of engineered human activity with identification and prediction of potential consequences. The occurrence of adverse natural phenomena, such as slope movements, collapse of ground, earthquake-induced damages, and failures of engineered works under the influences of deformations and stress accumulations also must be considered.

Engineering geology developed naturally as the result of building or excavating on or into the earth and placing heavy structural loads upon it, and from the need to construct dams and tunnels. In 1880 W. H. Penning published *Engineering Geology*, the first technical book on the subject. Four years later, U.S. Army Artillery Lt. Anthony Wayne Vodges published *Geology; Course of Sciences Applied to Military Art*, a manual containing the rudiments of engineering geology. By 1890 applied geologists were advising the architectural/engineering (A/E) firms of the day on a variety of construction projects involving excavation. By this time, the father of American engineering geology, William O. Crosby, was active as a consultant. Crosby served as professor of geology at the Massachusetts Institute of Technology from 1876 to 1916, and continued his private practice after retirement until his death in 1925. From about 1895 until his death in 1959, Charles Peter Berkey, professor of geology at Columbia University, was regularly retained by architectural/engineering firms and by federal construction agencies, such as the U.S. Army Corps of Engineers and the U.S. Bureau of Reclamation (USBR), to provide or to oversee geologic site characterization and to make recommendations for necessary structural and hydraulic design features.

By 1900, the University of California had become a place of practical instruction in engineering geology. Its Department of Civil Engineering, cooperating with the Department of Geology, produced what today are termed geological engineers, the most prominent of whom was Chester Marliave (class of 1907). From Stanford University came Hyde Forbes, protégé of Professor John C. Branner; and Professor Frederick L. Ransome, nominally an economic geologist (metallic ores), hailed from Throop Institute of Technology (now the California Institute of Technology at Pasadena, California).

The journal *Economic Geology* was created at Yale University in 1905. From its beginning, the journal solicited papers in engineering geology (seven papers appear in Volume 1, 1905–1906) and its pages, along with those of *Transactions* of the American Society of Civil Engineers (ASCE), were the most important forum for technological transfer in engineering geology until the 1960s. *Transactions*, initiated in 1872, began its coverage of engineering geology in 1894, with four papers, and in 1898, John C. Branner's classic paper on geology and topography appeared. It was not until 1964 that a distinct journal published exclusively for engineering geology appeared, the *Bulletin* of the Association of Engineering Geologists (AEG, known after 1994 as Environmental and Engineering Geosciences, a joint effort with the Engineering Geology division of the Geological Society of

America); it was followed in 1965 by *Engineering Geology*, published in Amsterdam; the *Quarterly Journal of Engineering Geology* (Geological Society of London) in 1968; and *Bulletin* of the International Association for Engineering Geology (IAEG) in 1970, the latter publishing at a rate of two numbers per year. These journals constitute the present-day body of serial publications for engineering geologists.

Upon U.S. entry into World War I, Alfred Hulse Brooks, chief of the U.S. Geological Survey (USGS) Alaskan branch, volunteered for service as Chief Geologist of the American Expeditionary Forces. The example of Brooks's volunteerism inspired other formidable engineering geologists such as OSCAR L. MEINZER, W. C. Lee, Douglas Johnson, Edwin C. Eckel, Sydney Paige, and Kirk Bryan to follow suit. By Armistice Day in 1918, officer-grade engineering geologists were assigned to every combat infantry division, and several were operating in the American sector producing water supply and sanitation studies. Others consulted on construction of fortifications, improvement of the road and rail networks necessary for troop and supply movements, and in the positioning of heavy artillery batteries and railway guns.

Federal and commercial waterpower development constituted the major involvement of engineering geologists in the 1920s. Bureau of Reclamation (USBR) projects burgeoned across the West, and state agencies were created and became operational for a variety of water supply, irrigation, and wastewater treatment facilities. Notable among private consultants were Chester Marliave, Hyde Forbes, the CalTech faculty, and Irving B. Crosby, son of William O. Crosby.

About 1925, annual meetings of the American Institute of Mining and Metallurgical Engineers (AIME) began to feature symposia on the newly developed field of engineering geophysics, including such methods as seismic refraction and electrical resistivity. In 1929 C. P. Berkey convened a landmark technical symposium on geology applied to dam construction. By the mid-1930s engineering geophysics had joined engineering geology in routine application to highway route surveys in some states. Engineering geophysics moved ahead with only moderate advances until World War II. Postwar advances in electronic circuitry resulted in shrinkage of the burdensome bulk and complexity of field equipment, beginning about 1955.

Federal New Deal programs after 1933 produced large water supply and hydroelectric projects, which were dominated by works of the Army Corps of Engineers, USBR, U.S. Soil Conservation Service, and the Tennessee Valley Authority (TVA). Major E. C. Eckel became the first chief geologist of the TVA and by 1934 produced a geologic overview of the TVA area of operations. Concurrently, the California Division of Water Resources (DWR) hired Chester Marliave and R. C. Eckis (later president of Richfield Oil Corporation) to begin to survey its area of operations for hydrogeologic resources and characteristics of sites for dams and other water resource projects. Eckel promoted engineering geology in the *Proceedings* and *Transactions* of ASCE, which began to feature papers and whole symposia on the subject of engineering geology.

With American rearmament, beginning in 1938, engineering geologists were hired by the Army Corps of Engineers as civilians and assigned to engineer districts, working both on military and civil works projects. Within a month of Japan's attack on Pearl Harbor on 7 December 1941, the chief of engineers had established the Military Geology Unit (MGU) of the U.S. Geological Survey. MGU first participated in the Allied invasion of Sicily and later operated in all but the European theater (geology being assigned to British Intelligence, with U.S. field armies retaining their own geologic personnel). The army itself acquired a large number of geologists in the course of its activation of reserve officers or from the selective service. As in World War I, many of these names became known as prominent members of the profession, including George A. Kiersch, Burton Marliave (son of Chester Marliave), and Reuben Newcomb (executive officer of an engineer water supply company assigned to General George Patton's Third Army), who were officers; and Manuel Bonilla and Clifford A. Kaye, who were sergeants. All served with high distinction as engineering geologists, generating combat terrain evaluation, trafficability assessments, and road and airfield construction.

Beginning in 1950, the rapid growth of the prewar federal flood control and state water supply projects (notably those of California, through its Central Valley Project) resulted in a frantic hiring of engineering geologists. For many engineering geologists the timing for entry to the profession was perfect, especially for those who had completed university degrees through the educational benefits of the G.I. Bill. By the early 1990s these engineering geologists, who had become leaders in the profession, were beginning to retire from formal practice.

In 1947, Geological Society of America past president and then executive secretary C. P. Berkey, along with other prominent practitioners of the day, formed the society's first operational specialty unit, the Engineering Geology Division (EGD). By 1950, Sydney Paige had edited a volume of papers titled *Engineering Geology in Practice*, dedicated to Professor Berkey. This volume was the cornerstone of engineering geology technology and, as such, is a timeless reference to the practice. The California Association of Engineering Geologists, formed in 1963 by federal agency geologists in Sacramento, became a national organization in 1967 as the Association of Engineering Geologists. On a worldwide scale, the International Association for Engineering Geology came into being on 21 December 1964 as a result of an approved motion made at the 22nd International Geological Congress, held at New Delhi, India.

Environmental concerns have, since the early 1950s, constituted a significant portion of the work of engineering geologists. In 1954, the Federal Housing Act was passed, with provisions (Sec. 701) for funded studies relating to urban planning (*see* URBAN PLANNING AND LAND USE). A large and diffuse body of publications and reports was completed under these provisions, mainly as a result of congressional amendments in the late 1960s and early 1970s. Most of the related formal publications were released by state geological surveys. State legislation, most notably the California Environmental Quality Act (1970), began to supplement the National Environmental Protection Act (1969). With this new funding, the breadth of engineering geology increased immeasurably, providing significant work for those engineering geologists who practiced in the subfield of hydrogeology.

Engineering geology is one of the most challenging areas of practice in the earth sciences, and, for that matter, perhaps in all of science. Each and every site is unique in terms of its geologic origin, made more complicated by the processes that acted during its geologic history and made doubly complex by the functional needs of the project to be built. Every field-practicing engineering geologist must have at his or her fingertips an array of technical knowledge and literature references just as broad as the science of geology, and a fundamental knowledge of the design processes and operational facets of all manner of engineered works. This requirement to know and retain codified technical knowledge has always been a basic necessity for practitioners. In addition to the individual papers in the specialty journals of engineering geology, the Berkey volume, the *Case History* and *Review* volumes of EGD, and the 1991 EGD *Heritage of Engineering Geology* as the sequel to the Berkey volume, there has been an ongoing need for development and harmonization of the standards, procedures, and guidelines of practice.

The first codification of engineering geologic field procedures was developed in the early 1970s by the Army Corps of Engineers, as *Engineer Technical Letters*, promulgated by former Chief Geologists Edwin Burwell, Gordon Prescott, Robert Nesbitt, and Lloyd Underwood, at the direction of the Office, Chief of Engineers. Various field-procedural guidelines also have been issued by various army engineer districts, USBR, California DWR, and as in-house documents by architectural engineering firms. In 1972, the U. S. Nuclear Regulatory Commission issued its first *Standard Format and Content for Preliminary Safety Analysis Reports* (PSARs). As of 1995, the *Bibliography of Engineering and Environmental Geology*, a comprehensive standards and procedural reference was in press.

Following the era of nuclear power plant siting and construction (which peaked from 1970 to 1984), the next major event involving new and emerging technologies has been the federal, legislatively driven remediation of uncontrolled hazardous waste sites, beginning with the Superfund law of 1980. The complexity of site characterization typical of engineering geology has again grown even more complex with the addition of the double dimensions of chemical contaminants and of their random and temporally variable disposal.

In the 1990s, engineering geology has been affected by the expanded capabilities of computer database storage, manipulation, and plotting of three-dimensional data. Most of the computer codes have been developed by applied mathematicians unskilled in geologic complexities; these codes are subject to gross simplification in order to meet the limitations of data arrays and computational matrices employed to smooth the differential calculus of dynamic processes such as groundwater flow and contaminant transport. Use of computer codes presents a challenge between those who adequately consider geologic complexities and those who must work with only the assumptions, data smoothing, and iterative alteration of variables so as to make the output analysis balance its own variables. Of the most serious potential value of computers to engineering geology is the Geographic Information Systems (GIS) technology, which can prove unmatchable in terms of storing, manipulating, and presenting subsurface databases for various forms of perspective-plotting, graphical analyses, statistical evaluations, and presentation of interpretations as perspective drawings (see GEOGRAPHIC INFORMATION SYSTEMS). As is true of all engineering geologic work, computer output can be only as accurate as the field work effort in collecting data and as sensible as the engineering geologist who approves of the input and who checks and verifies the results.

Bibliography

BRANNER, J. C. "Geology and Its Relations to Topography." *Transactions* 39 (1898):53–95.

BUCK, L. L. "A Few Remarks about the Niagara Gorge." *Transactions* 32 (1894):205–213.

HATHEWAY, A. W., comp. *Bibliography of Engineering and Environmental Geology.* Sudbury, MA, 1995.

KIERSCH, G. A., ed. *The Heritage of Engineering Geology: The First Hundred Years.* Geological Society of America. Centennial Special vol. 3, *Decade of North American Geology.* 1991.

MERRIMAN, M. "The Strength and Weathering Qualities of Roofing Slates." *Transactions* 32 (1894):529–543.

PENNING, W. H. *Engineering Geology.* London, 1880.

VOGDES, A. W. *Geology; Course of Sciences Applied to Military Art,* Part 1 *Geology and Military Geography.* Fort Monroe, VA, 1884.

WILSON, G. L. "The Sand Rock Sewers of St. Paul, Minn." *Transactions* 32 (1894):195–204.

ALLEN W. HATHEWAY

ENTHALPY

See Thermodynamics and Kinetics

ENTROPY

See Thermodynamics and Kinetics

ENVIRONMENT AND EARTH SCIENCE

Earth science or geoscience consists of at least twenty subdisciplines whose principal objective is the study of materials, processes, and terrain features that comprise Earth. Included within these are mineralogy, petrology, geochemistry, geoengineering, geomorphology, and geophysics, among others. Each subdiscipline has its own set of techniques and sphere of studies that seek to understand the composition and dynamics of Earth. These range from analysis of elemental particles to determine the radiometric ages of rocks and minerals to studies in mega-geology that deal with the movement of entire ocean basins, continents, and mountain ranges. It is difficult to find another major discipline with more importance and relevance to environmental matters than the earth sciences. Humankind lives, works, and plays on the surface of the Earth so human activities are always destined to change terrain features and processes in some manner. It is the earth scientist who is most closely tuned to societal interactions with the environment and habitations that are selected.

People have always been involved with and used skills that fall within the purview of earth science and environmental matters. The use of earth materials has been in the forefront of many different substantial advances in human history. From ancient times geological resources have been employed to sustain or enhance life, such as flint to cut meat, make weapons, and start fires, and earth pigments to paint caves and objects. The agricultural revolution depended on the proper use of soils and often required the importation of water by geoengineered structures. The industrial revolution necessitated plentiful iron, coal and limestone resources for an entirely new realm of manufactured products. The petroleum revolution needed sufficient oil supplies and the ability to discover new deposits and emplace deep wells. The success for all these required the skills of people who were performing earth science tasks.

Society has always required the materials and processes that constitute the geologic environment—the soil, rocks, minerals, and energy resources. However when these substances are stored, consumed, and manipulated, such changes can produce alteration, deterioration, and destruction in the land-water ecosystem. Thus the earth scientist's role in the environmental arena is more than just locating and cataloging Earth resources. Analysis is necessary to understand and predict the character and magnitude of such human changes to the natural condition. The context of such studies can provide a framework in the management of environmental affairs wherein human impacts become minimized and produce the least stress on environmental systems. Questions must be asked and resolved, such as: Will the proposed action produce deleterious feedback to other parts of the environment? Will the proposed actions cause degradation and irreversible changes? What are the geologic constraints that will affect human activities? Which type of managing ethic should be adopted for the proposed project?

Earth scientists have a keen awareness of the reciprocal relationship between humans and nature. Although people can greatly change natural materials and processes, nature can also produce grievous losses to society in the form of such geologic hazards as volcanoes, earthquakes, landslides, and floods. The geologist must understand and evaluate these hazards so that appropriate planning and management can occur in hazard-prone areas. Such features need to be located, mapped, and their frequency, size, and intensity determined. Advice can also be provided regarding such policy matters as which preventive and remedial actions should be taken or are possible to minimize loss of life and property.

Geologic Resources

Geologic or earth resources include minerals, rocks, fossil fuels, water, and soil. Earth is almost a completely closed system and there are finite limits to the geologic resources, which differ from such living resources as timber and crops because when mined out they cannot be replaced or grow again. Except for water, Earth resources are nonrenew-

able and must be mined in place. They also have high discovery costs and their extraction costs generally increase with time. It is the earth scientist who is called upon to locate and evaluate these materials for their use by society. To study these substances the geologist has devised special sets of investigative tools and techniques. Determination of the quality and quantity of such materials requires the mapping, testing, and analysis of the deposits. The character of the earth materials, their ease of removal or use, and which parts of the environment will be unduly damaged or destroyed must be assessed. Recovery, use, and removal of geologic materials can often produce undesirable changes that require remediation. Underground mining and withdrawal of fluids, for example, can lead to surface subsidence. Extraction of resources by surface mining can produce gaping holes that require reclamation procedures, and improper forestry and farming methods can cause enormous soil losses and unwanted sedimentation.

No other segment of environmental affairs has so many problems as those associated with the use and manipulation of water resources. This basic societal commodity is besieged by conflicts that arise in its quality, quantity, location, movement, and distribution. Although surface water can be considered a renewable resource, groundwater is very different because of the unusually slow movement of underground water. When "mined" out, it takes decades, centuries, and even millennia for recovery. Typical water problems that must be resolved include the apportionment of supplies among domestic, industrial, agricultural, and mining competitors. Additionally, water can be a two-edged sword when there is either too much, resulting in flooding, or too little during drought times or in dryland environments. Other problems associated with water include land erosion, chemical contamination, siltation, soil salinization, water logging, and earth subsidence.

Historically water resources policies have been organized around its occurrence, pattern of storage, distribution, and types of usage. All of these involve input from earth scientists. Water resource location, especially of groundwater, requires the expertise of hydrogeologists. Knowledge of such characteristics as permeability and porosity of the aquifer, and hydraulic conductivity and transmissivity (rate at which the water can pass through the materials in the aquifer), is needed to predict varying aquifer conditions accurately. The distribution of water via tunnels, canals, and aqueducts needs geoengineers for their design and installation.

Soils have provided the backbone of civilization and their appropriate stewardship is indispensable to society. Soils can be lost or degraded by many kinds of human activity including farming, deforestation, urban sprawl, and construction projects. Soil is also not a renewable resource so measures for its stability in ways to prevent erosion and maintain fertility are of vital importance and fall within the domain of the earth scientist. Expansive soils are especially vulnerable to extreme degradation by human installations. These are unusually fragile materials that can greatly shrink or swell depending upon the amount of available water. Each year they cause billions of dollars in damages to highways, building foundations, underground pipelines, pilings, retaining walls, and sidewalks in parts of the United States. Identification of such materials by geologic mapping can be crucial in preventing costly failures to engineered structures.

Geologic Hazards

Earth scientists are acutely aware of their responsibilities in dealing with such geologic hazards as volcanoes, earthquakes, landslides, and floods. These hazards result in significant loss of life and property. Compared with other natural damages, a geologic hazard is distinguished by its short duration. Erosion, sedimentation, and expansive soils all produce enormous losses but generally occur over periods of time measured in months or years. Only within the last few decades has society recognized the need to take action in attempts to abate, prevent, and protect losses from such processes. These measures take the form of investigations of hazardous conditions and recommendations for solving the problems. Such advice may lead to new laws and ordinances that prohibit or restrict development in hazard-prone terranes. Sometimes steps can be taken that can prevent or reduce potential damage. In other cases the hazard may be so overwhelming that no amount of money or geoengineered structures can prevent losses.

Volcanoes and earthquakes share some similar hazard characteristics and may occur only in certain Earth zones. They can also be of such magnitude as to be uncontrollable by human endeavors. It is of paramount importance to identify, map,

and restrict development in such zones. Structures already emplaced can be strengthened to resist earthquake damage, and legislation and ordinances can prohibit additional encroachment. Volcanologists are continuing to develop more sophisticated techniques that can provide early warning of possible volcanic activity. Similarly geophysicists are constantly studying new methods for prediction of potentially damaging earthquakes. Water level changes, increase of radon in wells, physiochemical and electrical resistivity changes in rocks, and differences in seismicity patterns have all been identified as precursors for some earthquakes.

Landslides and flooding phenomena have several common denominators. Generally they are more predictable than volcanic activity and earthquakes. They also are more common and occur throughout many different types of regions. In some instances it is possible to prevent, mitigate, or control the amount of devastation. Losses from these phenomena are increasing because of the world's exploding population. Lowlands and floodplains have always been the preferred habitat of humans, whose structures have changed river systems and increased damages from flooding. Conversely when developments must then be moved to hillslopes they can be in jeopardy because of instabilities created on earth materials in this environment.

The mechanics of rivers requires their periodic overflow and flooding. Debates have ranged for decades about whether to control flooding by such structures as dikes, levees, and dams. However during the devastating flood of the Mississippi River in 1993, which caused more than $15 billion in damages, the man-made structures failed to contain the onslaught of this massive flood.

Landslides occur on hillside terranes when the stability of soil and rock has been overwhelmed by downslope forces of gravity. Numerous land-use surveys since 1975 have mapped potential areas of hillside displacements. Strategies have been used that in some cases prove effective for increasing slope stability. These include dewatering of sensitive zones, hillside benching, emplacement of buttresses and retaining walls, and grouting or sinking of rock bolts. The geomorphologist is the scientist in the best position to determine and measure the possibility of slope movements, and the geoengineer can provide the best designs for stabilization of the material.

Wastes and Urban Areas

Physical problems in congested areas present earth scientists with some of their greatest challenges. As the world becomes increasingly urbanized, congestion in the urbanized terrane creates more constraints and less maneuverability for solution of Earth-related problems. Disposal of waste, both fluid and solid, places a difficult burden on the geologist. Waste disposal involves the location of landfills, excavation of earth materials, and the instrumentation of a network of monitor wells. Investigation reports, required by governmental agencies for landfill sites in order to determine the optimum location for emplacement of civilization's waste debris, often cost millions of dollars.

Urban areas create many other problems that must be addressed by the earth scientist. Typical of these are the increased incidence of floods and landslides, the possible paving over of mineral resources, the need to import large amounts of earth materials, and investigations to determine the strength of rocks to withstand stresses imposed by construction activities.

Environmental Policy and Management

The very character of environmental affairs assures that policies enacted will rarely receive unanimous approval. Conflicts arise from what legislation to enact, how to allocate funds, how to assign priorities concerning which structures to emplace, and which ethics style to adopt. Environmental issues are very complex and nearly always cross into several different disciplines. For example, phreatophytes (groundwater-using plants) in the Southwest consume prodigious amounts of underground water. Some hydrologists want them eradicated because such waters could be used by society. However, wildlife biologists want them saved for they provide shelter and food for animals and birds.

Three competing ethics govern environmental decisions. The business or utilitarian ethic favors commercialization of the environment and encourages economic growth for the maximum development of all resources. However, the preservation ethic wants nature and natural processes retained and held intact for the use of future generations. The conservation ethic provides a type of middle ground that champions the careful design and use of nature and earth materials that permit the maximum good for the maximum amount of time. The most important service the earth scientist can offer is to provide a sufficient database that contains cogent data in an understandable report. Determination should be made whether the proposed action is physically possible, economically justifiable, sociopolitically acceptable, legal and constitutional, and environmentally balanced.

To be effective in the public arena of policy-making, earth scientists must successfully defend their data and research methods. They have a unique understanding of the threshold concept, wherein a given action is most likely to produce an undesirable disequilibrium in the land-water ecosystem. Recognition and prediction of those actions that can trigger a threshold are based on such factors of knowledge as the physical history of the area, the use of threshold analogs from similar situations, and the development of an accurate database regarding earth materials and processes at the site. In addition to the involvement of geoengineers for construction projects, the word and subdiscipline of "geotechnology" has become an integral part of the environmental scene. In 1988 the National Research Council of the National Academy of Science established a Geotechnical Board for the purpose of applying "geotechnical input" for such national pressing problems as: waste management; infrastructure development and rehabilitation; construction efficiency and innovation; national security, resource discovery and recovery; mitigation of natural hazards; and frontier exploration and development.

It is difficult to overstate the importance of earth scientists in matters that concern environmental affairs. Their involvement at all levels of decision-making is often crucial to the success of human projects and in matters dealing with the natural world.

Bibliography

BELL, F. G., ed. *Ground Engineer's Reference Book*. Stoneham, MA, 1987.

———. *Engineering Geology*. Cambridge, MA, 1993.

COATES, D. R. *Environmental Geology*. New York, 1981.

———. "Perspectives of Environmental Geomorphology." *Z. Geomorphologie* 79 (1990):83–117.

FINKL, C. W. *The Encyclopedia of Applied Geology*. New York, 1984.

MORISAWA, M., ed. *Geomorphology and Natural Hazards*. New York, 1994.

MONTGOMERY, C. W. *Environmental Geology*, 3rd ed. Dubuque, IA, 1992.

WOLFE, J. A. *Mineral Resources A World View*. New York, 1984.

DONALD R. COATES

EROSION

See Weathering and Erosion

ESKOLA, PENTTI

Born on 8 January 1883, in Lellainen, Honkilahti, southwestern Finland, on the 300-year-old family farm, Pentti Eskola lived for nearly 81 years. He married Mandi Wiiro in 1914 and they had two children. Their daughter, Päivätär, became a chemistry teacher, but their son, Matti, was killed in 1940 during the Winter War between Finland and the USSR. T. Mikkola (1968) provided a warm memorial to the human qualities of his former teacher, Pentti Eskola, who emerges as an exceptional scientist, teacher, and mentor. Eskola died in 1964.

Eskola attended the University of Helsinki from 1901 to 1914. He obtained a degree similar to a master of sciences (M.S.) in chemistry in 1906 and completed his doctoral dissertation on the metamorphic rocks of the Orijärvi region, Finland, in 1914. Subsequently, he studied in 1920–1921 at the Mineralogical Museum in Oslo with Professor VICTOR MORITZ GOLDSCHMIDT (the first modern geochemist), who guided Eskola to evaluate rocks as chemical systems and minerals as thermodynamic phases. Eskola also worked at the Geophysical Laboratory at about this time and was strongly influenced by NORMAN L. BOWEN (the first modern experimental petrologist). Eskola was a geologist for the Finnish Geological Survey (1922–1024) and was Professor of Geology at the University of Helsinki for many years (1924–1953).

Pentti Eskola is one of the giants of petrology. He placed metamorphic reactions on a chemical and thermodynamic basis with pressure, temperature, and composition considered as important variables. He was the first to use composition space as a tool in predicting metamorphic assemblages and in unraveling chemical reactions that relate to different assemblages. In 1920 he introduced the concept of metamorphic facies, which relates metamorphic mineral assemblages found in common rock types to certain ranges of pressure and temperature that are encountered repeatedly within the crust and uppermost mantle. Although subsequently expanded and modified by Frank J. Turner and others, Eskola's concept remains in wide use today (*see* METAMORPHIC ROCKS).

Eskola wrote an important treatise on the eclogites of Norway in 1921. Eclogites are metamorphosed basaltic rocks comprised dominantly of garnet and sodic pyroxene; their origin(s) and significance have been the subject of continuing controversy. Eskola recognized the importance of the eclogite facies in terms of deep burial of crustal materials, a process that is now interpreted in terms of plate tectonics. Crustal eclogites and eclogite-facies rocks are thought to form deep in the roots of the giant mountain belts that originated during subduction and/or continent-continent collisions (*see* PLATE TECTONICS). Thus, occurrences of eclogites in metamorphic terranes around the world provide direct evidence of ancient mountain-building, or orogenic events. Norwegian eclogites formed in the roots of a major mountain belt named the Caledonides, which extended along the spine of Scandinavia through Scotland and northern Ireland. Plate tectonic reconstructions indicate that the Caledonian Orogen joined with North America and continued down the length of the Appalachian Mountains in mid-Paleozoic times.

Initially influenced by Jakob J. Sederholm, Eskola favored a metasomatic origin and solid state granitization for many rocks. However, he also identified (1933b, 1952) the important roles of partial melting and the in situ production of granitic magmas within the granulite facies. Granulites, which are dry rocks dominated by assemblages of pyroxenes and feldspars, formed at high pressures and temperatures deep in the continental crust. These two hypotheses for the origins of granulites survive in modified form to the present time. Most scholars favor one of two hypotheses: either granulites transformed in the solid state from more hydrous rocks during equilibration with CO_2-rich fluids from the mantle that pervaded and dried the lower crust, or alternatively they were associated with underplated mantle-derived basaltic magmas that heated the crust and produced hydrous granite magmas as partial melts, the removal of which would dry out the deep crust.

Altogether Eskola contributed greatly to the geology of Scandinavia and Finland in some 170 publications. Much of Eskola's life's research was presented in major chapters of two books, *Die Entstehung der Gesteine* (1939), and *The Precambrian* (1963). His lasting contribution was the consideration of rock and mineral assemblages as chemical systems that are tending toward a state of chemical equilibrium. As a result, most scholars who study metamorphic rocks today first treat them in terms of equilibrium concepts, with any deviations from equilibrium representing an anomaly worth investigating. Eskola also had a deep philosophical bent, as revealed in several of his popular books with a wider scope in which he explored the origin of Earth, the evolution of life, and the implications for the meaning of life. In his own lifetime Eskola received many awards and medals from various geological societies; in turn the Geological Society of Finland now presents the Eskola Medal to scientists once every five years in his honor.

Bibliography

AMSTUTZ, G. C. "Eskola, Pentti Elias." Vol. 4, *Dictionary of Scientific Biography*, ed. C. C. Gillispie. New York, 1971.

BARTH, T. F. W. "Memorial to Pentti Eskola." *Geological Society of America Bulletin* 76 (1965):117–120.

ESKOLA, P. "On the Petrology of the Orijärvi Region in Southwestern Finland." Helsinki, Finland, 1914.

MIKKOLA, T. "Memorial of Pentti Eskola." *American Mineralogist* 53 (1968):544–548.

————. "On the Relation Between Chemical and Mineralogical Composition in the Metamorphic Rocks of the Orijärvi Region." *Bulletin de la Commission Géologique de Finlande* 44 (1915):114–117.

————. "The Mineral Facies of Rocks." *Norsk Geologisk Tidsskrift* 6 (1920):143–194.

————. "On the Eclogites of Norway." *Skrifter Videnskabsselskabet Kristiana* 1 (1921):1–118.

————. "On the Origin of Granitic Magmas." *Mineralogische und Petrographische Mitteilungen* 42 (1932):455–481.

————. "On the Principles of Metamorphic Differentiation." *Bulletin de la Commission Géologique de Finlande* 97 (1933a):68–77.

————. "On the Differential Anatexis of Rocks." *Bulletin de la Commission Géologique de Finlande* 103 (1933b):12–25.

————. "The Problem of Mantled Gneiss Domes." *Geological Society of London, Quarterly Journal* 104 (1949):461–476.

————. "On the Granulites of Lapland." *American Journal of Science* 250A (1952):133–171.

————. "Precambrian of Finland." In *The Precambrian*, Vol. 1, ed. K. Rankama. New York, 1963.

ERIC J. ESSENE

EWING, WILLIAM MAURICE

William Maurice Ewing, son of Floyd Ford and Hope Hamilton Ewing, was born on 12 May 1906 in Lockney, a town in the Texas Panhandle. After graduating from Lockney High School at the age of sixteen he entered Rice Institute (now Rice University) in Houston, Texas. Ewing received his bachelor's degree with honors in mathematics and physics in 1926. He then won a fellowship in physics, from 1926 to 1929. In 1927 he completed his master's degree in physics and in 1929 he obtained a job as instructor of physics at the University of Pittsburgh. In 1930 he transferred to Lehigh University where he stayed as instructor, assistant professor and, finally, associate professor until 1944. While at Lehigh he was able to complete his Ph.D. in physics from Rice in 1931.

At Lehigh he started seismic refraction experiments on a limited scale using small explosive sources. At the encouragement of Richard Field, professor of geology at Princeton, and William Bowie, chief of the Geodesy Division of the Coast and Geodetic Survey, Ewing carried out seismic refraction experiments on the continental shelf, first aboard the Coast and Geodetic Survey ship *Oceanographer*, and followed by further experiments aboard the Woods Hole Oceanographic ship *Atlantis*. These were the first seismic refraction measurements made in the open sea, and they led Ewing to discover that a thick layer of sediments lay beneath the ocean floor.

The work on *Atlantis* also set a pattern for Ewing's life work, in which either he devised new experiments in the ocean or greatly improved existing instrumentation and methods. Not satisfied to simply demonstrate the success of these methods by obtaining a few measurements, Ewing sought to employ new instruments and methods on a large scale in order to learn about what lay beneath the oceans. Throughout his career Ewing engaged in seismic reflection and refraction experiments, but also devised or improved on a number of other instruments used for exploring the oceans.

One of the first instruments Ewing developed was the precision depth recorder (PDR), used to accurately measure the depth of the ocean. A crystal clock measured the time taken by sound to travel from the ship to the ocean bottom and back. One of the discoveries that came from using the PDR was that huge abyssal plains, with perfectly flat surfaces, formed large parts of the floors of the North Atlantic basins. This led in turn to the discovery (made jointly with his student Bruce Heezen) that turbidity currents brought vast amounts of unstable sand and silt from the edge of the continental shelf and spread them on the abyssal plain floors, gouging deep submarine canyons on the continental slope in the process. Ewing, with Edward Thorndike, developed the first underwater camera. Later, the same instrument was modified into a nephelometer, which measured particulate matter in the seawater. Ewing encouraged the development of the marine magnetometer and also the extensive use of the Vening Meinesz pendulum apparatus aboard submarines for making gravity measurements. He assisted the Allies during World War II by discovering a sound channel (named the SOFAR Channel), which allowed sound to be transmitted vast distances in the oceans. A network of monitoring stations with listening devices in the SOFAR Channel could accurately locate planes, ships, and life rafts in distress in the open ocean.

Ewing left Lehigh after the war and in 1948 became a professor of geophysics at Columbia University in New York. In 1949, Ewing became the founding director of the Lamont Geological Observatory. He left in 1973 to join the University of Texas in Galveston to become the Henry and Ida Green Professor in Geophysics.

Ewing's principal work was carried out while he was at Lamont Observatory. In 1953 he was able to acquire the *Vema*, a three-masted schooner, which was joined in 1961 by *Robert D. Conrad*, a research vessel provided by the office of Naval Research. From 1955 until the 1970s, these ships roamed the world's oceans, collecting geophysical data. They carried magnetometers and gravimeters to measure the magnetic and gravity fields, respectively. They also continuously carried out seismic reflection measurements to measure the thickness of sediments lying beneath the ocean floor. *Vema* and *Conrad* usually stopped once or twice a day to make sea stations. At these stations camera and nephelometer measurements were carried out. Early on Ewing had improved the Kullenberg piston corer, and it was deployed regularly at sea stations to obtain bottom sediment cores as much as one hundred feet in length. Ewing had also initiated the development at Lamont of the Ewing thermograd.

By fastening thermistors to the coring tube the temperature within the sediments could be calculated, thus leading to a determination of the amount of heat escaping from below into the ocean bottom. Ewing's achievement was not only that he initiated all these measurements, but that he was also able to routinely carry out these experiments throughout the world's oceans.

Vema and *Conrad* expeditions led to the collection of the largest library of sediment cores, bottom photographs, nephelometer readings, and seismic, gravity, and magnetic records ever assembled in the deep oceans.

Throughout his career Ewing retained his love of developing new instruments and collecting large volumes of data. After moving to Texas, he continued to lead oceanographic expeditions, devising new coring instruments, and collecting multichannel seismic reflection data in the manner in which industry collected such information.

When Ewing started making geophysical experiments at sea, it was not known that the geology of the ocean floors was much simpler than the geology of the continents. Discoveries in geology of the oceans, through a few decades of subsequent investigations, revolutionized the notions about geologic processes that shape Earth's surface and its interior. Although Ewing did not initially subscribe to the ideas about seafloor spreading and plate tectonics, it was the detailed examination of the data he so assiduously collected and stored at Lamont that led to the rapid establishment of the new theories (*see* OCEAN BASINS, EVOLUTION OF; PLATE TECTONICS).

Ewing also made a considerable contribution to earthquake seismology. Being in New York State, distant from most large earthquakes, he was more interested in the long-lasting surface waves than in the body wave phases. Together with Frank Press, he devised the Press-Ewing seismographs to record the longer-period surface waves. He then used the dispersion of the Love and Rayleigh waves to determine the broad characteristics of oceanic and continental crustal structures. The discovery that the crustal structure of the Atlantic was similar to that of the Pacific was a direct result of the study of dispersion curves. Later, again with Frank Press, he initiated the use of phase velocity rather than group velocity of seismic waves for the study of crustal structure and applied it to a determination of the variation of crustal structure in different regions of the United States.

Ewing's scientific interests were not limited to geophysics. With Bill Donn, he developed a theory about the periodic appearance and disappearance of ice ages. His interests included climate changes and the use of Foraminifera for detecting cycles of temperature changes in the oceans. He collaborated on papers in such diverse areas as marine geodesy and lunar seismology. He worked hard to initiate the Deep Sea Drilling Project (DSDP), and was chief scientist on its first leg.

Ewing was the recipient of a large number of honors, including eleven honorary degrees. He was a member of the U.S. National Academy of Science, and of a number of foreign academies. He served as the president of the American Geophysical Union and the Seismological Society of America. Ewing is perhaps remembered most for training graduate students and associates who went on to become scientific leaders in their own right. These include Frank Press, who was later elected president of the National Academy of Science; Al Vine, the force behind the submersible *Alvin*; Charles Drake, the first president of the International Geodynamic Project; Jack Oliver, who pioneered the exploration of the deep continental crust by seismic reflection methods; and his brother, John Ewing, who was a close collaborator in seismic work at sea. Another student, Bruce Heizen who, together with Marie Tharp, produced the Tharp-Heizen physiographic diagram of the ocean bottom, also collaborated with Ewing for many years.

Maurice Ewing died on 4 May 1974, in Galveston, Texas. To him, scientific research was not just an important undertaking—it was the most important undertaking.

Bibliography

BULLARD, E. "William Maurice Ewing, 1906–1974." *Biographical Memoirs of the Royal Society*, 21 (1975):264–311.

MANIK TALWANI

EXPANDING UNIVERSE

See Cosmology, Big Bang, and the Standard Model

EXTINCTIONS

Extinction in the evolutionary, paleontologic sense refers to the final demise of a particular species, genus, or higher biological category that left absolutely no direct descendants. For example, the famous saber-toothed tigers from Rancho La Brea in the Los Angeles area totally disappeared in the later Pleistocene: no saber cats exist anywhere today, nor any evolutionarily modified cats descended from saber cats.

The concept that the history of life has been interrupted many times by extinctions affecting many unrelated organisms at about the same time is an old idea. In the beginning of the nineteenth century Georges Cuvier had already learned enough about the geologic history of vertebrates to be able to recognize the existence of an age of reptiles, now called the Mesozoic, followed by an age of mammals, now called the Cenozoic (*see* CUVIER, GEORGES). His younger contemporary, Alcide d'Orbigny, soon recognized the same phenomenon among marine, bottom-dwelling invertebrates: d'Orbigny recognized far more faunal units than did Cuvier. Both Cuvier and d'Orbigny thought that the sporadic extinctions they recognized were total, or virtually total, and blamed them on major catastrophes that wiped out all life to be followed by the generation of new life. Since their day we have come to realize, based on a wealth of additional information gathered from many parts of the globe, that extinctions were not total but involved varying percentages of the previously living fauna that range from very low for minor extinctions to truly monumental for some in which the majority of organisms in most environments were eliminated.

How are extinction events recognized in the history of life? First, one needs to recognize the difference between a "true" extinction and other phenomena that may be confused with extinction. One such phenomenon that might be confused is the continuous evolution of organism lineages, species, and genera into other species and genera; with the result that the older ones may seem to have "disappeared." This may be referred to as "pseudoextinction" because it involved evolutionary change (and corresponding name change), not the true loss of an organism. At the higher taxonomic levels, this problem of one taxon (category) evolutionarily grading into another is not as pronounced, that is, the termination of family names in most groups commonly represents a relatively high percentage of extinctions, and an even higher percentage of extinctions as one climbs the hierarchical ladder. We have no fossil evidence for one phylum or another evolutionarily grading into another at the genus or species level; their disappearance represents an extinction, not a transition.

To recognize true extinction events from the fossil record it is common to plot first appearance and final disappearance of organisms based on their known fossil record. In principle such a procedure is adequate except that it assumes that the fossil record faithfully documents first and last appearances of all organisms. Unfortunately, that is not so for the majority of organisms. Soft-bodied organisms are only exceptionally fossilized, and their known time ranges thus are commonly restricted. Skeletonized organisms also pose problems. Many skeletonized organisms, for example, such important groups as the echinoderms, commonly fall apart after death into a mass of disarticulated, unidentifiable plates, organic gravel (*see* TAPHONOMY, FOSSILIZATION AND THE FOSSIL RECORD). Equally if not more difficult to determine are both beginning and ending of relatively uncommon-to-rare organisms that make up the great majority of fossil genera and species. These uncommon-to-rare organisms are hard to find while

collecting: their known geologic range is therefore commonly much shorter and unreliable as a guide to the time interval during which they lived. The only reliable organisms useful for recognizing extinction events are those abundant enough to be collected in the average sample.

A second way to recognize extinction events is to note major changes in the relative abundance of organisms through time. It is significant if an organism is abundant during one time interval but uncommon to rare during other time intervals. For example, rhynchocephalian reptiles are reasonably abundant during the Mesozoic, but they exit the fossil record at the end of the Cretaceous. One genus, *Sphenodon*, still exists on a few islets off the north coast of South Island, New Zealand: the major change in abundance of rhynchocephalians across the Cretaceous-Tertiary boundary is part of the evidence for recognizing this important extinction. A second example involves the changes in relative abundance of the rhipidistian crossopterygian fish that disappeared from the fossil record at the end of the Cretaceous (although present today in some deeper water areas off East Africa as *Latimeria*): a major abundance change bearing on the extinction event at the end of the Cretaceous that also affected the dinosaurs.

A third class of evidence useful in recognizing extinction events has to do with changes in community type. A community is a regularly recurring association of genera and species in which the relative abundances of the varied organisms remain relatively fixed: rare things remain rare, abundant things abundant. Major changes in community structure are a powerful means for recognizing extinction events because during every extinction and subsequent adaptive radiation there are changes in both presences and absences of taxa as well as changes in abundances of surviving taxa. These changes make for easily recognizable changes in the communities above and below the extinction horizon (and also help to minimize the sampling problems presented by uncommon-to-rare taxa).

Since the middle of the nineteenth century it has been obvious that there are less than a dozen truly major extinction events as well as far more minor events. At the present time for marine metazoans most scholars recognize the following major extinctions: Ediacaran-Early Cambrian; Early Cambrian-Middle Cambrian; Late Cambrian-Early Ordovician; Early Ordovician-Middle Ordovician;

earlier Late Devonian-later Late Devonian; Permian-Triassic; Triassic-Jurassic; Cretaceous-Tertiary. Additional to these major extinctions there are a much larger number of lower-level events, such as that between the Mississippian-Pennsylvanian and the one near the end of the early Middle Devonian. The chief point here is that a complete spectrum exists between relatively minor extinctions affecting a small percentage of organisms and truly major items affecting many.

It also needs to be emphasized that the major extinctions affecting the terrestrial ecosystem appear to be in large part decoupled in time from those involving the marine ecosystem. The most obvious evidence of this is that the eras (Paleozoic, Mesozoic, Cenozoic) based on marine benthic invertebrates do not have time boundaries that correspond with those based on higher land plants (Palaeophytic, Mesophytic, Cenophytic). For example, the beginning of the Cenophytic, the age of flowering plants, occurred in the mid-Early Cretaceous, a time when nothing very dramatic was occurring in the marine ecosystem. Extinction events affecting terrestrial vertebrates for most post-Devonian intervals remain to be worked out, the dinosaurs and some Permo-Triassic reptiles and amphibians notwithstanding.

In the search for cause(s) one must keep firmly in mind the fact that for global extinctions one needs to have explanations that permit both the extinction of certain groups and the survival of others. For example, it will not do to call on a Cretaceous-Tertiary "nuclear winter" scenario lasting some months, or even years, because this would not have permitted the survival of many of the mammalian groups, owing to their continuous requirements for food, although it would help to explain the disappearance of some other non-mammalian tetrapods, and would not explain the presence of angiosperms on both sides of the extinction boundary. Extinctions of localized, endemic species and genera are another matter. There is much current discussion about the cause(s) of extinction, based largely on the 1980 attention directed toward the presence of some beds at the Cretaceous-Tertiary boundary sufficiently enriched in trace amounts of iridium to suggest the possibility of an asteroidal or cometary impact at this time as the source of the iridium and possibly the cause of the extinction (*see* IMPACT CRATERING AND BOMBARDMENT RECORD). However, since 1980 it has become clear that trying to under-

stand the cause(s) of these extinction events is a very difficult area where limited circumstantial evidence does not lend itself to simplistic single-factor explanations. Most of the causes discussed in the decade of the 1980s have been physical (such things as climatic change, sea-level change, seawater compositional change) rather than biological. The interest in asteroidal-cometary impacts and their physical consequences has not been accompanied by a critical consideration of biological possibilities, such as disease, evolutionary phenomena, or ecostructure collapse possibilities. This situation needs to be remedied. Recall that we still are far from agreement about the cause(s) of the extinction of a number of large mammals (mammoths, mastodons, giant ground sloths, giant marsupials of varied types, glyptodontids, and so on) in North and South America plus Australia, while contemporary large mammals in Africa and Eurasia largely survived. Some scholars implicate human populations, whereas others appeal to climatic/environmental changes. This being the case we have an even greater problem trying to decide cause(s) of extinction for long-extinct organisms with no close relatives that might provide us with behavioral, physiological, or other clues.

It is not yet clear whether extinctions extend over only a geologically brief interval, say a few thousand years or less, or whether they may occur over a few million years. If the latter, it is not yet clear whether or not the more endemic, specialized organisms tend to disappear before the less specialized ones, followed ultimately by a community collapse of the most abundant, cosmopolitan non-specialists. It does seem clear, however, that the major extinctions are sporadic in time rather than periodic as some researchers have suggested.

We have no real understanding about why some organisms become extinct whereas others survive. There has been little curiosity about this basic question in terms of different morphologies or life habits. However, it is safe to say that the relative giants in any major animal group have much shorter time durations than average-size organisms (this does not bode well for us, as giant anthropoids).

The monumental extinctions, particularly on land and in the freshwaters, caused by humans in the past few millenia, possibly beginning as much as ten to twelve thousand years ago, will certainly leave a major extinction record for which we cannot be proud. Human-perpetrated extinctions will probably rank with the truly major extinctions of all time and may well exceed them.

Bibliography

GRAY, J. "Major Paleozoic Land Plant Evolutionary Bio-events." In *Palaeogeography, Palaeoclimatology, Palaeoecology* 104 (1993):153–169.

KAUFFMAN, E. G., and O. H. WALLISER, eds. *Extinction Events in Earth History.* Lecture Notes in Earth Sciences 30. Stuttgart, Germany, 1990.

WALLISER, O. H. *Global Bio-Events: A Critical Approach.* Lecture Notes in Earth Sciences 8. Stuttgart, Germany, 1986.

ARTHUR J. BOUCOT

EXTRA HEAVY OILS

See Tars, Tar Sands, and Extra Heavy Oils

EXTRATERRESTRIAL LIFE, SEARCH FOR

> To consider the Earth as the only populated world in infinite space is as absurd as to assert that in an entire field of millet, only one grain will grow.
>
> Greek philosopher Metrodorus,
> fourth century B.C.E.

The question is as old as it is profound: is Earth the only site of life in the universe? While this ancient inquiry remains unresolved, it may be only a matter of time until experimental evidence ultimately proves that planet Earth is one of many places in the cosmos where biological organisms have emerged and evolved. Several groups of radio astronomers are currently attempting to show that biology has given rise to technology on the planets of other stars by detecting interstellar communications transmissions or radar leakage from distant technologies. This entry will examine what we now know about the possibility of nonterrestrial biology—including intelligent life—that may have developed culture and technology. No one yet knows for sure whether life of any kind exists beyond the Earth's biosphere, so a certain amount of speculation is unavoidable when addressing this topic.

A Barren Backyard

Except for Earth, our solar system appears to be completely devoid of life. From July 1976 until November 1982, NASA's two unmanned Viking spacecraft explored Mars, our neighboring planet that is most often linked in the human imagination with thoughts of otherworldy life (*see* MARS). One major objective of the Viking mission was to look for indications of life on Mars. Photographs from orbit showed clear evidence that water had once flowed on the surface of Mars when that planet had a much denser atmosphere than it does today. This raises the tantalizing possibility that life may have started on Mars about the same time that it began on Earth. However, when the Viking landers tested the Martian soil for signs of biological activity, no convincing evidence for life was found. There remains a chance that either extant or extinct microorganisms await discovery at other locations on Mars. If native microbes are ever found on Mars, the idea that life has cropped up in other solar systems would gain a powerful new impetus. Other bodies in our solar system, such as Jupiter's moon Europa and Saturn's moon Titan, may have experienced enough chemical evolution to now possess some forms of microbial activity, although most scientists believe that this is unlikely. Such possibilities await further exploration. In the meantime, scientists can search for the technological manifestations of advanced life in other solar systems and continue to study the basic processes of life that have evolved on Earth.

The first modern search for radio-using technological civilizations on the planets of other stars was carried out in the spring of 1960 by astronomer Frank Drake, who was then working at the National Radio Astronomy Observatory in Green Bank, West Virginia. Drake aimed a 25-m-diameter antenna at two nearby stars, Tau Ceti and Epsilon Eridani, and listened for three months. The following year, Drake conceived an approach to identify and bound the factors to be considered in any effort to estimate the number of technological civiizations in our galaxy that might be expected to emit detectable radio signals at any given time. Known as the Drake equation, it encompasses contemporary theory about the evolution of stars and planets, the habitability of planetary environments, chemical and biological evolution, the evolution of intelligence and technology, and the longevity of civilizations. Although there is no unique solution to the Drake equation, it is a generally accepted tool used by the scientific community to examine the question of intelligent life elsewhere.

Beyond the Backyard

The Milky Way Galaxy is only one of hundreds of billions of galaxies in the presently observable universe (*see* GALAXIES). The local star we call the SUN

is just one of some 400 billion stars in our galaxy alone. Astronomy has shown that the universe contains huge numbers of stars like the Sun and that the chemical elements characteristic of life are widespread (Figure 1). More than seventy different molecules have been detected by radio astronomers in dense interstellar clouds of gas and dust and in the extended envelopes around stars. Most of the interstellar molecules discovered since the 1960s are carbon-based—that is, organic molecules. Amino acids have also been found in meteorites. This indicates the wide diversity of extraterrestrial environments where organic molecules can be formed. The surfaces of planets provide another rich environment for organic chemistry.

The chemical evolution model of the origin of complex molecules was first tested experimentally in 1951 by Stanley Miller and HAROLD UREY. Their University of Chicago laboratory flask contained

Figure 1. Spiral galaxy (photograph courtesy of NASA).

water, methane, hydrogen and ammonia to simulate a primitive Earth atmosphere then thought to be high in hydrogen content. When electrical discharges released energy into the mixture, compounds were produced that are important for living systems. The Miller–Urey experiment was criticized during the 1970s by scientists who argued that Earth's atmosphere never had as much hydrogen as the mixture of gases used in the flask. This critique resulted in a new generation of "early-Earth" experiments that incorporated lower hydrogen levels. Such studies also produced compounds that are important for living systems, and scientists continue to refine their understanding of origin-of-life processes with similar experiments.

Plentiful Planets

Just a few generations ago, astronomers thought that planetary systems were very rare—that our solar system and our Earth with its life-supporting environment might well be unique. Advances in astronomy and physics have given renewed support to the theoretical expectation that planets are normally formed when stars are born out of the gravitational collapse of rotating clouds of gas and dust. If this is the case, as most astronomers expect, planets may number in the hundreds of billions in our galaxy alone. On some of those worlds, water should exist in a liquid state, thus permitting chemistry to give rise to biology, as happened on Earth about 4 billion years (Ga) ago.

A Biological Universe?

Some astronomers and evolutionary biologists contend that the development of technological intelligence on Earth represents one of the only instances in nature where such a phenomenon occurred. In their view, the evolution of human intelligence was essentially a fluke—a freak accident of terrestrial evolutionary history that may not have analogs anywhere else.

Most scientists who have thought carefully about the possibility of life elsewhere in the universe believe that a multitude of evolutionary pathways are likely to exist for the emergence of tool-using organisms on suitable planets. In this view, anatomical and behavioral innovations that aid survival can spread so quickly in a population that

such factors as increased brain size and cultural activity could appear over and over again throughout the universe.

SETI, the Search for ExtraTerrestrial Intelligence, addresses the most fundamental hypothesis to be tested in the science of exobiology: that life is a natural product of cosmic evolution, governed by the same physical and chemical laws that operate throughout the universe. From this statistical perspective, living systems should be relatively common in the galaxy. Abundant indirect evidence exists supporting the hypothesis that given the right environment and enough time, a rich biota will emerge through chemical and biological evolution. Earth has been without life for only a fraction of its age. Individual fossilized single-celled microorganisms are found in rocks 3.5 Ga old on the 4.5 Ga Earth. All evidence gathered to date indicates that the main ingredients of life (the biogenic elements) and the processes that lead to the origin of life are widespread throughout the Galaxy and the universe. In at least one case, these processes led to complexity and intelligence.

Since the discovery of extraterrestrial intelligent life would validate these hypotheses, SETI researchers try to detect radio transmissions from technological civilizations in distant planetary systems. Many SETI workers maintain that if humanity had not evolved when it did, another species would have eventually emerged on Earth to fill the technology-dependent niche called "civilization" now occupied by *Homo sapiens*. Technology is largely a consequence of a species' innate problem-solving abilities, anatomical structures that can manipulate objects, competition, and the specialization of work during cultural evolution.

Early in its evolution, humankind developed culture as an adaptive strategy to control its environment and ensure an adequate food supply. With a secure environment, stable food sources, and higher population densities, cultural evolution—the transmission of information between individuals and generations—was then added to biological evolution. It was during the most recent phase of this process of cultural evolution that our communicative social species developed a technological civilization that is detectable over interstellar distances due to its activities in the radio spectrum. If other civilizations have also developed the ability to concentrate radiative energy in broadcast signals, the manifestations of distant technologies may be detectable if we point our antennas in the right direction in the sky and tune to the right frequency.

Several privately funded research groups are working to detect evidence of radio technology in other solar systems. The SETI Institute's "Project Phoenix" is designed to carry on the targeted search of NASA's defunct SETI program. The Planetary Society's "Project META" (Megachannel ExtraTerrestrial Assay) surveys northern and southern hemisphere skies, and preparations are underway to initiate "Project BETA" (Billion Channel Extraterrestrial Assay). Scientists at the University of California, Berkeley, are conducting "Project SERENDIP" (Search for Extraterrestrial Radio Emissions from Nearby Developed Intelligent Populations), and researchers at Ohio State University are using new generation SETI systems to continue the sky survey they have been carrying out since 1973.

The laws of physics are independent of the biology of any species that may exploit them. In other words, widely separated intelligent species with completely different biological and cultural histories could well have converged on similar technologies—such as radio—due to the universal nature of physical law. There is only one radio spectrum and its characteristics are the same everywhere in the universe. If there are many possible routes to technology-using behavior among the varied biologies of the galaxy, a significant number of other cultures may have discovered and utilized the radio spectrum at some point in their social and technological evolution.

If interstellar transmissions exist and can be detected with the tools of radio astronomy, we may eventually learn about the existence and nature of civilizations that may be millions of years older than ours. Moreover, we may even obtain, in some sense, a glimpse of a possible direction for our own future if information becomes available on what other intelligent species have accomplished. The successful detection of interstellar communications would settle the debate over the existence of extraterrestrial intelligent life once and for all. It would also end the cultural isolation of the human race.

Proof that biological and cultural phenomena are widespread would open a striking new chapter in humanity's intellectual development.

Bibliography

Bova, B., and B. Preiss, eds. *First Contact: The Search for Extraterrestrial Intelligence.* New York, 1990.

Drake, F., and D. Sobel. *Is Anyone Out There? The Scientific Search for Extraterrestrial Intelligence.* New York, 1992.

Goldsmith, D., and T. Owen. *The Search for Life in the Universe,* 2nd ed. New York, 1992.

BOB ARNOLD

F

FAMOUS CONTROVERSIES IN GEOLOGY

Geology emerged as a distinctive science only about two centuries ago. From its earliest days the subject has been associated with controversy. Focusing on controversy can be illuminating because issues tend to get dramatized and the underlying assumptions and attitudes of the protagonists often brought out into the open. Furthermore, attention is concentrated on the matters most critical to growth and development of a given subject. In this entry attention will be confined to four major controversies, which will be considered in historical sequence.

Neptunists Versus Vulcanists and Plutonists

ABRAHAM GOTTLOB WERNER taught mineralogy at the Freiberg Mining Academy in Saxony for forty years, but he was much more than a mineralogist because he outlined a stratigraphic scheme that was claimed to be applicable to the whole earth and in effect introduced the concept of universal correlation. Based on rock types rather than fossils, his scheme became discredited by the early nineteenth century, but this should not detract from his being recognized as a great pioneer in the study of historical geology.

In interpreting his rock sequence, from "primitive" igneous and metamorphic rocks to unconsolidated sediments, Werner adopted the widely accepted view that initially Earth had been enveloped by a primeval ocean covering even the highest mountains. All the older rocks we see today he perceived as being chemical precipitates from this ocean, even igneous ones such as granite. As the waters began to subside, rock formations were laid down that consisted partly of chemical precipitates and partly of mechanically deposited sediments. In the course of time these latter came to predominate, and the youngest, alluvial, deposits were laid down in lowland areas, having therefore only a local distribution; they were derived mainly from mechanical disintegration of older rocks. The emphasis on the role of water in Werner's theory led to its being called "neptunist." It experienced great popularity, especially on the European continent, because it appeared to account in a simple, satisfying way for a wide range of geological phenomena and because of Werner's eloquence in promoting it. It successfully bridged the gap between mere cosmogenic theories that postulated a series of episodes in Earth history without providing supporting evidence and the numerous but scattered sets of empirical observations by diligent scholars in several countries.

Nevertheless, there was no shortage of contemporary sceptics. For example, the Italian geologist Scipio Breislak argued that the volume of water now present on Earth was utterly insufficient to contain in solution or suspension all the solid materials of the crust. So where had all the water gone? Werner and his disciples were never able to give a satisfactory answer to this question. Breislak, along with others, thought that many of the geological phenomena in question could better be explained by advocating uplift of the land rather than fall of ocean level.

The major conflict was, however, to arise over another subject, the origin of basalt. Werner could not deny, of course, the existence of volcanoes, but he restricted their activity to very recent times. Since there was no "interior fire" to serve as a source heat, he argued that eruption of lava took place where basalt and other rocks were melted by the combustion of underlying seams of coal. He insisted that the basalt layers in hills in his native Saxony, interlayed with other strata, were chemical precipitates from the universal ocean. This interpretation was progressively undermined by the detailed researches of Nicolas Desmarest and a number of others in the Auvergne, France. It was possible to demonstrate an intimate relationship between the rocks comprising the young volcanoes of that region and layers of basalt nearby. In fairness to the Wernerians, Desmarest admitted that, confronted with the evidence of the Saxony hills alone, he could not have determined that basalt is volcanic in origin, and we must remember that petrographic study of thin sections was not established until many years later.

More fundamental opposition to the neptunist scheme was provided by Guy de Dolomieu and JAMES HUTTON. In 1789 Dolomieu argued that the granite which underlies the Auvergne volcanic rocks was not primordial but was underlain by rocks of a very different composition, which had penetrated the granite to give rise to basaltic lava. The volcanic hearth could not, therefore, be located in sedimentary strata containing combustible materials, and the heat source must lie at some considerable depth below the consolidated crust, in the realm of Pluto, the great god of the underworld. Contemporary research in Scotland by Hutton indicated that granite itself was igneous in origin, and in many cases had penetrated overlying sedimentary strata. Fire rather than water might be the key to a wide range of geological phenomena.

Hutton can be called the original plutonist,

and his position was more radical than that of the vulcanists, many of whom were happy to continue to use Werner's general stratigraphic scheme. He recognized intrusive granite veins in several places in Scotland, thereby conclusively establishing that some granite, at least, was younger than the enveloping country rock and hence could not be the primordial material of common belief. Hutton's work was largely ignored initially, but his disciple, JOHN PLAYFAIR, was much more influential. The first two decades of the nineteenth century witnessed progressive defections from the neptunist camp, and plutonism became firmly established.

Catastrophists Versus Uniformitarians

The terms "catastrophism" and "uniformitarianism" were coined in 1832 by William Whewell, but catastrophism fails to do justice to the cluster of beliefs that characterized the opposition to CHARLES LYELL's doctrine because what can be called "directionalism" was also an important component of their system. Uniformitarianism, as outlined by Lyell, is both a system and a method. The term has frequently been used as an exact equivalent of the continental term "actualism," and refers to the study of present-day processes as a means of interpreting past events. It is, however, perfectly possible to use actualistic methods and come to catastrophist conclusions.

Thus it was widely believed in the latter part of the eighteenth century that the most dramatic features of the landscape—high mountain ranges with strongly contorted rocks at their core, and the deep gorges and valleys that cut into them—could not be accounted for by present-day processes. Modern streams and rivers apparently contained too little water and did not flow fast enough to cut the valleys in which they flowed. It was far from obvious how the comparatively modest recent uplifts of land recorded, for instance, in the volcanic regions of Italy could have anything to do with the strongly tilted or contorted strata of the Alps and other mountain ranges. If Hutton deserves the epithet "founder of modern geology" it is less for his plutonism than for challenging head-on the widespread contemporary belief in such a fundamental decoupling of past and present, and for introducing the notion of immense, if not indeterminate, time.

The leading catastrophists of the early nineteenth century were the great French comparative anatomist GEORGES CUVIER and his leading geological disciple, Léonce Elie de Beaumont. Based primarily on his stratigraphic work in the Tertiary of the Paris Basin, Cuvier invoked a succession of catastrophes that not only disrupted strata and caused dramatic changes in relative sea level but resulted in mass extinctions of fauna. Elie de Beaumont followed Cuvier in arguing that folded and tilted strata implied sudden disturbance, and that one was not entitled to extrapolate to such "catastrophic" phenomena from the manifestly slow and gradual "causes now in operation." Long periods of quiescence were interrupted suddenly by relatively short-lasting upheavals of the land or torrential inundations of the sea.

A lively debate was engaged in England in the 1820s and 1830s. Leading catastrophists such as William Buckland and ADAM SEDGWICK promoted the so-called diluvial theory, which accounted for many geological phenomena by the action of the biblical flood. The detailed researches of George Scrope in the Massif Central demonstrated clearly, however, that a simple twofold classification into antediluvian and postdiluvian periods was untenable. All the geological agents required to account for the landscapes were still in operation, though with diminished energy. All that was needed was the invocation of immense periods of time for the agents to act. This became a principal argument put forward by Lyell, who saw no reason to invoke greater rates of geological activity in the past than observable somewhere at the present time.

While the diluvial theory was quickly abandoned a stubborn belief persisted that Lyell had overstated his case, and that there was indeed a direction to Earth history rather than a steady-state condition of the sort favored by Lyell. This was established most clearly from the fossil record, which indicated a kind of organic progression to more complex forms culminating in man, a fact that was later explained by Charles Darwin's theory of evolution by natural selection (*see* DARWIN, CHARLES). It is ironic that Darwin was one of Lyell's leading geological disciples, adapting Lyell's extreme gradualism to his own biological theory. In the last few decades Darwin's evolutionary gradualism has been challenged by a number of paleobiologists, while a school of "neocatastrophist" thought has emerged amongst geologists.

The Age of the Earth

By the middle of the nineteenth century the belief that Earth was a mere 6,000 years old was adhered to only by biblical fundamentalists. While the consensus of geologists, following Lyell, was that Earth was immensely older than that, there was a general reluctance to undertake even approximate estimates of age. The first serious attempt was made in 1860 by John Phillips, who adopted the cumulative thickness of strata as the best available measure of geological time. Though not a uniformitarian he accepted the practical necessity of assuming uniformity of depositional rate of sediment for his calculation. Using data from the Ganges Basin he arrived at an age estimate of nearly 96 million years (Ma) for the formation of Earth's crust. This calculation strongly challenged the Lyellian notion of virtually unlimited time.

Only a few years after Phillips published his age estimate the eminent Scottish physicist William Thomson, later elevated to the peerage as Lord Kelvin, attempted an estimate on a completely different basis. He had long been convinced of the notion, dating back to Descartes and Leibniz, that Earth was originally a hot, molten sphere that had cooled gradually. The assumption was made that the heat was transmitted solely by conduction and was derived ultimately from gravitational energy. The data required to apply the appropriate mathematics consisted of: (1) the internal temperature; (2) the thermal gradient at the surface; and (3) the thermal conductivity of rocks. Kelvin fully appreciated the uncertainties involved but believed that the data were good enough to provide a best estimate of 98 million years. This figure was remarkably close to that of Phillips, and Kelvin's work was initially well received by geologists. Darwin became very concerned, however, because of the limited time allowed for evolution, but "Darwin's bulldog," Thomas Henry Huxley, questioned Kelvin's underlying assumptions.

Later in the century a number of independent geological estimates were made. Most notable were those of Mellard Reade, who attempted to improve on Phillips's method and eventually ended up with a very similar result, and John Joly, who attempted to estimate the age of the oceans from their sodium content on the assumption that they were initially free of salt. Joly analyzed in detail a variety of complications to arrive at a final result of 80–90 million years, representing the time elapsed since Earth cooled below 100° C, assuming of course an initial molten condition; this was remarkably close to the estimates of Phillips, Reade, and Kelvin.

A growing opposition to Kelvin nevertheless emerged progressively, which became especially sharp when he became increasingly dogmatic, reducing his estimate by 1897 to a mere 24 million years. Osmond Fisher challenged Kelvin's assumption that the cooling of Earth took place by conduction through uniformly solid matter, pointing out that, if the interior were fluid, then convection currents might alter his results appreciably. John Perry agreed with Fisher and observed that if Earth's thermal conductivity was not homogeneous, as Kelvin assumed, but increased towards the center, then the age estimate must be increased. The American geologist Thomas Chamberlin speculated that there might be sources of energy locked up in atoms, of which nineteenth century physicists were completely unaware. The discovery of radioactivity in 1896 confirmed this prescient thought and within a few years Kelvin's assumptions had been completely undermined. With radiometric dating well established, it became generally agreed early in the twentieth century that Earth must be several thousand millions of years old.

Continental Drift

By the end of the nineteenth century a consensus had emerged among geologists that Earth had both cooled and contracted in volume through time, and many attempted to explain orogenic belts as the consequence of this contraction. Different European and American schools of thought can be recognized. The European school, typified by EDUARD SUESS and Emile Haug, believed that extensive sectors of ocean were underlain by subsided continent. Isostatic theory had more influence among the Americans, such as JAMES D. DANA and Thomas Chamberlin, who were impressed by the fact that gravity surveys appeared to indicate that the oceans were underlain by denser crust than the continents, implying that they were permanent features whose underlying crust was not interchangeable with continents. Both groups denied the possibility of any significant lateral movement of continental masses through the oceans, a phenomenon that would have been completely inconsistent with the stabilist Earth model.

Although he was not the first person to propose such lateral movement, the notion of continental drift is irrevocably associated with the name of the German meteorologist and geophysicist ALFRED WEGENER because he was the first to put forward substantial evidence for a coherent and logically argued hypothesis that took account of a wide variety of natural phenomena.

The basic idea seems to have come to him from being struck by the remarkable congruence of the coastlines on either side of the Atlantic, and finding paleontological evidence for a former land connection between Brazil and Africa. He challenged the cooling, contracting-Earth model on a number of grounds. It was not clear from the contraction hypothesis why the shrinkage "wrinkles" represented by fold mountains were not distributed uniformly rather than being confined to narrow zones. Further, some basic assumptions about Earth's supposed cooling had been undermined by the discovery of widespread radioactivity in rocks, leading to the production of considerable amounts of heat acting in opposition to thermal loss into space by radiation.

Wegener postulated that, commencing in the Mesozoic and continuing up to the present, a huge supercontinent, Pangaea, had rifted and the fragmented components had moved apart, creating the Atlantic and Indian Oceans. During the westward drift of the Americas, the western Cordilleran ranges had been produced by compression, as had the Alps and Himalayas as Africa and India had converged with Eurasia. The principal geological arguments deployed were the following:

1. There are some striking geological similarities on the two sides of the Atlantic, such as matching fold belts, that suggest a former continuity.
2. There are many late Paleozoic and Mesozoic fossils in common to the southern continents that are now isolated by ocean. Basic biological principles demand a former free land connection to account for this, but transoceanic land bridges that have subsequently subsided were held to be untenable on geophysical grounds because light "granitic" crust could not sink into the denser rock of the ocean floor. The only reasonable alternative was that the continents had once been joined and had since drifted apart.
3. At a later stage Wegener used arguments from paleoclimatology. For example, the convincing evidence of late Paleozoic ice sheets in South

America, southern Africa, Australia, and especially India was not consistent with the present dispersed continental configuration.

Wegener also put forward a variety of geophysical arguments in support of his revolutionary hypothesis, and thought he could prove significant horizontal movement of Greenland away from Europe by means of geodetic observations using radio time transmissions. He was unable, however, to propose a plausible mechanism to account for drift.

The initial reaction to Wegener's hypothesis was not uniformly hostile, but it became increasingly so in the years between the two world wars. A number of geologists challenged the accuracy of the supposed "jigsaw fit" of the Atlantic coastlines. Paleontologists persisted in arguing in favor of transoceanic land bridges despite the fact that the subsidence of such features was inconsistent with isostatic principles and the geodetic measurements were shown to be erroneous. The most formidable opposition came, however, from those geophysicists, most notably Harold Jeffreys, who insisted that Earth possessed too great a strength for continents to migrate across its surface. These scientists ridiculed Wegener's proposed mechanism to account for drift.

Wegener had a few supporters, however, the most notable of whom were the English geologist ARTHUR HOLMES, a pioneer in the establishment of a radiometric timescale, and the South African geologist ALEXANDER DU TOIT. Holmes indicated that the new knowledge of the heat-generating capacity of rocks with radioactive elements made subcrustal convection currents a distinct possibility, and these provided a plausible mechanism for drift. Du Toit demonstrated the striking similarities of late Paleozoic and early Mesozoic geology between the southern continents, which strongly suggested that they had formerly been adjacent. Despite their efforts, supporters of continental drift were generally dismissed as cranks, and by the middle of this century the hypothesis had been almost totally rejected by Earth scientists.

Research developments in two different fields transformed the situation. Thorough surveys of suboceanic topography revealed the existence of a mid-oceanic ridge system, consistent with tectonic tension, and surveys of the bottom, developing later into an extensive drilling program, revealed that the basalt underlying sediments of the ocean floor was surprisingly young, generally not older than Cretaceous. Work on rock magnetism was important in two different ways. Studies on the continents initially revealed polar wandering in the geological past, and then it was found that polar wandering paths differed between continents. The paths only came into concordance if the continental positions were reconstructed along the lines proposed by Wegener. The recognition of magnetic reversals allowed the direct testing of Harry Hess's seafloor spreading hypothesis, which is consistent with continental drift (*see* HESS, HARRY). This testing was first done in the Indian Ocean by Fred Vine and Drummond Matthews, and Hess's idea was well confirmed. These developments proved a great stimulus to further research, which led to the posthumous vindication of Wegener. The theory of plate tectonics, an outcome of continental drift, was put forward in the late 1960s and generally accepted by the Earth sciences community within a few years.

Bibliography

BURCHFIELD, J. D. *Lord Kelvin and the Age of the Earth.* London, 1975.

HALLAM, A. *A Revolution in the Earth Sciences: From Continental Drift to Plate Tectonics.* Oxford, Eng., 1973.

———. *Great Geological Controversies.* 2nd ed. Oxford, Eng., 1989.

IMBRIE, J., and K. P. IMBRIE. *Ice Ages: Solving the Mystery.* London, 1979.

LAUDEN, R. *From Mineralogy to Geology: The Foundations of a Science, 1650–1830.* Chicago, 1987.

LE GRAND, H. E. *Drifting Continents and Shifting Theories.* Cambridge, Eng., 1988.

RAUP, D. M. *Extinction: Bad Genes or Bad Luck?* New York, 1991.

RUDWICK, M. J. S. *The Great Devonian Controversy.* Chicago, 1985.

SECORD, J. A. *Controversy in Victorian Geology: The Cambrian-Silurian Dispute.* Princeton, NJ, 1986.

ANTHONY HALLAM

FAULTS, FOLDS, STRAIN, AND STRESS

See Earthquakes and Seismicity; Structural Geology

FERNS

See Pteridophytes

FERTILIZER RAW MATERIALS

Chemical fertilizers are additives whose purpose is to maintain or increase the fertility of soils. In a natural system, organic nutrients removed from soil by growing plants return to the soil when plant debris (dead leaves, stems, etc.) falls to the ground. However, when humans harvest plant material for food, animal feed, or timber and paper, the soil loses nutrients and requires addition of either chemical or organic fertilizer to remain productive. The most important nutrients that are depleted—nitrogen (N), phosphorus (P), and potassium (K)—are the major components of fertilizers. The three numbers on bags of commercial fertilizer (for example, 10-10-10) refer to the ratio of these three elements. Other important soil additives include sulfur (S), used to increase acidity, and lime (CaO), which reduces acidity. In addition, sulfuric acid is an important ingredient in the production of phosphate fertilizers.

Geological Sources of Fertilizer

The major geological sources for fertilizer were formed by chemical or organic precipitation from ocean (marine) or lake waters. Extreme evaporation of seawater forms potassium deposits, while evaporation of lake waters with unusual composition forms deposits of nitrogen minerals. Accumulation of plant and animal remains in oceanic and lake sediments is the largest source of the phosphorus deposits. The best quality sulfur deposits form through the alteration of gypsum ($CaSO_4$), which is also a product of seawater evaporation. Other important fertilizer raw materials include phosphorus deposits associated with special igneous rocks (called alkaline rocks) and nitrogen gas in the earth's atmosphere. Atmospheric nitrogen is the primary source of nitrogen.

Quantity of Fertilizer Used and How It Works

Total world production of the primary fertilizer components is as follows: potassium (as K_2O), 25–30 million metric tons; nitrogen (as contained N), approximately 100 million metric tons; and phosphorus (as P_2O_5), 46–51 million metric tons, derived from 150–160 million metric tons of rock. Of the total world production, approximately 95 percent of potassium, 94 percent of phosphorus, and 85 percent of nitrogen are used to make fertilizer.

Fertilizers must provide these elements to plants in a form that can be utilized. Usually, the geological source material must be converted into a more soluble form that can dissolve in soil water and be absorbed by plants. There are no substitutes for potassium, an essential ingredient in plant and animal diet. About 90 percent of the potassium added to any area is absorbed by plants, requiring fresh applications of K-bearing fertilizer after each harvest. Only a few plants can remove nitrogen gas from the atmosphere, so fixed nitrogen, chiefly as ammonium compounds, is an important soil additive.

Sources of Potassium (Potash)

Potash is a generic term used to describe all potassium ores and refers to any water soluble potassium additive for crops. Potassium deposits, primarily in the form of the mineral sylvite (KCl), form during the late stages of evaporation of seawater, much later than halite (NaCl, table salt) forms, and after almost all of the water has evaporated. Therefore, sylvite deposits are much less common and much smaller than those of halite. When separated from associated impurities (halite, clay minerals, and other salts), sylvite is applied directly to crops or converted to potassium-ammonium salts, which supply both potassium and nitrogen. Brines (extremely salty water) and seawater can also serve as sources of potassium. For example, Israel has developed technology to recover a number of elements, including potassium, from seawater. Other sources of potash include granite meal, green sand, kelp meal, and wood ashes.

Sources of Nitrogen

Nitrogen constitutes 78 percent of the earth's atmosphere, mostly as N_2 gas with lesser amounts of compounds such as ammonia, but it is rare in terrestrial materials. Notable exceptions are the large deposits of nitrate minerals that formed by evaporation of lake waters in the Atacama desert of Chile. Although large, these deposits are inadequate to meet world demand. The bulk of nitrogen used today in commercial agriculture comes from the atmosphere and is removed using the Haber-Bosch process. This process reacts nitrogen gas with hydrogen at temperature to produce ammonia by the following chemical reaction:

$$N_2(g) + 3\ H_2(g) = 2\ NH_3(g)\ \text{ammonia}$$

Natural gas serves as the source of hydrogen. Because natural gas accounts for approximately 75 percent of the cost of producing ammonia, changes in the price of natural gas affect nitrogen production. For example, the higher cost for deregulated natural gas contributed to a decrease in nitrogen production between 1980 and 1983.

Russia is the largest single producer of ammonia in the world, accounting for 21 percent of the total production, followed closely by China. The United States ranks third. Fertilizers account for 80 percent of ammonia consumption in the United States, as urea, ammonium sulfate, ammonium phosphate, and ammonium nitrate. Nitrogen and ammonia are also used in explosives, fibers, resins, and plastics. Organic sources of nitrogen include blood meal, fish meal, cottonseed meal, soybean meal, and animal manures.

Sources of Phosphorus

Guano deposits (accumulations of bird and bat droppings) were an extremely important source of phosphorus until early in the twentieth century. These deposits, found mostly on islands in the South Pacific, still account for about eight percent of world phosphorus production. The mineral apatite (Ca-phosphate with hydroxyl (OH), fluorine (F), or carbonate (CO_3)) contains 18 percent phosphorus and is the most important source today. Other sources of phosphorus include bonemeal and rock phosphate. The largest reserves of apatite are shales that formed in coastal oceans below areas of high planktonic activity. World reserves of phosphate rock were 12,285 million metric tons in 1991, almost two-thirds of which are in Africa, primarily in Morocco. Deposits located on the coastal plains of Florida and North Carolina and ancient coastal plains in Tennessee account for most (84 percent) of the phosphate production of the United States. In some cases, such as the Phosphoria Formation, which outcrops over a 643,000-square-km area in parts of Idaho, Montana, Wyoming, and Utah, phosphorus deposits also occur in the sediments of ancient lakes. However, deposits in the arid western United States lack the abundant water needed for ore processing.

Small amounts of apatite are usually present in igneous rocks, and apatite is especially concentrated in alkaline igneous rocks. Alkaline rocks contain relatively high amounts of sodium and potassium compared to silicon, resulting in the formation of low Si, and Na- and K-silicate minerals in place of quartz. Alkaline rock complexes on the Kola Peninsula, Russia; in the Palabora complex in Africa; and the Araxa and Jacupiranga complexes in Brazil are good examples. When these rocks are exposed to weathering at the earth's surface, the relatively resistant apatite is concentrated in the soils and in altered rock material overlying these alkaline complexes. Igneous apatites account for 17 percent of the world apatite production.

Recovery of the phosphorus begins with the separation of apatite from other minerals (mostly clay minerals, silica sand, and dolomite). Because apatite does not dissolve readily enough to make it useful as a direct plant additive, it is usually treated with sulfuric acid to form phosphoric acid (52% phosphoric oxide), which is in turn used to make ammonium phosphate. This process generates large quantities of gypsum as a by-product, some of which is packaged and sold but much of which is discarded, creating a disposal problem.

Sources of Sulfur

Sulfur is one of a few elements that can occur naturally in elemental form. This native sulfur most often fills fractures and cavities in the anhydrites ($CaSO_4$) associated with layered salt deposits that formed by evaporation of seawater. In some cases, the salt was forced upward through overlying sedimentary rocks to form salt domes. When crude oils came in contact with anhydrite, bacteria converted the anhydrite into native sulfur and calcite ($CaCO_3$). Another source of elemental sulfur is volcanic hot springs.

Native sulfur-bearing rock can be removed by

open-pit mining, or the native sulfur alone can be recovered using the Frasch process, which takes advantage of the fact that sulfur melts and is very mobile between 120°C and 180°C. Hot water enters the sulfur-bearing rock through the outer ring, producing molten sulfur that rises in the second ring. The innermost ring injects superheated steam into the rising sulfur so that it will stay molten. The resulting native sulfur is very pure.

Metal sulfide minerals are another important source of sulfur. Pyrite (FeS_2) is sometimes mined exclusively as a source of sulfur, and treatment of sulfide ores, mostly ores of copper, lead, or zinc, also produces by-product sulfuric acid. However, sulfuric acid from sulfide minerals is fairly impure (often containing significant amounts of arsenic, selenium, lead, zinc, mercury, etc.). In addition to mineral sources of sulfur, natural gas normally contains hydrogen sulfide and some crude oils contain sulfur.

World reserves of sulfur total nearly 1,400 million metric tons, with the largest deposits found in the former Soviet Union. Large deposits are also located in the United States, Canada, Iraq, and Poland. World production capacity for sulfur from all sources in 1991 was 68.9 million metric tons. Actual world production in 1991 was 55.6 million metric tons, with 22 percent from native sulfur (through both the Frasch and from mining), including 18 percent from pyrite, 14 percent from metallurgy (treatment of sulfide ores), and 43 percent from petroleum and natural gas.

Sulfur is used for many purposes, but production of sulfuric acid dominates, accounting for over 80 percent of sulfur consumption in the United States. Most of the sulfuric acid (67.5% of U.S. consumption in 1991) is used to produce phosphate fertilizer for agricultural purposes; in 1991, approximately 8.3 million tons of sulfur were used in the United States for production of phosphatic fertilizers. The remainder has a wide variety of industrial applications, including acid leaching of some copper and uranium ores, production of paper and plastic goods, petroleum refining, and chemical manufacturing.

Sources of Lime

Lime (CaO, calcium oxide) is formed when $CaCO_3$, in the form of limestone or other carbonate rocks or minerals, is heated to temperatures of 1,000–1,300°C. Because all developed nations, and most developing nations, have lime production facilities, world supply is virtually inexhaustible. World production in 1991 was 146 million short tons, with by far the largest production in the former Soviet Union and the United States.

Production and consumption of lime in the United States in 1991 was 17.3 million short tons. Agricultural uses consumed less than one percent of the total. Chemical and steel industries used about 66 percent, construction (soil and road stabilization) 8 percent, and most of the remainder in environmental uses (water purification, stack gas treatment for removing sulfur, sewage treatment, etc.).

Bibliography

Brookins, D. G. *Mineral and Energy Resources: Occurrence, Exploitation, and Environmental Impact.* Columbus, OH, 1990.

Craig, J. R., D. J. Vaughn, and B. J. Skinner. *Resources of the Earth.* Englewood Cliffs, NJ, 1988.

Kesler, S. E. *Mineral Resources, Economics and the Environment.* New York, 1994.

U.S. Department of the Interior, Bureau of Mines. *Minerals Yearbook.* Vol. 1, *Metals and Minerals.* Washington, DC, 1991.

U.S. Bureau of Mines. *An Appraisal of Minerals Availability for 34 Commodities.* Bulletin 692. Washington, DC, 1987.

JOSEPH L. GRAF, JR.

FIELD GEOLOGY

See Mapping, Geologic

FISH EVOLUTION

In 1992 the oldest undisputed fish fossils were identified from Lower Ordovician marine sediments in central Australia, dating about 480 million years old (480 Ma). These include scales and bone fragments of primitive jawless fishes (agnathans) having resemblances to other recently discovered more complete Ordovician fishes such as *Sacabambaspis* from Bolivia in South America. The oldest articulated fish fossils are from slightly younger rocks also in Australia, comprising up to four distinct species, including *Arandaspis* and *Porophoraspis*. These earliest agnathans were simple fishes up to 20 cm in length, had up to 15 paired gill openings along the sides of their simple bony shields, and had elongate trunk scales covering the tail. The earliest fishes appeared to have originated in the great southern supercontinent of Gondwana (Long, 1994).

The agnathan fishes radiated in the Silurian period, when five major groups appeared. All of these groups were largely marine, and most died out by the Middle-Late Devonian, with only lampreys and hagfishes, which probably evolved sometime in the Early Palaeozoic (but have no fossil record until the Carboniferous), surviving to the present. The Heterostraci (meaning "different shield") are a diverse group of extinct agnathans possessing a bony shield made up of several overlapping plates, similar to their Ordovician ancestors. Heterostracans thrived in the Silurian-Early Devonian, with only a few groups straggling on to the Late Devonian (e.g., *Psammosteus*). Osteostracans (meaning "bony shields") were another diverse group of jawless fishes that had a single-piece bony shield covering the head and a tail covered in thick plate-like scales. Well-known forms like *Cephalaspis* from Britain typify this extinct group that includes many strange Early Devonian forms, some species with well-developed rods of bone projecting from the front of the skull (e.g., *Boreaspis*). The Galeaspida (meaning "helmet shield") were a group of jawless fishes unique to China. They were similar to osteostracans in having a single-piece bony shield covering the head but are unique in having a large median opening in front of the eyes. Several forms evolved long tubular projections and broad wings on the armor (e.g., *Asiaspis*), while others evolved triangular-shaped shields (e.g., *Tridenaspis*).

Whereas most early fossil agnathans had thick bony armor there were two groups lacking extensive bony plates, the Anaspida and the Thelodonti. Anaspids (meaning "no shield") were eel-like, thin fishes sometimes bearing fragile trunk scales and sometimes having scaleless bodies. Thelodonts (meaning "nipple teeth," alluding to the shape of the scales) lacked body or head plates but had a covering of many tiny scales, each having a bony base and often pointed crown made of dentine, a dense mineralized tissue occurring in teeth. New finds of Silurian and Devonian thelodonts from Canada show that some forms had deep bodies with large forked tails, and a stomach was probably present (Wilson and Caldwell, 1993). Thelodonts inhabited marginal marine to inshore river and estuarine environments. The characteristic shapes and tissue types of the various scales, often abundantly preserved in sediment samples, enable geologists to use thelodonts for age dating of Silurian-Devonian sequences where other fossil groups are absent.

A new class of agnathan fishes, the Pituriaspida (named after the aboriginal hallucinogenic pituri plant), was erected in 1991 based on strange armored shields from the Lower-Middle Devonian of central Australia. They are characterized by a single-piece shield with long projections of bone at the front of the armor and paired openings near the eyes (e.g., *Pituriaspis*).

Current theories of agnathan evolution favor the heterostracans and hagfishes as the most primitive groups. Anaspids probably gave rise to the lampreys. The single-shielded forms, Osteostraci, Pituriaspida, and Galeaspida, are considered close relatives. Of all the agnathan groups, the Osteostraci are most widely regarded as the closest to jawed fishes as they share many advanced features of their anatomy. Thelodonts are still enigmatic as to their exact relationship (Forey and Janvier, 1993).

The jawed vertebrates (gnathostomes) make their first appearance in the latest Ordovician or earliest Silurian, based on primitive shark-like scales from Mongolia. However, as shark teeth do not appear in the sediments until the Early Devonian, it is possible that these earliest "sharks" may have been toothless or even jawless. The first definite occurrence of fishes with jaws and teeth is from the Early Silurian, when the spined acanthodians appear.

Acanthodians are characterized by fin-spines preceding all fins, and by many thousands of minute scales, each with a well-rounded, bulbous base. Early acanthodians, like the climatiids, evolved complex bony armor around the pectoral fins, while the last surviving group, the acanthodiforms, were toothless filter feeders with well-developed gill rakers. The acanthodians died out at the end of the Permian period. The interrelationships and evolution of acanthodians have been discussed by Long (1986).

The placoderms were a highly successful group of early armored jawed fishes that appeared in the Silurian and dominated the seas and rivers of the Devonian, becoming extinct at the close of that period. They are characterized by overlapping bony plates covering the head and trunk and usually articulating at a neck joint. They reached enormous sizes (up to an estimated 10 m for *Titantichthys*) and diversified into six distinctive major groups, the most successful being the arthrodires ("jointed necks"). One group, the ptyctodontids, had external clasping organs, as do sharks and rays, and this suggests a close relationship of placoderms to chondrichthyans.

Recent new finds of exceptionally well-preserved placoderms from the Gogo Formation in Western Australia are demonstrating other links to sharks, like the presence of annular cartilages in some arthrodires (Figure 1). *Bothriolepsis*, a highly

Figure 1. A perfectly preserved 370-million-year-old placoderm fish from Gogo, north Western Australia. Such fossil discoveries are providing much new information on the early evolution of fishes. (K. Brimmell, The Western Australian Museum, Perth.)

successful antiarch placoderm with external bony arms, is now known from over ninety species from every continent, spanning the Middle and Late Devonian, from marine to fluviatile deposits. Many species are known. They are widely used around the world for correlation of Devonian strata. New developments in placoderm studies include the recognition of major groups of antiarchs from China in Australia (family Sinolepididae), and new discoveries from northern Vietnam.

The sharks (Chondrichthyes) are cartilaginous fishes known from fossil teeth, with the earliest body fossils dating from the Middle Devonian of Antarctica and Australia (Figure 2). New discoveries of Early Devonian shark teeth from Spain, Saudi Arabia, Antarctica, and Australia point to an early radiation of the group occurring in Gondwana. By the Late Devonian there are some thirty different species of sharks known to us principally from their teeth and fin spines found all around the world. The appearance of major shark groups is continually being pushed back in geological time. The oldest representative of the Neoselachii, to which all modern sharks belong, is now thought to be Middle Devonian (Turner and Young, 1987). The largest predatory sharks were the giant *Carcharocles megalodon* that reached up to 15.9 m in estimated size. Their fossilized teeth, up to 15 cm high, are found around the world in Miocene and Pliocene Age deposits.

The bony fishes (Osteichthyes) are the most successful group of living fishes, with over 23,000 species of living teleost, their most advanced group. Primitive osteichthyans include several groups of ray-finned fishes (Actinopterygii) and lobe-finned fishes (Sarcopterygii), the latter group including the piscine ancestors of the first amphibians. The earliest fossil osteichthyans are from the Late Silurian of Europe and China and include primitive ray-finned fishes known largely from isolated scales. By the start of the Devonian period the lobe-finned groups had all appeared and flourished in the seas, invading freshwater habitats by the Middle Devonian.

The ray-finned fishes (Actinopterygii) were known from under ten species in the Devonian, radiating to over forty-seven families during the Late Paleozoic. The appearance of the first teleosteans, by the late Triassic, heralded the onset of modern fish faunas. Primitive teleosteans are recognized by many advanced characters, including a complex tail fin internal skeleton, as well as by hav-

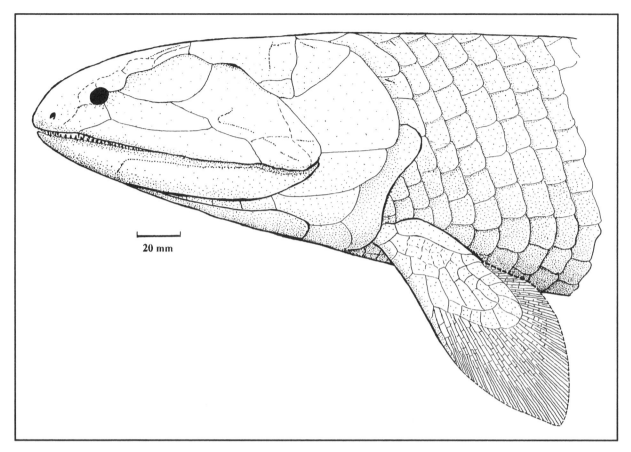

Figure 2. *Koharalepis jarviki,* a new crossopterygian described from the Devonian of Antarctica and representing a new family of osteolepiform fishes.

ing a unique feeding apparatus in which the jaws are freed from the rest of the cheek, enabling a wide range of feeding mechanisms to evolve.

The sarcopterygians include two major groups, the Dipnoi (or lungfishes) and the Crossopterygii (tassle-finned fishes). Fossil lungfishes are well-represented in the Early Devonian, with an ancestral form from China, *Diabolepis,* representing the link between lungfishes and other lobe-finned fishes. Lungfishes reached a peak of diversity in the Devonian, and by the Permian were restricted to terrestrial habitats and had achieved the ability to estivate—lie buried in the mud over dry summers to await the next wet season. Lungfishes all have powerful jaws with either hardened tooth-plates for crushing food, or denticle-lined palates for grasping food.

In recent years a number of new discoveries of perfect, three-dimensional skulls of Late Devonian lungfishes from the Gogo Formation, Western Australia, have been providing much new data on the early evolution of the group. Lungfish evolution slowed down dramatically after the Devonian period. The oldest known species of living vertebrate, the Queensland lungfish, *Neoceratodus forsteri,* on the evidence of its dentition, appears to have remained unchanged for over 100 million years.

Crossopterygians are characterized by a braincase ossified into two parts (ethmosphenoid and otico-occipital), and jaws with large conical fangs having a shiny enamel-like outer layer. The Actinistia, or coelacanths, represent the only living group, confirmed by a discovery of a living species, *Latimeria chalumnae,* in 1938, well after numerous fossil species had been described.

The relationships of the various lobe-finned fishes have been disputed in recent years by different scholars although most agree on a close relationship between osteolepiforms, rhizodontiforms and the tetrapods (four-legged land vertebrates), all of which have pectoral fins (the equivalent to the tetrapod "arm") with humerus, radius, and ulna bones present. The osteolepiform fishes radiated into many families during the Devonian and are typified by such well-studied forms as *Eusthenopteron* from Canada and Europe, and the Carboniferous form *Megalichthys*. New crossopterygians from the Middle Devonian Aztec Siltstone of Antarctica have been used to discuss crossopterygian interrelationships in detail. The Antarctic and Australian faunas include a new endemic family of osteolepiforms called the Canowindridae, containing the most primitive known members of the group.

Osteolepiforms probably gave rise to the first amphibians, and the transition from fish to land animals may have taken place in Gondwana, based on evidence from early amphibian fossils and trackways found in Australia. The panderichthyid fishes share a similar skull roof pattern, raised eyebrow ridges, and similar flat skull shape as seen in the earliest fossil amphibians. Furthermore, it has been shown that the retention of juvenile features into maturity (paedomorphosis) played a major role in the evolution of tetrapods from lobe-finned osteolepiform fishes (Long, 1990).

Bibliography

FOREY, P., and P. JANVIER. "Agnathans and the Origin of Jawed Vertebrates." *Nature* 361 (1993): 129–134.

LONG, J. A. "New Ischnacanthid Acanthodians from the Early Devonian of Australia, with Comments on Acanthodian Interrelationships." *Zoological Journal of the Linnean Society* 87 (1986): 321–339.

———. "Heterochrony and the Origin of Tetrapods." *Lethaia* 23 (1990): 157–166.

———. *The Rise of Fishes*. Sydney, Australia, and Baltimore, 1994.

TURNER, S., and G. C. YOUNG. "Shark Teeth from the Early-Middle Devonian Cravens Peak Beds, Georgina Basin, Queensland." *Alcheringa* 11 (1987): 233–244.

WILSON, M. V. H., and M. W. CALDWELL. "New Silurian and Devonian Fork-tailed 'Thelodonts' Are Jawless Vertebrates with Stomachs and Deep Bodies." *Nature* 361 (1993): 442–444.

YOUNG, G. C. "The First Armoured Agnathan Vertebrates from the Devonian of Australia." In *Early Vertebrates and Related Problems of Evolutionary Biology*, eds. M. M. Chang, Y. H. Liu, and G. R. Zhang. Beijing, 1991, pp. 67–85.

JOHN A. LONG

FISSION AND FUSION

See Energy from the Atom

FLOODS

Floods occur when runoff in a river exceeds the carrying capacity of the channel and overtops the banks, spilling water across the river floodplain. As natural phenomena, floods are important agents in transporting sediment, modifying the natural river channel, and creating floodplain adjacent to the river. The river floodplain is itself a natural safety valve for floods; in turn, floods are ecologically important in maintaining the complex assemblage of riparian wetlands along the river system. In stark contrast to this natural definition is the reality of damage caused by floods when rivers top their banks and destroy property and threaten lives.

The largest floods are caused by unusual meteorological conditions. For example, Colorado's Big Thompson Canyon flood of 1973 resulted from an extreme thunderstorm estimated to have produced 250 mm of rainfall over a single small area in just a few hours. The resulting flash flood that coursed down the narrow canyon caused 139 deaths and $35 million in property loss. In southern New England in 1955, two hurricanes within a nine-day period produced nearly 500 mm of rainfall in the foothills of the Berkshire Mountains west of Springfield, Massachusetts. The resulting floods were the greatest experienced in the region since the seventeenth century. The death toll exceeded 200 people and caused over $500 million in damages. The Mississippi River floods of 1973 and 1993 resulted from remarkably similar patterns of sustained rainfall that fell over a broad region for over half a year. In 1993, rainfall from January through June was up to two times the normal amount in the region of the upper midwest centered around the state of Iowa. These antecedent conditions saturated the soils within the river basin, filled the flood control reservoirs to capacity, and caused the river channels to reach their banks. Heavy rainfalls in July pushed the rivers over their banks and sustained the duration of this large regional flood. At Saint Louis, Missouri, the peak of the 1993 flood was the greatest since the flood of 1844, the historical flood of record. Damage estimates for the 1993 flood exceeded $10 billion.

To understand the risk of flooding, scientists attempt to calculate the probability that a flood of a given size will occur in any one year on a given river. The calculation is based on the historical record using the formula:

$$T_R = (n + 1)/m$$

where T_R is known as the return period, n is the length of record, and m is the rank of the flood based on its size. In the case of a ninety-nine-year record, the largest and number-one ranked flood would be the one-hundred-year flood. The inverse of the return period $(1/T_R)$ is the probability that the flood will occur in a given year. In the example of the one-hundred-year flood, the probability is 1/100 or .01 or a 1 percent chance. A common misconception is that the term "return period" implies that there is a waiting period between large floods. Rather, the chance that a one-hundred-year flood will occur is 1 percent each year, the chance of the fifty-year flood is one in fifty or 2 percent each year, and so on. In actuality, as Table 1 indicates, the chance of the one-hundred-year flood increases as time passes with its nonoccurrence such that there is a 68 percent chance of a one-hundred-year flood after an interval of one hundred years.

Because these statistical calculations are based on past history, they are very volatile and unstable estimates of the chance for big floods. For example, four of the largest floods on the Connecticut River at Hartford, Connecticut, have occurred during the past fifty years of an historical record that can be traced back to 1639, based on the diary accounts of the first European settlers.

Return period calculations are commonly graphed to illustrate the relationship between floods of different magnitude and their chance of occurrence for a given river. Figure 1 is the return period graph or flood frequency diagram for the Connecticut River at Hartford. In this example the magnitude of each flood is measured by the height or stage the water reached in Hartford. Prior to 1936, the largest flood of record on the Connecticut River was the flood of 1854. The graph illustrates that the 1854 flood was nearly equaled in 1927 and was exceeded in 1936, 1938, 1955, and 1984. The 1936 flood is now the historical flood of record. The effect on the flood frequency curve is to shift the graph upward; that is, statistically speaking, a flood of the magnitude of the 1854 flood now has a higher probability of occurrence because of the four larger floods that have all occurred in the last fifty years. Upward shifts in the slope of the flood frequency curve mean that the

Table 1. Chances of Occurrence of Floods of Different Magnitudes

Percentage Chance of Having One or More Similar or Bigger Floods in This Many Years					*Recurrence Interval of Flood*	*Relative Size of Flood*
100 years	50 years	25 years	10 years	1 year		
**	**	**	**	50%	2 years	smaller
**	**	**	**	20%	5 years	
**	99.3%	92%	63%	10%	10 years	
86%	63%	39%	18%	2%	50 years	
63%	39%	22%	9.5%	1%	100 years	
18%	9.5%	4.9%	2%	0.2%	500 years	
9.5%	4.9%	2.5%	1%	0.1%	1,000 years	larger

Note: adapted from Reiche (1973). Where values are not given, chances are greater than 99%.

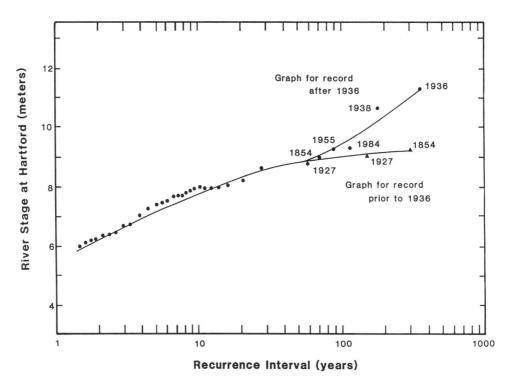

Figure 1. Flood frequency graph for the Connecticut River at Hartford, Connecticut. Flood magnitude is expressed as elevation or stage on the vertical axis and recurrence interval in years is on the horizontal axis. Each flood is plotted on the basis of a record that extends back to 1639. The flood frequency curve prior to 1936 is delineated by the plotting position of the 1854 and 1927 floods as shown by the triangles. After 1936 the curve was redefined by the magnitude of the 1936 flood. Note that prior to 1936, the return period for the 1854 flood was three hundred years. Today the 1854 flood is the fifth ranked flood in the historical flood series and has a return period of less than one hundred years.

risk of a given size flood has been underestimated. Changes in climate, and in land use and land cover within the river basin, also contribute to the highly variable character of flood runoff and make accurate knowledge of the risk of floods difficult to obtain. In some instances analysis of the geological record of flood deposits can be used to extrapolate and build a more complete flood history, a field of study known as paleohydrology.

Protection from flood hazards involves two distinct strategies: structural engineering works designed to prevent overbank flow of flood water and nonstructural land-management strategies designed to avoid the flood risk. Structural measures include the construction of dams, levees, and channelization projects. Dams are designed to increase the storage capacity for flood water within the river basin; levees are built to increase the capacity of

the river channel; and channelization projects are designed to speed the passage of the flood wave through the river system. The 1993 Mississippi floods revealed again that structural measures provide only a limited amount of flood protection and that it is not possible to replace entirely the natural storage of river floodplains with man-made structures. Flood control structures can be an additional liability when they fail, causing significant property damage and loss of life to the inhabitants of the river floodplain who placed their trust in the integrity of the structures.

Nonstructural approaches to flood risk are primarily accomplished through zoning and land management. In the United States, the National Flood Insurance Act of 1968 (Public Law 448) provides for subsidized insurance for individuals who live within flood hazard zones. The act requires that municipalities adopt a zoning program that regulates landuse within the limits of the one-hundred-year floodplain. Other governmental agencies have created zoning programs based on the elevation of the largest flood of record on a river. However, a line on a zoning map can cause the same false sense of security as a levee, causing development to crowd up against the line, and ultimately resulting in significant damage when a larger flood occurs.

How big can floods be? Long-term records in the United States indicate that the size of the largest floods, for a given river basin, continues to grow. Most hydrologists agree that this results from a longer and more accurate record of floods and not from a change to a climate more conducive to producing big floods. This upward trend in the size of floods warns us that the historical record is too short to have sampled the entire range of possible hydrologic conditions and that the upper limit of flood magnitude cannot be known.

Bibliography

BAKER, V. R., R. C. KOCHEL, and P. C. PATTON, eds. *Flood Geomorphology*. New York, 1988.

HOYT, W. G., and W. G. LANGBEIN. *Floods*. Princeton, NJ, 1955.

PARRETT, C., N. B. MELCHER, and R. W. JAMES, Jr. "Flood Discharges in the Upper Mississippi River Basin, 1993." *U.S. Geological Survey Circular 1120-A.* Washington, DC, 1993.

PAULSON, R. W., et al. *National Water Summary 1988–1989: Hydrologic Events and Floods and Droughts. U.S. Geological Survey Water Supply Paper 2375.* Washington, DC, 1991.

REICHE, B. M. "How Frequently Will Floods Occur?" *Water Resources Bulletin* 9 (1973):187–188.

PETER C. PATTON

FOSSIL FUEL USE, HISTORY OF

Fossil fuels—primarily biologically generated coal, petroleum, and natural gas—have served as primary energy sources for much of the world since the onset of the Industrial Revolution. They have served as the world's largest energy source for the past one hundred years and promise to do so well into the twenty-first century. The fossil fuels represent accumulations of organic matter that, as a result of geologic burial with resulting compression and heating, has been converted into combustible solids, liquids, or gases. Organic matter has been entombed in marine sediments at least since the beginning of the Paleozoic era some 570 million years ago and in terrestrial beds at least since the Devonian period 400 million years ago. Because natural geologic processes converted small proportions of this organic matter into fuels in periods of no more than a few million years, these fuels have been available throughout the history of humankind. Their use, however, dates back no more than about 4,000 years and large-scale purposeful combustion for heat no more than 1,000 years. Prominence for coal as a fuel came about with the beginning of the Industrial Revolution in the first half of the eighteenth century. Serious exploration of oil dates from the Drake Well (Pennsylvania) in 1859, but its large-scale usages awaited the development of the internal combustion engine in automobiles in the twentieth century. Local use of natural gas predates oil but large volumes were not available until after World War II, when pipeline systems permitted its transmission from the gas fields to the consuming homes and industries.

The different modes of occurrence, the different physical characteristics, and the different means of processing and distribution resulted in independent development histories for the various fossil fuels; these accounts are presented below. Although all the fossil fuels are biologically generated, they are considered as nonrenewable resources because they are presently consumed at rates much greater than they are being formed. The early histories of fossil fuel usage, in contrast to those of metals and some other mineral products, are difficult to reconstruct because the evidences no longer exist. Most of the evidences that were not burned away, washed away, or evaporated were consumed by oxidation and bacteria. There are no artifacts, carvings, or even broken fragments remaining as evidence; only a few charred camp fire remains.

Coal

There is general agreement that some of the Bronze Age peoples living in Wales 2000–1000 B.C.E. had discovered coal outcrops in the valleys and used some coal with wood in funeral pyres to cremate the dead. Along the remains of Hadrian's Wall, built across England in 121 C.E., there are evidences of coal having been used in camp fires. No doubt, there were scattered uses of coal in many places because coal beds crop out as a result of the natural weathering and erosion processes, and because coal is distinctive in its color and appearance.

The first significant traceable history of coal comes from the Anglo-Saxon Chronicle, which for the year 852 C.E. records a rent payment to an Abbott of "60 loads of wood, 12 loads of coal, 6 loads of peat." The use of these is not known, but it is probable that all were intended as fuel; peat, a nonfossil but potential precursor of coal, has long been burned as a fuel in England and Ireland. Little is known of coal consumption from that time through the Norman conquest in 1066, but it appears that its usage continued and expanded. By the thirteenth century coal, some gathered from the beaches of the east coast of England and some probably mined by hand from shallow pits, was being transported down to London by small ships. It reportedly burned inefficiently, giving off great smoke and terrible odors; the odors no doubt came, in part, from the burning of sulfur held in the coal as an organic constituent and as pyrite (FeS_2 or "fools gold"). The odors were so detestable that King Edward I banned the burning of coal as a fuel in London in the late 1200s.

Despite the rejection of coal in London, it continued to be dug and used at least locally in the coalfield areas of central England. Irregular records of coal shipments show that in 1377 more than 6,600 tons of coal were shipped from New Castle-on-Tyne; similar records exist for 1381 and 1465, so it is likely that coal was being mined and shipped continuously through this period. The Midlands coalfields developed irregularly, depending on the accessibility of coal seams, the proximity to water to facilitate shipping, and the desire of the landowner. By 1493, Sir Henry Willoughby of Nottinghamshire found it profitable to keep "goyng yerely five cole-pittes beside the level pit [i.e., an adit]." Between 1526 and 1547, these pits alone were producing 6,000 to 10,000 tons annually. Records remain scattered and incomplete but J. U. Nef in *The Rise of the British Coal Industry* estimates that total British coal production averaged 210,000 tons in the period 1551–1560, 2,982,000 tons in 1681–1690, 102,995,000 tons in

1781–1790, and 241,910,000 tons in 1901–1910. Britain's coal production dominated world output until the late 1800s; as late as 1868 Britain produced more coal than all other countries combined.

Coal production was spurred in part after about 1550 by the rising price and shortage of wood, the traditional fuel. Increased population, increased shipbuilding, and increases in many types of manufacturing placed great demands on Britain's forests and denuded great areas. Coal provided a cheap replacement that was a bit messy but gave greater heat per volume than did wood. As coal mining increased in the 1500s and 1600s the British mines grew deeper and more dangerous. The tragedies that killed countless miners were generally cave-ins resulting from weakened ground and methane explosions ignited by the torches and candles used by the miners to illuminate work areas. Children were employed in great numbers, especially to work in narrow seams where their small stature was an advantage. Groundwater flow was always a problem and frequently limited the depth to which mines could be driven; the development of the Newcomen steam engine in 1711 finally provided an efficient way to pump out water and allowed mining to greater depths.

The patenting of the blast furnace by Abraham Darby in 1735 marked a major shift in the use of coal from a fuel for heating to a fuel for iron production. This allowed for the cheap production of iron, gave a major impetus to the emerging Industrial Revolution, and spurred the demand for coal. All of this required the development of transportation and led to the development of the British canal system and later to railroads.

Coal was discovered in what would become the United States in 1673 by explorers near the site of Utica, Illinois, but no commercial mining for coal occurred until 1745, when a mine was opened near Richmond, Virginia. The great Pittsburgh coal seam in Pennsylvania was discovered around 1754 and became the major source of coal in America. Ohio coal beds were reported in 1755 and were being mined by 1770 when they were visited by George Washington. Little coal was used in the fledgling colonies because of the vast wood supplies. Most of the coal that was burned actually came from Nova Scotia and England. After the Revolutionary War, small coal mines were developed along the Appalachian coalfields from Pennsylvania to Kentucky and westward along the Ohio,

Illinois, and Mississippi rivers. The Americans addressed the transportation problems first with river barges and some canals and then with railroads. Railroads were a double boost for coal because they burned it as well as transported it. The American coal industry grew slowly but steadily, reaching one million tons per year in the 1840s. Even so, it was not until about 1900 that coal as an energy source surpassed wood in the United States; it remained the primary energy fuel in the U.S. until 1950, when it was surpassed by oil.

The American coal industry grew rapidly in the late 1800s as wealthy industrialists bought up mineral rights for the Appalachian fields and waves of immigrants arrived looking for employment. The mines were as dangerous as those in Britain and in the period from 1870, when record keeping began, until 1992, more than 100,000 miners were killed in U.S. mines. Methane explosions, though less common in the present day, still can kill many miners in a single blast. In fact, the worst mining disaster in history occurred when an explosion and fire killed 1,549 miners in Manchuria in 1942.

World coal production since the latter part of the nineteenth century has climbed steadily (Figure 1), the only downturns coming in the depression period of the early 1930s and during World War II. Production has climbed continuously since 1945 and shows no sign of changing. The early uses of coal, first as a means to heat homes and then as a fuel to power trains and industry, have disappeared in much of the world, but coal has become the principal means of generating electricity, a form of energy projected to continue to increase into the twenty-first century.

Petroleum

Petroleum, commonly called "oil," varies more widely in its physical characteristics and behavior than do the other fossil fuels. The earliest descriptions thus range from watery black seeps and films on water to nearly solid balls of asphalt. In nearly all of the major oil districts of the world there has been some seepage of oil to the surface where it may appear first as a low-viscosity fluid and then degrade by oxidation and loss of volatiles into more viscous oil, to tar, and ultimately to asphalt. Tar pits that served to trap prehistoric animals around the world attest to the presence of oil seeps long before humankind appeared.

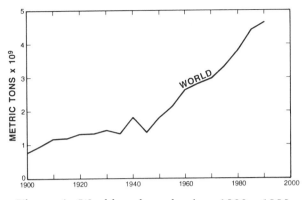

Figure 1. World coal production, 1900—1990.

The earliest documented reports of petroleum come from the Middle East about 2000 B.C.E. when writers described oil seeps, pools, and asphalt. The flammable nature of oil became well known, and the Roman writer Pliny is reported to have stated "its flammable nature, so closely related to fire, puts it far from being of any practical use." In fact, the Babylonians did find a practical use that did not make use of its flammable nature—they used it as mortar for bricks and as a water proofing. Even the biblical accounts of Moses floating in the reeds and of Noah's ark refer to the use of "pitch" (asphalt) to keep the vessels from leaking.

The native Americans recovered oil from seeps in western Pennsylvania and found that it had therapeutic properties for cuts and burns, information they passed on to the European settlers. The most important step in the history of oil came from those same Pennsylvania fields just before the American Civil War. Shortages of, and high prices for, whale oil, the most important illuminant for households, led a group of investors from New Haven, Connecticut, to search for oil as a possible substitute. Encouraged by laboratory distillation tests and overcoming several failures, they finally formed the Rock Island Oil Company and engaged Edwin Drake to drill on the bank of Oil Creek in Titusville, Pennsylvania. Drake used a converted steam engine and drilled at a rate of about 1 meter per day. In August 1859, oil seeped into the hole and the modern oil industry was born. The first barrels of oil were valued at $20 each but intensive drilling produced so much oil that within two years prices fell to $.20 per barrel. Distillation processes were successful in producing lamp oil, lubricating oils, tars, and gasoline. Gasoline had no practical use and proved to be dangerous because of its explosive nature; hence it was thrown away until the development of the internal combustion engine about forty years later.

The successes in Pennsylvania were soon matched in other parts of the United States with discoveries in West Virginia, Colorado, Texas, and California. Supply rapidly outpaced demand and the new industry soon found itself awash in oil. Many discoveries were significant but none more impressive than the discovery of the east Texas fields, ushered in by the gusher at Spindletop in 1901. It spouted 100,000 barrels a day, and rapidly constructed dikes held pools with more than three million barrels before the well was capped. Impressive dicoveries were also made in other parts of the world but foreign commercial production was very small relative to that of the United States until after 1900 (Figure 2). Coal was the dominant energy source for homes, factories, trains, and electric power plants, but oil gradually made inroads, especially in the young but growing automobile industry. The growing demand for gasoline and diesel fuel began to catch up with excess supplies so that a tight supply raised prices and stimulated exploration.

Oil had been known in the Middle East for millennia but the remoteness of the area and adequate supplies from other sources had limited exploration. The first significant discovery was made by William D'Arcy in Persia in 1908, but development was limited until the first of the big fields was found in 1938. With another world war threatening, the Allies recognized the region's potential and were determined to keep such a valuable resource out of Axis hands.

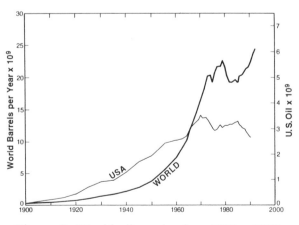

Figure 2. World oil production, 1900—1990.

Oil, of course, did play a major role in the history of World War II. Even before the war, Japan had moved into China and into Southeast Asia to procure oil supplies. In 1940 President Roosevelt placed an embargo on the sale of oil to Japan from the West. The Japanese military recognized their precarious situation—holding only eighteen months of oil supplies in reserve—and believed that a preemptive strike at Pearl Harbor would keep the United States out of the war and from menacing Japan's oil supply lines from Southeast Asia. Germany's fate was also strongly tied to oil. The Allies expended great efforts to destroy the oil supplies in Romania and, in the end, the famed North African desert campaign led by German Field Marshall Erwin Rommel ran out of gasoline and was defeated. Winston Churchill, recognizing the vital part played by oil, said, "the West floated to victory on a sea of oil."

The postwar boom led to a rapid expansion in the demand for oil products, and world production nearly doubled in the 1950s and more than doubled again in the 1960s. A historic seed was planted in 1960 when, amidst an oil glut, the major foreign oil producers in the Middle East cut the revenues paid to local governments. Representatives of Venezuela, Saudi Arabia, Iran, Iraq, and Kuwait met and formed OPEC (the Organization of Petroleum Exporting Countries) to protest the action and as a means of promoting their own oil revenues; subsequently membership grew to thirteen countries. OPEC was little noticed although its share of world oil production continued to rise until 1973 when, in response to the Arab-Israeli war, all OPEC countries followed the Arab lead by placing a boycott on oil shipments to the West and then tripling oil prices. This shocked the West and promoted conservation and the consideration of alternative energy sources and resulted in a short-lived drop in world oil production. When the shock wore off oil demand and production rose again, but this time with OPEC raising prices steadily to a maximum of nearly $40 per barrel (in comparison to $3 in 1972 and about $10 in 1974). This again brought about a reduction in demand and brought many other producers into the oil market; the subsequent glut resulted in a dramatic drop in oil prices to below $10 per barrel in December 1985. Subsequently, oil prices have remained in the $15–$20 range, OPEC has struggled for accord, and world demand has continued to rise. Iraq's invasion of Kuwait in 1991 and the uncertainty of oil flow from the Middle East briefly raised oil prices and concerns, but the Allied victory over Iraq rapidly alleviated concerns.

Proven oil revenues at the end of 1992 were approximately 1,000 billion barrels, with annual world production at about 23 billion barrels. This represents a 43-year supply; however, more than 75 percent of the reserves are held by OPEC members.

Natural Gas

Natural gas, composed almost entirely of methane (CH_4), is generated by methanogenic bacteria in swamps, bogs, and other surficial environments but also as a result of the same burial and heating processes that produce oil from organic matter. Hence, it was no doubt encountered by our ancestors thousands of years ago. The first known usage was in ancient China, where it was piped through bamboo tubes and burned to evaporate saline brines. By 600 C.E. eternal flames were kept burning near Baku on the west coast of the Caspian Sea by using gas issuing from cracks in the rocks.

The use of gas as an illuminant began not with natural gas but rather with distilled or manufactured gas prepared by heating coal, wood, and peat in Europe in the early 1600s. In the early 1800s William Murdock demonstrated the utility of gas lights by installing them in a cotton mill. This success led to the development of commercial gas light companies in London in 1812 and in Baltimore in 1816.

The use of natural gas traces its origins to French missionaries who reported seeing seeps of gas on fire in the northern United States area in 1775. George Washington reported "burning springs" near Charleston, West Virginia that same year. Commercialization started in the 1820s after William Hart, a gunsmith in Fredonia, New York saw fire set to rising bubbles in a well. He recognized the potential and constructed a small network of wooden pipes that serviced 66 gas lights. The success led to more wells and ultimately to the formation of a natural gas distribution company in Fredonia in 1865.

Meanwhile, natural gas was found also with the oil at Titusville, Pennsylvania, in 1859, but there was no local market or pipelines. Manufactured gas was more practical because it could be produced where needed. Nevertheless, a 40-km wooden pipeline was finally constructed in 1872 to

serve customers in Rochester, New York, and a 9-km steel pipeline carried gas to Titusville. The fledgling industry nearly died when Thomas Edison's electric light bulb appeared in 1879, but the use of natural gas for heating more than made up for the loss as an illuminant.

The usage of gas grew slowly but remained localized around the source areas. A major impetus came with the discovery of the large gas fields in Texas, Oklahoma, and Louisiana in the early 1900s, and by 1925 there were 3.5 million gas customers in the United States. The late 1920s and 1930s saw the development of a high-pressure seamless pipe that could be easily constructed and reliably used. Although it was known from the world's major oil fields, gas remained primarily a U.S. commodity with a slow but steady growth (Figure 3). The post-World War II economic boom and the rapid spread of suburbs provided a powerful stimulus that saw the rapid growth of gas consumption in the United States and worldwide starting about 1950. American growth began to falter in the early 1970s, but worldwide growth has continued with only a small hesitation in the early 1980s, when high prices resulted in some cutbacks.

The great oil fields of the Middle East also contain large quantities of natural gas, but its distance from markets made its use impractical until the development of liquid natural gas (LNG) facilities and ships in the late 1970s and 1980s. The cooling of the gas to below −162°C liquefies it and reduces its volume to 1/1600th of its gaseous state. This made the process economic, and now large LNG facilities are present in many Middle Eastern areas.

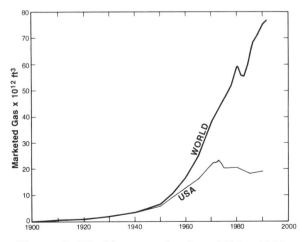

Figure 3. World gas production, 1900—1990.

Beginning in the early 1960s, the Soviet Union discovered great reserves of natural gas in Russia and developed pipelines to become a major supplier for Western Europe. The breakup of the Soviet Union slowed progress, but in the early 1990s Russia was working with Western companies to build additional infrastructure for increasing gas sales.

Coal bed methane began to develop as a major gas source in the United States in the late 1980s. It was recognized that the coal beds in many areas contained large quantities of extractable methane gas held in the coal structure. Consequently, what had always been the greatest threat to safety in coal mines has become an economic resource. Unfortunately, this does little to enhance coal mine safety because most coal bed methane is extracted from relatively low quality, deep seams that are probably not economic to mine.

Bibliography

CRAIG, J. R.; D. J. VAUGHAN; and B. J. SKINNER. *Resources of the Earth.* Englewood Cliffs, N.J., 1988.

NEF, J. U. *The Rise of the British Coal Industry.* Vol. 1. London, 1966.

PEEBLES, M. W. H. *Evolution of the Gas Industry.* London, 1980.

YERGIN, D. *The Prize: The Epic Quest for Oil, Money, and Power.* New York, 1991.

JAMES R. CRAIG

FOSSIL FUELS

Fossil fuels are naturally occurring substances such as coal, petroleum, or natural gas that are remnants of, or are derived from, organisms of a past geological age. These substances can be consumed or oxidized to produce energy. In contrast, synthetic fuels (synfuels) are produced through man-made distillation of fossil organic matter. Common sources of synfuels are oil shale and coal.

Fossil fuels share a common origin in the process of photosynthesis, which is the basis for mass production of organic matter (Tissot and Welte, 1978). Less than 1 percent of the organic matter produced through photosynthesis each year is buried in sediment (Durand, 1980). Photosynthesis is the process by which plants convert sunlight to chemical energy in chlorophyll-containing cells. In this process inorganic compounds, such as carbon dioxide and water, are first converted to simple organic compounds, such as glucose, and then to more complex biomolecules, such as cellulose and starch. Accordingly, fossil fuels can be considered to be ancient solar energy that has been locked away in deposits of coal and accumulations of petroleum and natural gas. Winchell (1886) writes, "in this flame [speaking of coal], the old warmth reappears; it is the warmth of the sun which shone in the Carboniferous Age."

Interrelationship of Fossil Fuels

Plant remains are the basis of fossil fuels, and differing types and amounts of vegetation produce various organic compounds in differing amounts. For example, terrestrial plants produce more oxygen-rich organic molecules, such as lignin in wood, which they require for structural support on land. In contrast, aquatic algae produce more hydrogen-rich organic matter for buoyancy in water. Thus differences in the type of plant producing the organic deposit result in differences in the type of fossil fuel produced.

Coal, petroleum, and gas can be found associated with or contained in sedimentary sequences. Depending on the type of organic materials in sediments, products from similar conditions that are subjected to increased temperature and pressure will yield either liquid hydrocarbons or gases. These changes result in an organic residue or high-rank coal that can be used as a fuel but may require a higher ignition temperature than lower-rank fuel. In all cases, fossil fuels can be used directly as fuels or indirectly as sources of other fuel-related products.

Organic matter formed from the remains of vegetation, either terrestrial or aquatic, can accumulate as sedimentary units such as peat or mud and, if buried, can be altered to coal or dispersed sedimentary organic matter. Subsequent chemical changes of the organic matter caused by increased temperature and pressure can follow one of two paths to the formation of either coal or carbonaceous rock. The process is called either coalification or catagenesis/metagenesis. The coal path leads to changes in the basic organic components (macerals) in peat to sequentially higher ranks of coal (such as lignite, bituminous coal, and anthracite) and can yield expulsions of gas and liquid hydrocarbons. The path to carbonaceous rock results in changes to organic components (kerogen, bitumen, macerlas) in the mud and to the expulsion of crude oil and natural gas; ultimately a residue of carbon and graphite remains in the rock. The products along each path are characterized as stages resulting primarily from incremental increases in temperature. In coal, the result is decreasing hydrogenation and increasing carbonization. With petroleum, the result is the thermal cracking of the large kerogen and bitumen molecules into smaller ones of lower molecular weight and, hence, lower viscosity.

The products of this evolution can be altered at any stage through a number of mechanisms: (1) biodegradation, whereby bacteria consume the lighter components, particularly the paraffin hydrocarbons; (2) waterwashing, which results in removal of the water-soluble components; (3) inspissation, the simple evaporation of the lighter components when oil is exposed at or close to the earth's surface; and (4) deasphaltening, whereby the very large asphaltene molecules are precipitated from the oil by natural gas, often collecting at the base of the reservoir as a tar mat.

Because coal, bitumen, and petroleum are effectively a hydrocarbon continuum, their classification is arbitrary, but it is based upon fundamental physical and chemical characteristics. In coal the hydrocarbon molecules are carbon rich, whereas in petroleum the molecules are hydrogen rich. The main cause of coalification or catagenesis for either coal or petroleum is the rising temperature resulting from the effects of depth of burial. The depth-temperature relationship that affects coalification may vary from place to place depending on geothermal gradient (the increase in temperature with depth). The stages of catagenesis and coalification are closely marked by the degree of reflectance (the measured amount of light reflecting from a flat, polished surface) of the coal maceral vitrinite (a plant-derived organic component), which predominates in coal and is commonly incorporated in sediments. The reflectance increases irreversibly with increasing temperature. Increasing coalification-like effects on organic-rich sedimentary rocks (containing hydrogen-rich macerals) can produce oil followed by the generation of natural gas, liquids (condensate), and methane gas.

Coal

Megascopic examination reveals the plant origins of coal. In some deposits, large coalified trees can be observed with the unaided eye (Plate 20).

Coal is defined as a brown to black combustible sedimentary rock composed principally of consolidated and chemically altered plant remains (ASTM, 1992); coal also can be yellow (brown coal). Coal is composed of water, minerals, and organic materials or macerals. Plant remains, such as wood, leaves, and spores, once preserved in peat, are primarily altered chemically in coal but retain suggestions of their plant origin when viewed microscopically (Plates 21–23). Water is expelled during coalification, and the organic chemical structure of macerals changes with the effects of coalification. Coal beds, or seams, are transformed deposits of peat that accumulated in environments such as wetlands, swamps, bogs, or mires. As a practical matter, in its natural state, coal contains more than 50 percent organic matter and water, thereby distinguishing it from carbonaceous shale, which contains less than 50 percent organic matter and water. Coal and carbonaceous shale are often interlayered, and the difference between the two in composition can be gradational.

As a fuel, coal is commonly burned in steam boilers that turn turbines for the generation of electricity. One product derived from coal is coke, which is produced by the destructive distillation of coal and is used in the manufacture of steel. By-products of coke making include gases, used during the late nineteenth century for lamps; crude light oils, sources of chemicals such as benzene, toluene, xylenes, and naphtha; tars, sources of acids, creosote oil, naphthalene, and pyridine bases; and ammonia, which can be used to make fertilizer. Gas, predominantly methane, can be extracted from coal or can be found naturally in great quantities in coal beds, where it is called coal bed gas, and can be extracted for use as a fuel. Light oils and tars can be extracted from coal and used directly as fuels or as feedstock in the manufacture of goods such as dyes. The industrial conversion of coal to liquid hydrocarbons suggests that coal could have been a source rock for some naturally occurring liquid hydrocarbons.

Petroleum

Although similar in color to coal, petroleum is a liquid, and contains no visibly recognizable primary components. Petroleum is a natural, brownish yellow-to-black, thick, flammable, liquid-hydrocarbon mixture found principally beneath the earth's surface and can be processed for gas, gasoline, naphtha, kerosene, fuel and lubricating oils, paraffin wax, asphalt, and various derivative products used in the manufacture of goods. Petroleum is best classified chemically on the basis of the amount of paraffinic, napthenic, or aromatic hydrocarbon compounds comprising the oil.

Bitumen, which occurs at or near the earth's surface in a number of forms, ranges from highly viscous liquid to semisolids and solids; it is a naturally occurring substance that is soluble in carbon disulfide and is composed principally of a mixture of hydrocarbons that are substantially free from oxygenated bodies.

Other bitumens include tars, natural mineral waxes, asphalts, and asphaltites. At least 99 percent of bitumen is in the form of natural asphalt, a very viscous liquid with a specific gravity approaching or exceeding that of water. Petroleum, a type of bitumen, is commonly found in reservoirs into which it has migrated from some source rock that was rich in dispersed, hydrogen-rich sedimentary organic matter. The insoluble substances, or kerogen, are fossilized insoluble organic material found in sedimentary rock that, through distillation, yield petroleum products. The kerogen in the source rock comprises mostly large, heteroatomic molecules (macromolecules) that break up with increasing temperature into small, lighter (lower molecular weight) molecules.

The paleoenvironments of the sediment that yields petroleum is usually marine or lake shores where rivers contribute organic matter and fine grained sediments—muds and silts—from land. Concurrently both algal and planktonic plant life were abundant in the ocean as a result of upwelling nutrients from the seafloor. As a result, organic matter was incorporated in the muds and silts. Rapid burial prevented much of the organic matter from being oxidized and destroyed. These organic matter–rich muds and silts were later consolidated into petroleum source beds. During the first stages of diagenesis, only nonthermal biogenic gas evolved (mostly CO_2 and CH_4). As the source beds were slowly buried under accumulating layers of sediment, the process of maturation advanced into the stage of catagenesis.

Gas

Gas can be found naturally in two forms: (1) natural gas, which is a mixture of hydrocarbon gases occurring with petroleum deposits, consisting principally of methane together with varying quantities of ethane, propane, butane, and other gases in which form it can be used directly as a fuel or in the manufacture of organic compounds; and (2) coal bed gas, which is primarily methane but also can contain minor amounts of ethane. In this case, the coal bed is both source rock and reservoir. Coal bed gas can be produced biogenically (through the action of bacteria) or by thermal processes (cracking). The latter process generally produces greater amounts of gas, but the coal must have been coalified at least to high volatile bituminous rank. Biogenic processes, which do not produce vast quantities of gas per volume of coal, can account for large resources of gas if coal beds are very thick. Generally, biogenic gas is predominantly methane, whereas thermogenic gas may include heavier hydrocarbons such as ethane, propane, and butane.

Bibliography

ASTM (AMERICAN SOCIETY FOR TESTING AND MATERIALS). *D 121: Standard Terminology of Coal and Coke.* 1992 Annual Book of ASTM Standards, vol. 5.05. Philadelphia, 1992, pp. 163–170.

DURAND, B. *Sedimentary Organic Matter and Kerogen. Definition and Quantitative Importance of Kerogen.* In *Kerogen: Insoluble Organic Matter from Sedimentary Rocks,* ed. B. Durand. Paris, 1980, pp. 13–34.

TISSOT, B. P., and D. H. WELTE. *Petroleum Formation and Occurrence: A New Approach to Oil and Gas Exploration.* Berlin, 1978.

WINCHELL, A. *Walks and Talks in the Geological Field.* New York, 1886, pp. 135–154.

RONALD W. STANTON

FOSSIL RECORD OF HUMAN EVOLUTION

Have you ever wondered how modern people have come to look and behave the way they do today? How did ancient pre-human ancestors live? What kind of food did they eat? And, what did they look like? These questions are fundamental to understanding how and where modern humans fit into the biological and physical world. The only direct evidence about earlier forms of life on Earth are fossils. They can reveal surprisingly detailed insights into the type of plants and animals that once lived on this planet (*see* FOSSILIZATION AND THE FOSSIL RECORD; TAPHONOMY; PALEONTOLOGY). Fossils can also reveal clues regarding the appearance, behavior, and lifeways of ancient human populations. Fossils are the preserved remains of once-living organisms, and scientists that study fossils are known as paleontologists. When the study of human fossils (human paleontology) is intimately coordinated with the study of cultural artifacts (tools, pottery, food refuse) by archaeologists, the interdisciplinary approach to human ancestry is known as paleoanthropology.

Two pieces of information crucial to the interpretation of any fossil include: (1) a specific description of the find site; and (2) a precise estimate of the antiquity of the fossil. The find site of a fossil is usually plotted on both topographic and geologic maps as well as positioned in a drawing of the stratigraphy of the find site. The antiquity of a fossil may be assessed by relative or absolute methods. Relative dating techniques are based on the law of superposition (*see* STRATIGRAPHY and GEOLOGIC TIME), and permit judgments of one fossil being older, because it is deeper stratigraphically than another, or younger, because it is found in higher levels of a stratigraphic column than another fossil. Absolute methods embrace a wide range of radiometric techniques that often rely on the decay rates of radioactive isotopes (*see* GEOLOGIC TIME, MEASUREMENT OF). These methods include potassium (K)/argon (Ar), argon/argon, uranium (U)/thorium (Th), radiocarbon, thermoluminescence, and electron spin resonance. Each of these methods permits a measurement of the antiquity of a rock or fossil in number of years before present and has an associated error of measurement factor expressed in $+/-$ years. A number of techniques are generally combined in assessing the age of a fossil deposit, some of which rely upon the integration of relative and absolute techniques, including magnetic signals preserved in rock, and associations of particular types of nonhuman fossil animals, for example. Since the discovery of Neanderthal fossils in 1856 in Germany, knowledge of the earlier stages

of human development has progressed episodically. Factors influencing our understanding of human evolution include the rate, location, and completeness of individual fossil discoveries. For example, the place of human origins has shifted from Europe (late nineteenth century), to Asia (early twentieth century), to Africa (mid to late twentieth century) as fossil discoveries of key stages in human ancestry were made in different continents. Some fossils are poorly preserved or incomplete, and they raise more questions than they answer. Other discoveries call well-established beliefs about our ancestors into question, or prove earlier views incorrect. The science of paleoanthropology is an exciting one because new fossil evidence is continually being discovered and because, over the years, you can see the impact fossil discoveries have for improving knowledge of our origins. This review of the fossil evidence for human evolution describes the stages of human ancestry and highlights key issues in the study of human evolution.

Stages of Human Evolution

This summary of the principal steps in the ancestry of modern humans is designed to provide a description of the physical appearance, behavior, and lifeways of each stage of development. While it is fascinating and enlightening to know these details, it is perhaps more interesting to consider questions about how and why these stages come into existence and why some, and not others, became extinct.

The four primary stages of human evolution are: (1) *Australopithecus;* (2) *Homo habilis;* (3) *Homo erectus;* and (4) *Homo sapiens.* These stages are named after the genus and species to which fossils have been allocated. A context for understanding the uniqueness and origins of the *Australopithecus* stage of human evolution requires some consideration of an antecedent phase known as the Miocene "ape" stage. Geographically, human origins and evolution occurred in tropical environments of the Old World during a period spanning the past 20 million years (20 Ma). This time frame includes three geological epochs known as the Miocene, Pliocene, and Pleistocene (Figure 1).

Miocene "Apes." During the seventeen million years of Earth history known as the Miocene (22–5 Ma), apelike animals were much more diverse and

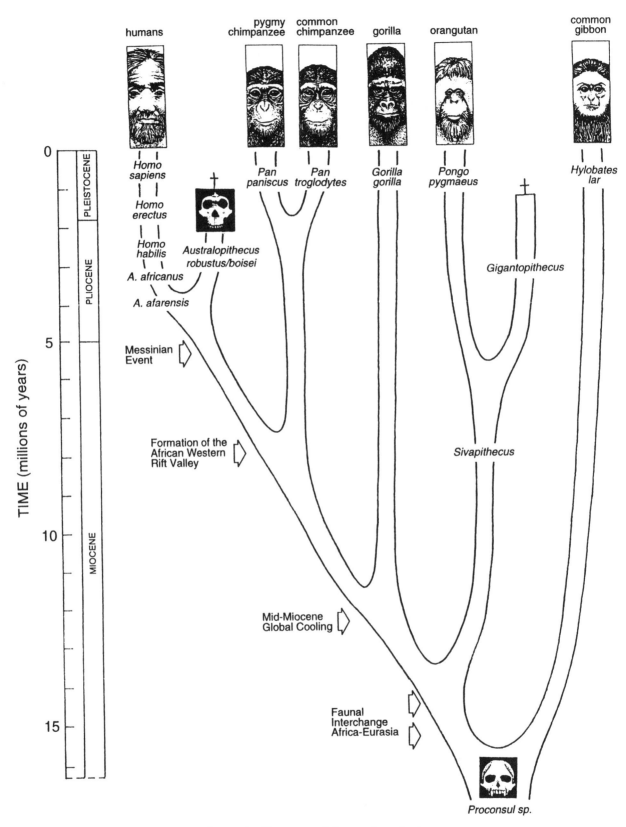

Figure 1. Human origins and evolution.

abundant than the apes alive today (chimps, gorillas, and orangutans). "Apes" of the Miocene, while apelike in general appearance, were not exact replicas of modern-day apes in anatomy, diet, movement, or behavior. Early Miocene apes are found primarily in East Africa between 22 and 17 Ma and are best represented by the genus *Proconsul*. Many genera and species of early Miocene apes are known to have inhabited the plains and forests of this volcanically active region. With a range of body sizes, locomotor patterns, and predominantly fruit diets, these early Miocene apes are thought to be the general ancestral source of a mid- to late-Miocene array of apes in Eurasia. The dispersion of apes from Africa into Eurasia was permitted by continental plate movements that established an Arabian bridge at about 17 Ma, resulting in animal migrations between Africa and Eurasia (Figure 1). These early apes were one small component of the complex intercontinental exchange of faunas. The predominantly Asian "ape" *Sivapithecus*, well known from Asia Minor (Turkey) to Pakistan to southern China, is regarded as the ancestor of the living orangutan. The genus *Sivapithecus* is also the probable ancestor of the very large apelike form known as *Gigantopithecus*. This giant ape displays many human features, lived between about 5 Ma and 500 thousand years ago (Ka), and is represented exclusively by fossil teeth and jaws from North India and southern China (Figure 1).

Australopithecines. While the exact Miocene "ape" stock from which *Australopithecus* developed remains obscure, our current picture of australopithecines is very vivid. As a group they are known from sedimentary deposits in south (caves) and east (lakeshore, riverine) Africa, between 5 and 1 Ma ago. An evolutionary mosaic, australopithecines combine some modern human features with traits that are typically apelike. Fully capable of walking in a human bipedal mode, the skull of *Australopithecus* is small-brained and large-toothed (apelike). This pattern is what would be expected of a "missing-link." The bipedal pattern of australopithecine locomotion is known from fossils of the pelvis; fossil hip, knee, and ankle joint surfaces; fossil hand and foot bones; and the impressions of footprints. *Australopithecus* fossils are generally grouped into two categories, robust and gracile, though by modern human standards both forms appear to have been muscular and capable of great

strength. Robust australopiths display massive jaws and cheek teeth, small anterior teeth (incisor and canine), bony crests for anchoring large chewing muscles, and brain cases the size of living apes. This anatomy, and evidence of microscopic dental wear and enamel chipping, suggests a vegetarian diet for the robust australopiths. Paleontologists disagree on the number of species of robust australopith, but four are often recognized: *A. boisei* and *A. aethiopicus*, in East Africa; *A. robustus* and *A. crassidens*, in South Africa.

The gracile australopith differs from the robust in the absence of cranial crests, smaller cheek teeth, larger anterior teeth, and different patterns of dental wear. By contrast with the robust form, gracile australopiths are interpreted to have a more omnivorous diet, one that included a greater portion of meat with the vegetarian base. Three species of gracile australopith are well established: *A. ramidus* and *A. afarensis* in East Africa and *A. africanus* in South Africa. *A. afarensis* is the species to which the fossil "Lucy" belongs. She is the most complete skeleton of an australopithecine ever discovered. Lucy stood about 1.05 m tall and lived in Ethiopia 3.25 Ma ago. Both robust and gracile forms were similar in body size, equally capable of bipedalism and manual dexterity, and undoubtedly used naturally occurring objects as tools in foraging for food. While the robust australopiths became extinct about 1.0 Ma ago, it is generally agreed that some form of gracile australopith, *A. africanus* or *A. afarensis*, gave rise to the human lineage.

Homo Habilis. Fragmentary fossils discovered by Louis Leakey at Olduvai Gorge in the early 1960s included hand bones interpreted as capable of toolmaking, and cranial fragments suggestive of a brain size greater than australopiths. These fossils were given the scientific name *Homo habilis*, a species now known to have inhabited East Africa between about 2.0 and 1.6 Ma ago. This earliest stage in the lineage leading to modern humans is characterized by a slight reduction in size of the cheek teeth, and a moderate increase in brain size over australopith ancestors. Habilines are regarded as tool-using omnivorous scavengers, fully capable of bipedal locomotion. Sites revealing an association of stone tools and animal bones, and animal bones with both carnivore tooth-marks and stone tool cut-marks, are interpreted as evidence that early human ancestors competed with carnivores for meat. Evidence of active hunting of large animals is absent, though small mammal, bird, and reptile bones suggest collecting of a variety of small animals for food.

Homo Erectus. Originally discovered and defined on the basis of fossils found in Java and near Beijing in China, this stage of human ancestry is now well known in Africa too. *Homo erectus* fossils first appear in Africa 1.7 Ma ago and are last seen in East Asia and northwest Africa between 0.4 and 0.25 Ma ago. With a long and low skull that displays conspicuous bony thickening over the eye sockets, *Homo erectus* also displays smaller cheek teeth and reduced facial and jaw bones. Thick bones of the braincase and bony thickening at the back of the skull are unique features of *Homo erectus* and may be related to heavy chewing stress and use of anterior teeth to hold or manipulate objects, or to aid in tool making. The broad range of distribution of *Homo erectus* in time and space may be responsible for physical differences in specimens attributed to this species. This stage of human evolution marks a number of significant achievements: expansion of pre-human populations out of Africa; colonization of nontropical environments; controlled use of fire; and cooperative hunting of large mammals.

Homo Sapiens. As the study of our ancestry approaches modern humans, fossils are better preserved—they become more complete and abundant—permitting a fuller picture of prehistoric humans and their lifeways. Our species is subdivided into three sequential substages: early or archaic *Homo sapiens;* Neanderthal; and anatomically modern *Homo sapiens.* Early *Homo sapiens* fossils are known from Africa and Eurasia, and from about 400 to 130 Ka ago. It now appears that humans, represented by late *H. erectus* or early *H. sapiens,* first settled Europe about 500 Ka ago and developed into a locally unique form known as Neanderthal. The question of whether Neanderthals, whose fossil remains are limited to Europe and southwestern Asia from 130 to 35 Ka ago, are ancestral to modern people is a highly controversial issue and is discussed below. In Asia, early *H. sapiens* are more modern looking than their *erectus* ancestors, primarily in the expansion of the braincase and continued reduction of cheek teeth. Interestingly, in Asia early *H. sapiens* retain some key features of their *erectus* predecessors, including a pos-

terior bony thickening of the skull, large browridges, and absence of prominent forehead development. While few early *H. sapiens* sites have yielded both human fossils and artifactual evidence of culture and behavior, a great deal is known of the Neanderthals and their successors—anatomically modern humans. The Neanderthal period of human evolution is thought to have taken place in a glacial period, and the biological and cultural features of Neanderthals are interpreted to reflect their adjustment to this harsh environment. Making more sophisticated stone tools than *H. erectus,* Neanderthals and their limited array of stone tools are often recovered from the caves they occupied, and in which they intentionally buried their dead. The question of altruistic, ritual, and symbolic behavior of Neanderthals is hotly contested and is closely linked to interpretations of their place in human ancestry. Scholars who see Neanderthals as an extinct side-branch of human evolution regard their culture as fundamentally primitive and of limited diversity.

The application of new dating methods of old fossils, and the discovery of new fossils, leads anthropologists to believe that anatomically modern *Homo sapiens* (modern humans) originated in Africa between 130 and 100 Ka ago. These first modern people dispersed into the Middle East at 100 to 80 Ka ago and are not known in western Europe until about 35 Ka ago. When modern humans first reached western Europe they encountered Neanderthals, who had been living in this region since at least 300 Ka ago. Fossils from western Europe strongly suggest that modern humans replaced Neanderthals in this region. The absence of fossils with an intermediate form, or mixture of Neanderthal and modern human traits, suggests to some anthropologists that each group was reproductively separated from the other. While this view of modern human origins is supported by some interpretations of modern human genetic diversity (such as the study of mitochondrial DNA), more complete fossils and more careful analysis of human genetic variation is needed to prove it conclusively.

Key Trends and Issues in Human Evolution

A broad overview of the physical evolution of humankind reveals several notable patterns of change. The trends include: (1) increasing bipedality and terrestriality; (2) increasing manual dexterity and tool making; (3) reduction in size of teeth and jaws; (4) dietary shift to include meat, and an increasing reliance on hunting; and (5) increase in brain size. Some trends are expressed and observable at different points in our past (bipedality, early; hunting, later), while others are gradual and occur over the course of our evolution (dental reduction, increase in brain size).

Questions regarding the stimulus for major events in human evolution generally include climatic and ecological factors, relationships between species, or movement of faunas between continents. Critical evolutionary events, such as the splitting of lineages, that appear to coincide with major climatic changes are noted in Figure 1, following Boaz (1993). First, the origin of *Sivapithecus* in Asia coincides with the continental collision of northwest Africa and Southwest Asia and associated faunal migrations. Second, mid-Miocene cooling may have triggered the bifurcation of the gorilla lineage. Third, formation of the western rift valley system in East Africa possibly caused the separation of chimpanzees from later prehuman ancestors. Fourth, climate changes toward greater aridity associated with the desiccation of the Mediterranean Sea (Messinian Event) could have precipitated the split of ancient prehumans from specialized robust australopithecines. And finally, the specialized features of Neanderthals in western Europe may be partly related to the cold ice age climate in which they lived.

A current controversy in human evolution surrounds the issue of modern human origins. Are the races of modern humanity of recent origin (100–200 Ka) or are the differences we observe today between people of separate continents very ancient (1.0–1.5 Ma)? Some investigators find the fossil evidence compatible with recent studies of human genetic variation (mitochondrial DNA), and advocate a recent "African Eve" hypothesis of modern human origins. Other human paleontologists see the evolutionary forces of natural selection and gene flow in ancient human populations resulting in the origin of modern humans from their ancient ancestors through continuous change in all regions of the Old World. More evidence, including genetic analysis and fossil discoveries, is essential to the resolution of this controversy.

To what extent did the extinction of robust australopithecines and *Gigantopithecus* involve competition with tool-using prehumans? Was their de-

mise due to competitive interaction with other mammals, ecological change in plant communities, or physiological factors about which we may never know? While our understanding of all stages of human evolutionary development has grown tremendously over the past two decades, many perplexing questions remain to be resolved.

Bibliography

BILSBOROUGH, A. *Human Evolution.* London, 1992.

BOAZ, N. T. *Quarry: Closing in on the Missing Link.* New York, 1993.

GAMBLE, C. *Timewalkers: The Prehistory of Global Colonization.* Cambridge, MA, 1994.

JONES, S., R. MARTIN, and D. PILBEAM. *The Cambridge Encyclopedia of Human Evolution.* Cambridge, England, 1992.

KLEIN, R. G. *The Human Career: Human Biological and Cultural Origins.* Chicago, 1989.

STRINGER, C., and C. GAMBEL. *In Search of the Neanderthals: Solving the Puzzle of Human Origins.* London, 1993.

THORNE, A. G., and M. H. WOLPOFF. "The Multiregional Evolution of Humans." *Scientific American,* April 1992, pp. 76–83.

WILSON, A., and R. L. CANN. "The Recent African Genesis of Humans." *Scientific American,* April 1992, pp. 68–73.

JOHN R. LUKACS

FOSSIL SOILS

Paleosols are soils of past landscapes. Some of these are preserved like fossils by burial in sedimentary and volcanic rock sequences. Paleosols also include soils at the surface that formed under conditions no longer prevailing, such as permafrost soils from the Ice Age preserved in temperate lowlands of Britain, or tropical Eocene soils preserved in the mountains of northern California.

The science of paleopedology was founded in 1927 by Boris Polynov. Polynov sought to address anomalies for pedology, the study of the geographic distribution and formation of soils. Buried soils in sedimentary sequences had been recognized well back into the eighteenth century, but little use was made of them until the emergence of pedology and paleopedology as sciences in the early twentieth century. A particularly influential theoretical framework for soil science was promoted by Hans Jenny in a 1941 book (*Factors in Soil Formation*) that urged the development of mathematical formulas for the five main factors of soil formation: climate, organisms, topographic position, parent material, and time. These ideas and formulas can be used to interpret paleosols as paleoclimatic indicators, as trace fossils of ancient ecosystems, as guides to paleotopography, as products of particular substrates, and as records of time elapsed between events of sedimentation. Paleosols are now of interest for interpreting the evolution and earthquake risk of landscapes as well as for more general problems of the long-term geological history of environments and life on land.

Recognizing Paleosols

Three main kinds of evidence are used to recognize paleosols in sedimentary and volcanic rock sequences. Among these are fossil root traces, although this criterion fails for paleosols older than Silurian. Soils and paleosols also are distinguished by horizons, which represent compositionally distinct kinds of alteration extending downward from land surface. Commonly the boundaries between horizons are gradational, whereas the top of the profile is sharply truncated (Figure 1). Finally, soils and paleosols have distinctive structures, such as the irregular system of cracks and coated surfaces produced by roots, burrows, drying out, and water movement in a soil. These modified internal surfaces are called cutans, and the soil clods they define are called peds.

Many fossil soils have been altered during burial to such an extent that they are difficult to recognize. Three kinds of burial alteration are particularly common: burial decomposition, chemical reduction of organic matter, and burial reddening of iron stain. Although coal seams are a kind of paleosol formed in peat swamps, many paleosols formed on dry land have much less organic matter than comparable surface soils. Much of their original organic matter has presumably been decomposed by microbes whose activity extends deep into the ground from soils forming in covering deposits. Some of these microbes are capable of metabolizing organic matter under conditions low in oxygen, and a by-product of their activity is the

Figure 1. These buried Mollisols in middle Miocene volcaniclastic deposits near Fort Ternan in southwestern Kenya show the sharp tops, gradational lower contacts, and unbedded ped structure characteristic of buried soils. The geological hammer for scale has a handle 25 cm long.

chemical reduction of natural iron stain in the original soil. Greenish-gray and bluish-gray haloes around fossil root traces are a common product of this burial gleization process. The brown to red color of most well-drained soils that forms by oxidation of iron weathered from minerals also can be reddened during burial as the water is driven off from pigmenting compounds during compaction, and they become brick-red iron oxide, or hematite.

Kinds of Soils

Eleven main kinds of soils are recognized by the classification of the U.S. Soil Conservation Service. Entisols are very weakly developed soils, showing little alteration of the initial material. As paleosols, Entisols typically show much sedimentary bedding penetrated by fossil root traces. Inceptisols are better developed with some weathering of mineral grains and accumulation of peat, clay, or carbonate, but not so much as to qualify as other kinds of soils. Andisols are similar to Inceptisols but have abundant volcanic ash remaining little weathered. Alfisols and Ultisols are soils with substantial accumulations of clay in subsurface horizons and are distinguished on the basis of chemical tests that reflect soil fertility. Alfisols are rich in nutrient cations such as K^+, Na^+, Mg^{2+}, and Ca^{2+}. They contain common, easily weathered minerals such as feldspar and calcite. Ultisols, on the other hand, are less fertile soils, with very few easily weatherable minerals. Oxisols also are deeply weathered clayey soils with virtually no nutrient cations or easily weatherable minerals. Spodosols are similarly infertile but are sandy and quartz-rich rather than clayey. Their most distinctive feature is a subsurface horizon rich in iron oxide and humus. Histosols are peaty soils forming in bogs and swamps. Vertisols are thick clayey soils that crack deeply in seasonally dry climates. Mollisols have a surface horizon dark with intimately admixed clay and organic matter. They have a distinctive structure of pellets of organic clay produced by earthworms and the abundant fine roots of grasses. Finally, Aridisols are soils with accumulations of calcite in nodules close to the surface. Aridisols form in desert climates, too dry to leach the soil of calcite and other easily weatherable minerals.

Soils in Space

Soils and paleosols discovered on the Moon and planets provide new challenges for Earth-bound classifications of soils and understanding of soil-forming processes. Conditions on our waterless and atmosphereless Moon are most unlike Earth. The most important soil forming process on the Moon is continual bombardment of the surface by sand-sized micrometeoroids. These micrometeoroids pulverize and locally melt the soil, darkening the surface with added metal. Dark metal-rich horizons found in lunar cores represent fossil soils buried by the debris of exceptionally powerful impacts.

On Mars there is a thin atmosphere of carbon dioxide and evidence of water in the distant geological past. Soils analyzed there by robotic landers are rich in iron and swelling clay (smectite), with subsurface hardpans of salts. These soils are similar to salty Aridisols now forming in the Dry Val-

leys of Antarctica. Such soils require warmer temperatures and more water than currently available on Mars, and they may be relict paleosols dating back to the time of fossil channels on Mars more than 2,000 Ma ago.

Some scientists have also suggested that certain kinds of meteorites may be parts of paleosols. Like Martian paleosols, carbonaceous chondrites also are rich in iron and smectite, and some are veined with salts. By this interpretation, some of these carbonaceous chondrites represent the oldest soils in the solar system, some 4,500 Ma old.

Fossil Record of Soils

The record of paleosols in rock sequences from Precambrian time provides another challenge to our understanding of soils and soil formation on Earth, because conditions on the early Earth were very different from those now prevailing. The most ancient known paleosols on Earth, some 3,000 Ma old, are thick, deeply weathered, and green gray in color. These peculiar soils do not fit easily within modern soil classifications and may reflect an early atmosphere rich in carbon dioxide but with very low amounts of oxygen. Red and oxidized paleosols, including Oxisols that indicate the rise of oxygen in the atmosphere, do not appear until about 2,000 Ma ago (Figure 2). At this time, with the appearance of the first large continents by amalgamation of smaller island arcs, came the first paleosols recognizable as Vertisols and Aridisols of dry climate. There is evidence of land plants and animals in Late Ordovician Entisols, Inceptisols, and Andisols. Histosols do not appear until the advent of more substantial land vegetation during the Early Devonian. Forested soils such as Alfisols are not known earlier than Late Devonian. Spodosols and Ultisols may be as old but are currently not known among paleosols older than Carboniferous. Mollisols of grasslands appear relatively late in geological history with the Tertiary rise of grasses and grazers.

This long fossil record of soils also is complementary to evidence of fossils and sediments for the history of life and environments on Earth in the geological past. Paleosols can be expected to provide evidence of the earliest microbial life on land, in the form of microfossils, microscopic trace fossils, or chemical anomalies. Indeed, some scientists envisage soils as sites for the origin of life itself. Burrows and other traces in paleosols indicate

Figure 2. Many Precambrian paleosols are known from major unconformities where sediments overlie metamorphic rocks, such as the thickly bedded sandstones some 800 million years old overlying gneiss and amphibolite with quartz veins on this rock platform near Sheigra, northwest Scotland. The tape extending down from the top of the buried soil is 2 m long.

that multicellular organisms, perhaps millipedes, lived on land as early as Ordovician. Root and rhizome traces in paleosols are evidence of early rhizomatous land plants (as early as Silurian) and forests (as early as Devonian). The emergence of grasslands can be inferred from the appearance of paleosols with fine root traces and granular peds during late Tertiary time. The role of grassland ecosystems in the evolution of humans and the emergence of agriculture also can be investigated using fossil soils. Soils and life have both diversified through geological time, and paleosols record this fundamental part of terrestrial ecosystems.

Bibliography

MARTINI, I. P., and W. CHESWORTH, eds. *Weathering, Soils and Paleosols.* Amsterdam, 1992.

REINHARDT, J., and W. R. SIGLEO, eds. *Paleosols and Weathering Through Geologic Time: Principles and Applications.* Special Paper of the Geological Society of America. Vol. 216. Boulder, CO, 1988.

RETALLACK, G. J. *Soils of the Past.* London, 1990.

———. *Miocene Paleosols and Ape Habitats in Pakistan and Kenya.* New York, 1991.

GREGORY J. RETALLACK

FOSSILIZATION AND THE FOSSIL RECORD

Fossils are any evidence of ancient life on Earth. The word fossil originally referred to any object that was dug from the earth, but because organisms that are living at the present time are not considered to be fossils when they become buried, paleontologists (the scientists who study fossils) arbitrarily exclude all organisms that have been buried since about 10,000 years ago. Fossils provide the only means to study directly the ancient record of evolutionary change in the biosphere. They are also useful for interpreting paleoecological relationships and solving a wide variety of geologic problems. As examples, they are the principal means of determining the relative ages of sedimentary rock formations; they are generally sensitive indicators of ancient climates and environments in which sedimentary deposits accumulated; they can be used to help determine the former positions of Earth's lithospheric plates at times during the geologic past; and they are useful in the search for economic mineral resources.

Commonly, the word "fossil" is used in an adjectival or figurative sense for nonorganic structures in rocks, such as fossil ripple marks, fossil mudcracks, and fossil fuels. However, this entry is restricted to a discussion of organic remains or organically produced structures.

Types of Fossils and Processes of Fossilization

Evidence of past life is preserved in two basic ways: as body fossils and as trace fossils. Body fossils are any direct evidence of prehistoric life. They include whole bodies of animals, plants, or other organisms, or any parts of bodies such as shells, bones, teeth, or leaves. Trace fossils, in contrast, are any indirect evidence of prehistoric life. They include tracks, trails, burrows, boreholes, bite marks, and coprolites (fossil excrement).

The study of the postmortem or post-molting history of fossils is called taphonomy; it includes all of the processes of fossilization as well as other factors relating to the preservation and eventual discovery of fossils. The two most important prerequisites for the fossilization of body parts are conditions that retard or prevent destruction of bodily remains; and, in most cases, the presence of some organic hard parts such as bone, teeth, shell, or wood. Traces in unconsolidated sediments require burial to become fossils, whereas traces on the bodies of other organisms require the same conditions for preservation as body fossils. Secondary taphonomic processes that can destroy fossils are recrystallization or metamorphism of their enclosing rock, and weathering and erosion.

Destruction of organic remains is reduced by enclosure in a protective medium; generally this involves burial in sediments. Burial is important because it seals organic remains from the destructive effects of bacterial decay, degrading enzymes, and scavenging animals. Because burial is so essential to preservation, organisms that live in water—particularly marine water where there is a net accumulation of sediments—have a much better chance of becoming fossils than do organisms that live in terrestrial environments, where there is a net loss of sediments. Burial is especially effective for preserving organic remains if anoxic conditions (conditions where oxygen is not present) are reached in the enclosing medium. Anoxia not only inhibits bacterial and enzymatic decomposition, it also inhibits scavengers, and it inhibits burrowing organisms, which can disaggregate buried organic remains. In marine muds, anoxic conditions are common just a few centimeters below the sediment surface. In terrestrial environments, burial or enclosure in volcanic ash, tar, glacial ice, or resin from pine trees is also effective for removing organisms from oxygenated conditions, but such conditions are rare. Enclosure of organic remains in very dry air, which rapidly removes the water from organic remains, can also retard decomposition; it is responsible for the preservation of rare natural mummies.

Most body fossils are remains of hard parts because they are more resistant to physical and chemical breakdown than are soft tissues or lightly skeletalized (or tough) remains. Hard parts are usually made of organically secreted calcium carbonate (such as shells of clams, snails, and other mollusks; see MOLLUSKS), calcium phosphate (such as teeth and bones of vertebrate animals), silica (such as the tiny spicules of some sponges and the skeletons of diatoms), or wood (such as tree trunks). Hard parts may be preserved in a chemically unaltered state, or they may be altered in some way. Common types of alteration are recrystallization, replacement, permineralization, dissolution, and carbon-

ization. Recrystallization involves the reorganization of the original skeletal remains into a different mineral (such as a change from aragonite to calcite) or into larger crystals of the same mineral. Most fossil shells and coral skeletons have been recrystallized. Replacement involves the removal of the original skeletal material and replacement of it by another mineral. Fossils of most marine invertebrates that have been preserved by silica or pyrite have been replaced from their original calcium carbonate composition. Permineralization involves the filling of porous materials by the precipitation of minerals from water. Petrified wood and dinosaur bones are common examples of permineralization. Dissolution involves the chemical removal of the original skeleton from rock, which leaves a hollow impression called a mold. Most fossils from strata composed of dolostone (calcium-magnesium carbonate) are molds. Carbonization involves the removal of volatile compounds (primarily oxygen, hydrogen, and nitrogen) from an organism, which leaves only a carbon film in rock. Leaves and other plant parts are commonly preserved as carbonized films.

Soft and lightly skeletalized parts are rare as fossils, so rare in fact that sedimentary deposits that contain extraordinarily preserved fossils are given a special designation, Lagerstätten (see ARTHROPODS). True soft parts are such things as skin, muscle, and internal organs. The best-known examples of soft-part preservation are remains of mammoths and rhinoceroses in glacial ice of Alaska and Siberia. Lightly skeletalized (or tough) parts can be defined to include the unmineralized external cuticle of insects, spiders, and some other arthropods, the external cuticle of some worms; and feathers of birds. Whereas true soft parts usually decay quickly or are quickly scavenged after the death or molting of an organism in a normally oxygenated environment, lightly skeletalized parts may last weeks to years longer before disaggregating if scavengers or sediment-burrowing organisms do not disturb them. As such, they stand a somewhat better chance of being preserved as fossils than do true soft parts. Excellent examples of lightly skeletalized marine fossils have been found in Cambrian rocks of Yunnan Province, China, and British Columbia, Canada, where arthropods have been preserved with appendages, gills, and internal organs intact, and worms have been preserved with the cuticle intact. In the Jurassic of Germany, marine arthropods have been preserved with their appendages intact, and some birds retain impressions of feathers. Lightly skeletalized terrestrial insects have been preserved in Tertiary amber from the Baltic Sea region and the Dominican Republic.

Trace fossils provide valuable information about the behavior of organisms and precisely where or how they lived. Some information that they convey would not be otherwise obtainable from body fossil evidence. Dinosaur trackways, for example, show that dinosaurs carried their tails in the air rather than dragging them on the ground. In addition, such trackways provide information about herding behavior in some species and how fast some animals could run. Bite marks on some marine animals give clues to which animals were their predators, and in some cases even provide information about the predators' attack strategies and the preys' evasive behavior. Trace fossils that were made in soft sediment that since has been turned to rock provide unequivocal evidence that an organism lived in that area. Unlike the bodily remains of an organism, which usually can be transported a great distance, movement of the sediment containing a trackway or burrow will destroy the trace.

Completeness of the Fossil Record

The fossil record extends back more than 3.5 billion years (Ga) of geologic time and preserves representatives of all living kingdoms of organisms, and representatives of numerous ecosystems. More than 250,000 fossil animal species have been described and named; estimates of the total number of animal and other species that have lived on Earth over the last 3.5 Ga range in the tens of millions.

The sample of organisms from the fossil record is both biased and nonrandom. Although the quality of preservation varies greatly from one setting to another, invertebrate animals that secreted robust skeletons generally have the best fossil record. This is true even though such organisms probably were far outnumbered by those composed solely of soft or lightly skeletalized parts.

The fossil record is more complete for marine depositional environments than for terrestrial ones. Even among marine settings, there is a substantial range of variability in the completeness of the fossil record. The record is much better for continental shelf areas than for deep-sea areas.

Among terrestrial environments, the record is better for swampy lowland areas near coastal rivers than for mountainous areas, hills, and deserts.

Our understanding of the fossil record is generally better for younger intervals of geologic time and for regions that were not at the edges of continents. Younger and more inboard sedimentary rocks are more likely to have survived the destructive effects of erosion, subduction, and metamorphism. Also, younger rocks have been better studied than older ones in the search for mineral resources.

Sampling of the fossil record by humans has been variable. Abundant, widespread organisms tend to be more frequently studied than are rare organisms. Exceptions are some organisms that had multielement skeletons and that readily disarticulated after death, such as the echinoderms (*see* ECHINODERMS). Organisms that are useful for stratigraphic correlation, such as trilobites, ostracodes, conodonts, foraminiferans, ammonoid cephalopods, graptolites, and pollen-producing plants, generally have been more intensively studied than organisms with limited utility for correlation such as sponges.

Uses of Fossils

One major use of fossils is for interpreting the course of biological evolution on Earth. The oldest fossils, which are found in 3.5-billion-year-old rocks of the Archean Eon, are primitive, single-celled, blue-green bacteria that are found in internally layered domal structures called stromatolites. Bacteria that lack cell nuclei and chromosomes were the major life forms in the oceans for approximately the first two Ga that life existed on Earth. Marine algae, which are the oldest known organisms to have true cell nuclei and chromosomes, are first found in strata deposited during the Proterozoic Eon, approximately 1.4 Ga ago. Multicellular organisms made their first appearance as marine algae about 1.2 Ga ago. Multicellular animals are first recorded by trace fossils indicating digging, feeding, and other behavior in formerly soft, unconsolidated marine sediments that are probably younger than 1.0 Ga. Imprints of soft-bodied or tough-bodied creatures have been discovered at numerous Late Proterozoic localities worldwide in rocks that were deposited approximately 650 Ma ago. Some of these creatures resemble modern cnidarians, including jellyfishes and sea pens. Others, however, defy easy placement within a known group.

The beginning of the Paleozoic era, the Cambrian period (approximately 570 Ma ago), is marked by an explosion in the number of preserved body fossils of marine animals. Plants first colonized the terrestrial environment in the Ordovician period, and animals (arthropods) followed, probably during the Silurian period. The Devonian period is notable for the appearance of terrestrial vertebrates (amphibians), the first forests, the first insects, and a group of marine cephalopods called ammonoids (*see* MOLLUSKS). Fully terrestrial vertebrate animals (reptiles) appeared during the Pennsylvanian period.

The terrestrial record of the Mesozoic era is dominated by dinosaurs and other reptiles, but this was also the time of the first appearance of mammals (Triassic period) and birds (Jurassic period). Flowering plants (angiosperms) evolved in the Cretaceous period. The seas were home to large reptiles and fishes as well as a wide variety of invertebrates and marine plants.

The fossil record of the Cenozoic era is characterized by the development of essentially modern biotas both in the oceans and on land. Mammals, insects, and flowering plants represent the most important terrestrial fossils, whereas marine ecosystems are dominated by mollusks, corals, arthropods, and fishes.

The fossil record is not a smooth, continuous progression through time but is punctuated by numerous mass extinction events. Major extinctions occurred near the end of the Proterozoic Eon, near the end of the Cambrian, Ordovician, Devonian, Permian, Triassic, and Cretaceous periods, as well as during the Pleistocene epoch of the Quaternary period. Various explanations have been offered for mass extinction events, and it is clear that one cause does not successfully explain all of these events. Some possible causes of extinction are dramatic changes in marine water temperature, loss of continental shelf space where many marine creatures lived, extensive volcanic activity, and events related to the impact of extraterrestrial bodies on Earth (*see* EXTINCTION).

Fossils are excellent indicators of the environments under which sedimentary deposits were laid down. Deposits that contain echinoderms, corals, brachiopods, cephalopods, and other exclusively marine organisms demonstrate that the enclosing sediments were deposited on the seafloor, even if those deposits are now far inland. Terrestrial environments, in contrast, are indicated by the presence of land plants and animals. Certain fossils indicate even more specific details about ancient sites of deposition, particularly when their enclosing sediments are also studied. Coral reef fossils, for example, strongly suggest that the enclosing deposits were laid down in warm, shallow, tropical marine water.

Paleoecologic relationships can be reliably inferred from body and trace fossils. For example, large Cretaceous fishes from Kansas that preserve smaller fishes as gut contents provide vivid details of a predator-prey relationship. Healed bite marks preserved on Paleozoic trilobites reveal that not all predatory attacks were entirely successful, and tumors on trilobites reveal parasitic relationships with microorganisms.

Fossils can provide important information about geography and climatic conditions during the geologic past. They can be used to readily identify the distribution of land and seas for an interval of geologic time, whether for a local region, a continent, or the entire Earth. Organisms are commonly sensitive indicators of temperature or climate. Leaves of deciduous trees preserved in Pliocene sediments of Antarctica, for example, show that this ice-covered continent was not always quite so cold; it probably experienced temperate conditions during parts of the Pliocene epoch. Temperature conditions are partly dependent on latitude, which means that temperature-sensitive organisms can be used to help ascertain the positions of continents or any small pieces of them during intervals of the geologic past. For example, the presence of reef-making animals in Cambrian rocks of North America indicates that the continent was located in the tropics during the Cambrian period (Figure 1).

Fossils are indispensable for correlating and classifying sedimentary formations throughout the world. The relative arrangement of hundreds of thousands of fossil species in strata worldwide is now well documented. Using guide fossils, which are the fossils most useful for interpreting the relative ages of rocks, it is possible to accurately determine the ages of most strata using either individual fossils or assemblages of them. When guide fossils in two or more areas can be matched, a close equivalence in the time of deposition of strata in those areas can be inferred. Gaps in the stratigraphic

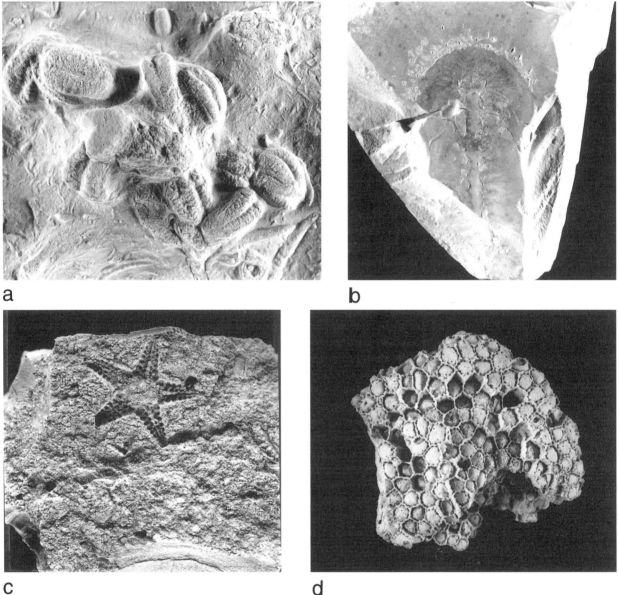

Figure 1. Trace fossils (a), and body fossils showing a range of preservational types (b–d).

a. Trace fossils, mostly *Rusophycus pudicum,* which are burrows produced by the trilobite. *Flexicalymene;* from the Ordovician of Ohio; ×1. The burrows were dug in soft mud and then filled in by limey muds that lithified to form limestone; the fossils are the natural casts in limestone.

b. *Naraoia longicauda,* a trilobite that lacked a calcified exoskeleton, showing exceptionally well preserved remains of the soft parts, particularly the ramifying intestines in the head; from the Chengjiang Lagerstätten (Cambrian) of Yunnan Province, China; ×1.5.

c. Natural mold left by the dissolution of a starfish, *Tretaster wyvillethompsoni;* from the Ordovician of Scotland; ×3.

d. A coral, *Favosites* sp., that was originally composed of calcium carbonate and later replaced by silica; from the Devonian of Ohio; ×1.

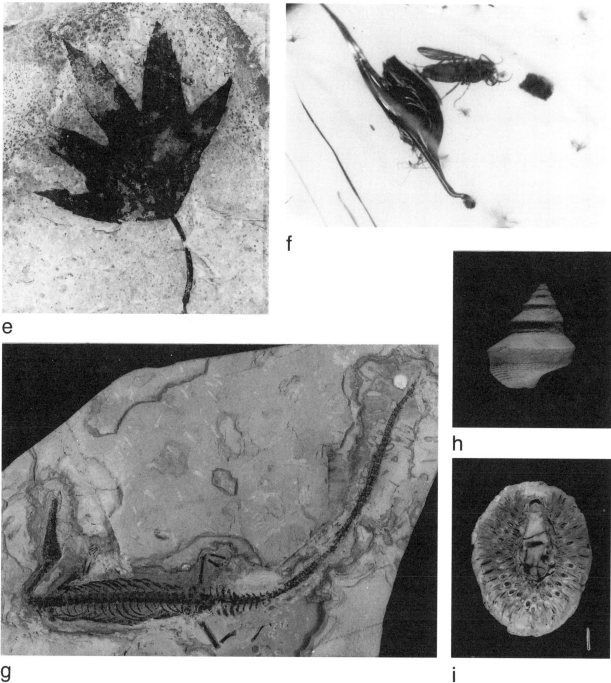

Figure 1. Continued

e. A sycamore leaf, *Plantanus wyomingensis,* preserved by carbonization; from the Eocene of Colorado; ×1.

f. An insect preserved in amber; from the Oligocene of the Baltic Sea area; ×16.

g. The skeleton of a reptile, *Mesosaurus* sp., preserved in a chemically unaltered state; from the Permian of Brazil; ×0.4.

h. A gastropod, *Worthenia tabulata,* which was originally composed of aragonite and later recrystallized to calcite; from the Pennsylvanian of Texas; ×1.

i. A pine cone, *Araucaria mirabilis,* that was permineralized by silica; from the Jurassic of Argentina; ×0.5.

Source: All photos by Loren E. Babcock.

succession of an area can be detected by noting the absence of guide fossils in the stratigraphic succession, assuming that the same sedimentary environments are preserved in the areas being studied. Correlation of sedimentary strata is an important concern because it is a necessary first step for many stratigraphically based investigations including the exploration for, and production of, oil, natural gas, coal, and other mineral resources.

Bibliography

ALLISON, P. A., and D. E. G. BRIGGS, eds. *Taphonomy: Releasing the Data Locked in the Fossil Record.* New York, 1991.

BOARDMAN, R. S.; A. H. CHEETHAM; and A. J. ROWELL, eds. *Fossil Invertebrates.* Palo Alto, CA, 1987.

BRASIER, M. D. *Microfossils.* London, 1980.

CARROLL, R. L. *Vertebrate Paleontology and Evolution.* New York, 1988.

DODD, J. R., and R. J. STANTON, JR. *Paleoecology, Concepts and Applications.* New York, 1981.

GOLDRING, R. *Fossils in the Field.* New York, 1991.

STANLEY, S. M. *Exploring Earth and Life Through Time.* New York, 1993.

TAYLOR, T. N., and E. L. TAYLOR. *The Biology and Evolution of Fossil Plants.* Englewood Cliffs, NJ, 1993.

LOREN E. BABCOCK

FRACTIONALISM

See Isotope Tracers, Stable

FROGS AND SALAMANDERS

See Amphibians

G

GALAXIES

The Sun is one of about 100 billion stars that are gravitationally bound together in a giant system called the Milky Way Galaxy. An observer outside our Galaxy looking back at it would see the stars arranged in a pattern that resembles a fried egg. That is, there is a thin disk that is 270–1,000 parsecs (pc) thick and that extends to a radius of at least 25 kpc (a parsec is 3.26 light-years; a kiloparsec is 3,260 light-years). Centered on the disk is a spherically shaped halo; half of the light in the halo is contained within a radius of 3 kpc. At a radius of 25 kpc the surface brightness of the stars in the disk has dropped to 25 magnitudes/arcsec^2 (the angular measure "arcsec" is 1/3600th of a degree). It is this disk of stars, whose light is blurred together, that is seen as the prominent band of light in the summer night sky. The Sun is located in this disk, 8.5 kpc from the center. The galactic center itself is in the direction of the constellation Sagittarius and is obscured from sight at optical wavelengths by dust.

The blue absolute magnitude, M_B, of the Milky Way is −20.3, which is a brightness that is 2×10^{10} times the brightness of our Sun in blue light. The total mass of the visible Galaxy, stars plus gas, is about 16×10^{10} times the mass of the sun, $M_\odot$.

The general picture of the formation of the Galaxy is one in which it began as a large, spinning ball of gas. This gas was mostly composed of hydrogen and helium atoms with a small percentage of slightly heavier elements that were formed in the "big bang." Under the force of its own gravity, the ball of gas began to collapse. As it collapsed, tiny pieces of it condensed to form stars and clusters of stars. The first generation of stars and the round, rich star clusters called globular clusters formed the halo and are among the oldest stellar objects in the Galaxy. There are several hundred observable globular clusters, each of which contains up to 100,000 stars; these clusters range in age from 13 to 18 billion years (18 Ga).

Because the gas of the young Galaxy was spinning, it continued to collapse gravitationally into the thin disk that forms the rest of the Galaxy. The oldest stars in the disk appear to be about 9 ± 2 Ga old, suggesting that there was a break of several billion years between the end of star formation in the halo and the begeining of star formation in the disk. Most of the gas settled into this disk, and it is here—in a zone ± 50 pc thick—where stars are continuing to form out of the gas today.

Most of the present gas in the Galaxy is still in the form of neutral hydrogen (HI) and helium atoms. There is a total of 4×10^9 $M_\odot$ of HI and it can be traced to a far greater distance from the galactic center than the stars. It falls steadily in density from the center to a radius of 35 kpc, where it has a density of 1 $M_\odot$/pc^2. However, the HI gas is not uniformly distributed. There are HI clouds up to 10^7 $M_\odot$ in mass, as well as loops, holes, and filaments, which are probably caused by the violent energy output of young massive stars in the form of stellar winds and supernova explosions.

The places where stars form are regions where the density of the gas is high, greater than 100 atoms/cm^3. These regions are called molecular clouds because they are places where the atomic hydrogen has bonded to form molecular hydrogen. Traces of molecules more complex than H_2 are found in these clouds, including carbon monoxide, alcohol, and formaldehyde. The molecular clouds range up to 10^6 $M_\odot$ in mass and 60 pc in diameter. About half of the $1-2 \times 10^9$ $M_\odot$ of molecular matter in the Galaxy is in the largest clouds, which dominate the star formation activity. The molecular clouds are not distributed evenly along the radius of the disk of the Galaxy. Within the inner 4 kpc only a small amount of molecular material is found. From 4 to 8 kpc radius there is a rich molecular ring. Beyond that the molecular material declines in density; at a radius of 15 kpc, it is down by a factor of 100. Molecular clouds have been traced to a radius of 20 kpc.

Star-forming regions come in a wide range of sizes, but smaller ones are more numerous than larger ones. Near one end of the range is the Orion Nebula (Figure 1), the closest star-forming region to Earth. It contains only a few massive stars (greater than 10 $M_\odot$). At the other end of the range are regions like the Tarantula Nebula in the Large Magellanic Cloud, a nearby galaxy. This star-forming region contains hundreds of massive stars in a dense cluster, and the ionized gas cloud itself, if placed at the distance of Orion, would cover 25 degrees on the sky. In most star-forming regions there is a large range in the masses of the stars that have formed—from stars of 100 $M_\odot$ down to 0.1 $M_\odot$ (100 times the mass of Jupiter), and the lower mass stars are much more common than the higher mass stars. However, the upper and lower stellar mass limits in different environments are still controversial.

As stars evolve, they convert the gas out of which they were formed into heavier elements, producing light in the process. Some fraction of the elements that are produced is returned to the interstellar medium. This return of material occurs gently in the form of stellar winds throughout a star's life as well as explosively at the end when the star casts off its outer envelopes. The elements that are returned to the interstellar medium enrich the

Figure 1. The Orion Nebula is a star-forming region that is located at 450 pc or 1,500 light-years from Earth. It can be seen with the naked eye as a fuzzy patch in the sword of the Constellation Orion. The image shows the cloud of gas that is left over after the stars have formed. Young, hot, massive stars are ionizing the gas, causing the different atoms to emit light which is seen in this photograph. The picture was obtained with the Kitt Peak National Observatory 4-m Mayall telescope.

Source: The National Optical Astronomy Observatories.

gas that remains, so that each successive generation of stars makes the gas increasingly richer in heavy elements. The stars in the Galaxy, therefore, have not all started with the same proportions of heavy elements. There is also a gradient in the proportion of heavy elements; that is, the stars in the outer part of the disk being two to three times less rich in heavy elements than the stars near the center. This implies that the central part of the Galaxy has evolved faster than the outer parts.

After ejecting the outer envelopes, the core of a dying star goes on to collapse into a dense object, either a white dwarf, a neutron star, of a black hole, depending on its mass. The dense object that is left permanently behind locks up 10 to 60 percent of the original gas mass that went into forming the star. Thus, with time the total amount of gas in the Galaxy decreases. The time will come when there will no longer be enough gas left in the Galaxy to make the clouds necessary to form stars, and then star formation will cease. At the current rate at which the Galaxy is forming stars, there is enough gas left to continue forming stars for approximately another billion years. However, the rate at which the Galaxy has formed stars has not been constant. The Galaxy formed stars at a much higher rate in its early history, and the rate has been slowly declining ever since. It is possible, therefore, that the Galaxy will form stars at a slower and slower rate in the future until eventually it stops doing so altogether.

The Galaxy formed out of a spinning ball of gas, and it continues to rotate. The Sun has an orbital speed of 220 km/s about the center of the Galaxy and takes about 240 million years (Ma) to go once around the galactic center. Outside of the center, objects in the Galaxy have the same speed at all radii. Since the circular path around the Galaxy is longer at larger radii, this means that objects at larger radii have longer orbital periods and slowly lag behind those closer to the center. The rotation has been measured to large radii, and nowhere does the rotation velocity drop as it would be expected to if the gravitational attraction is just that due to the mass that is seen in the stars and gas. This has led to the conclusion that the Galaxy contains much more mass than is visible. Estimates of the total mass of the Galaxy from rotation measurements and from the orbits of other satellite galaxies give 2×10^{12} $M_\odot$. This would mean that there is nearly thirteen times more matter than is visible. The nature of this mysterious "dark matter" is a topic of debate and of many observational searches.

The Milky Way is itself only one of billions of galaxies in the universe. However, it took several centuries after the invention of the telescope to resolve the controversy over the nature of the faint fuzzy patches in the sky that we now identify as galaxies outside of our own. In the mid-1700s Thomas Wright and Immanuel Kant suggested that these fuzzy patches were other systems like the Milky Way located at large distances from us. However, there were other people who argued that these were small nebulae of stars or gas, like the Orion Nebula, that were located nearby and within the Milky Way. The debate was settled in the 1920s when EDWIN HUBBLE used the periodic variation in brightness of Cepheid stars that he detected in the Andromeda Galaxy to measure the distance to that object. The distance he measured placed the Andromeda Galaxy well outside the Milky Way.

Galaxies generally come in a variety of sizes and shapes. Our galaxy is what is known as a spiral galaxy (Figure 2). An observer from the outside

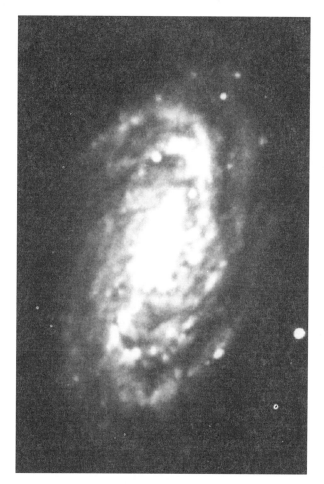

Figure 2. The spiral galaxy NGC 2903, which is shown here, is classified as type Sc. The picture shows the spiral arms coming out of the center of the galaxy and curving around the disk. The galaxy is 9 million pc or 29 million light-years from Earth. This blue-wavelength image was obtained with a digital Charged-Coupled Device detector by S. J. Bus using the 1.1-m Hall telescope of Lowell Observatory.

would see spiral-like arms superposed on the disk that originate from the center of the Galaxy. These spiral arms are highlighted by young stars and clusters, places of current star formation, and the larger molecular clouds. The Sun is located in a branch of the Orion–Cygnus spiral arm. Two other arms have been identified as well.

The two galaxies that are closest to the Milky Way are the Large and Small Magellanic Clouds, which are visible from the southern hemisphere. These galaxies are close enough that they are probably interacting gravitationally with the Milky

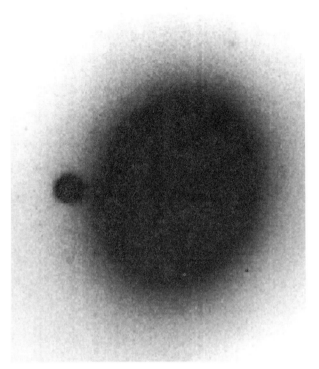

Figure 4. NGC 4472 is a bright elliptical galaxy in the Virgo Cluster of galaxies. The field of view that is shown is about 80 arcsec, and the object to the left of the Galaxy is a star. North is up and east is to the left. This image was obtained in April 1994 with an RCA CCD on the Lowell Observatory 1.1-m Hall Telescope by B. Keel (University of Alabama) and A. Zasov (Moscow State University).

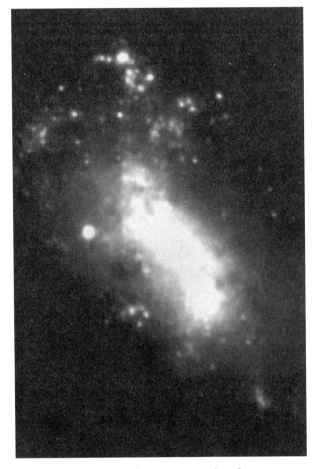

Figure 3. NGC 4449 is an example of an irregular-type galaxy. This green-wavelength image shows the lack of a symmetrical pattern that typifies the spiral galaxies. The Galaxy is located 5.4 million pc or 18 million light-years from Earth. The image was obtained by the author in February 1990 with the 1.8-m Perkins telescope of Ohio State and Ohio Wesleyan Universities at Lowell Observatory.

Way. There is a band of stars and gas called the Magellanic Stream that may have been pulled from the Large Magellanic Cloud by the gravitational field of the Milky Way. These two galaxies are much smaller than the Milky Way, having masses that are about 100 and 1,000 times smaller than our galaxy. They also do not have the symmetrical spiral structure of our galaxy but instead are irregular in appearance (Figure 3). Small irregular galaxies are in fact quite common in the universe. In addition to spiral and irregular galaxies there is a class of galaxies called ellipticals. These symmetrical, smooth looking galaxies appear to have used up most of their gas, converting it into stars when the galaxies were young (Figure 4).

The Galaxy, like other galaxies, is not static. The lag between objects at different Ladü changes the relative positions of objects with time. Spiral waves move through the Galaxy, compressing the gas and forming clouds. Gas clouds formed in this manner as well as through other mechanisms condense, produce new stars, and break up. The stars that are continuously forming slowly deplete the general gas supply. Stars are also continuously enriching the gas in the Galaxy with heavy elements and fading from sight when they cease to support internal nuclear reactions. In addition, the orbits of the stars, including that of the Sun, can be perturbed. These perturbations can result from infall of gas from the outer part of the Galaxy, gas clouds or spiral arms passing by, or chance encounters with other stars. Thus, the Galaxy is a dynamic place, home to many ongoing processes.

Bibliography

Bok, B., and P. Bok. *The Milky Way.* Cambridge, MA, 1981.

Gilmore, G., and B. Carswell, eds. *The Galaxy.* Dordrecht, Netherlands, 1987.

Hunter, D. A. "Regions of Recent Star Formation." In *Star Formation in Stellar Systems: III Canary Islands Winter School of Astrophysics,* eds. G. Tenorio-Tagle, M. Prieto, and F. Sánchez. Cambridge, 1992.

Mihalas, D., and J. Binney. *Galactic Astronomy: Structure and Kinematics.* San Francisco, 1981.

van Woerden, H., R. J. Allen, and W. B. Burton, eds. *The Milky Way Galaxy.* Dordrecht, Netherlands, 1985.

DEIDRE A. HUNTER

GALILEAN SATELLITES

The four largest, brightest satellites of Jupiter, known as the Galilean satellites, are named after the Italian astronomer Galileo Galilei, who is credited with discovering them in 1610. German astronomer Simon Marius also observed these satellites at approximately the same time (Burns, 1986). Chinese astronomers claim to have made naked-eye observations of these moons as early as the fourth century B.C.E. (Burns, 1986; Pang, 1983). In order of increasing distance from Jupiter, the Galilean satellites are Io, Europa, Ganymede, and Callisto. (*See* SATELLITES, SMALL; SATELLITES, MEDIUM-SIZE.)

In December 1973 *Pioneer 10* visited the Jupiter system but sent back little data on the surfaces of the satellites. In 1974 *Pioneer 11* visited the system with similar results. It was not until March 1979, when *Voyager 1* flew by Jupiter, that images of the surfaces of the Galilean satellites were taken. Four months later *Voyager 2* flew through the Jovian system and also sent back spectacular images of these satellites. The spacecraft *Ulysses* passed through the Jovian system in 1992 making observations of the magnetic field. The spacecraft *Galileo* is scheduled to rendezvous at Jupiter in December 1995. One of the instruments aboard *Galileo* is an imaging camera which will once again send back observations of the surfaces of the Galilean satellites.

The diameters of the inner two satellites, Io and Europa, are 3,630 km and 3,138 km, respectively, and thus are similar to the diameter of the Moon (3,476 km). The outer two satellites, Ganymede and Callisto, have diameters of 5,262 km and 4,800 km, respectively, which are closer in size to that of Mercury (4,878 km). Just like Earth's moon, each of these satellites rotates synchronously, always keeping the same face toward Jupiter. The inner three satellites are in a 1:2:4 orbital mean motion resonance: for each time Ganymede orbits Jupiter, Europa orbits Jupiter twice, and Io completes four orbits.

The density and bulk composition of these four moons are correlated with their distance from Jupiter. Io, Europa, Ganymede, and Callisto have densities of 3.57 g/cm^3, 3.04 g/cm^3, 1.94 g/cm^3, and 1.86 g/cm^3, respectively (Burns, 1986). Density in this system decreases with increasing distance from Jupiter and roughly spans the range between rock (3.5 g/cm^3) and water (1.0 g/cm^3). The bulk content of rock in these satellites thus decreases with increasing distance from Jupiter, while the bulk content of water ice increases. These correlations are similar to those seen in the solar system in general (planets closer to the sun are composed of more rock, and are therefore more dense, than those farther from the sun), so the Jovian system is often referred to as a "miniature solar system."

Io

Io is unique among objects in our solar system. It is one of only three planetary bodies known to have active volcanism. *Voyager 1* observed nine active volcanoes, and eight of these volcanoes were still active four months later, when *Voyager 2* imaged the surface. Volcanism has resurfaced Io so efficiently that impact craters, which dominate ancient surfaces like that of the Moon, have been obliterated. The energy source for all this volcanic activity is thought to be tidal distortion and heating driven by the gravitational interaction with Jupiter and its orbital resonance with Europa. The color and spectral properties of many of the volcanic deposits indicate that they consist of sulfur or sulfur dioxide. Unlike the other Galilean satellites, however, no water is present on the surface.

Io landforms and albedo (the fraction of incident light reflected from a surface) features seen by *Voyager* have been classified into three categories: volcanic vent regions (Figure 1), plains, and mountains. Some vent regions include dark, craterlike depressions, sometimes surrounded by bright halos, from which flows radiate. The depressions resemble terrestrial calderas in morphology (Nash et al., 1986). Other vent regions, however, have been identified simply as albedo markings or as sources of flows and deposits. Plains are the most extensive landform on Io. They show layering in several areas, with escarpments from 150 to 1,700 km high outlining smooth-topped, tabular plateaus (Nash et al., 1986; Schaber, 1982).

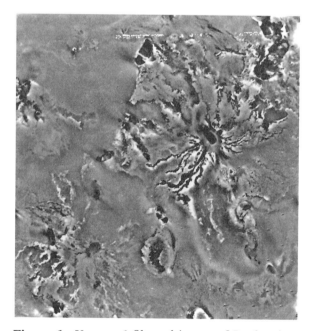

Figure 1. *Voyager 1* filtered image of Io showing a vent region and its associated flows (FDS 16390.06). Photo courtesy of the Lunar and Planetary Institute.

Mountains on Io range in size up to 9 km high (Nash et al., 1986). The presence of these features on Io is somewhat puzzling. Sulfur is not strong enough to support the topography of these surface features (Carr et al., 1979; Schaber, 1982), so that some amount of silicate rock material is required. However, silicate rocks have not been detected in the spectral data. Perhaps the silicate material on Io is covered with sulfur-rich material, mostly from volcanic activity. Only a thin veneer of sulfur-rich material would be needed to hide any silicate signature in the spectral data.

The internal structure is not known, but several model structures have been proposed to explain Io's volcanism and landforms. One model proposes that the interior of Io is molten and overlaid with a rigid crust. A second hypothetical model (Nash et al., 1986) of Io's interior consists of an outer quasirigid crust overlying a plastically flowing asthenosphere. This outer shell is separated from the interior by a global liquid layer, below which is a hot but solid mantle, possibly overlying an iron-rich core. Convection currents within the asthenosphere carry tidally generated heat from the lower to upper asthenosphere, from which volcanic eruptions and conduction carry heat to the surface (Nash et al., 1986).

Radiation from Jupiter's powerful magnetic field, in which Io is deeply embedded, has driven some of Io's surface material into space by sputtering (the geologic process of dislodging materials—usually atoms—by bombarding the surface with high energy particles). Some of this material, trapped in the magnetic field, forms a torus containing neutral atoms of sodium, potassium, sulfur, and oxygen, plus ions of sulfur, oxygen, and sulfur dioxide. Io is also surrounded by a banana-shaped cloud of neutral sodium vapor, the majority of which travels ahead of Io and interior to its orbit (Nash et al., 1986). The source of the sodium cloud is also believed to be materials driven from Io's surface.

Io also has a tenuous atmosphere, but its basic nature and the amount of gas present remain unknown. In 1973 the *Pioneer 10* radio occultation experiment detected an ionosphere. In 1979 the *Voyager 1* infrared spectrometer detected sulfur dioxide gas. These two observations suggest a relatively "thick" atmosphere with a pressure of 10^{-9} to 10^{-7} bar (Nash et al., 1986). Earth's atmosphere is 1 bar. However, the large flux of fast-moving sodium atoms within the torus and sodium cloud can be explained by surface sputtering only if the atmosphere is "thin," that is, if it has a surface pressure of less than 10^{-11} bar.

Europa

Europa presents an intriguing enigma. Its density indicates that it is approximately 85 percent rock by volume, yet Earth-based telescopic and spacecraft spectra show that its surface is composed of water ice and frost. Even though Europa has a greater density than Ganymede or Callisto, and therefore more rock in its interior, it has more water ice on its surface. The amount of water present on the surfaces of each of these three satellites decreases with increasing distance from Jupiter.

The lack of numerous impact craters on Europa indicates that its surface is the youngest among the Galilean satellites after Io. Europa's surface is also extraordinarily smooth, and topographic relief is only on the order of meters. Geologic units on Europa are therefore classified on the basis of color and albedo rather than topography. The two major units are plains and mottled terrains. The plains are thought to be older than the mottled terrains (Lucchitta and Soderblom, 1982). Linear albedo features, or lineaments, also crisscross the globe of Europa (Figure 2). These are thought to be cracks in the surface filled in with slightly darker material. They may be produced by fracturing of the crust by tidal stresses, expansion due to chemical and phase changes in the interior, or compression due to internal cooling (Malin and Pieri, 1986). In addition, four different types of landforms that have some topographic expression have been identified. These include (1) a few impact craters; (2) pits and irregularly shaped depressions; (3) mounds, domes, and irregular positive relief features; and (4) cyclodial ridges having a scalloped outline (Malin and Pieri, 1986). Structural models of Europa's interior describe a differentiated body. Europa has a rock interior overlain with an ice crust of undetermined thickness. In the early 1990s a debate was ongoing over evidence that liquid water may be present between the crust and interior; a theory that has been amplified, with life forms, in science fiction tales.

Like Io, Europa orbits Jupiter within that planet's large magnetic field. The magnetic field rotates more quickly than Europa orbits, so trapped ions traveling with the magnetic field sweep over Europa's trailing hemisphere. (Satellites that orbit syn-

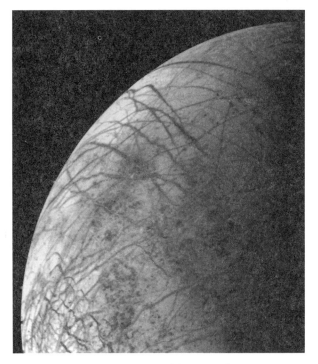

Figure 2. *Voyager 2* shaded image of Europa. Both plains and mottled terrains are shown. The lineaments which criss-cross the surface are also visible (FDS 20649.10). Photo courtesy of the Lunar and Planetary Institute.

chronously have a leading hemisphere, which faces the direction of motion, and a hemisphere that trails the direction of motion.) The trailing hemisphere of Europa is redder and darker than the leading hemisphere. This is attributed to sulfur ions, which are believed to have originated from Io's torus, been transported to Europa by the magnetic field, and implanted in the surface.

Ganymede

Ganymede is the most geologically diverse of the Galilean satellites. Its surface contains at least 50 percent water ice, and the albedo of the surface varies from about 30 to 70 percent. These brightness variations, as well as the fact that the albedo is lower than Europa's, are attributed to the presence of dark minerals like those composing carbonaceous chrondrite meteorites.

The terrains on Ganymede's surface are divided into two groups: darker, older terrain and lighter-colored, younger terrain (Figure 3). The dominant

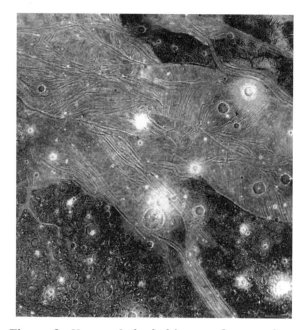

Figure 3. *Voyager 2* shaded image of cratered and grooved terrains on Gamymede (FDS 20637.17). Photo courtesy of the Lunar and Planetary Institute.

process that shaped the darker terrain is impact cratering. However, impact craters on Ganymede are distinct from those on the Moon. Large and old craters are relatively flat. The older craters are probably flat because the ice in Ganymede's crust is too weak to support high topographic relief features. Over time, these flow out and become flattened by a process called "viscous relaxation." The largest craters, called palimpsests, have lost nearly all their topography in this way and are recognized primarily as circular patches with a higher albedo.

Tectonism and volcanism have shaped dark terrain. Narrow troughs, called furrows, crisscross the surface and are one of the oldest features on Ganymede. Some are arranged concentrically around large palimpsests and may be impact fractures, but others appear unrelated and may be tectonic. The evidence for volcanism is smooth patches that have buried small topographic features and partly infilled large craters. However, unlike volcanism on the terrestrial planets, in which melted rock is erupted, on Ganymede the volcanic material was melted ice.

All of the lighter-colored terrain is less cratered than the darker terrain and is therefore younger. The lighter terrain can be divided into three units: grooved units, smooth units, and reticulate terrain.

The grooved units are covered by narrow troughs, or grooves, that occur singly or in parallel bands. Smooth units occur in discrete areas and resemble lunar plains, or maria (Shoemaker et al., 1982). Reticulate terrain is transitional in form between dark terrain and grooved terrain, exhibiting crisscrossing grooves like those in grooved terrain but a low albedo like the dark terrain. Both tectonic and volcanic processes appear to have been involved in forming the lighter-colored terrains. Three mechanisms for the formation of lighter-colored terrains have been proposed: (1) lithospheric spreading; (2) unconfined ice volcanism; and (3) ice-volcanic infilling of rifts. In the lithospheric-spreading hypothesis the surface cracked and spread, and material from below filled in the cracks. In unconfined ice volcanism, ice magma was extruded through fractures in the surface and covered wide areas. The third hypothesis involves the formation of downdropped rifts (geologic landforms where the crust spreads and the material at the spreading center drops due to gravity) that were subsequently filled in with ice-rich lavas. This hypothesis best explains the observed landforms, but some smooth units may have formed by unconfined ice volcanism. Grooves are formed by extensional deformation after emplacement of the ice magma.

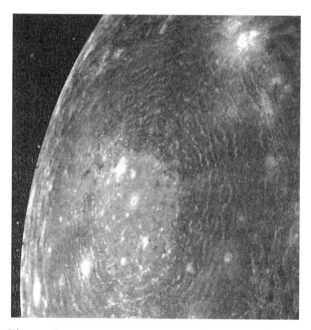

Figure 4. *Voyager 1* shaded image of Valhalla basin on Callisto (FDS 16422.11). Photo courtesy of the Lunar and Planetary Institute.

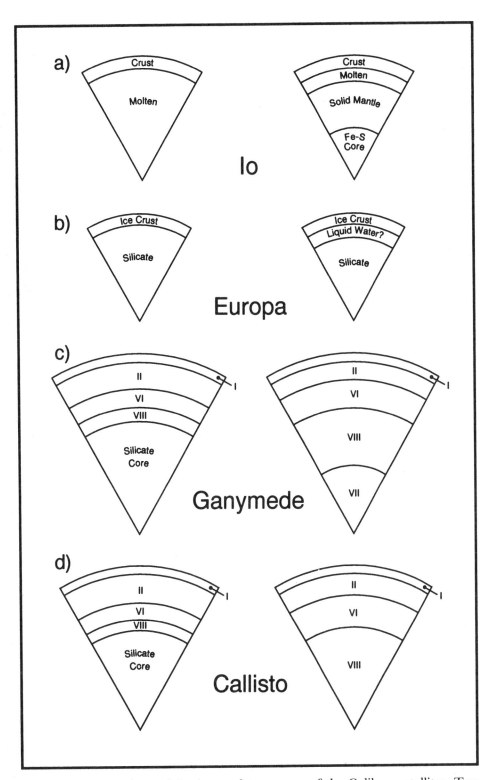

Figure 5. Comparison of the internal structures of the Galilean satellites. Two models are presented for each satellite: a. Io, b. Europa, c. Ganymede, d. Callisto. The roman numerals I, II, VI, VII, and VIII refer to different solid phases of water ice.

The internal structure of Ganymede is not known. Ganymede may have differentiated into a dense rock core, a lower density ice mantle, and a crust consisting of nearly pure ice. In this case the darker surface component would be meteoritic debris. Or Ganymede may have remained undifferentiated, and the crust and interior are simply a mixture of rock and ice.

Callisto

Callisto is considered the least surprising of the four Galilean satellites. It has the lowest density among the four, but spectral data indicate that it has less ice on its surface than Ganymede or Europa. As in the case of Ganymede, the non-ice component in the surface is probably similar to carbonaceous chondrite (see METEORITES).

The surface geology of Callisto is dominated by impact cratering. The large number and global coverage of impact craters make Callisto's surface the oldest of the four Galilean satellites, so that yet another trend seen with increasing distance from Jupiter is increasing surface age. Callisto's surface also exhibits extensive multiringed structures centered around palimpsests, not unlike the furrows on Ganymede. The largest of these are Valhalla (Figure 4) and Asgard. Minor ice volcanism has occurred along these rings.

One of the major controversies concerning the Galilean satellites is the origin of the Ganymede-Callisto dichotomy. Why are the surfaces so different geologically when the internal properties are so similar? Volcanism has been more active in the dark terrain on Ganymede than on Callisto, and Ganymede experienced emplacement of light terrain that has not occurred on Callisto. Tectonism has also been more active on Ganymede than on Callisto as seen in the formation of grooves on Ganymede.

Most hypotheses that attempt to explain this dichotomy include a warmer thermal evolution for Ganymede than Callisto. One possible cause is that Ganymede differentiated while Callisto's interior did not, although this is difficult to substantiate. Alternatively, because it is larger than Callisto, Ganymede may have crossed some "threshold" necessary for its greater geologic activity. Another plausible cause for Ganymede's divergent surface evolution is its 2:1 orbital resonance with Europa.

The evolution of this orbital resonance would have tidally heated Ganymede, causing the larger amounts of volcanism and tectonism seen on its surface, whereas Callisto did not experience such tidal forces (Figure 5).

Bibliography

BURNS, J. A. "Some Background About Satellites." In *Satellites*, eds. J. A. Burns and M. S. Matthews. Tucson, AZ, 1986.

CARR, M. H., H. MASURSKY, R. G. STROM, and R. J. TERRILE. "Volcanic Features of Io." *Nature* 280 (1979):729–733.

JOHNSON, T. "The Galilean Satellites." In *The New Solar System*, eds. J. K. Beatty and A. Chaikan. Cambridge, MA, 1982.

LUCCHITTA, B. K., and L. A. SODERBLOM. "The Geology of Europa." In *Satellites of Jupiter*, ed. D. Morrison. Tucson, AZ, 1982.

MALIN, M. C. and D. C. PIERI. "Europa." In *Satellites*, eds. J. A. Burns and M. S. Matthews. Tucson, AZ, 1986.

NASH, D. B., M. H. CARR, J. GRADIE, D. M. HUNTEN, and C. F. YODER. "Io." In *Satellites*, eds. J. A. Burns and M. S. Matthews. Tucson, AZ, 1986.

PANG, K. D. "Planetary Astronomy in Ancient and Modern China." *Bulletin of American Astronomy Society* 15 (1983):840.

SACHABER, G. G. "The Geology of Io." In *Satellites of Jupiter*, ed. D. Morrison. Tucson, AZ, 1982.

SHOEMAKER, E. M., B. K. LUCCHITTA, J. B. PLESCIA, S. W. SQUYRES, and D. E. WILHELMS. "The Geology of Ganymede." In *Satellites of Jupiter*, ed. D. Morrison. Tucson, AZ, 1982.

DEBORAH L. DOMINGUE

GARRELS, ROBERT M.

At the head of a small group of prescient students and colleagues, Robert M. Garrels led a revolution in the use of quantitative physical chemistry data for solving a host of low-temperature geochemistry problems, especially aqueous problems such as the importance of geochemical cycling and recycling. His half-century legacy in this field resulted in an abrupt and extremely productive new direction for low-temperature geochemistry and helped to bring about a new understanding of, and appreciation for, the use of physical chemistry methods and data in the earth sciences.

Robert Minard Garrels was born in Detroit, Michigan, on 24 August 1916, the second of three children of John Carlyle Garrels and Margaret Anne Garrels. Garrels spent his early years in the mountains of southwestern Virginia, where his father worked as a chemical engineer for a company that used salt and limestone as raw materials. In an unpublished autobiography he said that the three factors that most influenced his career were his father's trade, the rich natural setting of southwestern Virginia, and a gifted schoolteacher named James Moore. Garrels attended high school in Grosse Ile, Michigan, where he performed well in mathematics and science; his main interests, however, were athletics and reading. According to his autobiography, his first scientific thrill was the construction of a stroboscope for use in measuring the rate of rotation of an automobile wheel.

Garrels entered the University of Michigan in 1933, intending to follow his father's trade, chemistry, but instead he studied geology and obtained a B.S. degree in 1937. Garrels next enrolled at Northwestern University, and, after taking all the available geology courses, began to take courses in chemistry, a subject in which he found renewed fascination. Influenced by talented faculty, he began thesis work on the genesis of lead and zinc deposits in Newfoundland, and honed lifelong talents for hard work, high standards, and healthy scientific skepticism. His Ph.D. thesis, which was awarded by Northwestern in 1941, used electrochemistry techniques to determine complex ion formation between lead and chloride in aqueous solutions.

Garrels stayed at Northwestern as an instructor, but during World War II he joined the U.S. Geological Survey (USGS) in military affairs. Garrels returned to Northwestern and served until 1952, teaching and conducting research in ore deposits, as well as introductory courses in geology and physical science. He published *A Textbook in Geology* in 1951, but this groundbreaking work was not widely used, owing to its liberal use of quantitative chemistry and physics. He did, however, publish a series of important and widely read papers dealing with aqueous geochemical problems in fluid-inclusion geothermometry, crystal growth, complex ion formation, and oxidation reduction potentials (Eh). Included among these is the classic

"Origin and Classification of Chemical Sediments in Terms of Ph and Oxidation-Reduction Potentials," published with W. C. Krumbein in 1952, which put the indelible stamp of quantitative physical chemistry on understanding sediments and sedimentary rocks. R. A. Berner (1992), who was strongly influenced by Garrels and worked with him as a colleague, commented, "One couldn't think of Garrels without thinking of Eh-Ph diagrams."

Although he would return twice, Garrels left his faculty position at Northwestern to rejoin the USGS for three years as an administrator, dealing with experimental and theoretical aspects of uranium and vanadium geochemistry. Garrels returned to academia in 1957, assembling an energetic and productive laboratory group of students and associates at Harvard University. Garrels' associates, who made major contributions to their developing field, included R. A. Berner, O. P. Bricker, B. Hanshaw, H. C. Helgeson, and M. Sato. Garrels published two influential books in this period, *Mineral Equilibria at Low Temperature and Pressure* (1960), and its successor *Solutions, Minerals, and Equilibria* (1965), written with C. L. Christ (1965). The classic "A Chemical Model for Sea Water at 25 C and One Atmosphere Total Pressure," written with M. E. Thompson and published in 1962, showed how chemical activities (effective concentrations) of the main ions in seawater could be determined through the use of ion-pairing and complexing calculations.

Garrels returned again to Northwestern and began a series of productive investigations into mineral-equilibria controls of seawater composition, working with colleagues F. T. Mackenzie and R. Wollast, among others. Struck by the work of the Swedish chemist L. G. Sillen, Garrels and these colleagues developed hypotheses for the control of seawater chemistry by reaction (buffering) with silicate and carbonate minerals brought into the oceans by streams. These ideas and calculations led to models for the chemical mass balance between rivers and the oceans and the silicate-bicarbonate balance in the oceans. The central argument was that clay minerals, "degraded" by weathering, were reconstituted by reaction with cations and silica in seawater (reverse weathering), thus changing the chemical compositions of both. (Although the model helped to explain the processes whereby freshwater becomes seawater, later understanding of other influences, especially the widespread reaction of basaltic rocks with seawater at mid-ocean spreading centers, also become known as an important contribution to the development of seawater chemistry.)

A coming together of Garrels' ideas on chemical recycling was put forward in the important book *Evolution of Sedimentary Rocks* (1971), written with F. T. Mackenzie. Garrels and Mackenzie postulated that the formation of secondary materials (sediments, sedimentary rocks) from parental, primary igneous and metamorphic rocks through erosion, weathering, solution, transport, sedimentation, burial, and low-temperature alteration (diagenesis), is commonly followed by regional metamorphism, plutonism, and a resetting of the rock cycle. As R. A. Berner pointed out in his memoir on Garrels (1992), the fundamental concept of recycling, forcefully advanced by JAMES HUTTON 150 years earlier, was not well appreciated until work by T. F. W. Barth in the 1950s, and later and more quantitatively by Garrels and Mackenzie.

In 1969 Garrels moved to the Scripps Institution of Oceanography for two years, then to the University of Hawaii for another two years, returning once again to Northwestern for five years, and finally to the University of South Florida in 1979, remaining there until his death on 8 March 1988 in Saint Petersburg. Working with Mackenzie, A. Lerman, E. Perry, and R. Wollast, among others, Garrels continued the development of concepts on the temporal cycling and isotopic compositions of carbon, oxygen, and sulfur, work that continues to be useful in current models for climatic change. "The Carbonate-Silicate Geochemical Cycle over the Past 100 Million Years" (1983), an important paper written with Berner and A. Lasaga, argues that the greenhouse gas CO_2, affected by plate tectonic processes, could have produced global climatic change through geologic time.

R. M. Garrels was married to Jane Tinen and later to Cynthia A. Hunt, with whom he coauthored two books on water and the environment. He received many honors, among them membership in the National Academy of Sciences, the Penrose and Authur L. Day medals of the Geological Society of America, the Goldschmidt Award of the Geochemical Society, the Wollaston Medal of the Geological Society of London, and the Roebling Medal of the Mineralogical Society of America, as well as three honorary doctoral degrees.

Bibliography

BERNER, R. A. "Robert M. Garrels." *National Academy of Sciences Memoir* 61 (1992):195–212.
GARRELS, R. M., and C. L. CHRIST. *Solutions, Minerals, and Equilibria*. New York, 1965.

E. JULIUS DASCH

GEMS AND GEM MINERALS

Most gems are minerals and in the American gem trade the use of the word gem is restricted to stones of natural origin. A gem is defined as a mineral which, by cutting and polishing, possesses sufficient beauty to be used in jewelry or in personal adornment. Although beauty is a very important requirement, the more desirable and valuable gems are also both rare and durable. Several organic gem materials included in the term "gem" are:

- Pearl (natural and cultured), the result of deposition of a very fine grained form of calcium carbonate—aragonite crystals—as part of the nacre in a mollusk shell, which is formed in response to an irritation or a parasite.
- Amber, a fossil resin that was exuded from trees millions of years ago and which, through lengthy burial, oxidation, and loss of volatiles, has become a hardened material.
- Coral, which consists of another form of calcium carbonate (called calcite) that is deposited by very small marine animals—called coral polyps—as their exoskeletons. Precious coral ranges in color from white through light pink to dark pink, whereas black coral is brown to black. Both types of corals that are used are of present-day origin.
- Jet, a relatively soft, very dark brown to black, fossilized wood.

The great majority of well-known, durable, attractive, and valuable gems, however, are minerals of inorganic origin that show exceptional qualities of such properties as clarity, color, and color play. A mineral is defined as a naturally occurring homogeneous solid with a definite (but generally not fixed) chemical composition and a highly ordered atomic arrangement. It is usually formed by inorganic processes. There are approximately 3,500 known mineral species, but because a mineral must have certain qualifications to be placed in the special category of gem minerals, there are only about seventy minerals that meet these requirements. These are listed in Table 1, and of these there are only about fifteen mineral species that are considered important gem minerals (these are shown in bold type in Table 1).

Gem Qualifications and Cuts

Beauty is the most important of several attributes of gemstones. The factors that contribute to beauty are color, luster, transparency, clarity, the skillful cutting and polishing of stones, brilliance, and fire. Most gems possess several of these properties, but in some nontransparent stones, such as turquoise, beauty is expressed mainly by color. Opal, which owes its attractiveness to the flashes of spectral colors diffracted in the stone's interior, displays what is known as "play-of-color."

In addition to beauty a gem must exhibit durability, which means that it must be resistant to scratching and abrasion. Durability, which depends on the hardness and toughness of the material, is the second most important requisite of a gem mineral. A durable gem should have a minimum hardness value of 7 (that of quartz), as measured from 1 to 10 on the Mohs' scale. Only ten to twelve gems satisfy this requirement, but others cut from softer minerals retain their luster over time if worn with care. Gem materials with a hardness over 7 may still behave in a brittle fashion and a sharp blow may cause them to fracture or cleave. For example, a diamond with a hardness of 10 can be cleaved quite easily by a sharp impact.

Table 1. Gem Minerals

NATIVE ELEMENTS	TUNGSTATES	SILICATES *continued*
Diamond	Scheelite	Axinite
SULFIDES	PHOSPHATES	**Beryl**
Sphalerite	Beryllonite	**(emerald and aquamarine)**
Pyrite	Apatite	Cordierite
OXIDES	Amblygonite	**Tourmaline**
Zincite	Brazilianite	Enstatite-hypersthene
Corundum	**Turquoise**	Diopside
(ruby and sapphire)	Variscite	**Jadeite** (Jade)
Hematite	SILICATES	Spodumene
Rutile	Phenacite	Rhodonite
Anatase	Willemite	**Tremolite-actinolite**
Cassiterite	**Olivine (peridot)**	(nephrite jade)
Spinel	**Garnet**	Serpentine
Gahnite	**Zircon**	Talc
Chrysoberyl	Euclase	Prehnite
HALIDES	Andalusite	Chrysocolla
Fluorite	Sillimanite	Dioptase
CARBONATES	Kyanite	**Quartz**
Calcite	**Topaz**	**Opal**
Rhodochrosite	Staurolite	Feldspar
Smithsonite	Datolite	Danburite
Aragonite	Titanite	Sodalite
Malachite	Benitoite	Lazurite
Azurite	Zoisite	Petalite
SULFATES	Epidote	Scapolite
Gypsum	Vesuvianite	Thomsonite

Source: C. Klein and C. S. Hurlbut, Jr., *Manual of Mineralogy*, 21st ed. New York, 1993.
Important gem minerals are shown in bold.

In addition to beauty and durability, gems that are highly desirable are also rare and in fashion. Gems that are prized for their rarity are emerald, ruby, sapphire, and diamond. The type of gem that is highly desirable may change with the fashion of time. For example, during the Victorian era red pyrope garnets, mainly from Bohemia (in the western Czech Republic), were in much demand but are little used in jewelry today.

There are many ways in which rough gem minerals can be fashioned into gemstones. The two basic types, however, are the faceted cut and the cabochon cut. The faceted cut is achieved by carefully cutting and polishing various plane surfaces (facets) along the outside of the gem. Different names are given to these facets, depending on their position on the cut gem. When a faceted gem is viewed from above, the outline of the stone may be round, oval, rectangular, square, pear-shaped, and other shapes. Figure 1 shows the type of cut that is most common for diamond, the brilliant cut. The upper part of the stone is referred to as the

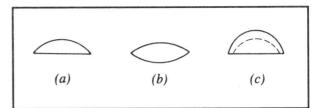

Figure 2. Cross section of cabochon cut stones. a. Simple cabochon with domed top and flat base. b. Double cabochon. c. Hollow cabochon. From C. Klein and C. S. Hurlbut, Jr., *Manual of Mineralogy*, 21st ed. New York, 1993.

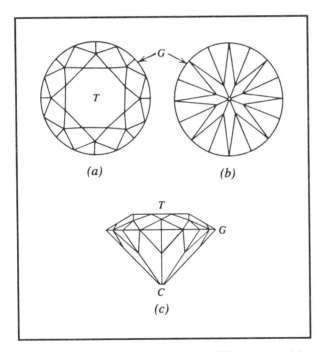

Figure 1. A faceted gemstone (brilliant cut) with 58 facets. a. The bezel showing the top facets. b. The pavilion showing the bottom facets. c. Front view showing the table (*T*), culet (*C*), and girdle (*G*). From C. Klein and C. S. Hurlbut, Jr., *Manual of Mineralogy*, 21st ed. New York, 1993.

bezel, crown, or top; the lower part is known as the pavilion, base, or back. The central top facet is the table and the small facet parallel to it (at the base) the cutlet. The cabochon cut generally consists of a smooth-domed top and a flat base (Figure 2). Generally, transparent gem material (with good clarity and without numerous small inclusions) is faceted, whereas translucent and opaque gems are cut as cabochons. Sometimes a desired optical effect dictates the cut. As such the cabochon cut is used for the best definition of the microscopic inclusions in the six-rayed star of star rubies and star sapphires.

The Most Important Gems

Diamond. Diamond is the hardest substance known, with a hardness of 10 on the Mohs' scale. It is composed of carbon (C), an element that also occurs naturally as graphite. Diamond and graphite are the two most common polymorphs of carbon. Diamond occurs in nature as single crystals of gem or of industrial quality and as very fine grained polycrystalline intergrowths known as boart. Although diamond is extremely hard, if it is struck in specific directions it will cleave. This cleavage property is used in the preparation of polished gem diamonds.

Natural diamond crystals occur most commonly in octahedral shapes (Figure 3) that reflect their internal isometric structure. Twinned crystals (Figure 3) and cleavage fragments are also common. The largest diamond found, the Cullinan, was a cleavage fragment.

Diamond is generally considered as the most highly prized of the gemstones. This is because it is

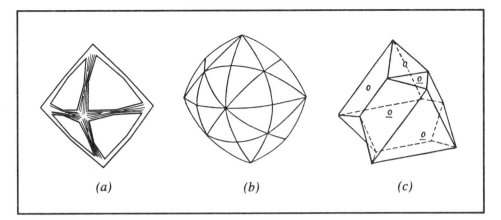

Figure 3. Examples of some diamond crystals. a. An octahedral crystal modified along the edges by hexoctahedral faces. b. A hexoctahedron. c. An octahedral crystal twinned according to the spinel law. From C. Klein and C. S. Hurlbut, Jr., *Manual of Mineralogy*, 21st ed. New York, 1993.

the hardest known mineral that also has a very high refractive index and dispersion. These optical properties account for the very high brilliance and fire of a well-cut diamond.

Faceted diamonds are graded on the basis of what is known as the four C's—carat, cut, clarity, and color. The carat is the unit weight used in the gem industry. It is standardized as 0.2 g (200 mg or 0.0071 oz) and is subdivided into 100 points. Therefore, a 20-point diamond weighs 0.2 ct (40 mg). The largest diamond ever found, the Cullinan, was the size of an average fist and weighed 3,106 ct. The cut is how well the facets and the shape of the faceted stone compare to a standardized model shape. A finely proportioned and finished diamond tends to be much more valuable than a stone of similar weight that is poorly cut. A diamond with a brilliant cut (see Figure 1 and Plate 8) is considered to be the best compromise between a high degree of brilliance and fire. The clarity of a cut diamond depends on the absence or presence of defects such as inclusions, cleavage planes, and other natural flaws. A flawless diamond is defined as one without such defects and with the greatest clarity. Although most diamonds are generally colorless or close to colorless, they can occur naturally in all colors. Such colored stones are known as "fancies," and they are highly prized when of good color and clarity. The most famous colored diamond is the blue Hope diamond (44.5 ct) on display in the Smithsonian Institution in Washington, DC.

Diamonds are formed in the upper part of the earth's mantle in a range of temperature of 900° to 1,200°C. They are transported to the earth's surface by explosive vents that are filled with rock types such as kimberlite and lamproite. The weathering and erosion of these rocks produce diamond occurrences in alluvial or placer deposits. The major diamond producing countries (including gem and industrial diamonds) are South Africa, Botswana, Zaire, Australia, Russia, Angola, the Central African Republic, and Ghana.

Diamonds can be produced synthetically by subjecting carbon to extremely high temperatures (in the range of 1,500°C) and simultaneously to very high pressures (in the range of 50 to 60 kilobars) in the laboratory. The General Electric Company of the United States has produced synthetic diamonds since 1955. The majority of these are used in industrial applications. Colorless gem-quality synthetic diamonds can be produced but their cost tends to be greater than that of the natural gem. The most common product that imitates diamond is cubic zirconia (ZrO_2 in composition; often abbreviated as CZ) with a hardness of 8 on the Mohs' scale, and with a refractive index and dispersion similar to that of diamond.

Ruby and Sapphire. Ruby is the red-colored gem variety of the mineral corundum (see Plate 9). Sapphire is the name for gem corundum of all other colors (see Plate 10), including the highly sought after cornflower-blue variety. Sapphire colors are specified as follows: purple sapphire, yellow sapphire, and so on.

The chemical composition of corundum is Al_2O_3, and crystals commonly show well-developed hexagonal or rhombohedral outlines. The internal structure of corundum is based upon the hexagonal closest packing of oxygen ions with the Al^{3+} ions in octahedral coordination between the oxygens. The external rhombohedral symmetry, in the hexagonal system, is a reflection of this internal structural arrangement. Corundum has a hardness of 9 and the common mineral variety ranges in color from some shade of brown, pink, or blue to colorless. The red color of gem ruby is attributed to the presence of small amounts of Cr^{3+} substituting (up to a maximum of about 4 weight percent) for Al^{3+}, which causes differential absorption of transmitted white light such that parts of the red spectrum are strongly enhanced. The blue color of gem sapphire is attributed to the presence of small amounts of Fe^{2+} and Ti^{4+} in substitution for Al^{3+}. Trace amounts of these two elements cause absorption of much of the white light spectrum, except in the region of the wavelength for blue light. In reflected light, both ruby and sapphire may show concentrations of microscopic fibrous inclusions that are arranged in three crystallographic directions at 120° to each other. A cabochon stone cut from such material will show the presence of a six-pointed star. This phenomenon, known as asterism, is present in star rubies and star sapphires.

Corundum is a common accessory mineral in some metamorphic rocks, such as crystalline limestone, mica-schist, and gneiss. It is also found as a primary constituent of silica-deficient igneous rocks such as syenites and nepheline syenites. The finest rubies have come from Myanmar (formerly Burma), where the most important locality is near Mogok, 145 km north of Mandala. Some stones are recovered from the metamorphic limestone that underlies the area, but most are found in the overlying soil and stream gravels. Rubies of fine quality are also mined in southern Kenya. Darker, poorer quality rubies have been found in alluvial deposits near Bangkok, Thailand, and Battambang, Cambodia. Sapphires are found in alluvial deposits near Bangkok, Thailand, and Battambang, Cambodia. Sapphires are found in alluvial deposits of Thailand, Sri Lanka, and Cambodia associated with rubies. The stones from Cambodia, of a cornflower-blue color, are most highly prized. Sap-

phires are also found in central Queensland, Australia, and Kashmir, India. In the United States small sapphires of good color are found in several localities in Montana.

Both ruby and sapphire can be produced by synthetic processes. The synthetically produced colors of red and blue can be strikingly similar to those of the natural gems. Star rubies and star sapphires are also synthetically grown, and cabochon cut stones of these materials rival the beauty of the natural stones they imitate.

Turquoise. Turquoise is a gem mineral with a distinctively strong sky-blue, or bluish green to apple green color (see Plate 11). It is a member of the phosphate group of minerals with a generalized composition of $CuAl_6(PO_4)_4(OH)_8 \cdot 5H_2O$. Considerable Fe^{2+} may substitute for copper and Fe^{3+} may substitute for part or all of the aluminum forming a complete chemical series from turquoise to chalcosiderite, $CuFe_6(PO_4)_4(OH)_8 \cdot 5H_2O$.

Crystals of turquoise are rare and are triclinic in symmetry. Most turquoise is massive, dense, and cryptocrystalline to fine-granular. It commonly occurs as veinlets or crusts and in stalactitic or concretionary shapes. It has a hardness of about 5 to 6 and a vitreous to waxy luster. The distinctive light blue coloration of much turquoise is the result of the presence of Cu^{2+}; limited substitution of the copper by Fe^{2+} produces greenish colors.

Turquoise is a secondary mineral, generally formed in arid regions by the interaction of surface waters with high-alumina igneous or sedimentary rocks. It occurs most commonly as small veins and stringers traversing more or less decomposed volcanic rocks. Since the times of antiquity turquoise of very fine quality has been produced from a deposit in Persia (now Iran) northwest of the village of Madèn, near Nishapur. It occurs also in Siberia, Turkistan, the People's Republic of China, the Sinai Peninsula, Germany, and France. The southwestern United States has been a major source of turquoise, especially the states of Nevada, Arizona, New Mexico, and Colorado. Extensive deposits in the Los Cerillos Mountains, near Santa Fe, New Mexico, were mined very early by Native Americans and were a major early source of gem turquoise. However, much of the supply of gem quality turquoise has now been depleted in the southwest.

Because of the great demand for turquoise at relatively low prices, and the scarcity of gem quality turquoise that is generally highly priced, many materials that are not completely natural or that are turquoise-colored imitations have appeared on the market. Some of these are:

1. Stabilized turquoise, which is a poor quality natural turquoise that has been chemically impregnated and hardened with organic resins to improve the color and the hardness of the final product. Such material becomes very workable, permanently hardened, stable, and can be very attractive.
2. Oil treated turquoise, in which the color of the natural turquoise has been enhanced by impregnation with oil, paraffin, or oil based polishes. Such treatment generally results only in the temporary improvement of color.
3. Treated turquoise, which means natural or stabilized turquoise that has been dyed to enhance the color.
4. Reconstituted turquoise, which means dust, turquoise particles, or nuggets bonded together with plastic resins.
5. Imitation turquoise, which means any natural or synthetic compound or mineral that is manufactured or treated (e.g., turquoise colored plastics, glass enamel, dyed magnesite, and dyed chalcedony) so as to closely approximate turquoise's appearance.

Peridot (Olivine). Olivine is a rock-forming silicate mineral that ranges most commonly in composition from Mg_2SiO_4 to Fe_2SiO_4. The Mg-rich end member is known as forsterite and the Fe-rich end member as fayalite. The members of this series are orthorhombic in symmetry and occur most commonly as embedded grains or granular masses in dark colored igneous rocks such as gabbro, peridotite, and basalt. The uncommon gem variety of the olivine series is known as peridot; it has a chemical composition in which approximately 10 atomic percent of the Mg in forsterite is replaced by Fe (the peridot composition therefore is commonly noted as 90% forsterite, 10% fayalite). It has a vitreous luster, is transparent to translucent, and ranges in color from yellowish green to greenish yellow and brownish green (see Plate 12). The color is the most diagnostic property of peridot, but distinguishing a cut stone from green chrysoberyl or garnet or synthetic green spinel may require special tests such as those for refractive index and specific gravity. The characteristic green color

of peridot is the result of Fe^{2+} in octahedral coordination in the olivine structure causing differential absorption of much of the red color component of white light, resulting in an overall green coloration.

The classic locality for gem peridot is the island of Zebarget in the Red Sea, about 80.5 km from the Egyptian port of Berenice. High-quality large stones are found in Myanmar. The largest commercial source of gem-quality peridot in the United States is from the San Carlos Apache Indian Reservation in Arizona, where the peridot occurs in irregularly shaped granular masses in a very dark grey to black olivine-rich basalt known as basanite. The granular, irregularly shaped olivine-rich masses within the basalt are considered to have formed originally in the upper part of the earth's mantle, and are called mantle xenoliths.

Garnet. The name garnet refers to a common rock-forming silicate group of minerals, all of which have the same isometric internal structure but show extensive ranges of chemical composition. The external form of garnet crystals is commonly highly diagnostic (Figure 4). Although the public associates the term garnet most commonly with a mineral that is red in color, the color of garnet (see Plate 13) is highly variable as a function of variations in chemical composition. The hardness of garnet ranges from 6.5 to 7.5.

The extensive variation in chemical compositions of garnet is expressed not only in chemical formulas but also by various end-member species names that apply to the garnet group as follows:

$$Pyrope—Mg_3Al_2Si_3O_{12}$$
$$Almandine—Fe_3Al_2Si_3O_{12}$$
$$Spessartine—Mn_3Al_2Si_3O_{12}$$
$$Grossular—Ca_3Al_2Si_3O_{12}$$
$$Andradite—Ca_3Fe_2Si_3O_{12}$$
$$Uvarovite—Ca_3Cr_2Si_3O_{12}$$

Among the above six end-member species there is extensive solid solution between pyrope and almandine and spessartine. That is, the elements Mg, Fe, and Mn substitute for each other extensively in specific atomic sites of the garnet structure. Similarly there is much solid solution among Al^{3+} and Fe^{3+} in the grossular-andradite series. Garnets most commonly are constituents of metamorphic rocks such as mica-garnet schists and gneisses. Some can also be of igneous origin.

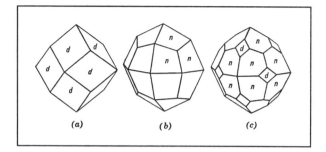

Figure 4. Examples of some garnet crystals. a. Dodecahedron. b. Trapezohedron. c. A trapezohedron modified by a dodecahedron. From C. Klein and C. S. Hurlbut, Jr., *Manual of Mineralogy*, 21st ed. New York, 1993.

Although pyrope would be colorless, substitution of some Fe^{2+} and also some Cr^{3+} in the pyrope structure gives rise to dark red and orange colors. The best-known occurrence of gem-quality pyrope is in the Czech Republic, formerly known as Bohemia, and it is called "Bohemian garnet."

Almandine and members of the pyrope-almandine series are the most common red, reddish orange, and brownish red garnets. Because they are common rock-forming minerals, gem localities are found in many countries including India, Tanzania, Madagascar, Sri Lanka, and Brazil.

Spessartine is less common than pyrope and almandine and can be of gem quality in beautiful yellowish orange to reddish orange cut stones.

Grossular may also be red but orange-yellow ("hessonite") and orange-brown ("essonite") material has been most widely used as gem material. The most important locality for "hessonite" is in gravels in Sri Lanka. An emerald green variety of grossularite, called "tsavorite," has recently been discovered in Tsavo National Park, Kenya. Such stones are generally small but highly valued. Their bright green color is due to the presence of small amounts of vanadium in the crystal structure.

Andradite ranges from yellow green to black in color but only the transparent green variety known as "demantoid" is an important gem. Its green color is due to the replacement of some of the iron in the crystal structure by chromium. Demantoid is a rare gem and stones of several carat size can be expensive. The main source of demantoid is in the Ural Mountains, Russia.

Uvarovite occurs as brilliant deep-green grains and crystals, in association with chromium mineralization in the Ural Mountains, Russia, and the Outokumpu area, Finland. The grains and crystals are too small for cutting, but if they were larger they would make lovely gems.

Topaz. Topaz is a relatively uncommon silicate mineral of composition $Al_2SiO_4(OH, F)_2$. It has orthorhombic symmetry (Figure 5) and a hardness of 8. It crystallizes from fluorine-rich vapors that are given off during the last stages of solidification of siliceous igneous rocks. It occurs in cavities in rhyolitic lavas and in granite, and it is a characteristic mineral in pegmatites, especially those that contain tin. Topaz has a wide range of colors, including yellow, orange, brown, pink, red, purple red, blue, light green, and also may be colorless. The blue color of natural or artificially irradiated topaz is the result of slight imperfections in the atomic structure of topaz, structural defects that are also known as a "color centers." The highly prized yellow orange to orange brown variety, known as "imperial topaz" (see Plate 14), occurs in the region of Ouro Preto, Minas Gerais, Brazil, in kaolinized veins. Fine blue colored topaz (see Plate 15) is also found in Minas Gerais, Brazil; in Mardan, Paki-

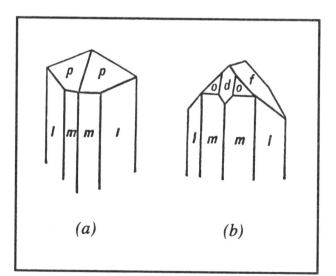

Figure 5. Examples of the orthorhombic form development in topaz. a. Vertical prisms (*l* and *m*) terminated by a dipyramid (*p*). b. Vertical prisms (*l* and *m*) terminated by several forms (*o*, *d*, and *f*). From C. Klein and C. S. Hurlbut, Jr., *Manual of Mineralogy*, 21st ed. New York, 1993.

stan; in Omi and Mino prefectures, Japan; in San Luis Potosí, Mexico; and in the gem gravels of Sri Lanka and Myanmar. The pale blue color of the highest prized natural topaz can be imitated by artificial irradiation, commonly followed by heat treatment, of originally colorless topaz. Such treated topaz is often referred to in the trade as "London blue," or "super blue."

Emerald and Aquamarine (Beryl). Emerald and aquamarine are two of several gem varieties of a relatively common rock-forming silicate—beryl. The chemical composition of beryl is $Be_3Al_2Si_6O_{18}$. It is the most abundant beryllium-containing mineral and is the principal source of this lightweight metal that is similar to aluminum in many of its properties. It has a hardness of 7.5–8, and it crystallizes commonly in prismatic crystals with a well-developed hexagonal shape (Figure 6). It occurs usually in granitic rocks, or in pegmatites. It is also found in mica schists and associated tin ores.

Emerald (see Plate 16), the deep green, transparent variety of beryl, is one of the most highly prized gems, and may have a greater value than, for example, diamond. The deep green color of emerald is the result of small amounts of Cr^{3+} replacing some of the Al^{3+} in six-coordinated (octahedral) sites in the beryl structure. When white light traverses such a Cr-containing beryl, most of the violet, blue, orange, and red components of the light spectrum are absorbed, resulting in a distinc-

tive emerald green color. Most natural emeralds contain abundant microscopic crystalline inclusions as well as fluid inclusions. They may also show extensive fine fractures. The most important localities for emerald mining are in Colombia, in the districts of Muzo and Chivor. Brazil is also a source of emeralds, as are Pakistan, Zambia, and Zimbabwe.

Emeralds have, for many years, been produced by synthetic processes. In order to distinguish between a natural and a synthetic emerald, the expertise of a gemologist is required. He or she can ascertain the origin of a stone on the basis of a combination of tests of refractive index, behavior under ultraviolet light, and the shape and size of possible inclusions.

The light to medium blue to blue green colored, transparent gem variety of beryl is known as aquamarine (see Plate 17). This color is the result of small amounts of iron replacing Al in the beryl structure. The iron can occur as Fe^{2+} or Fe^{3+} and charge transfer reactions (also known as molecular orbital transitions) take place in the beryl structure between adjacent ions of Fe^{2+} and Fe^{3+}; this involves the transfer of valence electrons between the two ionic states of iron, which causes absorption of many of the wavelengths of white light, except those belonging to the blue part of the spectrum. Gem-quality aquamarine occurs in several localities in Minas Gerais, Brazil.

Several other gem varieties of beryl are given specific names on account of their color. "Heliodor" or "golden beryl" (Plate 17) is light to medium to dark yellow. "Morganite" is a pink to orange pink variety and "goshenite" is colorless.

Tourmaline. Tourmaline is a fairly common rock-forming silicate with a very complex range of compositions. A general formula representation for its chemistry is $(Na, Ca)(Li, Mg, Al)(Al, Fe, Mn)_6(BO_3)_3(Si_6O_{18})(OH)_4$. As such, it is a complex hydrous silicate of B and Al, with substitutions of Ca for Na, of Mg and Al for Li, and of Fe^{3+} and Mn^{3+} for Al. The internal, atomic structure of this mineral consists of six-fold rings of Si_6O_{18} composition, which are cross-linked to each other by triangular (BO_3) groups and octahedral groups of $(Li, Mg, Al)O_4(OH)_2$ and $(Al, Fe, Mn)O_5(OH)$ composition. Well-formed crystals of tourmaline are not uncommon and if they are doubly terminated they exhibit different crystal forms at the top and bottom of the crystal (Figure 7). Their overall sym-

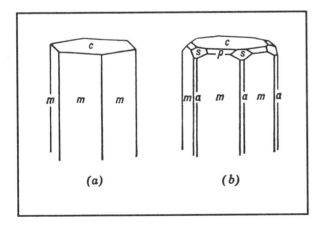

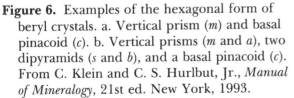

Figure 6. Examples of the hexagonal form of beryl crystals. a. Vertical prism (*m*) and basal pinacoid (*c*). b. Vertical prisms (*m* and *a*), two dipyramids (*s* and *b*), and a basal pinacoid (*c*). From C. Klein and C. S. Hurlbut, Jr., *Manual of Mineralogy*, 21st ed. New York, 1993.

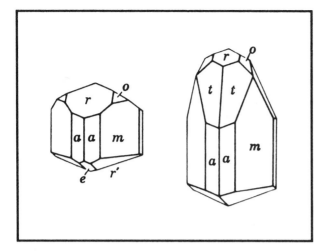

Figure 7. Examples of the complex hexagonal (trigonal) crystal form development in doubly terminated tourmaline crystals. From C. Klein and C. S. Hurlbut, Jr., *Manual of Mineralogy*, 21st ed. New York, 1993.

metry is part of the hexagonal system, with pronounced trigonal (three-fold) symmetry.

The most common and characteristic occurrence of tourmaline is in granite pegmatites and in rocks immediately surrounding them. It is also found as an accessory mineral in some igneous and metamorphic rocks.

Tourmaline is the gem mineral that occurs in the greatest variety of colors (see Plate 18). Single crystals may show distinct different color zones across the crystal (Plate 18). The many colors are the result of variations in the complex chemistry of the mineral. Gemologists and jewelers commonly use various varietal names to describe tourmaline of a specific color, although it is more informative to describe all colors of the mineral by an adjective to tourmaline, such as blue tourmaline, or pink tourmaline, and so on. Several of these varietal names are: "achroite," colorless; "indicolite," violet blue to greenish blue; "rubellite," brown to orange, or purplish pink to red; "verdelite," blue green to yellowish green. Cut stones or crystals showing more than one distinct color are known as "particolored" while those exhibiting two color zones are called "bicolored." "Watermelon" is a term used for tourmaline crystals or cut gems that are pink to red at the center and green at the periphery.

Artificial irradiation can produce pink to red colors in material that was originally colorless or light blue or light green.

Opal. Opal is a relatively common mineral in its nongem form. This is known as common opal and lacks the play-of-color for which gem or precious opal is known. All opal is of relatively simple chemical composition, $SiO_2 \cdot nH_2O$ in which n represents a variable amount of H_2O that is housed in the internal structural arrangement of the mineral. The water content generally ranges from about 4 to 10 weight percent but may be as high as 20 weight percent. The hardness of opal ranges from 5 to 6. The specific gravity and refractive index of opal decrease as a function to increasing H_2O content. Opal ranges from dense and glassy with a conchoidal fracture to quite porous such as siliceous geyserite, which occurs at the site of thermal springs and geysers. In addition to being formed at the site of hot springs, opal is deposited by meteoric waters, or by low-temperature hydrothermal solutions. It is found lining and filling cavities in rocks and may replace wood buried in volcanic tuff. Precipitation may result from evaporation or occur by organic action of organisms such as siliceous tests and diatoms (this is known as diatomaceous earth or infusorial earth).

Most opal is found to be amorphous, that is, apparently lacking an ordered internal, atomic structure, when studied by optical or X-ray diffraction techniques. Transmission electron microscope (TEM) studies have shown that the internal arrangement of opal is not made up of cations and anions in a regular ordered, structural array, but instead consists of closely packed spheres of silica in hexagonal and/or cubic closest packing. The sphere diameters vary from one opal to another and range from 1,500 to 3,000 Ångstroms (Å— $1Å = 10^{-8}$ cm). Air and H_2O molecules occupy the voids between the spheres.

The color of common opal ranges from transparent, glassy, and colorless to white and bluish white. Common pigmenting agents such as iron produce yellow, brown, red, green, and commonly several colors in a single specimen. Precious opal displays a play-of-color (see Plate 19) that is the result of white light being diffracted by the relatively regular internal array of silica spheres. That is, when white light passes through an essentially colorless opal, it strikes the planes of spheres and voids between them, and at these interfaces the white light is diffracted in such a way that certain wavelengths of the light flash out of the stone as nearly pure spectral colors. Some wavelengths may be completely internally reflected so the full color spectrum may not escape from a flat surface; however, the curved surface of a cabochon cut opal will allow most or all colors to escape.

Several trade terms are used to describe the appearance of opal based on transparency, body color, and the type of play-of-color. Some of these terms are: "black," translucent to almost opaque, dark gray to black body color, with play-of-color; "fire opal," transparent to semitransparent, with yellow, orange, red, or brown body color and with or without play-of-color; "harlequin" or "mosaic" opal in which the play-of-color occurs in distinct, broad, angular patches; "matrix opal," which consists of thin seams of high-quality gem opal in a matrix.

Because gem opal is commonly found in very thin veins and stringers inside a rock matrix, and because it tends to be fragile, much material may be unusable in gem and jewelry applications. Such material can, however, be mounted in composites known as opal doublets or triplets. Doublets consist of a thin slice of opal backed by a dark-colored material, which may be a dyed black chalcedony or black glass. Opal triplets are constructed the same way except that a colorless cap, commonly made of clear, transparent quartz, is cemented to the top of the opal section. This top cap provides added durability for the stone when set in jewelry. Because opal is a hydrous mineral, considerable care is required in the polishing and handling of the material.

Synthetic opal has been produced by Pierre Gilson in Switzerland since 1970. This product is identical to natural material in its chemical and physical properties, including a beautiful play-of-color. A lay person would be unable to distinguish this material from natural precious opal. Imitations such as colored man-made glass with some internal color reflections, and a few plastics, reproduce some aspects of precious opal to various degrees.

The largest source of precious opal is Australia, where it occurs in various districts as vein or void fillings in sedimentary rocks. Important districts are located in Queensland, New South Wales, South Australia, and Western Australia. In the Lightning Ridge area of the New South Wales, it occurs in voids formed by the leaching of fossils, woods, and various minerals.

Bibliography

AREM, J. E. *Color Encyclopedia of Gemstones*, 2nd ed. New York, 1987.

GEMOLOGICAL INSTITUTE OF AMERICA. *Gem Reference Guide*. Santa Monica, CA, 1993.

HURLBUT, C. S., JR., and R. C. KAMMERLING. *Gemology*, 2nd ed. New York, 1991.

KLEIN, C., and C. S. HURLBUT, JR. *Manual of Mineralogy*, 21st ed. New York, 1993.

LIDDICOAT, R. T. *Handbook of Gem Identification*, 12th ed. Santa Monica, CA, 1989.

CORNELIS KLEIN

GEOBAROMETRY

See Geothermometry and Geobarometry

GEOCHEMICAL TECHNIQUES

Earth scientists can use the chemistry of the rocks, minerals, liquids, and gases they study to help determine the origin and subsequent evolution of geological materials (*see* EARTH, COMPOSITION OF). Analytical methods in geochemistry have addressed problems of interest that range in scale from determining the growth history of small crystals to addressing the origin of the solar system. A wide variety of analytical tools that are available are described below.

Each element in the periodic table has unique physical and chemical properties that allow us to distinguish its respective concentration in any rock sample. Recall that an element is defined by the number of protons that provide a positive charge in the nucleus of the atom. Neutrons with similar mass accompany the protons in the nucleus but carry no electronic charge. Many elements have more than one stable nucleus, each with a different number of neutrons. These are the different isotopes of the element. Electrons of negative charge surround the nucleus by filling shells of progressively higher energy level and balance the positive charge of the nucleus. Each electron in the atomic structure of each element has a particular energy associated with it. Geochemical analytical techniques take advantage of these elemental properties to determine quantitatively the concentrations of each of the elements that are found on Earth.

There are two general types of analysis. Bulk analyses on whole rocks (or on assemblages of many grains of one mineral separated from a rock) were developed first, and many of these techniques are still popular today. Bulk analyses require samples that range in size from a few milligrams or more. Microbeam analyses were developed in the mid–1960s. They involve focusing a beam of electrons, ions, or X rays on a carefully selected "spot" on the mineral of interest. The interaction of the beam with the atomic structure of the sample generates a signal that is converted into a quantitative analysis. The amount of material analyzed ranges from a thousandth down to a billionth of a gram.

Bulk Analysis

Bulk chemical analysis of geological materials takes two general forms: (1) part of the rock is ground up and analyzed; and (2) after light crushing, specific minerals are selectively plucked from the rock for analysis. In both cases, the sample of interest is ground to a fine powder. Traditional wet chemical analyses are performed by dissolving the powder in acid and progressively titrating out successive precipitates of familiar chemical compounds. Modern bulk analyses of rocks are most commonly obtained using X-ray fluorescence, atomic emission spectroscopy, neutron activation analysis, or mass spectroscopy.

X-Ray Fluorescence. In X-ray fluorescence, a known mass of the powdered sample is combined with a flux (usually a lithium-boron oxide) and heated to a few hundred degrees centigrade. The rock dissolves into the low-melting flux material, and the resulting liquid is cooled to a glass disk. This disk is then bombarded with a beam of X rays (commonly Cu X-rays generated by the impact of an electron beam on a copper metal target.) The

incoming X rays have sufficient energy to "knock out" electrons from the individual atoms dissolved in the glass, and another electron in the atom commonly "falls" into the vacant spot, or orbit (Figure 1). The "falling electrons" jump from a higher energy state to a lower energy state, and thus release a quantity of energy in the form of X rays. Each element in the periodic table has a characteristic X-ray spectrum that shows up as a series of peaks of different intensities at certain energies. In X-ray fluorescence, the intensities of selected peaks are measured with an X-ray detector and compared to the signal observed from a sample of precisely known composition. X rays from light elements (H, He, Li, Be, B, C, N, O, F) are difficult to measure because of their very low energies. Thus the lithium-boron oxide flux is invisible to the detector. X-ray fluorescence analysis is sensitive, with levels of detection down to about 1 to 10 mg/g for many elements.

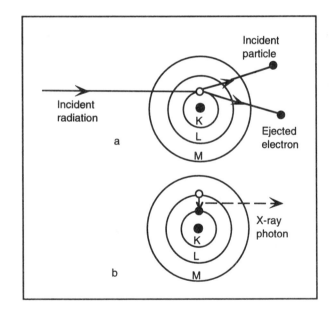

Figure 1. Production of X rays. a. An electron from an orbital close to the nucleus of an atom (the K shell) is ejected by incident radiation (electrons or X rays). b. An electron from an outer orbital (L or M shell) moves to the vacant orbital with lower energy, and an X ray is emitted. The energy of the X ray depends on the difference in energy between the two orbitals, and so depends on the element emitting it.

Inductively Coupled Plasma-Atomic Emission Spectroscopy (ICP-AES). When an element is heated to a very high temperature, it will emit light with frequencies that are unique to that element. Similarly, if a bulk sample is heated to high temperature, each of its component elements will emit light with its characteristic frequencies such that the intensity of the emitted light is proportional to the concentration of the emitting element. This is the principle behind the analytical technique of emission spectroscopy. Recent advancements in generating inductively coupled plasmas (ICP) have made this method the preferred technique for generating the high temperatures required for this analysis. ICP involves the generation of a stream of high-energy ionized gas (a plasma). The plasma is generated when ions of argon (Ar) couple to an oscillating magnetic field that is established by an induction coil that surrounds the gas stream. The motion of the particles within the oscillating field absorbs the energy from the coil, and gas temperatures reach 10,000 K. Small amounts of the sample are allowed to flow into the argon plasma stream where it quickly dissociates, ionizes, and its constituent elements become excited so that light is emitted. Diffraction gratings separate the emitted light into its component wavelengths, and photomultiplier tubes detect and record the relative intensities at each wavelength. In some instruments, the ions that are generated are steered into a mass spectrometer for analysis (see below). Most major and trace elements of geologic interest are amenable to ICP analysis with detection levels on the order of 1 mg/g.

Neutron Activation. A variety of analytical techniques utilize a stream of particles that interact with the nucleus of the atoms in a sample of interest. In the technique of neutron activation, the same is bombarded with a stream of neutrons that are absorbed by the nuclei of its constituent elements. The neutrons are generated by controlled fission of the radioactive isotope ^{235}U in nuclear reactors, slowed down to manageable velocities with the use of a moderator (typically heavy water; D_2O), and then directed toward the sample of interest. When a neutron enters the nucleus of an atom, the product nucleus is often transformed into an unstable isotope and becomes radioactive. The induced intensity of radioactivity in a sample is dependent on which elements are present, as well as the relative concentrations of these ele-

ments, and thus allows for accurate determination of the concentration of many trace elements in rock and mineral specimens.

Mass Spectroscopy. Analytical techniques involving mass spectroscopy take advantage of the fact that different elements (and different isotopes of the same element) have different masses. In this technique, the sample of interest is converted into a stream of charged ions that are accelerated down an evacuated flight tube. A magnet is situated around the tube so as to generate a magnetic field that interacts with the accelerated particles in order to bend their flight path; heavier ions are bent to a lesser degree than are ions of smaller mass. Ion collectors are situated at the end of the flight tube and, through electronic means, record the quantity of atoms that strike the detector. Variation of either the magnetic field or the accelerating voltage allows the investigator to record a mass spectrum of the sample, such that the peak height associated with each mass is proportional to the concentration of the isotope with that mass. This analytical technique is capable of measuring the concentration and isotope composition of many elements, including the light atoms such as hydrogen and oxygen as well as the heavier atoms such as uranium and lead (*see* ISOTOPE TRACERS, RADIOGENIC; ISOTOPE TRACERS, STABLE).

Microbeam Analysis

Microbeam analytical techniques allow earth scientists to obtain chemical information on specific areas of minerals within rocks. This is very useful for analyzing very small rock samples, as in the case of meteorites, or for analyzing experimental-run products synthesized in very small capsules. In addition, microbeam techniques allow for the study of small-scale variations such as those observed in zoned igneous and metamorphic minerals. The sample is typically a polished section of a rock about 2.54 cm in diameter ground to a thickness of 30 micrometers (μm). Microbeam techniques include the electron microprobe, synchrotron source X-ray fluorescence, secondary ion mass spectrometry, and laster ablation atomic emission spectroscopy.

Electron Microprobe. As for X-ray fluorescence (described above), the electron microprobe technique utilizes the X rays emitted from a selected mineral to obtain a chemical analysis. Within the electron microprobe instrument, however, the X rays are generated by the impact of an energetic (10 to 25 keV) electron beam. The electron beam is focused to a fine spot (typically from 1 to 20 μm in diameter). The electron microprobe is particularly useful at determining the concentrations of elements in minerals with abundances greater than 100 mg/g. This includes all the major mineral-forming elements and many trace elements of geologic importance. As is typical of microbeam techniques, the sample is not destroyed by the analysis.

Synchrotron Source X-ray Fluorescence. Physicists studying the structure of the atom have built large accelerators to obtain the products of collisions between atomic particles at energies of several MeV. X rays are also produced in these accelerators and are so intense that even if they are blocked by a plate with a 30 μm pinhole, the resulting 30-μm-diameter beam will generate X rays from the elements in a sample which are then detected and quantified in a manner similar to that for bulk X-ray fluorescence analysis. Because the X rays are at such a high energy, samples must be thicker than typical geological thin sections (30 μm), but the high intensity allows measurement of low levels of many elements, and the X-ray bombardment does not destroy the sample.

Secondary Ion Mass Spectrometry (SIMS). In SIMS, a beam of positive or negative primary ions is focused to a spot (from 1 to 100 μm in diameter) and directed at a mineral or glassy phase of interest. Ions strike the sample at impact energies of a few to 20 keV, and each impact knocks off, or "sputters," several atoms from the sample surface (Figure 2). Some of these atoms (less than 1 to 10 percent) are ionized during the sputtering process, and the resulting charged particles are steered into a mass spectrometer (described above) where they are detected according to their mass. The sample is consumed during analysis, but detection levels are typically so low that only a small amount (1 to 3 billionths of a gram is common) is needed to determine the concentration of the element of interest. The mass spectrometer method of detection is used to determine the variation in isotope ratios of certain elements. In some cases, mineral ages are determined on very small spots (approximately 30

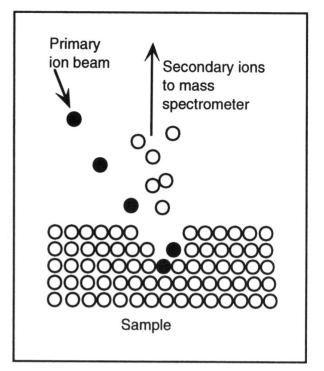

Figure 2. Generation of secondary ions in SIMS. A primary ion beam strikes a sample, knocking atoms out of the first two or three atomic layers. Some ejected atoms are ionized in the process and a positive or negative voltage on the sample pushes like-charged ions into a mass spectrometer for chemical identification and quantification.

μm diameter). In other cases, changes in the mineral environment during its growth are reflected by variable isotope ratios from the interior to the edge of a crystal.

Laser Ablation Inductively Coupled Plasma Atomic Emission/Mass Spectroscopy. With the use of a high energy laser, the analytical techniques of atomic emission spectroscopy and mass spectroscopy are now applied to geological samples with the advantages of a microbeam technique. A high-energy laser is directed at the area of interest (from tens to a few hundred micrometers in diameter) to blast (or ablate) material from the surface. The ejected material is transported into a plasma torch where chemical analysis is performed by either (1) the characteristic emission spectrum of the element, or (2) by accelerating the ions down an attached mass spectrometer flight tubes (see methods described above). Laser ablation consumes more of the sample than does SIMS, but the precision of the measurements can be significantly better than SIMS.

Bibliography

BENNINGHOVEN, A., F. G. RÜDENAUER, and H. W. WERNER. *Secondary Ion Mass Spectrometry: Basic Concepts, Instrumental Aspects, Applications, and Trends.* New York, 1987.

FAURE, G. *Principles of Isotope Geology.* New York, 1986.

GOLDSTEIN, J. I., D. E. NEWBURY, P. ECHLIN, D. C. JOY, C. FIORI, and E. LIFSHIN. *Scanning Electron Microscopy and X-Ray Microanalysis.* New York, 1984.

SMITH, K. A. *Soil Analysis.* New York, 1991.

PHILLIP DEAN IHINGER
RICHARD HERVIG

GEOCHEMISTRY, HISTORY OF

The science of geochemistry developed largely during the twentieth century; nevertheless, the concept of an autonomous discipline dealing with the chemistry of Earth is an old one, and the term "geochemistry" was introduced by the Swiss chemist Christian Schönbein (discoverer of ozone) in 1838. The history of geochemistry naturally includes much of the history of chemistry and geology. Because geochemistry is basically concerned with abundance and distribution of the chemical elements, their discovery and recognition are landmarks in the history of the subject. The modern concept of an element can be said to date from Antoine-Laurent Lavoisier's definition of his "Traité élémentaire de chimie" (1789), although some seventeenth- and eighteenth-century scientists certainly understood the distinction between elements and compounds. Lavoisier recognized thirty-one elements. This number was steadily increased during the nineteenth century, mainly through the discovery and analysis of the minerals containing them. The introduction of the periodic table by Dmitri Mendeleev in 1869 provided a limit on the total number of elements and enabled the prediction of missing ones—Mendeleev's eka-aluminum (gallium, Ga), eka-boron (scandium, Sc) and eka-silicon (geranium, Ge). In 1894 argon, the first of the noble gases, was discovered, followed rapidly by the remaining members of the group—neon (Ne), helium (He), krypton (Kr), and xenon (Xe). The discovery of radioactivity by Henri Becquerel in 1896 resulted in the recognition of polonium (Po) and radium (Ra) by the Curies in 1898 and actinium (Ac) by André Debierne in 1899.

By 1900, therefore, the periodic table was essentially complete, except for some short-lived radioactive elements and for europium (Eu, 1901), luketium (Lu, 1907), hafnium (Hf, 1923), and rhenium (Re, 1925). Proof of this completion, however, was first provided in 1914, when H. G. J. Moseley (1888–1915) demonstrated the correlation between X-ray spectra and the atomic numbers of the elements.

Throughout the nineteenth century geochemical data were mainly the by-product of general geological and mineralogical investigations and consisted of more and better analyses of the various units—minerals, rocks, natural waters, and gases—that make up the accessible parts of Earth. For many years this work was largely confined to European laboratories, but with the organization of the U.S. Geological Survey and the appointment of F. W. Clarke (1847–1931) as chief chemist in 1884, a center devoted to the chemical investigation of the Earth was founded on the American continent.

Clarke was chief chemist for forty-one years, until he retired in 1925. He was responsible for a vast and every-growing output of analyses of minerals, rocks, and ores collected by the field staff or submitted for examination; moreover, he was always concerned with the fundamental significance of the mass of factual data thus acquired. In 1889 he published a classic paper, "The Relative Abundance of the Chemical Elements," which represented the first attempt to use the accumulated rock analyses to determine the average composition of Earth's crust and the relative abundances of the elements. It is remarkable that even with the inadequate data then available, he was able to draw some notably prescient conclusions. Clarke wrote,

If . . . we assume that the elements have been evolved from one primordial form of matter, their relative abundance becomes suggestive . . . the process of evolution seems to have gone on slowly till oxygen was reached. At this point the process exhibited its maximum energy, and beyond it the elements forming stable oxides were the most easily developed, and in the largest amounts. On this supposition the scarcity of the elements above iron becomes somewhat intelligible; but the theory does not account for everything and is to be regarded as merely tentative.

However, in his great compendium, *The Data of Geochemistry*, which was first published as *U.S. Geological Survey Bulletin* 330 in 1908 and passed through five editions in less than twenty years, Clarke did not pursue the theory of the origin of the elements.

In many respects the publication of the fifth edition of *The Data of Geochemistry* in 1924 marked the end of an era. During the preceding century geochemical research was largely synonymous with the analysis of those parts of the Earth accessible to visual inspection and chemical assay. From the nature of things it could be little more. Interpretative geochemistry, the creation of a philosophy out of the mass of factual information, had to wait for the development of the fundamental sciences, physics and chemistry. A single illustration will do: the failure of all attempts to explain adequately the geochemistry of the silicate minerals before the discovery of X-ray diffraction provided a means for the determination of the atomic structure of solids.

The development of geochemistry in new directions was greatly advanced by the establishment of the Geophysical Laboratory by the Carnegie Institution of Washington in 1904. The laboratory's policy of careful experimentation under controlled conditions, and the application of the principles of physical chemistry to geological processes, was an immense step forward. Previously geologists and chemists had been skeptical of the possibility of applying the techniques and principles of physics and chemistry to materials and processes as complex as those on and within the earth.

At this time too, a new school of geochemistry was growing up in Norway. Founded by J. H. L. Vogt and W. C. Brøgger, it attained worldwide distinction through the work of Victor M. Goldschmidt (1888–1947) and his associates. Goldschmidt graduated from the University of Oslo in 1911, and his doctoral dissertation "Die Kontaktmetamorphose im Kristianiagebeit," was a basic contribution to geochemistry (*see* GOLDSCHMIDT,

VICTOR MORITZ). It applied the phase rule, codified by the work of Bakhuis Rooseboom, to the mineralogical changes induced by contact metamorphism in shales, marls, and limestones and showed that these changes could be interpreted in terms of the principles of chemical equilibrium. During the next ten years Goldschmidt's work was devoted largely to similar studies on rock metamorphism. These studies stimulated related research in other Scandinavian countries and led to the enunciation in 1920 of the principle of "mineral facies" by Pentti Eskola (1883–1964) in a paper published from Goldschmidt's institute (*see* ESKOLA, PENTTI).

In many ways 1912 can be considered a critical date in the development of geochemistry. In that year M. von Laue (1879–1960) showed that the regular arrangement of atoms in crystals acts as a diffraction grating toward X rays and thus made the discovery that enabled the atomic structure of solids to be determined. Since the geochemist is largely concerned with the chemistry of solids, the significance of this discovery can hardly be overestimated. However, some years elapsed before the impact of this new development was felt in geochemistry. In the 1924 edition of *The Data of Geochemistry*, no mention is made of it. It is a tribute to Goldschmidt's forward thinking that he not only realized the significance of crystal structure determinations for geochemistry but also devised a plan of research that led to a maximum of results in a minimum of time. Between 1922 and 1926 he and his associates at the University of Oslo worked out the structures of many compounds and thereby established the extensive basis on which to found general laws governing the distribution of elements in crystalline substances. The results were published in a classic series of monographs entitled *Geochemische Verteilungsgesetze der Elemente* (1923–1938); in these publications Goldschmidt's name is associated with T. Barth (1899–1971), W. H. Zachariasen (1906–1979), L. Thomassen (1896–1972), G. Lunde (1902–1942), I. Oftedahl (1894–1976), and others, all of whom had notable careers.

In 1929 Goldschmidt left Oslo for Göttingen, where he began investigations on the geochemistry of the individual elements, applying the principles he had established in the preceding years and making use of the current developments in quantitative spectrographic analysis for rapidly determining small amounts of many elements with a high degree of precision. The results were summarized in the final part of *Geochemische Verteilungsgesetze der Elemente* (1938); in this he presented the first comprehensive table of the absolute abundances of the elements, based largely on his analyses of meteorites. This table has been refined in more recent years, but it provided the database for theories of nuclear structure and for the origin of the elements.

During this period an important school of geochemistry developed in the Soviet Union. Its greatest adherents were VLADIMIR I. VERNADSKY (1863–1945) and his student and colleague A. E. Fersman (1883–1945). Its productivity has been immense. Geochemistry in the Soviet Union has been particularly directed toward the search for an exploitation of mineral raw materials, evidently with considerable success.

Developments during World War II resulted in a quantum leap in geochemical research in the postwar years, as many entries in this encyclopedia attest to. The discovery of atomic fission and the subsequent development of the atomic bomb led to intensive searches not only for uranium but also for other exotic elements. New techniques in instrumental analysis for minor and trace elements lowered detection limits and improved precision (*see* GEOCHEMICAL TECHNIQUES). High-resolution mass spectrometry (*see* NIER, ALFRED O.) opened up a new field, isotope geochemistry (*see* ISOTOPE TRACERS, RADIOGENIC and ISOTOPE TRACERS, STABLE). These powerful analytical methods continue to be improved and provide better data on smaller geochemical samples. This is important because it allows geochemical systems to be studied in even finer detail, leading to better control on geochemical hypotheses and theories.

The availability of lunar samples, remote probe data from planetary atmospheres and surfaces, and the acceptance of the concepts of plate tectonics in the 1960s and 1970s gave new perspectives to geochemical sampling and theory. The addition of extensive data from the Moon (*see* MOON) as well as from Earth and meteorites (*see* METEORITES) allowed many theories of planetology and cosmochemistry to be tested (*see* NUCLEOSYNTHESIS AND THE ORIGIN OF THE ELEMENTS). The concept of material recycling through subduction of ocean floor material requires reevaluation of the geochemical cycles of the elements (*see* EARTH, STRUCTURE OF). Today, geochemistry is undoubtedly on the threshold of new and exciting discoveries and changes.

Bibliography

CLARKE, F. W. "The Relative Abundance of the Chemical Elements." *Bulletin of the Philosophical Society of Washington* 11 (1989):131–142.

———. "The Data of Geochemistry." *U.S. Geological Survey Bulletin* 330 (1908), 491 (1911), 616 (1916), 695 (1920), 770 (1924).

GOLDSCHMIDT, V. M. "Die Kontaktmetamorphose im Kristianiagebeit." *Videnskaps Skrifter I*, 11. Oslo, 1911.

———. *Geochemische Verteilungsgesetze der Elemente*, I–IX. Oslo, 1923–1938.

———. *Geochemistry*. Oxford, Eng., 1954.

MASON, B. "Victor Moritz Goldschmidt: Father of Modern Geochemistry." *Geochemical Society Special Publication* 4, 1992.

BRIAN MASON

GEOGRAPHIC INFORMATION SYSTEMS

A geographic information system (GIS) is a computerized tool for characterizing landscapes that allows quantitative analyses of environments ranging from natural ecosystems to urban developments. The design of a GIS is based on data acquired from maps, and this provides unique capabilities that cannot be accomplished with traditional relational database techniques. Information from multiple maps can be synthesized in the GIS, creating new map products or derived summaries that address specific information needs. The analytical ability of GIS typically has been used in fields such as natural and water resource management, urban planning, and facility siting. Applications of GIS technology are numerous and the use of GIS has grown dramatically, creating a multibillion-dollar, international industry (*see also* EARTH SCIENCE INFORMATION).

Information in a GIS database is referenced by geographical coordinates. The data stored in a GIS may be envisioned as a collection of maps that overlay one another. In his 1969 book, *Design with*

Nature, Ian McHarg showed how environmental planning could be aided by using a set of shaded transparencies that corresponds to a common map base. Each transparent overlay would represent a different variable in the planning process, such as soil type, water availability, or protected areas. The shading pattern on each overlay would represent the suitability of different areas with respect to that variable. By overlaying the shaded transparencies, one could visually determine those areas on the map that satisfy planning constraints in the most acceptable manner. However, manually redrafting maps to a common scale and orientation is slow, overlays become unwieldy for large and complex analyses, and overlays cannot contain many shading patterns or layers before they become impossible to interpret. A digital GIS avoids many of these problems, providing much more power and flexibility for storage, analysis, modification, and presentation of spatial data. In fact, one of the first digital GISs, the Canadian Geographic Information System, was implemented in 1964 prior to the publication of McHarg's book. During the 1980s the use of GIS grew rapidly, and in the 1990s GIS software became available for systems ranging from portable personal computers to high-end workstations.

While the ability to handle features defined by map coordinates might also be provided by computer aided design (CAD) or automated mapping/facilities management (AM/FM) systems, GIS is distinguished by its ability to manipulate and cross-reference features from different map sources. Historically, CAD systems have been drawing tools that allow unconstrained placement of features in a graphic. Because of this, CAD systems have had limited ability to automatically derive relationships such as co-occurrence and proximity of features from different maps. These two AM/FM systems take CAD capabilities one step further by linking map features to descriptive records in a database. CAD systems generally rely on visual interpretation of computer displays. In contrast, the data in a GIS are organized into consistent data layers where locations are assigned to categories of information in a structured manner. This structure defines the geometric relationships *between* mapped objects, not just those of the objects themselves. This enforcement of spatial relationships between objects is referred to as maintaining topology, or topological relationships, and it allows the spatial analysis of maps to be automated.

Types of GIS

Several data structures have been developed for manipulating map-based data in a GIS. The structure of the database is an important factor in determining the ability of a GIS to accomplish a given task. A GIS using the raster data structure stores map information as coded numbers in rectangular grids. This can be envisioned as looking down at a landscape through a sieve and assigning what you see in each hole of the sieve to a predetermined number system (e.g., corn field = 1, wheat field = 2). The individual grid cells are referred to as pixels. The raster data structure allows rapid combinations between map layers because it is very easy to calculate the way in which two grids overlap. This data structure is most effective for storage and manipulation of data with continuously changing values, such as elevation, because the grid system creates a consistent sampling framework. However, the grid-based approach may create large data volumes and can cause geometric distortions, such as stairstep-shaped edges along map boundaries that should be smooth.

A second common data structure in GIS is referred to as the vector or polygon method. Rather than sampling every location in a map with a systematic grid spacing, the vector GIS stores the coordinates of those points that define the precise boundaries of mapped objects. Objects are represented as points (e.g., wells), lines (e.g., transportation networks), or polygons (e.g., political units). Because vector systems define shapes by using critical points that are connected to define boundaries, geometric representations are typically more accurate than raster systems. The vector data structure allows efficient storage of maps that depict discrete features such as buildings. However, analyses that combine different maps are generally slower since the geometric intersection of features in different maps is not based on a predetermined framework and can require many calculations. Also, this method does not allow for smooth transitions between adjacent features, so it is not useful for representing continuous variables like elevation.

More complex versions of these two systems have been developed, though the conceptual approach basically follows that of the raster or vector structures. In the quad-tree data structure an area is recursively divided into four equal-sized sub-rectangles until a single value can be assigned to each of the nested subunits. This approach is conceptually similar to the raster structure. A modification of the vector approach yields the triangulated irregular network (TIN), which can represent continuously varying surfaces by dividing the entire landscape into a set of triangular facets. Slope and aspect can be calculated and stored for each of these facets.

Vector-based systems may use a relational database management system (RDBMS) to help keep track of the topological relationships that join points into lines and lines into polygons. Such systems also allow you to link tabular data from other sources to map features by using the RDBMS. Raster-based systems are not generally linked to relational databases. This is due to potentially large data volumes and the types of applications from which raster systems evolved, rather than an inherent conceptual limitation. Some raster GIS packages have an improved ability to store multiple attributes for a given raster value (e.g., 1 = sandy loam soil, 0.25 porosity; 2 = silty clay, 0.15 porosity). However, this approach affords none of the power offered by relational databases. While early GIS software typically supported either the raster or vector approach, it has become common that a single package will allow both methods. This allows the best computerized representation to be chosen for a given type of map data and analysis.

Data Entry

Data may be entered into the GIS in a number of ways. A map may be entered into the system by laying it on a digitizing tablet and tracing features with an electronic pointing device. The digitizing tablet is a flat board that has a fine network of wires underneath the surface which can precisely detect where the pointing device is positioned. Methods have also been developed to input maps semiautomatically by using optical scanners, though these approaches are limited in the complexity of the maps they can process. Text or tabular data describing mapped units might be entered manually, input from existing computer files, or linked from tables in relational database systems. Field data collected using survey techniques or global positioning systems (GPS) can also be entered, either as text files or through direct hardware connections to the survey device.

Some GIS packages contain image-processing functions that allow imagery from satellite or air-

craft platforms to be used. Such imagery can rapidly provide information on land cover and land use over very large areas. Image processing functions are inherently raster-based, and many of the earlier raster GIS systems grew out of software that was developed for analyzing satellite imagery. Information can be extracted from these images by tracing land cover patterns on the computer screen with a pointing device, such as a mouse. Computer programs have also been developed that automatically divide the image into areas which are similar in appearance and then allow the analyst to label these areas as specific land cover types. Because of the complexity of image interpretation, manual approaches are still generally more accurate than automated methods.

The data entry required to crease a GIS database is often the most expensive part of creating a system, perhaps as much as 70 percent of the total cost. However, after the database is created the power of computerized analysis can save vast amounts of time and effort over manual analysis techniques. Since the landscape keeps changing, map data must be updated to remain accurate. Editing data in the GIS is quicker and cheaper than having to manually redraft the proofs for paper map sheets. For example, GIS systems with image-processing capabilities can display images from aircraft or satellites behind map features so that an analyst can quickly make changes on the computer screen. Figure 1 provides an example displaying roads from a vector database on top of high-resolution imagery from the French SPOT satellite.

Functionality

Maps that are input into a GIS will almost invariably be compiled with different projections, datums, and scales. In order to handle these data consistently, GIS software includes functions for converting between map projections and for precise realignment of mapped features based on selected control points identified in each of the map layers. Individual maps in the GIS may have their characteristics manipulated by renaming categories (e.g., forest and agriculture → land; lakes and streams → water) or using mathematical operations (e.g., feet to meters). Proximity may be tested using functions that calculate distances between features or which can create buffer zones around them. The relationship between proximate fea-

```
Selected Feature
----------------
ID             3546
Name           US 101
Direction      Northbound
Feature Type   Paved Highway
Lanes          2
Speed Limit    55
```

Figure 1. Road network superimposed on a SPOT satellite image.

tures in a map can also be calculated to provide new information. For example, pixels containing elevations in a raster dataset can be analyzed with respect to their neighbors to determine the local slope and aspect of the landscape.

The true strength of GIS, however, is in its ability to perform overlay operations between map layers. In cases where map features represent discrete categories, overlay operations can determine the intersection or union of fetures from different map sources. Maps representing numerical values may also be combined using mathematical relationships. As an example, a GIS may be used to find a good site for a power plant by recoding map layers for soils, slope, and proximity to cooling water and markets into suitability scores or cost estimates. These suitability maps could then be combined mathematically to create a derived map indicating

the relative costs and suitabilities for building a facility throughout an entire region.

An example in which decisions must be made regarding the health effects of a hazardous waste dump is portrayed in Figure 2. Substances potentially may leave the dump through runoff, leaching into the groundwater, or through airborne dispersal. There is also a risk of exposure near roadways which are used to transport waste to the dump. A GIS provides a consistent framework for integrating the variety of data sources which are required for informed decision making regarding these issues. Maps of streams and water bodies, geological characteristics of the aquifer, well locations, transportation networks, and land use may be digitized and converted to a common coordinate framework using a digitizing tablet and GIS software. Attributes can then be linked to map fea-

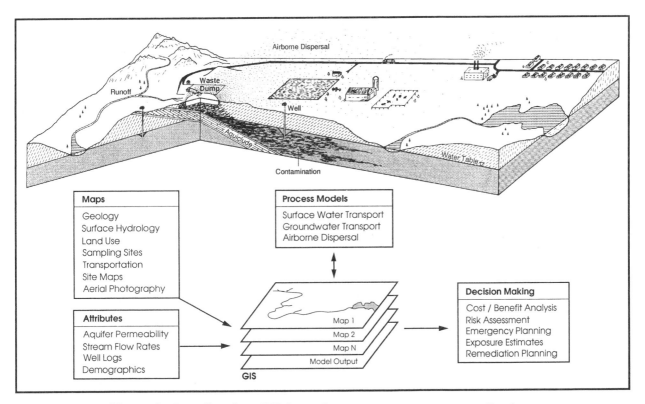

Figure 2. Data flow in a GIS hazardous waste management application.

tures, such as associating the chemical composition of groundwater samples with well locations. Map data that are required to predict contaminant transport would be exported to computerized models of environmental processes, such as a groundwater model. The output of such a model would be a map of the expected concentration of contaminants over time. This information would be input back into the GIS and spatially referenced to information on human activities, such as the locations of drinking water sources or residential areas. Buffer zones around roads which are used to transport waste can be calculated in the GIS and combined with demographic data as well. Automated routines could even generate form letters for addresses in potentially affected areas or voting districts. The result of this type of integrated analysis is improved information on the relationships between people and the environment.

GIS generally have well-developed capabilities for presenting derived information products. Data in the GIS may be displayed as maps, as views of the landscape from different vantage points, or even as animated "fly-through" videos. Statistical summaries may also be generated, such as the number occurrences, area covered, and mean value of selected features.

Despite the analytical power of GIS technology, GIS systems still are limited in their ability to perform complex analyses of three-dimensional phenomena over time, such as groundwater or atmospheric modeling. In these cases, data are typically exported from the GIS into a specialized computer program that simulates the natural process. The output of that program is then fed back into the GIS so that results can be interpreted with respect to concerns such as affected population centers or regulatory jurisdictions.

Bibliography

ANTENUCCI, J. *Geographic Information Systems: A Guide to the Technology.* New York, 1991.

BERRY, J. *Beyond Mapping: Concepts, Algorithms, and Issues in GIS.* Fort Collins, CO, 1993.

BORROUGH, P. *Principles of Geographical Information Systems for Land Resources Assessment.* Oxford, Eng., 1986.

CASSETTARI, S. *Introduction to Integrated Geo-information Management.* New York, 1993.

FOTHERINGHAM, S. *Spatial Analysis and GIS*. London, 1994.

MCHARG, I. *Design with Nature*. New York, 1969.

RIPPLE, W., ed. *GIS for Resource Management: A Compendium*. Bethesda, MD, 1987.

STAR, J., and J. ESTES. *Geographic Information Systems: An Introduction*. Englewood Cliffs, NJ, 1990.

TOMLIN, D. *Geographic Information Systems and Cartographic Modeling*. Englewood Cliffs, NJ, 1990.

WORRALL, L., ed. *Geographic Information Systems: Developments and Applications*. New York, 1990.

KENNETH C. MC GWIRE

GEOID

See Earth, Shape of

GEOLOGIC TIME

This entry addresses the enormity of geologic time and the central place that geologic time—its estimate, its measurement, and its uses—has held in the understanding and development of the earth sciences, especially the development and refinement of the geological timescale (see frontmatter).

Historical Overview

Ancient Greek and Hindu thought held that time was eternal, or cyclic with long-lived intervals; most theologically based estimates placed the age of Earth at five to ten thousand years. During the middle ages, the concept of time was considered mainly in terms of a closed-system Earth (no external influences), whose beginning was not too long ago and whose end was probably not far away. Much religious thought held that Earth was the center of the heavens, with the Sun, the Moon and the known planets revolving around Earth. The heavenly sphere of stars existed farther out. A widely accepted age for Earth, calculated through a literal interpretation of Hebraic writings, was

about 6,000 years. The beginning of the end of the Earth-centered universe hypothesis came with the separate discoveries by Johannes Kepler (1571–1630) and Galileo Galilei (1564–1642), who, using a new instrument, the telescope, brought forth the notion of a Sun-centered system, an idea championed earlier by Nicolaus Copernicus (1473–1543). By the mid–1600s, the concept of Earth was that of a dynamic body revolving in space, with almost limitless space beyond. This scientific revolution brought with it a different understanding of time as well as space.

For earth scientists, the clash between religious and scientific theories of time (a controversy still evident; *see* SCIENTIFIC CREATIONISM) was clearly defined just before the beginning of the nineteenth century by JAMES HUTTON, who is considered the father of geology. Hutton, a keen observer of rocks in his native Scotland, postulated that ongoing, albeit slow, "everyday" geologic processes such as rainfall, erosion, deposition, lava flows, and landslides could generate the thick sequence of rocks he studied, and account for the complex relations between them over time. This "uniformitarianism" view, which holds that the present is the key to the past, contrasted sharply with the most fundamental "catastrophist" interpretation that the Earth and the heavens were created in a dramatic stroke of divine creation. (Certainly, catastrophic processes such as explosive volcanic eruptions, asteroid impacts, and earthquakes also shape Earth and its history, but the term catastrophism is used in a scientific sense without implicating God as a causal agent.)

Hutton's carefully documented theory defied the prevailing views of a very young Earth, positing that enormous amounts of time—millions of years—had ensued during the formation of Earth's crust. Uniformitarianism, lucidly explained by JOHN PLAYFAIR and forcefully extended by CHARLES LYELL, became the scientific standard of interpretive geology, extending geologic processes over hundreds of millions of years. The concept of uniformitarianism, in concert with the law of original horizontality (layered rocks essentially are formed horizontal to Earth's surface), and the principle of superposition (in an undisturbed sequence of layered rocks, the oldest are at the bottom), formed the basis for the interpretation and correlation of rocks and the establishment of the stratigraphic record and the geological timescale (*see* GEOLOGIC MAPPING; STRATIGRAPHY).

The work of nineteenth-century evolutionist CHARLES DARWIN further buttressed the argument for a geologic timescale that extended well beyond that explained by conventional religious interpretation. Darwin showed that biologic evolution was slow, requiring long periods of time. Fossils of successions of extinct organisms that lay buried under thousands of feet of younger stratified rock obviously required millions of years to form. The stage was set for the first quantitative efforts to measure Earth's age.

Early Scientific Estimates for Earth's Age

Early scientific attempts to determine Earth's age were based on hypotheses, many of which, though seemingly reasonable at the time, were flawed. The various theories, the dates they provided, and the principal authors who proposed them include: (1) cooling rate for an initially molten Earth (Lord Kelvin, 1862: 98 Ma); (2) origin of the Moon by rotational wrenching from early Earth (George H. Darwin, 1898: 50–60 Ma or higher); (3) accumulation of salt in the ocean (John Jolly, 1909: 150 Ma or less); and (4) rates of sediment accumulation (CHARLES WALCOTT: 35–80 Ma). (The interested reader is referred to Dalrymple, B., 1991, for a more detailed account of early attempts to date Earth's age.) Although these calculations resulted in minimum ages for the Earth of many millions of years, they were still too young to satisfy the tenets of organic evolution and uniformitarianism. The seminal breakthrough in calculating Earth's ancient age, as well as the foundation for determining geologic ages ever since, was the discovery of radioactivity by Henri Becquerel in 1896.

Determination of Geologic Time

Geologic field study commonly begins locally, mapping rocks and geologic features in a local area (see MAPPING, GEOLOGIC). The use of the concepts of uniformitarianism, original horizontality, superposition, and other field relations, allows the scientist to construct a geologic map. A geologic map shows not only the areal distribution of the various rock types present, but also a legend or explanation of the mapped units, arranged from oldest to youngest, and perhaps also cross sections through the mapped area that interpret the relations of the

rocks to one another. Such maps are the first and commonly the best information not only for determining geologic history of the area, but also in determining reserves of ores, oil, or other materials of value which may be present. Field mapping may be supplemented by paleontology studies (see PALEONTOLOGY), if fossils are present, or by drilling and geophysical and geochemical testing (see DRILLING FOR SCIENTIFIC RESEARCH; GEOCHEMICAL TECHNIQUES; and GEOPHYSICAL TECHNIQUES). The best maps thus require teams of people, sophisticated instrumentation, money, and perhaps many years of study. As more and more quality maps become available for Earth's continents, ocean floors—even Earth's interior—the interpretation of Earth's ancient and complex history becomes more tractable.

Radiometric age determinations (see GEOLOGIC TIME, MEASUREMENT OF) are based on the uniform rates of decay of certain unstable (radioactive) isotopes of elements such as potassium (40_K), uranium (235_U, 238_U), samarium (147_{Sm}), and carbon (14_C, or radiocarbon), among others. These radioactive isotopes decay to stable (radiogenic) isotopes at different rates; if one knows the rate of decay of a certain radioactive isotope, the amount of that isotope in the mineral or rock, and the amount of radiogenic isotope that has accumulated, the quantitative, geologic age of that specimen—millions of years, or billions of years—may be computed. The most useful absolute age determinations are those which meet two criteria: (1) they are as analytically accurate as possible; and (2) they are carefully integrated with respect to the best understanding of the mapped field relations of the rocks or minerals that were analyzed.

Geologic mapping and the assignment of chronologic markers for thick sections of mapped rock and their contained fossils proceeded from local areas to larger regions. Correlation from regions to even more remote areas, even continents, proceeded by comparison of rock types, but, more importantly, by the paleontologic affinity of contained fossils, and, where available, radiometric ages. Correlation by rock type works well locally—a distinctive sandstone, for example, may be traceable across wide regions—but ultimately fails, for no strata or rock type of a given age are traceable around the world. (Two exceptions deserve mention: ash from an explosive volcano, ejected high enough into the atmosphere so that particles settle out all over the Earth, may form a unique and practical time marker; somewhat similarly, ejecta from massive asteroid impacts on Earth—*see* IMPACT CRATERING AND THE BOMBARDMENT RECORD—may provide a geologically instantaneous stratigraphic layer.)

An additional complication is that rocks of different types may be of equivalent age, and, conversely, a rock traceable over a significant region may not be of the same age in all of its exposures. Consider the types of sediments—later to be indurated into sedimentary rocks—across a shoreline from land to deep ocean. Although all are being deposited at the same time, and grade into one another, resulting rock types may be coarse-grained beach sandstone, finer-grained siltstone, biogenically formed limestone under quieter waters further offshore, and finally very fine-grained mudstone or shale deposited in the deep ocean. On the other hand, ocean water transgressing onto a continent may deposit coarse sandstone in a continuous deposit inland, though formed perhaps over a significant period of time; the resulting rock, though perhaps continuous in exposures (outcrops), is of different ages in different places perpendicular to the changing shoreline (*see* STRATIGRAPHY).

Correlation of rocks around the world is thus made possible through the addition of radiometric dates and age assignments made from diagnostic (guide) fossils. Such correlation indicates that: (1) no section on Earth contains rocks of all ages, from the oldest to the most recent; and (2) most sections contain large time gaps—that is, rocks of certain ages are missing. The gaps may represent enormous amounts of time. The most significant type of gap is the unconformity, a surface separating rocks formed at significantly different times. Consider again the sequence of sedimentary rock formed across the coastline previously described. Sediment accumulation may be interrupted by regional uplift and the deposition process replaced by erosion of the newly exposed rocks. Later subsidence may reinstate deposition, but on a surface (the unconformity) that might represent a significant period of time when no sediments accumulated. Such natural breaks in the stratigraphic record are commonly distinctive in the field, and are used to establish the base and/or top of a mappable rock unit.

Thick sections of rock, perhaps distinctive owing to rock type, the kinds of fossils contained, and major unconformities, have been given hierarchies

of names (see the geological timescale in frontmatter) that are understood by geologists around the world. Although boundaries between major subdivisions may change or be modified with additional study, and the correlation of rock sequence and/or chronologic interval may be refined, the resulting chart—the geological timescale—is the chart at the heart of our understanding of the earth sciences. The timescale indicates the formation of Earth and the Moon, Earth's oldest rocks (*see* OLDEST ROCKS IN THE SOLAR SYSTEM), the earliest record of life on Earth (*see* PALEOBOTANY; PALEONTOLOGY; PALYNOLOGY), the 100 Ma sequence of rock that contains the fossil record of the mighty dinosaurs (*see* DINOSAURS), and thin beds of distinctive sediment that record global catastrophes such as huge asteroid or comet impacts (*see* IMPACT CRATERING; IMPACT CRATERING AND THE BOMBARDMENT RECORD), or volcanic eruptions (*see* VOLCANISM).

Bibliography

ALLEGRE, C. J.; G. MANHES; and C. GOPEL. "The Age of the Earth." *Geochimica et Cosmochimica, Acta* 59(1995):1445–1456.

DALRYMPLE, G. B. *The Age of the Earth.* Stanford, CA, 1991.

EICHER, D. L. *Geologic Time*, 2nd ed. Englewood Cliffs, NJ, 1976.

FAURE, G. *Isotope Geology*, 2nd ed. New York, 1986.

HOLMES, A. "A Revised Geological Time-Scale." *Transactions of the Edinburgh Geological Society* 17(1960):183–216.

E. JULIUS DASCH

GEOLOGIC TIME, MEASUREMENT OF

The dimension of time is central to geology and understanding the history of the earth, including the processes that shape it. Geologists have long been fascinated with the timing and sequence of earth processes and events. The field of geochronology, which measures geologic time, determines precisely when these processes took place and over what time interval. Many different geochronologic methods have been developed because of wide variations in both the materials (e.g., sedimentary, volcanic, biological) of natural systems and the timescales (e.g., years to billions of years) of the processes affecting them (*see also* GEOLOGIC TIME).

The realization that geologic time is vast and that most geologic processes operate over long time periods are central to the development of modern earth science. In the eighteenth century, JAMES HUTTON established the principle of uniformity, a cornerstone of geologic reasoning which recognizes that ordinary processes operating for long periods of time can affect great changes. Even earlier, geologists recognized that the sequence of natural events is preserved many times in rocks and in the relative relations among deposits.

To decipher earth history, it is important to know the sequence of events, for example, the position of a particular plate (*see* PLATE TECTONICS) during a given point in geologic time. To reconstruct the history, it is essential to know the order of events with respect to one another; this is termed "relative dating." It is also important to know the time of an event in an absolute sense, that is, the specific point on the timeline that marks the formation of the earth and the interval separating that point from the present. The measurement of time in this sense is termed absolute dating and produces a numeric value between zero and approximately 4.55 billion years (Ga) ago when Earth formed (*see also* ISOTOPE TRACERS, RADIOGENIC). Many dating methods are used, including those that involve radioactive decay and dependent chemical reactions.

Relative Dating

Sequence. The order of geologic events or sequence is important not only to relative dating but also ultimately to the definition of absolute dates for geological periods. The appropriate logic expressed by the law of sequence holds that the record of an event or sequence of events surrounds, overlies, or is impressed upon the record of earlier events. This law, considered as common sense today, is an early geologic principle credited to Nicolaus Steno in the seventeenth century. Steno stated two other principles, the laws of original horizontality and superposition. The law of original horizontality recognizes that water-laid sediments are deposited in strata that are almost horizontal and parallel or nearly parallel to the surface upon which they accumulate. The law of superposition maintains that in a pile of sedimentary strata that has not been disturbed, the youngest stratum is at the top and the oldest is at the base.

Principles. The above laws are used in relative dating by applying several simple principles. According to the principle of superpostion, in stratified or layered rock formations, the oldest is on the bottom and the youngest is at the top. That is, younger deposits are laid upon older ones. The assumption, of course, is that the entire rock section has not been overturned after rock formation, which is generally testable by examining structures produced during deposition.

The principle of cross-cutting relationships holds that disrupted rock units or geologic patterns are older than the processes or events that produced the disruption. For example, a sedimentary bed that is cut by a fault is older than the fault. Likewise, a dike that intrudes the bed is younger than the bed. Similarly, the principle of inclusion states that a fragment of a rock incorporated or included in another rock unit is older than the host unit. A good example of this principle would be boulders in a sedimentary rock; obviously, the rock unit from which the boulders were derived is older than the sedimentary rock.

The principle of faunal succession is frequently applied to sedimentary rocks that contain fossils. This principle states that groups of plants and animals occur in a definite order with time and their fossilized remains can be used to recognize a geologic time interval. William Smith is credited with establishing the principle. A British surveyor in the early nineteenth century, Smith showed that fossil assemblages in England change systematically from older to younger beds and used the assemblages to correlate and define rock sequences. Smith's discovery preceded the scientific explanation for the changes in fossils over time, namely the theory of biological evolution that was formulated a half century later by CHARLES DARWIN. An important attribute of Smith's principle is that it can be used to correlate rock sequences worldwide.

Applications. The principles of relative dating are essential tools for virtually all areas of earth science. They are an indispensable part of sedimentary and historical geology. Very precise relative ages are commonly determined using fossil correlation for sedimentary rocks, often with a precision exceeding that provided by absolute ages. The standard geologic timescale was established using these methods of relative dating.

Absolute Dating

Principles and Assumptions. Absolute ages of natural materials are determined, or estimated, by a variety of methods that have various chemical and physical bases. These methods are based on processes that are time-dependent (e.g., chemical reaction or radioactive decay) or rely on specific markers (e.g., magnetic characteristics or isotopic compositions) that have time significance. All methods proceed from certain assumptions that must be satisfied to provide accurate dating. When evaluating a given absolute age, the validity of applicable assumptions must be evaluated.

Chemical, Physical, and Biological Methods. Methods of absolute dating have widely divergent chemical, physical, and biological bases. Most of these approaches rely on the time dependence of chemical, physical, or biologic processes or on the correlation of certain markers or imprints of these processes. Additionally, most of the methods rely on calibration with other absolute dating methods, generally those based on radioactive isotopes. These dating methods are most useful for geologically recent events and for surface or near-surface processes.

Chemical methods are based on changes that occur progressively with time, such as chemical reactions of minerals with water or the atmosphere. An example of a chemical method is obsidian hydration-rind dating. Obsidian glass hydrates with exposure to water, forming a hydrated rind or layer adjacent to a crack or surface. The thickness of the hydrated layer is determined and an empirical calibration is used to arrive at an age. Another example is the amino acid racemization method in which the conversion of amino acid of one form to another is time-dependent and the relative amounts present indicate the age. Chemical methods such as these assume a constant rate of the chemical reactions which, of course, may not be valid because reaction rates may vary with temperature and chemical environment.

Physical processes provide several methods of dating. The basic processes range widely from erosion and deposition on a large scale to subtle

changes in atomic structures of crystals. An example is thermoluminescence dating, reflecting a small-scale physical process. Natural radiation produces atomic structure changes or damage in solids that can accumulate over time so that minerals may serve as natural dosimeters. The total accumulated damage is ascertained from the amount of light emitted during laboratory heating. Since the damage is proportional to the radiation received (which is time-dependent), an age may be determined. A key assumption of this approach is that the damage is quantitatively retained until measured in the laboratory.

Biologically based dating methods are founded in the predictability of biologic processes such as growth rates or markers. A well-known method of this type is dendrochronology or tree-ring dating. Dendrochronology is based on the counting of tree growth rings that form annually and that also vary in size due to climate conditions. The patterns or sequences of ring thicknesses provide a method of correlation from one tree to another and extend the known total range to about 10,000 years ago or almost twice the age of the oldest known tree.

Processes that occur both continually and episodically leave imprints that provide methods of dating by correlation of specific features. Some examples of features that can be used for dating are the magnetism of rocks, the isotopic composition of stable and radiogenic isotopes, the occurrence of unique layers of volcanic ash, and the occurrence of glassy material produced during the impact of extraterrestrial masses. A good example of such an application is the dating of marine biogenic carbonates based on their $^{87}Sr/^{86}Sr$ ratios. The $^{87}Sr/^{86}Sr$ ratio of seawater is the same in all oceans at any point in time but has varied with time, with the last half of the Tertiary period showing a steady increase. If the carbonate has remained unaltered since its formation with the $^{87}Sr/^{86}Sr$ of seawater, it is possible to determine its age by measuring the $^{87}Sr/^{86}Sr$ ratio and by comparing the value of those empirically determined for seawater over geologic time.

Radioactive Methods. Radioactive isotopes of different elements provide important geological clocks that are essential to the absolute measurement of geologic time. This approach relies on the regular, predictable, and invariant radioactive disintegration of parent atoms with rates that depend on the half-live (the time required for half of the atoms of a radioactive substance to disintegrate) as described by the decay law. As such, these nuclear methods are preferred when feasible because they are typically more reliable than those based on physical, chemical, and biological markers or processes.

The approximately seventy-five naturally occurring radioactive isotopes have three main sources: long-lived (i.e., long half-life) isotopes that were present when the earth formed and still remain because their decay rates are slow (e.g., ^{40}K, ^{238}U); comparatively short-lived isotopes that are cosmogenic, that is, produced by the cosmic radiation that constantly bombards Earth (e.g., ^{14}C, ^{36}Cl); and similarly short-lived isotopes that are the radioactive daughters of long-lived radioactive U and Th parents (e.g., ^{234}U, ^{222}Rn). Methods of dating use radioactive isotopes from all three sources and can be described as one of two general types: decay clocks or accumulation clocks. Some important methods are listed in Table 1.

Decay clocks are based on the number of radioactive atoms that still remain in a sample. The age of a sample is determined from

$$T = (1/\lambda) \times \ln [N_0/N]$$

where T is "age"; N is the number of radioactive atoms now present; N_0 is the initial number of radioactive atoms; and λ is the decay constant for the radioactive isotope. Figure 1 illustrates the principle of a decay clock. An important assumption of decay clocks is that the initial amount of radioactive parent is independently known or determined. Another assumption is that the number of radioactive atoms changes only by decay or, in other words, the sample behaves as a "closed system." Radioactive parents with geologically short half-lives are used for decay clocks and the maximum age that can be determined is correspondingly short, typically up to several half-lives. Such dating methods are most useful for surface or near-sur-

Table 1. Some Important Methods of Absolute Dating Using Radioactive Isotopes

Method	Radioactive Parent	Half Life (years)	Radiogenic Daughter	Typical Age Range (years)	Commonly Dated Materials
Potassium-argon method	^{40}K	1.25 billion	^{40}Ar	50,000 to 4.6 billion	volcanic rocks and potassium materials (biotite, muscovite, hornblende, potassium feldspar)
Rubidium-strontium method	^{87}Rb	48.8 billion	^{87}Sr	10 million to 4.6 billion	igneous and metamorphic rocks and minerals (biotite, muscovite, potassium feldspar)
Samarium-neodymium method	^{147}Sm	106 billion	^{144}Nd	100 million to 4.6 billion	igneous and metamorphic rocks and materials
Uranium (and thorium)-lead methods	^{238}U	4.47 billion	^{206}Pb	10 million to 4.6 million	U (and Th) minerals (zircon, monazite, apatite, uraninite, thorite)
	^{235}U	703 million	^{207}Pb		
	^{232}Th	14 billion	^{208}Pb		
Carbon-14 method	^{14}C	5730	^{14}N	100 to 70,000	wood, charcoal, bone, CO_2-bearing waters, carbonate minerals
Uranium-234 (or disequilibrium) method	^{234}U	245,000	^{230}Th	5,000 to 1.5 million	marine carbonates (shells, corals) and sediments

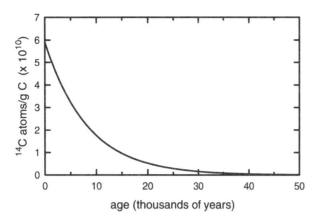

Figure 1. The principle of a decay clock using the ^{14}C method as an example. The age of a system, such as a piece of wood, may be found by determining the number of radioactive atoms (^{14}C) now remaining where the initial amount is assumed or known independently.

face events within about the past million years or less depending upon the isotope.

The carbon-14 method is a well-known example of a decay clock. This method employs ^{14}C, which is produced in the atmosphere by cosmic radiation, and is applied to C-bearing materials that have equilibrated with the atmosphere. For instance, a living organism exchanges carbon with the atmosphere and, therefore, contains ^{14}C. When the organism dies and no longer exchanges ^{14}C, the number of ^{14}C atoms simply diminishes by decay. The geological age, or in this case the length of time since the organism died, may be obtained by determining the fraction of original ^{14}C atoms remaining (Figure 1). Other decay clocks use various other cosmogenic isotopes or the radioactive daughters (such as ^{234}U) of the U and Th decay chains.

Accumulation clocks are based on the relative amounts of both parent and daughter atoms present in a sample. The age of a natural system may be determined using

$$T = (1/\lambda) \times \ln[(D/P) + 1]$$

where T is "age"; P is the number of radioactive parent atoms; D is the number of daughter atoms; and λ is the decay constant for the radioactive parent, which is, of course, directly related to its

half-life. Figure 2 illustrates this equation for the ^{238}U-^{206}Pb method. A key assumption of an accumulation clock is that the sample has been a closed system to loss or gain of both parent and daughter isotopes since initial isolation. The time since a given system formed or became dated is thus determined from the present-day ratio of the daughter to parent atoms. Parents with long half-lives are used for accumulation clocks. Therefore, the range of applicability of these methods is long, generally exceeding a million years, but extending up to the age of the solar system. Accumulation clocks (see Table 1) are most useful for igneous and metamorphic rocks and minerals. They provide the method for deducing the absolute timing of the vast majority of geologic time.

The U-Pb method is a well-known example of an accumulation clock. Here, both the natural, long-lived uranium isotopes, ^{238}U and ^{235}U, produce daughters ^{206}Pb and ^{207}Pb, respectively. Zircon, a common accessory mineral of igneous and metamorphic rocks, is frequently dated because it contains uranium that atomically substitutes for zirconium and because it initially contains very little lead. Over time, the number of parent atoms decreases while the number of daughter atoms produced increases. By measuring the ratio of the

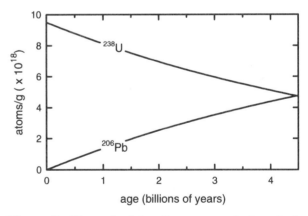

Figure 2. The principle of an accumulation clock for determination of absolute time using the ^{238}U-^{206}Pb method as an example. The age of geologic material, such as a crystal of the mineral zircon, may be found by determining the ratio of the number of radioactive parent atoms (^{238}U) now remaining to the daughter atoms (^{206}Pb) that have accumulated.

numbers of these atoms, the age may be determined (Figure 2).

Applications. The application of absolute dating methods is critically dependent on the type and age of the materials and processes to be dated. Typically, these factors dictate which method or methods may be used successfully. Absolute ages for recent surface processes are generally determined by methods with a physical, chemical, or biological basis or by radioactive methods of the decay-clock type. Radioactive accumulation clocks generally provide the times for most of the major large-scale crustal and mantle processes that have shaped the earth over its history. The rubidium-strontium, uranium-lead, samarium-neodymium, and potassium-argon methods in particular have been used to determine the ages for most of the earth's crust (and, indeed, Earth itself), meteorites, and the solar system. These methods have also provided the absolute timing or calibration of the geologic timescale and likewise provided the temporal framework for many processes. In addition, radioactive clocks may be "reset" during thermal events with different methods and minerals having different characteristic temperatures. In this manner, these clocks also provide simultaneously time and temperature information that permits the delineation of thermal histories of rocks and the crust. Relative and absolute dating complement one another and typically both are used together.

Bibliography

DALRYMPLE, G. B. *The Age of the Earth*. Stanford, CA, 1991.

DICKIN, A. P. *Radiogenic Isotope Geology*. Cambridge, Eng., 1995.

FAUL, H. *Ages of Rocks Planets and Stars*. New York, 1986.

FAURE, G. *Principles of Isotope Geology*, 2d ed. New York, 1966.

GEYH, M. A., and H. SCHLEICHER. *Absolute Age Determination*. New York 1966.

JÄGER, E., and J. C. HUNZIKER. *Lectures in Isotope Geology*. New York, 1979.

PIERCE, K. L. "Dating Methods." In *Active Tectonics*. Washington, DC, 1986.

STANLEY, S. *Earth and Life Through Time*, 2d ed. New York, 1989.

KENNETH A. FOLAND

GEOLOGIC WORK BY STREAMS

Rivers are ubiquitous on Earth's surface and form an important agent of erosion and deposition. (Rivers have also been active in Mars's earlier history [*see* MARS].) Wherever surface water is concentrated into discrete channels, some of the water's energy may be applied to eroding and transporting sediment, and thus shaping the landscape. As water flows downhill in response to gravity, potential energy is converted to kinetic energy. This kinetic energy may be expended in: (1) overcoming external frictional resistance produced by the channel boundaries; (2) overcoming internal resistance produced by turbulent mixing within the mass of flowing water; and (3) entraining (picking up) and transporting (carrying) sediment. The amount of sediment transported by a river is thus a function of total energy available for transport, which depends on the river's slope, discharge, and external and internal resistance, and the availability of sediment.

Sediment carried by a river may be eroded from the bed and banks of the river or may be introduced to the river by slope processes, such as debris flows or rockfall. Sediment may be carried as dissolved, suspended, or bedload. Dissolved load refers to minerals carried in solution. If a river flows over rocks that dissolve fairly readily under surface conditions, up to 90 percent of its sediment may be carried as dissolved load (Knighton, 1984). Suspended load refers to fine-grained sediment, generally silt and clay, that is carried in suspension within the water column. The amount of suspended load generally increases proportionally with discharge, and may constitute up to 100 percent of the river's total sediment load. The Mississippi River, for example, has an average annual suspended load of 350 million metric tons (Knighton, 1984). Bedload refers to sediment that moves downstream in contact with the channel bed. This coarser-grained sediment moves by rolling, sliding, or bouncing along the channel bottom. The proportions of sediment moved by each of these three methods depend on the stream's energy and on the nature of the bedrock within the stream's catchment that provides sediment. Fine-grained sedimentary rocks such as shale and siltstone, for example, weather to clay and silt-sized particles that generally constitute suspended sediment in streams. By contrast, limestone may weather to calcium (Ca) ions that form dissolved load, whereas conglomerate may weather to sand and cobbles that form bedload.

Although a river actually erodes a fairly narrow swath through the landscape—the width of its channel—the river may promote more widespread erosion. By gradually eroding downward, or incising, a river may create oversteepened slopes that are then modified by slope processes, gradually creating the v-shaped valley cross sections that characterize rivers. In a region with very resistant rocks, slope processes may do little to modify the river valley, with the result that rivers flow through deep, narrow gorges. Such gorges also characterize regions where the landmass is being uplifted, causing the rivers to incise rapidly in order to maintain their relation to baselevel.

Baselevel is a theoretical concept first proposed by American geologist JOHN WESLEY POWELL following his explorations of the Colorado River and its canyons in 1869–1871. Powell proposed that baselevel formed the lowest level to which a river would incise, and he described local baselevels of resistant rock along the river's channel as well as the ultimate baselevel into which a river emptied, which could be the ocean, a lake, or another, larger, river. Changes in a river's baselevel cause the river to incise or aggrade (deposit sediment) its channel in response so that baselevel may be thought of as a control point acting at the downstream end of a river. The ultimate baselevel for the world's rivers is global sea level. During periods of sea-level lowering, such as during the advance of the Pleistocene glaciers, the world's coastal rivers incised their lower portions, cutting canyons that were subsequently flooded when sea level rose during periods of glacial melting.

Rivers incise their channels through the processes of abrasion, dissolution, and cavitation. Abrasion refers to the grinding and scraping exerted on the channel boundaries by sediment in transport. Abrasion acts both on the channel boundaries and on the transported sediment itself, so that sediment in a channel generally becomes finer-grained downstream as a result of constant abrasion during transport. Dissolution refers to chemical weathering of channel boundaries by flowing water. Precipitation in many areas of the world is slightly acidic, allowing surface waters to gradually dissolve rock. Processes of dissolution are most effective on carbonate rocks such as lime-

stone, or in regions of warm, wet climate such as the tropics. Cavitation refers to the formation and subsequent implosion of small bubbles in flowing water as a result of abrupt changes in pressure associated with changes in velocity. The implosion of the bubbles generates shock waves that can gradually weaken rock, making it more susceptible to abrasion or entrainment.

In addition to incising vertically, a river may also move laterally, eroding and depositing sediment as the channel shifts back and forth across a valley floor. Reported rates of lateral movement range up to 750 m per year over a period of 150 years for the Kosi River in India (Ritter, 1986). The majority of lateral channel movement occurs during floods.

Most natural channels are always transporting sediment. If there is a net removal of material from a given portion of the river, the river is eroding. If there is a net gain of material, the river is aggrading. Sediment in transport is deposited along a channel if the flow energy decreases. This decrease in energy is often expressed as a decrease in velocity (the rate of flow) due to a decrease in channel slope or in discharge. Sediment may be deposited at many points along a channel for some length of time, only to be remobilized when flow increases, such as during a flood, or when the supply of sediment coming from upstream is decreased. These processes may be quantified in a sediment budget, which shows the relations among erosion of basin sediments, removal of sediment from the basin, and sediment storage within the basin. Because of storage, sediment production within the basin may not equal sediment yield (sediment actually discharged from the basin). Sediment budgets may be developed for different lengths of time, using such indicators as reservoir sedimentation, aerial photographs, or stratigraphic records of erosion and deposition.

A river may be simultaneously incising, aggrading, and maintaining an equilibrium condition along different portions of its length. This phenomenon is described in the idea of complex response. In order to understand complex response, consider the following hypothetical scenario: A small channel drains into a stockpond. When the stockpond is drained, the channel's baselevel is lowered. In response, the channel begins to incise. As the process of incision moves up the channel from the downstream end, sediment mobilized by the incision is carried downstream. Incision ceases in the lower part of the channel when the channel reaches its new baselevel at the bottom of the stock-pond. However, incision is still occurring in the channel's upper reaches, and the sediment flushed downstream as a result of this activity overwhelms the transport capacity of the lower channel and causes aggradation. Eventually incision in the upper channel ceases, causing the sediment supply to be reduced and initiating a new cycle of incision in the lower channel. Because of the lag time between some external change and the channel's response to that change, a channel may go through two or more cycles of incision and aggradation before finally adjusting to the new situation. Depending on the size of the channel and the erodibility of its boundaries, these adjustments may occur over a time period of weeks to tens of thousands of years.

In general, smaller channels respond to change faster than larger channels, and alluvial channels respond faster than bedrock channels. Alluvial channels are those formed in alluvium, or unconsolidated sediments, and their boundaries are generally more easily erodible than those of bedrock channels. Rates of incision for alluvial channels range up to several meters per year, while those for bedrock channels may be 7 to 130 cm per year.

As a group, rivers form the most important terrestrial agent of sediment transport. It is estimated that the world's rivers move more than 5,000 million metric tons of sediment annually (Knighton, 1984). (Nearly half of this sediment comes from Asia.) Equivalent calculations are not available for other agents of erosion and transport, such as glaciers and wind, but the latter often deliver sediment to rivers. In the great majority of contemporary terrestrial environments, rivers act as massive conveyor belts, transporting the sediment created by processes of physical and chemical weathering downstream to the oceans. Rates of erosion can be averaged over river basins and expressed as a rate of denudation, or general lowering of Earth's surface. Denudation rates generally reflect climate, human land use, topographic relief, and dominant agent of erosion. For example, denudation rates are generally high in regions of valley glaciers or badlands, and much lower in temperate and rain forest regions of low relief, where abundant vegetation limits sediment erosion from slopes. Denudation rates in the Himalaya Mountains are among the highest in the world (4.7 mm/yr) as a result of steep slopes, glaciated terrain, and intense land use (Press and Siever, 1986). Land use, in particular, may cause dramatic fluctuations in sediment movement from slopes. For example, sediment yield in a forested watershed in the eastern United States increased by a factor of ten when the forest was destroyed during urbanization (Ritter, 1986).

The role of rivers in creating their own valleys was not recognized until 1802, when the English geologist JOHN PLAYFAIR noted that channel and valley dimensions were proportional, and that tributaries joined larger channels at the level of the larger channels. John Playfair explained these observations by hypothesizing that each channel had created its own valley through gradual, prolonged erosion. This idea was initially controversial because other geologists attributed the formation of valleys to catastrophic events such as earthquakes, volcanic eruptions, or global floods.

Depositional Features

Rivers may also create depositional features if their flow energy decreases. At the small and relatively ephemeral end of the depositional scale are bars. Point bars form on the inside of channel meander bends as the speed of flow decreases going around the bend. As the channel migrates toward the outer portion of the bend, where erosion is occurring, the inside of the bend gradually builds a wide point bar. Lateral bars begin as arcuate masses of sediment attached to one bank, similar to a point bar. But lateral bars extend across the channel, meeting the opposite bank at an acute angle. Alternating bars, as their name implies, form along alternate sides of the channel in an offset pattern going downstream. Because of these bars, a channel with straight banks may have a sinuous central flow path. Central bars are teardrop-shaped, with their blunt end upstream. These bars form where an obstruction in the channel, such as a large boulder, creates a zone of quiet water in which sediment is deposited. Transverse bars form diagonal or perpendicular to the flow and may extend across the entire channel (a bar does not have to be exposed above the water surface but can be simply a topographically higher portion of the channel bed). Finally, channel junction bars develop where a tributary enters the main channel; as the flow of the tributary spreads into the main channel, it slows and deposits sediment. If a bar becomes stabilized by vegetation it is usually called an island.

When a channel floods over its banks, the floodwaters deposit sediment as they spread across the surrounding terrain. The coarsest sediment is deposited adjacent to the channel, forming mounded levees that parallel the channel. The finer sediment is spread in broad sheets called overbank deposits. Overbank and levee deposits, together with the sediments left behind as channels move laterally, create floodplains. A floodplain is the flat or gently sloping surface surrounding a channel that is flooded every few years. If the floodplain is isolated from the active channel—if, for example, channel incision is great enough that floods no longer regularly reach the floodplain—then the floodplain becomes a terrace. Terraces may be either fill/alluvial terraces of thick sediment accumulations, or they may be strath terraces consisting of a planar bedrock surface created through lateral erosion by a river, with a thin veneer of alluvium. Huge amounts of sediment may be stored in floodplains or terraces. The upper Mississippi River in Illinois, for example, has a floodplain approximately 9 km wide and 25 m deep (Strahler, 1963). Because of these sediment storage capacities, floodplains may play important roles in the movement of toxic materials, such as heavy metals, which may become concentrated in accumulations of silt and clay. For example, a mining operation may release heavy metal solutions into a channel. The metals are stored with floodplain sediments and may be subsequently eroded and remobilized decades after mining has ceased.

When a river confined in a narrow mountain canyon expands into a broad intermountain valley, some of the sediment that the river was carrying is deposited as an alluvial fan. An alluvial fan is a wedge of sediment, thicker but narrower at the mountain front, that spreads and thins toward the valley. Where several alluvial fans from adjacent channels coalesce along the base of a mountain range, they form a continuous apron of sediments called a bajada, or alluvial apron.

When a river enters a body of standing water, such as a lake or ocean, it forms a fan-shaped deposit known as a delta. If the river has a large sediment load and the tidal and wave currents of the receiving reservoir are relatively weak, the delta may be quite extensive, such as the delta of the Ganges River, which forms much of the country of Bangladesh. If the tidal and wave currents are strong they may transport the delta sediments away, leaving only a poorly developed delta, such as the one formed by the Columbia River in the Pacific Ocean. Unless they are artificially stabilized

like the Mississippi River, both alluvial fans and deltas are characterized by rapid and frequent lateral channel shifts due to aggradation . The Mississippi River delta, for example, has shifted 300 km to the east over the last 5,000 years (Ritter, 1986).

Bibliography

Knighton, D. *Fluvial Forms and Processes.* London, 1984.
Price, F., and R. Siever. *Earth,* 4th ed. New York, 1986.
Ritter, D. F. *Process Geomorphology,* 2nd ed. Dubuque, Iowa, 1986.
Strahler, A. N. *The Earth Sciences.* New York, 1963.

ELLEN E. WOHL

GEOLOGIC WORK BY WIND

Wind is capable of eroding, transporting, and depositing sand-, silt-, and clay-sized material in many geologic environments from sandy beaches to hot and cold deserts. Features resulting from wind action have also been seen on spacecraft images of Mars and Venus. Wind action is responsible for the formation of sand dunes and dust storms as well as the input of fine-grained material to desert soils and ocean sediments. Processes of desertification often involve wind action; ancient dune areas may be preserved as thick sandstones, some of which are important oil and gas reservoirs or sources of groundwater.

Wind Erosion

The wind must reach a critical threshold velocity dependent on particle size, shape, and density together with moisture content, vegetation cover, and cohesion before it can detach particles from the surface in a process known as deflation. Deflation involves a combination of lift and drag forces assisted by turbulence. It is a self-limiting process and stops when the surface becomes armored by particles too coarse for wind transport, forming one type of desert pavement. The impact of sand grains onto clay and silt aggregates in soils and playa surfaces also injects fine particles into the air. Areas of high wind erosion are often composed of uniform particle size or are subject to continual disturbance (e.g., by humans) or resupply of sediment (frequent flooding). Wind erosion of bedrock by a combination of abrasion by wind-driven sand and deflation forms grooved and faceted surfaces, or ventifacts. Large-scale (meters deep to kilometers long) versions of these grooves are known as yardangs and occur in materials as diverse as soft lake deposits and crystalline limestone.

Transport of Sediment by Wind

There are three distinct modes of sediment transport by wind that depend primarily on the grain size of the available sediment (Figure 1). Very small or dust-sized particles (smaller than 60–70 micrometers or μm) are transported in suspension and kept aloft for relatively long distances by turbulent eddies in the wind. Larger particles (approximately 60–1000 μm, or sand-size) move downwind by a series of hops or in saltation (after the Latin, *saltare,* to jump). The impact of saltating particles with the surface may cause short-distance movement of adjacent grains (reptation, or creeping). Larger (greater than 500 μm) or less exposed particles may be pushed or rolled along the surface by the impact of saltating grains. Interactions between saltating and reptating sand grains cause the sand surface to be "self-organized" in a series of regularly repeated wind ripples so as to maximize the flux of sediment. Strong winds are very important to sand and dust transport because the flux of sediment is proportional to the cube of wind shear velocity, while dust transport increases with the fourth power of wind shear velocity (Figure 2). For example, a wind blowing at 16 meters per second or m/s (35 mph) can move as much sand in a day as would be moved in three weeks by a steady 8 m/s (17.5 mph) wind.

Dust Storms

Dust storms can transport large amounts of material over long distances: major dust plumes from the Sahara have been identified over the eastern Atlantic Ocean (from satellite images and in deep sea sediments), while dust from central Asia reaches the northwest Pacific Ocean. Dust storms can have major impacts on humans by reducing visibility and air quality. Bare, loose sediments containing sand and silt, but little clay (which promotes crusting and cementation of particles), are favorable surfaces for dust production. Major sources of dust include ephemeral river flood plains, playas, alluvial fans, glacial outwash, and pre-existing aeolian deposits (especially where these are bare after agricultural operations). Dust storms occur as a result of strong winds associated with thunderstorms, the passage of weather fronts, and low-pressure systems. They occur most frequently in the Sahara (which is estimated to produce 300 million tons of dust a year), the Arabian Peninsula, the Indus Plain, Kazakhstan, and the Tarim Basin of China. In the United States, major dust sources occur in western Texas and the des-

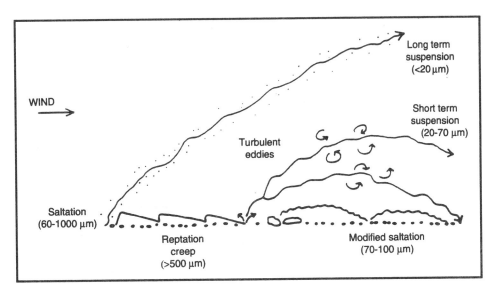

Figure 1. Modes of sediment transport by wind.

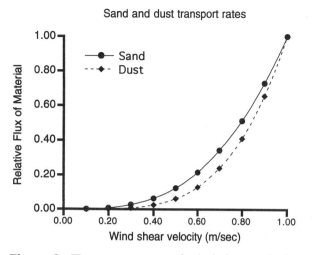

Sand and dust transport rates

Figure 2. Transport rate and wind shear velocity relations.

erts of southern California. Dust is transported from the Sahara by the Harmattan winds to West Africa and the eastern Atlantic Ocean. Deposition of dust occurs when wind velocity is reduced or particles become trapped by vegetation or rough surfaces. Alternatively, particles may aggregate or be precipitated by rain. Extensive deposits of silt-sized material known as loess were formed in the Midwest and Mississippi Valley, in central Europe, and China during glacial periods, when wind speeds were higher and large quantities of silt-sized material were available from glacial outwash plains. Much fine dust is deposited in ocean sediments downwind of arid regions, where it is an important nutrient source for marine ecosystems. Most medium and coarse dust (i.e., silt-sized) is deposited in desert margin areas where it is added to soils. Dust deposited on rocky desert surfaces infiltrates between stones and gradually raises and levels out the surface over periods of thousands of years. The stones "float" on the fine-grained material and remain on the surface to form a desert pavement.

Sand Dunes

Most wind-blown sand occurs in accumulations known as sand seas that comprise areas of dunes of varying morphological types and sizes as well as areas of sand sheets. Smaller dune areas are known as dune fields. Major sand seas occur in the deserts of the Sahara, Arabia, central Asia, Australia, and

southern Africa, where they cover between 20 and 35 percent of the area classified as arid. By contrast dunes cover less than 1 percent of the arid zone in North and South America. Sand seas are comprised of sand that has often been transported long distances by the wind. They accumulate where winds are checked by topographic obstacles, or where winds increase in directional variability and/or decrease in energy. Many sand seas show evidence of episodic accumulation and the strong effects of Quaternary climatic change, with lake deposits in inter-dune areas, and multiple generations of different dunes.

Dunes occur in a variety of morphological types (Figure 3), each of which displays a range of sizes (height, width, and spacing). There are four major dune types: crescentic, linear, star, and parabolic. Studies of the spatial distribution of dunes of different types show that the directional variability of the wind and sediment availability are major controls on their form (Figure 4). Wind speed, sand grain size, and vegetation play subordinate roles. Crescentic dunes occur where winds blow from one main direction, with barchans in areas of limited sand supply and crescentic ridges where sand is more abundant. Crescentic dunes are the simplest dune type and are the first dune type to form, approximately transverse to the wind. They are characterized by a gentle windward or stoss slope and a steep lee face. The wind accelerates up the stoss slope and erodes and transports an increasing amount of sand over the crest to the brink of the lee face. Here, the main part of the airflow separates from the surface. The saltating sand grains fall out and come to rest on the upper part of the lee face. This part of the slope soon becomes unstable, and the sand avalanches down the slope. Similar processes occur on all dunes. Over time, crescentic dunes migrate downwind as sand is eroded from one side of the dune and deposited on the other. Linear dunes are characterized by their great length (one dune in Australia is 200 km long), parallelism, and regular spacing, with wide flat areas between the dunes. They form in areas where the wind changes direction by 180° or less from season to season and are common in areas affected by the trade winds. Each wind approaches the dune at an oblique angle and forms an eddy on the lee side. Winds in this eddy transport sand along the dune. Linear dunes therefore extend over time rather than migrating. Star dunes are the largest of all dunes, and reach heights of sev-

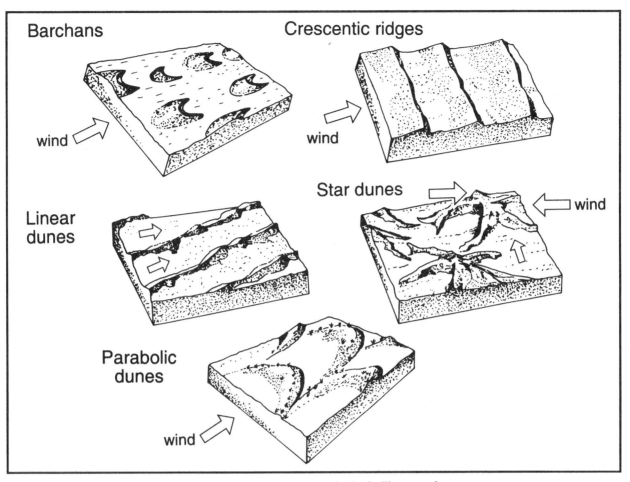

Figure 3. Dune types and winds illustrated.

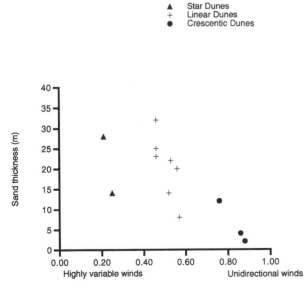

Figure 4. Dune types and wind regime relations.

eral hundred meters. They have a pyramidal shape and radiating arms. Star dunes form in areas where winds are from two or more opposed directions as in the Gran Desierto of Mexico. Once sand gets to a star dune, it rarely leaves, so the dune grows upward. Parabolic dunes occur in vegetated desert margin and coastal areas, where winds are unidirectional. Sand eroded from small areas known as blow outs forms a parabolic-shaped ridge with the apex pointing downwind. The bare center of the parabola moves more rapidly than the vegetated arms and the dune evolves to a hairpin shape.

As dunes grow, they create their own local wind patterns: as a result large dunes often have a complex form with smaller dunes superimposed on their lower slopes. Some of these dunes are often tens of thousands of years old, and their present form may be the result of both past and present wind patterns. In addition to dunes, many sand seas have large areas of sand sheets on their margins, which are dominated by coarse sand or a partial vegetation cover. They often form zones through which sand is transported to the adjacent dune areas.

Bibliography

GREELEY, R.; N. LANCASTER; S. LEE; and P. THOMAS. "Martian Aeolian Processes, Sediments and Features." In *Mars*, eds. H. Kieffer, B. M. Jakosky, C. W. Snyder, and M. S. Matthews. Tucson, AZ, 1992.

LANCASTER, N. "Dune Morphology and Dynamics." *Geomorphology of Desert Environments*, eds. A. D. Abrahams and A. J. Parsons. London, 1993.

LANCASTER, N., and W. G. NICKLING. "Aeolian Sediment Transport." In *Geomorphology of Desert Environments*, eds. A. D. Abrahams and A. J. Parsons. London, 1993.

PYE, K. *Aeolian Dust and Dust Deposits*. London, 1987.

PYE, K., and H. TSOAR. *Aeolian Sand and Sand Dunes*. London, 1990.

NICOLAS LANCASTER

GEOLOGICAL SURVEYS

Geological surveys are among the oldest science-based government agencies, with many tracing their origins to the early or mid-nineteenth century. They play an important role in the earth sciences because they generate much of the systematic documentation of the geology of Earth's surface and crust (*see* GEOLOGY, HISTORY OF).

This entry describes the origins of geological surveys and the general nature of their work. It uses mainly three organizations—the U.S. Geological Survey (USGS), the Geological Survey of Canada (GSC), and the British Geological Survey (BGS)—as illustrations.

These surveys have all made notable advances in geoscience applications. The U.S. Geological Survey pioneered digital topographic mapping, techniques in applying satellite imagery in geological interpretations, and new methods for assessing undiscovered mineral and fuel resources on public lands in the United States and elsewhere. The Geo-

logical Survey of Canada made major advances in interpreting the origin and evolution of ancient Precambrian rocks and it pioneered the use of aircraft and helicopters in developing rapid reconnaissance mapping techniques for Canada's extensive northern and arctic territories. The British Geological Survey from its beginnings established rigorous protocols in painstaking and accurate detail in classic geological mapping that were adopted throughout the world, and it was the first of the major surveys to incorporate contract research funded by external clients as an integral component of its programs.

Mission

Geological surveys are generally responsible for mapping and interpreting the geological framework of their jurisdictions and for documenting the mineral, fuel, and (in some cases) water resources they contain. Price (1992) put it succinctly: "The basic mission of any modern national geological survey is to ensure the availability of the geoscience information and expertise that are required for the wise use of the nation's mineral, energy and water resources, for the health and safety of its people and for the protection of the environment" (p. 98).

Functions

Most geological surveys are concerned primarily with geoscience mapping (geological, geophysical, geochemical), mineral and fuel resource studies, and research that supports such activities. Some larger surveys, such as the USGS, also conduct topographic mapping, hydrologic surveys, and the mapping of planetary surfaces other than Earth's (*see* MAPPING, PLANETARY). In many other countries aspects of water resources, notably groundwaters, are also the responsibility of surveys; commonly, however, this task is shared with other agencies.

Mapping and classifying surficial deposits (overburden, sand, gravel, clays, glacial deposits) are basic survey activities. Airborne geophysical surveys (magnetic, electromagnetic, and radiometric) are used as aids in geological mapping and in exploring for resources. Other geophysical techniques such as deep seismic sounding are used in investigating the properties of Earth's crust and mantle.

Seismograph networks are maintained for monitoring earthquake activity. Remotely sensed data from various Earth-monitoring satellites such as LANDSAT-TM (Land Satellite Thematic Mapper) are routinely used in reconnaissance mapping. Nearly all modern surveys use digital mapping techniques and GIS (Geographic Information Systems) to aid integrating, interpreting, and displaying data from multiple sources (*see* GEOGRAPHIC INFORMATION SYSTEMS; GLOBAL POSITIONING SATELLITES, GEOGRAPHIC INFORMATION SYSTEMS AND AUTOMATED MAPPING).

Geoscience information provided by geological surveys is used by industry in exploring for mineral and fuel (coal, uranium, oil, natural gas) deposits and by governments in formulating policies for land-use planning and management, including establishing parks and wilderness areas. It is used in developing regulations and protocols (building codes, engineering construction standards) for public safety and security in landslide, earthquake, and volcanic eruption protection and assessment. It is also applied in establishing baseline conditions (for example, the natural concentrations of potentially hazardous chemical elements) in many applications in environmental impact assessment.

Origins

Two general factors appear to have driven the establishment of formal geological survey organizations in the early and mid–1880s. First, as governments expanded into the frontier regions (western United States, northern and western Canada, much of Australia, colonial Africa), legislators and administrators realized that they needed better maps and systematic information, particularly about mineral resources, which could be used to spur economic development of these remote regions. In Europe, the colonial powers—Great Britain, France, Portugal, Spain—had long understood that organized knowledge about the resources of their colonies was important to their home economies.

The second factor was the scientific curiosity that spurred the great voyages of scientific discovery characteristic of the middle and late nineteenth century. CHARLES DARWIN's voyage aboard HMS *Beagle* was completed in 1841, and his *Origin of Species*, published in 1859, greatly influenced early geologists struggling to decipher the origin and evolution of sedimentary rocks on the basis of the fossils they contained. Later, Britain's famous HMS *Challenger* Expedition (1872–1876), with its fifty-volume documentation of the Atlantic, Pacific, and Antarctic ocean basins, laid the foundations for modern oceanographic studies. Everywhere in the natural sciences systematic explorations were being mounted to chronicle the natural history of Earth's little-known frontiers.

Chronology

The British Geological Survey (BGS), established in 1835, is credited with being the oldest national geological survey organization. Seven years later in 1842, the Geological Survey of Canada (GSC) was formed. There followed, throughout the nineteenth century, the formation of many of the world's surveys: the Service de la Carte Géologique de France in 1878 (later incorporated into the modern Bureau de Recherches Géologiques et Minières or BRGM), the U.S. Geological Survey (USGS) in 1879, and many other smaller European surveys (e.g., Hungary, 1869; Finland, 1885). Outside Europe and North America, the Geological Survey of India was established in 1851. In Africa, many survey functions were provided by the European colonial powers, but local surveys were formed in Egypt in 1896, Nigeria in 1911, and Morocco in 1912. By the onset of World War I, most industrialized countries had well-established geological surveys or geological institutes.

In some countries, state or territorial surveys predated national organizations. In the United States, for example, the oldest state survey, North Carolina (1823), was followed by a succession of some thirty-five other state surveys before the establishment of the USGS in 1879 (Socolow and Fakundiny 1994). In Australia a similar pattern developed, with colonial (state) surveys established in six territories (beginning with Victoria in 1852 and ending with Western Australia in 1888) long in advance of the national Commonwealth Bureau of Mineral Resources, Geology and Geophysics (now the Australian Geological Survey Organization) in 1946. In Canada, by contrast, the national survey (GSC) was established before a succession of provincial surveys was formed, beginning with New Brunswick in 1846 and ending with Saskatchewan in 1931.

British Geological Survey (BGS)

The establishment of geological surveys was commonly spurred by the efforts of individual organizations outside government. This was the case in Great Britain in the early 1830s when the Geological Society of London was pressing for the formation of a geological survey. In 1835, following the recommendations of the Society, the government formed the British Geological Survey, initially as a part of the existing Ordnance Survey. As its first director, the government appointed Henry de la Beche, a noted map-maker who had been producing geological maps of Ordnance Sheets for the west of England. The BGS put early emphasis on detailed geological mapping and documentation of Great Britain's mineral resources. It trained many young geologists who later occupied prominent positions in overseas and colonial surveys, including William Edmond Logan in Canada and Alfred Selwyn in Australia (Victoria), serving later in Canada.

The modern BGS stems from the 1965 merging of the original geological survey with the Overseas Geological Survey to form the Institute of Geological Sciences (Cook and Allen, 1994). Over the next twenty years a series of major reviews resulted in profound changes in the operations of government science agencies in Britain, including BGS (it resumed its original name following a period as the Institute of Geological Sciences). One of these studies specific to geological surveying needs was conducted by Sir Clifford Butler in 1985–1986 (Butler, 1987). It recommended that BGS implement a system that would combine long-term, "strategic" research in a "Core Programme" and shorter-term, client-specific research in a "Responsive Programme," the latter to be funded by both government and non-government clients.

Over the years this system has evolved so that BGS's scientific program has three components: a Core Program funded mainly through its internal (government) science budget; a Responsive Program that includes work funded by clients in both the public sector (other government agencies) and the private sector; and a research and development program contributing to both the core and responsive programs but funded through the internal science budget. The BGS budget varies because of the complex nature of the contract research funding relationships with client government departments, but in recent years it has been

of the order of 28 to 30 million pounds (British Geological Survey, 1993–1994). Within this structure, BGS conducts its programs with a staff of about 750 people from its headquarters at Keyworth near Nottingham and from regional offices in Edinburgh, Exeter, Belfast, and Wallingford.

Geological Survey of Canada (GSC)

In Canada the stimulus to form a geological survey occurred a few years later than in Great Britain, and its focus was initially centered on the east rather than the west of the American continent. In 1841 the Legislature of the Province of Canada granted "that a sum not exceeding 1,500 pounds sterling be granted . . . to defray the probable expense in causing a Geological Survey of the Province to be made" (Vodden, 1992). The Geological Survey of Canada was established the following year and given the task of exploring the territories of the "Province of Canada"—some twenty-five years before the official birth of the nation in 1867.

The fledgling GSC was profoundly influenced by the work of its first three leaders. Foremost of these was its founding director, William Edmond Logan. Logan was born in Montreal, educated in Scotland, and gained early apprenticeship in geological matters in coal mining in Wales. After his appointment as director, Logan, with a small band of assistants, began the immense job of mapping and documenting the resources of what is now eastern Canada. Logan was knighted for his efforts by Queen Victoria in 1856, and by the time he retired in 1869 at age seventy he had placed the GSC firmly on a course of scientific exploration and discovery.

The next two directors, Alfred Selywn (1869–1895) and George Dawson (1895–1901) continued in Logan's path, but with a shifting geographic emphasis. Following the confederation of Canada in 1867 and the incorporation of the new provinces and territories in the west and north (Manitoba, British Columbia, Northwest territories) in 1870 and 1871, much of the GSC's efforts became directed toward these regions. The next twenty-five years saw a period of remarkable geological journeys made on horseback, on foot, and by canoe as GSC's officers explored and mapped the immense territories of the new Dominion of Canada.

From these early beginnings GSC evolved to become the most senior federal scientific agency in

Canada. By 1992, when it celebrated its 150th anniversary, the GSC had an annual budget of about $120 million (Canadian) and a staff of about one thousand operating from its headquarters in Ottawa and from five regional offices distributed across the country from Dartmouth, Nova Scotia, on the east coast to Sidney, British Columbia, and Vancouver Island.

U.S. Geological Survey (USGS)

Following the end of the Civil War in 1865, four separate surveys (known collectively as the Territorial Surveys) mapped and explored the American West until toward the end of the 1870s. In 1879, following recommendations by the National Academy of Sciences, Congress enacted legislation to combine the geological functions of these territorial surveys into a new agency within the Department of the Interior to be called the U.S. Geological Survey.

As was the case in Canada, the early years of the USGS reflected the strong imprint of its first three directors. CLARENCE RIVERS KING, appointed the first director by President Rutherford B. Hayes, had planned and led the pioneer five-year Geological Exploration of the Fortieth Parallel (1867–1872), one of the original territorial surveys. From the beginning King focused the efforts of the USGS on the practical work of documenting and interpreting the West's mineral districts. JOHN WESLEY POWELL, appointed as the second director in 1881, headed the General Geology Division under King but had earlier gained widespread recognition through his explorations of the Colorado River and its regions, and for his influential "Report on the Land of the Arid Regions of the United States," which laid the groundwork for irrigation canal and reservoir surveys conducted by the USGS later in the 1880s. Under Powell's direction the USGS moved toward becoming a truly national organization with such bold projects as the beginning of a geological atlas of the entire United States. The third director, appointed after Powell's resignation in 1894, was CHARLES WALCOTT, a career geologist with the USGS and a widely known paleontologist. Walcott strongly entrenched the philosophy of sound scientific investigations as the basis of survey work and under his thirteen-year direction the agency expanded greatly.

By the turn of the century, the U.S. Geological Survey was firmly established as a national agency with responsibilities in geologic, topographic, and hydrographic surveying. In 1901 it had a budget of "more than $1 million . . . and a staff of more than 400" (Rabbitt, 1975). By the mid–1990s the USGS had an annual budget of about $585 million and a total staff of about nine thousand. Its principal programs are operated through five divisions (Water Resources, National Mapping, Geologic, Information Systems, Administrative) from its National Center in Reston, Virginia, and from regional centers in Denver, Colorado; Menlo Park, California; and various other field offices across the country.

Bibliography

BRITISH GEOLOGICAL SURVEY—NATURAL ENVIRONMENT RESEARCH COUNCIL. Annual Report, 1993–1994. Keyworth, Nottingham, Eng., pp. 46–47.

BUTLER, C. Report of the Study Group into Geological Surveying. Swindon, Eng., 1987.

COOK, P. J., and P. M. ALLEN. "The Example of the British Geological Survey: Past, Present and Future." In National Geological Surveys in the 21st Century, Geological Survey of Canada Miscellaneous Report 55. 1994, pp. 15–23.

PRICE, R. A. "National Geological Surveys: Their Present and Future Roles." Episodes 15, no. 2, (1992):98–100.

RABBITT, M. C. A Brief History of the U.S. Geological Survey. Washington, DC, 1975.

SOCOLOW, A., A. FAKUNDINY, and R. H. FAKUNDINY. "State Geological Surveys of the United States of America: History and Role in State Government." In National Geological Surveys in the 21st Century, Geologic Survey of Canada Miscellaneous Report 55. 1994, pp. 53–56.

VODDEN, C. No Stone Unturned—The First 150 Years of the Geological Survey of Canada. Ottawa, 1992.

D. CHRISTOPHER FINDLAY

GEOLOGY AND PUBLIC POLICY

The outcome of most human endeavors depends on appreciation of Earth's natural processes and resources. Public policy is the means that assures such endeavors have social and economic benefits for a nation and its citizens. Public policy—through laws and regulations—in the modern era, directs how and where human beings are to accommodate both natural and human-made conditions. The science of geology, the study of the earth, its processes, and its resources, is a basic component for useful public policy. Geology provides information essential to the outcome of any human endeavor that relies on Earth's natural processes and resources.

The public's interest is always well served when public policy decisions reflect an understanding of the geological aspects of a particular situation. The beneficial influence of geology on public policy is not, however, a modern invention. Our early ancestors instinctively connected the relationship between geology and public policy decisions, realizing, for example, that a geological resource such as water is essential to sustaining life, or that geological feature, such as caves, can afford shelter from harsh elements.

Early humans understood quickly that geological information can be the basis for hospitable and suitable environments. The decisions made by early humans to settle in certain areas, to explore beyond their known surroundings, to make tools, and to create complex social environments, all illustrate an instinctive comprehension that use of geological information improves the human condition.

When human activity became organized and structured, the important contribution that geological information makes to public policy decisions increased. Communities formed in areas where favorable environmental conditions, such as adequate water supply and arable soils, were found. Health, security, and well-being were assured when humans learned to make use of the earth's mineral resources. Commerce and the exchange of goods were established where suitable natural conditions for trade routes and ports existed. Just as in today's world, geological conditions often were key elements in shaping the policy decisions that caused or solved hostile territorial disputes.

The use of geological information to effect public policy decisions remained generally a matter of chance or circumstance until nations formed and government structures became established. As modern civilizations developed, geology became a key factor in public policy development. The advent of complex social and cultural environments required greater and not always compatible multiple uses of natural resources. For example, rivers that once were used only to support the most basic needs for human subsistance—food and water—were now also needed for transportation or as an energy source. Population centers developed even in areas where geologically unsuitable conditions could not be avoided. The long history and vitality of the city of Venice, Italy, provides ample testimony to the human ability to make prudent policy decisions to accommodate a geologically unsuitable natural environment.

Nations developed economies based on the absence or presence of certain natural resources. Every developed nation recognized the necessity for specific rules and procedures to protect its citizens' well-being and to assure wise use of its natural resources. Among the first formal laws established throughout modern history have been those that guided the use of a country's natural resources in support of that country's social interests.

Geological information greatly influenced early public policy decisions on education, for example. Geology and public policy both benefited, in the sixteenth century, from publication of the first "textbook" on economic geology. *De Re Metallica* defined the geological settings and engineering suitability of the extensive mineral resources found in Bohemia, now part of the modern Czech Republic. The mineral wealth of Bohemia also provided the impetus, in 1716, for the first school to educate and train mining engineers to effectively utilize the mineral resources.

The industrial revolution, in the latter part of the eighteenth century, stimulated social and cultural changes that forced decision makers to consider more carefully the relationship between human activity and natural conditions. For the first time the relationship between geology and public policy began to address formally concerns about the quality of the human environment, problems associated with pollution, waste management, land use, and resource conservation. The procedures were simplistic by contemporary standards, but the results were effective. Safe municipal water supplies and sanitary sewage systems were developed. Prohibitions were enacted to prevent random disposal of both domestic and industrial waste. It was also during this period that concepts of land-use planning and resources management began to emerge.

As a nation, the United States has been the consummate beneficiary of the close relationship that developed between geology and public policy throughout the rise of western civilization. The territorial formation of the United States, as well as its laws and traditions, originated out of the backgrounds and experiences of its mainly European founders. American explorers and settlers formed the new nation by applying old-world geological insight and public policy experience to accommodate new-world conditions. The ill-fated colonists at Jamestown, Virginia, who were killed by Native Americans in 1622, undertook the first search for economic minerals in North America. After an initial false discovery when the glitter of mica was mistaken for gold, economic-grade iron ore was found, and ironworks were active for about four years until the colony's tragic demise (Rabbitt, 1979).

During colonial times a geomorphic oddity guided policy decisions about where to locate some of the first American cities. The presence of a fall line, a series of then impassable rapids, in the major rivers that differentiate the Piedmont plateau from the mid-Atlantic coastal plain, established the setting for the cities of New York, Philadelphia, Baltimore, Richmond, and Columbia, along the rivers, but downstream of the fall line. After independence was achieved and the United States became a nation, its founders realized that economic sufficiency was essential for development. The U.S. Constitution, in Article 1, specifically authorizes Congress to "promote the progress of science" for the public good. And so here again, to the citizens' benefit, wise decision makers understood the inherent relationship that must be maintained between geology and public policy.

The developing federal government of the nineteenth century also recognized the importance of basing policy decisions on sound scientific information. Each state of the United States justified its formation and subsequent admission to the union with information developed through geological organizations established to support state or territorial interests. A special geological organization is considered an important part of the government structure in all fifty states today.

During the formative years of the United States, elected officials developed policies to further geoscientific inquiry in a number of ways. At the time of the Louisiana Purchase in 1803 almost nothing was known about the land area involved or its attendant resource potential. Numerous laws were enacted that encouraged westward expansion and settlement. These laws all included provisions that addressed prudent use and wise management of the immense natural resources found within the borders of the Purchase territory—an area that lies between the Mississippi River and the Rocky Mountains and that extends from the Gulf of Mexico to the U.S.-Canadian border.

Scientific excursions to investigate the unknown parts of the nation were organized and funded. The scientific success and national benefits derived from several of the western trips, called "surveys," led to the establishment in 1879 of the U.S. Geological Survey to focus federal geological research in a national organization. The use of geology to effect U.S. public policy, although economically and socially beneficial, has not been entirely free of controversy. In 1854 a group of business promoters engaged one of America's first and foremost geologists, Professor Benjamin Silliman of Yale University, to determine the resource potential of a substance known as "rock oil" that bubbled up in creeks in northwestern Pennsylvania. Until Professor Silliman's assessment, "rock oil" was used principally as a medicine for human consumption. Professor Silliman's findings, which demonstrated the energy fuel potential for "rock oil," are considered to be the basis for the development of the U.S. petroleum industry (Yergin, 1991). The late twentieth-century ambivalence and controversy about the appropriate uses of fossil energy resources were already in place in 1854.

U.S. public policy derives its authority from federal legislation, and many important legislative actions of the twentieth century illustrate the wisdom of engaging geological information to support public policy decisions. The record of public benefits derived through the use of geological information to shape legislation in the twentieth century is impressive, but of course unfinished. The Mineral Leasing Act of 1920, the Outer Continental Shelf Lands Act of 1953, the Atomic Energy Act of 1954, the National Environmental Policy Act of 1969, the Coastal Zone Management Act of 1972, the Safe Drinking Water Act of 1974, the Federal Land Policy and Management Act of 1976, the Clean Water Act of 1977, the Comprehensive Environmental Response Compensation and Liability Act of 1980, and the National Energy Strategy Act of 1992 are only a sample of the numerous federal laws that affect both present and future lifestyles in the United States (EESI, 1994). Every one of these influential statutes (as well as many others) owes its conceptualization and execution to an appreciation of geological knowledge and applications.

Historically, the geological contributions made to policy decisions have been the results of efforts of individuals or small informal groups. Many early policy decisions were influenced by "geological" data because the decision makers themselves were either educated in, or serious observers of, natural events or phenomena. Thomas Jefferson (1801–1809) and Herbert Hoover (1929–1933) were two notable presidential examples. Jefferson, who as President negotiated the Louisiana Purchase, was an avid earth scientist and enjoyed a reputation as a first-rate "naturalist." A paper published by Jefferson on fossil vertebrates remains in the paleontological literature. President Herbert Hoover, a member of the National Academy of Sciences, was a strong advocate of applying science to support effective natural resources management and conservation. A mining engineer of world renown, Hoover received the first degree in geology granted by Stanford University and served as a summer field assistant with the U.S. Geological Survey during his undergraduate years (Rabbitt 1989). (Mrs. Lou Henry Hoover, who also graduated in geology from Stanford, gained her own scientific acclaim for preparing the first modern translation of De Re Metallica.)

The practice of employing geological advice to shape public policy has been shown to be as old as human history. It is the present and the future, however, that offer both the greatest opportunity and the obligation for a connection between geology and public policy. If human society intends to remain in harmony with the planet, it is essential that the link between geology—a hard science—and public policy—a soft art—be strengthened.

Public policy decisions must be designed to address the risks involved with such geological-related issues as climate change, natural-hazard management, health effects of a number of naturally occurring substances (lead, asbestos minerals, radon, as well as a whole array of other useful but potentially harmful minerals), scientific literacy, land use and natural resources management. Elected officials of every developed nation, particularly those of the United States, must be attuned to their constituents' interests.

The future of the relationship between geology and public policy depends on the ability of the geological professions to be effectively responsive to the interests of both public policy makers and the concerned public.

Bibliography

CAMERON, E. N. At the Crossroads: Mineral Problems of the United States. New York, 1986.

ENVIRONMENTAL AND ENERGY STUDY INSTITUTE (EESI). 1994 Briefing Book on Environmental and Energy Legislation. Washington, DC, 1994.

NEWTON, E. G. "Geology in the Public Arena." San Antonio, 1986.

———. "Geology in the Public Arena." Pennsylvania Geology 18, no. 2 (1987):2.

———. Ethical Obligations in Public Policy. Geology for 21st Century Penrose Conf. Keystone, CO, 1987.

RABBITT, M. C. Minerals, Lands, and Geology for the Common Defense and General Welfare. Vol. 1, Before 1879; Vol. 2, 1879–1904; Vol. 3, 1904–1939. Washington, DC, 1979, 1980, 1986.

YERGIN, D. The Prize: The Epic Quest for Oil, Money & Power. New York, 1991.

ELISABETH GUERRY NEWTON

GEOLOGY, HISTORY OF

Approaches to Studying Earth

All through the development of geological thought, two philosophical approaches have competed in establishing the working model. These are uniformitarianism and catastrophism. The former maintains the concept of uniformity in nature, that the present is the key to the past—no surprises. Catastrophism, on the other hand, emphasizes sudden, catastrophic episodes to effect changes. In classical times Aristotle (384–322 B.C.E.) believed that Earth is eternal and that its processes have a uniformitarian cyclicity, where decay of the land is followed by rejuvenation. Plato (427–347 B.C.E.), in contrast, had taught that Earth was created and therefore not eternal but punctuated with sudden calamitous events like the disappearance of the legendary island of Atlantis.

The astute observations of the early Greeks and Romans concerning Earth were fortunately preserved by Arab scholars while scholarship lay dormant in the Western world during the Dark Ages. With the revival of scientific scholarship around the thirteenth century, a third approach, embodied in the Bible, had to be incorporated because the few scholars of the day were mostly ecclesiastics. To them the philosophy of Plato was more palatable.

The Foundation of Geology

Practical interest in mining produced the first handbook in mineralogy, written by Georg Bauer (1494-1555), better known as Agricola. Leonardo da Vinci (1452–1519) was among the earliest, after the early Greeks, to recognize the organic origin of fossils. But into the seventeeth century, the prevailing opinion on "fossils" or "figured stones," terms that also include minerals, was that they were objects possessing a certain "plastick vertue" generated in situ in the soil, *lapides sui generis*. The resemblance of fossils to shells or leaves was considered simply a sport of nature, *lusus naturae*. Two men stand out as contributors to the foundation of geology in the seventeenth century. They are the Dane Niels Stensen, known as Steno (1638–1686), and the Englishman Robert Hooke (1635–1703). Steno expressed a number of simple laws, such as the Law of Superposition (in a layered, undisturbed sequence of rocks, the oldest are at the bottom), establishing a systematic approach to studying strata. While Steno emphasized Noah's Flood and sudden collapse of strata as a result of formation of caverns, Hooke documented both a uniformitarian concept of cyclic processes and a history of disastrous earthquakes and volcanic collapses. In crystallography, Steno's diagrams expressed the Law of Constancy of Interfacial Angles, while Hooke's showed that he believed the external form of crystals is an expression of the internal arrangement of particles, illustrating both Steno's Law and Haüy's Law of Rational Intercepts. Both men recognized the organic origin of fossils.

Steno was guided in his ideas by the Mosaic account of Creation; thus fossils were relics of the Deluge. Hooke was largely unhampered by religious constraints. He placed the dynamic processes that shape terrestrial features within the context of celestial mechanics and theorized that Earth's shape is an oblate spheroid because of its rotation. He was also the first to describe polar wandering. Among his many startlingly on-target statements were those that foreshadow the developmental or evolutionary theory in biology, including the concept of extinction of species because of environmental changes and generation of new species.

By the time JAMES HUTTON (1726–1797), of Scotland, widely acclaimed as the "Father of Modern Geology," arrived on the scene, the major ideas in the pre-continental-drift geological paradigm had been expressed. Hutton's ideas on the cyclic process of sedimentation, consolidation, uplift, erosion, and so on were essentially those of Hooke; but while Hooke would allow either heat/fusion or aqueous solution for consolidation of loose sediments, Hutton argued against aqueous solution and in favor of only heat/fusion.

The fundamental difference between the seventeenth-century Hooke and eighteenth-century Hutton is in the dimension of time. While Hooke undoubtedly wished for a much longer time than the Scripture allowed, he had no notion of the vast geological time frame. Hutton, however, authored the famous quote, "We find no vestige of a beginning—no prospect of an end." In making this statement, Hutton precipitated a profound revolution in the history of geological thought.

Neptunism and Plutonism

Ironically, Hutton's opposition to Hooke's aqueous solution idea for consolidation of loose sediments became the rallying point around which the Plutonists (Huttonians), associated with uniformitarians, gathered in their controversy with the Neptunists (Wernerians), linked to the Flood and catastrophism. The latter were disciples of ABRAHAM GOTTLOB WERNER (1749–1817), who was a talented and influential teacher at the Freiberg Mining Academy in Germany. He devised an Earth system called Geognosy using a stratigraphical-chemical/mineralogical approach, claiming granites, schists, and basalts are chemical precipitates from solution in opposition to Hutton's heat/fusion process. Werner's most loyal admirer was Robert Jamieson (1774–1854) of Scotland who founded the Wernerian Society. JOHN PLAYFAIR (1748–1819), in the Plutonists camp, skillfully publicized the Huttonian Theory, which received a boost from the experiments of Sir James Hall (1761–1832). The latter fused "whinstones" (basalts) in an ironfoundry furnace and observed them revert to a crystalline state when slowly cooled.

During this period, the idea of uniformity was expressed in France by Comte de Buffon (1707–1788) and in Italy by Giovanni Arduino (1714–1795). Catastrophism was advocated by GEORGES CUVIER (1769–1832) of France and his associates, among them ALEXANDRE BRONGNIART (1770–1847). Cuvier recognized that transgressions and regressions of the sea, alternating with periods of continental deposition by freshwater, caused catastrophic extinctions of organisms. He believed, however, in the fixity of species.

Unconcerned with the feuding academic world, the practical William "Strata" Smith (1769–1839), working as a surveyor and engineer for coal mines in England, developed a significant advance in stratigraphy. He recognized the succession of strata and that some rocks, especially limestone, could be distinguished from each other by the fossils they contained. Accordingly, he constructed maps based on this knowledge.

The Glacial Theory

Religion was still a strong influence on a group of distinguished geologists, including William Buckland (1754–1856) of Oxford and ADAM SEDGWICK (1785–1873) at Cambridge, who were called the diluvialists. They attributed the cause of all surface features to Noah's Flood and pointed to areas covered with a jumble of muds, sands, and gravels as evidence. Huge glacial erratics, whose provenance were many miles away, were more difficult to explain by the Flood. It was known at that time that icebergs could drift for great distances and upon melting dump their unsorted load. The diluvialists quickly accepted the drift theory, especially because the presence of marine shells in the "drift" deposits proved transgression by the sea—in accordance with Cuvier's catastrophic episodes. Eventually, through the work of a Swiss engineer Ignatz Venetz (1788–1859), and Swiss geologist Johann de Charpentier (1786–1855), and the latter's ability to convince JEAN LOUIS RODOLPHE AGASSIZ (1807–1873), the glacial theory became accepted and diluvialists were then either convinced of the theory and the reality of the Ice Age, or they died out.

The Age of Earth

Whether they were uniformitarians or catastrophists, geologists by now were convinced of a long geological time frame. CHARLES DARWIN (1809–1882), who took CHARLES LYELL's *Principles of Geology* with him on his voyage on the *Beagle*, was impressed by how Lyell's principles were evident in many foreign geologic sites. Darwin's evidence of gradual ongoing incremental uplifts of the Andes, for example, argued against catastrophic explanations of mountain-building like those of Elie de Beaumont (1798–1894), who attributed mountain-building to the cooling and contraction of Earth's crust and its sudden collapse. Lyell calculated that 240 million years (Ma) had elapsed since the Cambrian period. Darwin's biological evolution required at least as much.

This reliance on a long timescale was shattered by the pronouncement of physicist Lord Kelvin, William Thomson (1824–1907), that Earth was less than 100 Ma old, based on the finite age of the Sun and assuming its energy is dissipated according to the laws of thermodynamics, and that no other *then unknown* source of energy existed. He thus dealt a blow against both the concept of uniformity in geology and the concept of evolution: in the first instance, processes cannot cycle eternally; in the second, there is inadequate time for evolution. Darwin's staunch defender, Thomas H. Huxley, courageously questioned Kelvin's calculations but to no avail—the laws of physics are inviolate. The controversy ended when radioactivity was discovered. Other sources of heat *were* available. Radioactive isotopes became a powerful tool in dating the age of rocks. Scientists eventually arrived at the consensus that Earth is about 4.6 billion years (Ga) old (*see* OLDEST ROCKS IN THE SOLAR SYSTEM).

Geology in America

Religion pervaded the thinking of most American geologists in the nineteenth century. Important names include William Maclure (1763–1840), who constructed the first geologic map of the eastern United States, and Benjamin Silliman (1779–1864), who taught geology at Yale and founded *The American Journal of Science*, the oldest scientific journal in America. His pupils included Amos Eaton (1776–1842), Edward Hitchcock (1793–1864), and JAMES D. DANA (1813–1895). Eaton's students included James Hall (1811–1898) of New York, who recognized the important concept of "facies" (lateral variations in lithology and fossils in stratigraphy), and showed them to Lyell during the latter's American trip.

American interest in exploration and western expansion spurred the government to sponsor great western surveys such as the CLARENCE RIVERS KING (1842–1901) survey in 1867 along the Fortieth Parallel; the Ferdinand Hayden (1829–1887) geological survey of the Territories in 1869; the JOHN WESLEY POWELL (1834–1902) survey of the Colorado River canyon starting in 1869; and the George M. Wheeler (1842–1905) survey in 1874 with GROVE KARL GILBERT (1843–1918) and Clarence Dutton (1841–1912). The privately sponsored expeditions in search of vertebrate fossils by Othniel Charles Marsh (1831–1899) and Edward Drinker Cope (1840–1897) and their intense rivalry are legendary. The amount of vertebrate fossil material these men collected and the number of new genera and species named were phenomenal and afforded many years of study at various American museums. The Canadian Geological Survey is the oldest federally sponsored survey in North America. Its first director, Sir William Logan (1798–1875), published *A Report on the Geology of Canada* in 1869. In 1890 the Survey sent Lawrence Morris Lambe (1863–1919), in close and friendly association with Henry Fairfield Osborn (1857–1935) of the United States, to study and describe fossil finds that led to the fruitful exploration of the dinosaur fields in Alberta. Many other Canadian geologists, such as Elkanah Billings (1820–1876), George Mercer Dawson (1849–1901), and Joseph Burr Tyrrell (1858–1957), contributed to the coming of age of American geology. The massive publications of the great surveys of North America were monumental volumes of new information.

Continental Drift

Geology in the first half of the twentieth century became a period of "normal science," as described by historian of science Thomas Kuhn in his analyses of scientific revolutions. The established geological principles were synthesized by the Swiss geologist EDUARD SUESS in his classic four-volume *Das Antlitz der Erde* ("The Face of the Earth," 1885–1909). The central idea is that the geological phenomena on the earth are the result of a cooling and contracting planet. Vertical movements of the crust are allowed, governed by the principle of isostasy applying to the gravitational balance over a broad region of different altitudes and topographical relief.

The German climatologist and Greenland explorer ALFRED WEGENER (1880–1930), however, advocated large-scale horizontal movements. Paleon-

tologists had conceived huge land bridges spanning the oceans to explain the similarities of many fossils found on different continents. Wegener showed that these bridges were unnecessary if all the continents had been one landmass he termed *Pangaea*. Starting at the end of the Mesozoic era, *Pangaea* broke up into its present configuration. He collected a massive amount of geologic, biostratigraphic, paleontologic, and climatologic evidence. He invoked Hooke's concept of polar wandering in conjunction with continental drift for the interpretation of the distribution of sediments and climatic belts at various geologic periods. While polar wandering and continental drift are two distinct concepts, the sediment and climatic distribution at various periods could not be interpreted coherently unless the continents were assembled more or less the way Wegener had proposed in his theory. Although the South African geologist ALEXANDER DU TOIT (1878–1948), having seen the evidence in the southern hemisphere, supported Wegener, the reaction to Wegener elsewhere was one of almost total rejection.

Plate Tectonics

The main criticism of Wegener's theory was that there was no acceptable mechanism for the sialic continents (made of lighter rocks containing such elements as silicon and aluminum) to sail through the denser sima (made of rocks with greater specific gravity containing such elements as iron and magnesium). In 1962, Robert S. Dietz and HARRY HESS (1906–1969) independently conceived of the seafloor spreading concept in which convection currents rising at a mid-oceanic ridge spread out on either side and act like a conveyor belt carrying continental masses. Mantle convection had actually been postulated by ARTHUR HOLMES (1890–1965) and, ironically, cited by Wegener as a possible mechanism in the fourth edition of his book *The Origin of Continents and Oceans*.

Gravity anomalies, heat-flow measurements, magnetic surveys, and studies on seismicity and volcanicity all contributed to confirmation of seafloor spreading. Specifically, magnetic surveys dramatically demonstrated the linear anomalies symmetrically distributed parallel to the axis of the ridge system corresponding to alternately "normal" and "reversed" magnetism depending on the polarity of Earth's magnetic field when the rock was solidified.

In 1968 the *Glomar Challenger,* a unique research vessel, satellite-navigated and dynamically positioned to operate in deep waters, was built to the specification of the Joint Oceanographic Institutions for Deep Earth Sampling (JOIDES), financed by the U.S. National Science Foundation. Samples of the sedimentary layers of the oceans documenting the history of the ocean basins were collected in cores drilled by this ship, which are yielding volumes of data and information.

After the confirmation of the Dietz–Hess hypothesis, it was a matter of synthesis and extension to evolve the ideas into a new paradigm—plate tectonics. The new concept shows Earth's surface to be divided into major and minor plates bounded by oceanic ridges, transform faults, or trenches. The plates are in constant relative motion with one another as evidenced by earthquakes along the boundaries. It is a unifying theory satisfying many previously puzzling research results, integrating a massive amount of evidence and data collected by scientists from around the world.

Geologists have settled in the paradigm of plate tectonics in a renewed period of "normal science." In doing so they have made exciting new discoveries such as the existence of life at sites of undersea hot springs. But what causes plates to form and to move? What causes hotspots? The debate over the demise of the dinosaurs at the end of the Cretaceous period has prompted the advance of the meteorite impact theory. This idea that giant meteorite strikes at various times in geologic history have caused extinctions has been supported by the discovery of an iridium layer at the Cretaceous-Tertiary boundary. Iridium occurs in meteorites whereas its occurrence is rare in Earth's lithosphere because of its early differentiation. Some geologists speculate that giant meteorites also may have caused the cracking of Earth's crust and the formation of hotspots. Time will tell whether terrestrial catastrophic events caused by extraterrestrial objects have a major role in geologic history. Both approaches in the study of geology, however, uniformitarianism and catastrophism, have contributed profoundly to knowledge of our planet.

Bibliography

DRAKE, E. T. "The Hooke Imprint on the Huttonian Theory." *American Journal of Science* 281 (1981):963–973.

DRAKE, E. T., and W. M. JORDAN, eds. *Geologists and Ideas: A History of North American Geology.* Boulder, CO, 1985.

FAUL, H., and C. FAUL. *It Began with a Stone.* New York, 1983.

GOHAU, G. *A History of Geology.* New Brunswick, NJ, 1991.

HALLAM, A. *Great Geological Controversies.* Oxford, 1983.

HOOKE, R. *Micrographia.* London, 1665.

———. "Lectures and Discourses of Earthquakes and Subterraneous Eruptions, etc." In *The Posthumous Works of Robert Hooke,* ed. Richard Waller. London, 1705.

HUTTON, J. "Theory of the Earth." *Royal Society of Edinburgh Transactions* 1 (1788):209–304.

STENO, N. *De solido intra solidum naturaliter contento dissertationis prodromus.* English version with notes by J. G. Winter. New York, 1916.

WEGENER, A. *Die Entstehung der Kontinente und Ozeane,* 4th rev. ed. Brunswick, Germany, 1929.

ELLEN TAN DRAKE

GEOMORPHOLOGY

See Landscape Evolution

GEOMORPHOLOGY, HISTORY OF

Geomorphology literally means the form of the surface of Earth. Water moving in streams, waves striking the coast, wind blowing across barren land, glaciers moving down mountain slopes, and landslides are some of the processes that contribute to perpetual change of the surface of Earth. Evidence indicates that geomorphic processes have been operating for billions of years. Some of the oldest known rocks (nearly 4.0 billion years or Ga old), for example, contain a pattern of sediments that required deposition by moving water. As geologic forces create mountain ranges and high plateaus, geomorphic processes slowly but effectively erode landforms into sediments that are gradually moved toward the ocean.

The study of landforms and related processes has a history that reaches far back in time, as some of the earliest known records attest. Only the events in the last two hundred years, however, have been well documented. This brief history of geomorphology includes references to key individuals, the different approaches employed to study geomorphic processes, and how the present technological revolution helped geomorphology achieve worldwide distribution and universal acceptance.

Before Recorded History

The history of geomorphology really begins as people successfully adapted and thrived on a surface constantly changing. As knowledge about the environment was acquired, generation after generation probably heard stories about processes and landforms to aid in survival. Extreme events, such as large floods, debris flows, or landslides, initiate major landscape changes and are easy to remember. Most geomorphic agents, however, operate so slowly that the amount of change may be very small over an individual's lifetime. Small changes would have little impact on people focused on acquiring the necessities of life (food, water, and shelter). Today, in contrast, the general population, whether residing in high-density urban centers or small towns, often have little knowledge of the processes of surficial change. But knowledge of such change has been growing since the Greek scholar Herodotus wrote about surficial changes in the fifth century B.C.E. What other cultures knew about surficial processes is slowly being revealed as Chinese and Iraqi literature becomes known to Western scholars (Walker and Grabau, 1993).

In Western tradition, thoughts about Earth and how it changed over time were dominated for centuries by religious teachings. But in the late eighteenth century, scholars of geology began to reject church concepts in favor of ideas supported by field evidence. This departure signifies the beginning of geomorphology although the actual term "geomorphology" was not used until the late nineteenth century (Tinkler, 1985).

Beginning of Modern Geomorphology

Research on the history of geomorphology acknowledges JAMES HUTTON, an English geologist (1726–1797), as the modern founder of geomor-

phology (as well as the father of geology more generally) because he was the first individual who broke from church tradition. Excellent descriptions of his contributions to geomorphology can be found in Tinkler (1985, 1989). Hutton clearly identified running water as a dominant agent of surficial change. He noted that many years were required to create the changes that he observed in the rock record. In the two hundred years since Hutton, the discipline spread throughout the world, as documented by H. Jesse Walker and Warren E. Grabau in *The Evolution of Geomorphology: A Nation-by-Nation Summary of Development* (1993).

The Nineteenth Century

During the nineteenth century, exploration and landform descriptions dominated the interest of many geologists and physical geographers. Whereas the search for precious metals spurred many explorers, scholars recorded landscapes controlled by geologic structure or geomorphic processes. In the United States, the federal government sent numerous parties into the West, including scientists such as JOHN WESLEY POWELL (1834–1902) and GROVE KARL GILBERT (1843–1918). Powell became the director of the U.S. Geological Survey (USGS) in 1880. In contrast, Gilbert performed field research and wrote insightful reports about how processes shaped the surface. His concepts of how the surface changed were revisited by modern scholars, including John T. Hack (1913–1991), who credited Gilbert's ideas as contributing to the theory of dynamic equilibrium.

JEAN LOUIS RODOLPHE AGASSIZ (1807–1873) was a European scholar who investigated glaciers in the Swiss Alps before moving to the glaciated landscape of North America. Scientists with state geological surveys mapped surficial geology, including the impact of continental glaciers in many northern states. Field investigations helped explain that boulders found in England, once attributed to the biblical flood, were associated with glaciation. The great progress made in recognizing surface form and process in the nineteenth century laid the foundation for the growth of the discipline.

Before World War II

The Geological Society of America (1888) and the Association of American Geographers (1904) are professional societies that promoted interaction among geomorphologists employed in academia or government service. Whereas G. K. Gilbert was an important geomorphologist with the USGS, WILLIAM MORRIS DAVIS (1850–1934) was the academician who dominated the transition into the modern era. His simplified model of how the landscape developed, the cycle of erosion, was taught to many students, who then applied this model to various geomorphic processes that shaped the surface. This qualitative model was easy to understand and apply because it stressed the evolution of forms from youth—such as initial mountain formation—to maturity—the dissection and erosion of the original form—to old age—erosion to a landscape of gentle relief. The impact of William Morris Davis on geomorphology has been well documented by Chorley, Beckinsale, and Dunn (1973) as Part Two of a three-volume effort about the history of geomorphology (see also Chorley, Dunn, and Beckinsale, 1963; and Beckinsale and Chorley, 1991).

Description of the surface was the dominant theme in the first part of the twentieth century. Physiography, the classification and description of surface form without consideration of the process of formation, occurred worldwide. The leader in this effort was N. M. Fenneman (1865–1945), who described the physiography of the United States. Artistic renderings of the landscapes of states, of the United States generally, and of all countries on Earth by A. K. Lobeck (1886–1958) and Erwin Raisz (1893–1968) focused attention on how the surface exhibited different landforms at a time when vertical perspectives of such differences were not easily obtained. Aerial photography, satellite imagery, and space shuttle photography now provide regional views of Earth, a remarkable change from hand-drawn perspectives. A spectacular digital shaded-relief map of the United States by Gail P. Thelin and Richard J. Pike of the USGS contains over 12 million elevations. Such modern techniques have renewed interest in regional landforms.

Although William M. Davis dominated the qualitative, descriptive period before World War II, some scholars opted to focus on processes of surficial change and resultant forms. Douglas W. Johnson (1878-1944) described coastal processes in 1919. Concepts about mass movement as a process were developed by C. F. S. Sharp in 1938. The role of the wind in the formation of sand dunes was described by RALPH ALGER BAGNOLD (1896–1990) in 1941. These scholars recognized the importance of various geomorphic processes in shaping Earth.

Various activities during World War II required application of geomorphic knowledge. Troop landings on beaches, construction of ports, bases, airfields, and roads, and movement of large armies over various terrains required detailed knowledge of form and process. Measuring and mapping techniques were enhanced by aerial photography and detailed analyses of soil, bedrock, the ocean, and the atmosphere. Scientific knowledge was critical to the war effort. The educational boom generated by the war continues to expand in response to the technological revolution.

Since World War II

The postwar period was marked by population growth, educational boom, and technological revolution. With prosperity came the need for more natural resources, fossil fuels, and a greater knowledge of the surface to aid in resource acquisition and safe development. Descriptive studies of landforms gradually acquired a quantitative perspective spurred in part by Robert Horton's efforts (1945). Early quantitative innovators included Arthur N. Strahler (1992), his students, and fluvial geomorphologists associated with the USGS, including Luna B. Leopold (1915–), Walter B. Langbein (1907–1982), and Thomas Maddock (1907–). A conceptual development in biology probably helped geomorphologists understand the complexity of surficial change. Ludwig von Bertalanffy (1950) described the biologic aspects of the environment as a system in which every element has an impact and is affected by every other element in the system. He emphasized that people were part of rather than apart from the system. People could have an impact on nature but nature was not controlled by human action. The application of system theory to geomorphic processes, promoted by A. N. Strahler (1952) and Richard J. Chorley (1962), plus the quantitative revolution, established the basis for modern geomorphology.

As the academic enterprise grew in the United States and around the world, geomorphology also expanded. Professional organizations created divisions, such as the Quaternary Geology and Geomorphology Division with the Geological Society of America in 1955, and the British Geomorphological Research Group (BGRG) in the United Kingdom in 1960, to promote geomorphology. Informal field trips, organized as the Friends of the Pleistocene by Richard J. Flint (1902–1976) and J. Walter Goldthwait (1880–1947) in 1934, are currently offered regularly by eight regional groups. These field meetings were designed to generate discussion of research topics at field sites (Vitek, 1989). Another type of informal gathering, the Binghamton Geomorphology Symposium, was established in 1970 by Donald R. Coates and Marie Morisawa. This annual gathering has established a tradition of excellence by attracting prominent geomorphologists to address a single topic without the distractions present at professional conventions.

Most research in geomorphology follows established paradigms, that is, stable patterns of scientific activity that have slowly evolved. Specific paradigms include process, technology (computer modeling), remote sensing and regional geomorphology, planetary geomorphology, and theoretical geomorphology. In addition, the need for energy to drive a system to exceed a threshold, that is, a point of irreversible change, has been recognized although the capability of precisely measuring energy input and expenditure is lacking at this time. A brief review of important paradigms, including key individuals, establishes the current status of geomorphology.

The process paradigm, linked by John T. Hack to Grove K. Gilbert, is also found in Douglas W. Johnson's work on coastal processes, Kirk Bryan's investigations of hillslopes and fluvial processes, Luna B. Leopold and Thomas Maddock's work at the USGS on fluvial geomorphology, Richard F. Flint's research into glacial processes, Arthur N. Strahler's work on fluvial processes, and recently in Stanley A. Schumm's field research and laboratory models of the fluvial system. Tinkler (1985) and Beckinsale and Chorley (1991) are valuable references from which to acquire more information about the contributors to the process paradigm. The academic perspective in North America is dominated by process studies, in contrast to the efforts in France and Germany that emphasize climatic geomorphology. But emphasis on process has not prevented development of other paradigms.

Although interest in physiography waned after World War II, acquisition of photographs and satellite images covering large regions, plus the the-

ory of plate tectonics, revitalized the regional perspective. Hence, a paradigm developed in conjunction with tectonic geomorphology. Cliff Ollier (1981) summarized the status of this paradigm and contributed the first chapter on the morphotectonics of continental margins in Morisawa and Hack (1985). William B. Bull, Theodore M. Oberlander, and Milan J. Pavich have also been major contributors to the tectonic geomorphology paradigm.

As space probes provided images of planets, moons, and asteroids, surficial phenomena on these images were often explained through analogy to similar forms on Earth. Victor R. Baker (1993) is one of the dynamic leaders of the planetary geomorphology paradigm. His research on fluvial geomorphology with respect to catastrophic and historic events was extended to phenomena observed on planetary bodies. He noted that new landscapes present the same problems of explanation as did the landscapes on Earth. Theories must be built and tested; scientific inquiry proceeds at its finest when faced with issues raised on other planets. Geomorphologists have contributed to the research literature on the craters of the moon, canals of Mars, and fluvial-like forms of Venus. When the resolution of images improves, the quality of explanation for extraterrestrial phenomena will also improve.

Probably the most important changes in geomorphology have been associated with the technological revolution in information processing, from data collection, including remote sensing, to data manipulation such as that embodied in the geographic information system (GIS). Sophisticated mathematical models have been developed and manipulated by software in high-speed computers in efforts to understand and predict change. Knowledge of processes, from detailed studies of weathering of rocks and detritus formation to understanding the mechanics of the processes that move debris toward the ocean, has been incorporated in these models. Contributors to developments in theoretical geomorphology include Michael J. Kirkby (1994), Frank Ahert, Adrian Scheidegger, Richard Craig, Larry Band, and Jonathan Harbor. Although computer research may seem far removed from process-oriented field studies, in actuality, field data for form and process are necessary to calibrate each model. Results of the models can only approach reality if equations and software reflect reality. Predicting surficial change is critical because any change impacts human activity.

Conclusion

Understanding how the surface was created and how it changes has interested people for millennia. Great strides have been made toward these goals but the future holds incredible promise. Technology is changing how we search for answers. From satellite imagery of Earth and planets to the ability to observe the surface of atoms, geomorphologists have acquired tools to assist with data collection. Data are necessary to test hypotheses and confirm or change what we believe. The integration of scientific knowledge from geology, chemistry, physics, biology, mathematics, and computer science aids geomorphologists in their primary goal—description of landforms and explanation of how processes create surficial change.

Bibliography

BAKER, V. R. "Extraterrestrial Geomorphology: Science and Philosophy of Earthlike Planetary Landscapes." In *Geomorphology: The Research Frontier and Beyond*, eds. J. D. Vitek and J. R. Giardino. Amsterdam, 1993.

BECKINSALE, R. P., and R. J. CHORLEY. *The History of the Study of Landforms or the Development of Geomorphology*. Vol. 3, *Historical and Regional Geomorphology 1890–1950*. London, 1991.

CHORLEY, R. J. "Geomorphology and General Systems Theory." *U.S. Geological Survey, Professional Paper 500-B*. Washington, DC, 1962.

CHORLEY, R. J., R. P. BECKINSALE, and A. J. DUNN. *The History of the Study of Landforms or the Development of Geomorphology*. Vol. 2, *The Life and Work of William M. Davis*. London, 1973.

CHORLEY, R. J., A. J. DUNN, and R. P. BECKINSALE. *The History of the Study of Landforms*. Vol. 1, *Geomorphology Before Davis*. London, 1963.

HORTON, R. E. "Erosional Development of Streams and Their Drainage Basins: Hydrophysical Approach to Quantitative Morphology." *Bulletin Geological Society of America* 56 (1945):275–370.

KIRKBY, M. J. *Process Models and Theoretical Geomorphology*. West Sussex, Eng., 1994.

MORISAWA, M., and J. T. HACK. *Tectonic Geomorphology*. Boston, MA, 1985.

OLLIER, C. D. *Tectonics and Landforms*. London, 1981.

STRAHLER, A. N. "The Dynamic Basis of Geomorphology." *Bulletin Geological Society of America* 63 (1952):923–938.

———. "Quantitative/Dynamic Geomorphology at Co-
lumbia 1945–60: A Retrospective." *Progress in Physical Geography* 16 (1992):65–84.

TINKLER, K. J. *A Short History of Geomorphology*. London, 1985.

———. *History of Geomorphology: From Hutton to Hack*. Boston, 1989.

VITEK, J. D. "A Perspective on Geomorphology in the Twentieth Century: Links to the Past and Future." In *History of Geomorphology: From Hutton to Hack*, ed. K. J. Tinkler. Boston, 1989.

VON BERTALANFFY, L. "The Theory of Open Systems in Physics and Biology." *Science* 111 (1950):23–28.

WALKER, H. J. and W. E. GRABAU, eds. *The Evolution of Geomorphology: A Nation-by-Nation Summary of Development*. Chichester, Eng., 1993.

JOHN D. VITEK

GEOPHYSICAL TECHNIQUES

Geophysical methods used in investigations of Earth's interior involve physical measurements at and near Earth's surface, which can be related to properties of rocks, minerals, and fluids in Earth or to processes occurring within Earth. Mostly we make passive measurements. In some instances we generate signals that travel through the earth, and then we detect them when they return to the surface. In every case it is necessary to carefully interpret the data that we collect in order to draw conclusions about the internal structure of Earth or the nature of geological processes. The limited accuracy of the data, and the generality of some of the physical laws we use to interpret the data, often allow more than one, perhaps many, possible models to explain the measurements.

The most important techniques include: seismology (which in effect measures the elastic properties), magnetism (which depends upon magnetic properties and upon electromagnetic processes), gravity measurement (which is affected by density), heat flow (which is determined from measurements of temperature and thermal conductivity), and various electrical methods (which measure natural currents or the electrical conductivity of the earth). Earth's magnetic field—and its measurement—is covered in a separate entry (*see* EARTH'S MAGNETIC FIELD).

Seismology

Seismology is one of the most powerful methods for obtaining information about the internal structure of Earth, either within the shallow parts of the crust as in exploration for oil, or through investigating the large-scale structure of the deep interior. In both cases we measure the time it takes elastic waves to travel through various parts of the earth. Elastic waves are similar to everyday sound waves in that they travel through matter with speeds depending upon the properties of the particular medium. However, since most of Earth is solid, seismic waves (as they are called) are more complicated. There are four main types of seismic waves, two of which travel through the interior of the earth (called body waves), and two of which only travel along the surface (called surface waves).

One of the body waves is very similar to normal sound waves since it travels by means of alternate compression and decompression of the medium through which it travels. It is the fastest of any of the seismic waves, and therefore is the first to arrive at a seismic wave detector, whether produced artificially as by an explosion, or naturally, as in an earthquake. For this reason such waves are called Primary or "P" waves. The other body wave travels by distorting the transport medium in a shearing motion. It is the second wave to arrive at a seismic recorder and is therefore called the Secondary or "S" wave. S waves differ from P waves in that they cannot travel through liquids, because liquids will not support shear stresses.

Rayleigh waves are one type of surface wave and closely resemble the surface waves seen when an object is thrown into still water. Love waves, the other type of surface wave, resemble the waves that can sometimes be observed when wind blows across a field of tall grass. Although surface waves may be useful in investigating the earth, they are generally far less important than P and S waves.

When body waves encounter a boundary between two different materials they may either be reflected back from the interface or transmitted into the second medium, almost always with some bending (refraction). In either case the waves obey Snell's Law:

$$\frac{\sin \Theta_1}{V_1} = \frac{\sin \Theta_2}{V_2}$$

in which Θ_1 is the angle of the incident ray at the boundary between two layers, and Θ_2 is the angle of the refracted ray. V_1 is the speed of sound in layer 1 and V_2 is the speed of sound in layer 2.

Most studies of Earth's crust make use of reflected P waves to determine the shape of layers or surfaces in the crust. This is especially important in oil exploration, since the shape of subsurface structures is often essential in forming containers or "traps" in which oil has collected and been preserved. In simple terms seismic prospecting involves generating waves using explosives or other means and measuring the times it takes the waves to reflect from a buried surface and return to a set of seismic detectors (called geophones) set out of the earth's surface. Careful analysis of the different arrival times at various detectors and using different source ("shot") locations can result in a model of the depth and shape of one or more underground surfaces. This is similar to trying to determine the shape and distance of an irregular wall by using the timing of echos at various locations. Very large amounts of money and computer time are spent in collecting seismic reflection data and analyzing the reflection signals to find oil.

The refraction, or bending, of seismic waves passing through the earth's deeper interior is very important in defining the major subdivisions of the earth. The path, and average speed, of body waves depends upon the properties of the layers of the earth through which the waves pass. A global network of seismic recording stations allows us to measure the time of arrival of earthquake waves at many locations at different distances from the source of the earthquake (Figure 1). The travel time graphs constructed from carefully timed earthquake recordings can be used to determine the epicenter (surface geographic location) of subsequent earthquakes, using a simple triangulation procedure and at least three recording stations. The travel time graphs can also be interpreted in terms of the properties of Earth's interior, allowing seismologists to determine accurately the changes in both compressional and shear wave velocities as a function of depth within the earth. These results have been used to divide Earth into four spherical "shells": crust, mantle, outer core, and inner core. The subdivisions are based upon striking changes in the velocities of body waves at the boundaries between the layers.

The crust is a thin outer region about 10 km thick beneath ocean basins and about 40 km thick beneath the continents. At the base of the crust both compressional and shear wave velocities in-

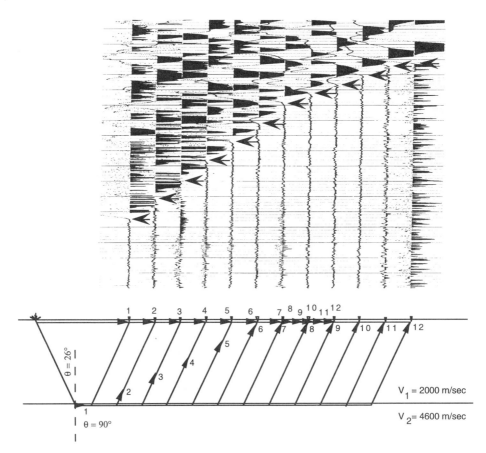

Figure 1. Twelve-station shallow seismic refraction profile. The cross section at the bottom of the figure shows the paths followed by the direct wave (which travels horizontally from the source explosion to the geophones) and the refracted wave (which follows a path down to the deeper layer, along the interface between the two layers, then back to the surface). At each geophone the two signals generally arrive at different times. The numbered arrows show the positions of each wave at the time that the first one arrives at each numbered geophone. Note that for geophones near the source the refracted wave arrives much later than the direct wave, while for more distant geophones the higher wave velocity in the deeper layer allows the refracted wave to catch and then pass the direct wave. The seismic records shown at the top of the figure show signals arriving at each geophone position over the geophone location. On these records time increases in the vertical direction on the page. The arrows on the figure show the point in time where the first signal arrives at each geophone. The change in slope of the arrows, which occurs close to geophone seven, is due to the higher speed of the refracted wave in the deep layer.

crease suddenly by about 10–20 percent. The mantle is a solid layer that extends from the base of the crust to about half way to the center of the earth (nearly 3,000 km down). The mantle consists of minerals that are rare at Earth's surface. These minerals have higher densities than most common minerals of the crust. Some of them may be found in unusual rocks that have been explosively erupted onto the surface of the earth. The mantle comprises the largest part of the earth, both in terms of volume and mass. At the base of the mantle there is a profound change in the properties of the material of the earth. There is a sharp increase in density and the material is known to be molten because it does not transmit shear waves. This latter fact is evident from recordings made at seismic stations sufficiently far from an earthquake source. Beyond a certain critical distance seismographs do not record a shear wave signal at the expected time. The outer core, as this region is called, consists mostly of molten iron-nickel alloy, based upon the speed of compressional waves that travel through it, and upon its density. The outer core extends to a depth of just over 5,000 km. The inner core is also composed of iron-nickel alloy, but it is solid. It is capable of transmitting shear waves in particular a wave that travels through the outer core as a compressional wave but is converted to a shear wave at the boundary between outer and inner core. When the wave reaches the other side of the inner core, it is converted back to a compressional wave and can thus travel through the outer core and eventually be detected at Earth's surface.

Because of the large number of layers with very different properties, the seismic recordings made around the earth can be bewilderingly complicated. In fact, within the mantle there are additional sublayers defined by more subtle changes in the properties of the rock. In the upper part of the mantle there is a layer known as the low velocity zone (LVZ), where seismic wave velocities are unusually low and even decrease with increasing depth. This is thought by many geophysicists to be the result of the formation of a tiny amount of melt within the mass of rock. Whatever the cause of the LVZ, its upper boundary closely corresponds to the change from rather rigid rock, the material of the "plates" of plate tectonics, to the much more plastic (though still very much solid) material of the asthenosphere, a region where the rock flows very slowly but does not fracture like the plates.

Indeed, it is the brittle fracturing of the plates that gives rise to most of the earthquakes that have been used to study the interior of Earth. The precise location of earthquakes has demonstrated that they are not randomly distributed over the surface or in the interior of Earth, but occur in narrow bands mostly in the outer 60–70 km of the earth, and never at depths below about 700 km. The narrow lines of earthquake distribution define the boundaries of the rigid plates, which generate earthquakes as they scrape against each other

while slowly sliding on the plastic asthenosphere below. While this picture seems obvious to most geoscientists today, and although the basic distribution of earthquakes was known for decades, it was another branch of geophysics that provided the crucial evidence for the plate tectonic model of the earth by establishing the reality of continental drift (*see also* SEISMIC LINES, COCORP, AND RELATED GEOPHYSICAL TECHNIQUES).

Seismic Tomography

In the last two decades seismologists have used techniques similar to those used in medical tomography (one type of which is commonly known as a CAT-scan) to look for lateral variation in the seismic properties of Earth's interior. Ideally one could imagine sending seismic waves diametrically through the earth and comparing the time it takes the signal to reach the other side. Variations in the travel times for many such measurements could then be used to produce a three-dimensional picture of variations of seismic velocity within Earth. Unfortunately we cannot actually do such an experiment. For one thing most of Earth is under water, which makes it difficult to place seismic recorders. Second, we would need too many such experiments to cover the entire planet. In addition, seismic waves in the earth generally do not follow straight-line paths the way X rays do when they travel through tissue. This latter fact makes the analysis of the data more difficult than for X-ray imaging. On the bright side, the curved paths and the large number of seismic recorders spread around the earth provide a wealth of data from waves which traverse virtually all of the interior of Earth. The increase in seismic data and computer power over the last ten years has provided us with a good first "image" of variations in seismic properties in 3D within Earth. Among the most intriguing features identified to date are low seismic velocities under ocean ridges, and high seismic velocities that seem to be related to ancient "cold slabs," which are the remains of subducted ocean lithosphere (*see* SEISMIC TOMOGRAPHY).

Magnetism

Most magnetic investigations of Earth's interior depend on the presence of a predominantly dipolar magnetic field (the familiar arrangement with "north" and "south" magnetic poles), the influence of magnetic minerals in Earth's crusts on the local strength of that field, and the ability of rocks to record Earth's magnetic field at the time they form (*see* EARTH'S MAGNETIC FIELD).

Earth's magnetic field changes with time on many timescales, from daily variations due to solar magnetic storms, to decadal changes in pole position and field strength, to changes in the polarity of the field (north and south magnetic fields "flipping") on a scale of 100,000 years or so. Historical and archaeomagnetic observations of Earth's field strongly suggest that the magnetic poles correspond to the geographic poles on the average (over a period of thousands of years) even though at any given time the location of the magnetic poles may be many hundreds of kilometers away from the geographic poles. Overall, the dipolar aspect of the field appears to have been a feature of Earth throughout its history.

The presence of a strong magnetic field immediately tells us something interesting about the interior of Earth. Out knowledge of Earth's internal temperature (see below) and the magnetic properties of the minerals present in the earth preclude the possibility that Earth's main field is due to permanent magnetism. In addition, the known time variations in the field also argue against permanent magnetism. The only plausible mechanism for producing Earth's magnetic field is a "self-regenerating dynamo." This is a natural system that is similar to an electric generator, with a conductor moving in a magnetic field, and which produces a self-maintaining magnetic field. Such a dynamo requires the presence within the earth of a conductor capable of large-scale motion and a power source to sustain the motion. It is easy to envision the molten metal outer core of Earth as the conductor, and Earth's internal heat resources as the energy source. The details of such a source for the magnetic field are poorly understood, but no better alternative has been suggested as yet.

Unusual concentrations of magnetic minerals in crustal rocks are occasionally related to the presence of valuable ore minerals. A magnetometer, which measures the intensity of the local magnetic field, can be used to survey an area. This may be done on a regional scale using aircraft-borne magnetometers, and at a local scale using a handheld magnetometer. By recording the magnetic field at different locations, unusually high and low values of the field may be used as indicators of "interesting" rocks beneath the surface. In some cases this may lead to the discovery of ore deposits. This approach is becoming more important because most of the ore deposits that are exposed at the surface of the earth have now been found. In order to locate deposits beneath the surface it is necessary to drill holes, which is quite expensive. Using magnetic surveys, geophysicists can locate some of the most promising locations for drilling at modest cost (Figure 2).

Studies of the past magnetic field of the earth, as preserved by the permanent magnetism in rocks formed at various times, have been crucial in establishing the current model of how the earth behaves on a planet-wide scale: the plate tectonic theory. These results come in two main forms. First there are numerous examples in which the permanent magnetism of ancient rocks from different continents gives different locations for the earth's magnetic poles. In fact, rocks of different ages from the same continent give pole positions different from the present pole. These curves are different for different continents and were produced not by the magnetic pole moving with respect to the continents, but by the movements of continents relative to the magnetic (and by implication the geographic) pole. In other words, the continents have drifted with respect to the pole, and they have also moved relative to one another (*see* PLATE TECTONICS).

The other discovery, which finally established continental drift and led to the development of plate tectonics, involved the careful measurement of the magnetic field around previously known features on the seafloor called oceanic ridges. The slight increases and decreases in magnetic field caused by the magnetized rocks in the ocean crust (called marine magnetic anomalies) were found to form bands nearly parallel to one another and to the topographic axis of the ridge itself. Moreover, the bands were found to be symmetrical about the ridge crest. At about the same time studies of numerous layered lava flows indicated that Earth's magnetic field has changed its polarity (i.e., the north and south poles had switched) on a number of occasions over the past few million years.

The bands are produced as new seafloor is created at the ridge crest by intrusion and extrusion of basaltic rock. The new rock is magnetized by Earth's field and preserves this magnetization.

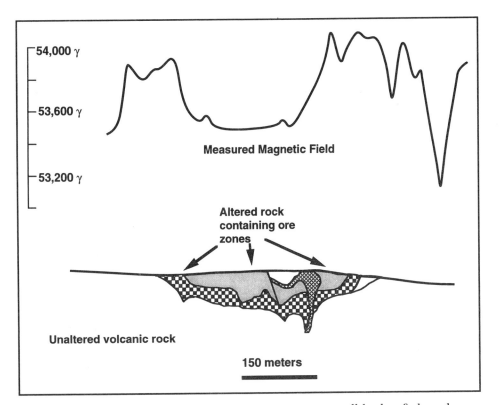

Figure 2. Total magnetic field profile (top) across a small body of altered volcanic rock containing gold ore (cross section at bottom). The unaltered volcanic rock is relatively strongly magnetic because it contains a small amount of magnetite. The alteration process destroys the magnetite, producing less strongly magnetic minerals. The alteration process is also responsible for the deposition of the ore. In this case a relatively lower magnetic field is a guide to the presence of ore. (Modified from R. Van Blaricom, "The Geophysical Response of the Buckhorn Mine—Case History," in *Practical Geophysics for the Exploration Geologist II—1989.* Northwest Mining Association, Spokane, WA, 1989.)

When more rock is added it pushes older rock to each side. When the polarity of Earth's field reverses, the rock formed thereafter is magnetized in the opposite direction. The repetition of this process over many reversals of the magnetic field produces symmetrical bands of rock that are oppositely magnetized. By combining this model with the chronologic pattern of magnetic reversals determined from lava flows, it is possible to determine the rate of seafloor spreading (rate at which the two sides of the ridge are moving apart as the result of new crust created at the ridge crest). This rate turns out to be about 5–10 cm/yr, and varies somewhat from ridge to ridge. The seafloor spreading model is a key part of plate tectonics.

Gravity

The attraction between Earth and objects dropped near its surface is obvious to everyone. The acceleration produced on these falling objects (9.8 m/s^2) is familiar to many. The use of a constant number for g (the acceleration due to Earth's gravity) is justified for most purposes by the nearly spherical shape of the earth. In fact g varies from about 9.78 m/s^2 at the equator to about 9.83 m/s^2 at the poles. While this does not seem like very much, the sensitivity of modern gravimeters is so high that differences in g of less than .0000001 m/s^2 can be readily measured. This ability to measure very small differences in g allows us to determine variations in mine variations in the composition of parts of the earth, either within the crust or at greater depths, which are related to variations in the density of the rock.

A gravity survey involves careful measurement of gravity at a number of stations over an area where geophysicists hope to find an interesting geological feature which has unusually high or low density. Since measured gravity depends on elevation (the further from the center of the earth the lower the gravitational attraction), geophysicists adjust the measured value of gravity at each station to account for effect of elevation. Usually this involves adjusting the measured value to what it would be if the gravity meter could be placed at sea level. Because the rock between sea level and the actual elevation at which the measurements are made also affects measured gravity, it is also necessary to adjust the data to account for this effect. In making this adjustment it is necessary to make an estimate of the density of the rock under the station. Usually some average value of density is chosen, based upon the rocks found at the surface. The adjusted value for the measured gravity is compared to the value of gravity which an ideal model of Earth's gravity would have at sea level at the station. This ideal model of gravity takes into account the shape and rotation of Earth, and incorporates the expected change in gravity with latitude. It is very important in making gravity surveys to know the elevation of each measurement location, normally to better than 30 cm. The difference between the measured (and adjusted) gravity and the ideal gravity is called the Bouguer gravity anomaly. A positive value of the Bouguer gravity anomaly indicates that there is extra mass below the station compared to average Earth, probably due to material of unusually high density (say a deposit of iron ore). A negative Bouguer anomaly indicates low density material (such as a salt dome, see Figure 3).

Measurements of gravity can also tell us about important differences between major geologic features such as ocean basins, continental platforms, and mountain ranges. Gravity studies led to the development of the important concept of ISOSTASY.

Electrical Methods

Electrical conductivity of rocks depends upon mineral composition and the presence or absence of

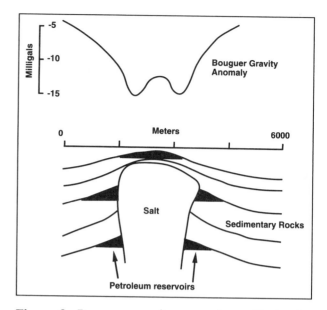

Figure 3. Bouguer gravity anomaly profile (top) across a salt dome (cross section at bottom). The Bouguer gravity profile shows a distinct gravity low over the salt dome. The lower density of the salt compared to the surrounding sedimentary rocks leads to less gravitational attraction in the vicinity of the salt. The small increase in gravity over the center of the dome is a common feature often due to higher density cap-rock at the top of the salt. As the diagram shows, salt domes have been found to form important traps for petroleum. In fact the development of precise gravity meters was initially for purposes of exploring for salt domes in the Gulf Coast region of the United States. (Modified from R. E. West, "The Land Gravity Exploration Method," in *Practical Geophysics for the Exploration Geologist II—1989.* Northwest Mining Association, Spokane, WA, 1989.)

conducting solutions in the pore spaces between the mineral grains. Many ore minerals have relatively high conductivity (often shown by a metallic luster), which makes conductivity measurements a valuable tool in mineral exploration. The most common method for making measurements involves inserting electrodes into the ground, passing a current through the ground, and then measuring the voltage associated with the current flow. There are numerous specific techniques for doing these measurements depending upon the goal of the investigation, but all are directed at determining the variation in earth conductivity as either a function of depth or lateral position.

Because the earth is generally a poor conductor of electricity, it is possible for charges to build up within the earth when currents are forced to flow, as in the method just described. When the current is shut off, these charge buildups produce a short-term current flow that can also be measured. The ability of rocks to hold a charge is dependent upon electrical properties of the rocks and is an additional factor that can be measured to study the subsurface environment.

Magnetotellurics is an additional electrical technique that can provide information about both shallow and deeper layers of the earth. Ionospheric electrial currents produce short-term variations in Earth's magnetic field that, in turn, drive current flows within the earth (much like the currents produced in electromagnetic dynamos or generators). These currents can be detected by measuring voltages at the surface. Because of the regional scale of the ionospheric currents that force the terrestrial current flows, these measurements often are most suited to studying regional-scale structures and properties of deeper layers of the earth.

Ground penetrating radar is an electromagnetic method that has become popular recently for very shallow studies. High frequency waves are transmitted into the earth and reflections from good electrical conductors, such as fuel tanks and barrels, are detected by a sensor at the surface. This technique is especially valuable in environmental investigations of potentially contaminated sites.

High Pressure Geophysics

Most of Earth is inaccessible to us and is at high temperatures and pressures. The boundary between Earth's mantle and core, at a depth of about 3,000 km (nearly halfway to the center of the earth), is at a pressure of nearly 1,300 million atmospheres and a temperature of perhaps 5,000 K. Most of the geophysical data from Earth's deep interior is in the form of seismic velocities and densities calculated using those velocities. In order to interpret these data it is necessary to make measurements on likely samples of minerals that we think will be found at great depths within the earth. However, to be meaningful (and even to make some of the minerals thought to exist there) experiments must be carried out at high pressure.

(It turns out that high temperature is not as important as high pressure.)

The most important device for carrying out high pressure experiments is the diamond anvil press. This is a surprisingly small but very ingenious device consisting of two gem-quality diamonds with carefully ground flat faces. The diamonds are set in a framework of screws or levers by which the faces can be carefully aligned and forced together. Because diamond is very hard, the faces can be pushed together to generate very high pressures, and because the faces are small, the amount of force required can be produced by a wrench turning a screw. The highest pressures yet generated by such a device are equal to pressures well inside Earth's core. Obviously the samples tested must be very small, typically less than .1 mm in diameter. Various microscopic techniques have been developed to study such small samples. High pressure geophysicists can now make measurements of optical, electrical, magnetic, and elastic properties on small pressurized samples. Among the first measurements made were X-ray diffraction observations that allowed the calculation of density at high pressure. The use of these high pressure techniques is steadily improving knowledge of the likely composition of different parts of Earth's deep interior.

Bibliography

BROWN, G. C., and A. E. MUSSETT. *The Inaccessible Earth.* London, 1981

BURGER, H. R. *Exploration Geophysics of the Shallow Subsurface.* Englewood Cliffs, NJ, 1992.

DZIEWONSKI, A. M., and J. H. WOODHOUSE. "Global Images of the Earth's Interior." *Science* 236 (1987):37–48.

JAYARAMAN, A. "The Diamond-anvil High-pressure Cell." *Scientific American* 250 (1984):54–62.

PARASNIS, D. S. *Principles of Applied Geophysics*, 4th ed. New York, 1986.

GEORGE H. SHAW

GEOPHYSICS, HISTORY OF

Geophysics is the study of Earth by quantitative physical methods and measurements. In addition to the solid earth, geophysics includes the study of the atmosphere and hydrosphere and components of geodesy, meteorology, oceanography, astronomy, engineering, and other earth and planetary sciences. This entry focuses on solid earth geophysics, which includes the fields of seismology, potential fields, chemical-thermal, and laboratory investigations.

Although many civilizations were necessarily concerned with geophysical topics and likely wrote or communicated on these topics, one of the earliest writings that has survived is that of Aristotle (384–322 B.C.E.) called "Meteorologica," which dealt with weather, earthquakes, oceanography, astronomy, and meteors (Bates et al., 1982). One of the earliest tools of geophysics that has survived came from China, where Chang Heng (132 C.E.) developed a simple seismograph that signaled the occurrence and origin direction of an earthquake. We have few records that relate to solid earth geophysics until the sixteenth century, when Leonardo da Vinci (1452–1519) studied the mechanics of faulting, wave propagation, fossils, gravitational attractions, and acoustical reflections.

From the sixteenth century through the nineteenth century, recorded advances in solid earth geophysics steadily increased with the works of numerous scholars including: William Gilbert (1540–1603), magnetism and electricity; Viscount Francis Bacon (1561–1626), gravitation and magnetism; Galileo Galilei (1564–1642), gravity and thermometry; Christian Huygens (1629–1695), reflection and diffraction of waves; Isaac Newton (1642–1727), laws of motion; Pierre Bouguer (1698–1758), gravity; John Mitchell (1724–1793), earthquakes and gravity; C. F. Gauss (1777–1855), magnetism; Robert Mallet (1810–1881), seismic methods and earthquakes; and Roland von Eotvos (1848–1919), gravity.

The earlier works primarily involved applying principles of mathematics and physics to calculate and deduce various observations. Observation was the key since early geophysical investigations were limited by what could be observed with the human senses. In the nineteenth century solid earth geophysics began to rapidly grow when devices that permitted better observation were developed. The major fields of solid earth geophysics that have since developed include seismology, potential fields, thermal/chemical, and laboratory methods.

Seismology

The basis of seismic theory is the wave equation of elementary physics. Seismology is the most widely used tool in geophysics, as it is used to investigate phenomena involving earthquakes, the structure of Earth's subsurface, the location of subsurface resources, and the location of nuclear tests and other warfare weapons. Most seismic measurements can be grouped into components of three areas of seismology: earthquake, exploration, and tomography.

Earthquake seismology was the earliest form of seismological investigation. It uses seismic waves to locate and understand seismic sources, and to determine physical characteristics of Earth's deep interior. Earthquake seismology also provides abundant information about the earthquake process, including the amount of stress released, the direction of movement, the orientation of the earthquake fault, the location of the source, and the magnitude of the earthquake. Late-twentieth-century research in earthquake seismology focuses on forecasting and mitigating the effects of large-magnitude earthquakes. A successful example of progress toward forecasting the location of major earthquakes is the 1989 Loma Prieta earthquake near San Francisco, California. Precise prediction of major earthquakes, however, remains an elusive target. Mitigation efforts involve understanding and compensating for strong shaking from seismic waves, the effects of local geology, types of seismic waves, and the earthquake rupture process.

Reflection and refraction seismology, sometimes referred to as exploration seismology, is used to explore the subsurface of Earth for its resources and to determine the planet's internal composition and structure. Seismic sources include explosions, vibrating machines, earthquakes, and other impact sources. Exploration seismology dates back to the time of Robert Mallet (1860s) but was not widely used until the 1920s and 1930s. Reflection seismology, which uses seismic waves that reflect from subsurface horizons, has been used most extensively since the 1950s in petroleum exploration. Due to numerous data acquisition and data processing improvements, reflection seismology is an indispensable tool in oil and gas exploration. One of the most recent developments is the use of three-dimensional seismic reflection techniques, whereby images of the subsurface are depicted with respect to area and depth. Three-dimensional images allow seismic stratigraphy, a method of identifying stratigraphic horizons on the basis of seismic reflections, to be more readily used. Seismic refraction techniques, which measure the velocity of seismic waves in the earth, became widely used in the early days of oil and gas exploration, but since the 1960s the academic geophysical community has used seismic refraction techniques to identify structures and properties of the earth's crust and mantle for purposes of unraveling tectonic histories, mineral and resource exploration, and understanding wave propagation.

Seismic tomography methods use multiple observations of the travel times of seismic waves in a particular volume of the earth to determine relative velocity and structure. Sources are as diverse as earthquakes, explosions, vibrators, and random noise. Exploration targets include oil and gas deposits (borehole tomography), Earth's crust around volcanoes and earthquake epicenters, and the entire lithosphere beneath continental areas. As in most geophysical disciplines, there is considerable overlap of methodology. For example, seismic tomography uses components of earthquake seismology and exploration seismology.

Potential Field Methods

Potential field geophysical measurements include gravity, magnetics, electrical, and electromagnetic methods, all of which involve use of equations derived from Gauss's divergence theorem, a mathematical description of flux outward across the surface of a bounding region. Gravity and magnetics are among the oldest of geophysical methods, with major contributions dating back to the 1500s (see above list of scholars).

Gravity methods involve measurements of variations in Earth's gravitational field due to changes in the density of rocks below and at the earth's surface. The foundation of gravity methods rests on Newton's law of gravitation. Gravity methods are frequently used in conjunction with seismic methods for exploring the subsurface because both measurements relate to the density of the rocks and provide complementary information.

Magnetic methods involve measuring the magnetic field associated with rocks to identify structures resulting from deposition, displacement, and deformation of various rock layers. The magnetic field varies with the age and composition of the rocks. The magnetic properties of magnetite were first recognized by the Chinese between the fourth and sixth centuries C.E. One of the late twentieth century developments in gravity and magnetic methods involves the use of airborne measurements and new data processing techniques to investigate large areas over short periods of time.

Electrical methods include self-potential, telluric and magnetotelluric, resistivity, and induced polarization techniques. Each of the electrical methods involves using natural currents or inducing electrical currents into the subsurface at a given locality or localities and measuring those currents as they propagate through Earth. Because of differences in electrical conductivity of various rocks and minerals, careful attention to the resistivity, conductivity, and decay of the electrical currents yields a great deal of information about the composition and physical state of the subsurface rocks. Most electrical methods were developed in the middle-to-late twentieth century by numerous individuals or companies. A major advancement in electrical prospecting is the use of three-dimensional profiling and modeling, which better define structures laterally in the Earth.

Thermal/Chemical Methods

Heat flow refers to the transfer of thermal energy from Earth's hot interior to its cooler surface. The interior heat is largely due to radioactive decay of various isotopes and the remaining heat is believed to have originated during the formation of Earth. The amount of heat that reaches the surface in any given locality is a function of the amount of heat at any given depth in Earth and the thermal conductivity of the rocks at that locality. Heat flow is measured within deep boreholes (greater than 100 m) or below the ocean floor and yields information about the physical state, particularly temperature, of rocks deep with Earth. The first systematic study of heat flow was undertaken by British committees in the late nineteenth century.

Measurements of the radioactivity of various rocks are used principally to explore for uranium, thorium, and potassium, used as nuclear fuels and in weapons of warfare. Antoine Henri Becquerel (1852–1908) originally discovered radioactivity in

1896, but it was not until the twentieth century that it was used in geophysical applications. In addition to geophysical exploration for radioactive elements, radioactivity methods also include measuring gamma rays in boreholes to determine stratigraphic correlations.

Laboratory/Theoretical Methods

Most geophysical field measurements are remote measurements of rocks and conditions within Earth. Laboratory measurements are necessary in order to correlate known substances and conditions with the field measurements. For example, mineralogical phase transformations in the mantle are experimentally determined in the lab using high-pressure devices, such as the diamond anvil, and representative samples of various minerals. The diamond anvil produces pressures equivalent to those at 700 km depth in Earth by squeezing small samples between diamond tips. Phase and compositional changes in the sample are correlated with variations in seismic velocity and other geophysical measurements to estimate composition and physical state deep within Earth.

Bibliography

BATES, C. C.; T. F. GASKELL; and R. B. RICE. *Geophysics in the Affairs of Man.* New York, 1982.
BATES, R. L., and J. A. JACKSON, eds. *Dictionary of Geological Terms*, 3rd ed. Garden City, NY, 1984.

RUFUS D. CATCHINGS

GEOTHERMOMETRY AND GEOBAROMETRY

Geothermometry and geobarometry are terms used to describe a method in petrology in which the compositions of minerals found in a rock are used to calculate the temperature and pressure at which the minerals crystallized. The method is based on the thermodynamic principle that the equilibrium constant (K_{eq}) for a chemical reaction among minerals is a function of both temperature

and pressure. Consider the reaction among the components grossular ($Ca_3Al_2Si_3O_{12}$) in garnet, anorthite ($CaAl_2Si_2O_8$) in plagioclase, sillimanite (Al_2SiO_5) in sillimanite, and quartz (SiO_2) in quartz (*see* MINERALS, SILICATES):

$$3CaAl_2Si_2O_8 = Ca_3Al_2Si_3O_{12} + 2Al_2SiO_5 + SiO_2$$
$$\text{anorthite} \qquad \text{grossular} \qquad \text{sillimanite} \quad \text{quartz}$$
$$(1)$$

for which the equilibrium constant may be written:

$$\Delta G° = -RT \ln K_{eq}$$
$$= -RT \ln \left(\frac{(^a\text{anorthite})^3}{(^a\text{grossular})(^a\text{sillimanite})^2(^a\text{quartz})} \right)$$
$$(2)$$

[$\Delta G°$ is the change in the standard state Gibbs free energy of the reaction, T is the temperature in degrees Kelvin, R is the gas constant (8.314 J/mol-K), and ^{a}anorthite, ^{a}grossular, ^{a}sillimanite, and ^{a}quartz are the activities (thermodynamic concentrations) of the anorthite, grossular, sillimanite, and quartz components in plagioclase, garnet, sillimanite, and quartz, respectively.] $\Delta G°$ is itself a function of pressure and temperature:

$$\Delta G° = \Delta H° - T\Delta S° + P\Delta V° \qquad (3)$$

($\Delta H°$, $\Delta S°$, and $\Delta V°$ are the enthalpy, entropy, and volume changes for the reaction; see standard texts in physical chemistry.) Equation (2) can thus be written:

$$\Delta H° - T\Delta S° + P\Delta V° =$$
$$-RT \ln \left(\frac{(^a\text{anorthite})^3}{(^a\text{grossular})(^a\text{sillimanite})^2(^a\text{quartz})} \right)$$
$$(4)$$

$\Delta H°$ can be thought of as a standard energy change for the reaction at a set of reference conditions, and $\Delta S°$ and $\Delta V°$ can be thought of as the temperature and pressure sensitivity of the reaction. Values of H, S, and V are tabulated for many minerals so that it is possible to compute $\Delta H°$, $\Delta S°$, and $\Delta V°$ for a large number of reactions among minerals.

The activities of the components in Equation (4) are proportional to the composition of the mineral. The exact relationship between composition, which is a measured quantity, and activity, which is a thermodynamic quantity related to concentration, can be quite complex, and is not perfectly known for many minerals. The simplest activity model is that of ideal solution among components, in which case the activity is equal to the mole fraction of the component raised to the power of the site multiplicity of the cation. For the anorthite component in plagioclase, this is

$$^a\text{anorthite} = {}^x\text{anorthite} = \frac{Ca}{Ca + Na}$$

and for grossular component in garnet the activity is

$$^a\text{grossular} = \left(^x\text{grossular}\right)^3 = \left(\frac{Ca}{Ca + Mg + Fe + Mn}\right)^3$$

The activities of sillimanite and quartz both have values of 1 because they are pure phases that display no solid solution. If we make the substitutions of composition for activity, Equation (4) becomes

$$\Delta H° - T\Delta S° + P\Delta V° = -RT \ln\left(\frac{(^x\text{anorthite})^3}{(^x\text{grossular})^3}\right) \quad (5)$$

The compositions of minerals can be measured using analytical instruments such as the electron microprobe. For a particular rock, therefore, it is possible to measure the composition of garnet and plagioclase, substitute the values into Equation (5), and arrive at an expression that defines a line on a temperature-pressure diagram (Figure 1).

Reaction (1) is an example of a geobarometer (i.e., a reaction that is useful for determining pressure) because it is more sensitive to pressure than to temperature. Good geobarometers typically have a large ΔY and small $\Delta S/\Delta V$ of reaction, because it is ΔV that determines the pressure sensitivity and $\Delta S/\Delta V$ is the P-T slope of the reaction.

Reactions that are more sensitive to temperature than pressure are useful as geothermometers and for a reaction to be temperature sensitive, the ΔH and $\Delta S/\Delta V$ of reaction must be large. The most common geothermometer reactions are those that involve exchange of cations such as iron (Fe) and magnesium (Mg) between two minerals, for example, garnet and biotite:

$$\underset{\substack{\text{almandine} \\ \text{in garnet}}}{Fe_3Al_2Si_3O_{12}} + \underset{\substack{\text{phlogopite} \\ \text{in biotite}}}{KMg_3AlSi_3O_{10}(OH)_2}$$

$$= \underset{\substack{\text{pyrope} \\ \text{in garnet}}}{Mg_3Al_2Si_3O_{12}} + \underset{\substack{\text{annite} \\ \text{in biotite}}}{KFe_3AlSi_3O_{10}(OH)_2} \quad (6)$$

If we assume ideal solution among components of garnet and biotite, the thermodynamic expression that relates mineral composition to temperature is

$$\Delta H° - T\Delta S° + P\Delta V°$$

$$= -RT \ln\left(\frac{^a\text{pyrope} \, ^a\text{annite}}{^a\text{almandine} \, ^a\text{phlogopite}}\right)$$

$$= -RT \ln\left(\frac{^x\text{pyrope} \, ^x\text{annite}}{^x\text{almandine} \, ^x\text{pyrope}}\right)^3 \quad (7)$$

Applications

Figure 1 (a–c) shows an example of the application of this technique to a sample of schist, a metamorphic rock (see ROCKS AND THEIR STUDY), from central New Hampshire. Reaction (1) defines a line on the P-T diagram with a gentle positive slope and reaction (6) defines a line with a steep positive slope. For every measured set of plagioclase, garnet, and biotite there will be a unique line on a pressure-temperature diagram, and it is a fundamental premise of geothermobarometry that the pair of minerals crystallized in equilibrium at conditions somewhere along the line. Note that two equilibria are needed to uniquely constrain both temperature and pressure, in this example 600°C and 3.8 kbar. This corresponds to a crystallization depth of approximately 13 km. There is also an uncertainty associated with these calculations, as illustrated by the error ellipsoids in Figure 1c. This error propagates from uncertainties in the analysis of the compositions of the minerals in the sample and the measurement of the thermodynamic properties of the reaction.

Equations such as (4) and (7) form the basis for geothermobarometry (geothermometry and geobarometry). Every reaction such as (1) or (7) can be expressed as a function of temperature and pressure in a manner such as described above, and for reactions that involve minerals with variable composition (i.e., solid solutions), the temperature and pressure at which the reaction occurs are a function of the compositions of the minerals.

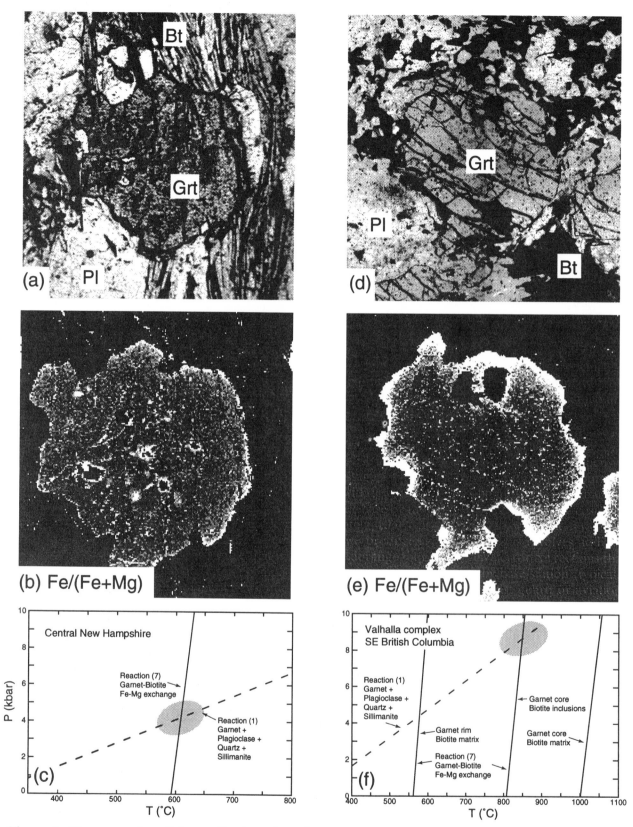

Figure 1. Two examples of the application of geothermometry and geobarometry. a–c. From central New Hampshire (from Spear et al., 1995). d–f. From southeastern British Columbia (from Spear and Parrish, 1996). a and d show photomicrographs of garnet (large crystal) + biotite (dark crystals) + plagioclase (light crystals) from each sample. b and e show the distribution of Fe/(Fe + Mg) in each garnet. Lighter regions are higher Fe/(Fe + Mg). Note that the garnet from New Hampshire (b) is

410

There are many such reactions encountered in rocks that are useful for geothermobarometry. Many reactions involving garnet and plagioclase are useful for geobarometry because of the large volume difference between calcium (Ca) in plagioclase (which has a low density) and garnet (which has a high density). Similarly, many reactions involving the exchange of Fe and Mg between two minerals (e.g., garnet, biotite, amphibole, pyroxene, olivine) are useful for thermometry because this exchange is temperature sensitive and pressure insensitive.

Practical Considerations and Pitfalls

The application of geothermometry and geobarometry would be quite simple if all minerals in a rock reacted and equilibrated perfectly at the peak of metamorphism and then cooled instantaneously so that the compositions at the peak were perfectly preserved (*see* MINERALS AND THEIR STUDY). However, reactions in rocks are complex phenomena and not all minerals equilibrate at the same rate. Minerals that do not display solid solutions have, of course, always the same composition, but some minerals may develop complex chemical zoning patterns as a result of reaction and growth processes as well as diffusional reequilibration during cooling. Plagioclase, for example, reequilibrates very slowly so that rocks often display an array of plagioclase grains each with a different composition, or individual plagioclase crystals that are chemically zoned from the core to the rim.

Garnet is another mineral that typically displays chemical zoning because diffusion rates are relatively slow in this mineral. Figure 1 (d–f) shows an example of the type of problem that may arise because chemical zoning is preserved in garnet. The compositional zoning (in this case Fe/(Fe + Mg) is obvious in the X-ray composition map (Figure 1e) and shows that the core of the garnet is lower in Fe/(Fe + Mg) than the rim. Although not apparent

in the figure, the composition of biotite is also quite variable. As a consequence, the temperatures inferred from examination of the compositions of garnet and biotite (Equation 6) vary from 570°C to more than 1,000°C. This enormous range in inferred temperature reflects the inhomogeneity of the sample and illustrates the care required in the application of this method.

Conclusion

By systematic application of geothermometry and geobarometry to metamorphic and igneous rocks it is possible to calculate the temperatures and pressures of crystallization of rocks that have recrystallized in the cores of mountain belts during mountain forming episodes (*see* MOUNTAINS AND MOUNTAIN BUILDING). Petrologists have been able to determine that rocks now exposed on the surface of Earth have, in some instances, been buried to enormous depths. For example, it has been discovered that rocks from the cores of many mountain belts, such as the Alps, the Himalayas, and parts of the Appalachians have been subjected to depths of 30–40 km and subsequently exhumed to the surface. This knowledge has contributed to our understanding of Earth as a dynamic planet capable of large crustal movements.

Bibliography

SPEAR, F. S., and R. PARRISH. "Petrology and Petrologic Cooling Rates of the Valhalla Complex, British Columbia, Canada." *Journal of Metamorphic Geology* (1996).

SPEAR, F. S., M. J. KOHN, and S. PAETZOLD. "Petrology of the Regional Sillimanite Zone, West-Central New Hampshire, U.S.A. with Implications for the Development of Inverted Isograds." *American Mineralogist* 80 (1995):361–376.

FRANK S. SPEAR

very nearly homogeneous with only minor zoning on the garnet rim, whereas the garnet from British Columbia (e) displays substantial zoning on the rim. c and f are P-T diagrams showing the results of geothermobarometry calculations. The New Hampshire sample (c) yields a unique P-T estimate (600°C, 3.8 kbar), whereas the British Columbia sample (f) yields a wide range of possible temperatures, depending on where in the sample the analysis of garnet and biotite was collected. The ellipsoids outline the preferred peak P-T conditions for each sample and illustrate the analytical uncertainty in the method.

GILBERT, GROVE KARL

Gilbert is recognized as the finest geologist to have been employed by the U.S. Geological Survey (USGS) and perhaps the greatest American geologist of his generation. This reputation is based on the significant contributions he made to a variety of fields in geology. Gilbert was born on 6 May 1843 in Rochester, New York, and died on 1 May 1918 in Jackson, Michigan.

Like many of his era he did not have advanced degrees, but when he graduated at nineteen from the University of Rochester, Gilbert had excellent grounding in mathematics and classical languages. His first occupation was with Henry Ward's natural history establishment, preparing specimens for museums and universities. He participated in the excavation and assembly of the Cohoes mastodon. At age twenty-six he began a two-year apprenticeship as a volunteer on the Ohio Geological Survey. Much of his investigations were involved with ancient lake levels and other glacial effects. In later years he studied in detail the retreat of Niagara Falls, and, more generally, fluctuation of the Great Lakes.

Gilbert's chief in Ohio, John Strong Newberry, was impressed with his efforts and recommended him when army exploration of the Grand Canyon was planned. Gilbert had his first western experience in May 1871 when he joined Lt. G. M. Wheeler's U.S. Geographical Survey west of the 100th meridian. The exploration included crossing Death Valley in the summer and boating up the Colorado River during his first season. He spent three field seasons in the west and a fourth year writing reports in Washington. He emphasized the significance of vertical uplift in the Basin and Range Province compared to the folding of rocks of the Appalachian Mountains.

Late in 1874, Gilbert joined John Wesley Powell's U.S. Geological and Geographical Survey of the Rocky Mountains, which later became a division of the F. V. Hayden territorial survey (see PO-WELL, JOHN WESLEY). During his first field season he explored the Henry Mountains and developed the concept of lacoliths, the upbowing of strata by injection of magma from below. During other field seasons he examined ancient Lake Bonneville in Utah, speculated on past climates, and studied the

effects of erosion on the Colorado Plateau in both Utah and Arizona, in addition to conducting a great deal of topographic work. His observations and interpretations in these seemingly unrelated areas relied heavily on his mathematical background, and he set a standard of more rigorous and less speculative writing.

In 1879 Gilbert became one of the original members of the new U.S. Geological Survey and remained a member of the organization until his death. Under Clarence King he was transferred to Salt Lake City as head of the division of the Great Basin, where he continued his studies of Lake Bonneville (see KING, CLARENCE RIVERS). Monograph 1 of the USGS (published in 1890) details his investigations of the ancient great lake of the desert. In addition to studies of climate and physiography, his investigations led him into the concept of isostacy as he studied the uplift of lake terraces after the water drained out.

Shortly after John Wesley Powell became director of the USGS, Gilbert was appointed Chief Geologist, an administrative task for which he seemed less suited. In addition to overall supervision of geologic studies, however, he was also directly responsible for supervising field investigations in the Appalachians and later for preparation of a series of major bulletins on the correlation of stratigraphic units.

In 1891, Gilbert investigated Coon Butte (now Meteor Crater); because he could not find meteoritic iron on the site, however, he concluded that the feature was volcanic in origin. In 1892, Gilbert spent eighteen evenings at the U.S. Naval Observatory examining the Moon's surface. He supplemented this with laboratory attempts to model impact craters, and for this work he deserves credit as the first astrogeologist.

In 1892 Gilbert served as president of the Geological Society of America. He was reelected in 1909 and is the only person to have served twice in that office. The Geological Society of London awarded him the Wollaston Medal, only the third American to be so honored.

Beginning in 1892, the Geological Survey underwent budget cuts. To help restore the organization, in 1894 Director Charles D. Walcott assigned Gilbert to study water resources in eastern Colorado (see WALCOTT, CHARLES). He made useful contributions to groundwater studies, but he also became interested in the cyclic sedimentation of the Cretaceous and attempted to correlate it to astronomic cycles.

T. C. Chamberlin and Gilbert shared an interest in interpreting the scientific method of study, culminating in the general concept of multiple working hypotheses. If there is a philosophical base to American geologic investigations, it stems from the writings of these two men.

In 1899, Gilbert was chief scientist on a cruise to Siberia and Alaska organized by E. H. Harriman. Fortunately, at this time major changes were occurring in the tidewater glaciers and once again Gilbert made significant observations.

Gilbert spent several more seasons in the Great Basin, but found the west coast more to his liking. In 1903 he investigated the high Sierra of eastern California. The destruction caused by hydraulic mining in California and the effects of the sediment load throughout the river course became the next major emphasis in his career, beginning in 1904 and continuing for a decade. His studies in both humid and arid regions place him among the premier ranks of geomorphologists.

As a result of his sediment studies, Gilbert was in California when the 1906 San Francisco earthquake occurred. He immediately began investigations of movement along the San Andreas fault and later became a member of the state earthquake commission. H. F. Reid is properly given credit for the elastic rebound theory of earthquakes, but he was aided materially by Gilbert in developing this concept.

Despite ill health during the last years of his life, the earthquake rekindled his interest in vertical uplift. In 1917 he again visited the Wasatch Mountain front in Utah. His final observations were assembled in the posthumous Geological Survey bulletin *Studies of Basin-Range Structure* (1928).

Bibliography

PYNE, S. J. *Grove Karl Gilbert: A Great Engine of Research.* Austin, TX, 1980.

YOCHELSON, E. L., ed. *The Scientific Ideas of G. K. Gilbert: An Assessment on the Occasion of the Centennial of the United States Geological Survey (1879–1979).* Geological Society of America, Special Paper No. 183. Boulder, CO, 1980.

ELLIS L. YOCHELSON

GLACIAL AGES THROUGH GEOLOGIC TIME

In 1837, JEAN LOUIS RODOLPHE AGASSIZ presented his famous and startling *Discours de Neuchâtel* (Discourse at Neuchâtel) to the Swiss Society of Natural Sciences, and later in 1840 he published his monumental *Études sur les glaciers* (Studies on Glaciers). These works together provide the basis for his revolutionary conclusion that the earth had experienced an ancient ice age in the not too distant past. The works of Agassiz provided for the first time a holistic view of an ice age that was to become the basis for modern thinking and research.

More than 150 years later, we now know that Earth has experienced not just one, but at least three, ice ages in the last billion years during times when the earth cooled to an extent that allowed glaciers of continental proportions to form on one or more continents at high and midlatitudes.

The possible mechanisms of climate change—especially those that drive ice ages—have long been considered and fall into two categories, extraterrestrial and terrestrial. The former includes the shading effect of passing dust clouds in space or dust generated by meteoric impacts on Earth, variations in solar heat output, and the orbital positioning of Earth relative to the Sun. Terrestrial mechanisms include positioning of the continents by continued drift mechanisms, dust generated by volcanic eruptions, and changes in global ocean circulation patterns and intensities.

Clearly all of these mechanisms are operative, but on different time scales, and many of them, such as those caused by volcanic eruptions, meteoric impacts and cosmic dust clouds, are apparently random events, while the others have been quantified with increasing precision toward the present time.

There is growing evidence that the strong global climate signals are driven primarily by orbital positioning, as predicted by Milutin Milankovitch in 1930, coupled with changes in the global thermohaline ocean circulation elucidated by W. S. Broecker and G. H. Denton in 1990. These two climate forces are undergoing severe scrutiny by the scientific community, and general understanding suggests that together these may be the dominant cause of the present glacial interglacial cycle, and that the anthropogenically induced "greenhouse" warming currently underway will only play a minor role in dampening the longer term "natural cycle."

Examining the geological record back in time from our present Late Cenozoic ice age, we find that there were at least two earlier long intervals of global cooling identified as ice ages. During each of these ice ages, continental ice sheets repeatedly developed on continents that were, in part, located in polar latitudes, according to paleomagnetic analyses of the rocks.

The earliest of these two glacial ages occurred during Precambrian time and is dated as having occurred between approximately 800 and 600 million years (Ma) ago, while the second major glacial period occurred during the Permo-Carboniferous period between about 350 and 250 Ma ago, with continental glaciers then occupying parts of the supercontinent of Pangea. The processes of continental drift have subsequently broken and distributed parts of Pangea on the earth's surface and along with them the rock sequences containing the ancient glacial record (Hambrey and Harland, 1981). Pieces of the depositional and erosional record for this glaciation are found on all of the continents of the Southern Hemisphere and resemble parts of the jigsaw puzzle that was formerly joined as Pangea. The first evidence for the Permo-Carboniferous age glaciation was reported from tropical India by W. T. Blanford et al. (1856), just twenty-four years after Louis Agassiz shocked the scientific world with his holistic concept of an ice age.

This evidence consisted of a boulder bed exposed at Talchin containing rock types transported from considerable distances, many of which displayed surface faceting and striating that Blanford attributed to glacier transportation. This conclusion was not generally accepted at the time. However, twenty years later when Fedden reported an extensive bedrock surface at Irai in 1875, which displayed abrasion features only attributed to glacial action, this find, coupled with the evidence at Talchir, was generally accepted as proof of paleoglaciation in tropical central India. Growing geological evidence from this classic area of India called Gondwana now suggests that there were at least two phases of continental ice sheet glaciation on Pangea. Those glaciations may have been initially centered on what now is North Africa and ended centered in Antarctica, perhaps spanning the time from the Ordovician through Permian periods, a time from 400 to 200 Ma ago, rather than being restricted to the conventional and shorter Permo-Carboniferous time as a single-phase glaciation. The current or late Cenozoic glacial age began about 55 Ma ago with the initiation of a general cooling of Earth, a trend that has persisted until today. The reasons why Earth experienced at least these three glacial ages, which are well recorded in the geological record, are largely conjectural, especially for the Precambrian event, because the very distant geological record is not only sparse but also dim. However, it appears as if these climate cooling events, recorded by extensive proxy geological evidence of continental glaciation, primarily resulted from a random combination of variations in orbital positioning within our solar system, continental drift, and elevation changes of the continents.

Continental drift occasionally moves continental-sized fragments into polar positions and at the same time the arrangements of the continents may change oceanic thermohaline circulation so as to restrict equatorial-to-pole heat transport, thus promoting polar cooling. The orbital positioning, coupled with variations in axial tilt and wobble, periodically places Earth in positions relative to the Sun that result in the general cooling of Earth. This promotes temperature variations between the equational and polar zones and between hemispheres as predicted by the analyses and predictions of Milutin Milankovitch in 1930, known as the astronomical theory of ice ages. However, orbital changes, although seemingly a fundamental cause, cannot alone be the cause of ice ages. These orbital changes demand that the cold periods alternate between the two polar hemispheres, but inphase mountain glacial expansions and contractions in both polar hemispheres appear to not support this speculation. Although the Milankovitch prediction of out-of-phase temperature changes between the hemispheres is correct, the fact of the inphase glacial events indicates that there are, in addition to the Milankovitch changes, yet unknown additional factors driving climate changes between the hemispheres.

The combination of all or several of these factors may result in a long-term general cooling of the earth that is of sufficient intensity to allow the development of ice sheets over substantial areas of those continents positioned by continental drift in high to midlatitudes. The randomness of achieving the proper combination of factors has apparently resulted in the random distribution of the position, lengths, and nature of the ice ages recorded in the geological history of the earth.

Such a confluence of conditions may have accounted for the Precambrian ice age; however, the record is too fragmented, sparse, and dim to test this hypothesis.

The record of the younger Permo-Carboniferous ice age—composed of glacial deposits and erosional features preserved on all of the continents of the Southern Hemisphere now dispersed by continental drift—is considerably clearer. During Permo-Carboniferous time, the present continents of the Southern Hemisphere were joined as the supercontinent Pangea whose center straddled the equator while its southern tip reached into the south polar region. This situation placed the landmass, which is now South Africa, India, southern South America, Australia, and Antarctica, in the high latitudes of the Southern Hemisphere and, hence, in an ideal position to support a single great continental ice sheet. The center of this ice sheet moved and eventually dissipated as continental drift shifted southern Pangea northward away from the South Pole and started to fragment and move pieces of the supercontinent into the present configuration of new continents (Hambrey and Harland, 1981).

The glacial regime that prevailed during the Permo-Carboniferous time was followed by an approximately 200-Ma-long period of warmth. This ended with the beginning of the Late Cenozoic cooling trend at 55 Ma ago that eventually brought Earth into the present ice age which began about 2 to 5 Ma. Global cooling progressed and by about 10 Ma ago mountain glaciers had formed at several locations in the Northern Hemisphere. At the same time, cooling in the Southern Hemisphere was more intense as reflected by the formation of an ice sheet over Antarctica about half the size of the present ice sheet. At about 5 Ma ago, the Antarctic ice sheet expanded to approximately its present size, covering the entire continent.

By about 3 Ma ago, global cooling had progressed to the extent that allowed ice sheets to form in the Northern Hemisphere on landscapes adjacent to the North Atlantic in North America, Greenland, and northwestern Europe; and it was a time when even in the warmer interglacial periods extensive masses of permanent ice persisted in both polar regions as they do in the current interglacial phase on Antarctica, on Greenland, and on much of the surface of the Arctic Ocean. During the colder phases, these bodies of ice enlarged significantly and new ice sheets formed, expanding to cover nearly all of Canada and the northern half of the United States, and in northwestern Europe covering Scandinavia, western Siberia, northern Germany and Poland, and most of the United Kingdom and Ireland.

Contemporaneously with the development of these continental ice sheets, valley glaciers and ice caps formed over the high mountain areas of the world, including the Alps, the southern Rocky Mountains, the Andes of Chile and Argentina, and the Southern Alps of New Zealand, as a result of the global cooling and consequent lowering of the snowline.

Since their initial formation, the ice sheets of late Cenozoic time have expanded and contracted periodically, reflecting their apparent sensitivity to astronomical variations predicted by Milankovitch in 1930: (1) eccentricity of the orbit (period 91,800 years); (2) inclination of the axis to the ecliptic plan (period 40,000 years); and (3) shifting of the perihelion (period 21,000 years). The combination of terrestrial and marine records shows that Earth, over the last million years, has experienced at least eight glacial phases lasting on the order of 100,000 years each, separated by intervening nonglacial or interglacial phases that lasted about 10,000 years. Earth is presently in another warm phase, and there is no compelling reason to believe that this is not a precursor of another cold glacial phase.

The last maximum extent of these ice sheets of the Northern Hemisphere was achieved by about 20,000 years ago and at that time the area covered by the glaciers, including that of Antarctica, was approximately 43.73×10^6 km^2 compared to the area covered today of only 14.90×10^6 km^2. The volume of ice on land at the last glacial maximum was approximately 76.97×10^6 km^3 as compared to today's volume of about 26.25×10^6 km^3. As a consequence of this disruption of the hydrologic cycle, where water from the world's oceans was stored on land as ice, global sea level was lowered approximately 197 m below present levels at the last glacial maximum.

The documentation of late Cenozoic ice-age events has primarily resulted from the interpretation of geologic, paleobotanic, oceanographic, and ice-core records. The rapid escalation of research in this area has largely been in response to demands during the last two decades for more precise records of ice-age events, required for understanding the problem of causes with the aim of ultimately being able to understand and predict future global changes of climate.

Early studies employing modern science essentially began in the mid-nineteenth century and were focused primarily on the terrestrial erosional and depositional features left by former mountain and continental glaciers. Although a great deal has been learned since then as a result of new techniques, especially radiocarbon (^{14}C; *See* GEOLOGIC TIME, MEASUREMENT OF) and other absolute dating methods, the ability to resolve the terrestrial record is still limited by the fact that successive glacier expansions and recessions generally destroyed or masked the record of earlier glaciations.

As oceanographic science progressed, it was realized that the sedimentary records that could be obtained from deep-sea cores underwent a continuous and relatively high temporal resolution proxy of the terrestrial glacial events integrated on a global basis, and also could provide direct records of changes in the environmental parameters of the ice-age marine environments such as sea surface temperatures, microfaunal changes, and salinity changes, reflecting addition and subtraction of water from the oceans and ocean circulation patterns. In turn, it has been possible to correlate many of these changes with those on land using radiometric or paleomagnetic techniques.

The increasing number of techniques and quality of analyses, and the expanding geographic distribution of lake sediment cores containing sequences of macro and micro vegetation remains, are yielding high-precision records of vegetation distributions, plant migration patterns, and changes in the composition of plant communities during the last glacial phase and also during the following Holocene times. These records largely reflect changes in the distribution of atmospheric temperatures and precipitation through time, and the availability of newly exposed land surfaces as deglaciation progressed.

In recent years there has been an increasing interest in and focus on obtaining and analyzing ice cores from Greenland and Antarctica because they hold the promise of yielding annual to ten-year resolution proxy records of atmospheric changes recorded at those locations spanning the ages of up to 250,000 years ago. This is the case with deep ice cores extracted from Greenland by the U.S.

Greenland Ice Sheet Project (GISP) and the European Greenland Ice Sheet Project (GRIP).

In summary, the rapidly accumulating high-precision multidisciplinary records, especially documenting the last glacial phase of the late Cenozoic glacial period, where the records are the clearest and most available, demonstrate that the cycles predicted by the astronomical theory closely coincide with the major, well-documented Northern Hemisphere glacial/interglacial cycles of the Quaternary Period; however, the cause and effect relationships are not yet clear and are the subject of intense research. In addition, from a global perspective, factors in addition to those of orbital origin must be involved to account for the totality of events of the late Cenozoic ice age.

One research area of great importance to solving the problem of causes now appears to be related to the strong possibility that the great global heat conveyor, the oceanic thermohaline circulation system, can abruptly start and stop. This activity can in turn rapidly affect the heat distribution in the oceans and consequently the closely coupled global atmospheric circulation that controls climate.

Rapidly accumulating evidence suggests that the transitions between major glacial and interglacial phases are synchronous with, and are somehow caused by, rapid reorganizations of the oceanic-atmospheric circulation system. However, this hypothesis is still being tested and, in addition, there is an escalation of research directed toward finding links between the ocean, the atmosphere, climate, and glacial fluctuations on other timescales.

Bibliography

AGASSIZ, L. *Discours de Neuchâtel.* Neuchâtel, Switzerland, 1837.

———. *Etudes sur les Glaciers.* Neuchâtel, Switzerland, 1840.

ANDERSEN, B. G., and H. W. BORNS, JR. *The Ice-Age World.* Oslo, Norway, 1994.

BLANFORD, W. T., M. F. BLANFORD; and W. TEHOBOLD. "On the Geological Structure and Relations of the Talchir Coalfield in the District of Cuttack." *Memoir of the Geological Survey of India* (1856):33–38.

BROECKER, W. S., and G. H. DENTON. "What Drives Glacial Cycles." *Scientific American* 262 (1990):49–56.

CHORLTON, W. *Ice Ages.* Alexandria, VA, 1983.

DENTON, G. H., and T. J. HUGHES, eds. *The Last Great Ice Sheets.* New York, 1981.

FEDDEN, F. "On the Evidence of Ground Ice in Tropical

India during the Talchir Period." *Record of the Geological Survey of India* 8 (1875):3–26.

HAMBREY, M. J., and W. B. HARLAND. *Earth's Pre-Pleistocene Glacial Record.* Cambridge, Eng., 1981.

IMBRIE, J., and K. P. IMBRIE. *Ice Ages, Solving the Mystery.* Cambridge, MA, 1986.

MILANKOVITCH, M. *Mathematische Klimalehre und Astronomische Theorie der Islimaschwankungen, in Handbuch der Klimatologie, I(A).* Berlin, Germany, 1930.

HAROLD W. BORNS, JR.

GLACIERS

See Earth's Glaciers and Frozen Waters

GLAESSNER, MARTIN F.

Martin Glaessner's career was remarkable for its numerous strands, each sustained through the decades with profound and significant scientific achievements. Glaessner's career balanced basic research and service to the profession with a commitment to education and to the promotion of geology's role in national development, especially in petroleum exploration. He made major contributions to the use of microfossils in correlating sedimentary strata and in clarifying the geology of young mountain belts, work that anticipated the modern discipline known as paleoceanography. Subsequently, Glaessner pioneered the study of the first known animals on Earth.

Martin Fritz Glaessner was born in Aussig, Bohemia, Austria, on 25 December 1906 to Luise and Arthur Glaessner, a chemical engineer and industralist. Glaessner was educated during the "golden autumn" of Vienna in the 1910s and 1920s. At age sixteen, he became a research associate at the Natural History Museum in Vienna, embarking on what would become a sixty-six year career in research and publication. He studied both law (for job security, at his family's insistence) and geology and palaeontology (at his own insistence), and was

awarded the L1.D (1929) and the Ph.D. (1931) from the University of Vienna. By the age of twenty-six, he had made an impact on the three major divisions of palaeontology (vertebrate, invertebrate, and micropalaeontology) as well as on stratigraphy and tectonic geology. In his education and early research Glaessner developed a profound respect for unified geology and paleontology, based on but never restricted to rigorous specializations. His keenest interest became the power of the fossil record to reveal the history of life and organic evolution, the history of the earth's crust, and concentrations of resources such as petroleum and natural gas. Glaessner cultivated a three-way balance between his own tools (fossils and strata), a broad and eclectic approach needed in the study of earth and life history, and the comparable breadth needed in economic geology.

After a period of travel and study in Europe, including two appointments as Visiting Research Worker at the British Museum (Natural History), Glaessner was invited in 1932 by the director of the State Petroleum Research Institute in Moscow to organize research work in micropalaeontology, for the purpose of correlating the zones and strata of the oil fields in the Caucasus and Crimea. In 1934 he continued this task in Moscow's new Institute of Mineral Fuels, where he demonstrated that the fossil succession of planktonic foraminifera (single-celled organisms with calcareous shells that fossilize) could be used to subdivide and correlate strata. Because these organisms live in the billions in the open ocean, are readily incorporated during sedimentation, and are recovered in great numbers in drilling, they have a range and versatility of application far beyond the macrofossils that had been in use for more than a century. Glaessner's work in the USSR in the 1930s, and parallel studies that began in the Caribbean region soon after, were driven by two, overlapping geological demands: understanding the architecture of sedimentary basins for petroleum exploration and development, and unravelling the complexities of the geologically young (Alpine) mountain belts. (Thirty years later, micropaleontology became one of the central pillars of deep-ocean drilling and the new science of paleoceanography.)

In 1936 Glaessner married Tina Tupikina, but the couple were forced to leave the USSR because of new restrictions on foreign workers. They returned to Vienna but had to depart after the Anschluss (1938), which joined Austria with Nazi

Germany, before World War II. He was invited to join the Anglo-Iranian Oil Company (forerunner of British Petroleum) and then asked to organize the joint company's laboratory in Port Moresby, New Guinea, where he served as Chief Micropalaeontologist of the newly formed Australian Petroleum Company. Glaessner spent the war years mostly in Melbourne working on a geological map of the territory of Papua and New Guinea and on other tasks for the Australian Army and in consulting for the Irak Petroleum Company. During this period, Glaessner completed a book that he had begun in Moscow, *Principles of Micropalaeontology*, which became highly influential in establishing modern micropalaeontology and was reprinted several times.

Glaessner responded to an invitation from Sir Douglas Mawson to join the department of geology at the University of Adelaide, South Australia, where he remained until his retirement as professor emeritus in 1971. In Adelaide, Glaessner made the most marked change of his career. Until the mid-twentieth century, scientists had found almost no record of life on Earth (other than "algae") below the trilobites and other shelly fossils of the Cambrian period. Discovery of the Ediacara fauna in South Australia in the 1940s dramatically altered the existing record. Glaessner interpreted those fossils as animals belonging to still-living groups such as cnidarians (jellyfish and polyps) and annelids (proto-arthropods and others), concluding that phylogenetically advanced animals were among those appearing suddenly in the fossil record before mineralized skeletons were discovered. Through Glaessner's efforts and those of his students, the Ediacara fauna has taken its place as the most important record of early animal evolution (*see* HIGHER LIFE FORMS, EARLIEST EVIDENCE OF). Prior to his death in Adelaide on 22 November 1989, Glaessner prepared a second edition of his 1984 book, *The Dawn of Animal Life*.

Martin Glaessner's other areas of expertise included the systematics and evolution of the decapod crustaceans, in stratigraphy and in tectonics. At various times he acted as the honorary editor of the *Journal of the Geological Society of Australia*; served as Chairman, National Committee of Geological Sciences; was seated on the South Australian Underground Waters Appeal Board and then the S.A. Waters Resources Tribunal; and was a consultant on geology related to International Law of the Sea for the governments of the Bahamas

and Greece. Numerous awards from several countries included a trio of outstanding trophies: the Lyell Medal of the Geological Society, the Walcott Medal of the U.S. National Academy of Sciences, and the Suess Medal of the Geological Society of Austria. He was honored by the Australian Government "for services to geology, particularly micropalaeontology."

Bibliography

GLAESSNER, M. F. *Principles of Micropalaeontology*. New York, 1963.
———. *The Dawn of Animal Life; a Biohistorical Study*. Cambridge, MA, 1984.

BRIAN MCGOWRAN

GLASS

Glass, both natural and manufactured, is a supercooled, inorganic liquid that exhibits rigidity. That is to say: (1) its constituents have a more or less random arrangement similar to the atoms of liquids, rather than the ordered arrangement characteristic of crystalline substances; (2) it is inorganic in composition; and (3) it is rigid except at highly elevated temperatures. Some people also apply the term amorphous to glass, thereby recognizing the random arrangement of its constituents.

Although many inorganic elements and compounds can be converted into glass, most commercially manufactured glass is made up mostly of silica (SiO_4^{-4}) tetrahedra. The diversely modified silica glasses are the subject of this entry.

Properties and Production

Common silica glass is colorless, transparent, lustrous, relatively hard, and brittle. When heated sufficiently, however, it can be blown, cast, drawn, pressed, or rolled into diverse forms. In addition, glass has small electrical and thermal conductivities that vary only slightly with temperature.

Silica glasses are produced by fusing silicon oxide (SiO_2) in the presence of a flux (to lower the melting temperature) and a stabilizer (to make the melt cool to a rigid mass without crystallizing).

Soda ash (Na_2CO_3) is commonly used as the flux. Lime (CaO) and alumina (Al_2O_3) are the common stabilizers. Glasses produced for many purposes, however, have rather complex compositions.

Virtually all properties of glass can be altered by modifying the composition of the parent melt. Consequently, melt compositions are closely controlled to obtain desired characteristics. For example, several elements, compounds, and combinations thereof produce diverse colors and tints (Table 1); lead increases luster; lithium improves strength; and potassium increases thermal expansion.

Several other additives give properties with special applications. Three of these follow:

Feldspar [(Na, K)$AlSi_3O_8$] is often added to glass-producing melts because of its alumina and alkali contents. Alumina increases resistance of glass to both physical (e.g., impact and bending) and thermal shock and inhibits any tendency the glass has to devitrify. Alkalis, as vitrifying agents, lower the temperature of fusion of melts and increase workability, which facilitates shaping of, for example, jars and bottles.

Boron (B) is added to produce borosilicate glasses, such as Pyrex, that withstand severe temperature changes without cracking. Boron is also added to glass-melt formulae used to produce glasses to be converted into glass fiber; the boron facilitates formation of fiber by reducing the viscosity of the heated glass. Glass fibers have wide-scale utilization as fiberglass, as reinforcement for plastics, as glass fiber for communication, and in chemically resistant and noncombustible products.

Strontium (Sr) additions produce radiation-shielding glass that blocks out X rays such as those used for the faceplates of color television picture tubes.

Cerium- and titanium-oxide (CeO_2 and TiO_2) additions have been shown to produce virtually colorless glass that absorbs ultraviolet radiation. These glasses have the prospect of being used for containers to prevent the loss of certain vitamin contents of fruits and other foodstuffs that are sensitive to ultraviolet rays and also for protecting fabrics, photographs, and other perishable art objects from deterioration caused by exposure to sunlight.

History of Use

Stone Age humans used natural glass (i.e., the volcanic rock obsidian and tektites) for tools, weapons, and ornaments (see TEKTITES). Pliny, in his *His-*

Table 1. Elements and Compounds That When Added in Small Amounts to Glass-Melts Give Glass Certain Colors*§

Color	Added Constituent
Red	CdS + Se; Au; Cu or Cu stain; or Sb_2S_3; U in high-lead glass
Pink	Se†
Orange	CdS + Se
Yellow	CdS; CeO_2 + TiO_2; Ag stain; or UO_3‡
Green	Cr_2O_3; Fe_2O_3 + Cr_2O_3 + CuO; V_2O_3; or CuO
Blue	Co_3O_4; or Cu_2O + CuO
Purple	Mn_2O_3; or NiO^+
Brown	MnO (reduced); MnO + Fe_2O_3; TiO_2 + Fe_2O_3; MnO + CeO_2; or NiO^+
Amber	Na_2S (reduced)
Black	Co_3O_4 + Mn, Ni, Fe, Cu, Cr oxides; MnO + Co_3O_4 in high-lead glass
Opaque	CaF_2 (fluorite)

* Whereas most coloring agents go into solution in glass, some of them become fine, dispersed particles within the glass.

† Selenium in relatively large amounts colors glass pink or red; in smaller amounts it is used to decolorize glass by neutralizing the often objectionable green hues produced by the almost ever-present traces of iron (Fe).

‡ The yellow color thus produced has a green fluorescence.

§ Table based largely on data compiled by Kreidl (1974).

toria naturalis, written in the first century C.E., recorded the story that manufactured glass was first made by accident by some Phoenician sailors who were cooking over open fires on a beach near Acre, on the Bay of Haifa:

> Once a ship belonging to some traders in natural soda put in here and scattered along the shore to prepare a meal. Since, however, no stones suitable for supporting their cauldrons were forthcoming, they rested them on lumps of soda from their cargo. When these became heated and were completely mingled with the sand on the beach a strange translucent liquid flowed forth in streams; and this, it is said, was the origin of glass.
>
> Translated by D. E. Eichholz

However true that tale of serendipity, archaeologists working in Egypt have discovered manufactured glass beads and amulets that date to the middle of the sixteenth century B.C.E., fragments of opaque glass vessels made during the early fifteenth century B.C.E., and fine-quality glass jewelry in the fourteenth-century tomb of the pharaoh Tutankhamen. Also, recipes used for glazes made in Mesopotamia were recorded by 1700 B.C.E. (Weiss, 1971).

Today, glass is used in such diverse ways as optical lenses, glass abrasives, and simulated gemstones ("paste"). The major use categories, however, include container glass for bottles, jars, and so forth; pressed and blown glass for fiberglass, light bulbs, art and novelty ware; and flat glass for windows, plate glass, and the like.

A few evolutionary steps in the use of glass for windows are especially noteworthy. Romans used glass left over from their mosaics to close openings in walls of their dwellings; that use of glass, most of which was nearly opaque or only translucent, very likely led to glass windows during the third century C.E., when recipes for clear glass production were discovered. Stained glass windows, especially the magnificent rose windows, served not only to decorate but also to lighten the rather somber interiors of the Renaissance cathedrals of western Europe. More recently, glass became the standard for windows in vehicles, and since World War II, glass windows have assumed a major role in construction. Indeed, many buildings, especially skyscrapers, have become glass-clad "steel skeletons" (i.e., their curtain walls consist largely of heat absorbing, highly reflective glass).

In addition, fiberglass is gaining increased utilization. It is used for such things as pipes for plumbing, electrical and thermal insulation, and acoustic absorption materials. Part of this use stems from the fact that unlike asbestos, fiberglass is nontoxic. In addition, fiberglass plus resin has become the preferred material for fabrication of several things such as small boat hulls and some automobile body parts.

The history of the uses of glass in the arts and crafts is another subject of great interest. These uses, which span the ages, include the fashioning of diverse decorative as well as useful artifacts made of pressed and blown glass, and such use has perhaps reached its apogee with the fabulous creations of the artist Dale Chihuly. For information about the history of these uses, see, for example, Wakefield (1957) and Weiss (1971).

Geological Deposits Suitable for Glass Production

As already noted, several elements and compounds are used in the manufacture of glass. As also mentioned, for most glass-melt formulae, the basic ingredients are silica (SiO_2), soda ash (Na_2CO_3), and lime (CaO) and/or alumina (Al_2O_3).

Sand and lascas (i.e., natural nonelectronic-grade quartz) are used as the sources of silica by most producers of silica glass. Synthetic quartz is used only rarely. High-quality silica sands (i.e., those with less than 0.03 percent iron) are suitable for most uses. Most of these sands are derived from sandstones—for example, those in West Virginia and Illinois. Recently published figures show that approximately 10 million short tons of glass sand with an average sales value exceeding $12 per ton (f.o.b. quarry) were produced during the mid–1980s. Lascas is used primarily as a feedstock for the manufacture of certain large lenses, mirrors, and windows. Most of the lascas quartz comes from pegmatites in Brazil and is made into the fused-quartz feedstock in Germany.

Most soda ash used in the glass industry is currently produced from the sodium-containing mineral trona or from sodium carbonate-bearing brines. In the past, it was manufactured by the Solvay process from common salt, limestone, and coal. Trona is mined from ancient alkaline lake deposits in southwestern Wyoming. Sodium carbonate brines are pumped to the surface from be-

Table 2. Elements Sometimes Incorporated in Glass-making Melts*

Material	Geological Source	Geographic Source	Reference
Boron			
(borax and kernite)	lake fed by hot springs	Boron, California	Lyday
(colemanite and ulexite)	marine evaporite	Emet, Turkey	Lyday
Lead			
(galena)	stratabound cavity fillings	southeastern Missouri	Woodbury
Lithium			
(spodumene)	pegmatite masses	Spruce Pine, North Carolina	Ferrell
(LiCl brine)	subsurface salars	Silver Peake, Nevada	Ferrell
Potassium			
(sylvite)	layered evaporite	Carlsbad, New Mexico	Searls
Selenium			
(chalcopyrite† and bornite†)	copper sulfide ores (e.g. porphyry coppers)	Ajo, Arizona; Noranda, Quebec; Sudbury, Ontario; Flin Flon, Manitoba	Jensen
Strontium			
(celestite)	lenses in limestone strata	Paila district, Mexico	Fulton

* The geologic and geographic sources for the raw materials are only examples.
† Small percentages of selenium commonly substitute for sulfur in these minerals, and the selenium is recovered by refining electrolytic copper slimes or slags.

neath Searles Lake, California. Much of the salt used in the Solvay process was recovered from salt beds as an artificially produced brine; the limestone was quarried, in most cases at some nearby location; and the coal was mined.

Lime is usually introduced as limestone, but dolostone, burnt lime, or hydrated lime are used in some furnaces. Limestone and dolostone, both rocks, are quarried or mined; burnt lime and hydrated lime are produced from limestone and dolostone.

Alumina is usually added by way of feldspar concentrates; in some cases, however, nepheline syenite rocks are used instead. Either source serves also to add a desired alkali content to the melt. Most feldspar concentrates so-used in the United States are recovered from pegmatite masses—for example, those in North Carolina, Connecticut, Georgia, and South Dakota; minor amounts are derived from feldspar-rich sands recovered in California and Oklahoma. Several nepheline syenite masses occur as igneous intrusives in the Blue Mountain Region north of Toronto, Canada.

Recycled glass is also used as part of the feedstock by some producers. In fact, previously produced glass powder of the same composition as the glass being made is often added to the melt as a matter of procedure.

A few of the other constituents frequently incorporated into glass-melts and their sources are given in Table 2.

Bibliography

BOYD, D. C., and D. A. THOMPSON. "Glass." In vol. 11 of *Encyclopedia of Chemical Technology*, eds. R. E. Kirk and D. F. Othmer, 3rd ed. New York, 1980, pp. 807–880.

DAVIS, L. L., and V. V. TEPORDEI. "Sand and Gravel." In *Mineral Facts and Problems*. Washington, DC, 1985. U.S. Bureau of Mines, Bulletin 675 (1986):689–703.

EICHHOLZ, D. E. *Pliny Natural History*. Vol. 10. Cambridge, MA, 1962.

FERRELL, J. E. "Lithium." In *Mineral Facts and Problems*. Washington, DC, 1985. U.S. Bureau of Mines, Bulletin 675 (1986):461–470.

FULTON, R. B., III. "Strontium." In *Industrial Minerals and Rocks*, 5th ed. New York, 1983, pp. 1229–1233.

JENSEN, N. L. "Selenium." In *Mineral Facts and Problems*. Washington, DC, 1985. U.S. Bureau of Mines, Bulletin 675 (1986):705–711.

KOSTICK, D. S. "Soda Ash and Sodium Sulfate." In *Mineral Facts and Problems*. Washington, DC, 1985. U.S. Bureau of Mines, Bulletin 675 (1986):741–755.

KREIDL, N. J. "Optical Properties." In vol. 2 of *Handbook of Glass Manufacture*, ed. F. V. Tooley. New York, 1974, pp. 957–997.

LYDAY, P. A. "Boron." In *Mineral Facts and Problems*. Washington, DC, 1985. U.S. Bureau of Mines, Bulletin 675 (1986):91–102.

POTTER, M. J. "Feldspar." In *Mineral Facts and Problems*. Washington, DC, 1985. U.S. Bureau of Mines, Bulletin 675 (1986):255–263.

SEARLS, J. P. "Potash." In *Mineral Facts and Problems*. Washington, DC, 1985. U.S. Bureau of Mines, Bulletin 675 (1986):617–633.

WEISS, G. *The Book of Glass*. Translated by J. Seligman. New York, 1971.

WOODBURY, W. D. "Lead." In *Mineral Facts and Problems*. Washington, DC, 1985. U.S. Bureau of Mines, Bulletin 675 (1986):433–452.

RICHARD V. DIETRICH

GLOBAL CLIMATE CHANGES, HUMAN INTERVENTION

Human activities are substantially increasing the atmospheric concentrations of the greenhouse gases, most notably carbon dioxide, methane, chlorofluorocarbons, and nitrous oxide. Essentially, greenhouse gases are "selective" absorbers, being relatively transparent to the short wavelengths of incoming solar energy but selectively absorbing the longer wavelength outgoing terrestrial radiation from the earth. The selective absorption of outgoing terrestrial radiation will promote global warming if no other compensating factors occur.

The increase in the atmospheric concentration of the greenhouse gases is unmistakably due to human activities. For 1990, the emissions of carbon due to the burning of fossil fuels (petroleum, natural gas, and coal) were estimated by the Intergovernmental Panel on Climate Change (IPCC, 1992) as approximately six gigatons of carbon. The rate of carbon emissions continues to increase. The

IPCC also estimated the carbon emissions from tropical deforestation at about 1.6 gigatons for the decade of the 1980s, although this number is considerably more uncertain. About one-third of these emissions are taken up by the ocean. Much of the remainder remains within the atmosphere. The 1990 level of carbon dioxide in the atmosphere was 353 parts per million by volume (ppmv) compared to a pre-industrial level of 280 ppmv.

Current methane concentrations are almost 1.72 ppmv, nearly double the preindustrial value. The major sources of methane can be categorized as natural and anthropogenic. The largest natural sources are from wetlands and termites. The largest anthropogenic sources include natural gas, coal mining and petroleum production, rice paddies, fermentation in the stomachs of cows and cattle, landfills, and biomass burning. The total methane emissions are estimated by the IPCC at approximately 500 terragrams of methane per year, with about 70 percent of the emissions from anthropogenic sources.

Chlorofluorocarbons, used as aerosol propellants, solvents, and in refrigeration and air conditioning, continue to grow in background concentration in the atmosphere, although consumption has decreased markedly over the last decade. Chlorofluorocarbons are one example of halogenated gases produced by humans. Nitrous oxides are an important long-lived greenhouse gas that have increased about 8 percent from preindustrial levels to present concentrations of 310 parts per billion by volume (ppbv). The major natural sources from oceans and forests currently exceed the major anthropogenic sources from cultivated soils and biomass burning.

Human activities have clearly altered the chemical composition of the atmosphere. The greenhouse gases described above should, barring other compensating factors, promote global warming. The real question is not whether Earth will warm due to increases in greenhouse gases associated with human activities, but rather what will be the magnitude and the rate of warming. The only feasible mechanism to estimate the magnitude and rate of global warming is through models of the climate system. The primary tools available to predict the climate impact of human activities are General Circulation Models (GCMs). GCMs are time-dependent global models with a horizontal resolution approaching 2.5° by 2.5° in latitude-longitude dimensions and tens of vertical levels.

GCMs incorporate atmospheric motions, the hydrologic cycle, and radiative transfer, following conservation laws for mass, momentum, and energy on a rotating planet. These models have improved dramatically over the last decade with respect to physical realism and their ability to simulate the present-day climate.

Most model predictions are designed to assess the importance of the dominant greenhouse gas resulting from human activities, carbon dioxide. Two types of simulations are performed with GCMs, equilibrium and transient simulations. In an equilibrium simulation, carbon dioxide levels in the model are increased (CO_2 doubling experiments are the most common) and the climate is computed that would be in balance, or in equilibrium, with that level of carbon dioxide. In transient experiments, the CO_2 levels in the model are steadily increased, matching scenarios of future carbon emissions to the atmosphere.

A variety of GCMs have been employed to assess the role of carbon dioxide doubling. The development of GCMs is an intensive effort, and the majority of the models are developed in national laboratories—for example, the National Center for Atmospheric Research, the National Oceanic and Atmospheric Administration (NOAA), Geophysical Fluid Dynamics Laboratory, and the National Aeronautics and Space Administration's Goddard Institute for Space Studies in the United States, and internationally at facilities such as the United Kingdom's Meteorological Office. The thirty or so GCMs that have been applied to this important problem yield estimates of 1.5 to 5.3°C globally averaged temperature increase for a CO_2 doubling. To place this value in perspective, during the last major ice age, 18,000 years ago, in which ice caps covered much of northern Europe and northern North America including the northeastern United States, a 3 to 5°C globally averaged temperature decrease occurred. In short, these models predict a climate change of possibly ice age magnitude over a period of decades in response to human activities. The ice age, in contrast, was a climate change that occurred over thousands of years.

In the first decades of the transient model simulations, the CO_2 response is difficult to separate from the natural variability of the climate in the model. However, after the first decades these simulations produce results that are largely similar to the equilibrium experiments. In addition, the transient model experiments give a "best" estimate of

the rate of global warming, 0.3°C per decade. The range of estimates is from 0.2 to 0.5°C per decade.

Both equilibrium and transient experiments yield consistent results for the following aspects or regions:

1. The largest model sensitivity is at high latitudes, associated with the poleward retreat of sea ice and snow cover. Tropical regions illustrate less sensitivity, although even small increases in tropical temperatures can have dramatic impacts.
2. Surface temperatures increase more over land than over the ocean areas.
3. Precipitation tends to increase at high latitudes, in midlatitudes during winter, and in monsoon areas.
4. In many midlatitude regions, the models predict soil moisture decreases. Even in some areas with higher precipitation rates, the increased evaporation associated with warmer temperatures results in a greater moisture deficit or a smaller moisture surplus.

The model simulations described above focus on the role of carbon dioxide; however, a number of greenhouse gases have been identified that are increasing as a result of human activities. The IPCC defines the Global Warming Potential (GWP) of all the major greenhouse gases as an aid for policy makers in comparing the relative importance of each of these radiatively active gases. The direct GWP (Table 1) takes into account two factors, the lifetime of a specific molecule of each gas in the atmosphere and the change in radiative forcing, each in relation to carbon dioxide. The table demonstrates that on a per molecule basis the chlorofluorocarbons, nitrous oxides, and methane have a greater global warming potential on a per molecule basis than does carbon dioxide. Fortunately, the concentrations of these gases are lower than for CO_2. The net climate forcing for each of the greenhouse gases is also given in Table 1 based on a "business as usual" scenario for projecting the concentrations of greenhouse gases to the year 2100. These data indicate that the climate forcing of all the human-induced increases in greenhouse gases is approximately 1.4 times the role of carbon dioxide alone.

Many issues remain. For example, how accurate are the models' predictions? Accurate simulation of the present-day climate with GCMs does not guar-

Table 1. Important Greenhouse Gases

Gas	Direct Global Warming Potential (GWP)	Changes in Radiative Forcing 1765–2100 (W m^{-2})
Carbon dioxide	1	6.84
Methane	11	1.09
Nitrous oxide	270	0.47
CFC-11	3400	0.14
CFC-12	7100	0.39
HCFC-22	1600	0.59

antee a capability to simulate a climate which is very different from the present day, like for conditions of increased carbon dioxide. In fact, several inadequacies of current GCMs have been identified.

First, because of their complexity and scale, clouds are highly parameterized in coarse spatial resolution GCMs. For example, clouds may be parameterized as a function of relative humidity. When the relative humidity exceeds a specific threshold, the grid box becomes covered or partially covered with clouds. Clouds have a two-fold role in climate. Clouds are highly reflective and hence cloud increases should result in a cooling effect. But clouds also radiate energy to space at lower rates (because the cloud top is cooler than the surface since temperature decreases with height in the troposphere) than the surface to the earth. An increase in low cloud may result in a net cooling, while an increase in high cloud may result in a net warming. The primary question is whether the simple cloud parameterizations in GCMs accurately capture these important feedbacks. As evidence of the importance of clouds, if the GCMs described above are executed in a clear sky mode (clouds parameterizations are deleted) the models produce very similar results. This suggests that much of the discrepancy between GCMs is a product of different cloud-climate feedback parameterizations. This is clearly significant as the difference between a 1.5 and a 5.3°C globally averaged temperature increase is substantial in terms of potential impact.

Second, fully resolved ocean models are only now being coupled to the atmospheric GCMs. In many cases, the GCMs lack any ocean currents (they provide moisture to the atmosphere and heat storage). Yet, ocean heat transport is a substantial percentage of the total poleward heat transport by

the climate system. It is unlikely that the oceans will remain static in the face of climate change and we must question the degree to which future climate may be sensitive to changes in the ocean. Recently, a number of fully coupled ocean-atmosphere models have been completed. However, to date these models have required unreasonable adjustments at the ocean-atmosphere interface in order to produce reasonable simulations of the present-day climate.

Third, the land surface is crudely parameterized in most models. Vegetation is either incorporated to ensure that surface albedos are correctly specified or vegetation is modeled as a "big leaf" over the grid square to incorporate parameterizations of evapo-transpiration. None of the models incorporate vegetation as a dynamic quantity that changes distribution as a function of the climate change. Some experiments suggest that vegetation changes, if large enough, can result in a substantial regional or even global climate response.

Finally, the ice caps are fixed in most experiments because the time scale of ice growth and decay is so long compared to the timescale of the atmosphere, making coupled experiments prohibitively expensive in terms of computational time required to reach equilibrium. Yet, much of the sensitivity exhibited by GCMs is at high latitudes. Therefore, the climate predictions may be suspect if we cannot incorporate the growth or decline of the major ice areas of Antarctica and Greenland.

Because of the uncertainties, the IPCC has adopted a "best" estimate of global warming of about 2.5–3°C, but also states that the sensitivity of the global mean surface temperature to a doubling of CO_2 is unlikely to lie outside of the range of 1.5 to 4.5°C.

A second major question concerns whether other factors may compensate for or accentuate the climate response to the greenhouse gases. Human sulfur emissions from industrial activity, biomass burning, subsonic aircraft, and volcanic eruptions contribute to the formation of stratospheric and tropospheric aerosols. Human activities have resulted in increased concentrations of atmospheric aerosols, especially in the industrialized Northern Hemisphere. Interestingly, the sulfur emissions, which reflect combustion processes, have a history that is similar to the production of anthropogenic CO_2. The primary role of these aerosols appears to be a global cooling because sulfate aerosols reflect solar energy. Some scientists speculate that the increase in aerosols may have compensated for the increases in carbon dioxide over the last decades. However, aerosols have a short residence time in the atmosphere, their distribution in the atmosphere is far from globally uniform, and aerosol loading of the atmosphere (also related to acid rain and air quality issues) is expected to decline with the growing emphasis on clean air.

An additional human-induced climatic forcing factor is land-use changes and deforestation. Much of the vegetation and character of the land surface has been modified by human activities. Vegetation plays an important role in the surface energy balance and has the potential to be an additional regional or global climatic forcing factor.

With the steady increase in carbon dioxide concentration over the last decade, considerable research is being applied to detect global warming. The anomalously high temperatures of the late 1980s and very early 1990s promoted considerable debate about whether global warming had occurred. Careful analysis of the surface temperatures recorded over the last century reveals an increase in global mean air surface temperature of 0.3 to 0.6°C. The size of the global warming is broadly consistent with transient climate model predictions, and the correspondence is even better if the role of aerosols is included. However, it is also not outside of the range of plausible natural climate variability. Unequivocal detection of human-induced global warming is not likely for a decade or more, because the temperature increase has not yet risen above the "noise" of natural climate variability.

An analysis of the temperature records indicates several interesting trends. The greatest increases are associated with nighttime temperatures rather than the maximum daytime temperature. The lower atmosphere has warmed but the surface warming is not uniform. The largest temperature increases are associated with higher latitudes and some continental interiors.

What should the societal response be to the prospect of global warming? Large changes are predicted by climate models, but the models clearly have specific limitations. An analysis of the temperature records indicates a warming over the last century but it is as yet too small to isolate from what might be natural variability. Three courses of action are possible: (1) wait for model improvements and for another decade or so of temperature records to determine the "actual" rate and magnitude of global warming, but perhaps by waiting it will be too late to adapt without considerable hardship; (2) act now to mitigate global warming, but perhaps this will be at too high a cost if the climate system proves to be less sensitive to CO_2 than currently projected by models; or (3) examine our vulnerability to climate change and seek to limit the areas of costly vulnerability (e.g., the availability of water resources) while waiting for improved predictions and observations. The strategies for mitigation and adaptation may well prove to be some of the most important issues of the coming decades.

Bibliography

BARRON, E. J. "Earth's Shrouded Future: The Unfinished Forecast of Global Warming." *Sciences* (Sept./Oct. 1989):14–20.

INTERGOVERNMENTAL PANEL ON CLIMATE CHANGE (IPCC). *Climate Change: The IPCC Scientific Assessment*, eds. J. T. Houghton, G. J. Jenkins, and J. J. Ephraums. Cambridge, 1992.

———. *The Supplementary Report to the Scientific Assessment*, eds. J. T. Houghton, B. A. Callander, and S. K. Varney. Cambridge, 1992.

ERIC J. BARRON

GLOBAL ENVIRONMENTAL CHANGES, NATURAL

Planet Earth, the "water planet," is unique in the solar system because it: (1) has a continuous history of global environmental change throughout its 4.6

billion years (Ga) of existence; (2) a dynamic global hydrologic cycle; and (3) a biosphere. The solar system includes the Sun, nine known planets, an asteroid belt—mainly fragments of a protoplanet that orbit the Sun between Mars and Jupiter, and some separate families of asteroids which have "Earth-crossing" orbits—and numerous comets. The four innermost planets, presumably the outermost planet Pluto, and Earth's moon are composed primarily of silicate rocks; the Earth-Moon system can be considered as a double planet on the basis of comparative size and composition. The outermost planets, except for Pluto, have numerous moons (satellites) orbiting them, and are composed of hydrogen and other gases. Of the terrestrial planets, only Earth and Pluto have single moon; Mars has two captured asteroids.

The innermost or terrestrial planets are quite different from one another. Mercury and the Moon ("Cratered Planets") are geologically inactive rocky spheres, with no significant atmospheres (although Mercury has a tenuous atmosphere of sodium and potassium). Their surfaces show the effect of intense meteor bombardment during the early history of the solar system. A few really extensive effusive lava flows from fissures, and some discrete volcanoes have been documented (see MERCURY; MOON). Venus, whose surface temperature reaches 467°C (873°F) under runaway "greenhouse" conditions, lies shrouded in a dense atmosphere that is 99 percent carbon dioxide (see VENUS). The radar imager on the *Magellan* spacecraft revealed a surface fractured with faults and fis-

sures, vast areas covered by effusive lava flows from fissures, numerous large volcanoes, and countless small volcanoes, and especially lava shields. The surface of Venus shows some deformation features but only about one thousand impact craters. Venus is the "Volcano Planet," largely dominated by volcanic processes and products, but despite its similarity to Earth in size, it clearly has had a very different planetary history than Earth. Venus' crust is presumed to be thicker and composed of a brittle upper layer, with no evidence of horizontal plate tectonics.

Mars, the "Desert Planet" and presently a dry planet (see MARS), has an atmosphere that is 98 percent CO_2, but it is far colder than Earth because of its thin atmosphere (1/1,000th that of Earth)—Mars loses its atmospheric gases because of low gravity—and its greater distance from the Sun. The surface of Mars reflects a dynamic past, with landform evidence recorded on *Viking Orbiter* and *Mariner 9* images of past epochs of flowing surface water, extensive glaciation, prolonged volcanic activity (lava flows and numerous volcanoes, including four huge composite shield volcanoes) and large-scale rifting episodes; there is no evidence of plate tectonics. Intense winds continue to modify its surface, and ephemeral polar caps exist. Whether life is or ever was present on Mars must await intensive scrutiny of surface deposits for evidence of fossil or living organisms.

When we compare the systems of the terrestrial planets, only Earth has a biosphere and active biogeochemical cycles (Figure 1). Earth and Mars

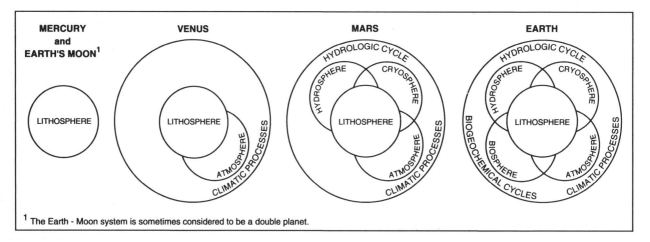

Figure 1. Comparison of Earth system with those of other terrestrial planets.

have complete geospheres (lithosphere, atmosphere, hydrosphere, and cryosphere), with hydrologic cycles and climatic processes; but on Mars, despite aeolian processes, these are weak compared to the intensely active processes and dynamic interaction of the components of the Earth system. An environment conducive to liquid in the atmosphere and on and within the lithosphere seems to be the key to the major differences between Earth, Mars, and Venus. Mercury and the Moon have only a lithosphere; Venus has a lithosphere, atmosphere, and climatic processes. Alone among the terrestrial planets, only Earth has an active lithospheric crust, which is composed of sixteen major plates, all moving relative to one another (*see* PLATE TECTONICS). The planetary uniqueness of Earth is derived from the presence of: (1) a dynamic hydrologic cycle with water in its three phases (solid, liquid, and gas); (2) a fractured crust with sixteen major crustal plates currently in motion; and (3) the early appearance and eventual diverse evolution of life on the planet, now represented by at least five to ten (perhaps as many as one hundred) million species within five kingdoms that occupy a countless variety of interactive ecosystems. In fact, James Lovelock, author of the controversial Gaia hypothesis, believes that organisms formed and continue to maintain the unique composition of Earth's atmosphere (*see* SOLID EARTH-HYDROSPHERE-ATMOSPHERE-BIOSPHERE INTERFACE).

Throughout the 4.6 Ga history of the planet, sudden and gradual changes have taken place within the Earth system. Some of these were environmental, biological, and climatological changes, which are recorded in its rocks and sediments, and glacier ice. Deciphering this record of change in Earth's past environments is the purview of geologists in collaboration with scientists from many other disciplines. The geologic record of our planet's environmental history becomes more difficult to interpret the farther back in time we look because of gaps in the record or due to inherent ambiguities of the rock, sediment, fossils, or ice samples (*see* FOSSILIZATION AND THE FOSSIL RECORD).

Modern humans (*Homo sapiens*) began to alter the global environment about 10,000 years (10 Ka) ago. Before this time, all global environmental changes were natural. Human influence on the global environment remained relatively small until the onset of the age of industrialization (beginning about 300 years ago). Since then, human impact has accelerated because of rapid population growth (from an estimated 200 million to around 6 billion in 2,000 years) and increased per capita consumption of natural resources. Human activities now have a measurable impact on the Earth system, including climatic processes, biogeochemical processes, and the hydrologic cycle. Humans also are altering the surface of the lithosphere, changing the composition of the atmosphere (by way of increased CO_2, CH_4, etc.), changing the hydrosphere (through river diversion, groundwater depletion, higher sea levels, etc.), producing change in the cryosphere (such as reductions in glacier ice volume), and causing displacement, geographic extirpation through habitat loss, and extinction of species throughout the biosphere. Humans are causing changes in many elements of the Earth system. Changes due to human intervention are occurring much faster than would be expected from natural processes as interpreted from the geologic record.

External agents also can produce changes in the Earth system. Examples of such external agents include variations in the energy received from the Sun (suspected cause of the "Little Ice Age," about 300 to 100 years before present). Impact on the Earth is by asteroids or comets (suggested by some as the possible cause of the five major species extinction events in the geologic past). Natural changes in the atmosphere occur both rapidly (movement of fronts and storms, volcanic eruptions) and slowly (changes in composition of the atmosphere); natural changes in the hydrosphere proceed at a slow pace (changes in sea level); natural changes in the lithosphere are also generally slow (movement of plates), but can be sudden (large volcanic eruptions or earthquakes); natural changes in the cryosphere (waxing and waning of "ice ages") and extinction of species in the biosphere (except for catastrophic events such as asteroid and comet impacts) occur more slowly. Importantly, these natural changes have produced global environmental changes far greater in magnitude than any produced by humans to date.

Because the list of natural global environmental changes is so large, the following discussion will address only those processes or changes that have had the greatest global impact on the geosphere and the biosphere during geologic time.

Geosphere

Lithosphere. The surface of Earth (the outer skin of the lithosphere) is continually changing in response to natural processes operating in the geosphere and biosphere. There is a continuum of time scales under which these processes operate, some quite evident to casual human observation, others only visible from the microscopic, macroscopic, or time-lapse view. The movement of the sixteen major crustal plates is slow on the scale of human life times; in Iceland, the North American and Eurasian plates are moving apart at a relative rate of 4 cm per year; on Hawaii, the Pacific plate is moving about 10 cm per year. Plate boundaries are either moving apart (spreading), such as in Iceland between the North American and Eurasian plates, colliding, one beneath the other, such as along the western coast of South America and in the Aleutian trench, slipping past one another, as along the San Andreas fault, or colliding into each other as with India and the Asian continent. Volcanic and earthquake activity, subaerial and submarine, is concentrated along plate boundaries. Volcanism occurs within plates over "hot spots," where plumes of magma are upwelling from Earth's mantle, such as currently in Hawaii and possibly in Yellowstone National Park. Most of Earth's volcanism—about 80 percent—is submarine. Volcanic activity also can reshape the land surface through effusive lava flows, lahars (mud flows), and the deposition of tephra (ash and other airborne particles) during explosive eruptions. Explosive volcanic activity also can temporarily depress global annual temperatures. The geologic record contains evidence of very large effusive fissure eruptions (e.g., Columbia River basalt flows in the northwestern United States) and of very large explosive eruptions from either rifts or volcanoes that had global effects (creating tephra layers, such as occur in glacier ice cores, far from their sources and sometimes with a global distribution).

The recognition that Earth's surface is presently composed of sixteen major plates, known as crustal or tectonic plates, that are moving horizontally with respect to one another, and that the present configuration of the continents is ephemeral, provided an explanation for why certain organisms, now separated geographically by ocean basins, are so similar. Plate movement over geologic time also explained how ancient tropical sedimentary deposits could now be found in high latitudes. The climate did not have to change latitudinally; rather, the deposits originally formed at a different latitude in a different climatic zone. Climatic zones can also expand and contract latitudinally. Throughout Earth's history, the crust has been broken into plates, sometimes joined together as a supercontinent, sometimes separated widely as individual continents with the surrounding oceans. The continents might be located near the Equator, near the poles, or spread out from pole to pole. Throughout Earth's history different configurations of continents and oceans would produce a different global climate because of the changing amounts of glacier ice, distribution of mountain ranges, and orientations and temperatures of ocean currents. One widely accepted theory holds that a single supercontinent called Pangaea existed from 300 to 200 million years (Ma) ago and included ancestral parts of all of the present-day continents. This supercontinent broke up into two large continents, one in the north (Laurasia), another in the south (Gondwana); the present-day continents are dispersed fragments ("continental drift") of these two large continents. During the course of Earth's history, agglomeration and fragmentation of continents and crustal plates must have occurred many times.

Other natural changes that occur on or near Earth's surface result from severe storms, such as tornadoes and hurricanes (typhoons). Such storms can cause significant destruction along their paths, from tree falls in tornadoes to coastal erosion and flooding and injury or death to vegetation and other organisms during hurricanes (*see* HURRICANES). River floods can also cause severe erosion and deposition of sediment along floodplains. Glaciers slowly erode and modify Earth's surface, eventually depositing vast amounts of sediment downstream from their termini. Glacier-outburst floods (known as *jökulhlaup*) are the largest floods ever recorded. A good example are the floods that created the channeled scablands topography of the Pacific Northwest in the region of the Columbia River basalts. Earthquakes can trigger landslides and either uplift or depress Earth's surface in limited regions. When earthquakes occur on the seafloor, they (and volcanic eruptions) often cause seismic sea waves called tsunamis, which can severely flood coastal regions. Natural forest fires, ignited by lightning or volcanic activity, can burn extensive areas of grasslands and forests. Insect infestations can decimate susceptible plant species and animals that depend on the plants. Plants respond rapidly to changes in either soil moisture or temperature. Vast changes in the distribution and types of flora and fauna took place during the past 20 Ka in North America, Eurasia, South America, and Africa in response to warmer climate, altered precipitation patterns, and change in sea level following the end of the latest Ice Age.

Hydrosphere. The single largest reservoir of water on our planet is the oceans, with 97 percent of the total volume. Glaciers contain 2 percent, and all other water (e.g., atmospheric water vapor, the most important greenhouse gas; rivers and lakes; soil moisture; and groundwater) makes up the remaining 1 percent, although soil moisture and surface water are crucial to the survival of biota in terrestrial ecosystems. The dynamic ocean currents transport energy, nutrients, and marine and terrestrial organisms throughout the nearly three-quarters of the surface of Earth that are covered with water. Ocean basins also are the final repositories for sediments, naturally occurring chemicals, and organic debris discharged into the sea by rivers or produced by oceanic organisms. The tropical oceans are the sites of origin for giant low-pressure storms, called hurricanes (in the Atlantic) or typhoons (in the Pacific), that can cause considerable changes (erosion and redeposition of sediment) in barrier islands and low-lying coastal regions.

In response to variations in the volume of glacier ice on the continents, sea level has repeatedly risen and fallen within a range of about 200 m. Approximately 20 Ka, for example, sea level was about 125 m lower than at present. If all of the present glacier ice were to melt, sea level would rise an additional 80 m (260 ft). During the past 11 Ka, sea level has risen at a rate of 24 m over two 1-Ka intervals. Organisms must adapt to these natural changes. For example, reefs of hermatypic corals (those that live in the wave-spray zone) must be able to grow fast enough to keep pace with sea level rise. Another consequence of changes in sea level is a change in global coastlines and outlines of the continents. At a time of lower sea level, for example, what is now Chesapeake Bay would have been occupied by a surface-stream network; today Chesapeake Bay is an estuarine arm of the ocean. The distribution of oceanic currents also changes over geologic time in response to changes in continental configurations, changes in sea level, and tec-

tonic changes (including movement of crustal plates). The uplift of the Isthmus of Panama 3 Ma stopped circulation of ocean currents between the western Caribbean and eastern Pacific. This phenomenon isolated organisms on either side of the isthmus and created a land bridge for the migration of flora and fauna between North and South America. According to a recent theory, even the largest ocean currents, such as the Gulf Stream, can suddenly shift position or diminish in transport volume and cause significant changes in sea-floor deposition and climate.

Surface water and groundwater change relatively slowly over time in response to changes in climate (temperature and precipitation), sea-level change, and tectonism. During the last Ice Age, regional precipitation/evaporation patterns were different than today, and large pluvial lakes existed in the Great Basin of the western United States. The largest ancient lake was Lake Bonneville of which Utah's Great Salt Lake is a small remnant. Scientists have found evidence that the present-day rain forest of the Amazon River basin may have grown from isolated remnants of an older rain forest, which survived in refugia since the end of the last Ice Age. Much of the Amazon basin may have been a savanna during the Pleistocene, similar to many regions in central and southern Africa today.

Atmosphere. Earth's atmosphere contains 78 percent nitrogen, 20 percent oxygen, and 1 percent water vapor; the remaining 1 percent is made up of carbon dioxide (CO_2), methane (CH_4), and other gaseous compounds such as the inert gases argon and neon. Water vapor, CO_2, and CH_4 are the principal, naturally occurring "greenhouse gases." These greenhouse gases maintain an average global temperature that makes Earth far warmer than would be expected considering its distance from the Sun (150 M km). Water-vapor content in the atmospheric column varies seasonally and regionally; CO_2 and CH_4 have varied over geologic time by processes not yet understood. From analyses of ice cores extracted from the Antarctica and Greenland ice sheets, we know that both CO_2 and CH_4 were reduced in the atmosphere (to about 180 parts per million [ppm] for CO_2) during ice age temperature minima. During interglacials, atmospheric concentrations of CO_2 and CH_4 increased (to about 280 ppm for CO_2). This was the concentration in the atmosphere at the start of the industrial revolution (300 years ago). Today the concentration of CO_2 is above 360 ppm, about 30 percent higher than at any time during the past 160 Ka (based on glacier ice-core analyses) and is about to surpass the maximum thought to have occurred during the past 3 Ma.

The amount of energy Earth's atmosphere (and surface) receives from the Sun varies latitudinally during the course of the year. Over many thousands of years, predictable changes in Earth's astronomical elements (degree of inclination, seasonal distance from the Sun—Earth's orbit around the Sun is elliptical—and seasonal tilt toward the Sun), if they coincide, result in climatic cooling or warming. Short-term variations in the Sun's energy output also occur and may be related to sunspot activity. Between 500 and 150 years ago there was a period of reduced sunspot activity, referred to as the Maunder Minimum, that coincided with and perhaps caused an extended period of cooler climate, especially in the Northern Hemisphere. The cooling was so pronounced that the historical period is called the Little Ice Age. Just before the beginning of the Holocene Epoch (10 Ka), climate briefly returned to ice-age conditions; paleoclimatologists and geologists refer to this interval as the Younger Dryas (about 10,500 Ka). Recent analyses of ice cores from Greenland have shown that climate can suddenly shift from cold to warm in a matter of decades in the waning stages of an ice sheet's history. The mechanisms that produce such "sudden" shifts in climate are not known, but it is a cause for concern because a rapid change in our global climate system today would cause many serious problems with food supplies, energy needs, and possibly produce regional political instability (*see* EARTH'S GLACIERS AND FROZEN WATERS).

Volcanic eruptions can also reduce the amount of energy received at Earth's surface for several years, especially if the eruption ejects a large quantity of sulfuric acid aerosols into the stratosphere. The 1991 eruption of Mount Pinatubo in the Philippines, for example, caused a worldwide suppression of temperature for a few years after the explosive outburst. Another feature of the atmosphere that produces changes on Earth's surface is the ozone layer. Photochemical reactions in the upper atmosphere (stratosphere) create a layer of ozone that filters out most of the energy emitted by the Sun in the ultraviolet part of the electromagnetic spectrum. Sulfur-rich eruptions can deplete ozone in the upper atmosphere by injection of sulfur dioxide into the stratosphere, which oxidizes to sulfate aerosols whose surfaces catalyze ozone-reducing reactions, much as ice-crystal surfaces do in the pronounced ozone "hole" over the Antarctic peninsula. This is thought to be a natural process for reducing the amount of ozone in the ozone layer of the stratosphere. Synthetic chemicals—such as chlorofluorocarbons that were used in aerosol sprays, as a cleaning agent, and in refrigeration systems—have been implicated as an agent that reduces upper-atmosphere ozone. An international treaty is in place that will eventually reduce the amount of ozone depleting chemicals to zero.

Cryosphere. The cryosphere is frozen water in all its forms: glaciers; snow cover; frozen ground (permafrost); and sea, lake, and river ice (*see* EARTH'S GLACIERS AND FROZEN WATERS). Frozen water is part of the global hydrologic cycle that stays sequestered until it melts, except for loss due to sublimation (change in phase from solid to gas without going through a liquid state). The cryosphere is a dynamic part of the Earth system and is very sensitive to global and seasonal changes in temperature. Today, the cryosphere is primarily located in high latitudes (polar regions) or on high mountains where the annual mean temperature is low enough to permit ice to persist from year to year. The temperate regions sustain ice during the winter months in the form of snow, lake and river ice, and frozen ground, but it disappears by spring or early summer. At higher latitudes and on high mountains snow pack can persist from one year to the next if the summers are cool, leading over time to the formation or expansion of glaciers during prolonged periods of cooling. During the Little Ice Age about 300 years to 100 years ago temperate glaciers expanded on a global scale. During the last 100 years most of these glaciers have been in recession, and some glaciers that appeared on old maps no longer exist. Polar sea ice expands significantly during the winter months and shrinks during the summertime, so there are great annual changes in the ice-covered area of polar seas. Permanently frozen ground exists over the northern polar regions in North America and Asia as areas of discontinuous (southern margin) and continuous permafrost. At some sites in northern Canada, permafrost is more than 1,000 m thick and is a relict from the Ice Age of the Pleistocene epoch. The upper 1 m or so of permafrost, the so-called active layer, thaws during the summer months, but it quickly refreezes when colder temperatures return in the fall. Most terrestrial organisms adapted to polar regions migrate, go into hibernation (some animals), or become dormant (most plants) to survive the polar winters.

Glaciers are the dominant component of the cryosphere and have waxed and waned throughout the history of Earth whenever global climate, latitudinal position of the continents, and latitudinal position and elevation of mountain ranges have been conducive to their formation. From a geological viewpoint, glaciers are an ephemeral landform, a monomineralogical metamorphic rock, that, if large enough, can modify Earth's surface through erosion and deposition of a variety of landforms. At present, glaciers cover 15.8 million km² of Earth's land surface and contain a volume of 33 million km³. The water that comprises glacier ice is derived from the oceans. When the last ice age was at its maximum, about 18 Ka, sea level was 125 m lower than today. At that time an additional 50 million km³ of ice was located on the North American, Eurasian, and South American continents as three huge ice sheets. The ice sheets displaced northern flora and fauna far to the south in both North America and Asia. After the ice sheets disappeared, the organisms that survived reoccupied the deglaciated landscape. Many of the Pleistocene megafauna that lived along the ice-sheet margin either could not adapt to the climate change, the vegetation changes that resulted, or were hunted to extinction by nomadic hunters. Nearly complete specimens of extinct mammoth and steppe bison have been recovered in the permafrost of Russia and Alaska. The geologic record of the latest of Earth's ice ages, the Pleistocene—one of the two epochs that form the Quaternary period (the other is the Holocene)—is best known. The most recent major glaciation, the Wisconsinan, is relatively well studied and understood. Traditionally, the Pleistocene has been divided into four major glacials and four interglacials. Work in Iceland, however, suggests that there were ten glaciations. Further analyses of sediment cores from the ocean floor will probably provide a definitive answer to how many glaciations occurred during the Pleistocene. Some cores indicate more than ten glacials. There is also evidence of glaciation during earlier periods in Earth's history, as, for example, during the Permian, the last period of the Paleozoic era. Whenever circumstances are favorable for an ice age to occur on Earth, the effects are planetwide, espe-cially disruption of regional hydrologic cycles and the distribution of flora and fauna. Glacial induced changes include cooler climates at all latitudes; cooler oceans; changes in temperature of oceanic currents; dislocation, extinction, and changed speciation rates of terrestrial and marine biota (flora and fauna); changes in temperature and precipitation on the continents; compression of climatic belts and changes in their latitudinal locations; and changes in surface water discharge and groundwater levels. Conversely, an absence of glaciers on the planet produces a much warmer global climate and its attendant corollaries, although some components of the cryosphere probably were present even during an equatorial location of the continents. These components would include seasonal sea ice in polar regions because of the extended seasonal periods of darkness, and snow pack and glaciers on high mountains. Permanent snow and glaciers occur today on the Equator in the high mountains of Africa, South America, and Indonesia (Irian Jaya). Polar ice sheets and the colder temperatures produced in continental air masses in polar regions have a profound effect on the Earth system. They create conditions for a very energetic global climate system through exchange of energy between the polar and tropical regions via atmospheric (fronts and storms) and oceanic (currents) circulation.

Biosphere

The biosphere of Earth today includes an estimated 5 to 30 million marine and terrestrial species (about 100 phyla) in five kingdoms: single-celled prokaryotes, nucleated single-celled eukaryotes, and the three multicellular kingdoms: fungi, plants, and animals (*see* BIOSPHERE). The present profusion of life on Earth began about 3.5 Ga when the first prokaryotic organism appeared. It was not until about 1.5 Ga later, or about 1 Ga, that two prokaryotes apparently combined to create a single cell with a nucleus. Soon thereafter, the first multicelled organisms evolved; a great diversity of marine life developed by about 600 Ma. Earth's flora and fauna have undergone many extinction episodes and divergent evolutionary paths since multicellular life first appeared. Species diversity in the biosphere has a definite geologic and geographic bias. For example, there are more tree species in 10 hectares of a tropical rain forest in Borneo than in the entire North American temperate and sub-arctic forests. Endemism of species is related to prolonged geographic isolation, such as is the case with the diverse marsupial fauna of Australia or the unusual flora and fauna of Madagascar. The present-day geographic distribution of terrestrial flora and fauna, and the degree of endemism represented in each biogeographic province, can be explained, in part, through dispersal by migration or movement of crustal plates. Other factors include past changes in sea level, mountain-building episodes, and past and present regional climate, among others.

Each major episode of faunal and floral extinctions has had losers and winners, and the surviving species radiated rapidly and evolved into new taxa to fill vacant ecological niches. The earliest fish appeared about 500 Ma, land plants by about 400 Ma. Terrestrial vertebrates (amphibians) were present by 360 Ma; early reptiles by about 320 Ma. The first mammals appeared about 240 Ma, the first birds by about 180 Ma. Flowering plants spread over the landscape about 140 Ma, and the first hominids appeared by about 3.0 Ma. Natural extinctions have many causes, but the principal ones appear to be changes in Earth's climate, movement of crustal plates into different latitudes and hence new climatic zones, and impact by asteroids or comets. One of the most successful faunas or faunal groups, the dinosaurs, radiated into a very large number of taxa around 225 Ma to fill a wide variety of terrestrial (including aerial) and marine niches during a 160-Ma reign, but then became extinct at the end of the Mesozoic era (65 Ma)—the so-called K-T boundary between the Cretaceous period of the Mesozoic era and the Tertiary period of the Cenozoic era—when a large asteroid may have impacted with Earth, causing a mass extinction of taxa. Paleontologists have identified four other mass extinctions at 439 Ma, 362 Ma, 245 Ma, and 208 Ma, and still more may be recognized. Trilobites, ammonites, and brachiopods were all very successful organisms for hundreds of millions of years; today, only a few species of brachiopods survive; ammonites and trilobites are extinct. "Living fossils" that have managed to avoid the fate of their brethren include horseshoe crabs, coelecanths, sharks, Metasequoias, stromatolites, and so on.

What is unique about these organisms that enabled them to survive one or more major extinction events? This is a scientific mystery, because for most species in the animal kingdom the average time before extinction or speciation is 5 to 10 Ma. The long persistence and evolution of life has been one of the most remarkable aspects of Earth's history. Each extinction event has eventually resulted in a greater diversity of species than before but with a decidedly different mix of multicellular life. For example, whereas the Mesozoic era was dominated by dinosaurs, the Cenozoic era is dominated by mammals. The progenitor species for both of these groups survived a major extinction event and then rapidly radiated into relatively empty terrestrial and marine ecosystems.

Bibliography

DE BLIJ, H. J., ed. *Earth '88: Changing Geographic Perspectives*. Proceedings of the Centennial Symposium, National Geographic Society. Washington, DC, 1988.

BRETHERTON, F. P. *Earth System Science: A Closer View. A Program of Global Change*. Report of the Earth Systems Science Committee, NASA Advisory Council. Washington, DC, 1988.

ELDREDGE, N. *Fossils: The Evolution and Extinction of Species*. New York, 1991.

GOULD, S. J., ed. *The Book of Life. An Illustrated History of the Evolution of Life on Earth*. New York, 1993.

HOUGHTON, J. T., G. J. JENKINS, and J. J. EPHRAUMS, eds. *Climate Change: The IPCC Scientific Assessment*. Cambridge, Eng., 1990.

LEVENSON, T. *Ice Time: Climate Science, and Life on Earth*. New York, 1989.

LOVELOCK, J. E. *Gaia: A New Look at Life on Earth*. Oxford, Eng., 1979.

RAVEN, P. H., L. R. BERG, and G. B. JOHNSON. *Environment*. New York, 1993.

SILVER, C. S., and R. S. DEFRIES. *One Earth. One Future. Our Changing Global Environment*. Washington, DC, 1990.

WILSON, E. O. *The Diversity of Life*. Cambridge, MA, 1992.

RICHARD S. WILLIAMS, JR.

GLOBAL POSITIONING SATELLITES, GEOGRAPHICAL INFORMATION SYSTEMS, AND AUTOMATED MAPPING

The Global Positioning System (GPS) consists of twenty-four Earth-orbiting satellites, developed by the U.S. Department of Defense (DoD) for purposes of navigation and positioning. By tracking four or more global positioning satellites simultaneously, users from any region in the world can quickly determine their coordinates (longitude, latitude, and ellipsoidal height) using portable, battery-powered receivers. Although GPS was originally designed to deliver positions with an accuracy of a few meters, a number of important technological advances have enabled accuracy within a few millimeters. With such precision, earth scientists can investigate a variety of geophysical phenomena, including deformation around faults in the Earth's crust that eventually rupture, causing earthquakes.

The Global Positioning System (GPS)

GPS consists of three segments: the space segment, the control segment, and the user segment. The space segment consists of the global positioning satellites. The control segment is operated by the DoD, and consists of a network of ground stations, and a system to communicate with the spacecraft. The user segment includes anyone who uses the satellite signals, including earth scientists.

The global positioning satellites are deployed in six evenly spaced orbital planes, at an inclination of 55 degrees to the equator (Figure 1). These orbits are nearly circular, with a radius of 26,600 km (approximately four Earth radii), and with an orbital period of approximately 11 hours and 58 minutes. As a result, the satellites repeat their positions to an observer on the ground for each complete turn of Earth (1 day minus 4 minutes). The satellites are about the size of a bathroom and have two sets of solar panels for power. They have a redundant set of atomic clocks, which are used for timing and for providing a reference frequency with which to generate the radio signals. The signals provide the user with the satellite position, the time when the signal was emitted from the satellite, and the error in the satellite clock.

Using several GPS receivers around the globe, the DoD's control segment monitors the satellite orbits, predicts their future positions, and communicates this information to the satellites. In addition, the satellites are told how much their atomic clocks are in error. The control segment also monitors the status of the satellites, and sends commands (e.g., to deactivate a clock that appears to be failing).

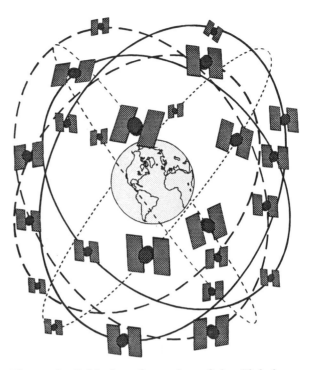

Figure 1. Orbital configuration of the Global Positioning System.

There is no limit to the number of simultaneous users. Using a portable, omni-directional antenna (often mounted on a surveyor's tripod) a receiver can seek out and lock onto the GPS signal from several satellites simultaneously (with as many as ten in view). The receiver then processes the signal to produce observations (described below), which are recorded and subsequently downloaded to a personal computer. Alternatively, some receivers allow for remote control and data transmission over phone lines. The data can then be processed in the laboratory together with data from other simultaneously observing receivers.

GPS satellites transmit microwave carrier signals at two frequencies (L1 = 1.57542 GHz and L2 = 1.22760 GHz), modulated with a code (P-code) that identifies the time of signal emission according to the satellite's atomic clock. Using knowledge of this code, a GPS receiver can observe the difference in the time of signal reception (according to its own clock) and the time of emission. Multiplying this time difference by the speed of light gives a distance observation, called "pseudorange" because the measured distance is biased by timing errors in the receiver and satellite clocks. Another source of error is the delay of the signal in the Earth's ionosphere. L1 and L2 coded signals are delayed by different amounts, but dual-frequency receivers can use this knowledge to correct for this error.

The GPS signals are also encoded with the navigation message, which tells the receiver the positions of the GPS satellites and their clock errors. Using the navigation message and pseudorange observations from at least four satellites, a receiver can estimate its three coordinates of position, and its own clock error, with a precision of the order of 10 m and 0.1 microseconds or μs, respectively. This process, known as point positioning, is used extensively for real-time applications, such as navigation, but is inadequate for earth scientists.

GPS Technology for Precise Positioning

Earth scientists are interested in positioning accuracy within a few millimeters. The key to this level of accuracy lies in the following three concepts: (1) track the carrier phase signal; (2) estimate the relative positions and clock errors of at least two receivers operating simultaneously; and (3) use an ultraprecise version of the navigation message.

The carrier phase can be tracked with few-millimeter level precision. Multiplying the phase by the appropriate wavelength (approximately 19.0 cm for the L1 signal, 24.4 cm for L2), we produce a distance observable that is like a pseudorange, but with an additional complication. Carrier phase is biased by an unknown number of cycles, known as cycle ambiguity. Cycle ambiguity can be inferred in several ways by a procedure known as ambiguity resolution. This effectively produces a very precise pseudorange observable, which can be used for precise positioning. The carrier phase is actually speeded up by Earth's ionosphere. Dual frequency phase observations can be used to correct for this effect almost perfectly, and to enhance ambiguity resolution techniques.

The relative positions of simultaneously operating receivers can be estimated very precisely due to approximate cancellation of common errors (primarily clock errors). Moreover, if one of the receivers is at a location whose coordinates are known very accurately in a global reference frame, another receiver's position can also be determined within that frame. Some software packages difference the data between pairs of receivers and satel-

lites to cancel errors, and some estimate the clock biases explicitly. Both approaches work well and are practically equivalent for most purposes.

The above approach does not exactly cancel errors from the navigation message, and therefore a civilian organization called the International GPS Service for Geodynamics (IGS) produces its own version of this information for the earth science community (with a delay of about two weeks). Using its own network of high-precision GPS receivers, IGS produces very precise tables of orbit positions, satellite clock errors, and even the orientation of Earth's axis of rotation, as the pole wanders by several centimeters from day to day.

There are numerous other technology advances which have enabled positions with few-millimeter accuracy. For example, we must account for propagation delay in the lower atmosphere (troposphere). For this purpose, scientists have developed algorithms which use the GPS data themselves to estimate tropospheric water vapor content and how it fluctuates in time.

GPS accuracy increases in proportion to the sophistication and expense of the hardware, software, and field measurements. At the most basic level, a relatively inexpensive single-frequency receiver, costing a few hundred dollars, can be used for point positioning to within a few tens of meters. At a slightly higher cost, users can position themselves to within a few meters using "differential GPS." This involves additional radio hardware to receive error corrections that are broadcast from a nearby base station (typically operated by a country's coast guard or aviation authority). For a few thousand dollars, relative positions to within a few centimeters can be obtained using a pair of single-frequency GPS carrier phase receivers and relatively simple post-processing software that double-differences the carrier phase data. For a few tens of thousands of dollars, the earth scientist can produce positions to within a few millimeters using dual-frequency geodetic receivers, IGS products, and post-processing software of the type developed by universities and research institutions. It should also be emphasized that increased accuracy also demands longer times for data collection and processing.

The DoD intentionally degrades the GPS signals in one of two ways. "Anti-spoofing" (often abbreviated A/S) is the easiest of the two to understand. The DoD encrypts the signal, thus denying access to the code which allows the receiver to form a precise pseudorange measurement. Advances in signal-processing technology have largely rendered A/S ineffective, as there are now civilian GPS receivers that can produce precise pseudoranges on both L1 and L2 channels even in the presence of A/S. The other type of signal degradation, "selective availability" (often abbreviated S/A), has two components. The first component is an intentional dithering of the satellite clock. Civilian receivers do not have access to the precise knowledge of how the clock is behaving. The induced error can be removed by differencing data between two receivers, hence it is not a problem for earth scientists who, in any case, difference the data to reduce other systematic errors. The second component of S/A is the intentional degradation of the satellite position as broadcast in the navigation message. This type of S/A is not a problem if a precise version of the navigation message is used (for example, from IGS).

Earth Science Experiments Using GPS

Geodesy is the science of measuring the shape and deformations of Earth. Space geodesy achieves this end by using extraterrestrial sources of electromagnetic radiation, such as GPS satellites, laser-reflecting satellites, laser reflectors on the moon, or distant quasars that emit radio waves (*see* SATELLITE LASER RANGING AND VERY LONG BASELINE INTERFEROMETRY). GPS gives earth scientists ready access to high-precision space geodesy with relative economy, portability, and ease of operation.

Earth scientists can observe many geophysical phenomena, by making repeat GPS measurements of a benchmark several times over an extended period of time. Areas of active research include motion of Earth's tectonic plates, regional deformation of Earth's crust in plate boundary zones, changes in the orientation of Earth's axis of spin, and in Earth's angular speed of rotation. GPS can measure changes in Earth's shape over a broad scale of time and space.

For example, over the past 10,000 years, Earth's surface in Scandinavia has been slowly rising at a rate of several millimeters per year. This is in response to the melting of thick glaciers following the last ice age. Earth is slowly resuming its original shape before the time that glaciers weighed down on Earth's crust. The rate of "glacial rebound" is a function of the viscosity of mantle deep below

the crust. Experiments using GPS are underway to measure the uplift, and thus infer mantle viscosity.

On much shorter timescales, GPS can be used to measure the near-instantaneous permanent deformation caused by large earthquakes. These measurements, together with seismometer readings, have been used to construct a more complete picture of the rupture details following the Landers earthquake in southern California in June 1992.

As well as directly measuring the position of Earth's crust at various points, GPS has also been used to precisely position low Earth-orbiting satellites. One example is the Topex/Poseidon satellite, which has used GPS together with altimeters to measure sea-surface topography (*see* SATELLITE LASER RANGING AND VERY LONG BASELINE INTERFEROMETRY). In addition, GPS can be used to precisely measure velocity and acceleration, which can be used together with accelerometers in boats, aircraft, and satellites to measure Earth's gravity field. Other more exotic uses of GPS include measuring total electron content in the ionosphere, tropospheric water vapor estimation for climate studies and weather prediction, and estimation of temperature in the stratosphere (using occultation techniques).

Developments to increase positioning accuracy have revolutionized the field of geodesy and have added GPS to the earth scientist's toolkit. The wide variety of scientific uses of GPS continues to grow as accuracy and precision have improved beyond expectation. The exotic applications show that we have reached a level where "noise" to the geodesist is being viewed as "signal" to other earth scientists. This is a familiar theme in the development of new technology, leading to unforeseen scientific applications.

Bibliography

BILHAM, R. "Earthquakes and Sea-Level: Space and Terrestrial Metrology on a Changing Planet." *Reviews of Geophysics* 29 (1991):1–29.

BLEWITT, G. "Advances in Global Positioning System Technology for Geodynamics Investigations: 1978–1992." In *Contributions of Space Geodesy to Geodynamics: Technology*, eds. D. E. Smith and D. L. Turcotte. Geodynamics Series, vol. 26, American Geophysical Union. Washington, DC, 1993.

DIXON, T. H. "An Introduction to the Global Positioning System and Some Geological Applications." *Reviews of Geophysics* 29 (1991):249–276.

HAGER, B. H.; R. W. KING; and M. H. MURRAY. "Measurement of Crustal Deformation Using the Global Positioning System." *Annual Reviews of Planetary Science* 19 (1991):351–382.

LAMBECK, K. *Geophysical Geodesy: The Slow Deformations of the Earth.* Oxford, Eng., 1988.

LEICK, A. *GPS Satellite Surveying.* New York, 1990.

VANICEK, P., and E. KRAKIWSKY. *Geodesy: The Concepts.* Amsterdam, The Netherlands, 1986.

GEOFFREY BLEWITT

GOLD

Gold, symbolized as Au (from Latin *aurum*, meaning "golden dawn"), is a deep yellow metal whose chemical element number is 79 and atomic weight 196.967. It is classified as a heavy, noble, or precious metal consisting naturally as a single isotope Au^{197} with a specific gravity of 19.3. It has a melting point of 1,064.43°C and a boiling point of approximately 2,807°C. Twenty-four radioactive isotopes of gold have been prepared; their half-lives range from 1.35 seconds (s) for Au^{177} to 183 days for Au^{198}. The crystal structure of gold is face-centered cubic with a unit cell dimension of 4.078 Å. The extreme malleability of gold has led to its widespread use in gilding, a process of applying films as thin as 0.00008 mm as decoration to metal, ceramic, wood, paper, or other substances.

History

Along with copper and silver, gold was one of the first metals used by humans. As early as 5000 B.C.E., these metals had been discovered in their native states and fashioned into amulets and jewelry. The first use of gold for coinage began in Asia Minor and Greece about 600 B.C.E. Gold coinage has been used widely in many countries, including the United States from 1795 through 1933, but its use is mostly limited today to commemorative and bullion coins.

Throughout history, gold has played an important role in the migrations of peoples and the prosperity of societies. It was cast into medallions, jewelry, amulets, and vast varieties of ornamental objects; several societies even interred the dead with large quantities of gold. The annual world production of gold was very low (less than 5 metric tons per year [mt/yr]) throughout history until the discovery of the New World in 1492 (Figure 1). Columbus's encounter with native Americans who possessed gold provided a powerful stimulus for the Spanish conquistadors and led to large (for the time) shipments of gold and silver back to Europe. Spanish records reveal shipments of 181 metric tons of gold and 16,000 metric tons of silver in the period 1520–1650. From this period through 1850 world annual production of gold increased to 20–30 mt/yr.

Clearly the principal historical event for gold in the United States was its discovery at Sutter's Mill, California, by James Marshall on 24 January 1848. This led to the 1849 gold rush that lured thousands to the gold fields of California and was the major stimulus in populating the American West. Subsequent major rushes in the next fifty years (Australia, 1851; South Dakota, 1874; South Africa, 1886; Klondike and Yukon, 1896) raised world average annual rates of production to the level of hundreds of metric tons per year. Rates continued to increase reaching 1,000 mt/yr by 1936. In the post–World War II period, and especially since 1968 when the price of gold was allowed to rise and fall in response to world market forces, the production of gold has continued to rise, reaching 2,000 mt/yr in 1990. The total amount of gold produced throughout all of human history was believed to have been approximately 128,000 metric tons by the end of 1994. Of this, less than 20 percent has been lost, used in unrecoverable industrial materials, or unaccounted for. Of the balance, approximately two-thirds is privately owned as coin, bullion, or jewelry, and about one-third is held by central banks.

Occurrences

The average geochemical abundance of gold in Earth's crust is approximately 0.0025 parts per million (ppm) (=2.5 parts per billion or 0.00000025 percent) but various geological processes, primarily involving hot water, may concentrate the gold to levels of 1 ppm or more where

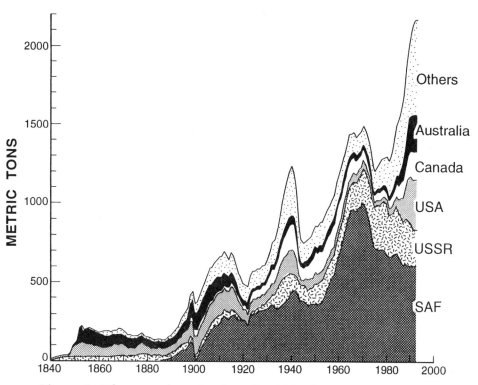

Figure 1. The annual production of gold in the world since 1840.

extraction may be economic. Gold occurs primarily as the free metal that usually also contains some silver (the term electrum is used if the silver content is 20% or more) and rarely as telluride minerals such as calaverite (Au Te$_2$) and sylvanite (Au Ag Te$_4$). Native gold and, to a lesser extent, the tellurides occur in trace quantities in the rocks of many types of ore deposits; hence, gold is mined not only from gold deposits but also as a by-product from many mines operated primarily for copper, lead, and zinc.

Primary gold occurrences are broadly categorized as placer or lode deposits. Placer deposits are sands and gravels containing grains or nuggets of gold that have been eroded from preexisting rocks. The very high density (15–19 g per cm^3 for most natural gold and electrum grains) relative to common stream sand and gravel constituents (mostly quartz with a density of about 2.6 g per cm^3) results in concentration of gold particles into areas where the water can effectively separate the minerals by winnowing out the lighter grains. Small placer deposits occur in many parts of the world and their discovery has generally been the key in locating the world's major gold districts. In

fact, the world's largest known deposit, the Witwatersrand in South Africa, is interpreted as a fossil placer that accumulated in a basin about 2 billion years (Ga) ago.

There are many types of lode deposits but the most important are veins and disseminations where waters heated by igneous intrusions or merely by deep circulation have moved along fractures and precipitated gold and other minerals during cooling or as a result of chemical reaction with enclosing rocks. Many of the earliest mined and famous deposits (e.g., California's Mother Lode) were open faults in which there was deposited quartz, gold, and pyrite ("fool's gold"). Although such deposits are still mined, most lode gold production is from massive base-metal sulfide ores, from large but very low-grade porphyry deposits, and from so-called invisible gold deposits. These latter deposits, now mined on a very large scale in Nevada, were formed by relatively low-temperature (less than 300°C) fluids that circulated along fracture zones leaving very finely dispersed gold in the adjacent rocks, or by types of hot-spring activity (similar to that observed in Yellowstone National Park).

Recovery of Gold

Gold may be recovered by physical or chemical means. Gold's high density allows it to be relatively easily separated from the nonvaluable, low-density minerals in a deposit through the use of gold pans, sluices, or jigging tables. In each case flowing water, sometimes combined with a vibrating motion, permits the more dense gold grains to separate from the lighter waste minerals. Physical recovery techniques may be effectively employed when the gold occurs as loose grains in placer deposits, or as free grains after crushing and grinding of lode ores. Amalgamation, the addition of mercury to gold pans, or other recovery practices that absorb the gold were widespread in the past. Today the use of mercury has been banned for environmental reasons in most areas but is still commonly used by small-scale, and often illegal, operations.

Chemical solution, in which the gold is dissolved in dilute cyanide solutions, is the primary means used today in the recovery of gold from large, low-grade deposits. This method generally is carried out as heap-leaching in which the crushed ore is piled on large plastic-lined flat areas. The cyanide solution is sprayed or dripped on the tops of the heaps and allowed to percolate downward, dissolving out the gold (and silver). The solutions are collected, treated to recover the gold, and then returned again to the leach piles. This technique is very simple and inexpensive but very efficient if the ore has been crushed finely enough that the solutions can contact the small and dispersed gold particles. Ores are termed "refractory" if the gold is trapped within other minerals, usually quartz or pyrite, which shield the gold from the cyanide solutions and hence prevent its solution and recovery.

Gold commonly occurs in small amounts in sulfide ores mined for copper, lead, and zinc. In such ores physical or chemical separation of the gold alone is not practical, but the gold usually occurs with one or more of the sulfide minerals. The sulfide minerals are usually separated by selective flotation techniques; thus the gold usually concentrates with the copper minerals. The gold is recovered at the time those minerals are smelted.

Usage of Gold

Gold has been used throughout history for jewelry, ornamentation, amulets, and coinage. Because it does not react with common corrosive elements, gold artifacts are readily preserved unless they are physically abraded. Gold coinage, once widely used, is now primarily restricted to special commemorative and bullion coins. The most common use of gold throughout the world today is as jewelry, where the purity is measured in terms of fineness or caratage. Caratage denotes the gold content such as 24 carat represents 100 percent pure gold, 18 carat is 75 percent, 12 carat is 50 percent, and so on. In North America jewelry is usually 14 to 18 carat but in Europe 20 and 22 carat is relatively common. Other metals, such as silver, copper, and tin, are alloyed with the gold to enhance color and strength and to reduce costs. Fineness is the ratio of gold over gold plus silver content in weight percent multiplied by 1,000. Hence, if 50 percent is gold and 50 percent is silver, the fineness is 500. Fineness, however, does not take into account other metals; thus an alloy of 50 percent gold, 25 percent silver, and 25 percent copper is 12 carat but has a fineness of 667.

Gilding, the application of extremely thin sheets of gold as a decoration for stationary, jewelry, book decoration, and so on, is still widely used. The great malleability of gold allows it to be pounded into sheets as thin as 0.00008 mm; thus, very small amounts of gold can cover large areas. Throughout history there have been local traditions of using gold gild to cover foods and to be placed in certain types of liquors (e.g., Goldwasser). These are unusual examples, but the small amounts of thin gold appear to have no medical effect.

Gold also has widespread usage in the electronics industry because it is an excellent conductor of electricity and does not tarnish. It is an excellent reflector of infrared radiation and is used as a very thin coating on heat-insulating windows and special mirrors. Gold is largely inert in terms of human body reaction and has long been used in dental amalgams; it has also been successfully employed as Auranofin, a complex organic molecule, in the treatment of certain severe cases of rheumatoid arthritis.

Chemistry of Gold

Gold in the natural realm is quite noble (unreactive), occurring only as a free metal, an alloy with one or more of several metals, and as tellurides. In the laboratory it may be synthesized into a large number of compounds as Au^{+1} or Au^{+3}. Thus it readily forms a variety of halides, ammonia complexes, thiosulfate complexes, thiocyanate complexes, and a variety of organometallic compounds.

How Much Gold Is There?

The average rock gold content of 2.5 parts per billion means that the crust of Earth contains 62.5 billion metric tons of gold. If this were concentrated in a square slab of gold 1 m thick, it would be 57,358 m on each side (that is, more than 57 km on each side). The gold content of the oceans is difficult to measure but appears to be on the order of 0.001 parts per billion. This means that the oceans contain 1,320,000 metric tons of dissolved gold, enough for a 1-m-thick square slab that would be 264 m on a side. All of the gold produced in human history (estimated through 1992) would be equal to a 1-m-thick square slab that is 80.5 m on a side or a cube about 15 m on a side (a block about one-tenth the height of the Washington Monument). All known world reserves today total about 44,000 metric tons. At the present world rate of production of 2,170 metric tons (in 1992), known world reserves will last about twenty years. The world's largest producer has been South Africa (600 metric tons in 1992) for many years; the United States (300 metric tons in 1992) is second with most of the gold being mined in Nevada.

Bibliography

BOYLE, R. W. *Gold: History and Genesis of Deposits*. New York, 1987.
CRAIG, J. R., D. J. VAUGHAN, and B. J. SKINNER. *Resources of the Earth*. Englewood Cliffs, NJ, 1988.
JASTRAM, R. W. *The Golden Constant*. New York, 1977.
MARX, J. *The Magic of Gold*. Garden City, NY, 1978.

JAMES R. CRAIG

GOLDSCHMIDT, VICTOR MORITZ

Victor Moritz Goldschmidt was born in Zurich, Switzerland, on 27 January 1888, the son of Heinrich Jacob Goldschmidt and Amelie Goldschmidt (née Koehne). His father was a prominent chemist who eventually received an appointment as the chair in chemistry at the University of Oslo. His mother came from an established, well-educated and widely traveled Danzig (now Gdansk, Poland) family. Victor Goldschmidt was named after a well-known colleague of his father, the chemist Victor Meyer.

During World War I, V. M. Goldschmidt began his research in geochemistry as a result of his investigation of the mineral resources of Norway. Of those years, Goldschmidt later said of his work: "Among the problems proposed there was one of outstanding importance, i.e., to find the general law and principles which underlie the frequency and distribution of the various chemical elements in nature—the basic problem of geochemistry. I proposed to attack the problem from the viewpoint of atomic physics and atomic chemistry and to find out the relationships between the geochemical distribution of the various elements and the measurable properties of their atoms and ions." No one better defined the field nor contributed more to it. He is the father of geochemistry.

Goldschmidt's earliest schooling was in Amsterdam and Heidelberg. (In Amsterdam, his father worked with J. H. van't Hoff, perhaps the foremost physical chemist of his day.) In 1905, Goldschmidt continued his schooling in Oslo and later enrolled in the university, where he studied chemistry from his father and from Thorstein Hiortdahl, and mineralogy and geology from the eminent Waldemar Christopher Brøgger.

Brøgger inspired Goldschmidt and influenced him to engage in research on metamorphic rocks near Oslo, research that led in 1911 to the publication of the nearly 500-page monograph "Die Kontaktmetamorphose im Kristianiagebiet" (Contact Metamorphism in the Kristiana Region), a classic in geology. Based on the careful field and mineralogic studies that had been carried out by Brøgger and coworkers, Goldschmidt described mineral assemblages of specific metamorphic rock zones formed at the contact of sedimentary rocks with intrusive igneous rocks. He established that these assemblages were controlled completely by three factors—the composition of the original sedimentary rock and the temperature and pressure reached during their metamorphism by the igneous contact. In his doctoral thesis, Goldschmidt used these data to construct the Mineralogical Phase Rule, a simplification of J. Willard Gibb's

Phase Rule ($P + F = C + 2$) that relates, at chemical equilibrium, Phases (in this case minerals), degrees of Freedom (composition, temperature, pressure), and Components (elements or simple oxides). This fundamental geologic tool was instrumental in the widespread application of thermodynamics to rock study.

Goldschmidt's careful, physical chemistry approach to the understanding of contact metamorphism, augmented by his later, exhaustive studies of progressively metamorphosed rocks over large regions, was pivotal in establishing and extending the fundamental criteria for their study—distinctive indicator minerals, and grade, or intensity, of metamorphism.

A cornerstone of geochemistry—a genetic assignment for the partitioning of elements based on their affinity to metal (siderophile, i.e., iron-loving), silicate (lithophile, rock-loving), and sulfur (chalcophile, sulfur-loving) phases, and later to vapor (atmophile) and organic (biophile) phases—was postulated by Goldschmidt as a result of his chemical analysis of iron meteorites, stony meteorites, and troilite, the main sulfide mineral of meteorites. By a metallurgy analogy with raw materials (ore) and resulting metals and slag—analogous to Earth's primordial composition and its later metallic core and silicate mantle—along with numerous and careful analyses of terrestrial minerals and rocks, Goldschmidt's work was instrumental to our understanding of the composition of Earth and its major domains. A measure of Goldschmidt's creativity was his postulate that the composition of glacial flour (clay), scoured from all types of rocks in the Fennoscandian shield (ancient granitic rocks of Norway, Sweden, and Finland), should represent the average composition of Earth's crust (excepting the most soluble elements).

Goldschmidt's foremost contribution was to the field of crystal chemistry, a field that did not exist prior to his work and one to which he contributed its fundamental laws. In 1929 he accepted the mineralogy professorship in Gottingen, along with an arrangement to build a first-rate laboratory in X-ray and carbon arc spectrographic analysis, a position he held until 1935. For the next six years, with outstanding coworkers, including the renowned T. F. W. Barth, he performed an extraordinary sequence of quality analyses on almost two hundred prepared chemical compounds of geologic significance. These results were published in the classic eight-part series "Geochemische Verteilungsgesetze der Elemente" (Geochemical Principles Concerning the Distribution of the Elements).

Seminal results from this work were: (1) the realization and documentation that atomic structure was more important in explaining mineral similarities than their chemical composition; (2) that binary crystal structures were explicable simply in terms of the ratio of the two ions; and (3) that the most important parameters governing crystal chemistry were ionic size, charge, and ion polarizability. Collectively, these studies provided a clear and effective model for the explanation of the periodic table of the elements. Comparing ionic size with ionic strength, Goldschmidt also defined "ionic potential," which usefully distinguishes major groups of elements during sedimentary processes—those which remain in true ionic solution, those which precipitate by hydrolysis, and those which form soluble anionic complexes.

Goldschmidt received many honors, including the Wollaston Medal, the highest honor given by the London Geological Society; Foreign Member of the Royal Society; member of the Royal Norwegian Order of St. Olaf; the Fridtjof Nansen Prize; and three honorary doctorate degrees.

V. M. Goldschmidt never married and left few close relatives; as a result, little is known of his personal life. His health was poor throughout his life. As a Jew, he was persecuted and uprooted several times before and during World War II, and, on one occasion, narrowly escaped transfer to a Nazi death camp. He was kind and generous to strangers as well as family and was admired and respected by his colleagues. He could, however, be suspicious and bad-tempered with colleagues; though these tantrums rarely lasted, this trait resulted in a few lifelong enemies. Goldschmidt died on 20 March 1947 in Oslo, Norway.

Bibliography

MASON, B. *Victor Moritz Goldschmidt: Father of Modern Geochemistry.* The Geochemical Society, Special Publication No. 4. San Antonio, TX, 1992.

ROSBAUD, P. "Victor Moritz Goldschmidt (1888–1947)." In *Great Chemists,* ed. E. Farber. New York, 1961.

TILLEY, C. E. "Victor Moritz Goldschmidt (1888–1947). In *Obituary Notices of Fellows of the Royal Society* 6 (1948–1949):51–66.

E. JULIUS DASCH

GREAT DARK SPOT

See Uranus and Neptune

GREAT RED SPOT

See Jupiter and Saturn

GREENHOUSE EFFECT

See Planetary Atmospheres

GROUNDWATER

Earth is a water-rich planet, with immense quantities of water stored in the oceans, the atmosphere, the polar ice caps, and as streams and lakes on the continents. Many people do not know that significant quantities of water are also stored below the land surface in the form called groundwater. This subsurface water has been used by humans for thousands of years, as is evidenced by references to wells in ancient writings. Groundwater was once thought to exist primarily as underground rivers or lakes. Although some caves contain underground rivers, large underground water bodies of this type are quite rare. The vast majority of groundwater is held in networks of microscopic openings in sediment or rock. These openings, or pores, include the irregular spaces between grains of sand as well as cracks and fractures in rocks.

The ratio of the volume of pores to the total volume of sediment or rock is the porosity. The amount of water that can be stored beneath the land surface depends directly on the porosity. Be-

cause porosity varies from less than 1 percent in some igneous and metamorphic rocks to over 50 percent in some types of sediments, it is difficult to estimate the volume of groundwater with the same precision as one might estimate the volume of water in a lake. Nevertheless, we know that the total volume of groundwater far exceeds the total volume of water in lakes and streams. Because ice caps and glaciers, the only other major sources of fresh water, are not readily available for human consumption, groundwater makes up over 95 percent of the world's utilizable fresh water. Although groundwater supplies only about 20 percent of the total water used in the United States (not including water for hydroelectric power), it makes up almost 50 percent of water used for drinking and other household purposes (see U.S. Geological Survey, 1990). Even in areas of abundant surface water, groundwater is often preferred as a drinking water source because it is generally less susceptible to contamination by pathogenic micro-organisms (*see* POLLUTION OF GROUNDWATER AQUIFERS).

Not all pores in the ground are completely filled with water. Except in areas occupied by wetlands or immediately adjacent to some streams or lakes, pores near the land surface contain both air and water (Figure 1). This zone in which air and water coexist is called the zone of aeration. Water in the

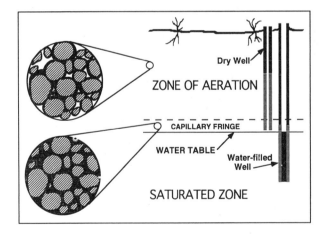

Figure 1. Water content in the zone of aeration and the saturated zone. In magnified diagrams, striped areas are sediment grains, white areas are air-filled pores, and black areas are water-filled pores. Pores in the capillary fringe and below the water table are entirely water-filled, but water flows into a well only below the water table.

441

zone of aeration, known as soil moisture, is held on the sediment by a force known as capillary attraction. Because pressure of the water held by capillary attraction is lower than atmospheric pressure, soil moisture will not flow into a well in the zone of aeration. Plants can overcome this capillary attraction to make use of soil moisture but humans cannot extract it economically. In regions where rainfall is abundant, the zone of aeration is generally less than 30 m thick. In some desert regions, however, the zone of aeration may extend to many hundreds of meters below the land surface.

No matter how little rainfall there is in an area, at some depth below the ground all pores are entirely filled, or saturated, with water. (This fact explains the reported success of "water witches" in locating underground water; if you drill deep enough anywhere, you will find groundwater.) Water-filled pores at the base of the zone of aeration, in a relatively thin zone known as the capillary fringe, still hold water by capillary attraction, preventing flow into a well. Once the pores become saturated, however, the pressure of the water increases with depth due to the weight of water in overlying pores. Water pressure reaches atmospheric pressure at a surface called the water table (Figure 1). Water from pores at or below the water table will flow into a well. The water table forms the top of the saturated zone, and water in the saturated zone is groundwater.

Water in the saturated zone can flow not only into wells but also through the pore network. Most groundwater is in constant motion although groundwater velocities may be only a few meters per year compared to stream velocities that can be greater than a meter per second. On a regional scale, groundwater flows from areas of high elevation such as hills, where it is "recharged" by rain infiltrating through the zone of aeration, to lower elevations where it flows out, or "discharges," into lakes, streams, or springs. This general direction of flow is similar to that of streams, which flow downhill under the influence of gravity. At a more local scale, however, groundwater flow is quite different from that of surface water. Between the locations where water enters and exits the saturated zone (the recharge and discharge areas, respectively), groundwater may take a very irregular path, flowing upward, downward, or horizontally.

The physical law governing groundwater flow was discovered in the 1800s through experiments conducted by Henri Darcy, a French engineer

(Price, 1985). These experiments showed that groundwater flows in response to both gravity and pressure. A combined measure of these two forces is the hydraulic head. Darcy's law states that groundwater flows from higher to lower hydraulic head. Fortunately for those who would like to determine the direction of groundwater flow, hydraulic head is relatively easy to measure. The hydraulic head at some point in the saturated zone corresponds to the level to which water rises in a well at that location.

Darcy also discovered that the rate of groundwater flow depends on a property of the rock or sediment called permeability. For a given change in hydraulic head over some distance, groundwater flows more quickly through high permeability materials than through low permeability materials. Permeability depends in part on the porosity, because the greater the volume of pores, the greater the volume through which flow can occur. Permeability also depends on the size of pores. Water will flow more easily through a few large pores than through many small pores, even if the total volume of the pores is the same in both cases. A final factor affecting permeability is the connectedness of the pores. Coarse-grained sediments such as sand and gravel with large, well-connected pores have high permeabilities, whereas igneous and metamorphic rocks containing few fractures have very low permeabilities. Clay-rich sediments and rocks such as shale also have low permeabilities due to the small size and poor connection of pores. High permeability rocks or sediments make up water-bearing units called aquifers; low permeability zones are known as confining units.

If the water table occurs in high permeability material, the saturated zone below it is a water table aquifer. In most cases, a confining unit occurs at some depth and forms the lower boundary of the water table aquifer. If another permeable zone exists beneath the confining unit, that aquifer is called an artesian, or confined, aquifer (Figure 2). Water enters an artesian aquifer either in a recharge area where the confining unit is absent or by downward flow through the confining unit. Water exits the artesian aquifer by discharge to some lower elevation, either directly or by upward flow through a confining unit. The water level in a well drilled into an artesian aquifer rises above the top of the aquifer into the zone occupied by the confining unit, the water table aquifer, or in some cases even above the land surface. If the stable water

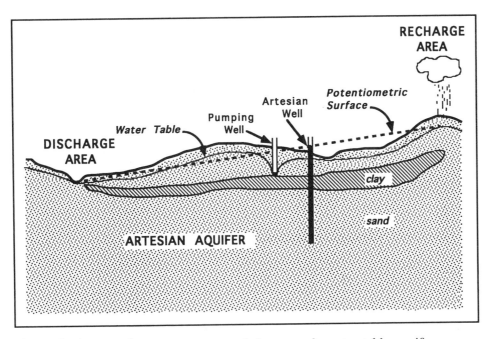

Figure 2. A groundwater system containing a sandy water table aquifer, a clay confining unit, and a sandy artesian aquifer. Note the cone of depression in the water table surrounding the pumping well.

level, or potentiometric surface, of an artesian aquifer is above the land surface, wells open to the aquifer will flow without being pumped and are called flowing artesian wells. Flowing wells do not always require the presence of a confining unit because they may also occur in areas of upward flow near streams or lakes that serve as discharge areas for a water table aquifer.

Variations in permeability in the saturated zone can create a complex arrangement of aquifers and confining units leading to complex patterns of groundwater flow. Permeability variations can also have important effects on flow in the zone of aeration. A particularly interesting situation is one in which an extensive layer of low permeability sediment or rock occurs below the surface in a recharge area. The low permeability layer slows down the infiltrating water and can cause pores in and above it to fill with water. A local zone of saturation is called a perched water table. Perched water tables may persist over long periods of time or may be temporary features that develop only during periods of high rainfall and rapid infiltration. Perched water tables can be a source for springs and seeps on cliffs and hillslopes and can contribute to landslides.

Groundwater can be brought to the surface for human use by pumping. As water is removed from the aquifer, pressure decreases, causing a decline, or drawdown, of hydraulic head. Drawdown is greatest near the well and decreases rapidly with increasing distance. This creates a cone-shaped depression in the water table or potentiometric surface (Figure 2). If several pumping wells are located near each other, their cones of depression can overlap, causing even greater drawdown. Extensive drawdown can significantly increase the cost of producing groundwater because lower water levels require deeper wells and greater energy for pumping. Drawdown occurs rapidly when pumping begins but slows down as water flows to the well from farther away in the aquifer. Water levels may eventually stabilize if the rate of water removed is balanced by natural recharge to the aquifer from rainfall or by decreased discharge to lakes, springs, and streams. The lowered hydraulic heads in the aquifer may also induce water to flow in from other aquifers through the confining units or from surface water. When a pump is shut off, water levels begin to increase and may eventually recover to the pre-pumping levels, given sufficient time and recharge.

Stable water levels during pumping indicate that the groundwater can be considered a renewable resource. In some parts of the world, particularly where groundwater is used for irrigation or urban development in arid areas, groundwater levels may continue to decline over many years without stabilizing. When pumping rates exceed the rate at which the water is replenished by recharge, the groundwater is being "mined." The Ogallala aquifer, which extends over the High Plains from South Dakota to Texas, was extensively mined from the 1940s through the 1970s. Improved water management practices in the 1980s, coupled with several years of heavy rain and snow, reduced the rate of drawdown and even caused water levels to rise in some areas (Zwingle, 1993). In parts of California, surface water from the mountains is sent to the drier areas, where it is allowed to infiltrate through streambeds or ponds as "artificial" recharge.

Excessive groundwater use can have some other important consequences for human activities in addition to increased pumping costs and decreased availability of water for human consumption and agriculture. One of these is subsidence, a lowering of the land surface that can cause damage to buildings and increases the risk of flooding in coastal areas. Subsidence is the result of compaction of clays. Before pumping, these sediments are supported in a relatively loose arrangement by the pressure of groundwater. As pressure decreases during pumping, the clays begin to shift into a more compact arrangement. The resulting compaction cannot be reversed by shutting off the pumps and allowing water levels to rise or even by injecting water. In areas containing thick beds of clay, such as California's San Joaquin Valley or Mexico City, the land surface has dropped up to 10 m as a result of groundwater pumping.

Groundwater is also important in a variety of geologic processes. Water flowing through hot rocks, as are found in areas of recent volcanic activity, may reach boiling temperatures and escape from the ground as a geyser. Hot groundwater and steam are sources of geothermal energy. Hot groundwater can also dissolve a variety of metals, moving them from a source rock to a cooler area where they may precipitate to form ore deposits. Some rocks, such as limestone, dissolve even in cool groundwater. As the rock dissolves, the pores through which the water flows become enlarged, increasing the permeability. The increased permeability allows even higher rates of groundwater flow and more dissolution. Caves, sinkholes, and the irregular topography known as karst are the end products of this dissolution.

Bibliography

PRICE, M. *Introducing Groundwater.* London, 1985.
U.S. GEOLOGICAL SURVEY. *National Water Summary 1987—Hydrologic Events and Water Supply and Use.* U.S.G.S. Water Supply Paper 2350. Washington, DC, 1990.
ZWINGLE, E. "Ogallala Aquifer, Wellspring of the High Plains." *National Geographic* (March 1993):80–109.

JEAN M. BAHR

GUTENBERG, BENO

Beno Gutenberg was one of the most prominent of the early seismologists. His career spanned the whole development of the field from its beginnings to its emergence as an organized discipline. Gutenberg was born on 4 June 1889 and died on 25 January 1960. He was active in determining the structure of Earth, from the crust to the core, and also contributed to the study of the structure of the atmosphere. He is best known for discovering the low-velocity zone in the upper mantle and for determining an accurate value for the radius of the core. Gutenberg was an early advocate of the theory of continental drift. Along with Charles Richter he developed quantitative methods for assigning magnitudes to earthquakes (*see* RICHTER, CHARLES). He is well known for his catalogs of earthquakes.

Beno Gutenberg was born and educated in Germany, receiving his Ph.D. from the University of Göttingen in 1911 under the supervision of Emil Wiechert, a pioneer in the new science of seismology. At the time Göttingen was a noted center of research in many branches of science and mathematics. Gutenberg took the opportunity to hear lectures by Born, Hilbert, Minkowski, Klein, Runge, Schwarzschild, Voight, Weyl, Wiechert, and others. Emil Wiechert had just set up a new geophysical institute. Gutenberg worked there with Wiechert, Karl Zoeppritz, and Ludwig Geiger on problems in this new discipline. Gutenberg's thesis was on microseisms. He then proceeded to collect seismograms from around the world and identified and characterized most of the various kinds of seismic waves that are used today, both routinely and in special studies. In 1923 he determined the radius of the core to be 2,900 km. From 1913 to 1929 he worked at the International Seismological Association in Strasbourg, the army meteorological service, the University of Frankfurt, and to supplement his income, as a factory executive.

In 1929 Gutenberg attended a conference sponsored by the Carnegie Institution of Washington to chart future directions of the Seismological Laboratory in Pasadena, California. He became a member of the laboratory in October 1930 and also taught at the California Institute of Technology (Caltech). In 1936 the Seismological Laboratory was transferred to Caltech, and in 1947 Beno Gutenberg was officially given the title of director, which he retained until 1958.

The foundations of modern deep-Earth seismology were laid in a series of papers between 1931 and 1939, written with Charles Richter, entitled "On Seismic Waves." Earth structure models were derived from the travel times of seismic phases reported in these papers.

A long period wave recorded by horizontal seismographs was called the G wave, for Gutenberg. It was later identified as a Love wave. Gutenberg discovered, identified, and cataloged more seismic waves than any other person. His detailed research notes are still being used by active researchers. The low-velocity zone in Earth's upper mantle is now widely referred to as the asthenosphere and is regarded as a low-viscosity layer. The channels, or wave guides, discovered by Gutenberg in the atmosphere and ocean have many practical applications.

Beno Gutenberg and Charles Richter published the influential *Seismicity of the the Earth* in 1941. It was repeatedly revised and enlarged. Various magnitude scales were devised to accommodate deep-focus earthquakes and the various kinds of seismic waves. His geography of earthquakes was influential in developing the theory of plate tectonics. His publications range over all aspects of the physics of the deep interior and the descriptions of earthquakes. His views were summarized in the book *Physics of the Earth's Interior,* published in 1959.

Beno Gutenberg was a prolific author and edi-

tor. He published almost 300 articles and three major books. He was a member of the National Academy of Sciences and president of the Seismological Society of America. He received the Bowie Medal of the American Geophysical Union (1933), the Lagrange Prize of the Royal Belgian Academy (1950), the Weichert Medal of the Deutsche Geophysikalische Gesellschaft, and an honorary Ph.D. from the University of Uppsala (1955). His name is attached to the boundary between the mantle and the core and to the wave guide in the upper mantle.

Gutenberg's associates at the Seismological Laboratory of Caltech included Charles Richter, Hugo Benioff, and Frank Press. Under their guidance the Seismological Laboratory became a world center in deep-Earth and earthquake studies.

Bibliography

GUTENBERG, B. *Internal Constitution of the Earth.* New York, 1939.
———. *The Physics of the Earth's Interior.* New York, 1959.
GUTENBERG, B. and C. F. RICHTER. "On Seismic Waves." *Beiträge zur Geophysik* 43 (1934), 45 (1935), 47 (1936), and 54 (1939).
———. *Seismicity of the Earth.* New York, 1941.
JEFFREYS, H. "Beno Gutenberg." *Quarterly Journal of the Royal Astronomical Society* 1 (1960):239–242.
RICHTER, C. F. "Memorial to Beno Gutenberg." In *Proceedings of the Geological Society of America for 1960.* Boulder, CO, 1962.

DON L. ANDERSON

GYMNOSPERMS

Unlike the seeds of flowering plants (*see* ANGIOSPERMS), the seeds of gymnosperms are always naked. This means that gymnosperm seeds may be borne on small stems, directly on a main stem, on the surfaces or edges of leaves, or within cupules or cones, but they are never completely enclosed inside a sealed structure called a carpel, which is part of an angiosperm flower or fruit. Naked seeds may have evolved independently in several Paleozoic lineages, or in only one lineage that rapidly diversified.

Living gymnosperms include conifers, taxads, cycads, gingko, and the gnetales. Although modern gymnosperm diversity is relatively low, several of the extant families may be only distantly related to one another. In groups with such long evolutionary histories (for gymnosperms, at least 350 million years or Ma), fossils can provide crucial information regarding the rate and direction of character change. At most systematic levels, the ancestral relationships of gymnosperms are still quite open to question. Cladistic and other systematic analyses require clear recognition of which similarities are shared ancestral traits, and which are more recently derived, in order to decipher the pattern and degrees of relatedness.

Conifers

The most common gymnosperms in the northern hemisphere temperate zone are the conifers, which include pines, cedars, and possibly taxads (trees or shrubs of the family Taxaceae). While, like gymnosperms, the conifers may be a polyphyletic (multi-origin) group, the Pinaceae is widely accepted as a natural family. The pines have an excellent fossil record extending back 100 Ma into the Cretaceous. Pines are always trees or shrubs. Pine leaves are narrow and needle-like and, like the woody stems and seed cones, contain numerous resin canals. The seed cones of fossil pines are much like those of modern pines. By comparison with the most ancient conifer seed cones, however, pine seed cones are reduced and simplified.

In contrast to the robust seed cones, the pollen cones of modern and fossil pines are ephemeral structures that generally shrivel and drop from the tree after the pollen has been wind-dispersed. The structure of conifer pollen cones is very simple in organization and may have changed relatively little through geologic time. The pace of evolution within the pines may have been mosaic (uneven) and comparatively slow, with organs such as the seed cones revealing much more rapid change than the woody trunks or the pollen cones.

Taxads

The taxads are rather widely dispersed geographically, occurring today in America, Europe, and Asia, including Malaysia. As their modern diversity is low, this broadly disjunct distribution may be due to dispersal of their seeds by birds. Although the geologic record of taxads begins in the Lower Jurassic, it is limited to very few, rather poorly preserved specimens. As a result, there is much disagreement as to taxad origin and their systematic relationship to other gymnosperms, including the cone-bearing conifers. Taxads bear their seeds as individual units, with only a few scale-like leaves and a fleshy aril that extends around the seed at maturity. Taxad wood is generally comparable to that of conifers, but the pollen cones are not like those of conifers.

Two quite different views of conifer and taxad relatedness exist at present. One view is that the taxads and conifers are separate at the class level, that is, the Taxopsida and the Coniferopsida. This interpretation assumes that they have been distinct lineages since their separate origins, because the earliest conifer fossils all have seeds borne on leaves in cones, whereas the earliest taxad fossils all have individual seeds on small stems. A contrasting view recommends placing the Taxaceae within the order Coniferales and class Coniferopsida, since the leaves and wood are comparable and the seed-bearing shoots of taxads may have been derived by genetically controlled suppression of early developmental stages in an ancestral conifer lineage, resulting in one terminal seed per leafy shoot. Ontogenetic study of taxad seed development could provide insight into this controversy.

Cycads

The cycads are widespread throughout the tropical and warm temperate regions of both hemispheres in small disjunct populations. They reached maximum diversity during Mesozoic time and, despite the presence of several neurotoxins, are eaten today and probably provided food for the dinosaurs. Fossil cycad cones, leaves, and the unique woody trunks are known from Upper Paleozoic coal and shale deposits, especially in North America, China, and Antarctica. Although most cycads bear large cones, the trunks are generally unbranched and resemble palms or tree ferns. Cycads are not closely related to either group, however, and cycads have no close living relatives. Their rich fossil record includes many extinct lineages of pteridosperms and cycads. The thick, pithy trunks and persistent leaf bases may provide protection from fires. Cycads may be the only modern gymnosperms with insect and algal symbionts and insect assistance in pollination, and their large colorful seeds attract many animal dispersers.

Ginkgo

Ginkgo biloba is often considered a living fossil because of its resemblance to tree-forming Carboniferous pteridosperms. It is the only genus and species in its class, and has no closely related living relatives. Fossil and modern *Ginkgo* leaves are broadly triangular with dichotomously parallel veins. These attractive leaves are easily recognized worldwide in Mesozoic rocks, but fossils of the wood and seeds are not common. As an ornamental tree with edible seeds, *Ginkgo* is widely planted throughout the temperate climate zones, but natural populations occur only in southeastern China. The large seeds have a stony inner layer and a fleshy outer layer that develops a strong odor when ripe. Seeds are borne on small unbranched axillary axes on spur shoots; each seed has a basal collar of tissue, and it is not clear from the fossils or from developmental studies whether the seeds are borne on modified leaves or on modified stems. This is thought to be an important distinction that would reveal whether *Ginkgo* is more closely related to cycads or to conifers. *Ginkgo* and the cycads are the only gymnosperms that exhibit the apparently archaic trait of swimming sperm cells, but their leaves and wood are quite dissimilar.

Gnetales

The gnetales comprise three genera, *Gnetum*, *Welwitschia*, and *Ephedra*, that are by far the most unusual assemblage of gymnosperms alive today. The gnetums are mostly tropical lianas, the welwitschias are restricted to coastal deserts in southwestern Africa, and the ephedras occur in cool arid regions in both the eastern and western hemispheres. The features they share include several that are usually restricted to angiosperms but also can be interpreted as evidence of convergence or uniqueness to these particular gymnosperms. Unlike angiosperm vessels, the vessels in gnetalean wood develop by increased broadening of circular bordered pit fields in tracheid endwalls. The reproductive structures are variously referred to as compound cones or flowers because of the presence of tissue layers around the seeds and pollen-bearing units. Understanding the significance of these features will be greatly assisted by recognition of appropriate well-preserved fossils as well as ontogenetic studies of early growth and development.

Gymnosperm Origins

All gymnosperms recognized today have naked seeds and simple wood composed of ray cells and tracheids. In the Devonian there were plants with gymnospermous wood, but no seeds. We call these plants progymnosperms because we think gymnosperms originated within this group of woody plants that had free-sporing reproduction.

Among the most ancient gymnosperm seeds are Archaeosperma and Elkinsia. These seeds are from Upper Devonian (Famennian) sediments deposited 350 Ma. They occur as isolated specimens enclosed loosely in enveloping cupules with basal portions only millimeters wide. We do not know if these cupulate seeds were borne on plants that had gymnospermous wood. Recent study of such seeds suggests that they may be more correctly interpreted as pre-seeds. If so, they may represent a lineage that led to one or more of the richly diverse and geologically persistent gymnosperms.

Bibliography

BECK, C. B., ed. *Origin and Evolution of Gymnosperms*. New York, 1988.

GIFFORD, E. M., JR., and A. S. FOSTER. *Morphology and Evolution of Vascular Plants*, 3rd ed. New York, 1989.

GILLESPIE, W. H., G. W. ROTHWELL, and S. E. SCHECKLER. "The Earliest Seeds." *Nature* 293 (1981):462–464.

JONES, D. L. *Cycads of the World*. Washington, DC, 1993.

NIKLAS, K. J. "Airflow Patterns Around Some Early Seed Plant Ovules and Cupules: Implications Concerning Efficiency in Wind Pollination." *American Journal of Botany* 68 (1981):635–650.

NORSTOG, K. J. "Cycads and the Origin of Insect Pollination." *American Scientist* 75 (1987):270–279.

STEWART, W. N., and G. W. ROTHWELL. *Paleobotany and the Evolution of Plants*, 2nd ed. New York, 1993.

TAYLOR, T. N., and E. L. TAYLOR. *The Biology and Evolution of Fossil Plants*. Englewood Cliffs, NJ, 1993.

GENE MAPES

H

HAZARDOUS WASTE DISPOSAL

The disposal of waste material containing substances hazardous to human health has become a problem of large and increasing magnitude. In the early 1990s more than 300,000 metric tons of such waste were being generated in the United States alone each year, and the amount grows larger as additional materials are recognized as potentially harmful. For most of these substances the common method of disposal is burial under the earth's surface, so that geology plays a pivotal role in the selection and preparation of disposal sites. The geologic factors important in determining the feasibility and safety of underground disposal are discussed herein.

Hazardous materials contained in waste can be grouped in three principal types: (1) metals poisonous to human health like mercury, lead, and cadmium; (2) organic chemicals, both naturally occurring and those produced by the chemical industry; and (3) radioactive substances, especially those used in and generated by nuclear reactors. These toxic substances vary greatly in the concentrations that constitute a hazard and in the length of time they will remain dangerous if released to ordinary human environments. Some of the organic compounds and some radioactive elements decay rapidly, so that disposal for a few years or decades is enough to render them harmless, while the metals

and a few of the more stable organic compounds remain a potential hazard indefinitely. Some organic compounds need to be present in substantial amounts to be dangerous, but metals like mercury or plutonium are safe in the environment only if they occur in tiny concentrations of a few parts per billion or trillion. The disposal problem is complicated by the fact that permissible concentrations change from time to time as more is learned about the effects of various substances on human health. A further complication comes from recognition that some waste constituents—selenium and chromium are good examples—are poisonous in large concentrations but are essential components of the human diet in small amounts.

The object of hazardous waste disposal is not the elimination of every atom or molecule of the offending substances but to make sure that concentrations in any effluent from a disposal site do not exceed levels recognized medically and legally as acceptable in the environments of everyday life. If disposal is to be accomplished by burial underground, this generally means that concentrations in enveloping groundwater must be kept below these established levels. Nearly everywhere beneath the earth's land surface, rocks at depth are saturated with water; the dissolving of material from buried waste by this water is the chief mechanism by which toxic substances can escape to surface environments. The level of the top of the saturated zone in rock or soil—the water table—varies with topography, climate, and removal by water wells (*see* GROUNDWATER). It may be near the surface in humid areas of low relief or at depths of hundreds of meters in deserts. The water nearly everywhere is moving slowly, and the danger is that the movement may carry dissolved hazardous materials to places where the groundwater comes to the surface in springs or seepages or where it may be tapped by wells for domestic or agricultural water supplies. Geologic problems of hazardous waste disposal are largely concerned with controlling the amounts of waste constituents that may be dissolved and transported away from a disposal site by moving groundwater. Escape of volatile material may be an additional problem with some kinds of organic and radioactive waste, but generally groundwater is the source of greatest worry in underground disposal.

A method commonly used for the subsurface disposal of toxic waste is the same as that used for much ordinary industrial waste: the technique of shallow landfill (*see* WASTE DISPOSAL, MUNICIPAL). This method means putting the waste in trenches a few tens of meters deep excavated at sites well above the regional water table and ensuring minimum contact with rainwater by providing good drainage and a cover of impermeable material. Arguments have arisen about the effectiveness of this procedure, and some of the early disposal sites have had to be abandoned when contamination of groundwater above permissible levels was noted. But shallow landfill has proved generally satisfactory for much hazardous waste, provided that extra precautions are taken to shield the waste from contact with groundwater and surface water, including rainwater.

Most serious are the problems posed by high-level waste—waste containing high concentrations of toxic material—and to this issue most attention is devoted herein. Major emphasis will be on the underground disposal of radioactive waste, because it well illustrates the most difficult problems encountered in waste disposal and because it is currently a subject of intense popular concern. High-level radioactive material, it is generally agreed, will require disposal at deeper levels than those at landfill sites. No disposal at deep levels has yet been undertaken for either radioactive material or other kinds of toxic waste, but the geologic requirements for such an operation have been and are being thoroughly explored. This continuing study is the principal focus of this entry.

There are two principal varieties of high-level radioactive waste: the spent-fuel elements from nuclear reactors, discarded after most of their fissionable uranium-235 has been exhausted in the generation of energy, and the liquid wastes resulting from the production of plutonium for nuclear weapons. Both kinds contain radioactive elements in great variety and high concentrations, in part coming from the fission of uranium-235 (for example strontium-90 and cesium-137), and in part from neutron capture by atoms of uranium-238 to form elements heavier than uranium (for example neptunium-237 and plutonium-239). The spent-fuel elements are solid waste in the form of clusters of thin metal tubes kept immersed in large basins of cooled water that are maintained at reactor sites. The waste from weapons production ("reprocessing waste") is a liquid-solid mixture stored in huge steel tanks at the two places in the United States where plutonium has been generated in large amounts: Hanford in southern Washington and the Savannah River plant in South Carolina. Both kinds of waste require handling by remote control, not only because of the intense radiation they emit but also because they are thermally hot from the release of energy by radioactive decay.

With time, of course, both heat and radioactivity will diminish so that the waste becomes progressively less hazardous on standing. Some of the radioactive elements, however, have half-lives in the tens of thousands to hundreds of thousands of years, meaning that by human standards they will remain dangerous for a very long time. Putting a definite number on the time during which these wastes must be kept isolated from surface environments is difficult, but it is certainly well over 10,000 years.

The wastes at present are doing no harm to their surroundings. Radiation is absorbed by their containers, and heat is removed by circulating air and water. But to preserve this happy state requires careful maintenance: the wastes will remain harmless only as long as someone is at hand to repair leaks, to keep the water cold and circulating, to fix damage from accidents, and to build new tanks as needed. The difficulty is that maintenance over a span of 10,000 years can hardly be expected, and it is therefore agreed that some more permanent means of disposal must be found. Of the many methods that have been proposed, the one uniformly regarded as safest and most feasible is mined geologic disposal—putting the waste in a cavity excavated in rock deep underground. This is the procedure being studied intensively both in the United States and in most other countries that face a similar problem of managing high-level radioactive waste.

The plan is essentially simple. A repository for the waste in the form of a shaft with tunnels branching from its base will be excavated, much as an underground mine is excavated, at a depth of several hundred or a few thousand meters. The liquid waste will be solidified, presumably in the form of glass made by heating with silica, soda, lime and boron oxide, and both glass and spent-fuel rods will be enclosed in metal containers. Holes will be drilled in the floors or walls of the mined tunnels, the waste containers will be inserted, and both holes and tunnels will be filled with crushed rock and clay. These procedures can be accomplished with well-known mining techniques. The waste will be far enough from the surface to ensure that radiation will be absorbed; it will also be far enough from the surface to make it unlikely that geologic accidents like earthquakes or deep erosion will expose the waste. At first glance, safe disposal seems assured.

But a skeptic would object immediately that groundwater has not been considered. Surely at some time in the future water will seep into the tunnel fillings and collect around the metal containers. The metal will eventually corrode, and water will be in contact with the waste. Some of the waste will dissolve, and radioactive elements will then become part of the moving groundwater. Ultimately, the water will reach the earth's surface, and whoever lives nearby will be exposed to damaging radiation. A confirmed skeptic could suggest other possible calamities—a major earthquake that might crack the rock and bring waste to the surface, or an invasion of the repository by hot liquid rock and gas from below, carrying waste to the surface in the form of a volcanic eruption. But the groundwater scenario is the one most likely to cause a hazard.

In answering the skeptic, a geologist would have no trouble pointing out places for repository construction where the more extreme possible disturbances could be avoided, places, for instance, where earthquakes and volcanic activity are very unlikely. A geologist would also, obviously, suggest avoidance of sites where future exploitation of mineral or oil and gas resources, or future concentrations of population, seem at all probable. But

can a geologist find sites where the entrance of groundwater can be long delayed and where its capacity to transport radioactive elements to surface environments is limited? This is a query that has occupied, and is still occupying, the attention of many scientists and engineers throughout the world. The possibility of geologic disposal of high-level waste depends on finding a satisfactory answer.

The geologist's first concern would be to look for a suitable kind of rock, at the proper depth, in which a repository could be excavated. Ideally, it would be a rock strong enough to maintain an opening, at least for the time during which waste is being emplaced, a rock with low permeability and few cracks through which water could move easily, and a rock with high heat conductivity to help in keeping temperatures low. One likely candidate would be rock salt, a rock that has the virtues of easy mining, unusually high head conductivity, impermeability to groundwater, and an ability to flow plastically, over a period of years, so that it would ultimately collapse around the waste containers and hold them firmly in place. Another good repository rock is granite, which is less easy to mine than salt but better able to maintain repository openings, generally impermeable, and containing little groundwater unless it is cut by many cracks. A third suggested rock is volcanic rock basalt, attractive only if it can be found in unusually thick lava flows without much cracking. A more promising volcanic rock would be a thick bed of ash or tuff, consolidated enough to maintain repository openings but not enough to destroy its capacity for sorbing materials from solution. All of these alternatives have been given serious study in the United States, and in the early 1990s attention was concentrated on volcanic ash at Yucca Mountain in southern Nevada. Salt has been the first choice in Germany, and granite has had most appeal to Canadian, French, and Swedish geologists.

Next on the geologist's agenda would be a detailed study of the amount and composition of present-day groundwater in the region of a proposed repository, with emphasis on estimating its capacity for dissolving and transporting the different radioactive elements. Of particular interest would be the length of the path the water would follow on its way from the repository to the ground surface, and the ability of the rock along this path to precipitate or adsorb ions of the various radioactive elements dissolved in the water. In a desert area where the groundwater table is deep, an attractive possibility would be to locate the repository well above the water table so as to prevent long-term filling of the tunnels with water—as the Department of Energy is hoping to do at Yucca Mountain. Study of the groundwater also makes possible some suggestions about the best metal to use for the waste containers, the best solid form in which to put the waste, and the best material to use for filling tunnels and other openings so that groundwater would be delayed as long as possible in making contact with waste and would then dissolve no more than a minimum amount.

Following such studies of rock and water at a given site, the geologist's next move is to set up a reasonable model of the future behavior of an imagined repository from which predictions can be attempted regarding the amounts of radioactive material that will escape at any time in the future, up to at least 10,000 years. For this purpose the various properties of rock and waste and water can be assigned numbers—solubilities, amounts of adsorption, rates of water movement—and from the numbers the probable behavior of each radioactive species can be estimated. Refining the estimates also requires consideration of the long-term effects of heat and radiation on the rock along the path of the groundwater. The calculation is obviously complex and cannot pretend to give precise answers. But at least limits can be set: if numbers in a range are always chosen conservatively, to give, for example, the most rapid groundwater movement and the highest solubilities, the predictions will surely err on the high side in their estimates of release of radioactivity to the surface and hence will give a comfortable margin of safety. The many such predictions that have been made, despite their differences, indicate clearly that a repository can be built in a well-chosen site that will almost certainly keep concentrations of radioactive species that reach surface environments below regulatory limits for at least one hundred centuries.

This conclusion seems evident at least to the great majority of geologists and of other scientists and engineers who have concerned themselves with the question. Inevitably there are some who disagree: the number of variables is so large, and some of the numbers are so uncertain, that ample room exists for differences of opinion. The differences can be narrowed by further research, but how much additional research is warranted? Or to ask the question differently, how much consensus of opinion is needed before actual disposal of high-level waste can begin? There will never be complete agreement. No geologist, and no other scientist or engineer, can ever claim that mined geologic disposal of high-level radioactive waste is completely without risk. So the question of where to undertake disposal has become truly difficult, in a sense more sociological than technical. How much difference of opinion among experts can society tolerate before making a decision, and what degree of risk will society accept? We have no answers as of the early 1990s, and no high-level waste is being disposed of in a mined geologic repository, either in this country or abroad. The technical questions about managing the geologic disposal of high-level toxic waste are in large part answered, but social and political questions remain.

Bibliography

CHAPMAN, N. A., and I. G. MCKINLEY. *The Geological Disposal of Nuclear Waste*. Chichester, Eng., 1987.

OFFICE OF TECHNOLOGY ASSESSMENT, U.S. CONGRESS. *Managing the Nation's Commercial High-level Radioactive Waste*. Washington, DC, 1985.

YUCCA MOUNTAIN SITE CHARACTERIZATION PROJECT, U.S. DEPARTMENT OF ENERGY. *DOE's Yucca Mountain Studies*. Office of Civilian Radioactive Waste Management. Washington, DC, 1992.

KONRAD B. KRAUSKOPF

HEAT BUDGET OF THE EARTH

Before 1896, when Henri Becquerel of France discovered radioactivity, the earth's internal heat presented an unsolvable scientific problem. Geologists realized that the sedimentary rocks that cover large areas of the earth required hundreds of millions of years to accumulate, but the British physicist Lord Kelvin showed by a mathematical argument that the flow of heat from the earth, as observed in mines, was inconsistent with an age greater than about 25 million years (Ma). Kelvin's argument agreed well with an earlier calculation by the German physicist H. Helmholtz on the heat source of the Sun. The only known energy available to the

Sun was the gravitational energy released by shrinking and this could not have produced the present heat output of the Sun for more than 20 to 40 Ma. Both of these calculations were seen to be irrelevant as soon as radioactivity was discovered and the enormous energy released by nuclear processes was recognized. As we now know, radioactivity accounts for most of the heat that is flowing steadily from Earth's hot interior. But it does not account for all of it. Earth is slowly cooling. This means that it was hotter when it was young and that geological processes, such as volcanic eruptions, were more vigorous than they are now.

The important elements producing heat in the earth by radioactive decay are uranium, thorium, and potassium. The rate of heat production per kilogram of each of these elements is given in Table 1, together with an estimate of the total mass of each of them in the earth. This allows the total heat generated by radioactivity to be calculated, as in the fourth column of Table 1. We can put these large numbers into perspective by comparing them with a familiar heat source, kilowatts (2,000 W or 2×10^3 W). We see that the radioactive heating of the earth is equivalent to 14 billion hot plates. But the concentration of radioactivity in the earth is very low. The average concentration is such that it would take nearly 80 km^3 of rock to produce the heat of one hot plate.

The essential reason why Earth is gradually cooling is that its radioactivity is decreasing with time. As radioactive atoms decay, they become fewer and the disintegrations become correspondingly fewer. The decays have been very carefully measured and so we know how much stronger the radioactivity was in the past. When we calculate back to the origin of Earth, 4.5 billion years (Ga) ago, we obtain the numbers in the final column of Table 1. They show that when Earth was very young, the radioactive heat was four times as great as it is now.

Although heat is transported through most of the depth of the earth by convection, the final stage is by conduction, that is, the simple flow of heat through the crust to the cool surface, where it escapes to the atmosphere and oceans. The rate is estimated by measuring the variation of temperature with depth (about 20°C/km) in mines and boreholes and the thermal conductivities of crustal rocks. It varies from place to place over the earth, but many measurements have been made and the average is 0.063 W m^{-2}. This conducted heat flow is augmented by the convective circulation of water in the crust, especially on the flanks of the ocean ridges where mantle convection is producing fresh ocean floor. The total heat flow, averaged over the whole earth, is about 0.083 W m^{-2}. This means that heat equal to that from a single hot plate (2×10^3 W) flows from an area of 24,000 m^2 (2.4 hectare or nearly 6 acres).

Such a small heat loss is not noticeable at the surface. It is dwarfed by the heat from the Sun, which amounts to more than 1 kW/m^2, or the equivalent heat from one hot plate for each 2 m^2 at midday in tropical areas. However, when we multiply the average heat loss by the total surface area of the earth (5.1×10^{14} m^2) we obtain the total surface heat loss, 42×10^{12} W, which is substantially greater than the heat generated by radioactivity.

We now have the two most important numbers to use in drawing up a heat budget for Earth (Table 2). There are two minor items that we also need to take into account. Since we conclude that there is a net loss of heat by Earth, it is contracting thermally. This means that gravitational energy is lost because each element of Earth's mass is falling, and this energy appears in Earth as heat. The rate of contraction is slow, about 2 mm in radius per 1,000 years, but the energy release, 2.1×10^{12} W, is not negligible.

Table 1. Heat Produced by Radioactivity in Earth

Radioactive Element	Heat Produced (mW/kg of element)	Total Earth Content (kg)	Heat Produced (10^{12} W)	Heat 4.5 Ga Ago (10^{12} W)
Uranium	98.4	13.25×10^{16}	13.04	70.2
Thorium	26.6	47.2×10^{16}	12.56	15.7
Potassium	0.0035	7.14×10^{20}	2.50	30.2

Source: Data from F. D. Stacey. *Physics of the Earth*, 3rd ed., Brisbane, Australia, 1992.

Table 2. The Heat Budget

Income		
Radioactive heat	28.1	
Thermal contraction	2.1	
Gravitational segregation and latent heat	1.8	
		32.0
Expenditure		
Surface heat loss		42.0
Net loss of heat		10.0

All values are given in units of 10^{12} W.
Source: Numbers taken from Stacey, F. D. *Physics of the Earth*, 3rd ed., Brisbane, Australia, 1992.

Table 3. Heat Losses from the Major Regions of Earth

Crust	8.2
Mantle	30.8
Core	3.0
Total	42.0

All values are given in units of 10^{12} W.

The other entry in Table 2 is a sum of several contributions that are individually small but arise from processes that are important for other reasons. Like the rest of Earth, the fluid metal core is cooling and as it does so it is slowly solidifying, producing, at the center, the solid inner core. There is, therefore, a release of latent heat of solidification. More important, the composition of the inner core is not identical to that of the outer core. The outer core is a solution in liquid iron of light elements that are probably a mixture of oxygen, sulfur, carbon, and perhaps silicon or hydrogen, but these elements are rejected by the solid in the same way as salt-free ice is formed when salty water freezes. The inner core is therefore more dense, and an excess of the light constituents is left at the bottom of the fluid outer core. Such a layer of lower density is gravitationally unstable. It mixes with the rest of the fluid, stirring the outer core as it does so, and the stirring action generates Earth's magnetic field. Ultimately all of this energy appears as heat. There is a small contribution also by the chemical segregation of Earth's crust from the denser mantle.

Now we can see how far Earth's heat budget is out of balance (Table 2). The difference between loss of heat and the heat generation amounts to 10×10^{12} W. This figure is a measure of the rate at which Earth is cooling. The cooling is not uniform throughout Earth. For the mantle the estimated rate is 70°C per Ga, but for the core it is closer to 20°C per Ga.

The heat loss from the three principal regions of Earth (Table 3) occurs by different mechanisms, with different geological and geophysical effects. The crust is a thin veneer and heat can escape from it by conduction. The crustal entry in Table 3

does not include the heat transfer to the oceans from the ocean floors, which must be regarded as mantle heat. Although it is finally transferred to the oceans by conduction from the cooling rigid surface layer, known as the lithosphere, it is convectively transported to the surface from the deeper interior.

Convection is the only possible mechanism of heat transfer from the deep parts because Earth is too large to have cooled appreciably by conduction, even in its entire lifetime, 4.5×10^9 years. Thus the mantle heat loss, 30.8×10^{12} W by the estimate in Table 3, is accounted for by convection, the bodily motion in which hot material rises and is displaced by sinking of the cooled and therefore more dense, rigid lithospheric slabs. Mechanical energy is generated in this process and we can refer to a thermodynamic efficiency, that is, the ratio of mechanical energy to the heat transported. Its value is estimated to be 0.156, that is, the mechanical energy generated by mantle convection is 0.156 times the mantle heat flux of 30.8×10^{12} W. So, the mechanical power of mantle convection is 4.8×10^{12} W. This is the power available for plate tectonics. It corresponds well to the power required to drive the plates at the observed rate of a few centimeters per year against stresses of the magnitudes seen in earthquakes. Thus thermal convection of the mantle is a convincing explanation for tectonics.

Loss of heat from the core occurs by conduction into a boundary layer at the base of the mantle, known in the jargon of geophysics as the D layer (*see* PHYSICS OF THE EARTH). This is a layer of hot and therefore softened and thermally expanded material that is convectively unstable and generates thermal plumes that rise through the mantle. They have surface expressions as isolated hot spots of volcanism. Thus the mantle has two superimposed and quite different patterns of thermal convection, one driven by buoyancy generated at each of its boundaries. Cool rigid slabs (plates) sink from the surface and hot soft plumes rise from the base.

Bibliography

LOPER, D. E. "A Simple Model of Whole Mantle Convection." *Journal of Geophysical Research* 90 (1985):1809–1836.

STACEY, F. D. *Physics of the Earth,* 3rd ed. Brisbane, Australia, 1992.

FRANK D. STACEY

HERSCHEL FAMILY

Few families have been such prolific contributors to astronomy as were the Herschels: William (originally Frederich Wilhelm), his sister Caroline Lucretia, and William's son John Frederick William. These three astronomers are renowned for the discovery of Uranus, the building of large telescopes, the discovery of comets and moons, the compilation of a catalog of star clusters and nebulae, numerous solar studies, and extensive observations of the southern hemisphere sky, among other accomplishments.

Sir William Herschel was born in Hanover, Germany, on 15 November 1738. His father, Isaac Herschel, was the regimental bandmaster in the Hanover Guards Band, and at age fifteen William joined his father and older brother as a musician in the band. At the start of the Seven Years' War in 1757, William moved to England and held various teaching and musician positions in several English cities over the next twenty-five years. After moving to Bath in 1766, William began an in-depth study of mathematics, optics, and astronomy. He rented a 60 cm focal length Gregorian reflector to begin his study of the skies, progressively making several larger telescopes of his own (reflectors of 1.5 m, 3 m, 6 m, 9 m, and 12 m focal lengths as well as over two hundred mirrors of 2.10 m focal length) as his interest in astronomy intensified. Using these telescopes (the best of the day), William began a methodical and ambitious program to survey the heavens in an attempt to determine the three-dimensional structure of the Milky Way. In the course of his observations, he studied over 2,500 nebulae and star clusters, discovered over eight hundred double stars and showed that some of these stars revolved around each other (the first proof that the law of gravitation applied beyond the solar system), and was a pioneer in the new area of stellar photometry. William is best known, however, for his observation of a greenish blue disk on the night of 13 March 1781. At first he thought the object was a comet, but he soon realized that it was a planet, the first to be discovered telescopically. William named the object Georgium Sidus (George's Star), after England's King George III, but this name was rejected in favor of Uranus (God of the Heavens). William was elected a fellow of the Royal Society in 1781 and received the Copley Medal that same year in recognition of his discovery of Uranus. He was granted a pension by King George III, which enabled him to retire from his music career and practice astronomy fulltime. Following his discovery of Uranus, William discovered two of the planet's moons (Titania and Oberon) in 1787 as well as two moons around Saturn (Enceladus and Mimas) in 1789. He was the first to suggest that the Sun moves around the center of the Galaxy, and he discovered the infrared region of the electromagnetic spectrum. He continued an active observing and publishing schedule in spite of declining health. William died in Slough, Buckinghamshire, England, on 25 August 1822.

Caroline Herschel was born 16 March 1750, in Hanover, Germany, the younger of two girls in the Herschel household. Against the desires of her mother, Caroline actively participated in philosophical and scientific discussions with her father and brothers and pursued instruction in a musical career. Following her father's death in 1767, Caroline performed the household duties for her mother, Anna, and older brother, Jacob, until William brought her to Bath, England, in 1772 to encourage her singing career. She sang in oratorios in Bath, Bristol, and other English cities until 1782, when she began to devote herself full-time to assisting William with his astronomical observations. She served as recorder of his observations, executed the extensive calculations necessary for his work, prepared the star cluster and nebulae catalog for publication, and edited his papers prior to their submission for publication. William gave Caroline a small refracting telescope in 1782, and she began independent studies of the heavens. After she discovered three new nebulae in 1783, William gave her a larger reflecting telescope, which she used when William was away from home and opportunities for her to conduct independent research occurred. Caroline discovered eight comets between 1786 and 1797, a record by a woman astronomer, which was not surpassed until 1987 (by Carolyn Shoemaker). Caroline was granted a stipend by King George III in 1787 for serving as William's assistant. Shortly after the discovery of her last comet in 1797, Caroline compiled a cross-index of Flamsteed's Star Catalogue, making it much more useful for astronomical observations. This index was published by the Royal Society in 1798 as the *Catalog of Stars*. Following William's death in 1822, Caroline returned to Hanover, where she compiled a new catalog of nebulae arranged by zone. Her astronomical contributions were acknowledged with gold medals from the Royal Astronomical Society (1828) and the King of Prussia (1846), and she was awarded honorary membership in the Royal Astronomical Society (1835), since women were not permitted to be full members at the time. Caroline died in Hanover on 9 January 1848.

The Herschel legacy was continued by William's only son, John, who was born on 7 March 1792, in Slough, England. Sir John Herschel contributed to many areas of science and mathematics, including terrestrial magnetism, photography, differential calculus, finite differences, summation of series, calculus of operations, and optics. John devoted much of his life to studies of the Sun, pioneering the use of chemical analysis of the solar spectrum in 1819, recording the first satisfactory measurements of the amount of solar radiations in 1836, and studying the structure of solar floccules in 1864. During the 1820s John continued and revised his father's studies of binary star systems, nebulae, and star clusters in the northern hemisphere. From 1834 to 1838, he observed the southern hemisphere sky from the observatory at the Cape of Good Hope in South Africa, concentrating on observations of nebulae and star clusters and revising the nomenclature of the southern stars. John was a cofounder of the Analytical Society of Cambridge in 1813, the same year he was elected a fellow of the Royal Astronomical Society. He served as president of the Royal Society between 1827 and 1832. He received numerous medals and awards from a variety of scientific societies throughout his lifetime, including a gold medal and the Lalande prize from the Royal Society in

1825. He also served in several nonscientific positions, including lord rector at Marischall College in Aberdeen (1842) and master of the mint (1850–1855). John died on 11 May 1871, in Collingwood, Kent, England.

Bibliography

LEY, W. *Watchers of the Sky*. New York, 1963.

LUBBOCK, C. A. *The Herschel Chronicles: The Life-Story of William Herschel and His Sister Caroline Herschel*. Cambridge, Eng., 1993.

OGILVIE, M. B. "Caroline Herschel's Contributions to Astronomy." *Annals of Science* 32 (1975):149–161.

SIDGEWICK, J. B. *William Herschel, Explorer of the Heavens*. London, 1953.

NADINE G. BARLOW

HESS, HARRY

Earth scientists remember Harry Hammond Hess primarily for developing in the early 1960s the hypothesis of seafloor spreading, which was a key element in the forthcoming plate-tectonic revolution. Long before seafloor spreading, however, Hess had secured his reputation as an earth scientist of unusual breadth for his contributions to the diverse fields of mineralogy, petrology, tectonics, and marine geology. Elected to the prestigious National Academy of Sciences in 1952, Hess soon became a ranking civilian advisor to the federal government on scientific policy. He was also the long-time chairman of the Department of Geology at Princeton University and a vigorous promoter of interdisciplinary research within the earth sciences. After active duty with the U.S. Navy in World War II, Hess remained in the naval reserve and eventually attained the rank of rear admiral. In view of this remarkable range of achievement, Hess's teacher and colleague at Princeton, Arthur F. Buddington, would eulogize him as a model statesman-scientist who had "accomplished the impossible" in living "five lives contemporaneously."

Harry Hess was born in New York City on 24 May 1906 and grew up in suburban New Jersey. At Yale University he was one of two undergraduates in a geology department devoted primarily to graduate studies and received his B.S. degree in 1927. After a brief stint as a working geologist in Rhodesia, Hess resumed his formal education at Princeton University and completed his Ph.D. in 1932. Hess's dissertation concerned the hydrothermal alteration of peridotite, a magnesium-rich igneous rock common in the earth's mantle but only rarely intruded into the crust. This project was originally suggested by Edward Sampson, the economic geologist at Princeton, because of its relevance to the genesis of talc, soapstone, and asbestos deposits. However once Hess realized that altered ("serpentinized") peridotite was exposed in only a handful of geologic settings involving profound crustal disturbance, such as in island arcs and alpine mountain belts, the future course of his research was largely set.

Although Hess credited Sampson for the "not too gentle push" into his initial study of altered peridotite, three other professors at Princeton probably had a greater impact on his career. Hess learned petrology in the field from Buddington and mineralogy in the laboratory from Alexander H. Phillips, which gave him the tools to investigate a wide range of problems in addition to the formation of serpentinized peridotite. For instance, Hess's meticulous studies of the pyroxene group of minerals and the igneous rocks in which they commonly occur were fundamental contributions to a specialized field, still regarded as classics. The third professor was Richard M. Field, an early promoter of marine geology and geophysics whom Hess described fondly as "brilliant and erratic." Field taught Hess the importance of "thinking big" and in 1932 arranged for the young graduate student to accompany the renowned Dutch geophysicist, Felix A. Vening Meinesz, on a gravity survey of the Caribbean island arcs.

Vening Meinesz had developed an extremely sensitive instrument for measuring minute variations in the force of gravity, which he installed in submarines to escape the disturbing effect of surface waves. Beginning in the East Indies in the mid 1920s Vening Meinesz discovered unexpectedly large deviations from "normal" gravity in the vicinity of island arcs, which permitted him to infer the hidden crustal structure and speculate upon its origin. Hess's unique perspective on island arcs, based on his studies of serpentinized peridotite, provided a perfect complement to Vening Meinesz's highly theoretical "remote sensing." Although Vening

Meinesz and Hess would never again collaborate directly, their continuing research over the coming decades contributed to a major new synthesis on the origin and evolution of island arcs, which remained viable until the advent of seafloor spreading and plate tectonics in the 1960s.

After completing his dissertation Hess left Princeton briefly but returned in 1934 to join the faculty. At Field's suggestion he entered the naval reserve to provide easier access to submarines for follow-up studies in the Caribbean. Shortly after the Japanese attack on Pearl Harbor in 1941, Hess reported for active duty with the navy. Ironically, his first assignment was a desk job predicting the movements of German submarines. Eventually, however, he received command of a transport vessel in the Pacific, the U.S.S. *Cape Johnson*, and in the interest of science kept the ship's depth recorder in constant operation to chart submarine topography. In this manner Hess discovered those curious, flat-topped submarine mountains in the western Pacific he would later call "guyots." Hess interpreted these features as drowned volcanic islands whose tops had been planed off by wave action prior to submergence. Guyots provided clear evidence of profound vertical movements of the seafloor, much as Charles Darwin had postulated a century earlier in his theory of coral reefs.

Following the war Hess established the famed Caribbean Research Project at Princeton, which would yield nearly three dozen Ph.D. dissertations in geology and geophysics by the time of his death in 1969. He also broadened his inquiry into vertical movements of the seafloor to include possible mechanisms for the origin and disappearance of submarine ridges. By 1960 the seemingly disparate threads of Hess's research on island arcs, serpentinized peridotite, and submarine ridges had come together in his crowning achievement, the hypothesis of seafloor spreading. In this model Hess proposed that new oceanic crust was created along the axis of a mid-ocean ridge and then transported like a conveyor belt above the convecting mantle, only to be destroyed by plunging beneath the island arcs at the periphery of the ocean basin. In 1963 Fred J. Vine and Drummond H. Matthews proposed that the curious seafloor magnetic stripes represented the imprint of Earth's reversing magnetic field on Hess's spreading crust, and this powerful corollary hypothesis received confirmation in 1966. Seafloor spreading also implied continental drift, and by the end of the decade the entire framework of the plate-tectonic revolution had fallen into place.

Toward the end of his life Hess was increasingly consumed by advisory and committee work on behalf of the National Academy of Sciences and other organizations. He was instrumental in proposing the ill-fated Mohole project in 1957, with the goal of drilling through the oceanic crust to sample the upper mantle. Initially managed by a small group of scientists and engineers under the auspices of the National Academy, the first phase of the project successfully demonstrated the feasibility of deep drilling from a floating rig. The second phase required an outside contractor who began planning for the "ultimate" hole, but after charges of political favoritism in the bidding process, frequent cost overruns, and bitter internal conflicts over management of the project, Congress killed the funding in 1966. Nevertheless, the early technological achievements of the Mohole project would facilitate the highly successful Deep Sea Drilling Project (1968–1983) and subsequent drilling programs (*see* DRILLING FOR SCIENTIFIC RESEARCH).

As the National Aeronautics and Space Administration (NASA) aimed for the Moon and beyond, the National Academy appointed Hess, the geologist who "thought big," as chairman of its influential Space Science Board. Hess, the pyroxene expert, also advised NASA on the handling of the anticipated lunar samples. A lifelong smoker, Harry Hess died on 25 August 1969 of a heart attack in Woods Hole, Massachusetts, while presiding over a meeting of the Space Science Board.

Bibliography

BUDDINGTON, A. F. "Memorial to Harry Hammond Hess, 1906–1969." *Geological Society of America Memorial* 1 (1973):18–26.

JAMES, H. L. "Harry Hammond Hess, May 24, 1906–August 25, 1969." *Biographical Memoirs of the National Academy of Sciences* 43 (1973):109–128.

MOORES, E. M., and F. J. VINE. "Alpine Serpentinites, Ultramafic Magmas, and Ocean-Basin Evolution: The Ideas of H. H. Hess." *Geological Society of America Bulletin* 100 (1988):1205–1212.

RUBEY, W. W. "Presentation of the 1966 Penrose Medal to Harry Hammond Hess." *Geological Society of America Proceedings Volume for 1966* (1968):83–86.

ALAN O. ALLWARDT

HIGHER LIFE FORMS, EARLIEST EVIDENCE OF

Orthodoxy states that life, and by implication intelligence, pervades the Galaxy, but to date our only certain knowledge of life is here on Earth. The earliest evidence for life is equivocal, but the consensus accepts that it had evolved by at least 3.8 billion years ago, or 3.8 Ga (giga annum). Interestingly, this age coincides with the amelioration of the major episode of meteorite bombardment that probably included impacts of such severity so as to evaporate entire oceans and sterilize the planet's surface. However life originated, the time available for it to arise may have been geologically short (*see* LIFE, ORIGIN OF).

The common possession of various organic compounds in all living organisms, such as RNA, suggests that life had a common ancestor. But there are some very deep divisions in life and in principle we need to know: (1) times of origination; (2) history of diversification in changing environments, and (3) the likelihood of direct inspection from the fossil record. The present concept of life has a basic division into the archaebacteria and eubacteria, comprising the prokaryotes and the eukaryotes (protistans, plants, fungi, and metazoans [animals]), respectively. This framework of understanding is largely based on molecular biology, which continues to reveal new surprises and qualifications of older views. First, although the simpler condition of prokaryotes suggests an earlier origination, the origin of eukaryotes is also very ancient. Second, the nearest living approximation to earliest life (or at least to the last common ancestor) may be the eocytes, a group of thermophilic archaebacteria that live in acidic hot springs. Third, archaebacteria and eukaryotes share a common ancestor. Fourth, during their history eukaryotes have acquired symbiotic prokaryotes that are now represented by organelles such as mitochondria (from purple bacteria) and chloroplasts (from coccoid cyanobacteria). Fifth and finally, eubacteria, archaebacteria, and eukaryotes have all undergone separate diversifications of which a major episode in the last group is particularly noteworthy, not least because it gave rise to the animals and so ourselves.

The basic structure of life needs to be placed in the context of Earth history (Figure 1). The oldest well-preserved sediments are from Isua in west

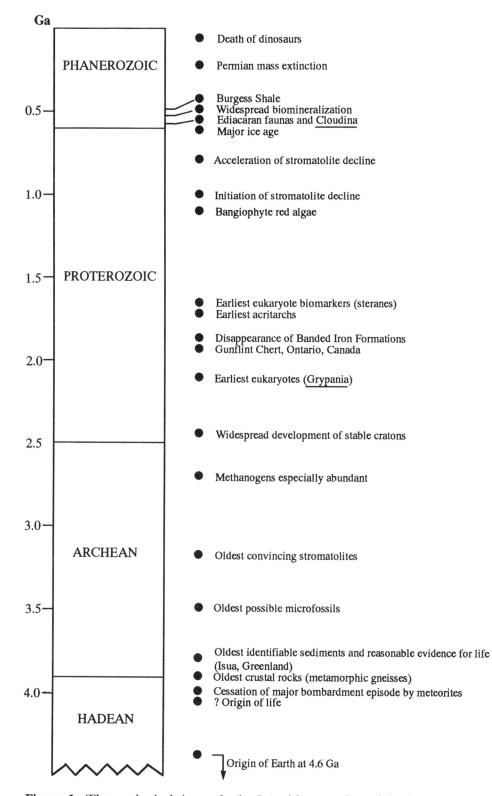

Figure 1. The geological timescale (in Ga) with a number of the key events during the Precambrian and early Phanerozoic. Note the dates of many of these events are tentative.

Greenland (ca. 3.8 Ga old). This marks the effective beginning of the Archaean (3.8–2.5 Ga); the preceding Hadean (4.6–3.8 Ga) is largely shrouded in obscurity owing to a minimal rock record. The Archaean was characterized by vigorous volcanism and very active tectonics, a reflection of the higher heat budget of Earth. The transition to the Proterozoic (2.5–0.54 Ga) was neither instantaneous nor synchronous, but over hundreds of millions of years there was a shift to a more stable tectonic regime and to larger continents rimmed by epicontinental seas. Oceans and atmosphere formed very early, and the single most important change was the introduction of oxygen by photosynthesis. How fast and to what levels oxygenation proceeded is very contentious, but appreciable quantities probably were present by 2 Ga.

Palaeontologists recognize three types of fossil: body, trace, and chemical. This classification can readily be applied to the evidence for Archaean and Proterozoic life. Most informative are actual cellular remains. The oldest are dated as 3.5 Ga and occur in cherts from the Warrawoona Group near a locality known as North Pole (northwest Australia) and the Onverwacht Group of the eastern Transvaal (South Africa). The fossils are filamentous, while cocooids are also reported from North Pole. But their interpretation remains very problematic and by no means all palaeontologists accept them as genuine. The subsequent record for Archaean cells is also very contentious, and some experts still argue that it is not until the 2 Ga Gunflint Chert (Lake Superior region) that absolutely unequivocal cells can be identified. Gunflint fossils include various bacteria, but this is exceptional because overall the Proterozoic record is biased toward the cyanobacteria, whose tough gelatinous sheaths confer a significant preservation potential.

Fossil cells are exceedingly rare, and the principal evidence for life in the Archaean and Proterozoic is based on stromatolites. These are a special type of trace fossil, formed by the binding and trapping of sediment of microbial mats, especially phototactic cyanobacteria. The characteristic laminations in a stromatolite reflect episodic burial of the mat and its upward migration to the sediment surface so that photosynthesis may resume. There are, however, many complexities, at least if recent stromatolites are a reliable guide. For example, the microbial ecology of mats is very complex with pronounced vertical stratification. Moreover, movement of the microbes is controlled by several factors, not only light intensity, and may be downwards as well as upwards.

Because stromatolites trap sediment and are aggregative they readily fossilize, although the component microbes very seldom fossilize. The oldest stromatolites are from North Pole and the Transvaal, but here, too, these claims are strongly contested. Nevertheless, unequivocal stromatolites occur as far back as 3.2 Ga and are abundant from about 2 Ga ago.

Chemical fossils hold considerable potential, although avoiding contamination and obtaining rocks of a low enough metamorphic grade to enable the organic molecules to survive remain problematic. Broadly, there are two approaches. Fractionation of carbon isotopes in favor of C^{12} over C^{13} (a ratio notated as δC^{13}) by photosynthesis, such as in the cyanobacteria, results in the carbon of organic matter being isotopically light. Subsequent metamorphism of the sediments can alter δC^{13}, but using correcting factors it has been suggested that carbon from 3.8 Ga sediments in Isua is probably organic in origin. Thereafter, measurement of δC^{13} in Archaean and Proterozoic sediments averages −27 per mil, the signature typical of photosynthetic activity. But there are some interesting anomalies. These include significant excursions (δC^{13} of up to −60 per mil) about 2.8–2.5 Ga ago. These may represent unusually high activities of methanogens, a group of archaebacteria that live in sediment and produce isotopically very light carbon. In addition, a significant shift in δC^{13} about 2 Ga ago may mark a major increase in productivity.

The other line of investigation is via biomarkers. These are molecules that are only produced by specific groups of organisms: for example, the former presence of protistan dinoflagellates can be inferred from dinosterols (molecules similar to cholesterol). The study of biomarkers is a fast-moving field because the biomarkers derived from many organisms are still imperfectly known, as are their diagenetic alterations after burial in the sediment.

The fossil record of early life remains patchy and frustrating. For many important microbial groups there is no direct evidence whatsoever, although data from molecular biology suggest ancient originations and these phylogenies also provide a framework into which future discoveries can be placed. The present consensus remains that the

Archaean fossil record is exclusively prokaryotic, notwithstanding some evidence of a very early eukaryotic origin (see above). Much of the subsequent fossil record of cells and stromatolites is reasonably attributed to cyanobacteria, with circumstantial evidence for methanogens (from very negative δC^{13}) and other bacteria (from biomarkers).

What of eukaryotes? The earliest evidence, compelling but not unequivocal, is of spiral fossils (*Grypania*) from 2.1 Ga old sediments in Michigan. *Grypania* is better known from younger strata (in Montana, India, and China), where it grew at least 60 cm in length, presumably anchored to the seafloor and coiling upward. Other presumed multicellular algae occur from about 1.9 Ga onward, while possible encystment bodies (known as acritarchs) occur from about 1.8 Ga. This evidence broadly accords with recovery of sterane biomarkers from 1.7 Ga old Australian sediments that almost certainly derive from eukaryotes.

Evidence from molecular biology suggests that many of the higher groups within the eukaryotes arose during a major diversification. Calibrating this event against geological time is difficult, but a sensible figure is about 1 Ga ago. Among the exquisitely preserved algae material from approximately 1 Ga old sediments in Somerset Island (northern Canada) are clearly red algae (bangiophytes), while equally spectacular remains were described in 1992 from somewhat younger Proterozoic strata exposed in the Yangtze Gorges, China.

It is possible animals originated at least 0.8 Ga ago, and some palaeontologists ascribe the coincident decline in stromatolite diversity to the grazing and burrowing activities of primitive worms. But the earliest acceptable evidence is from the Ediacaran faunas, globally distributed assemblages of soft-bodied animals of latest Proterozoic (Vendian) age (ca. 0.6–0.56 Ga). The most spectacular occurrences are from South Australia (Flinders Ranges), southeast Newfoundland (Mistaken Point), and north Russia (White Sea). Many Ediacaran fossils are difficult to interpret, but the consensus identifies a predominance of Cnidaria, a phylum that today includes the sea anemones and jellyfish. The Ediacaran fossils include pennatulaceans related to the living sea pens (anthozoans), floating hydrozoans and cubozoans, and less certainly true jellyfish (scyphozoans). More complex animals might be represented by segmented

forms, possibly stem arthropods and stem annelids, and early echinoderms. But all these assignments are controversial because a radical alternative has been promulgated by A. Seilacher (Vendozoa, 1989). He argues that the Ediacaran fossils are multicellular eukaryotes (termed by Seilacher the Vendobionta) with a distinctive body architecture of internal canals and tough integument. The latter feature is regarded as conferring a high fossilization potential and hence their widespread distribution. One problem in deciding whether Ediacaran organisms are either genuine vendobionts or early animals is that critical features of anatomy are not readily preserved or are open to more than one interpretation. Either way most Ediacaran species apparently disappeared before the beginning of the Cambrian (ca. 540 Ma), possibly in mass extinctions. Some forms, however, cross the Proterozoic-Cambrian boundary and have been found in the Burgess Shale. Study of these fossils appears to refute the vendobiont hypothesis.

The "Cambrian explosion" refers to the extraordinary irruption of metazoan life between 540–500 Ma ago. Evidence for this key evolutionary event is most obvious from the widespread appearance of animal skeletons and a coincident development of calcareous deposits in some algae. Calcified algae are known from the late Proterozoic, but they are rare. Only one calcified genus of animal occurs in Ediacaran strata: a distinctive worm tube known as *Cloudina*. In contrast, the Lower Cambrian teems with skeletal remains. Some are relatively familiar (e.g., mollusks and brachiopods), whereas many others are currently enigmatic. These latter include tubes, sclerites derived from complex cataphract arrays, grasping teeth, as well as a considerable number whose entire biology is highly problematic.

Also integral to the "Cambrian explosion" is a spectacular radiation of trace fossils, clear evidence that soft-bodied animals were diversifying. The burgeoning record also unequivocally shows a rapidly expanding repertoire of behaviors. Trace fossils present problems, however, because very rarely is the actual animal found in situ. To make matters worse, unrelated animals may make very similar traces. It is fortunate, therefore, that we have remarkable insights into soft-part preservation of animals via the famous Burgess Shale. This deposit is about 525 Ma old (Middle Cambrian) and is located in the Rocky Mountains of eastern British

Columbia, Canada. The fauna consists of a bewildering array of species. Many are arthropods, but also present are worms (priapulids, polychaete annelids), sponges, chordates, and a variety of seemingly more problematic species. Similar faunas are now being discovered elsewhere, of which the Lower Cambrian Chengjiang (south China) and Sirius Passet (north Greenland) assemblages are the most significant. Burgess Shale-type faunas have attracted particular attention for four reasons. First, the fossils show exquisite preservation, such as gut contents and musculature. Second, they give an unrivaled glimpse into Cambrian diversity and demonstrate not only the magnitude of the "Cambrian explosion" but the relative insignificance of skeletal taxa. Third, it is rich in evolutionary information, such as the earliest known chordate (*Pikaia*). Fourth, supposedly bizarre species have attracted attention because of claims they represent extinct phyla, which, had they survived, would have radically altered the subsequent history of life. Such a view is naive. It is more important to realize that there is a coherent scheme of metazoan phylogeny still to be worked out in which the presently enigmatic Burgess Shale species will play a crucial part.

The study of the early history of life is moving rapidly. Significant developments are inevitable. It is an exciting area in which to work, not least because it is interdisciplinary, involving groups as disparate as palaeontologists, molecular biologists, and geochemists. From Archaean archaebacteria to Cambrian metazoans vast changes occurred in complexity over enormous lengths of geological time. But even if the outline is clear, much remains to be discovered.

Bibliography

BROWN, G. C., C. J. HAWKESWORTH, and R. C. L. WILSON, eds. *Understanding the Earth: A New Synthesis.* Cambridge, Eng., 1992.

GLAESSNER, M. F. *The Dawn of Animal Life: A Biohistorical Approach.* Cambridge, Eng., 1984.

NISBET, E. G. *The Young Earth: An Introduction of Archaean Geology.* Boston, 1987.

SCHOPF, J. W., and C. KLEIN, eds. *The Proterozoic Biosphere. A Multidisciplinary Study.* Cambridge, Eng., 1992.

VENDOZOA. "Organismic Construction in the Proterozoic Biosphere." *Lethaia* 22 (1989):229–239.

SIMON CONWAY MORRIS

HOLMES, ARTHUR

Arthur Holmes, one of the outstanding geologists of the twentieth century, has aptly been called the genius of geological age-dating.

Holmes was born at Hebburn on Tyne in the northeast of England on 14 January 1890. His father was a cabinetmaker, his mother a schoolteacher. Holmes was taught locally at Gateshead High School, where he was introduced by his physics master to Lord Kelvin's work, especially his calculations on the age of the earth. He was awarded a scholarship in 1907 to study physics at Imperial College, London, under the instruction of R. J. Strutt (later Lord Rayleigh). He also attended lectures on geology given by W. W. Watts. Holmes gained his bachelor of science degree in 1909 and started postgraduate work at Imperial on the age-dating of rocks. In 1911 he undertook an expedition to Mozambique as a member of a mineral exploration party. This fieldwork in Africa lead him to his second great interest, petrology, but bad attacks of malaria and blackwater fever took a toll on his health. He was appointed demonstrator in geology at Imperial College in 1912, a post he held until 1920. (Poor health precluded his serving in the military during World War I.) Financial need forced him to seek a job with an oil company in Burma, but by 1924 he was back again in England as reader in geology at the University of Durham, where he was promoted to professor in 1925. In 1943 he successfully applied for the Regius Chair of Geology at the University of Edinburgh, attracted by the much larger and more modern department and by a higher salary. He retired because of ill health in 1956. Among the many honors bestowed on him were the premier awards of the geological societies of London (Wollaston Medal) and America (Penrose Medal), both in 1956, and the Vetlesen prize, the equivalent of the Nobel prize in geology, in 1964.

The discovery by Madame Curie of radioactivity and the detection by R. J. Strutt of radioactive elements in the rocks of the earth's crust in 1906 led to Holmes's lifelong interest in the geochronology of Earth's rocks. His first paper, published by the Royal Society of London while he was still only twenty-one, examined the association of lead with uranium and its application to the measurement of geological time. Holmes's first book, *The Age of the Earth* (1913), was published when he was twenty-three. In it Holmes concluded that Earth was approximately 1,600 million years or Ma old—an increase over Lord Kelvin's 1897 calculation of 20–40 Ma, which had been based on the assumption that the earth lost heat at a constant rate. With increasing refinement of laboratory techniques, Holmes concluded in 1964 that Earth was 4,550 Ma, a finding by C. Patterson two years earlier (*see* GEOLOGIC TIME, MEASUREMENT OF). He also calculated a timescale for the Phanerozoic geological periods.

For his distinguished petrographic work Holmes relied on little more than a petrological microscope. His books *The Nomenclature of Petrology* and *Petrographic Methods and Calculations* were published in 1920 and 1921, respectively. He was especially interested in alkaline rocks (igneous rocks with higher than average concentrations of the alkali metals, such as sodium and potassium) and Precambrian rocks, and he became increasingly attracted to the theory of metasomatism in which emanations from the mantle were believed to change preexisting rocks. Spurred on by his second wife, Doris Reynolds, a distinguished petrologist herself, he supported granitization, a theory of rock formation zealously overpursued by its adherents in the 1940s and 1950s.

Holmes was a strong supporter of continental drift when it was an unfashionable hypothesis, and in a paper as early as 1928 he considered the possibility that convection currents within the mantle might be the means of moving continents. This paper, unobtrusively published in the *Transactions of the Geological Society of Glasgow*, helped sow the seeds of the later theory of plate tectonics. Above all, Holmes is remembered for his brilliant student textbook, *The Principles of Physical Geology* (1944). Translated into many languages and in its fourth edition in 1994, it revealed its author's breadth of vision and a beauty of English prose unmatched by any other geologist of the twentieth century.

Arthur Holmes was a kind but shy man who avoided committees, academic politics, and unnecessary traveling. In his later years in Edinburgh he walked to his department—he lived less than a quarter of a mile away—gave his lectures lucidly and with a calm authority, did what little administration was necessary, and returned home at lunch time. For the rest of the day he read, wrote, played the piano, or listened to music. He much enjoyed investing in the stock market and was always willing to pass on good advice—geological and other-

wise—to his junior staff. In those days the geological faculty was small and senior students few. Holmes's major influence on the geological world came from his papers and books, and his worldwide correspondence. He wrote mostly as sole author, a form of communication now rarely possible in the greatly expanded literature of the earth sciences. Holmes died in London on 20 September 1965 at the age of seventy-five.

Bibliography

DUNHAM, K. C. "Authur Holmes, 1890–1965." *Biographical Memoirs of the Fellows of the Royal Society of London* 12 (1966):290–310.

HOLMES, A. *Holmes' Physical Geology,* 4th ed., ed. P. McL. D. Duff. London, 1992.

GORDON Y. CRAIG

HOT SPOTS

See Mantle Convection and Plumes

HUBBERT, M. KING

Marion King Hubbert often spoke of having the good fortune of being able to contribute productively (many of his colleagues might say dominantly) to diverse scientific fields of research—exploration seismology, tectonics, fluid flow in rocks, and oil- and gas-resource assessment, for which he is best known. Hubbert also pointed to the fact that he had worked in every major type of research institution—academia at Columbia University, industrial research at the Shell Oil Company, and governmental research at the U.S. Geological Survey (USGS). For his contributions to science and society he won many awards, prizes, memberships, distinguished lectureships, and medals. In 1981 Columbia University awarded him the Vetlesen prize, which is the highest honor for contributions in the earth sciences and is often equated with the

Nobel prize by the scientific community. Hubbert also was a fellow in the American Academy of Arts and Sciences and a recipient of the Arthur L. Day and the Penrose medals of the Geological Society of America.

Hubbert, who was born on 5 October 1903 in San Saba County, Texas, grew up in a rough-and-tumble fashion, working on a farm and attending school when he could. He loved to tell stories about his early life and his struggle for a college education; many of these have been widely cited as a tribute to the grit and determination of someone who would become one of the true giants of twentieth-century geology.

In his late teens, Hubbert realized that he needed far more academic stimulation than was available at the small junior college where he had been a student for two years. Following the advice of the college faculty, he applied to the University of Chicago and was accepted. In the era before college loans, this penniless youth set out with little more than the clothes on his back and headed north for the beginning of the academic year. Hubbert began his odyssey as a field worker following the 1924 summer wheat harvest across Texas and Oklahoma. In Kansas, down to his last dollar, he joined a railroad work gang laying heavy steel track.

He reached Chicago in time to matriculate for the fall semester. As a student, he continued to work to support himself with jobs at the post office, the phone company, and a restaurant. He was always strong-willed and sought only the broadest possible academic education. It was only after the dean ordered him to declare a major that Hubbert chose a joint major in geology and physics. By this time, he had almost certainly developed the iconoclastic manner that would serve him so well.

Hubbert took a break from the University of Chicago in 1926–1927 and joined a geophysical crew that was doing early work on calibrating reflection seismology. He always remembered this work fondly—seeing inside the earth by watching sound waves returning to the surface was exciting in 1927.

Hubbert was surprised when he was awarded his doctorate in 1937 for his fundamental work, *The Theory of Scale Models as Applied to the Study of Geologic Structures,* because he had gone to the University of Chicago only to learn and not necessarily to gain advanced academic degrees. He often said that he had chosen to work in this area because he

had to know why the earth could be as strong as steel and as soft as putty at the same time.

From 1937 until his retirement in 1976, Hubbert was a productive and controversial research scientist. He jolted the scientific community during the 1940s with his work *Theory of Ground-Water Motion.* With the courage of convictions that were based upon his own theoretical and experimental work, he battled with his many critics and, in the process, made much of what they had written in the scientific literature obsolete. A decade later, Hubbert would publish one of his most famous papers in which he adapted this work to the search for petroleum. In it, he explained the counterintuitive observation that oil-water water contact in an oil field need not be horizontal. This was explained by variations in the hydraulic head across the bottom of the oil fields.

Hubbert left Columbia University in 1941 after serving as an assistant professor in geophysics, having tried in vain for ten years to introduce more mathematics and physics into the geology curriculum. He spent the next two years at the Board of Economic Warfare, where he analyzed the supply of mineral resources. After joining the Shell Oil Company in 1943, he worked with D. G. Willis to determine how fluids fracture rocks. This research led to the development of the applied field of hydraulic fracturing, which has been used to discover large volumes of petroleum that may otherwise not be available from reservoir rocks. At this same time, he also worked with WILLIAM W. RUBEY, a USGS geologist, on an explanation of thrust fault mechanics.

When Hubbert retired from Shell in 1963, he joined the U.S. Geological Survey. However, retirement meant little to him other than having a new place to continue his research. At the USGS, he accelerated his work on petroleum-resource assessment, which he had pursued to a modest degree during the late 1940s and the 1950s. Later in his life, he would often refer to this work as his most important research; it was also the work that brought him notoriety. Hubbert, who always gave proper credit, picked up the kernel of the idea that he would later develop into his theory of petroleum-resource exhaustion from D. F. Hewett's 1929 paper *Cycles in Metal Production,* which, according to Hubbert, was one of the best ever written at the USGS.

Hubbert used a bit of differential calculus to formalize the idea that Hewett had developed for metal mining districts. In a 1948 speech before the American Association for the Advancement of Science, Hubbert argued that by using these ideas, he could predict the year in which petroleum production would peak in the United States. The petroleum industry never formally reacted to the speech when it was published a year later. During the early 1950s, Hubbert collected additional data and, in 1956, predicted that U.S. crude production would peak in ten to fifteen years (1966 to 1971). This forecast and subsequent, more refined forecasts published between 1962 and 1967 were later borne out. The actual date of peak crude petroleum production was in 1971.

With the publication of the 1956 prediction, Hubbert was immersed in a controversy that did not subside until well into the 1970s. The overall reaction to Hubbert's prediction was negative. Rather than taking note of Hubbert's prediction that the United States was running out of oil, his critics worked hard to reverse his conclusion. The tide did not really turn until production began to fall. With the Arab oil embargo in 1973, Hubbert, who had been reviled as a doomsayer, was proven to have been a prophet.

It should be noted that although history has proved Hubbert's prediction correct, the methods he used have been heavily scrutinized by the scientific community. Some flaws have been pointed out—his data-analysis techniques were a bit crude and were (and are) not acceptable to the data-analysis community. However, to quote David Root, who was Hubbert's last coworker, "To appreciate Hubbert's contribution, you must look at when he did his work. He had it all figured out in the 1950s when he saw the first derivative change in the production curve. Just remember that date because that is what is important about Hubbert."

Marion King Hubbert died in Bethesda, Maryland, on 11 October 1989.

Bibliography

HUBBERT, M. KING. "Theory of Scale Models as Applied to the Study of Geologic Structures." *Geological Society of America Bulletin* 48 (1937):1459–1520.
———. "The Theory of Ground-Water Motion." *Journal of Geology* 48 (1940):785–944.
———. "Degree of Advancement of Petroleum Exploration in the United States." *American Association of Petroleum Geologists Bulletin* 51 (1967):2207–2227.

LARRY J. DREW

HUBBLE SPACE TELESCOPE

See Telescopes

HUBBLE, EDWIN

Edwin Powell Hubble was one of the leading figures in astronomy in the twentieth century. His work in the field of observational cosmology was unrivaled and led to a new understanding of our place in the universe. His major accomplishments were his proof that there were other galaxies outside our Milky Way and that their motion away from us was consonant with the concept of an expanding universe.

Hubble was born in Marshfield, Missouri, on 20 November 1889. In 1898 he and his family moved to Evanston, Illinois, and by 1900 they had settled in Wheaton, Illinois, where Edwin demonstrated excellence in both academics and athletics. At the age of sixteen he entered the University of Chicago and received his bachelor of science (B.S.) degree in 1910. Upon graduation Hubble was awarded a Rhodes Scholarship to study at Oxford University in England, where he matriculated at Queens College and studied law while also participating in track, water polo, and rowing.

Hubble left Oxford in 1913 with second-class honors in jurisprudence and returned to his family, who by then resided in Louisville, Kentucky. He had originally planned to practice law in Kentucky but ended up teaching high school across the Ohio River from Louisville in New Albany, Indiana. After one year of teaching, Hubble decided to continue his education in the sciences and in 1914 returned to the University of Chicago as a graduate student in astronomy.

As an astronomy graduate student at Chicago, Hubble did his observational work at the university's Yerkes Observatory in Williams Bay, Wisconsin. At the observatory, Hubble spent most of his independent research time with the 60 cm reflecting telescope. With it, Hubble began studying the hazy patches of light in the sky known as nebulae, and this research grew into his Ph.D. dissertation, "Photographic Investigations of Faint Nebulae."

Although his dissertation was not an important work, it demonstrated Hubble's abilities to frame critical questions and his skill in getting the most information out of the instruments available to him.

Hubble proved to be an excellent student at Yerkes and he soon came to the attention of George Ellery Hale, director of the world-class Mount Wilson Observatory near Pasadena, California. The largest telescope in the world at that time, the 25 m Hooker reflecting telescope was under construction at Mount Wilson and Hale was looking for bright, young astronomers to use it when it was completed. Hale offered Hubble a job at Mount Wilson upon completion of his Ph.D., but just before he received his degree, the United States entered World War I. Hubble chose to postpone his appointment to Mount Wilson, and he joined the U.S. Army on May 15, three days after passing his Ph.D. examination. Hubble entered the infantry but never made it to the front by the time the war ended. Hubble spent some time after the war in England, where he had the opportunity to study and meet with a number of British astronomers. He returned to the United States and was discharged from the army with the rank of major in September 1919. Hubble was proud of his military service and many of his friends and colleagues always referred to him as Major Hubble.

Immediately following his discharge Hubble went to the Mount Wilson Observatory, where a job was still waiting for him. The 25 m telescope was just entering regular service and Hubble began his research program on the nature of the nebulae. Of particular interest were a class of spiral-shaped nebulae that many astronomers believed might be distant galaxies of millions of stars, similar to the Milky Way. Hubble concentrated on the larger, apparently nearby spiral nebulae for his research. The best known of these was the Andromeda nebula, or M31. Studying M31, Hubble discovered a number of stars of varying brightness in the nebula. After investigating the rise and fall of luminosity in these stars, he concluded that they were a special class of stars known as Cepheid variables. The period of the light variation of a Cepheid was known to relate directly to its luminosity. Therefore, by calculating the Cepheid's actual luminosity and comparing this to its apparent brightness, Hubble determined that M31 was nearly one million light years distant (the modern value is over twice that amount). Hubble's work on the distance of nearby spiral nebulae proved decisive in showing them to be galaxies of stars.

Hubble's other significant discovery was in showing that we live in an expanding universe. While a few astronomers had obtained marginal evidence that the more distant galaxies were receding from us faster than nearby galaxies, their arguments were not convincing. Hubble's access to the 25 m telescope gave him the opportunity to study galaxies in greater detail. With his colleague Milton Lasell Humason, Hubble measured the speed of recession (the "radial velocities") of many distant galaxies. They were able to determine these velocities by measuring the amount of displacement of certain lines in the galaxies' spectra (the "redshift"). Using his calculated values for the distances to the nearby galaxies, Hubble extrapolated from these to determine how far away the more distant galaxies were. In 1929, Hubble and Humason published their initial results showing that there was a relationship between the velocity and distance of galaxies. This relationship, now known as Hubble's Law, is expressed in modern notation as:

$$v = Hd$$

where H is referred to as the Hubble constant. Hubble's law implies that the universe is expanding: velocities of galaxies increase as an observer looks progressively farther outward from any point in the universe.

The value of the Hubble constant—a subject of great continuing interest—is important as it represents the expansion rate of the universe. Hubble determined its value to be 500 km per second per megaparsec (megaparsec being 1,000,000 parsecs, or approximately 3,260,000 light-years), but astronomers today believe it to be approximately 50–60 km per second per megaparsec (although some have argued for a value nearer to 100). The reciprocal of the Hubble constant in turn yields the age of the universe. A value for the Hubble constant of fifty-five would indicate that the universe is 18 billion years (Ga) old. However, if the Hubble constant is nearer to one hundred, as some have claimed, the age of the universe would be considerably less. All of this, of course, assumes that the universe has been expanding at a constant rate, which is not at all certain. In any event, Hubble's Law is very important in setting the time constraints within which earth scientists can place their theories regarding the age and evolution of Earth and the solar system.

Hubble continued to work in the field of observational cosmology by refining the velocity-distance relation and extending it to more distant galaxies. In 1949, the 50 m Hale telescope on Palomar Mount, the largest telescope in the world at that time, entered service, and Hubble could use it to study even more distant galaxies. A heart attack that same year prevented him from pursuing his research right away, and he only had a few years of work with the 50 m telescope before he died 28 September 1953. By the time of his death, Hubble was recognized as the leading figure in observational cosmology. The general public regarded him as a major celebrity, and he became a media icon and spokesperson for science. Hubble regarded himself as a renaissance man and had a great interest in a wide variety of subjects and enjoyed socializing with great artists, humanists, and philosophers as well as scientists.

Bibliography

BERENDZEN, R.; R. HART; and D. SEELEY. *Man Discovers the Galaxies*. New York, 1976.

HETHERINGTON, N. S. "Hubble's Cosmology." *American Scientist* 78 (1990):142–151.

NORTH, J. D. *The Measure of the Universe: A History of Modern Cosmology*. Oxford, Eng., 1965. Repr. New York, 1990.

OSTERBROCK, D. E.; J. A. GWINN; and R. S. BRASHEAR. "Edwin Hubble and the Expanding Universe." *Scientific American* (1993): 84–89.

SHAROV, A. S., and I. D. NOVIKOV. *Edwin Hubble, the Discoverer of the Big Bang Universe*. Cambridge, Eng., 1993.

SMITH, R. W. *The Expanding Universe: Astronomy's "Great Debate," 1900–1931*. Cambridge, Eng., 1982.

RONALD S. BRASHEAR

HURRICANES

Hurricanes belong to a class of weather systems called tropical cyclones. Each year approximately eighty tropical cyclones with maximum sustained surface winds of at least 20 meters per second (m s^{-1} or sometimes abbreviated as m/s) form over the tropical oceans (Gray, 1979). The impacts of intense tropical cyclones can be catastrophic. The surge of ocean water accompanying a 1970 tropical cyclone killed at least 300,000 persons in Bangladesh, and the strong winds of Hurricane Andrew in 1992 destroyed more than $20 billion of property in south Florida. Fortunately, few tropical cyclones are so destructive, and rainfall in weaker cyclones often benefits drought-stricken areas.

Forecasters in the Western Hemisphere refer to tropical cyclones with maximum sustained surface winds from 18–33 m s^{-1} as tropical storms and assign names to them. They become hurricanes (typhoons in the western North Pacific Ocean) when the maximum sustained surface winds are greater than 33 m s^{-1}.

Formation

Tropical cyclones develop over the tropical oceans (Figure 1), except south of the equator in the Atlantic and eastern Pacific Oceans where meteorological and oceanic conditions do not favor their development. They form at latitudes of 5° or more from the equator and usually move westward and poleward.

In the western North Pacific, tropical cyclones are observed throughout the year. In the other oceans, they occur during the summer and early fall. The official Atlantic hurricane season begins on 1 June and ends on 30 November, but most of the tropical cyclone activity occurs in August and September. Although the average annual numbers of all named Atlantic tropical cyclones and those reaching hurricane intensity from 1950 to 1994 were 9 and 5.5, respectively, there is considerable interannual variability. Landsea and Gray (1992) identified rainfall over western Africa as one factor in these interannual variations.

Tropical cyclones form in areas with sea surface temperatures warmer than 26°C, weak or moderate large-scale winds having little vertical shear, and high humidity in the middle troposphere. A preexisting weather disturbance with an organized area of deep convective clouds serves as the seedling for development. In the lower and middle troposphere above the tropical Atlantic and eastern Pacific Oceans, the seedling is usually an easterly wave that develops near 10°N over Africa. Easterly waves have east–west wavelengths of 2,500 km, periods of about 2.5 days, and maximum winds of

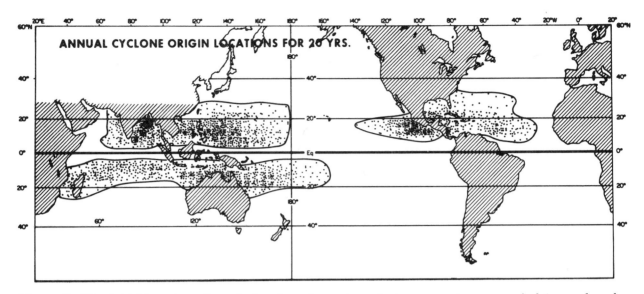

Figure 1. The initial location of tropical cyclone formations during a twenty-year period (reproduced from Gray, 1979, with permission from the Royal Meteorological Society).

5 m s^{-1}. A small fraction of these waves become named Atlantic storms as they move westward.

Storm Structure and Hazards

A hurricane is a nearly circular vortex that extends vertically throughout the troposphere and horizontally about 1,000 km from the storm center. The winds rotate counterclockwise to the north and clockwise to the south of the equator. Maximum sustained surface winds approach 80 m s^{-1} in the strongest hurricanes.

As a tropical cyclone strengthens from a tropical storm to a weak hurricane, a rain-free area, called the eye, forms at the center of the wind circulation. The eye is surrounded by a nearly circular ring, or eyewall, of deep cumulus convection with heavy rainfall (Figure 2) and the strongest winds. The eyewall is about 10 km across and may be located 10–50 km radially outward from the center of the eye. Outside the eyewall, the organization of the precipitation is in rainbands that spiral in toward the eye. The rainbands have a few isolated areas of intense precipitation and extensive areas of lighter, more uniform rain.

The sea-level pressure in the eye of a very strong hurricane is as much as 10 percent lower than that normally observed in the tropics. The low pressure results from warm temperatures in the eye that may be up to 20°C above normal in the middle and upper troposphere. The warm air results from the release of latent heat in the deep eyewall convection and descent in the eye.

Tropical cyclones are mainly coastal hazards and dissipate rapidly after landfall. The surge of ocean water above the normal tide in strong storms may reach heights greater than 6 m at the coastline. The surge is responsible for the largest loss of life and destruction of property along the coast. The damage from heavy rains and strong winds, including a small contribution from tornadoes, is at its greatest near the coast and decreases to zero about 250 km inland. Moisture-laden air in decaying tropical cyclones over land, however, can produce extensive flooding in mountainous terrain, even greater than 1,000 km from landfall.

Atlantic Hurricane Hugo

Hugo is a typical example of a late summer hurricane that developed from an easterly wave. On 10 September 1989, forecasters categorized the area of disturbed weather as a depression (Case and Mayfield 1990), a closed low-level cyclonic vortex with maximum surface winds less than 18 m s^{-1}. As it moved westward (Figure 3), Hugo became a

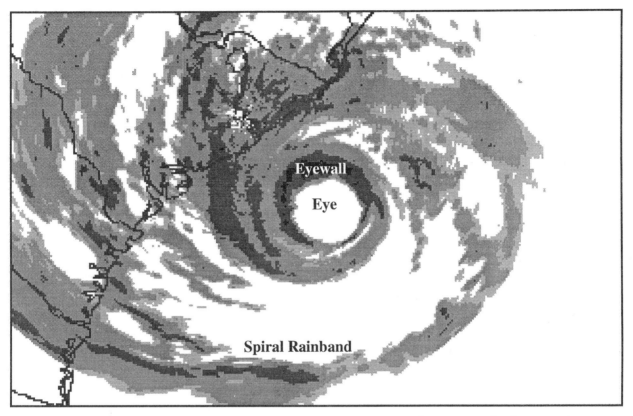

Figure 2. The radar structure of Hurricane Hugo late on 22 September 1989, a few hours before Hugo's eyewall struck Charleston, South Carolina. The rainfall rates increase with darker gray shades. White represents little or no rainfall and black indicates areas with rain rates greater than 50 mm h^{-1}. The labels show the location of the eye, eyewall, and a spiral rainband.

tropical storm and hurricane on the 11th and 13th of September, respectively. Tropical cyclones with similar tracks are referred to as Cape Verde hurricanes after the eastern Atlantic islands.

Hugo was a major hurricane (maximum surface winds greater than 50 m s^{-1}) and had surface winds exceeding 70 m s^{-1} just before it crossed through the northeastern part of the Caribbean Sea. The strong winds devastated several Caribbean islands. Hugo weakened slightly as it passed through the islands and then headed rapidly northwestward. Reintensifying just before landfall in Charleston, South Carolina, on 22 September, Hugo had surface winds that reached 62 m s^{-1} and a maximum storm surge of 6 m at the coastline to the right (northeast) of the storm track where the counterclockwise winds were onshore. Wind damage in South and North Carolina exceeded $5 billion.

Forecasting

A tropical cyclone moves in the general direction of the broad-scale wind patterns in which it is embedded. Forecasters predict the motion of tropical cyclones with the aid of computer guidance models and their subjective analyses of surface, aircraft, and satellite observations. The average error for a twenty-four hour track forecast is 185 km in the Atlantic. Forecasters issue hurricane warnings about twenty-four hours before landfall. The goal is to provide coastal residents twelve hours of daylight in which to complete preparations.

Operational forecasts of intensity change are not very accurate. Researchers are identifying the dominant physical processes responsible for intensity change and developing experimental models. The observing network over the tropical oceans is sparse, however, and the development of accurate

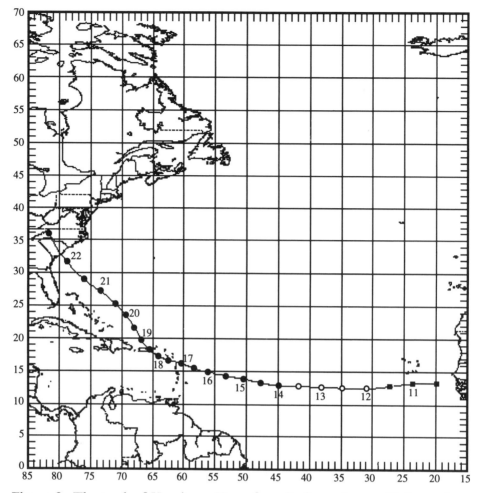

Figure 3. The track of Hurricane Hugo from its formation on 10 September to its landfall on 22 September 1989. The darkened squares, open circles, and darkened circles are Hugo's positions at twelve-hour intervals and indicate those parts of the track when Hugo was a depression, storm, and hurricane, respectively. The numbers are the day of the month and are next to the storm's 0000 universal time positions.

intensity change models based upon observations has been slow. The lack of oceanic observations also limits forecasters' capabilities to predict hurricane-related rainfall.

Conclusion

During the twentieth century, more accurate forecasts and warnings, better communications, and increased public awareness reduced the loss of life associated with Atlantic hurricanes. From 1970 to 1994, the improvement in forecast tracks averaged about 1 percent per year, while the U.S. population in hurricane-prone areas increased 3–4 percent per year. A few communities with limited escape routes now require much more time for preparation and evacuation than can be accurately provided by present warnings. Government agencies, however, are constantly improving the observations, analyses, and forecasts, optimizing procedures for evacuations, and accelerating post-hurricane recovery efforts. These programs should minimize the loss of life during future hurricane landfalls.

Bibliography

CASE, B., and M. MAYFIELD. "Atlantic Hurricane Season of 1989." *Monthly Weather Review* 118 (1990):1165–1177.

GRAY, W. M. "Hurricanes: Their Formation, Structure and Likely Role in the Tropical Circulation." In *Meteorology Over the Tropical Oceans*, ed. D. B. Shaw. Bracknell, Eng., 1979.

LANDSEA, C. W., and W. M. GRAY. "The Strong Association Between Western Sahelian Monsoon Rainfall and Intense Atlantic Hurricanes." *Journal of Climate* 5 (1992):435–453.

ROBERT W. BURPEE

HUTTON, JAMES

James Hutton was one of the first geologists to establish a general theory of the earth. His uniformitarian theory, which stressed the uniformity of geological change, was later accepted by much of the geological community and has been central to our understanding of the earth's processes.

James Hutton was born in Edinburgh, Scotland, on 3 June 1726. He was the only child of a wealthy merchant family. When he was three his father died and left him enough money that he was from then on independently wealthy. Freed from having to make a living, he spent most of his life pursuing a number of interests. He did his undergraduate work at Edinburgh University, where he developed a strong interest in chemistry. Later he studied medicine in Edinburgh and Paris and received his M.D. degree from Leyden University in 1749. Hutton did not intend to practice medicine. At the time medicine was one of the few advanced subjects in universities that was studied by those interested in knowledge of natural science.

For twenty years after he graduated in medicine, Hutton spent his time in developing a farm he inherited from his father. He was interested in the new farming methods that were revolutionizing agriculture in the eighteenth century. Apparently he was successful, and in 1768 he rented his well-run farm and settled in Edinburgh for the rest of his life.

Hutton belonged to a group of Scottish Enlightenment intellectuals who lived in Edinburgh. Among his friends were Joseph Black, Adam Smith, and Lord Monboddo. They were individuals who were interested in many subjects, especially history, technology, philosophy, economics, and natural science. Hutton spent the rest of his life pursuing experimental chemistry, geology, agriculture, meteorology, and philosophy. He was an active member of the Royal Society of Edinburgh and corresponded with many English and European scientists. He died in Edinburgh on 26 March 1797.

Hutton is remembered today primarily for his theory of the earth. He first presented his theory in 1785, and after further research he published his *Theory of the Earth: With Proofs and Illustrations* in 1795. Central to Hutton's geology was a set of highly controversial ideas, which were not accepted for many years.

Among the ideas that Hutton stressed was the great antiquity of the earth. Although some earlier French naturalists, among them Georges-Louis Leclerc, Comte de Buffon, had ventured the opinion that Earth was millions of years old, in Britain geologists still stressed the biblical creation story for geological history. This biblical account depicted Earth as several thousand years old and explained much of the diversity on the earth's surface by reference to Noah's flood. Hutton fundamentally broke with the biblical tradition by stating that knowledge of geology had to come from an investigation of nature and not from sacred documents. His view was that a study of Earth showed that it was so ancient that we cannot even detect its beginning.

Biblical geology had described a young earth whose appearance could be understood in terms of a set of unique historical stages. Hutton suggested a different approach to the understanding of the earth. Instead of describing discrete stages, Hutton emphasized the uniformity and continuity of the processes that have shaped the planet. These processes he thought were cyclic and resulted in the birth and decay of surface phenomena. Historians often label his approach to geology as "uniformitarianism" to stress the constant and uniform nature of geological change. Hutton also followed the "actualism" method of using only present-day processes to explain past events, a concept still widely used by geologists and embodied in the phrase "the present is the key to the past."

Hutton was impressed by the constant sources

471

of weathering and erosion that worked on Earth. He also believed that most of the rocks that currently make up the crust of Earth originally had been deposited on the bottom of the oceans. What we see on the surface of Earth, therefore, is not the original appearance of the planet. Instead, what today is the surface of Earth was formed under the ocean. At the time, a different part of the planet was above water and subject to forces of erosion and weathering. These processes produced large quantities of sediment. But what consolidated the products of erosion that were deposited in the oceans along with the remains of organic life that settled to the oceans' bottom? Hutton believed that subterranean heat was the principal factor.

Hutton claimed that the enormous heat that existed inside the earth was capable of fusing the sediment that was deposited on the ocean floor. Not only did this great heat alter and create rock, but it also was responsible for elevating the ocean bottom to produce the vast land masses that we see.

The process of sedimentation and elevation was not confined to small regions of the globe, but according to Hutton was a general process that accounts for all the current land masses and the relics of former ones. The earth was a dynamic system, and for that reason we did not see its beginning, nor was it likely that we could predict its end.

Hutton's ideas did not have an immediate influence on the geological thoughts of his day. Some writers have claimed that his difficult writing style discouraged people from reading his work and that even those who did had trouble understanding it. More significant, however, was the fact that his ideas contradicted many accepted notions. Many of his contemporaries held that his belief in the great antiquity of the earth was blasphemous. Many of the geologists of the time also doubted that heat could be such a major factor. And the cyclical picture of Earth's history did not accord well with the more traditional view that the surface of the earth could be explained by reference to a set of catastrophic events (like Noah's flood).

In spite of the immediate lack of attention, Hutton's geology came to be widely known and later accepted. His writings were popularized by John Playfair's *Illustrations of the Huttonian Theory of the Earth* (1802), and later, Charles Lyell's *Principles of Geology* (1830–1833) extended Hutton's ideas and gained widespread acceptance for them (*see* LYELL, CHARLES, and PLAYFAIR, JOHN). Hutton is now widely recognized as one of the pioneers of modern geology.

Bibliography

BAILEY, E. B. *James Hutton: The Founder of Modern Geology*. Amsterdam, 1967.

DOTT, R. H. "James Hutton and the Concept of a Dynamic Earth." In *Toward a History of Geology*, ed. C. J. Schneer. Cambridge, 1969.

PLAYFAIR, J. *Illustrations of the Huttonian Theory of the Earth*. 1802. Reprint New York, 1964.

PAUL LAWRENCE FARBER

HYDROTHERMAL ALTERATION AND HYDROTHERMAL MINERAL DEPOSITS

Virtually all rocks in the earth's crust contain small amounts of useful metals and minerals, but normally at concentrations far too low to be of practical interest. Hydrothermal mineral deposits are above-average concentrations of economically valuable materials within the earth's crust, formed by mineral precipitation from heated natural waters. Most commonly, the deposits contain sulfide minerals, such as chalcopyrite ($CuFeS_2$), sphalerite (ZnS), and galena (PbS), which can be processed to extract metals such as copper, lead, and zinc, or native metals such as gold (Au), silver (Ag), and mercury (Hg). Although the metal or mineral of interest is highly concentrated relative to average crustal rocks, it is still generally only a small percentage of the total rock mass, and extensive processing is required to extract it (Table 1).

Origins of Hydrothermal Fluids

Because hydrothermal mineral deposits are defined by their formation from hot water, the origins of this water have been a central focus of scientific study. For many decades, particularly in the United States, many scientists believed that virtu-

Table 1. Average Crustal Abundance of Elements versus Economic Concentrations

Element	Average Crustal Concentration (wt %)	Minimum Ore Grade (wt %)	Concentration Factor
Zn	0.0070	3.0	429
Cu	0.0055	1.0	182
Pb	0.0013	3.0	2,308
U	0.00018	0.5	2,778
Sn	0.0002	0.6	3,000
W	0.00012	1.4	11,667
Mo	0.00015	0.3	2,000
Hg	0.000008	0.6	75,000
Ag	0.000007	0.02	2,857
Au	0.0000004	0.001	2,500

Source: Data compiled from several sources including Skinner (1986) and Guilbert and Park (1986).

ally all hydrothermal mineral deposits were formed from magmatic fluids. However, in the past twenty years studies of the isotopic composition of mineral deposits have permitted direct analysis of the sources of hydrothermal fluids, and indicated a much greater role for meteoric and formation waters than previously recognized. Four principal types of hydrothermal fluids have been recognized (Figure 1).

Meteoric hydrothermal fluids are produced when groundwater, surface water, or seawater is heated by interaction with hot igneous rocks, or by the earth's normal geothermal gradient. Formation waters are produced when groundwater combines with aqueous fluids derived from the compaction and diagenesis of sediments within a sedimentary basin. Metamorphic hydrothermal fluids are derived by dehydration or decarbonation of minerals at high temperature and pressure deep within the earth. Magmatic hydrothermal fluids are composed of water and dissolved gases that escape from crystallizing magmas.

Sources of Metals

Two principal mechanisms are thought to be responsible for the presence of metals in hydrothermal fluids: concentration of incompatible elements in magmatic hydrothermal fluids, and dissolution of trace metals from rocks or sediments.

Concentration of Incompatible Elements in Magmatic Hydrothermal Fluids. Magmas are heterogeneous mixtures of molten rock, crystals, and dissolved water and gases. As a magma crystallizes, most of its chemical components are gradually incorporated into the structures of the mineral crystals being formed. However, some elements do not readily fit into the structures of common rock-forming minerals; these "incompatible elements"

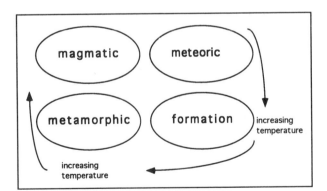

Figure 1. Meteoric, formation, metamorphic, and magmatic water represent the four idealized end-member sources of hydrothermal fluids. In nature, meteoric waters become formation waters through interaction with rocks; formation waters grade into metamorphic waters with increasing temperature and pressure.

include many economically important metals, such as copper, lead, zinc, gold, silver, tin, tungsten, and molybdenum. In addition, volatile components of the magma, such as carbon dioxide (CO_2), hydrogen sulfide (H_2S), hydrochloric (HCl) and hydrofluoric acid (HF) gas, are not incorporated into the solidifying rock mass. If dissolved water separates from a magma due to pressure release during the final stages of crystallization, incompatible elements and magmatic volatiles will be concentrated into the exsolving fluid phase. The result is a metal-rich hydrothermal fluid.

Dissolution of Trace Metals from Rocks or Sediments. When meteoric, formation, or metamorphic waters move through rocks or sediments within Earth's crust, trace metals may be gradually dissolved. Although the concentrations of trace metals in average crustal rocks are very low, the volumes of rock available within Earth's crust are more than adequate to account for observed mineral deposits. However, the solubilities of "raw," or uncomplexed, ions of copper, lead, zinc, and other metals are extremely low. For an aqueous fluid to carry metals in quantity, the metals must be complexed with anions, or anionic groups, known as complexing ligands. The most important anionic complexes in hydrothermal fluids are thought to be the chlorides.

Where does the chlorine come from? In magmatic hydrothermal fluids, chlorine is an incompatible element that concentrates in the residual fluid along with the metals of interest. In seawater, chlorine is a natural part of oceanic salinity as a chloride ion. In freshwater, formation water, and metamorphic water, chlorine must be dissolved out of minerals. For example, in the common mica muscovite [$KAl_3Si_3O_{10}(OH)_2$], chlorine ions (Cl^-) may substitute for the hydroxyl group (OH^-) in the mineral structure. However, some metals do not readily form highly soluble chloride complexes and may be carried by other complexing ligands. The most important of these is gold, which forms a stable bisulfide complex [$Au(HS)_2^-$] in the presence of sulfur. Sulfur may be present in sedimentary environments in association with organic matter, in hot springs with sulfur fumaroles, and in metamorphic rocks where sulfur is released during the metamorphism of organic matter. In these environments, fluids may be generated that selectively dissolve and transport gold.

Mechanisms of Transport and Deposition

The solubilities of metal chloride and bisulfide complexes are highly temperature-dependent and generally increase with increasing temperature. Therefore, the hotter the fluid, the more likely it is to carry metals in solution. But how do hydrothermal fluids get hot? In magmatic systems, fluids are generated from melts at temperatures in excess of 600–700°C. In metamorphic systems, fluids are generated from rocks at temperatures mostly between 350–500°C. In contrast, meteoric and formation waters are initially cool and are heated by Earth's natural geothermal gradient (about 20°C/km). In the upper portions of Earth's crust, where permeabilities are sufficient for fluids to flow in quantity, this commonly results in temperatures up to 50–100°C. However, most hydrothermal mineral deposits appear to have formed at higher temperatures (generally greater than 120°C). Therefore, many geologists believe that meteoric hydrothermal fluids and formation waters require an additional heat source, such as above-average heat flow in areas of high geothermal gradient, interaction with igneous rocks, or the addition of a magmatic component to the system.

If heat is necessary to dissolve and transport metals in solution, then cooling must be an important mechanism of precipitation, and mineral deposits may form by simple cooling of hydrothermal fluids. However, in most geological environments, cooling is accompanied by changes in solution acidity (pH), oxidation state (fO_2), and/or mixing of the metal-bearing hydrothermal fluid with other fluids, which also affect mineral stabilities. If dissolved metals are carried by aqueous chloride complexes (abbreviated as $MeCl_2$), and deposited as metal sulfide minerals (MeS), then precipitation can be described by the following general reactions:

$$MeCl_2 + H_2S \Leftrightarrow MeS + 2H^+ + 2Cl^-$$
$$\text{(neutralization)}$$

$$MeCl_2 + SO_4 \Leftrightarrow MeS + 2O_2 + 2Cl^-$$
$$\text{(reduction)}$$

These reactions suggest that oxidized and acidic solutions will tend to dissolve metal sulfide minerals; reduction and neutralization will cause mineral precipitation. However, if metals are carried to solution as bisulfide complexes, then a different set of reactions apply:

$$Me(HS)_{3-} + H^+ \Leftrightarrow MeS + 2H_2S$$
(acidification)

$$Me(HS)_{3-} + 4O_2 \Leftrightarrow MeS + 2SO_{4=} + 3H^+$$
(oxidation)

In this case, mineral precipitation will be caused by acidification and oxidation—the reverse of the metal chloride case (Figure 2). Thus, mineral precipitation is not caused by a single process operating in all geological environments but by a variety of processes depending on the nature of the complexing ligands.

The role of acidity and oxidation state has been especially stressed in the formation of gold deposits because of the sensitivity of gold bisulfide complexes to solution conditions. Figure 3 illus-

trates the stability fields of the gold bisulfide complex $[Au(HS)_{2-}]$ with respect to acidity, oxidation state, and temperature. Gold is soluble only under a relatively narrow range of geological conditions, and its solubility is particularly sensitive to oxidation state. Thus, precipitation could be produced by mixing a hydrothermal fluid with cool, oxidized, surface water. However, precipitation could also be achieved by simple cooling without mixing and oxidation. How can one discriminate between these two possibilities? Simple cooling is likely to be a gradual process resulting in gold dispersal over a large geographic area. Mixing is likely to be focused by structural controls and therefore better concentrate gold in a relatively small area. Both processes no doubt occur, but perhaps only the latter leads to the formation of hydrothermal gold deposits.

Two other mechanisms may be important in mineral sulfide precipitation: dilution and sulfur

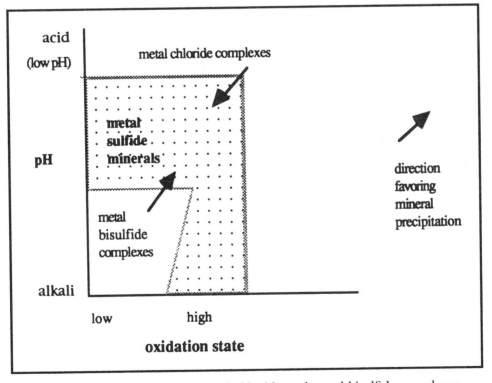

Figure 2. Relative stabilities of metal chloride and metal bisulfide complexes with respect to the acidity and oxidation state of a hydrothermal fluid. The bold arrows indicate directions favoring mineral precipitation. Sulfide minerals are precipitated when chloride complexes are reduced or neutralized, or when bisulfide complexes are oxidized or acidified.

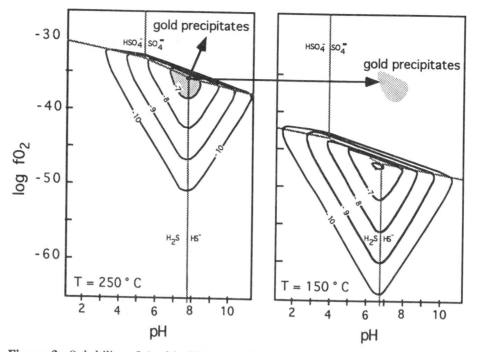

Figure 3. Solubility of Au bisulfide complexes with respect to pH and fO2, at 250 and 150°C. Contours are labeled by the logarithm of the activity of gold bisulfide in solution. Gold is highly soluble only in a narrow "bull's-eye" of solution conditions (stippled region in left-hand diagram). Precipitation may be caused by a variety of mechanisms but most readily by oxidation (short arrow). At lower temperature (right-hand diagram), the solubility field has shifted to lower pH and much lower oxidation state. Solution conditions that were previously favorable for gold solubility are no longer. Therefore, simple cooling can also cause gold precipitation (long arrow). Diagram is drawn for sulfur activity = 10^{-3}. After Sander and Einaudi (1990).

supply. Because the solubilities of most metals depend on the presence of chloride, dilution of a hydrothermal fluid—for example, by mixing with freshwater—may result in precipitation of the metals carried in that solution. This may seem counterintuitive: everyday experience suggests that a concentrated solution is more likely to precipitate materials than a dilute solution, and this would be the case if metals were carried as raw ions. However, if metals are carried as chloride complexes, then dilution decreases the concentration of chlorine in solution and, therefore, decreases the solubility of the metal. Of course, this presupposes the availability of sulfur, and in many ore deposits the availability of sulfur may be the controlling factor in mineral precipitation. If sulfur is supplied by a cool fluid, for example, a low salinity groundwater, then cooling, dilution, neutralization, and sulfur supply may occur simultaneously, and the result is a high-grade mineral deposit.

Hydrothermal Alteration

The reaction of hydrothermal fluids with wall rocks results in modification of both the chemistry of the fluid and the mineral composition of the rock. Most hydrothermal mineral deposits display distinctive patterns of wall rock alteration surrounding the central zone of mineral deposition. Because hydrothermal fluids can rarely, if ever, be observed directly, the alteration patterns produced in the rocks with which they interacted are one of the most important lines of scientific evidence in understanding the formation of hydrothermal mineral deposits.

One of the most commonly observed alteration patterns involves the transformation of K-feldspar to K-mica (sericitic alteration), and K-mica into clay (argillic alteration):

K-feldspar + hydrogen
$$= \text{muscovite} + \text{quartz} + \text{potassium}$$

$$3KAlSi_3O_8 + 2H^+$$
$$= KAl_3Si_3O_{10}(OH)_2 + 6SiO_2 + 2K^+$$

muscovite + hydrogen
$$= \text{kaolinite} + \text{potassium} + \text{oxygen}$$

$$2KAl_3Si_3O_{10}(OH)_2 + 2H^+$$
$$= 3Al_2Si_2O_5(OH)_4 + 2K^+ + 3/2\ O_2$$

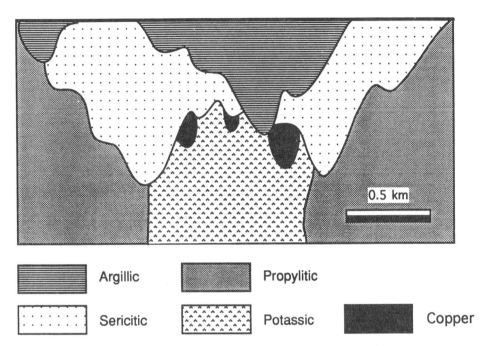

Argillic   Propylitic

Sericitic   Potassic   Copper

Figure 4. Hydrothermal alteration patterns surrounding porphyry copper deposits. In the central portions of the stock, high-temperature, high-grade copper sulfide mineralization (black) is associated with abundant K-feldspar and biotite, which commonly replaces or forms overgrowth on igneous Na-feldspar and hornblende (potassic alteration). Beyond the margins of the stock, the nature of wall rock alteration generally changes due to the contrasting composition of the surrounding rock, and a feldspar-(albite or adularia)-calcite-epidote-chlorite-pyrite assemblage is commonly observed (propylitic alteration). Both potassic and propylitic assemblages are overprinted by muscovite + quartz + pyrite (sericitic or phyllic alteration), particularly in the higher portions of the system associated with low-grade disseminated copper ore. All three alteration types may be overprinted by kaolinite (argillic alteration). These patterns reflect changing mineral stabilities as magmatic hydrothermal fluids cool and mix with meteoric water. Although the details of alteration patterns may be quite complex, the overall consistency of the general pattern makes recognition of wall rock alteration an essential exploration tool.

These reactions—known as hydrolytic alteration—consume hydrogen and result in neutralization of the hydrothermal fluid. Therefore, in areas where extensive hydrolytic alteration is observed in direct association with metal sulfide minerals, we can infer that neutralization was an important mechanism of mineral precipitation.

In addition to their scientific significance, patterns of hydrothermal alteration are critical in the exploration for new, undiscovered mineral deposits. Most hydrothermal mineral deposits are relatively small features—rarely more than 1 or 2 km across and often as little as a few hundred meters. The search for such small targets is an extremely difficult one, and many technologies have been developed to aid in it. However, the most effective technique in any search is to increase the size of the target, and this can be done by looking for the characteristic patterns of wall rock alteration that extend beyond the limits of the zone of metal enrichment. Perhaps the most famous and well-documented example is the hydrothermal alteration associated with porphyry copper deposits. These deposits generally occur in shallow-level igneous stocks, in which both the stock and the adjacent rocks are strongly altered in a distinctive and relatively consistent pattern (Figure 4). The central portions of the stock are characterized by K-rich mineral suites, while the outer portions are characterized by intense hydrolytic alteration. These patterns have been interpreted as the result of early, high temperature alteration of the stock by the magmatic hydrothermal fluid derived from it, overprinted by later, lower temperature alteration caused by cooling of the magmatic hydrothermal fluid and influx of meteoric water.

Outstanding Issues and Future Trends in Research

Scientific understanding of hydrothermal mineral deposits has increased dramatically in recent decades, but many important questions remain unanswered. How are meteoric hydrothermal fluids heated to temperatures adequate to generate mineral deposits, and to what extent might magmatic fluids be responsible for the metals in systems in which the fluid source is dominantly meteoric? How important is vapor-phase transport in high-temperature hydrothermal systems? What role do organo-metallic complexes play in mineral transport, and how important are adsorption and kinetics in controlling mineral precipitation?

From a practical standpoint, geologists would like to know why the world is filled with many small, low-grade mineral deposits but only a small handful of giant, high-grade deposits. What are the geological factors that control the formation of giant mineral deposits? Why do some portions of the world have so many more mineral deposits than other areas? Are there fundamental heterogeneities in the composition of the earth's crust or upper mantle that control the availability of metals? These questions demonstrate that the study of hydrothermal mineral deposits is not only necessary for ensuring the world's supply of essential material resources, but that it can also lead to a greater understanding of the fundamental nature of the earth.

Bibliography

BARNES, H. L., ed. *Geochemistry of Hydrothermal Ore Deposits.* New York, 1979.

BEANE, R. E., and S. R. TITLEY. "Porphyry Copper Deposits: Hydrothermal Alteration and Mineralization." *Economic Geology, 75th Anniversary Edition* (1979):235–269.

GUILBERT, J. M., and C. F. PARK, JR. *The Geology of Ore Deposits.* New York, 1986.

SANDER, M. V., and M. T. EINAUDI. "Epithermal Deposition of Gold during Transition from Propylitic to Potassic Alteration at Round Mountain, Nevada." *Economic Geology* 85 (1990):285–311.

SKINNER, B. *Earth Resources.* New York, 1986.

WHITE, D. E. "Diverse Origins of Hydrothermal Fluids." *Economic Geology* 69 (1974):954–973.

NAOMI ORESKES

I

IGNEOUS PROCESSES

The crusts and interiors of Earth and its neighboring rocky planets, including some asteroids, owe their overall character largely to the behavior of molten rock or magma in producing the igneous rocks. The vast diversity of igneous rocks is itself abundant testimony to the central role of magma in Earth history. The very origin of continents, which themselves provide a freeboard from the oceans for establishment of nonmarine life, rests with understanding magma and igneous processes.

That rocks, all rocks, can be melted simply with a welder's torch is surprising, but true. Lavas freely flowing or exploding from volcanoes are fairly commonplace, but the fact that all igneous rocks, even those casually found along roads and beaches, formed through the slow growth of minerals at very high temperatures, often many kilometers deep within Earth, is almost incomprehensible. Earth and Moon each may well have been jacketed early in their histories by thick oceans of magma. Cooling of magma spawns the nucleation and growth of crystals that move with and against the magma and often collect, like snowflakes, to form layers. Both the layering and the detailed nature of the crystals themselves represent pages in the book of magmatic evolution that records the intimate details of igneous processes. Just as chil-

dren resemble their parents, igneous crystals reflect the chemical composition of the magmas from which they grow, but they are also, like children, always chemically distinct from their parental liquids. Separation and sorting of crystals from magma chemically fractionates or differentiates magma to yield a wide diversity of igneous rocks. Here we explore the form and behavior of igneous systems, both large and small, to discover the fundamental controls on the evolution of igneous rocks and the crusts of the rocky planets. In many ways this is a story of the physical and chemical fractionation of crystals and liquids in solidification fronts.

Solidification Fronts

The formation and evolution of magma under almost all circumstances is controlled by solidification fronts. The word "front" is used much in the same sense that cold or warm front is used in weather forecasting; it represents a strong variation in temperature over a relatively short distance. Because rocks do not melt at a single temperature but over a range of temperature, across a solidification front magma changes from wholly liquid to wholly solid. The temperature of the first appearance of melting is called the solidus, and the temperature of disappearance of the very last tiny crystal is called the liquidus. The interval between the liquidus and solidus defines the solidification front, which commonly spans a range of about 200°C. Solidus and liquidus temperatures themselves depend strongly on the exact chemical composition of the rock being melted (McBirney, 1993).

Granite-like rocks are rich in the minerals quartz and feldspar, are light in color, and begin melting near 700°C; basalt-like rocks, poor in quartz and rich in dark ferromagnesian minerals, begin melting near 1,000°C. The basalt magmas of Hawaii, for example, undergo complete crystallization in cooling from about 1,200°C to 980°C as long as they are near Earth's surface, where the confining pressure is low. In going deeper in Earth, pressure increases systematically with depth; increasing by 1,000 times atmospheric pressure for every one kilometer of depth (1,000 m). This increasing pressure forces the atoms and molecules in magma closer together, allowing them less freedom to rush about and escape solids for melting. The net result is a general increase in both the solidus and liquidus temperatures with increasing depth in Earth, but the overall relation to one another remains essentially the same as at the surface. We can thus expect solidification fronts to look and behave similarly at depth as they do at or near Earth's surface. The only difference is that solidification fronts advance outward in source regions, causing partial melting, and in the near distance they advance inward upon the magma, causing progressive partial solidification within the front.

Although no one has ever trapped and sampled a live magma buried at depth, fairly large bodies of magma have accidentally ponded in Hawaii to form lakes of lava. The 1959 fire fountain eruption of Kilauea volcano, an example of hot spot volcanism, flowed into a large circular, washtub-like depression, filling it to a depth of 100 m. Within a few weeks a rocky crust formed on the surface much as a layer of ice forms on a winter pond. This crust is a solidification front. Scientists from the U.S. Geological Survey (USGS) used diamond-studded drill bits to penetrate the crust, obtain samples on the way through, and reach the inner region of magma. They found the crust of the lava lake to be rigid and drillable even when only partially solid as long as the solid crystals made up at least 50 percent (by volume) of the rock. At crystal contents below 50 percent, which we call the critical crystal-

linity, the drill stem can be pushed downward by hand. The magma beyond this point no longer acts rigid but is instead thick and gooey or viscous, like cold syrup.

Just as cold syrup pours slowly, viscosity or resistance to flow increases with decreasing temperature. But a much more profound influence on the viscosity of magma is the presence of crystals. As crystals grow they essentially lock or immobilize a small halo of liquid around themselves making the magma more viscous. As the magma cools and increases in crystallinity, viscosity systematically increases, but when there are so many crystals that they begin to touch each other, a dramatic effect sets in. The viscosity increases astronomically, perhaps by a factor of ten billion, over a relatively small drop in temperature of about 15–20°C. This is another result of the attainment of critical crystallinity. But just what is this point of critical crystallinity?

The critical crystallinity marks the point of maximum packing of crystals in the magma. At maximum packing the crystals touch, and no more crystals of the same size can be added to the system. The remaining liquid resides between the crystals, and further crystal growth occurs in these remaining pockets of liquid. But since the crystals are touching and are sticky—because they are still growing—they weld to one another and form a haphazard crystal network or mesh, much like a chicken-wire fence, which has strength. Once the magma reaches this point of maximum packing of crystals it is no longer mobile. A dramatic reflection of this process is the recognition that lavas never contain more than about 50 percent crystals; with more than 50 percent crystals not only is magma too viscous to erupt, but it has a new behavior. It behaves as a dilatant solid. That is, it swells or expands if it is moved, forming a plug in any conduit it attempts to ascend. This is the same effect we observe when walking along the beach at the water's edge. As we step, a halo of dry sand appears around each foot. The sand grains are at maximum packing and by squeezing them with our weight they roll out away from each other, expanding or dilating, making a new volume of sand too large to be filled by the original amount of intergranular water. The expanded sand now appears dry. This effect has yet a further meaning for the differentiation of magma.

Because the maximum packing for magmas is near a crystallinity of 50 percent, once a parcel of magma contains 50 percent solids, it is fully choked with crystals. Not only is the magma immobile, but the crystals cannot easily be separated from the accompanying liquid and the magma cannot be differentiated through fractionation (i.e., separation by fractional crystallization) of the crystals. A similar process is common to our everyday lives. A cup filled with crushed ice holds only half as much soda as the same ice-free cup, and the ice cannot be squeezed down to free any liquid. Only if the ice can be deformed and repacked to make a solid mass can ice-free liquid be partitioned to its own horizon of the cup. We will see presently that this process does indeed operate within Earth to collect drops of magma from slightly molten rock.

Having considered the fundamental personality of magma as a mixture of liquid and crystals (and also sometimes gas bubbles, which act much like crystals), it is appropriate to consider the life cycle of magma from generation to eruption.

The Magmatic Life Cycle and Ocean Ridges

The life of magma is its heat. The life cycle of magma in essence can be tracked by tracking one thing: magma crystal content. But of course crystal content is a reflection of magma temperature relative to the liquidus and solidus. A knowledge of temperature alone, however, is meaningless without a phase diagram, which links temperature to crystallinity (i.e., liquidus and solidus) and crystal composition.

Magma is almost invariably produced in Earth by a process of convection through a phase or melting boundary. Just as weather systems form due to warm or cold air moving into, respectively, cold and warm regions, often by rising or sinking, so it is also within Earth. Earth's mantle is forever turning over and shifting in response to global cooling and convection associated with plate tectonics (Kearey and Vine, 1990). Hot rock flows like glacier ice, and as the mantle rock ascends the pressure drops fast relative to its rate of cooling. The solidus is encountered and melting sets in. The deeper melting begins, the greater the amount of melt produced as the rising rock approaches Earth's surface.

Beneath ocean ridges, for example, melting begins at a depth of about 50 km below the surface. The entire region over the depth range of 10–50 km is a mushy, partially molten source region of magma. The most easily melted, low-temperature differentiate or distillate enters the earliest-formed melt. But how is this liquid extracted from the rock when the overall crystallinity is still much larger than that at maximum packing (i.e., about 50 percent)? Because the mush column is so tall and the rock is undergoing melting, the crystals are relatively soft and can be squeezed and deformed by the weight of the overlying column. The lowermost crystals undergo compaction and squeeze out the intercrystalline liquid, forcing it to flow upward. Magma collects at a location higher in the column to form a chamber of magma.

The magma chamber beneath ocean ridges has been detected by using seismology and is thought to be thin (100–200 m), wide (2–4 km), and very long, extending, perhaps discontinuously, along the entire 40,000 km of ridge (Sinton and Detrick, 1992). In every respect, it is a ribbon of magma. As the plates move apart the oceanic crust cracks open, and magma is pulled from the ribbon and erupts onto the seafloor. Magma also freezes onto the walls of the conduits connecting the magma chamber to the surface. The net result is the formation of the oceanic crust, which covers 60 percent of Earth's surface. The oceanic crust thus has a layered structure beginning with lavas, underlain by dikes (i.e., a plexus of vertical, sheetlike conduits called the sheeted-dike complex), then coarse-grained solidified magma (i.e, gabbro rock), and finally even coarser-grained, almost mantle-like rock, reflecting the accumulation of old crystals originally entrained from the mush column by the ascending magma. The grain or crystal size increases downward, reflecting the length of time allowed for crystal growth. The lavas typically have less than about 20 percent crystals and are about 1 mm in size, whereas the deep cumulate crystals attain sizes of 5–8 mm.

The oceanic crust overall contrasts strongly in chemical composition with the underlying mantle. This reflects the strong controls on composition by the prevailing igneous process of melt extraction during partial melting by crystal deformation and compaction in the resulting mush column. But once the extracted magma, carrying perhaps some crystals, reaches the ribbonlike chamber, cooling and crystal growth occurs principally in the solidification fronts that border the chamber. If the chamber is not resupplied, it solidifies in a few hundred years, much as a Hawaiian lava lake does. Although lateral solidification and continual spreading produces a solid oceanic crust, continual resupply of magma from the underlying mush column keeps the ridge chamber alive. But does much chemical differentiation occur? No.

Extensive drilling and dredging along ridges the world over shows mainly the same rock types: basalt lava and occasionally its coarse-grained equivalent, gabbro. Why is there so little chemical fractionation? Because in the ridge chamber crystal growth and settling are controlled by the solidification fronts. The only crystal separation is through settling of old crystals entrained from the mush column in the ascending magma. New crystals nucleated on the inwardly advancing, leading edge of the solidification front attain sufficient size for settling at a time when they are too deeply ensconced in the solidification front to escape. Thus, highly refined or differentiated, quartz-rich rocks, so common on continents, are never found as lavas along ocean ridges. Then, how are continents produced?

Continental Growth and Solidification Front Instability

Sheets of basaltic magma significantly larger than those beneath ocean ridges are not uncommon in continental regions. These sheets reflect the rifting of continents in the plate tectonic cycle and the rising and emplacement of magma much as along present-day ocean ridges. When well exposed, sheets as thick as 300–700 m and as expansive as 5,000–10,000 km^2 reveal an intricate history of solidification. If the magma is emplaced carrying no large, old crystals (called phenocrysts), very little chemical differentiation occurs during solidification in the liquid portion of the magma. These sheets cool symmetrically from the top and bottom; the last portion to solidify is near the center, and here the magma composition hardly changes over what it was upon emplacement. This process again reflects the strong controls of the solidification fronts on the chemical and physical evolution of the sheet of magma. But something drastic and extreme does happen within the upper solidification front.

To be sure, extreme chemical differentiation occurs within the solidification fronts, but the liquid or melt is trapped among the interconnected crystals and cannot be collected into an eruptable body of liquid. There is no upward force to compact the

upper front, and the lower front is generally much too thin to undergo compaction. Instead, once the upper solidification front becomes sufficiently thick, it begins to sag and tear away from the overlying rigid part of the front, where the crystal content is greater than about 60 percent. This sagging or delamination draws the refined, intercrystalline melt into the tears, which form 1–2 m thick interdigitating lenses of highly differentiated silica-rich melt. Sagging of the solidification front ceases after forming a wide horizon of tears, and only in the thickest bodies does the front entirely fall away to free and allow collection of the lenses themselves.

Once formed, these siliceous lenses can never be homogenized back into the original magma even if the entire sheet were to be remelted, which certainly can happen after prolonged burial. The reason for this is that siliceous magmas are much too viscous relative to the host basalt to be assimilated during any natural mixing process. These siliceous bodies are much like lumps of melted cheese in water.

Siliceous lenses or segregations are found in many sheetlike bodies, including the oceanic crust. These lenses are the seeds of continental crust. Early in the history of Earth, before continents existed, huge expanses of sheetlike bodies of magma, perhaps even oceans of magma, produced vast quantities of such lenses. Remelting freed the lenses to float to the top and collect; continual, long-term remelting steadily formed the early continental crust. Further processing due to erosion, chemical weathering, and further remelting eventually produced continental crust.

Disruption of Solidification Fronts in Island Arc Diapirs

Beneath island arcs thick tectonic plates descend deep into the mantle, initiating sharp bands or arcs of volcanism (Wilson, 1989). The magma itself ascends in discrete, bulbous masses known as diapirs. The ascent is slow and deliberate and is regulated by the enormous viscosity of the surrounding mantle. Each body or diapir moves much as a small bubble rises in thick shampoo. As the diapir ascends, solidification fronts form around the perimeter of the body. But because the body is moving slowly upward (perhaps at rates of 10–100 m/yr), the solidification front is continually disrupted. Some crystals are mixed back into the main body of magma, and others are lost in a trail of crystals through the floor of the body; this loss is another example of fractional crystallization. In ascending some 100 km to the surface, this process has the potential to produce strongly differentiated magmas. By far the most common magma of island arcs is still basaltic. To be specific it is high-alumina basalt, but also erupted are some highly differentiated andesitic and dacitic lavas containing as much as, respectively, 60 percent and 65 percent SiO_2 (by weight). It is these refined compositions that eventually collect, along with the silicic lenses of magmatic sheets, to further add to the growth of continents. And it is these continents floating on the denser mantle that provide an oasis for nonmarine life.

Bibliography

KEAREY, P., and F. J. VINE. *Global Tectonics*. Boston, 1990.

McBIRNEY, A. R. *Igneous Petrology*, 2nd ed. Boston, 1993.

SINTON, J. M., and R. S. DETRICK. "Mid-ocean Ridge Magma Chambers." *Journal of Geophysical Research* 97 (1992):197–216.

WILSON, M. *Igneous Petrogenesis*. London, 1989.

BRUCE D. MARSH

IGNEOUS ROCKS

Igneous rocks are one of the three major types of rocks (*see* SEDIMENTS AND SEDIMENTARY ROCKS, CHEMICAL AND ORGANIC; SEDIMENTS AND SEDIMENTARY ROCKS, TERRIGENOUS; and METAMORPHIC ROCKS). By definition, igneous rocks form when molten or partially molten rock, called magma, solidifies. There are two groups of igneous rocks. Extrusive igneous rocks include lava and volcanic ash. They form when magma erupts from a volcano at the surface of the earth. Intrusive igneous rocks, such as granites, form when magma cools and crystallizes at depth, often in large ovoid to irregular masses called plutons or thin sheets called dikes. Intrusive and extrusive rocks have distinctive textures that are related to the cooling rate of the magma (*see* INTRUSIVE ROCKS AND INTRUSIONS).

Extrusive Igneous Textures

Extrusive rocks cool quickly. Crystals that form during this rapid cooling tend to be small, generally too small to be seen by the naked eye. If cooling is very rapid, crystals may not form at all and the magma quenches to a supercooled liquid or glass. The rock called obsidian consists almost entirely of volcanic glass. Much volcanic ash is made up of tiny shards of volcanic glass. Lavas generally consist of mixtures of crystals and glass. Usually some slow crystallization of the magma occurs before the extrusive rock is erupted. The crystals formed during this pre-eruptive period of crystallization are commonly large enough to be seen with the unaided eye. These early formed crystals are called phenocrysts. The texture of a few large crystals set in a matrix of fine-grained crystals and glass is termed porphyritic.

Other extrusive rock textures are related to the presence of gas in the magma. Pumice and scoria form when gases (most commonly water vapor and/or carbon dioxide) that are dissolved in a magma when it is under pressure come out of solution as pressure is released during eruption. The process is analogous to the release of carbon dioxide bubbles when a bottle of soda is opened. The bubbles in the magma expand, forming a frothy mixture of liquid and gas. The froth solidifies to a porous, glassy rock with lots of bubbles. Pumice contains so many trapped bubbles that it actually floats on water. Scoria is denser and resembles coal cinder. When smaller amounts of dissolved gas are present in the magma, isolated bubbles or pockets of gas may form. These bubbles are called vesicles. Over time, mineral-rich solutions may deposit minerals within volcanic vesicles. Many geodes form in this way.

Intrusive Igneous Textures

Intrusive rocks cool very slowly. In the case of large bodies of magma deep within the earth, cooling may take millions of years. This slow cooling results in the formation of crystals large enough to be seen with the unaided eye. Glass is almost never found in intrusive rocks (except for those formed near the surface, such as in conduits feeding volcanos); hence they are termed holocrystalline, entirely crystalline. Grain size in intrusive rocks varies with a number of factors. In water-rich magmas, very large (centimeters to meters) crystals may form, producing a rock called pegmatite. More commonly, grain sizes are on the order of millimeters to a centimeter. Most intrusive rocks are approximately equigranular, containing mineral grains of more or less equal size. Intrusive porphyries contain large crystals set in a matrix of smaller, but still visible, crystals. Seriate texture, with a wide range of grain sizes, is relatively rare.

The textural relations among minerals in intrusive rocks can provide information about the crystallization history of the rock. Early-formed minerals can grow unimpeded and may have well-developed (euhedral) crystal shape. Later formed minerals must compete for space and, because of this, have their crystal outlines partially (subhedral) or completely (anhedral) distorted. Sometimes late-formed, anhedral minerals grow to large size, filling available void space and enclosing smaller, preexisting mineral grains (poikolitic texture).

Igneous Rock Types

Even though intrusive and extrusive rocks are very different in form and overall appearance, both form by solidification of magma. Hence, with a couple of exceptions, their compositions are similar. In the most commonly used classification scheme (LeMaitre, 1989), extrusive igneous rock types are defined by the abundances of the element silicon (Si) and the alkali metals sodium (Na) and potassium (K). Intrusive rock types are defined by the abundances of minerals whose presence generally reflects Si and alkali abundances (Figure 1).

Most rocks having SiO_2 concentrations less than 53 percent are termed basalt (extrusive) or gabbro (intrusive). If the concentration of alkalis is low, they are termed subalkaline or tholeiitic basalt/gabbro; if high, they are called alkaline basalt/gabbro. Characteristic minerals in these rocks include

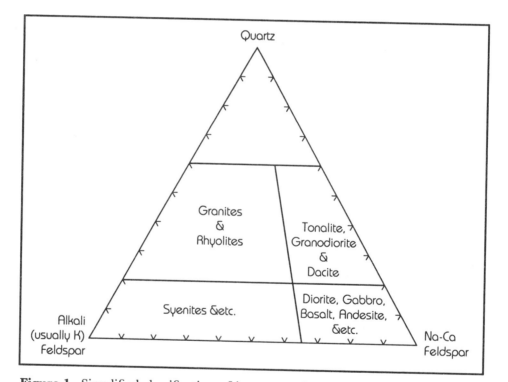

Figure 1. Simplified classification of igneous rocks modified after LeMaitre (1989). Most gabbros and basalts contain no quartz and plot along the feldspar sideline, often at or near the Na-Ca feldspar corner. Many common igneous rocks can be plotted on this diagram. However, most alkaline rocks and rocks with unusual compositions (e.g., carbonatites) cannot be classified using this diagram. See LeMaitre (1989) for a more detailed discussion of igneous rock classification.

olivine, pyroxenes, calcium-rich plagioclase, and, in intrusive rocks, amphibole. Quartz is rare or absent.

Rarely, igneous rocks will have less than 45 percent SiO_2. These unusual rocks include carbonatities (the same name is used for intrusive and extrusive varieties), which consist almost entirely of the minerals calcite and dolomite, and komatiites, very magnesium-rich lavas that are found primarily in Archaean rocks (see EARTH'S CRUST, HISTORY OF).

Andesite (extrusive) and diorite (intrusive) contain 53–63 percent SiO_2 and have slightly higher alkali contents than typical basalts. Trachy-andesite (extrusive) and syenite (intrusive) are the alkali-rich varieties. Typical minerals are plagioclase, pyroxene, and amphibole. The more Si-rich varieties may contain quartz, the more alkali-rich varieties potassium feldspar or feldspathoids.

Rocks with greater than 63 percent SiO_2 include the extrusive rocks dacite, rhyodacite, and rhyolite and their intrusive equivalents tonalite, granodiorite, and granite. Granites (by definition) and rhyolites (generally) contain more potassium than dacites and tonalites. They are usually (but not always) richer in SiO_2 as well. Typical minerals are quartz, Na-rich plagioclase, biotite, and, in the K-rich rocks, potassium feldspar.

A few intrusive rocks do not represent liquid magma compositions. Instead, they represent accumulations of crystals that have settled out or otherwise been separated from a crystallizing magma. These accumulations of crystals are called cumulates. Common cumulate minerals include olivine, pyroxene, and plagioclase. Many gabbros are actually crystal cumulates. Some cumulate rocks are strikingly layered. The study of cumulates can provide important clues to the crystallization history and chemical evolution of a magma body.

The Origin and Occurrence of Igneous Rocks

Processes by which magmas are produced and modified include partial melting of mantle and crust, fractional crystallization (the removal of minerals during the crystallization of a magma), mixing of magmas of different compositions, and contamination of magmas by surrounding rocks.

Magmas produced by partial melting of the mantle are hot, relatively fluid, and enriched in the elements iron (Fe), magnesium (Mg), and Ca. These magmas are generally basaltic in composition. Basalts are the most abundant extrusive rocks on Earth. The ocean floor consists of basalt overlain by a thin veneer of sediment. These ocean floor rocks are termed mid-ocean ridge basalts (MORB). They are produced at mid-ocean ridges where hot mantle ascends due to convection and begins to melt as pressure decreases. Perhaps 60 percent of all extrusive magmatism on Earth occurs at mid-ocean ridges. As voluminous as they are, MORBs are only rarely preserved on the continents; most are destroyed during subduction of the oceanic lithosphere back into the mantle (see PLATE TECTONICS). Gabbros are the primary constituent of the deep crust of the ocean basins. Some mid-ocean gabbros are the intrusive equivalents of MORBs; others are cumulates.

During the early history of Earth, the mantle was hotter than it is today and melting of the mantle at mid-ocean ridges was more extensive. This very hot melting produced the Mg-rich komatiites.

Other basalts and gabbros form as a result of plume (or "hot spot") magmatism. The best-known examples of "active" plume volcanoes are found on the island of Hawaii. Plumes are thought to form as a result of melting very deep within the earth, perhaps at the core-mantle boundary. Most plume-related magmas are chemically distinct from MORBS. Many are enriched in many trace elements and, especially, alkalis. These alkaline basalts (and other alkaline rocks, including carbonatites, syenites, and feldspathoidal rocks) are also common in continental rifts.

Finally, some basalts and gabbros occur in subduction-related volcanic arcs. In subduction zones a hydrous fluid, rising from the subducted slab, causes the overlying mantle to melt. Thus, subduction zone basalts tend to be richer in water and other volatiles than basalts from other tectonic settings.

Andesites and diorites occur in association with subduction zones in continental or oceanic arcs. They form when low-Si minerals, such as olivine and, especially, magnetite, are removed from a basaltic magma (crystal fractionation), thereby raising its Si content. Alternatively, andesite and diorite can form when Si-poor basalt and Si-rich crustal melts mix together in a magma chamber, producing a magma of intermediate Si content. Andesites are rare in ocean basins. One possible explanation is that magnetite readily crystallizes from water-rich arc basalts but not from water-poor MORBs.

The Si-rich dacites, rhyodacites, and rhyolites and their intrusive equivalents occur most commonly in arcs, particularly continental arcs. They can form by extreme fractional crystallization of basalt, by partial melting of the relatively Si and alkali-rich rocks of the continental crust, by contamination of basalt or andesite by silicic crust, or (as is probably true in many cases) by a combination of these processes. However, the fact that these rocks are most abundant today in continental arcs suggests that crustal melting must be at least partially responsible for their formation in most cases. In a typical continental arc, the heat required to melt the crust probably comes from the emplacement of a mantle-derived basalt within or at the base of the crust.

A special type of tonalite termed adakite is particularly abundant in the Archaean. It is thought to have formed when ocean crust partially melted, rather than dehydrated, during subduction into the hot Archaean mantle.

Bibliography

FRANCIS, P. *Volcanoes.* New York, 1976.
LeMaitre, R. W., ed. *A Classification of Igneous Rocks and Glossary of Terms.* Boston, 1989.
Press, F., and R. Siever. *Earth.* New York, 1978.
Raymond, L. A. *Petrology: The Study of Igneous, Sedimentary and Metamorphic Rocks.* Boston, 1995.

JAMES S. BEARD

IMPACT CRATERING

Even a cursory look at a photograph of the Moon, Mercury, or Mars is enough to impress one with the significance of craters in the geologic history of those planets. Although the extremely active nature of Earth's lithosphere has destroyed most of the craters formed here, there is still abundant evidence attesting to the importance of impact cratering throughout Earth's history (Grieve, 1991). Impact craters result from high-speed collisions between solid bodies, and they are exemplified by features ranging from submicron pits on individual grains of lunar soil to enormous structures such as the Hellas Basin on Mars, which is 2,000 km in diameter. While acknowledging the importance of small craters in the evolution of planetary and asteroidal surfaces, this section will concentrate on large (multikilometer) craters. Most of the mechanisms associated with the early stages of the cratering process occur regardless of scale, and many of the phenomena cited here are as valid for ten-micrometer craters as they are for the largest impact basins. The interested reader is directed to Melosh (1989), which is a much more extensive treatise on impact-cratering mechanics and its myriad effects and manifestations (*see* IMPACT CRATERING AND THE BOMBARDMENT RECORD).

Basic Impact Mechanics

A "shock," in the sense used here, is a mechanical wave across which exist often extreme discontinuities in pressure, velocity, and temperature. It travels faster than the speed of sound in, and imparts high velocities to, the medium in which it travels, although the material's velocities are somewhat lower than its own. It also raises the internal energy of the medium, often to the point of melting or vaporization. Virtually all motion induced by an impact can be attributed to the effects of shock waves propagated through both the target and the projectile. (It is often thought, incorrectly, that the formation of an impact crater is due to the explosion of the impacting body. While the energy deposited into the impactor is often sufficient to induce its vaporization, the resulting "explosion" occurs well after the shock front has already penetrated some distance into the target.)

The shape of the shock wave generated by a high-velocity impact, for descriptive purposes, is a portion of a sphere, with its center lying near the surface of the target. The initial motion imparted to the target material by this shock front is radial to the center of the sphere—a configuration that is soon changed, however, by "rarefaction waves," which perform the process of decompression. The shock front forces the material to a state of very high pressure; since the surface of the target is effectively at zero pressure, however, this high-pressure state cannot be maintained. The rarefaction front acts as a decompression wave, relaxing the compressed material much as a spring relaxes when a compressing force is removed. Neither the compression nor the decompression is instantaneous, however, and the target material set into motion can travel some distance before it is completely decompressed. This combination of compression and subsequent decompression forces the bulk of the affected target material downward and then upward and outward, resulting in a crater. (For a more detailed description of this process, see Gault et al., 1968.)

The Crater

The growing crater is often called the "transient cavity" because its shape and dimensions are only temporary. Various forces, associated predominantly with the rarefaction process and gravity, in-

variably modify the transient cavity. When these modification processes are only mild—as in the case of small craters (that is, those typically less than about 5–10 km in diameter)—"simple craters" result; they are commonly described as being bowl-shaped because they are relatively deep, have smooth walls, and are circular in plan (Figure 1). Larger cavities suffer more extensive modification through wall failure (slumping) and floor rebound, the latter bearing most of the responsibility for the formation of central peaks (Figure 2). These large, "complex craters" are relatively shallower than simple craters and, as their name implies, possess much more intricate morphologies.

The smaller cavities are modified less than larger ones because the strength of the target is more important than, or at least comparable to, the forces due to gravity. The range in diameter over

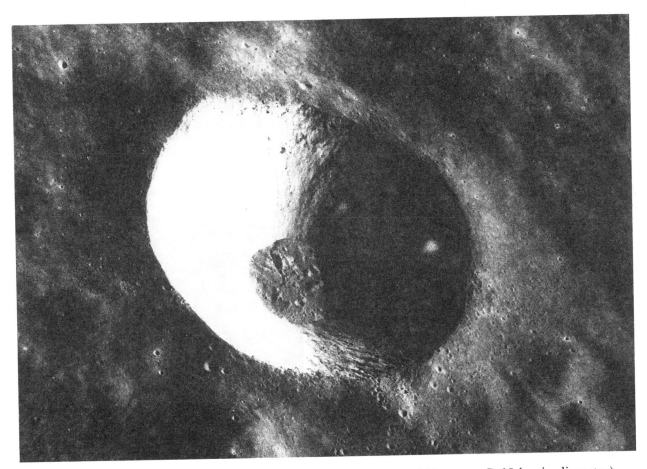

Figure 1. An example of a fresh, simple crater. This lunar crater (Alfraganus C, 10 km in diameter) illustrates the smooth walls, the large depth-diameter ratio, and the small, flat floor characteristic of bowl-shaped craters.

Figure 2. One of the best-known lunar craters, the 85-km Tycho, is an excellent representative of complex craters. Note the ubiquitous, flowlike deposits of impact melt inside and on the exterior of the crater, the central peaks, and the somewhat shallower aspect of the crater and its complex internal morphology relative to those of Alfraganus C.

which the transition from simple to complex craters occurs is both planet-dependent and relatively restricted and has been interpreted as defining the range over which gravitational forces become more important than the strength of the target during modification of the transient cavity.

Factors Affecting Crater Dimensions

Given two identical projectiles and identical targets, the faster projectile will create the bigger crater upon impact. While this may seem obvious, it is

not as straightforward a conclusion as it appears on the surface—faster projectiles waste more energy by transforming kinetic energy into heat, so an impact at 40 km s^{-1} (sometimes abbreviated as km/s) will not create a crater four times larger than that formed by an identical impactor at 10 km s^{-1}. If the projectiles and impact velocities were identical but the targets were in different gravity fields, the smaller crater would be formed in the target with the higher gravitational acceleration. Different results can be obtained, for instance, by varying the density of either the projectile or the target, by changing the angle of impact, or by making one target more porous than the other. Predicting the size of a crater resulting from a planetary-scale impact is an exercise fraught with uncertainty, and much effort has gone into the study of such "scaling relationships," which, not surprisingly, often take on complex forms (Schmidt and Housen, 1987).

The Ejecta

It is important to note that the entire volume of a transient cavity is not due to removal of material by ejection. Indeed, very detailed calculations indicate that a maximum of perhaps 50 percent of the cavity's volume is due to ejection of material, with plastic deformation of the target accounting for the remainder. Ejection composes a greater fraction of the final crater volume—that is, the crater after all modification processes have been completed—for complex than for simple craters because the latter are less modified versions of the transient cavity and are therefore more voluminous relative to their diameters.

The great bulk of the ejecta from a growing cavity travels on ballistic trajectories; a small fraction is fast enough to escape the gravitational field of the target planet. The remainder reimpacts the planet at distances proportional to the ejection velocity. The ejecta leaving the cavity last is the slowest and lands near the final crater's rim. This fraction, often mixed with local target material, makes up the thickest deposit, often called the "ejecta blanket." The curtain of ejecta, expanding its areal extent as it moves away from the crater, becomes more of a traveling assemblage of discrete fragments or clumps compared to its behavior as a continuous sheet closer to the crater. Upon impact, these clumps and fragments create clusters, chains, and isolated patches of "secondary craters." Deposition of such energetic ejecta mixes the target and can destroy or significantly degrade any relatively shallow, preexisting stratigraphy.

Shock Metamorphism

The intense pressures and high temperatures that are associated with shock waves are easily capable of causing metamorphism in the target rocks. Shocks generated by impacts at velocities characteristic of interplanetary encounters (10–20 km s^{-1} if the impactors are derived from asteroids, and several times higher for cometary projectiles) are easily capable of melting the target rock; higher velocities will even vaporize it. The conservation of energy and the second law of thermodynamics, however, dictate that the shock pressure decrease with distance from the point of impact. At some distance, the pressures and temperatures will be sufficiently low that melting cannot occur. Nevertheless, quantities of melt many times greater than the volume of the impactor can result. These melts are one means by which otherwise nondescript terrestrial structures have been identified as impact craters. While some impact melts are extremely difficult to distinguish from indigenous igneous rocks, others are so laden with clasts of different, spatially separated target rocks that there is little doubt as to their origin. In some cases, trace-element studies can identify the projectile type by its compositional signature of the impactor in the melt. (When the target rock is melted by shock, most of the projectile suffers the same fate. Mixing of the two liquids then occurs, and this can be studied with geochemical techniques.) When used as samples for isotopic age-dating techniques, impact melts are also ideal indicators of the age of the impact event.

At greater distances, pressures remain high enough to cause diagnostic features in the target rocks and their constituent minerals. Examples include shatter cones (poorly understood features that are nevertheless exclusively associated with impact craters), shock lamellae in quartz and other minerals, diaplectic glasses (glass-like phases apparently formed without melting the minerals in question), and high-pressure, high-density phases of various minerals, among which are coesite and stishovite (derived from quartz) and diamond (derived from graphite). Many mineralogical effects of shock compression are described in considerable detail by Stöffler (1972).

Almost every lunar sample and virtually every meteorite studied has suffered some shock damage. Indeed, shock melting is ubiquitous in the lunar-sample collection, and it is probable that most meteorites were ejected from their parent bodies (asteroids or asteroid-like objects) by impacts. Perhaps the most spectacular examples of this class are the SNC (shergottite, nakhlite, and chassignite) meteorites, which appear to be samples of Mars. Impact is the only known natural process that could remove material from a planet as large as Mars and place it into an eventual Earth-crossing orbit.

Craters as Tools

Because impacts destroy preexisting features, blur boundaries, and even obliterate each other, it is an unfortunate fact that many geologists consider craters to be a hindrance to their interpretive efforts. Unless samples are available, however, craters often hold the only key to determining the ages of planetary surfaces. Since all parts of a planet's surface should be subjected to the same rate of bombardment, the oldest terrains should have accumulated the greatest density of craters. Younger surfaces, on the other hand, should have fewer superposed craters. Counting craters on different geological units, therefore, is an important procedure in discriminating the ages of planetary surfaces, particularly when there is little other information to use in making a distinction between the units.

An example of such crater counts for two lunar terrains—the Apennine Mountains, created during the formation of the 1,200-km Imbrium Basin, and the central portion of Mare Serenitatis, one of the largest expanses of smooth basalt deposits on the Moon—is illustrated in Figure 3. (This is a cumulative plot, in which the number of craters larger than a given diameter are plotted against that diameter. There is a wide variety of other types of crater-density plot, each of which is tailored to a specific purpose.) Samples from the Apollo 15 mission yield ages for the Imbrium Basin of about 3.9 billion years (Ga). There are no samples identified as being from the center of Mare Serenitatis, but various stratigraphic arguments, among which are studies of crater statistics, lead to the conclusion that these basalts are about 700 million years (Ma) younger. The consequent difference in crater density is apparent in the figure. The different shapes of the two distributions contain an abundance of information regarding crater degradation and other processes. A powerful approach to photogeologic interpretation couples examination of planetary photography with crater-density plots specifically designed to enhance such differences in crater-size distributions.

There is a panoply of applications of crater analysis to topics as diverse as the generation of regolith (the layer of rubble that constitutes the surfaces of all airless bodies of appreciable size), the thermal evolution of planets, the disruption of asteroids and small satellites, and the early accre-

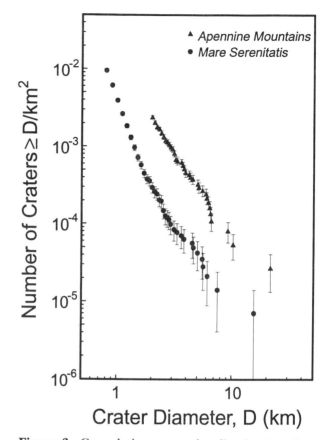

Figure 3. Cumulative crater-size distributions for the Apennine Mountains (the ejecta blanket of the Imbrium Basin) and the interior of Mare Serenitatis (extensive basalt flows). The greater age of the Imbrium Basin is reflected in the much greater density of craters in the Apennines. The deviations of both from smooth distributions can be due to such factors as changes in the size distributions of impacting objects, impact degradation of the two surfaces, or flooding by different sequences of lava flows in the case of Mare Serenitatis. The statistics for this figure were taken from Neukum et al. (1975).

tion of the solar system. In addition to the references cited above, others that the interested reader might find helpful are French and Short (1968), Roddy et al. (1977), and Schultz (1988).

Bibliography

FRENCH, B. M., and N. M. SHORT, eds. *Shock Metamorphism of Natural Materials.* Baltimore, 1968.

GAULT, D. E., W. L. QUAIDE, and V. R. OBERBECK. "Impact Cratering Mechanics and Structures." In *Shock Metamorphism of Natural Materials,* eds. B. M. French, and N. M. Short. Baltimore, 1968, pp. 87–99.

GRIEVE, R. A. F. "Terrestrial Impact: The Record in the Rocks." *Meteoritics* 26 (1991):175–194.

MELOSH, H. J. *Impact Cratering: A Geologic Process.* New York, 1989.

NEUKUM, G. B. KONIG, and J. ARKANI-HAMED. "A Study of Lunar Impact Crater Size-Distributions." *The Moon* 12 (1975):201–229.

RODDY, D. J., R. O. PEPIN, and R. B. MERRILL, eds. *Impact and Explosion Cratering.* New York, 1977.

SCHMIDT, R. M., and K. R. HOUSEN. "Some Recent Advances in the Scaling of Impact and Explosion Cratering." *International Journal of Impact Engineering* 5 (1987):543–560.

SCHULTZ, P. H. "Cratering on Mercury: A Relook." In *Mercury,* eds. F. Vilas, C. R. Chapman, and M. S. Matthews. Tucson, 1988, pp. 274–335.

STOFFLER, D. "Deformation and Transformation of Rock-forming Minerals by Natural and Experimental Shock Processes. 1. Behavior of Minerals under Shock Compression." *Fortschift Mineralogie* 49 (1972): 50–113.

MARK CINTALA

IMPACT CRATERING AND THE BOMBARDMENT RECORD

Impact craters are the dominant geologic features on most solid-surface bodies in the solar system, Earth being one of the notable exceptions. These craters are produced when fast-moving space debris strikes an object, creating an explosion that destroys the impacting object and creates a crater in the target material. The prevalence of impact craters on the solid-surfaced planets, moons, and asteroids indicates that impact cratering has been a common process throughout the history of the solar system (*see* IMPACT CRATERING). Earth likely has received its share of impact craters as well, but the active geologic environment on our planet has destroyed evidence of most of these events (*see* WEATHERING AND EROSION, PLANETARY; PLATE TECTONICS).

The number of impact craters on a surface provides important information about the age of that surface. If we assume that impact cratering has been a continuous process throughout solar system history, an area with more impact craters will be older than an equivalent area with few or no impact craters. This is the way in which scientists determine the ages of different geologic units on solid-surfaced objects in our solar system.

Relative Ages

Most planets display a variety of surface units with differing geologic characteristics. One area may display a large number of impact craters while another area of equivalent size contains only a few craters. If impact cratering has been a continuous process on a particular planet over time, the older surface will display more craters than the younger surface since statistically there is a better chance that an impact event will have occurred for an area that has been exposed for a longer period of time. Thus, an area with many craters is older than an area with fewer craters. This is the basis of relative age-dating using impact craters. This technique is relative since one talks about the age of a unit relative to that of another unit. Scientists determine the relative ages of terrane units through use of crater density, the number of craters greater than a certain size per unit area. Older units have greater crater density than younger units.

Scientists also use graphical representations to determine if the frequency of craters varies by crater diameter. These representations are called crater size-frequency distribution graphs. Two major techniques are utilized: (1) cumulative size-frequency distribution plots; and (2) relative size-frequency distribution plots.

The cumulative size-frequency distribution technique is the most common way in which crater data are displayed. In this method the total (or cumulative) number of craters greater than or equal to a specified diameter per unit area is plotted against crater diameter (Figure 1a). The result for most surfaces is an approximately straight line, which can be represented mathematically by a power law function with a −2 slope ($N = KD^{-2}$, where N is the total number of craters of size D and larger, K is a constant, and D is the crater diameter). Crater density is indicated by vertical placement of each line, with areas of greater crater density (i.e., older age) lying above the plots for areas of lower crater density (i.e., younger age).

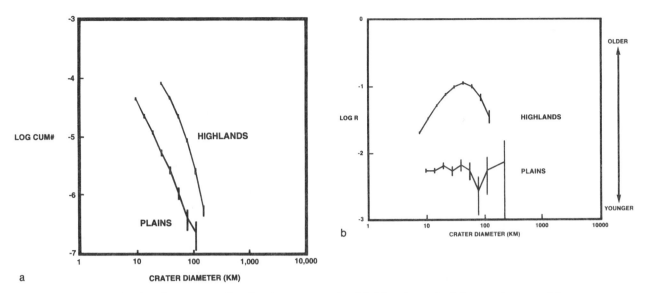

Figure 1. Example size-frequency distribution curves for highlands and plains regions on Mars. a. Cumulative crater size-frequency distribution curve. b. Relative crater size-frequency distribution curve for same data plotted in a.

A problem with the cumulative size-frequency distribution technique is that it tends to smooth out frequency variations within particular diameter ranges. These frequency variations provide information about differences in the size-frequency distribution of the impacting populations and therefore are important to the interpretation of cratering records. For this reason, a second size-frequency distribution technique, called the relative plotting technique (or R-plot), is used by some crater specialists. The R-plot technique differs from the cumulative technique in two ways: (1) the crater size-frequency distribution data are normalized to a power law distribution with a −2 cumulative slope; and (2) only craters within a specified diameter range are included in the computation of each frequency value. If crater size-frequency data follow a straight line of −2 slope, as suggested by the cumulative plot results, they should appear as a horizontal line on an R-plot.

Figure 1b shows the R-plots of the same cratered areas displayed in Figure 1a as cumulative plots. Two results are apparent: (1) crater density is still indicated on the plot by vertical placement, with older units higher than younger units; and (2) not all areas exhibit a crater size-frequency distribution that can be described by a single-sloped power law function at all diameters. The different shapes of the curves for heavily cratered and

lightly cratered units in Figure 1b have been ascribed to three possible causes: (1) the heavily cratered areas are "saturated" (regions so completely covered with craters than an existing crater is obliterated for each new crater formed); (2) erosion, whereby a single-sloped size-frequency distribution patter is altered by preferential destruction of small craters; and (3) actual differences in the size-frequency distribution of the impacting populations with time. For various reasons (see Strom et al., 1992, for discussion), the preferred explanation for the different curve shapes is changes in the impacting populations with time.

The Solar System Record

What do the cratered surfaces of the planets and moons in our solar system tell us about the impacting objects? Some interesting results are found by looking at the R-plots of cratered surfaces throughout the solar system.

Crater size-frequency distribution analyses of cratered surfaces within the inner solar system (Mercury, Venus, Earth, Moon, and Mars) suggest that two populations of impacting objects have dominated the cratering record in this region of the solar system (Figure 2). Heavily cratered regions of Mercury, Mars, and the Moon all display multi-sloped size-frequency distribution curves

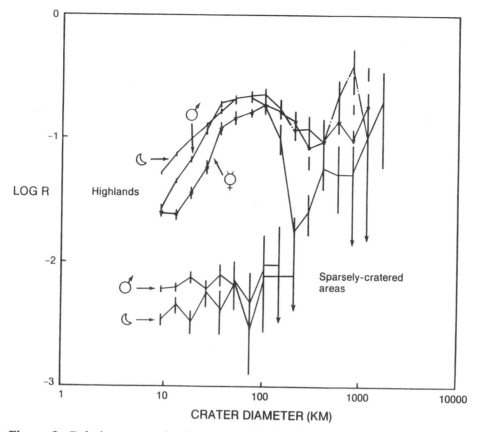

Figure 2. Relative crater size-frequency distribution curves for heavily cratered highlands and sparsely cratered areas on the Moon (☾), Mercury (☿), and Mars (♂).

that are statistically identical except for small variations which can be attributed to erosion and/or differences in impact velocity. These areas probably were impacted by material left over from the accretion of the planets. Lightly cratered regions of the Moon and Mars, as well as the cratered surfaces of Venus and Earth, can be described by a single-sloped size-frequency distribution function that probably is the result of impacts by asteroids and comets.

In the outer solar system, scientists look at the cratering record preserved on the icy satellites of the gaseous giant planets. In the Jupiter system, the cratering records of Ganymede and Callisto indicate only one distinguishable population of impacting objects, different from either of the two populations in the inner solar system and thus indigenous to the Jupiter system. The Saturnian moons suggest two populations of impacting objects endemic to the Saturnian system. Two popu-

lations of impacting objects also are suggested in the Uranian system, both different from the populations observed elsewhere in the solar system. The cratering record for the Neptunian system (based on craters on Triton) is too sparse to draw firm conclusions, and no information is yet available for the Pluto–Charon system. The cratering records on the two asteroids so far visited by spacecraft (Gaspra and Ida) indicate complex histories often reset by collisional disruption of larger objects to the current fragments.

Absolute Ages

The relative ages of terrane units are useful to scientific understanding of the geologic evolution of planetary surfaces. However, information about the thermal evolution of such bodies requires actual (or absolute) dates for the geologic events of

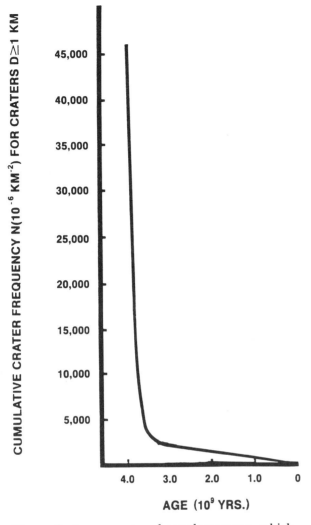

Figure 3. Lunar crater chronology curve, which relates the cumulative crater density to the formation age of the terrane. This curve is derived from comparison of the surrounding crater density to the radiometric ages determined for the lunar samples returned by the Apollo landings.

interest (*see* GEOLOGIC TIME; GEOLOGIC TIME, MEASUREMENT OF). Such absolute dating of a region generally can be obtained only through radiometric dating of rock samples. This type of dating technique has been applied only to terrestrial rocks, the limited selection of lunar samples returned to Earth by the Apollo astronauts and Lunokhod rovers, and meteorites (*see* METEORITES; METEORITES FROM THE MOON AND MARS). However,

absolute ages for geologic units on other planets and moons can be estimated by combining the lunar radiometric data with crater density information.

For each lunar sample dated by radiometric techniques, scientists are able to determine the crater density of the area surrounding the site from which the sample was collected. Putting this information together for all lunar samples results in a lunar crater chronology curve (Figure 3), which relates the crater density of any region to an estimated absolute age. Thus the lunar highlands, with their very high crater density, date from around 4×10^9 years ago, whereas the lightly cratered maria formed closer to 3.2×10^9 years ago. The transition from the high impact rates recorded in the lunar highlands to the lower impact rates recorded in the maria appears to have occurred around 3.8×10^9 years ago, a time believed to mark the end of heavy bombardment.

The lunar chronology curve can be extrapolated to other solid-surfaced bodies provided the following three conditions are met: (1) the craters were produced by the same impacting population; (2) the timing of the end of the heavy bombardment period is known; and (3) the impact rate relative to the lunar impact rate is known. The first condition is met within the inner solar system for each of the two impacting populations, as indicated by the similar shapes of the size-frequency distribution curves among the planets and the Moon. Most models of early solar system evolution suggest that the end of heavy bombardment was essentially simultaneous throughout the inner solar system, so the second condition can be met by assuming the end of heavy bombardment was at 3.8×10^9 years ago. Satisfying the third condition requires knowledge of the sources of the impacting populations, but the expected impact rates at the inner solar system planets with respect to the Moon are shown in Table 1. These assumptions then produce the often quoted absolute ages for different geologic features on the inner solar system planets. However, small changes in the basic assumptions can result in large changes in the absolute ages derived for various features, so these ages must be taken as estimates only. For the outer solar system, the impacting populations are different from the impact populations at the Moon, so absolute ages for terrane features on the icy satellites are even less well constrained than for the inner solar system planets.

Table 1. Estimated Impact Rates Relative to the Moon

Source Objects	Mercury	Venus	Earth	Moon	Mars
Asteroids	0.8	1	0.9	1	2
Comets	5	2	0.7	1	0.3
40% comets/60% asteroids	2	1	1	1	2
Mars crossers favored	2	1	2	1	4
Earth-crossing asteroids only	0.7	0.8	1.2	1	0.2
Maximum likely	5	2	2	1	4
Minimum likely	0.8	0.8	0.9	1	1
Most likely	2	1	1.5	1	2

Source: After *Basaltic Volcanism on the Terrestrial Planets*, p. 1080.

Cratering Rates

How often do impacts occur on Earth and Moon? We can estimate the current crater production rate from a number of sources, depending on the size of the impacting material. For very small material, we utilize measurements of impacts on spacecraft such as the Long Duration Exposure Facility, which was in Earth orbit for five years (*see* INTERPLANETARY MEDIUM, COSMIC DUST, AND MICROMETEORITES). Intermediate-sized impacts can be detected by seismic disturbances on the Moon and Earth, and the largest material can be telescopically observed as Earth-crossing asteroids and comets. Observational evidence leads scientists to estimate that the Moon has a current cratering rate of approximately 2×10^{-14} craters/km^2-yr for craters greater than 4 km in diameter. The observed cratering rate for craters greater than 22.6 km in diameter on Earth is 1.8×10^{-15} craters/km^2-yr. Cratering rates during the period of heavy bombardment were many orders of magnitude higher. Obviously impacts of segmented projectiles, such as the July 1994 impact of the twenty pieces of Comet Shoemaker-Levy 9 into Jupiter, can dramatically affect the general estimates of cratering rates given above. Recent calculations suggest that Earth is struck, on average, every hundred thousand years by objects of approximately 1 km in diameter. An impact 65×10^6 years ago of an approximately 10-km-sized object on Mexico's Yucatan peninsula is believed to have created the mass extinction of more than 50 percent of all species living at the time, including the dinosaurs (*see* EXTINCTIONS). Although such large impacts are rare, they can produce catastrophic results for life on Earth. As a result, astronomers have instituted the Spacewatch Survey, which uses telescopes to discover and identify Earth-crossing asteroids and comets that could potentially collide with Earth.

Bibliography

BASALTIC VOLCANISM STUDY GROUP. *Basaltic Volcanism on the Terrestrial Planets*. New York, 1981.

CHAPMAN, C. R., and D. MORRISON. *Cosmic Catastrophes*. New York, 1989.

GEHRELS, T., ed. *Hazards due to Comets and Asteroids*. Tucson, AZ, 1995.

STROM, R. G., S. K. CROFT, and N. G. BARLOW. "The Martian Impact Cratering Record." In *Mars*, eds. H. H. Kieffer, B. M. Jakosky, C. W. Snyder, and M. S. Matthews. Tucson, AZ, 1992.

NADINE G. BARLOW

INDUSTRIAL MINERALS

Several chemical elements and a number of minerals are used primarily in a variety of industrial applications to make products such as special glasses and insulation, metals and alloys, soaps and detergents, and chemicals.

Sodium Sulfate

Sodium sulfate is produced from both natural and synthetic (or by-product) sources, with approximately half of total production from each category. Natural deposits include anhydrous Na_2SO_4 (thenardite), hydrous sodium sulfate (mirabilite), and mixed sodium and calcium sulfates (glauberite). Dissolved sodium and sulfate are liberated during weathering of sodium-rich volcanic rocks and shales containing sulfide minerals; dissolved in surface water, the sodium and sulfate may enter young (postglacial) lake basins. When the lake water evaporates, sodium sulfate minerals start to precipitate. In addition, the concentrated brines (salty waters) in some lakes, such as Great Salt Lake in the United States, serve as a source of sodium sulfate. The world reserve base of natural sodium sulfate from evaporite sources is 4.6 billion metric tons; approximately half of this amount is in the countries of the former USSR and approximately one-third is in the United States.

Synthetic sodium sulfate is a by-product in a number of chemical industries, including those that manufacture rayon, hydrochloric acid, lithium carbonate, and sodium dichromate. World production of sodium sulfate in 1991 totaled approximately 4.8 million metric tons, of which 2.5 million tons were natural and 2.3 million tons synthetic. The United States consumed about 750 thousand metric tons of sodium sulfate in 1991, in the form of detergents (53 percent), paper and pulp (27 percent), and other products or processes.

Soda Ash (Sodium Carbonate)

The main mineral source for soda ash, trona (a compound of sodium carbonate, sodium bicarbonate, and water), forms through the evaporation of lakes fed by material weathered from sodium-bearing volcanic rocks and from carbonate rocks. Other sources of chemicals for such lakes may include carbonate-rich springs, atmospheric carbon dioxide, and carbon dioxide released by decay of plant and animal remains.

The world reserve base of soda ash is approximately 40 billion metric tons, of which almost 39 billion tons occur in the United States. The largest single deposit of trona is ancient Lake Gosiute in southwestern Wyoming. The deposits there are adequate to meet U.S. demand for over three

thousand years and world needs for at least seven hundred years, based on current levels of consumption. Smaller evaporitic lakes found in California (e.g., Searles Lake, Owens Lake) also contain appreciable quantities of trona. World production of soda ash in 1991 was about 31 million metric tons, including 9 million tons from the United States, 4 million tons from the countries of the former USSR, and 4 million tons from China. About 6.5 million metric tons of soda ash were consumed in the United States during 1991 in the following areas: glass and related industries (50 percent), sodium-based inorganic chemicals (24 percent), soap and detergents (13 percent), and miscellaneous uses, including water treatment, and pulp and paper (13 percent).

Boron

Boron is another element concentrated in economic deposits found only in lakes. Certain volcanic rocks in the western United States have relatively high boron contents, which are released during weathering and transported to lakes by streams. Springs may also feed dissolved boron into lakes. Evaporation of these lake waters causes precipitation of boron minerals, usually hydrous Na and/or Ca borates (tincalconite, colemanite, and ulexite). The large Kramer deposit at Boron, California, consists of a 15- to 80-m-thick accumulation of boron minerals, occurring over a square-mile area. Other large deposits of this type are located at Searles Lake and at Furnace Creek in Death Valley, both in California. Other large deposits are found in Turkey, Russia, Argentina, Bolivia, and Chile. Boron reserves of the United States are adequate to meet demand to the year 2000. Annual worldwide production capacity of boron oxide in 1991 totaled 1.4 million metric tons, with 53 percent produced in the United States and 40 percent in Turkey. World production of boron minerals that same year totaled almost 3 million metric tons, with the United States and Turkey each producing about 40 percent.

U.S. consumption of boron minerals totaled 261 thousand metric tons, with glass applications, including borosilicate glasses and fibers for insulation and textiles, accounting for 62 percent; soap and detergents for about 9 percent; and fire retardants account for 5 percent. Other important uses are in metal industries and agriculture.

Bromine

Bromine is present in seawater in small amounts, but it is usually recovered from evaporated brines that have higher bromine content than ocean water. World production capacity for bromine was 532 million kilograms (kg) in 1991. Most of this capacity is in the United States, Europe, and the Middle East (Israel). Annual production in 1991 was 401 million kg, primarily from the United States and Israel, with lesser amounts from Japan, the United Kingdom, and the countries of the former USSR.

Ethylene dibromide (EDB) is a bromine compound that is added to lead-bearing gasolines to keep lead from precipitating on cylinders and other engine parts. EDB is also used in the dye and pharmaceutical industries. Methyl bromide is a fumigant, and other bromine compounds are used for fire retardants, dyes, pharmaceuticals, photography, and plastics. Bromine is replacing chlorine as a water purifier in swimming pools, hot tubs, and spas. Bromine use in fire retardants accounted for 30 percent of annual production in 1988; drilling fluids accounted for 25 percent; EDB in gasoline (18 percent, down from 27 percent in 1983, reflecting the phase-out of leaded gasolines); and agricultural applications (for use as a soil fumigant), 15 percent.

Fluorine

Fluorite (CaF_2, commonly referred to as fluorspar) is the most important source of fluorine. Fluorite deposits are found in fissure veins (fracture fillings) and replacements (fluorite grains take the place of preexisting minerals) in limestones and other carbonate rocks, in such places as southern Illinois, South Africa, and Mexico. Fluorspar also occurs in replacement deposits in carbonate rocks at contacts with igneous bodies, veins in fracture zones in and around igneous bodies, and gangue (non-ore) minerals in base metal deposits. Fluorine is also an important by-product of phosphate rock processing.

Worldwide reserves of fluorspar totaled 239 million metric tons of contained CaF_2 in 1991. Largest reserves occur in the countries of the former USSR, Mongolia, South Africa, China, and Mexico. Of the worldwide production of 4.4 million metric tons of fluorspar in 1991, China's share

was 36 percent, Mongolia's 12 percent, Mexico's 8 percent, and South Africa's 6 percent.

Major uses of fluorine, based on U.S. figures for 1991, are in chemicals (68 percent), steel production (15 percent), and treatment of aluminum ores, ceramics, and other uses (18 percent). The purest ores, called acid-grade fluorspar, are used in chemicals. Ceramic-grade and metallurgical-grade fluorspar, used in ceramics and steel production, respectively, are progressively less pure.

Hydrogen fluoride (HF, hydrofluoric acid) is made by reacting fluorite with sulfuric acid, which forms calcium sulfate (gypsum or anhydrite) as a by-product. Much of this hydrogen fluoride is used to form fluorocarbons. Hydrogen fluoride is also used to make aluminum fluoride (AlF_3) or cryolite (Na_3AlF_6) for production of aluminum metal from bauxite ore. Iron and steel foundries add about 2.5 kg of fluorite per ton of steel produced to help remove sulfur and phosphorus and to promote effective slag formation. Hydrogen fluoride is also used to make uranium hexafluoride for separation of ^{235}U from ^{238}U, and to prepare sodium fluoride used in water fluoridation.

Graphite

Graphite is a form of carbon that is usually soft and black; the other polymorph (mineral of the same composition but with different structure and properties) of carbon is diamond. Graphite occurs in nature as fissure vein fillings (so-called lump graphite) and disseminated in metamorphic rocks (so-called flake graphite). Graphite veins probably formed from hydrothermal fluids (hot waters that circulate through the rocks and dissolve and precipitate carbon). Flake graphite formed from metamorphism of carbon compounds originally present in the rocks. So-called amorphous graphite (microcrystalline is a better name) is the product of metamorphism of coal. World graphite reserves are approximately 21 million metric tons. Annual production in 1991 totaled 629 thousand metric tons, with over half the total from China and Korea. Other major producers include India and the countries of the former USSR. The United States contains no workable deposits of graphite.

In 1991, the United States consumed 36,000 metric tons of graphite. Major uses were foundry facings and refractories (fire bricks, etc., that can stand very high temperatures), brake and clutch linings, and lubricants. Other uses include batteries, pencils, rubber, and a variety of other carbon products.

Barite

Barite, $BaSO_4$, occurs in sedimentary layers, along with cherts and siltstones; in veins in many rock types; and as replacement bodies in limestones. So-called residual deposits are formed when any of these deposits are weathered and the more resistant barite is concentrated at the surface. In the United States, bedded deposits are the most important, and those in Nevada produce about 78 percent of the U.S. total. World production capacity in 1991 was estimated to be 8.2 million metric tons, with the largest capacities in China, the United States, India, Mexico, the countries of the former Soviet Union, and Turkey. In contrast, actual production in 1991 was only 5.3 million metric tons, with approximately one-third from China. Other major producers included India, the United States, the former USSR, Morocco, and Mexico. Estimated consumption of barite in the United States in 1991 was 1.25 million metric tons, or about one-fourth of world production.

Because of its high specific gravity (4.5 for pure barite), barite is an important weighing agent in drilling muds, used primarily in oil and gas wells. The mud lubricates and cools the drill bit, serves as a transporting agent for cuttings (chips of rock produced by the drill bit), maintains formation pressures, and keeps the drill hole from collapsing. In 1991, approximately 46 metric tons of barite were used in each well drilled. Because 92 percent of its use is in drilling muds, barite consumption varies directly with the level of drilling for oil and gas. Another 7 percent of barite is used as a filler in paints, plastics, and rubber. Minor uses include production of numerous Ba chemicals, special papers, inks, and glass.

Bibliography

BROOKINS, D. G. *Mineral and Energy Resources: Occurrence, Exploitation, and Environmental Impact.* Columbus, OH, 1990.
CRAIG, J. R.; D. J. VAUGHN; and B. J. SKINNER. *Resources of the Earth.* Englewood Cliffs, NJ, 1988.
KESLER, S. E. *Mineral Resources, Economics and the Environment.* New York, 1994.
U.S. BUREAU OF MINES. *An Appraisal of Minerals Availability for 34 Commodities.* Bulletin 692. Washington, DC, 1987.
U.S. DEPARTMENT OF THE INTERIOR, BUREAU OF MINES. *Minerals Yearbook*, Vol. I, *Metals and Minerals.* Washington, DC, 1989.
————. *Minerals Yearbook*, Vol. I, *Metals and Minerals.* Washington, DC, 1991.

JOSEPH L. GRAF, JR.

INSECTS, HISTORY OF

The insects are a group of predominantly terrestrial arthropods (*see* ARTHROPODS), and are the most numerous of any animals on Earth. Nearly one million species of living insects have been described; estimates of the number of undescribed, living species range from one to four million. Thus, insects account for between about 80 percent and 95 percent of the living species of animals on Earth! Insects are an important source of food for other animals, and have economic importance for humans as pollinators of plants, or as destructive or disease-carrying pests.

The full importance of insects during the geologic past is difficult to evaluate because their soft, unmineralized exoskeletons, combined with their terrestrial habitat, have resulted in a generally poor fossil record (*see* FOSSILIZATION AND THE FOSSIL RECORD). Most fossil species (Figure 1) are from a relatively small number of localities of various geologic ages where organisms have been preserved with soft or lightly skeletalized parts intact. In most places where ancient insects are preserved, the wings are the parts most likely to be fossilized. This is fortunate for the purposes of categorization because characteristics of the wings are important in classifying living insects. Some of the most spectacularly preserved insects are ones preserved in their entirety in amber (fossil pine tree sap) from the Mesozoic and Cenozoic eras (*see* FOSSILIZATION AND THE FOSSIL RECORD). In spite of a poor overall fossil record, more than seven thousand species of ancient insects have been described. The total diversity of insects during the geologic past must have exceeded that number by many times, and probably ranged in the tens of millions of species.

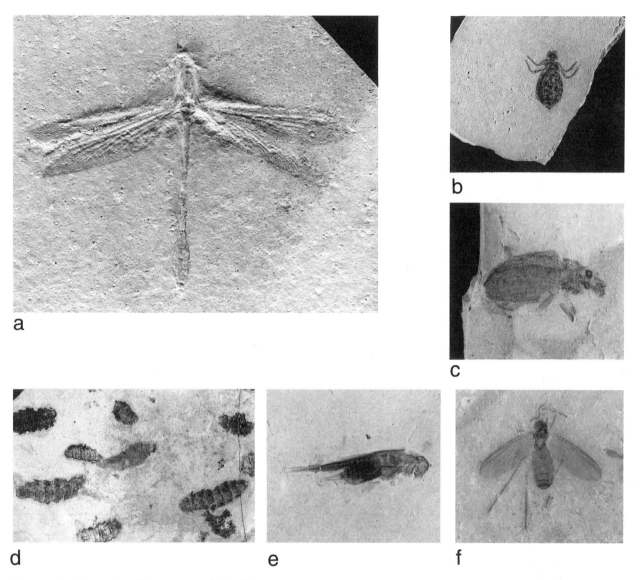

Figure 1. Exceptionally preserved fossil insects.
a. An adult dragonfly, *Tarsophlebia* sp.; from the Jurassic of Germany; ×0.75.
b. A larval dragonfly (or naiad), *Libellula doris;* from the Oligocene of Italy; ×1.25.
c. A coleopteran (beetle); from the Eocene of Colorado; ×5.
d. Botfly larvae, *Lithohypoderma ascarides;* from the Eocene of Colorado; ×0.75.
e. An orthopteran, *Pronembius* sp.; from the Eocene of Colorado; ×1.5.
f. A dipteran (fly); from the Eocene of Colorado; ×5.
All photos by Loren E. Babcock.

Insects are familiar to all people, but the question of what exactly constitutes an insect is still unsettled. Most specialists classify the Insecta as a class within the superclass Hexapoda, and three small groups that are primitively wingless and have covered mouth parts, springtails (Collembola), proturans (Protura), and diplurans (Diplura), are regarded as separate classes. That is the procedure followed here. A less common scheme involves the classification of these three small groups as orders, along with about twenty-five other orders, in the class Insecta. In the classification followed here, hexapods are arthropods that respire by means of tracheae (air tubes), and that have a well-developed head, thorax, and abdomen. The head generally has a pair of antennae and a pair of unjointed limbs that are modified into jaws in front of the mouth, as well as two other limbs behind the mouth. The hexapod thorax has three pairs of walking legs, and dorsal wings are commonly present. The hexapod abdomen usually contains eleven segments. Hexapods assigned to the class Insecta have exposed mouth parts. Most insects have wings but a few (such as silverfish and bristletails) are primitively wingless, and others (such as ants, fleas, and lice) have secondarily lost their wings through evolution. Insects are overwhelmingly terrestrial but some live in aquatic environments. Most insects feed on plant juices and tissues although some feed on dead animals, some catch and eat live animals (including other insects), and some are parasitic. Some ants and termites even cultivate fungus as a source of food.

Classification and Evolution of Insects

The oldest and most primitive insects are wingless forms such as bristletails and silverfish (Thysanura), which date from the Devonian period. These early insects seem to have been terrestrial, and must have evolved from some marine ancestor. Available fossil evidence indicates that insects became dominant on land by at least the Pennsylvanian period, and probably more likely the Mississippian period.

Representatives or common ancestors of most modern insect groups have a known stratigraphic record that begins in the Pennsylvanian period, and it seems likely that insects underwent one of their most important phases of evolutionary radiation shortly before this time, during the Mississippian period. Unfortunately, Mississippian insects are not known, a problem that is due largely to the widespread covering of continents by shallow seas during this time and the lack of deposits conducive to preserving terrestrial arthropods.

By the end of the Paleozoic era, representatives of the major groups of winged (flying) insects, paleopterans and neopterans, had appeared on Earth. Paleopterans, which include dragonflies (Odonata) and mayflies (Ephemeroptera), are insects that cannot fold their wings against their bodies and have a hemimetabolous development. Hemimetabolous insects have immature forms (nymphs if terrestrial, naiads if aquatic) that are often voracious predators and bear little resemblance to adults. Paleopterans evolved by at least the Pennsylvanian period, and some early forms achieved enormous size. One species of Pennsylvanian dragonfly had a wingspan of about 75 cm.

Neopteran insects, which can fold their wings against their bodies, have a fossil record that begins in the Pennsylvanian period. The ability to fold their wings enabled neopteran insects to move more quickly in search of food or to escape predators when they were on the ground. Two types of development, paurometabolous and holometabolous development, evolved among the neopterans. Cockroaches (Blattodea), praying mantises (Manteodea), termites (Isoptera), earwigs (Dermaptera), grasshoppers (Orthoptera), stoneflies (Plecoptera), booklice (Psocoptera), thrips (Thysanoptera), and aphids, bugs, and cicadas (Hemiptera) all have a paurometabolous development. In these forms, the immature forms (nymphs) resemble adults but are smaller and lack well-developed wings and genitalia. The oldest known cockroaches and the ancestors of grasshoppers and crickets are from Pennsylvanian rocks. The oldest known booklice, stoneflies, cicadas, and aphids are from the Permian period. So-called true bugs appeared during the Triassic period.

Neopterans having a holometabolous development have a fossil record that extends from the Permian period. Ants, wasps, and bees (Hymenoptera), beetles (Coleoptera), scorpionflies (Mecoptera), fleas (Siphonaptera), flies and mosquitos (Diptera), caddis flies (Trichoptera), and butterflies and moths (Lepidoptera) have a holometabolous development. In these forms, the immature forms are wormlike or caterpillar-type larvae that lack compound eyes. The larvae have feeding habits that are different from adults, which permits them to avoid competing with adults for food. Following the larval phase, the insect enters into a phase called a pupa, during which time it undergoes intense physiologic activity and morphologic change. It is during this time that the animal is transformed into an adult. With a few notable exceptions (such as ants and fleas), adult holometabolous insects have wings. Beetles, scorpionflies, and caddis flies are some of the holometabolous insects that appeared during the Permian period. Ants, wasps, bees, fleas, flies, butterflies, and moths first appeared during the Mesozoic era.

The rise or diversification of warm-blooded animals and flowering plants during either the Mesozoic or Cenozoic era seems to have been accompanied by coevolutionary radiations of a variety of insects belonging to the orders Diptera, Siphonaptera, Lepidoptera, and Hymenoptera. Blood-sucking parasites of mammals, notably mosquitos, horse flies, and fleas, have fossil records extending from the Jurassic or Cretaceous periods, but it is possible that these groups originated as early as the Triassic period. The earliest blood-sucking insects may have fed on warm-blooded dinosaurs, birds, or mammals. At present, it is uncertain whether genetic evidence about prey species will be extractable from preserved blood in the bodies of blood-sucking insects. Accompanying the great evolutionary diversification of mammals in the early Tertiary period, fleas seem to have undergone a similar adaptive radiation. Butterflies and moths made their first appearance during the Cretaceous period, as did the flowering plants (angiosperms). Butterflies and moths are important pollinators of angiosperms, and their caterpillar larvae feed primarily on the tissue of flowering plants. Angiosperms, butterflies, and moths seem to have undergone major adaptive radiations during the Cretaceous and Tertiary periods. Another group of pollinators, the bees, seem to have undergone evolutionary diversification beginning in the Cretaceous and coinciding with the rise of flowering plants. Angiosperms, which rely on bees for pollination, provide bees with pollen and nectar that is needed for the production of honey.

Bibliography

BARNES, R. D. *Invertebrate Zoology*. Philadelphia, 1987.

CARPENTER, F. M. "Superclass Hexapoda." In *Treatise on Invertebrate Paleontology, Part R, Arthropoda 4, Volumes 3, 4*, ed. R. L. Kaesler. Lawrence, KS, 1992.

HENNIG, W. *Insect Phylogeny*. New York, 1981.

HICKMAN, C. P., JR.; L. S. ROBERTS; and F. M. HICKMAN. *Biology of Animals*. St. Louis, MO, 1982.

LABANDEIRA, C. C., and B. S. BEALL. "Arthropod Terrestriality." In *Arthropod Paleobiology*, ed. D. G. Mikulic. Knoxville, TN, 1990.

POINAR, G. O., JR. *Life in Amber*. Stanford, CA, 1992.

ROBISON, R. A., and R. L. KAESLER. "Phylum Arthropoda." In *Fossil Invertebrates*, eds. R. S. Boardman, A. H. Cheetham, and A. J. Rowell. Palo Alto, CA, 1987.

LOREN E. BABCOCK

INTERPLANETARY MEDIUM, COSMIC DUST, AND MICROMETEORITES

The space between planets in our solar system, or interplanetary medium, is not empty. This region is host to numerous forms of energy, gas, and solid materials. The Sun continually sheds ionized gases and their associated magnetic field, collectively called the solar wind. The solid material ranges in size from small asteroids (less than 10 km), to rock-sized meteoroids, to micrometeoroids (less than 1 mm), down to interplanetary dust particles (less than 100 micrometers or μm).

The solar wind consists of an initially hot (10^5 K) plasma (a collection of ions and electrons), originating in the solar corona (outer atmosphere of the Sun). The plasma flows outward from the Sun in the form of a spiral, at a velocity of approximately 450 km/s. The density of this material varies with the inverse square of its distance from the Sun; at Earth the density is approximately 5 particles/cm^3 under ambient conditions. The ionized portion of the plasma carries an embedded magnetic field; at Earth the field strength is approximately 5×10^{-5} gauss. While at present the solar wind has been studied only near Earth, in the 1990s the *Ulysses* spacecraft (of the European Space Agency) will investigate it at high solar latitudes.

There is a constant flow of atomic nuclei (principally protons) and electrons called cosmic rays into Earth's atmosphere. The elemental ratios of nuclei in cosmic rays are similar to those for the Sun (this composition is called chondritic), which means that they are dominated by the nuclei of low atomic number elements (iron and lighter elements, especially hydrogen and helium). Those particles believed to originate outside the solar system, called galactic cosmic rays, are probably directed and accelerated on their travels by solar flares and supernovas. These particles penetrate into the solar system to a degree that varies inversely with the strength of the eleven-year solar cycle. When solar activity (and the solar wind) is at a peak, galactic cosmic rays are most efficiently excluded from the inner solar system, but during the solar minimum the cosmic rays can more easily penetrate; at Earth, galactic cosmic rays can be twice as intense during the latter period. The edge of the solar system is considered to be the point at which the solar wind merges with the interstellar medium (called the heliopause), probably located at 50–100 AU, depending on solar latitude. *Pioneers 10* and *11* and *Voyagers 1* and *2* are now being used to search for the heliopause.

The larger solid bodies in the interplanetary medium, asteroids and meteoroids, are described in other articles (*see* ASTEROIDS: COMETS). A cone of light is visible on the western horizon immediately after sunset and on the eastern just preceding sunrise in favorable viewing conditions. This zodiacal light arises from sunlight scattered off interplane-

tary dust particles lying within the orbit of Mars and concentrated on the ecliptic plane. An especially bright portion of the zodiacal light, directly opposite the Sun, is called the Gegenschein. Interest in this dust has been intense because the local cloud is a convenient place in which to examine such astrophysical processes as grain evolution and destruction, sputtering, magnetic fields, and radiation pressure from the Sun. The latter process (called Poynting Robertson Drag) acts to expel the finest dust (generally smaller than 1 μm) from the inner solar system and drag larger grains into the

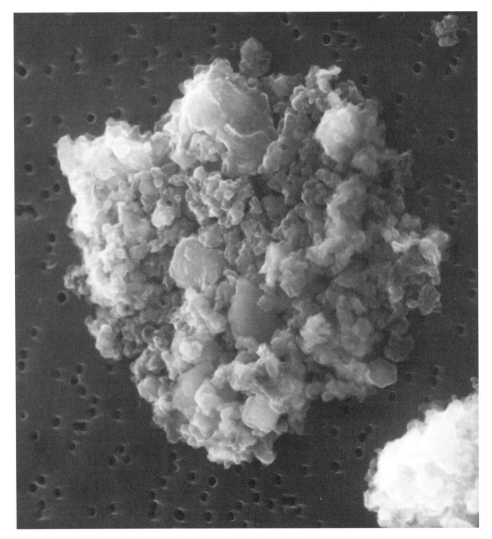

Figure 1. A scanning electron microscope image of a chondritic interplanetary dust particle collected in the stratosphere. Although it measures only 15 μm across, it is observed to be an aggregate of millions of far smaller grains. Note the high porosity of this potentially cometary particle. (Photo courtesy of Michael E. Zolensky.)

Sun, both with time scales on the order of 10^3 to 10^5 years. Thus, these particles must be continuously replenished by active comets, colliding asteroids, large impacts on planets and moons, and interstellar sources. About 10^8 kg of these particles enter Earth's atmosphere annually, whereupon they are referred to as micrometeorites.

Smaller solid bodies entering Earth's atmosphere at typical velocities of 12–72 km/s lose mass by ablation (fragmentation, melting, and vaporization), ionizing the air in the process; these incandescent materials can be observed from the ground as meteors or fireballs. Systematic observation of visible meteors has shown there to be two classes: (1) sporadic, which occur randomly and come from all directions; and (2) shower meteors, which come at certain dates, from one direction (called the radiant), and which appear to originate from meteoroid streams. Examples of these meteor showers are the Perseids (12 August, which appear to originate from comet Swift-Tuttle), Eta Aquarids (4 May, probably related to comet Halley), and Orionids (20 October, also associated with comet Halley). Unfortunately, cometary materials larger than dust do not survive atmospheric entry, due to low internal strength.

Interest in interplanetary dust has also been spurred by the realization that their impacts onto spacecraft with typical velocities in excess of 7 km/s are a hazard, resulting in the degradation of space-exposed surfaces and even the breaching of spacecraft hulls and spacesuits. Consequently, there have been numerous attempts to measure the flux of interplanetary dust particles in space,

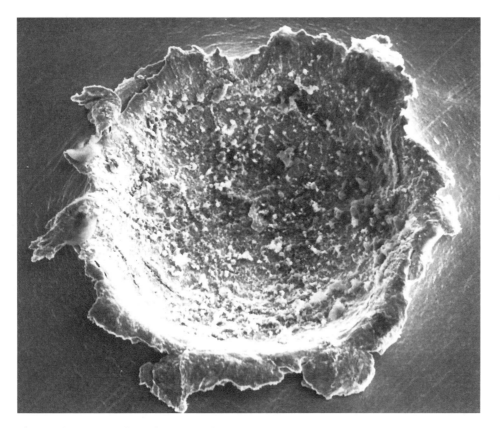

Figure 2. A scanning electron microscope image of a crater resulting from the impact of an interplanetary dust particle into aluminum on the Long Duration Exposure Facility Satellite. The bowl-shaped impact crater measures 24 μm across and contains clearly visible residue from the impacting grain (which probably measured only 5 μm across and was traveling at $\sim$10 km/s). (Photo courtesy of Ron Bernhard, Lockheed ESCO.)

involving the HEOS (Highly Eccentric Orbiting Satellite), *Helios* and *Pioneer* spacecraft, and, more recently, the Long Duration Exposure Facility (LDEF), which monitored the space environment in low-Earth orbit for 5.7 years. In spite of these efforts, the distribution and detailed sources of dust in the solar system are imperfectly understood.

Whether collected by spacecraft, in Earth's stratosphere, or from the oceans or polar ice caps, micrometeorites and interplanetary dust particles provide scientists with a unique sample of asteroids, our only samples of comets and, potentially, interstellar materials. The distinction between dust derived from asteroids versus comets is important in that comets are believed to preserve unaltered interstellar grains and solar nebular condensates, whereas most asteroids experienced significant parent-body processing (from low-grade metamorphism to complete melting) early in solar system history. Micrometeorites and interplanetary dust particles are collected and curated, similar to meteorites and lunar samples, at NASA's Johnson Space Center in Houston, Texas (Figures 1 and 2).

Studies have established that micrometeorites and interplanetary dust are widely variable in composition, mineralogy, and structure, although only particles with refractory and chondritic compositions (the latter with approximately solar abundances of elements heavier than He) have received detailed attention. The chondritic grains, in particular, are the most interesting, and consist predominantly of ferromagnesian silicates (variously hydrous, anhydrous, and amorphous), oxides, sulfides, carbonaceous materials, and Fe-Ni metal. These chondritic particles generally have bulk densities ranging from 0.3 to 6.2 g/cm^3, averaging 2.0 g/cm^3. Some of the lowest density particles, which are generally anhydrous and porous, are probably cometary grains, with the pores resulting from sublimation of ices. These particles can contain very high amounts of carbonaceous materials, up to 50 percent carbon by weight. A revolution in our understanding of carbon chemistry in the early solar nebula will occur when organic analyses are sufficiently developed to permit analysis of these nanogram-sized samples. Chondritic micrometeorites also contain micrometer-sized aggregates of much smaller mineral grains (down to nanometers in size), which appear to record the earliest stages of grain condensation and accretion in the solar nebula.

Bibliography

BRANDT, J. C. "Comets." In *The New Solar System*, 3rd ed., eds. J. K. Beatty and A. Chaikin. Cambridge, Eng., 1990.

McDONNELL, J. A. M., ed. *Origin and Evolution of Interplanetary Dust*. New York, 1992.

VAN ALLEN, J. A. "Magnetospheres, Cosmic Rays and the Interplanetary Medium." In *The New Solar System*, 3rd ed, eds. J. K. Beatty and A. Chaikin. Cambridge, Eng., 1990.

MICHAEL E. ZOLENSKY

INTRUSIVE ROCKS AND INTRUSIONS

There are two groups of igneous rocks. Extrusive rocks (*see* IGNEOUS ROCKS), such as lava and volcanic ash, form when magma is erupted or extruded onto Earth's surface (*see* VOLCANISM). Intrusive rocks are formed when magma that has intruded (insinuated its way into, around, between, or through) rocks beneath the surface of Earth cools and crystallizes without reaching the surface.

Types of Intrusions

The shapes of intrusive bodies range from pipelike and tabular to domed, pluglike, or irregular.

Dikes are tabular intrusions ranging in width from under a centimeter to a kilometer or more. In map view, they appear as narrow, nearly straight bodies that are much longer than they are wide. By definition, dikes are discordant, that is, they cut across the fabric of the surrounding rocks.

Sills are also tabular, but they are concordant intrusions, intruding between rock layers rather than cutting across them.

Laccoliths are another type of concordant intrusion. Unlike sills, laccoliths are domed, forcing the overlying layers of rock to warp upward. They are relatively rare.

Pipes, as their name implies, are roughly cylindrical, discordant, tubelike intrusions. Many pipes are near-surface intrusions. Volcanic necks, for example, are a type of pipe that forms when magma freezes in a conduit that once fed a volcano.

Technically, any mass of intrusive rock can be termed a pluton. In practice, however, this term is usually reserved for relatively large intrusions that are rounded, ovoid, or irregular in map view. Plutons are essentially frozen magma chambers. They can be very large; many are tens of kilometers across. A few, such as the Bushveld intrusion of South Africa, are hundreds of kilometers in maximum dimension. Plutons over 100 km² in area are called batholiths; plutons smaller than this are termed stocks. The shape of plutons in three dimensions is often poorly known. Most are discordant, but many are at least partially concordant.

Intrusive Rock Types

Most intrusive rocks have compositions similar to lavas, volcanic ashes, and other extrusive rocks. This fact is not surprising since intrusive and extrusive rocks form when the same magmas are emplaced in different ways. True granite, for example, is the intrusive equivalent of the extrusive rock rhyolite. Intrusive and extrusive rocks look different because they cool at different rates. Extrusive rocks cool quickly, forming many small crystals, most of which are too small to see without a microscope. Most intrusive rocks cool very slowly, forming relatively large crystals, which are easily visible to the naked eye. Some intrusions cool more slowly than others, so there is a range of mineral grain sizes in intrusive rocks. Rocks intruded as small bodies near the surface may look like volcanic rocks and be given names that reflect this (e.g., basalt dike or andesite volcanic neck). Unless otherwise noted, the rock types discussed below refer to slowly cooled rocks.

Intrusive rock types are defined by mineralogy, and, less commonly, texture. Granite, granodiorite, and tonalite (see IGNEOUS ROCKS) usually occur in plutons. They are often collectively referred to as "granitic" rocks. Granodiorite is probably the single most common rock type in plutons. The Sierra Nevada Batholith of California, for example, consists largely of granodiorite.

Most gabbros and diabases (see IGNEOUS ROCKS) are the intrusive equivalents of basalt. Gabbro is texturally similar to the granitic rocks and commonly occurs in plutons. Diabase contains a mixture of fine- and coarse-grained minerals and is a common rock type in dikes, sills, and other relatively small or near-surface intrusions. Some gabbros are crystal cumulates, collections of crystals that have been separated from crystallizing magma in a magma chamber (see IGNEOUS PROCESSES). Cumulate gabbros and related rocks are sometimes layered, much like sedimentary rocks. Diorites and quartz diorites (see IGNEOUS ROCKS), the intrusive equivalents of andesites, are intermediate in composition and mineralogy between tonalite or granodiorite and gabbro.

There are literally hundreds of other named varieties of intrusive rocks, but all are less common than those named above. Many of these rock types belong to the family of alkaline rocks that are rich in alkalis and/or Ca and relatively poor in silica. The most common of these are the syenites, rocks rich in potassium feldspar, but containing little or no quartz. Perhaps the most unusual of all intrusive rocks are the carbonatites. These rocks consist almost entirely of the minerals calcite and dolomite.

Geologic and Tectonic Setting

Intrusive rocks form in the same geologic and tectonic settings as extrusive rocks (e.g., magmatism, subduction, rifting, hot spots). However, intrusive rocks form at depth and are exposed only as the landscape is eroded away. Long after volcanos and other surface expressions of magmatic activity have been destroyed by erosion, intrusive rocks remain as indicators of the tectonic and geologic history of an area.

For example, large-volume, tonalite–granodiorite–granite batholiths usually represent the underpinnings of arcs established on continental crust (such as continental or Andean arc). The presence of true batholithic intrusions from Alaska to southern Caiifornia establishes that a continental arc was active over much of western North America during the middle to late Mesozoic. Chemical analysis of intrusive rocks can provide clues to paleo-tectonics. The presence of syenites or other alkaline plutons, for example, may indicate rifting or hot spot activity.

Dikes are another useful indication of tectonic and geologic setting. A dike can be thought of as a crack in the earth that has filled with magma. These cracks form when local or regional stresses pull Earth's crust apart, as during rifting (see RIFT-

ING OF THE CRUST). A large number of parallel dikes occurring over a wide area (dike swarm) may reflect the orientation of regional stresses that are pulling an area, sometimes a whole continent, apart. A dike swarm in eastern North America formed approximately parallel to the present coastline 200 million years (Ma) ago. The dikes record the initial stages of rifting that split Pangaea apart and ultimately led to the formation of the Atlantic Ocean.

Bibliography

CLARKE, D. B. *Granitoid Rocks.* New York, 1992.
PRESS, F., and R. SIEVER. *Earth.* New York, 1978.
RAYMOND, L. A. *Petrology: The Study of Igneous, Sedimentary and Metamophic Rocks.* Boston, 1995.

JAMES S. BEARD

INVERTEBRATES

Invertebrates are animals without backbones, a category that includes the majority of animals (*see* ARTHROPODS; BRACHIOPODS; COLONIAL INVERTEBRATE FOSSILS; ECHINODERMS; INSECTS; MOLLUSKS). Their history in the sea probably spans more than 600 million years (Ma), and from about 400 Ma ago they have also been a significant part of the land biotas.

The origin of animals is obscured by the poor fossil record of the late Proterozoic, a time when microbes dominated the biosphere. Comparisons of cell structures and molecular sequences suggest that animals all evolved from one lineage of unicellular protists and are closely related to fungi. None of the first animals appear to have been fossilized, however, and would probably be difficult to recognize as animals even if they were found.

The Ediacara biota, about 600 Ma old, gives the first unequivocal evidence for the presence of animals in the sea. The biota contains large disk-, frond-, and leaf-shaped organisms reminiscent of jellyfish and seapens, as well as traces of animals moving in the sediment (*see* GLAESSNER, MARTIN F.). Because of their preservation as impressions in rock, the interpretation of these organisms is con-

troversial: a number of them may represent extinct evolutionary side branches, but at least some genuine animals appear to have been present.

The main burst of animal evolution came about 550 Ma ago, at the transition to the Cambrian period (the earliest division of the Phanerozoic). It was a very rapid ecological reorganization of the biosphere: the almost exclusively microbial biotas of the Proterozoic were expanded by the appearance of various animal groups representing different lifestyles and feeding modes. The food chains rapidly attained a complexity comparable to those of the modern world (*see* HIGHER LIFE FORMS, EARLIEST EVIDENCE OF).

By the end of the Cambrian, almost all major groups of now living animals seem to have appeared in the marine realm (if the record of fossilizable animals is considered representative; we have no record of a number of groups with low potential for preservation, such as roundworms and flatworms). Arthropods were particularly diverse and abundant, trilobites being the major players (Figure 1). Although they contained easily recognizable representatives of modern groups, the early fossil biotas also counted a number of forms with more obscure relationships to known groups. Recent discoveries of exquisitely preserved specimens have allowed us to identify the relationships of some such problematic fossils, whereas others still are of enigmatic nature. Because extinction of lineages is in the long run more common than survival, it is to be expected that some early appearing forms represent groups that, had they survived, would be quite distinct from the ones actually living today (*see* LIFE, EVOLUTION OF; EXTINCTIONS).

The subsequent evolution of invertebrates underwent a series of quantitative and qualitative changes. The Ordovician radiation, about 500 Ma ago, was almost as dramatic as the Cambrian one and involved a number of groups (such as mollusks, brachiopods, echinoderms, and chordates) that would dominate the seas for part or all of the following times. The invasion of the land habitat by various invertebrate groups (e.g., gastropod mollusks and various kinds of arthropods) mainly took place during the Ordovician–Devonian interval, though the fossil evidence for this transition is very poor. The Permian–Triassic boundary, 245 Ma ago, saw the extinction of most of the marine animal biota, and the ancestors of now living invertebrates stem from the small group of survivors of

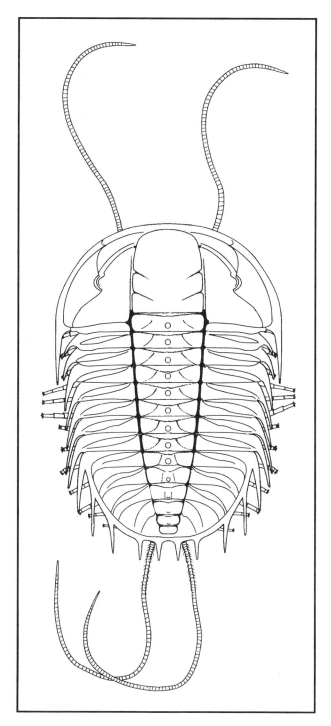

Figure 1. Trilobite.

that event. The late-Cretaceous extinction event (now thought to have been related to a major meteorite impact—*see* IMPACT CRATERING AND THE BOMBARDMENT RECORD) that ended the reign of the dinosaurs had profound effects on the whole biosphere, but it did not have the same bottleneck effect on the marine invertebrate life as the Permian–Triassic extinction. In addition to these major events, invertebrates have throughout their history been subject to a series of more or less profound extinctions and subsequent rebounds.

The pattern of invertebrate life and evolution through time is exceedingly complex, and even features that are reasonably well understood cannot be dealt with in detail in this entry. It is useful to look at the various invertebrates from the point of view of their ecological roles during their life history, but most statements on the ecology of specific groups should be seen as generalizations that do not apply to each individual case.

It is important to recognize that even the earliest appearing animals in the fossil record were fully functional organisms that most likely were no less viable than later evolving ones: many extinctions are due to chance events (for example, destruction of biotopes or major impacts of astronomical bodies) that affect "primitive" and "advanced" animals alike. Most evolutionary changes appear to consist of modifications of existing basic body plans during ecological replacements of extinct groups or during mutual adaptations between organisms dependent on each other, for example parasites and hosts, or predators and prey.

The water-filtering sponges (phylum Porifera) have a rich fossil record, though most sponge fossils consist of isolated skeletal elements (spicules) that may be difficult to identify. The cup-shaped Cambrian archaeocyathans are an important component of many algal reefs in the early Cambrian Seas. They were long regarded as an extinct phylum (Archaescyatha), but the recent discoveries of various living sponges with basal calcareous skeletons have made it quite likely that they too are a type of sponge. Similarly, the laminated stromatoporoids, particularly common in Silurian–Devonian reefs and on soft bottoms, are now generally interpreted as sponges.

In the fossil record cnidarians (phylum Cnidaria) are mainly represented by stony corals; the mostly solitary tetracorals in the Paleozoic; and from the Mesozoic the hexacorals, today a major component of coral reefs. Other cnidarians with a significant fossil representation are the octocorals (sea pens, gorgonians, and soft corals) and, probably, the peculiar conulariids, a mostly Palaeozoic group of animals that have a phosphatic skeleton shaped like an elongate pyramid. Most cnidarians capture small floating organisms by means of their stinging cells. Sponges and cnidarians are prominent reef-builders throughout the Phanerozoic, though other organisms, especially calcareous algae, are often the main contributors to the reef fabrics.

Worms, a variety of invertebrates that tend to be small and soft-bodied, have for obvious reasons a poor fossil record, but a couple of exceptions should be mentioned. TRACE FOSSILS (preserved traces of animal activities) include a lot of traces of wormlike animals. Tubes of various composition are abundant as fossils, and although their makers often cannot be identified, the tube-dwelling habit is common among annelids and other worms today. An example is the giant "rift worms" (vestimentiferans), which were recently discovered living around submarine hot springs in the deep sea (in symbiosis with chemosynthetic bacteria) and which have now also been discovered in similar fossil environments from various places.

Animals feeding on particles suspended in the water tend to be common throughout the fossil record, because they occupy the lower positions in the food chain. In addition to the sponges and many "worms," the bivalved brachiopods (lamp shells) were prominent during the Paleozoic (550–245 Ma ago), whereafter their ecological role appears to have been taken over by the bivalved mollusks (clams, mussels, etc.), which have steadily grown in importance since the middle Paleozoic. Bryozoans, minute colonial suspension feeders, have been common since the Ordovician. Colonial hemichordates (pterobranchs) have generally been less frequent, but during the Ordovician–Devonian one group, the graptolites, developed floating colonies that were a prominent component of the oceanic plankton and have been important as index fossils for age determination of, in particular, shales.

In addition to the suspension-feeding bivalved mollusks, other mollusk groups have been equally or even more successful. The gastropods (snails and slugs) are mostly grazers and predators. They have increased in numbers and forms since the early Paleozoic and have also occupied land habitats. The cephalopods (squids, octopuses, etc.) are fast-swimming predators that have a rich fossil record of shelled forms. Of these, the nautiloids include the Paleozoic straight-shelled orthoceratites and the well-known living coiled nautilus. The usually coiled ammonoids had a nautilus-like shell

and attained great size and diversity in the Mesozoic seas (they are important as index fossils). The coleoids include the Mesozoic belemnites, squidlike forms with a heavy internal skeleton, and all living cephalopods except the nautilus. As most living cephalopods are without shell, their fossil history is not so well known, but the presence of other hard parts, in particular jaw tips, has filled in some of the gaps in our knowledge.

Arthropods continued to flourish after the Cambrian bloom and are today among the most diverse and successful animal groups, having all conceivable life modes and living almost anywhere. The most prominent arthropods are the crustaceans in the sea and the insects and spiders on the land. The crustaceans include the bivalved ostracodes, of great value as index fossils throughout the Phanerozoic. The fossil record of insects goes back to the Devonian. It is possible that the arthropod invasion of land dates back to the Ordovician, as suggested by finds of trace fossils in soils that appear to have been made by myriapods.

Echinoderms (sea stars, brittle stars, sea urchins, sea cucumbers, sea lilies, etc.) include suspension feeders, deposit feeders, grazers, and predators, all living in the sea. They have an excellent fossil record due to their calcareous skeleton. Although all living echinoderms have a characteristic five-rayed symmetry, the earliest fossil echinoderms did not. The bizarre early Paleozoic carpoids have generally no symmetry plane at all. They have been suggested to be a kind of chordate, demonstrating evolutionary affinity between echinoderms and chordates (and, with them, the vertebrates), but this interpretation is coming under increasing query. The dominant echinoderms during the Paleozoic were the stalked crinoids (sea lilies), suspension-feeding sessile animals that are now mostly confined to the deep seas—the exception being certain stalkless crinoids (feather stars) in modern shallow seas. Since the early Mesozoic, the dominant echinoderms are the echinoids (sea urchins), asteroids (sea stars), and ophiuroids (brittle stars).

Chordates (other than vertebrates) have a very poor fossil record, owing to their general lack of hard parts. An amphioxus-like animal is known already from the Cambrian, however, and the appearance of vertebrates in the Ordovician supports the early origin of chordates. In particular, the toothlike microfossils called conodonts, for a long time of enigmatic affinity, are now generally interpreted as early chordates, perhaps even vertebrates, on the basis of recent discoveries of fossils with preserved soft parts resembling fins, V-shaped muscular segments, and possible eye capsules. Conodonts are important index fossils from their appearance in the Cambrian to their extinction in the early Mesozoic.

Each period in the history of animal life has its own signature, which makes it possible to determine the relative age of sedimentary rocks by means of the fossils in them (*see* STRATIGRAPHY). This does not mean that life forms were the same all over Earth at any given time. Just as today, there were geographical and ecological differences in the distribution of animals, though during some periods the biotas were more homogenous than during others. Also, the picture of invertebrate history is hardly one of "lower" organisms gradually being replaced by "higher" ones, as may be the impression gained from schematic accounts. After the initial Cambrian radiation, most of the complexity was already there, though subsequent evolutionary change involved substantial modifications of the original forms, and the invasion of land opened up new modes of life that led to the splendid diversity of land-living invertebrates, most particularly the insects.

Bibliography

BOARDMAN, R. S., A. H. CHEETHAM, and A. J. ROWELL. *Fossil Invertebrates*. Palo Alto, CA, 1987.

CLARKSON, E. N. K. *Invertebrate Palaeontology and Evolution*, 2nd ed. London, 1986.

STEFAN BENGTSON

IRON DEPOSITS

Iron is the most widely used metal in industrial societies. It is also the most abundant element in Earth, constituting 34.63 percent of Earth as a whole. Most of the iron is in Earth's core, however, and iron forms only 5.8 percent by weight of rocks in Earth's crust. Deposits from which iron can be economically recovered represent a minuscule portion of crustal rocks. These deposits, formed by a number of different geological processes, include deposition of extensive iron-rich sedimentary rocks by chemical or biochemical processes, called iron formations; segregation (separation) of molten iron minerals in igneous rocks, especially gabbros; and hydrothermal deposits formed adjacent to igneous intrusions. What processes produce these various types of deposits?

A vast majority of iron deposits formed in sedimentary environments were precipitated by chemical and biochemical processes. Iron formations consist mainly of iron minerals (hematite, magnetite, siderite, and several types of iron silicate minerals) interlayered with chert (fine-quarried silica). Because the interlayerings are strikingly banded, the term "banded iron formation" is commonly applied. Iron formation deposits may be very extensive. For example, iron formations in the Hamersley Range of Western Australia are blanketlike deposits with a cumulative thickness of approximately 1,000 m and cover an area approximately 500 by 400 km. Elsewhere, such as in the "Labrador trough" of eastern Canada, iron formation extends for approximately 1,000 km and has a thickness of 150 to 300 m. Deposits of similar size are present in South Africa, Brazil, and the Lake Superior region of North America.

Some banded iron formations formed in close association with volcanic rocks in rock sequences more than 2,500 million years (Ma) old. Iron formations are also interbedded with typical sedimentary rocks, sandstone, dolomite, and shale, in rock sequences that range from 2,500 Ma to about 2,000 Ma old. Iron formations are rare in sedimentary sequences younger than 1,800 million years old. Iron formations in the older volcanic sequences are referred to as "Algoma-type" by Gross (1980) because of their prominent development in the Algoma district north of Sault Sainte Marie, Ontario. Algoma-type iron formations are common in volcanic sequences older than 2,500 Ma (2.5 Ga) throughout the world. They are generally believed to have formed in submarine basins between volcanic islands where hot springs added iron- and silica-rich fluids to the seawater during periods of little, or no, volcanic activity. Algoma-type banded iron formations are typically overlain by rhyolitic volcanic rocks, suggesting that renewal of volcanic activity buried the iron formations.

Iron formations in sedimentary sequences, as opposed to volcanic sequences, are called "Superior Type" by Gross (1980) because of their widespread development in the Lake Superior region.

Superior-type iron formations were deposited on broad platforms in relatively shallow water.

Deposition of large quantities of iron as a chemical sediment requires that iron be present in solution in seawater as ferrous iron (Fe^{+2}) (Holland, 1973). Because the solubility of iron is strongly influenced by the presence of oxygen in the atmosphere and seawater, most geologists conclude that the large sedimentary iron deposits formed at a time when Earth's atmosphere was deficient in oxygen. It is widely postulated that iron in iron formations was precipitated as a consequence of luxuriant growth of photosynthetic (oxygen-producing) bacteria and algae (Cloud, 1973; LaBerge and others, 1987). Oxygen produced by the organisms presumably reacted with iron in seawater, converting it to insoluble ferric iron (Fe^{+3}). Development of an oxygenated atmosphere on Earth about 1,800 Ma ago converted most iron in surface environments to insoluble ferric iron. As a result, iron was not present in seawater after that time.

The sedimentary rock, banded iron formation is generally not mined as an iron ore because it contains only 20–40 percent iron. Chemical weathering, especially in tropical climates, dissolves silica out of the original rock and enriches the iron content to 65–70 percent Fe (nearly pure hematite). Extremely large deposits of high-grade hematite are being mined in the Hamersley Range of Western Australia and at Carajas, Brazil.

Extensive areas of iron formation have not been enriched by chemical weathering to form natural iron ores. In places such as the Lake Superior region unenriched iron formation (called taconite) can also be utilized for iron ore. The taconite is mined in large open pits, crushed to a fine size, and the iron oxide minerals, magnetite and hematite, are concentrated mechanically. The fine magnetite or hematite powder is then bonded together by clay to form "pellets" approximately 1–3 cm in diameter. The pellets are roasted to make them hard enough to withstand transport to smelters.

Sedimentary iron deposits that consist of concentrically layered iron-rich oolites in a calcareous or clay-rich matrix are present in some Paleozoic and Mesozoic sequences. These deposits, called "Clinton" or "Minette" type ores, lack the cherty layers typical of the older banded iron formations, and evidently form in shallow marine environments. Although they are much smaller than banded iron formations, these oolitic ores have

been important sources of iron in Newfoundland, the British Isles, and the Alsace-Lorraine area of France.

Iron may also be concentrated in gabbroic igneous rocks that are relatively rich in iron. As the magma crystallizes, oxygen may combine with iron to form magnetite crystals that settle out of the magma due to their high density. Convection currents within the crystallizing magma allow the heavier magnetite to accumulate as a layer while the lighter silicate minerals are kept in suspension in the magma. Alternatively, as the magma crystallizes, the oxygen and iron may combine to form a liquid of iron oxide that settles to the floor of the magma chamber. The process is referred to as "liquid immiscibility" because liquid iron oxide separates from the remainder of the magma. Layers of iron oxide liquid may accumulate within the crystallizing rock. Eventually the iron oxide layers cool and crystallize. The most common mineral in these igneous deposits is magnetite, which usually contains significant amounts of titanium and vanadium that were concentrated with the iron in the oxide liquid that separated from the magma. The titanium and vanadium may limit the use of these deposits as iron ores, but may render them important ores of titanium or vanadium. Examples of titanium (ilmenite) deposits are found at Allard Lake, Quebec (Bergeron, 1972), and vanadium-bearing magnetite bodies are found in the Bushveld igneous complex in South Africa (Willemse, 1969).

Hydrothermal iron deposits are produced locally, adjacent to intermediate composition of igneous rocks, especially where the magma came in contact with carbonate rocks. These deposits, called "skarns," are generally rather small and quantitatively unimportant except in countries such as Chile and Peru that have no sedimentary iron deposits. The skarn deposits are composed mainly of magnetite with variable amounts of metamorphic minerals that were partially replaced by iron-rich fluids. The iron was carried from the igneous intrusion into the surrounding rocks in hot, aqueous solutions. The solutions reacted with the surrounding rock, were cooled, and precipitated the iron minerals in the carbonate rock.

In small, isolated occurrences, limonitic iron ore, called "bog ore," is produced at the earth's surface. Iron-bearing groundwater may reach the surface in springs or in marshes. Bacterial action precipitates the iron by converting the relatively soluble ferrous iron in the groundwater to insoluble ferric-iron in the form of limonite. In earlier times these small deposits were mined for iron ore.

Bibliography

BERGERON, M. "Quebec Iron and Titanium Corporation Ore Deposits at Lac Tio, Quebec." *Twenty-Fourth International Geological Congress Guidebook.* Montreal, Canada, 1972.

CLOUD, P. E. "Paleoecological Significance of Banded Iron Formations." *Economic Geology* 68 (1973):1135–1143.

GROSS, G. A. "A Classification of Iron Formations Based on Depositional Environments." *Canadian Mineralogist* 18 (1980):215–222.

HOLLAND, H. D. "The Oceans: A Possible Source of Iron in Iron Formations." *Economic Geology* 68 (1973): 1169–1172.

LABERGE, G. L.; E. I. ROBBINS; and T.-M. HAN. "A Model for the Biological Precipitation of Precambrian Iron Formations—A Geological Evidence. In *Precambrian Iron-Formations*, eds. P. W. U. Appel and G. L. LaBerge. Athens, Greece, 1987.

WILLEMSE, J. "The Vanadiferous Magnetite Iron Ore of the Bushveld Igneous Complex." *Economic Geology* monograph. Lancaster, PA, 1969.

GENE L. LA BERGE

ISOSTASY

The concept of isostasy originated in the 1850s when pendulum gravity surveys near the Andes and Himalayas demonstrated that the pendulum bob was deflected toward the mountains by a smaller amount than what one would predict were the mountains an excess mass resting on an otherwise unperturbed Earth. J. H. Pratt (1855) and G. B. Airy (1855) proposed that the deficit in the gravitational attraction of the mountains could be explained by low-density material at depth supporting the excess mass of the surface load through buoyancy forces. A region would therefore be in the state of isostasy (from the Greek, literally "equal standing") if the excess mass of the topography is exactly supported by the buoyancy forces of the buried mass deficiencies so that stresses are hydrostatic below some depth, called

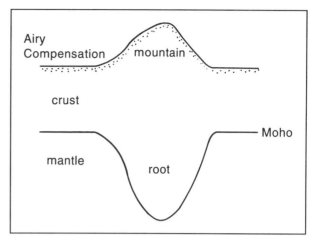
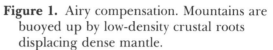

Figure 1. Airy compensation. Mountains are buoyed up by low-density crustal roots displacing dense mantle.

the compensation depth. Topographic features that have attained the state of isostasy are said to be isostatically compensated.

Airy and Pratt developed two different models to explain the nature and distribution of this compensating material. According to Airy's model, the mountains are underlain by thickened crustal roots that displace denser mantle material, such that the mountains float on the mantle in the same manner as icebergs floating in the ocean (Figure 1). The compensation depth in Airy's scheme would be the crustal-mantle boundary, the "Moho." Pratt hypothesized that the density of material in a vertical column is inversely proportional to the elevation of that column (Figure 2). Thus high regions are composed of low-density material and vice versa.

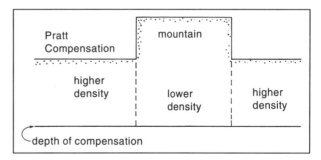

Figure 2. Pratt compensation. The density of the vertical column of rock above the depth of compensation is inversely proportional to the elevation of the column.

The compensation depth would be defined as the depth below which there are no longer any lateral density changes.

The Airy and Pratt models are of more than historical interest, because there do exist features on Earth that have been shown to be supported by both of these mechanisms based on the results of gravity and seismic surveys. For example, many continental mountain ranges are indeed underlain by thicker crust compared to the adjacent plains, in the manner of Airy isostasy. Mid-ocean ridges stand high on account of the fact that their geologically young lithospheric material is hotter, and therefore less dense, than that beneath the older abyssal ocean plains, in the manner of Pratt isostasy. The difference in elevation between continents and oceans is a combination of the two effects: continents stand high because their 30- to 40-km-thick granitic crust is both thicker and lighter than the 5- to 7-km-thick basaltic crust beneath the oceans.

The two models of Pratt and Airy are both examples of local compensation schemes in that the support for the elevated region is confined to the vertical column immediately beneath the topography. Thus the state of isostasy in the case of local compensation is that there exists an equal amount of mass in each vertical column. Any changes to the amount of mass in that column (e.g., on account of erosion at the surface or changes in temperature at depth) would have to lead to an adjustment in the height of that vertical column, for example, through movement along vertical faults, in order to maintain the condition of isostasy.

More recent models of isostasy take into account the fact that Earth's lithosphere is known to have

some finite strength (Vening Meinesz, 1941), such that a vertical column cannot adjust its elevation independently from that of adjacent columns. The strength of the lithosphere would thus cause compensation for surface loads to be distributed laterally as well as vertically beneath the elevated region. Models of isostasy that take into account finite strength of the lithosphere are called regional compensation, the most familiar of which is the elastic plate model (Gunn, 1943) which assumes that Earth's surface bends as a thin elastic plate overlying a fluid asthenosphere (Figure 3). The rigidity of the plate is measured by its flexural rigidity, D, which is defined as

$$ D = \frac{ETe^3}{12(1 - \nu^2)} $$

in which E is Young's modulus, Te is the effective elastic plate thickness, and ν is Poisson's ratio.

The elastic plate model has very successfully explained the isostatic compensation of volcanoes erupted on oceanic lithosphere. For example, the Hawaiian Islands are surrounded by a bathymetric moat and arch caused by elastic flexure of the Pacific seafloor under the weight of the volcanoes. Numerous studies of flexure of the oceanic lithosphere have demonstrated the validity of a remarkably simple relationship: Te of the oceanic lithosphere increases as the square root of the age of the plate when flexed (Watts et al., 1980). Because the temperature of the oceanic lithosphere decreases as the square root of its age, this result is interpreted as evidence that the base of the elastic plate corresponds to an isotherm (surface of equal

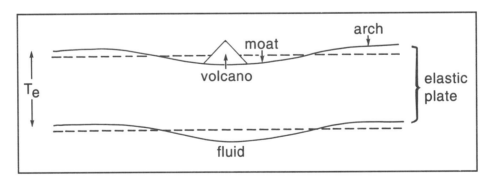

Figure 3. Elastic plate compensation. The load of a volcano depresses the lithosphere in the manner of a thin elastic plate to form a topographic moat and arch surrounding the feature.

temperatures) that marks the boundary between rocks at shallower depth that are cold enough to sustain elastic stresses for many millions of years, and rocks below that depth that are so hot that they flow to relieve the elastic stresses. The fact that the maximum value for Te even for very old (greater than 100 Ma) oceanic lithosphere is only 40 to 50 km indicates that the isotherm controlling the elastic/ductile transition is about 450°C. When one accounts for the weakening of the elastic plate by brittle processes near the surface, the estimate for the value of this isotherm is somewhat higher (McNutt and Menard, 1982).

Elastic plate models have also been applied to the compensation of topographic features on the continents (Karner and Watts, 1983), but with less success. The magnitude of gravity anomalies over narrow mountain belts and the existence of foreland basins in front of thrust sheets (slices of crust that have been stacked on top of adjacent crust via motion along compressional faults) are clear evidence for elastic behavior of continental lithosphere, but the variations in Te, from more than 100 km to less than 5 km, are less systematic and not easily interpreted in terms of the thermal age of the plate. The great diversity in composition of continental crust and the long, complicated geologic history of the continents are thought to be significant factors in preventing a simple interpretation of Te in terms of the depth to an isotherm (McNutt et al., 1988).

During the 1970s and 1980s the definition of isostasy was extended to include the concept of dynamic isostasy, for which variations in the elevation of Earth's surface are supported by convection in the mantle (Hager et al., 1985). It is predominantly buoyancy forces (caused by chemical or thermal differences driving convection) that produce the surface topography, because the viscous stresses from typical mantle flow velocities are much smaller. However, the viscosity of Earth's mantle causes the "compensation" for the buoyancy forces driving the flow to be partitioned between the surface and any internal density interfaces, such as the core/mantle boundary, deformed by the flow (Figure 4).

On account of the high viscosity of Earth's mantle, some regions may be temporarily out of isostatic equilibrium as the surface slowly adjusts to changes in loads. The best-known example of this phenomenon is the postglacial rebound that is presently occurring in Fenno-Scandia (Finland,

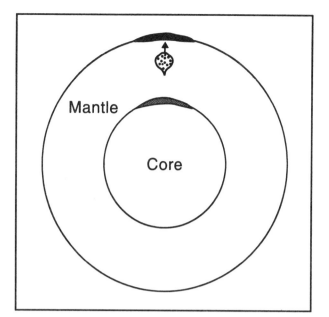

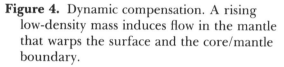

Figure 4. Dynamic compensation. A rising low-density mass induces flow in the mantle that warps the surface and the core/mantle boundary.

Sweden, and Norway). The weight of the Pleistocene ice sheets depressed Earth's surface at high northern latitudes such that the ice load was compensated by displacement of denser mantle material. The last major ice sheet began melting about 40,000 years ago, and was largely melted 10,000 years ago. Some of the topographic depression still remains as the slow flow velocities of the underlying mantle prevent rapid rebound. Observations of the rate of uplift of the land surface (as high as 1 cm per year) demonstrate that the viscosity of Earth's mantle must be of the order of 10^{21} Pascal seconds (Pa · s) (Cathles, 1975).

Bibliography

AIRY, G. B. "On the Computation of the Effect of the Attraction of the Mountain-masses as Disturbing the Apparent Astronomical Latitude of the Stations of Geodetic Surveys." *Philosophical Transactions of the Royal Society* 145 (1855):101–104.

CATHLES, L. M., III. *The Viscosity of the Earth's Mantle.* Princeton, NJ, 1975.

GUNN, R. "A Quantitative Study of Isobaric Equilibrium and Gravity Anomalies in the Hawaiian Islands." *Journal of the Franklin Institute* 236 (1943):373–390.

HAGER, B.; R. W. CLAYTON; M. A. RICHARDS; R. P. COMER; and A. M. DZIEWONSKI. "Lower Mantle Heterogeneity, Dynamic Topography, and the Geoid." *Nature* 313 (1985):541–545.

KARNER, G. D., and A. B. WATTS. "Gravity Anomalies and Flexure of the Lithosphere at Mountain Ranges." *Journal of Geophysical Research* 88 (1983): 10449–10477.

McNUTT, M. K., and H. W. MENARD. "Constraints on the Yield Strength of Oceanic Lithosphere Derived from Observations of Flexure." *Geophysical Journal of the Royal Astronomical Society* 71 (1982):363–394.

McNUTT, M. K.; M. DIAMENT; and M. G. KOGAN. "Variations of Elastic Plate Thickness at Continental Thrust Belts." *Journal of Geophysical Research* 93 (1988):8825–8838.

PRATT, J. H. "On the Attraction of the Himalaya Mountains, and of the Elevated Regions Beyond Them, upon the Plumb Line in India." *Philosophical Transactions of the Royal Society* 145 (1855):53–100.

VENING MEINESZ, F. A. "Gravity over the Hawaiian Archipelago and over the Madeira Area." *Proceedings of the Netherlands Academy Wetensia* 44 (1941):1–12.

WATTS, A. B.; J.H. BODINE; and N. M. RIBE. "Observations of Flexure and the Geological Evolution of the Pacific Ocean Basin." *Nature* 283 (1980):532–537.

MARCIA MCNUTT

ISOTOPE TRACERS, RADIOGENIC

Most elements are composed of more than one isotope, atoms with the same number of protons but different numbers of neutrons. For example, the oxygen humans need for survival consists of three isotopes, ^{16}O, ^{17}O, and ^{18}O, differing only in the number of neutrons each contains (8, 9, and 10, respectively). Most elements in nature have essentially the same isotopic composition no matter where they are found. The iron dissolved in seawater is the same as that in your mother's Ford, the same as that in Moon rocks, and the same as that in meteorites from the asteroid belt. There are, however, three cases where the isotopic compositions for a given element can be distinct. The very lightest elements can vary in isotopic composition because thermodynamic and/or kinetic effects operate slightly differently on lighter than heavier isotopes (*see* THERMODYNAMICS AND KINETICS; UREY, HAROLD). This causes mass-dependent fractionations (changes in isotope abundance due only to differences in mass) in cases where the relative

mass difference between two isotopes is large (*see* ISOTOPE TRACERS, STABLE). For example, ^{16}O is about 12 percent heavier than ^{16}O, a substantial difference. A second case is isotopic anomalies, which are common in some fractions of meteorites. Some of these anomalies are simply extreme forms of mass-dependent isotope fractionation, but others are due to nuclear reactions that occurred when the elements were formed. These anomalies persisted because of incomplete mixing of the matter that formed the solar nebula. The last cause of isotopic variations is accumulation of specific isotopes as daughters of radioactive decay of a parent element (*see* GEOLOGIC TIME; GEOLOGIC TIME, MEASUREMENT OF). Because of their mode of formation, these daughter isotopes are commonly known as radiogenic isotopes. This article is about the radiogenic isotopes and what we can learn about geologic processes from them.

There are two general classes of radiogenic isotopes, those whose parents are long enough lived to be extant, and those with short-lived, extinct parents (*see* NUCLEOSYNTHESIS AND THE ORIGIN OF THE ELEMENTS). The short-lived isotopic systems are used mostly to determine how long after the formation of the solar system a given object, such as a meteorite, was formed, and are not used much for radiogenic tracer studies. Hence, they will not be discussed further here; the focus will instead be the long-lived radiogenic systems. Table 1 lists long-lived radioactive nuclei, their stable daughter isotopes, and the isotopic ratios usually measured

Table 1. Radiogenic Isotope Systems Useful or Potentially Useful in Earth Sciences

Parent Isotope	Daughter Isotope	Half-life (Ga)*	Isotope Ratio Measured
^{40}K	^{40}Ar	11.9	$^{40}Ar/^{36}Ar$
^{40}K	^{40}Ca	1.40	$^{40}Ca/^{44}Ca$
^{87}Rb	^{87}Sr	48.8	$^{87}Sr/^{86}Sr$
^{138}La	^{138}Ce	269	$^{138}Ce/^{142}Ce$
^{147}Sm	^{143}Nd	106	$^{143}Nd/^{144}Nd$
^{176}Lu	^{176}Hf	35.7	$^{178}Hf/^{178}Hf$
^{187}Re	^{187}Os	42.3	$^{187}Os/^{188}Os$
^{232}Th	^{208}Pb	14.05	$^{208}Pb/^{204}Pb$
^{235}U	^{207}Pb	0.704	$^{207}Pb/^{204}Pb$
^{238}U	^{206}Pb	4.47	$^{208}Pb/^{204}Pb$

* Ga equals giga annum, or 10^9 years.

for the more commonly studied radiogenic isotopic systems.

The long-lived radioactive systems were first used to determine the ages of rocks. In the past few decades, geoscientists have made increasing use of simple measurement of the ratio of radiogenic daughter isotope to a stable isotope of the same element to investigate the large-scale structure of the earth's mantle, the evolution of the earth, and other geological processes. The simplest measurements are those on recent volcanic rocks. Here, decay of the parent nuclide has not significantly changed the amount of radiogenic daughter in the rock in the short time since it crystallized, so correction to the measured isotope ratio is not needed. The common assumption made is that the measured isotope ratio in the volcanic rock represents that of the region of the mantle that was partially melted. This is a good assumption in theory, as the high temperatures of mantle melting processes and the small mass difference between different isotopes of the daughter element minimize any mass-dependent fractionation. However, in practice the assumption is often difficult to satisfy because the magmas can be contaminated with continental material during ascent through the crust, or through interaction with seawater, or with groundwater. We will focus on the isotopes of strontium (Sr), neodymium (Nd), and lead (Pb) as applied to mantle evolution because these systems have been utilized the most, but will briefly mention other systems and processes.

Strontium (Sr) and Neodymium (Nd) Isotopes

Processes acting in the earth's interior, such as melting, subduction, and the stabilization of continental lithosphere, have fractionated the ratios of parent to daughter isotopes in different portions of the earth from the values they had in the primordial, undifferentiated earth. This has allowed different reservoirs to evolve distinct radiogenic isotope signatures over time. For example, rubidium (Rb) is more incompatible in melting processes than is Sr. Early in the earth's history, when the continental crust first began to form out of the primordial mantle, the Rb/Sr ratio in the crust was higher than that of the primordial mantle, and the mantle became depleted. Over time, this led to

growth of more radiogenic Sr in the crust than in the mantle. In the samarium (Sm)-Nd system, the opposite fractionation occurs; Nd is the more incompatible element and the crust has a lower Sm/Nd ratio than the depleted mantle. With time then, the isotopic composition of Nd in the mantle became more radiogenic than that of the crust. This is illustrated in Figure 1, a plot of present-day Nd isotopes versus Sr isotopes for various earth materials. The contrasting behavior of the radiogenic growth of Sr and Nd isotopes has been especially useful in constraining petrologic hypotheses.

Figure 1 can be divided into four quadrants by the intersection of the lines marking the ideal bulk earth Sr and Nd isotope compositions. These are the isotopic compositions that all earth materials would have today if chemical fractionation of Rb/Sr and Sm/Nd ratios had never happened. The upper left quadrant is occupied by depleted materials; materials with time-averaged Rb/Sr ratio lower and Sm/Nd ratio higher than the primordial earth. Occupying this quadrant are mid-ocean ridge basalts, or MORB, from all over the earth, plus some ocean island basalts (OIB). These basalts are formed from the upper mantle, and their position on Figure 1 is good evidence that the upper mantle is generally depleted in its low-melting fraction.

Continental crust, represented by analyses of loess, plot in the lower right quadrant of Figure 1. Loess is wind-borne dust that forms thick deposits found all over the earth, and it is believed to be fairly representative of the composition of the upper continental crust in the region where it is found. Loess contains very radiogenic Sr, but nonradiogenic Nd, a signature of material with high, time-averaged Rb/Sr and low, time-averaged Sm/Nd. This is signature of the low-melting fraction of the primordial earth. Also shown in Figure 1 are various basalt types from around the world from different tectonic settings. As can be seen, there is considerable variation in the isotopic signature of recent volcanic rocks, and this variation, coupled with other geochemical and petrological data, can be used to understand how the present-day earth works, and also something about the history of earth processes. The array of data from basaltic rocks is often referred to as the mantle array, because basalts should have the isotopic composition of their source regions in the mantle. Note that the intersection of this mantle array, with the Nd isotopic composition of chondrites, gives an estimate

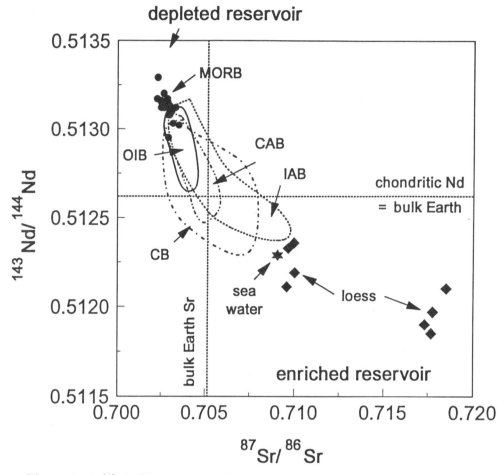

Figure 1. A ^{143}Nd/^{144}Nd versus ^{87}Sr/^{88}Sr diagram for various earth materials. Earth reservoirs depleted by partial melting (upper mantle) occupy the upper left quadrant, while reservoirs enriched in these mantle partial melts (continental crust) occupy the lower right quadrant. Lines for the bulk earth Nd and Sr isotopic composition are shown. The various basalt types labeled are: MORB, mid-ocean ridge basalt; OIB, ocean island basalt; CAB, continental arc basalt; IAB, island arc basalt; and CB, continental basalt. The diagonal field defined by MORB, OIB, and some IAB is commonly referred to as the mantle array, because these samples are believed to have the isotopic signature of their source regions in the mantle. Loess is representative of the local upper crust.

of the Sr isotopic composition in the bulk earth, and therefore the earth's Rb/Sr ratio.

The data shown in Figure 1 are the result of isotopic variations that developed with time. But when did the chemical fractionations occur that allowed these isotopic variations to grow? This question can be addressed by determining the ages and initial isotopic compositions of old rocks and searching for time-related isotopic variations. This process is much more difficult than simply determining the radiogenic isotopic character of recent volcanic rocks. First, several measurements of both the parent and daughter elements in the old rock need to be made so that its age and initial isotopic compositions can be determined. Second, old rocks have generally survived one or more episodes of weathering or metamorphism or both, with the possibility of disturbances to the radiogenic systems being investigated. Hence, the age and initial isotopic composition of old rocks are not always straightforward to determine or interpret. Nevertheless, the evidence from ancient Precambrian mafic rocks suggests that even early in the earth's history the mantle was already depleted in its low-melting fraction by partial melting and removal of the melt by volcanism.

Lead (Pb) Isotopes

The uranium (U)–thorium (Th)/Pb systems are unique in that three parent isotopes, two of U and one of Th, all decay to different isotopes of a single element, Pb. Lead has one stable, non-radiogenic isotope (^{204}Pb) with which the radiogenic daughter isotopes can be compared. The possibility of comparison of three different Pb isotope ages has been particularly useful in geochronology. Additionally, this makes the isotopic evolution of Pb a particularly powerful tool for investigations of the geochemical history of the earth, and indeed, Pb isotopic studies of volcanic rocks were the first to address the large-scale geochemical evolution of the earth. Particularly powerful are the ^{235}U/^{207}Pb and ^{238}U/^{206}Pb systems because a single parent element decays to different isotopes of a single daughter element. Hence, there are two decay equations, one for ^{235}U-^{207}Pb decay and one for ^{238}U-^{206}Pb decay, with only two unknowns, the time of chemical fractionation and the U/Pb ratio. Hence, the measurement of only one parameter, the Pb isotopic composition of a suite of rocks, allows one to infer not only the time-averaged U/Pb ratio, but also the time when this ratio was established. Figure 2 shows a so-called Pb-Pb diagram for recent volcanic rocks and models of Pb isotope growth that illustrates how geoscientists can use Pb isotopes to understand the history of the earth.

First let us consider models for Pb isotope growth. Based on studies of the least radiogenic Pb in meteorites, geochemists can infer the initial isotopic composition of Pb in the earth at the time of its formation, shown in Figure 2. If different parts of the earth had different initial U/Pb ratios, these reservoirs would evolve different radiogenic Pb compositions with time, as indicated by the various

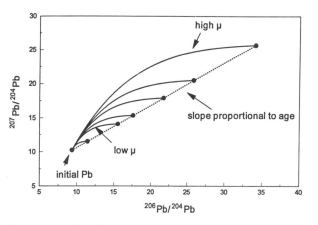

Figure 2. Lead isotope models for the earth, with an example from oceanic basalts. The initial Pb is assumed to be the least radiogenic Pb measured in meteorite samples with very low U/Pb ratio. The change in Pb isotopic composition with time depends on the U/Pb ratio. Commonly, geochemists use the ^{238}U/^{204}Pb ratio, which they denote μ. ^{238}U is the longer-lived U isotope, while ^{204}Pb is the only non-radiogenic Pb isotope. The spots represent model Pb isotope compositions as might be measured today. The slope of a line drawn through these points is proportional to the age of the samples.

curves in the figure. (The growth lines in the diagram are curved because ^{235}U has a much shorter half-life than ^{238}U. This leads to rapid growth of the ^{207}Pb/^{204}Pb ratio early in the earth's history before ^{235}U became depleted through decay. Present-day U is mostly ^{238}U because most of the ^{235}U has decayed, and therefore in earlier times, the ^{206}Pb/^{204}Pb ratio grew more rapidly than the ^{207}Pb/^{204}Pb ratio.) If samples from these different reservoirs were measured today, their Pb isotope compositions would plot at the solid circles on the diagram. A line passing through all of these analyses would have a slope proportional to the age of the earth, and pass through the initial Pb isotope composition as measured in meteorites.

The growth of Pb isotopes in the earth's mantle did not follow a single-stage history as outlined above. Shown in Figure 3 are analyses of oceanic basalts made in the 1970s by M. Tatsumoto of the U.S. Geological Survey. He showed that the array of Pb isotope analyses suggested that a two-stage growth of radiogenic isotopes is a closer match to

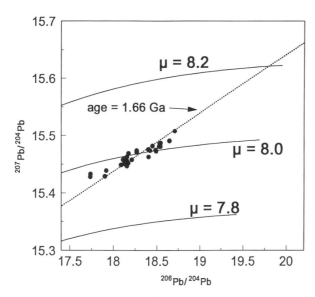

Figure 3. An example for oceanic basalts. A reservoir with $\mu = 8$ passes through the data, but gives an unrealistic age for the earth. The slope of the dotted line passing through the data defines an age of 1.66 Ga, and suggests that a two-stage history for Pb evolution is a better match for the data.

the data. He found that the first stage of Pb growth occurred in a reservoir with μ of about 8 (where μ represents $^{238}U/^{204}Pb$) from the time of formation of the earth until about 1.66 Ga. At that time, the uniform reservoir was fractionated into regions with different μ values, which evolved radiogenic Pb until sampled by the oceanic basalts. This two-stage model is a better match to the data, but is undoubtedly still too simple to explain the Pb isotope evolution of the earth. The age 1.66 Ga is probably better thought of as a mean age of fractionation of U from Pb, rather than as a discrete event that happened for the entire earth. Nevertheless, Pb isotope geochemistry is a powerful tool for unraveling the history of the earth.

Hafnium (Hf) Isotopes

The lutetium (Lu)/Hf system is a fairly recent addition to the radiogenic isotope toolbox, largely as a result of improvements in mass spectrometers used in isotopic analysis (*see* GEOCHEMICAL TECHNIQUES). This system is conceptually like the Sm/Nd system in that the daughter element Hf is more

incompatible during melting processes (more strongly partitioned into the magma) than is the parent. Hence, the crust has a low Lu/Hf ratio compared to the primordial mantle, and its Hf is relatively unradiogenic. The depleted upper mantle has a higher Lu/Hf ratio, and commensurably more radiogenic Hf. A plot of Hf isotopic compositions of rocks versus Sr isotopic composition would look very much like Figure 1 of Nd versus Sr.

Osmium (Os) Isotopes

Rhenium (Re) and Osmium (Os) are unlike the other isotope systems discussed so far in that both of these elements are siderophile (iron loving) elements. The bulk earth inventory for these elements undoubtedly resides in the core, not the mantle or crust (*see* EARTH, COMPOSITION OF). Nevertheless, a small amount of Re and Os are present in the mantle and crust. Rhenium is more incompatible than Os during silicate melting to form basalts, so the crust of the earth has a higher Re/Os ratio than the depleted mantle. Osmium isotope systematics on the earth will then mimic to some degree Sr isotope systematics; the crust will contain more radiogenic Os than does the mantle. Rhenium and Os are abundant in chondritic and iron meteorites, and the Os isotopic composition in meteorites is distinct from that of the earth's crust. Because of this, Os isotopic studies can be used to help identify a "meteoritic signature" in breccias and melt rocks recovered from suspected impact craters on the earth.

Other Systems, Other Processes

The radiogenic isotope systems described above are those in most common use. Other radiogenic isotopes may find use in the future, but now are little used owing to their low abundances, difficulties in measurement, or long half-lives of their parents. Radiogenic calcium (^{40}Ca) is formed by decay of potassium (^{40}K). Because Ca is more abundant in the earth than K, and because ^{40}Ca is one of the major isotopes of Ca (about 97 percent of Ca is ^{40}Ca), radiogenic Ca is not a sensitive system except in rocks with high K/Ca ratios such as granite. Radiogenic argon (^{40}Ar) is also formed by decay of ^{40}K. Because Ar is a gas, it is of limited use in studies of recent volcanic rocks. The abundance of ^{40}Ar in the earth's atmosphere, on the other hand, can be used to estimate the K content of the bulk earth. A final isotope system to mention is the lanthanum/cerium (La/Ce) system. This system is akin to the Rb/Sr system in that La is more incompatible than Ce. Hence, the crust should contain more radiogenic Ce than the upper mantle. In practice though, in most geologic systems fractionation of La from Ce is slight, and coupled with the long half-life, this limits the utility of radiogenic Ce isotope studies to understanding the evolution of the earth. Cerium can be fractionated from La during weathering, so there is a potential to use Ce isotopes to study ancient weathering processes.

Summary

Study of the distributions of radiogenic isotopes in the earth provides a powerful means to unravel the history of the earth and the geochemical fractionations that occurred as it was transformed from its primordial condition to the chemically stratified state of today. These studies, coupled with information gleaned from geophysics and petrology, have allowed geoscientists to probe the inaccessible interior of the earth in amazing detail, and have greatly contributed to our overall understanding of how all planetary bodies evolve through time.

Bibliography

DICKIN, A. P. *Radiogenic Isotope Geology*. Cambridge, MA, 1995.
FAURE, G. *Principles of Isotope Geology*, 2d ed. New York, 1986.

DAVID W. MITTLEFEHLDT

ISOTOPE TRACERS, STABLE

Any population of atoms can be grouped according to the number of protons and neutrons in the nucleus. Isotopes of an element are atoms that have an identical number of protons but differing numbers of neutrons. Inherently stable configurations exist only for those atomic particles that contain a narrow range of neutrons in the nucleus; outside this range atoms will spontaneously decay. Of the approximately 1,700 isotopes that can exist, only about 260 have stable configurations. For each element, there usually exists a single stable isotope that is common, while all others are relatively rare. It is the distribution of these rare and common stable isotopes in nature that provides insightful information about how the universe formed and evolved through time.

Isotopic Fractionation

Stable isotope distributions are not random but rather a consequence of reactions that are governed by kinetic theory and the laws of thermodynamics (*see* THERMODYNAMICS AND KINETICS). As such, the direction, magnitude, and temperature dependence of isotope partitioning can be used to trace the path of an element from reactant to product, determine whether a system is at isotopic equilibrium, or predict the temperature at which a reaction takes place.

Equilibrium Fractionation. The driving force that governs stable isotope distributions among molecules at equilibrium is the minimization of free energy. Energy minimization is quantum mechanical in nature and depends on the difference in zero point energies for isotopically substituted molecules. The isotope of an element that has the greater atomic mass (the sum of neutrons and protons) tends to substitute in the molecule that produces the greatest shift to lower vibrational energy, with the magnitude of isotope partitioning being related to the reduction in free energy. There is a greater separation between energy levels for isotopically substituted molecules at lower temperature; therefore, the magnitude of equilibrium partitioning decreases as reaction temperature increases (*see also* UREY, HAROLD).

The constant used to describe a system at isotopic equilibrium is the fractionation factor, or α, which is defined as the ratio of the rare to the common isotope (R) for an element in species a to that in species b: $\alpha_{a-b} = R_a/R_b$.

At equilibrium, α is related to the thermodynamic quantity K, or the equilibrium constant, for an exchange reaction involving two isotopes. The fractionation factor for a particular system can be calculated at any temperature using the methods of statistical mechanics or it can be measured empirically. Fractionation factors are a powerful tool that can be used to predict the distribution of isotopes among molecules in a system and determine the temperature at which isotopic exchange takes place.

Kinetic Fractionation. Kinetic fractionation of stable isotopes occurs in unidirectional reactions or those which are controlled by diffusional processes. For the case of unidirectional reactions, reaction rate controls isotopic composition. Since rates of reaction proceed faster for a molecule bearing the isotope of lower atomic mass, reactions that are influenced by kinetic fractionation tend to concentrate the lighter isotope in the product. Several intermediate or transition steps may occur as reactant is converted to product, and if one step is both rate-limiting and mass-sensitive, then kinetic fractionation will be expressed in the product. Almost all of the important biological reactions that occur in nature are characterized by kinetic effects. Diffusional processes fractionate stable isotopes according to Graham's Law, with the rate of diffusion being inversely proportional to mass.

Isotopes and Conventions

The most important stable isotopes to display variation in natural abundance belong to elements of low atomic mass ($\sim$ amu < 40) because the magnitude of isotopic fractionation is proportional to the relative differences between isotopes for a given element. The list of elements for which significant variation exists include: hydrogen (H), carbon (C), nitrogen (N), oxygen (O), and sulfur (S). It is not fortuitous that these elements also participate in most all of the important physicochemical and biological reactions that occur in nature.

The convention generally used to report isotopic composition is delta notation (δ). It is defined as: $\delta^m E(\%_o) = \{\{[R_a/R_b]_{sa}/[R_a/R_b]_{std} - 1\} \cdot 1000\}$, where m is the nominal atomic mass of the rare isotope for element E, and $[R_a/R_b]_{sa}$ and $[R_a/R_b]_{std}$ are the ratios of the rare to common isotope for a sample and an international standard, respectively. The symbol $\%_o$ represents the unit parts per thousand, or as it is most commonly referred to, parts per million.

Isotopic Fractionation in Earth Materials

The isotopic composition of earth materials depends on many factors associated with bonding environment, including oxidation state, ionic potential, and atomic mass. At equilibrium, the heavier isotope is preferentially incorporated in molecules having a higher ionic potential and lower atomic mass because such combinations produce the greater shift to lower free energy for a system.

Igneous and Metamorphic Rocks. Systematic regularities in the isotopic composition of many igneous and metamorphic mineral assemblages attest to the thermodynamic basis for isotopic fractionation. The rare isotope of oxygen (^{18}O) is generally concentrated in minerals where oxygen-cation bonds are stronger and more highly polymerized. Such equilibria processes are responsible for the general descending order of ^{18}O enrichment in the minerals, quartz > feldspar > pyroxenes and olivines > oxides.

If unaltered, the minerals that comprise igneous rocks generally show small variations in $^{18}O(-2$ to $16\%_o)$ due to the high formation temperatures of crystallization. Isotope composition provides insight into the source region (e.g., mantle or crust), temperature of formation, and evolutionary history of a magma during mineral formation. Igneous processes that produce isotopic variability include fractional crystallization, assimilation of country rock, and interactions with hydrothermal plumes/cells that are generated during magma ascent.

Metamorphic rocks are usually ^{18}O-enriched and have a wider range of isotopic compositions (5 to $25\%_o$) due to greater variability in source material and lower formation temperature. Hydrogen and oxygen isotopic compositions of metamorphic minerals can be used to determine thermal and fluid histories, time-temperature paths of metamorphism, physical flow paths, dehydration and decarbonation reactions, and the scale and extent of water–rock interactions.

Sedimentary Rocks. Since sedimentary rocks and minerals form at lower temperatures than their igneous and metamorphic counterparts, they are usually characterized by higher $\delta^{18}O$ values (up to $40\%_o$). Among the clastic sedimentary rocks, those that have a high percentage of igneous and metamorphic components (e.g., sandstones) are generally ^{18}O-depleted compared with those rocks composed primarily of authigenic minerals (e.g., mudstones). Most authigenic clays and chemical precipitates (e.g., carbonates, phosphorites, and cherts) have $\delta^{18}O$ values that reflect the precipitating fluid and temperature of crystal growth.

Sulfur-bearing Rocks and Minerals. The chemistry of sulfur, with its variety of bond-types and oxidation states, contributes greatly to the wide range of sulfur-isotope compositions found in nature (-65 to $90‰$). Variations in sulfur-isotope composition can be used to trace the source of sulfur-bearing fluids responsible for the precipitation of sulfides and sulfates, characterize the physicochemical conditions of mineral crystallization/precipitation (temperature, pH, and oxygen fugacity—corrected vapor pressure), characterize the pathways and duration of fluid infiltration events, and determine the mechanisms of fluid transport. Sulfur isotopes are also strongly influenced by kinetic effects during biologically mediated sulfate reduction reactions that occur at low temperature ($\sim<60°C$).

Stable Isotopes in the Biosphere. Most biochemical reactions are governed by kinetic fractionation. During photosynthesis, there are two main isotope discriminating pathways that plants utilize to transform inorganic carbon into biomass. Organic matter generated by way of the Calvin (or C_3) pathway has a range in $\delta^{13}C$ of -33 to $-23‰$, while that produced through the Hatch–Slack (or C_4) pathway ranges between -16 to $-9‰$. Most higher plants fix carbon using the C_3 pathway, while tropical grasses and warm climate plants use the C_4 pathway. Succulent plants that grow in water-limited environments display a bimodel distribution in $\delta^{13}C$ because they utilize the Crassulacean Acid Metabolism pathway (CAM), which fixes carbon by both the C_3 and C_4 pathways.

Waters. The hydrogen and oxygen isotope composition of seawater does not vary far from $0‰$ ($\pm1‰$) because of the tremendous mass, efficient mixing, and buffering of marine waters at mid-ocean ridges. The concordance of H and O isotopic compositions in seawater at approximately $0‰$ is no accident. The standard for isotopic comparison for both H and O is Standard Mean Ocean Water, or SMOW in isotopic jargon.

This does not hold true for most meteoric waters as they follow a systematic trend in isotopic composition, known as the "meteoric water line." Equilibrium fractionation during the process of evaporation and condensation produces water vapor that is enriched in the light isotopes ^1H and ^{16}O, with the magnitude of fractionation being dependent on ambient air temperature. The isotopic composition of meteoric waters decreases from the equator (δ^2H and $\delta^{18}O \sim 0‰$) to the poles ($\delta^2H \sim -350‰$ and $\delta^{18}O \sim -35‰$) as the isotopes of greater mass are preferentially removed by precipitation during the poleward movement of air masses. Similar patterns of light isotope enrichment are also observed for waters from coastlines to continental interiors and from lowlands to the mountains.

Water masses that are subjected to kinetic fractionation (<100 percent humidity) will display isotopic compositions that deviate significantly from the meteoric water line because of a preferential increase in water vapor ^{16}O relative to ^1H during evaporation. These kinetic and equilibrium processes produce distinct isotopic heterogeneities on a local to global scale that can be used to trace water sources, apportion mixing events, characterize fluid flow and transport dynamics, and distinguish various types of rock–water interaction.

Rock–Water Interactions

Almost all crustal rocks have undergone some form of interaction with aqueous fluids. The interaction of fluid with rock tends to shift the oxygen isotopic composition of both components, depending on the mass of each component (or water–rock ratio) in a system. The temperature of exchange plays a critical role as fluid and rock attempt to reequilibrate under different physical conditions. Other factors that influence water–rock interactions include dehydration–decarbonation reactions, mineral exchange kinetics, and time.

Paleoclimatology

Paleoclimatology has at its foundation the use of stable isotope geochemistry to determine the character and history of oceanic/atmospheric circulation and how it relates to climatological change. With the development of the "paleotemperature equation" in 1953, past temperatures could be inferred for the first time from the $\delta^{18}O$ of fossil shells composed of calcite. The shells of marine planktonic (surface) and benthic (bottom water) foraminifera were used to determine ocean temperature over the geologic past. The observed variability in isotopic records was eventually related to small but significant changes in the isotopic composition of seawater (0.5 to 1.8‰) that occur when excess ice (^{18}O-depleted) is stored at the poles during glacial periods. Subsequent isotopic records have been used to document at least forty glacial cycles for the last 1.5 million years (Ma). Such in-

formation is vital to a better understanding of the factors that drive global climatic change.

Bibliography

ARTHUR, M. A., T. F. ANDERSON, I. R. KAPLAN, J. VEIZER, and L. S. LAND. *Stable Isotopes in Sedimentary Geology.* Dallas, TX, 1983.
FAURE, G. *Principles of Isotope Geology.* New York, 1986.
HOEFS, J. *Stable Isotope Geochemistry.* Berlin, 1987.
VALLEY, J. W., H. P. TAYLOR, JR., and J. R. O'Neil. *Stable Isotopes in High Temperature Geological Processes.* Chelsea, MI, 1986.

CHRISTOPHER S. ROMANEK

J

JUPITER AND SATURN

Jupiter and Saturn are the two largest planets in the solar system, 318 and 95 Earth masses, respectively. They are the fifth and sixth planets from the Sun, orbiting at 5.20 and 9.54 astronomical units (AU). One astronomical unit is the average distance between Earth and the Sun, equal to 149.6 million km. Both planets are visible to the naked eye: Jupiter is brighter than Sirius, the brightest star, while Saturn is about as bright as Antares in the constellation Scorpius. The most striking features on Jupiter are alternating bright and dark cloud bands (Figure 1). Saturn's cloud features are muted compared with those of Jupiter, as is evident in Plate 25. Saturn's ring system is much more prominent; however, this entry will focus on the planets themselves rather than on their satellites and rings.

Remote Sensing

Our knowledge of these far-off worlds comes, thus far, from remote sensing. Remote sensing consists of measurements of the properties of an object (like a planet) conducted at a distance. Planetary astronomers use optical telescopes on mountain tops in dry locations, such as Arizona and Hawaii, to observe changing cloud features on Jupiter and Saturn. The Hubble Space Telescope and the International Ultraviolet Explorer are two examples

of observatories in Earth orbit which have observed the giant planets without interference by Earth's atmosphere. Radio telescopes, such as the Very Large Array in New Mexico, are also used to study Jupiter and Saturn remotely. The biggest advance in our knowledge of these planets came from NASA spacecraft that flew by Jupiter and Saturn. *Pioneer* 10 and *Pioneer* 11 blazed the trail between 1973 and 1979, followed by *Voyager 1* and *Voyager 2* between 1979 and 1981. The *Galileo* spacecraft will begin orbiting Jupiter in December 1995. A probe will be dropped into Jupiter's atmosphere near the equator. A parachute will slow the probe so that instruments can measure cloud particles and gases scooped up during the descent. The probe will make direct, or "in situ," measurements while the orbiter conducts remote observations. The next visitor to the outer solar system after *Galileo* will be the *Cassini* spacecraft, planned for a launch in 1997. *Cassini* is scheduled to begin orbiting Saturn in 2004 for a four-year study of the planet and its rings and satellites. *Cassini* will carry and release an atmospheric probe built by the European Space Agency, known as Huygens, that will land on Saturn's cloud-covered satellite Titan.

The Interior

Jupiter and Saturn are gas giants composed mostly of hydrogen and helium. There is no surface and, therefore, no geology. Physical properties such as

Figure 1. Jupiter as viewed by the *Voyager 1* spacecraft in February 1979 from a distance of 29 million km. The Great Red Spot is seen prominently in the southern hemisphere. (Photo courtesy Lunar and Planetary Photographic Services.)

radius and gravity are specified at a level in the atmosphere where the pressure is 1 bar (approximately the pressure at sea level on Earth). Models of the deep interior rely on knowledge of bulk properties of the planet such as mass, density, and rotation rate (Table 1). The pressure inside the giant planets increases with depth until, at the center, it is estimated to be 100 million bars for Jupiter and 50 million bars for Saturn. The corresponding temperatures are about 20,000 K and 10,000 K, respectively. Jupiter and Saturn have a layered interior structure, as shown in Plate 26. At the center is a core containing iron and silicates as well as an envelope of methane (CH_4), ammonia (NH_3), and water (H_2O). Above this envelope lies an unusual substance known as liquid metallic hydrogen. Under normal conditions hydrogen is a gas, but when squeezed at a pressure of 3 million bars or more it behaves like a metal. Electrons are free to roam in this layer. These electrical currents are believed responsible for generating Jupiter's strong magnetic field. Saturn has a more modest magnetic field because its interior has less of this electrically conducting material. From this level outward to the top of the atmosphere, Jupiter and Saturn consist mostly of gaseous molecular hydrogen (H_2, rather than atomic hydrogen, H) and helium (He). These light gases are responsible for the low density of the giant planets. Hydrogen and helium are very compressible. Jupiter has more of these compressible gases than Saturn; therefore it has a higher density despite the fact that Saturn has a larger amount of heavy elements. There is no abrupt boundary between the interior and the atmosphere of a gas giant like Jupiter or Saturn. We simply have more information about the outer envelope of these planets than we have about the deeper levels.

Table 1. Orbital and Physical Properties of Jupiter and Saturn

Characteristics	Jupiter	Saturn
Orbital distance (Earth = 1)	5.203	9.539
Orbital eccentricity	0.048	0.056
Orbital period (years)	11.86	29.46
Perihelion date	July 1987	July 2003
Northern winter solstice	Aug. 1994	Oct. 2002
Inclination of equator to orbit	3.12°	26.73°
Rotation period	9h55m30s	10h39m24s
Mass (Earth = 1)	317.94	95.18
Equatorial radius, R_e (km)	71,492	60,268
Polar radius, R_p (km)	66,854	54,364
Oblateness $(R_e - R_p)/R_e$	0.0649	0.0980
Mean density (g/cm³)	1.33	0.69
Gravity at equator (Earth = 1)	2.34	0.93

The Atmosphere: Visual Appearance and Weather

Jupiter and Saturn have bands of clouds parallel to the equator, as is evident in Figure 1 and Plate 25. Jupiter's dark cloud bands are called belts and brighter ones zones. From the equator northward these bands are named as follows: Equatorial Zone, North Equatorial Belt, North Tropical Zone, North Temperate Belt, and North Temperate Zone. There are similar names in the southern hemisphere. Saturn's cloud bands have much lower contrast than those on Jupiter. This is due to Saturn's colder temperature, which results in its clouds forming at deeper levels. Sunlight reflected from these clouds traverses a longer path through the atmosphere, which absorbs some of the reddish colors. The cloud patterns on each planet are organized by winds powered by sunlight as well as by heat from the interior. Winds on Jupiter and Saturn generally blow east-west, changing speed and sometimes direction with latitude. On Jupiter's equator a strong jet blows from west to east at about 100 meters per second (m/s, sometimes abbreviated as m s^{-1}); its counterpart on Saturn has a speed of 500 m/s. Jupiter's bright zones are usually found on the equatorward side of an eastward jet; dark belts are located on the poleward side. Winds on Saturn are hard to measure from Earth because there are very few cloud features to track. However, about once every saturnian year (29.5 Earth years) a spectacular storm occurs. An outburst occurred in September 1990, when a "great white spot" appeared just north of the equator. It expanded into an oval and then stretched eastward into a ribbon-like feature. It soon expanded all the way around the planet. Each month it became more difficult to see, until after a year it faded away altogether.

The atmospheres of Jupiter and Saturn are much colder than Earth's atmosphere. The average temperature at 1 bar is 288 K on Earth, versus 165 K on Jupiter and 134 K on Saturn. Weather on Jupiter and Saturn is quite different from that on Earth because there are no topographic features (like mountains) to affect the flow of winds. In addition, there are two sources of energy to drive atmospheric motion—sunlight and heat from the interior. Surprisingly, Jupiter and Saturn are warmer than one would expect based on their distance from the Sun. Each planet radiates about twice the amount of heat as is absorbed from the

Sun. This extra heat comes from the deep interior, and it is believed to be left over from the time the giant planets formed 4.6 billion years (4.6 Ga) ago. Both planets receive more sunlight at the equator than at the poles, as does Earth. However, heat radiation emitted at infrared wavelengths is fairly constant at all latitudes on Jupiter and Saturn. Thus, heat must be transported across latitude circles at some unknown depth inside these planets. This north-south motion is much slower than the observed east-west winds. We do not understand completely how large a role the internal heat source plays in determining weather on Jupiter and Saturn.

Gas Composition and Clouds

Jupiter and Saturn are believed to be largely unchanged since their formation 4.6 Ga ago. These massive planets hold their atmospheres tightly, in contrast to the terrestrial planets, which have lost a large part of their atmospheres to space. Measurements of the gas composition of Jupiter and Saturn provide important clues to how these planets formed. Both planets are expected to have an elemental composition similar to that of the Sun. Since the outer planets are much colder than the Sun, these elements exist in molecular form, not as atoms. The most common elements in the Sun after hydrogen and helium are oxygen (O), carbon (C), and nitrogen (N). In the hydrogen-dominated, or reducing, atmospheres of the giant planets, these atoms will reside in the form of molecules of H_2O, CH_4, and NH_3. Measurements of CH_4, for example, suggest that Jupiter and Saturn are enriched in carbon by factors of about 2.5 and 5 compared with the Sun, as shown in Table 2. Other hydrocarbons have been detected in the upper atmospheres of Jupiter and Saturn. These are manufactured by ultraviolet light from the Sun, which breaks CH_4 into fragments that combine to form heavier hydrocarbons. These, in turn, form a haze or smog layer. Larger amounts of this haze on Saturn may partially explain the more muted, low-contrast appearance of Saturn's clouds when compared with those of Jupiter.

Chemical models predict three cloud layers on Jupiter and Saturn. The deepest is believed to be an H_2O ice cloud. In the middle is a cloud formed by a chemical reaction between NH_3 and hydrogen sulfide (H_2S) to form ammonium hydrosulfide (NH_4SH). The uppermost cloud, which is largely

Table 2. Chemical Composition of the Atmospheres of Jupiter and Saturn

Element	Molecule	Jupiter Above Clouds		Jupiter Below Clouds		Saturn Above Clouds		Saturn Below Clouds		Sun or Meteorites	
Hydrogen	H_2	90	%	90	%	96	%	96	%	84	%
Helium	He	10	%	10	%	3	%	3	%	16	%
Carbon	CH_4	2000	ppm	2000	ppm	4000	ppm	4000	ppm	700	ppm
Nitrogen	NH_3	30	ppm	300	ppm	50	ppm	500	ppm	200	ppm
Oxygen	H_2O	4	ppm	3000?	ppm	1?	ppm	6000?	ppm	1600	ppm
Phosphorus	PH_3	0.4	ppm	0.7	ppm	1.4	ppm	7	ppm	0.7	ppm
Germanium	GeH_4	0.7	ppb	0.7	ppb	0.4	ppn	0.4	ppb	8	ppb
Arsenic	AsH_3	0.7	ppb	0.7	ppb	0.4	ppb	3	ppb	0.5	ppb

responsible for the visual appearance of Jupiter and Saturn, is thought to contain NH_3 ice. It is difficult to verify this model using remote sensing because none of these ices have ever been detected on Jupiter or Saturn as of 1995. The non-detection of NH_3 ice in the highest cloud layer is quite baffling. Small particles of pure NH_3 ice absorb strongly at several wavelengths in the infrared, but Jupiter and Saturn do not show any of these features. This may be because the ice is "dirty," that is, coated with impurities which give a yellowish color to the clouds. Another possibility is that the ice particles are large compared to the wavelength of light. Either process may make the ice absorption features too weak to be detectable.

The best way to measure the chemical composition of Jupiter and Saturn is to choose wavelengths that probe the warm, deep atmosphere where most molecules are in the gas phase rather than frozen out as ices. Studies of radio waves emitted from the deep atmosphere of Jupiter and Saturn indicate that gaseous NH_3 is more abundant below the clouds, as is expected if the top two clouds contain compounds derived from ammonia. Observations of Jupiter at thermal infrared wavelengths show small amount of H_2O vapor. The small amount of H_2O is puzzling. Either Jupiter is much drier than expected or these observations do not probe deep enough to account for all of Jupiter's water inventory. One possibility is that the infrared data pertain to a region immediately above a massive water cloud. If so, much larger amounts of H_2O should exist below the cloud. In Table 2, notice that many of the molecules indicated are more abundant below the clouds, where it is warmer and where they are protected from harmful ultraviolet sunlight.

An important objective of the *Galileo* probe is to measure the cloud structure as well as the composition of Jupiter's atmosphere beneath all of the clouds.

The Great Red Spot

The most famous cloud feature on Jupiter is the salmon-colored Great Red Spot (GRS), as shown in Figure 1. One could place two Earths, side by side, within this oval feature, which measures 26,000 km east–west by 14,000 km north–south. The GRS moves in longitude, yet it does not move in latitude. The GRS has survived at least three centuries, whereas much smaller cloud features last for only a few days or weeks. Material within the spot moves counterclockwise at varying speeds; the average period is about six days. The GRS is a high-pressure feature; thus, it is probably not a jovian hurricane. An important unanswered question is: why is the spot red? All of the proposed cloud layers for Jupiter and Saturn are colorless. Sulfur (S) and phosphorus (P) are red under certain conditions. In the giant planets these elements reside in gaseous hydrogen sulfide (H_2S) and phosphine (PH_3). Hydrogen sulfide is predicted to be present but it has never been detected on either Jupiter or Saturn. The absence of gaseous H_2S may mean, but does not prove, that sulfur is chemically tied up in the clouds. Phosphine is relatively abundant, especially on Saturn (Table 2). However, there does not appear to be any more PH_3 in the GRS than in surrounding regions. Furthermore, the spectrum of laboratory phosphorus at visible wavelengths does not match the GRS very well. We suspect that the colors in the spot and perhaps in other regions on Jupiter may be due to impurities in the ammonia cloud, rather than in either of the deeper clouds. However, after many years of observations from ground-based telescopes and several spacecraft encounters, we still do not know for certain which chemicals are responsible for the yellow cloud colors on Jupiter and Saturn or why the Great Red Spot is red.

Origin of Jupiter and Saturn

In Table 2 the chemical composition of Jupiter and Saturn is compared with that of the Sun. Measurements of the solar abundance of certain elements come from studies of meteorites (*see* METEORITES). Interestingly, Jupiter and Saturn are both enriched in most of the observed elements compared with the Sun. The enrichment factor is larger for Saturn than for Jupiter. This evidence supports a particular model for the formation of the giant planets. Some 4.6 Ga ago, the cores of Jupiter and Saturn were built up by accretion of small fragments orbiting the young Sun in a cloud of dust and gas called the solar nebula. This core contained silicates and ices composed of H_2O, CH_4, and NH_3. Once they reached a size of about 10 Earth masses, these protoplanets had enough gravity to grab all of the gas in the neighborhood. Since Jupiter was closer to the Sun, where more gas was present, it grew to become the largest planet. Less material was present at Saturn, so it did not grow as big as Jupiter. Mixing took place between the core, which is enriched in the elements oxygen, carbon, and nitrogen, and the outer envelope of gases, containing mostly hydrogen and helium. Since Saturn's core is a larger fraction of its total mass (10/95 versus 10/318), Saturn should be more enriched in heavier elements than Jupiter. This seems to be the case for most elements, except helium. Helium is thought to be concentrated in the deep interior of Saturn, leaving the outer, observable shell depleted.

Gravitational Effects

One consequence of the enormous size of Jupiter and Saturn is the effect their gravity had in sweeping the solar system free of debris. Comets are believed to be condensates from the solar nebula that have changed little since the solar system was formed. Jupiter, and to a lesser extent Saturn, ejected many of the billions of comets originally traveling in the plane of the solar system into a spherical shell known as the Oort cloud. This helped to spare Earth and the other terrestrial planets some of the early devastation that might otherwise have occurred. A spectacular reminder of the power of cometary impacts occurred in July 1994 as twenty pieces of comet Shoemaker-Levy 9 smashed into Jupiter's southern hemisphere. The largest pieces produced dark spots in Jupiter's upper atmosphere that were prominent when viewed through small telescopes (*see* COMETS). Jupiter's gravity also affects the motions of asteroids. Several thousand asteroids orbit the Sun between Mars and Jupiter in the main asteroid belt. A nineteenth century astronomer named Daniel Kirkwood noted that there are very few asteroids with periods one-half, two-fifths, one-third, and one-fourth the period of Jupiter. Such asteroids would undergo repeated tugs by Jupiter and would be kicked into a different orbit, creating a "Kirkwood gap." Curiously, the effect is reversed in the outer asteroid belt. Asteroids prefer to have periods two-thirds, three-quarters, and equal to Jupiter's period, rather than avoiding those orbits. Thus, Jupiter's gravity creates both stable and unstable regions for asteroids to orbit the Sun.

Bibliography

ALLISON, M., and L. TRAVIS, eds. *The Jovian Atmospheres.* NASA Conference Publication 2441. Washington, DC, 1986.

BEATTY, J. K., and A. CHAIKIN, eds. *The New Solar System,* 3rd ed. New York, 1990.

HARTMANN, W. K. *Moons & Planets,* 3rd ed. Belmont, CA, 1993.

HUBBARD, W. B. *Planetary Interiors.* New York, 1984.

INGERSOLL, A. P. "Jupiter and Saturn." In *The Planets,* ed. B. Murray. San Francisco, 1983.

MORRISON, D., and T. OWEN. *The Planetary System.* New York, 1987.

GORDON L. BJORAKER

K

KINETICS

See Thermodynamics and Kinetics

KING, CLARENCE RIVERS

Clarence King was born on 6 January 1842 in Newport, Rhode Island. The King family had been in trade with China, but shortly after Clarence's birth, his father died and financial reversals occurred. His mother remarried when he was seventeen; King withdrew from school and went into trade but later attended the Sheffield Scientific School at Yale and completed the three-year postbaccalaureate course in only two years.

In the spring of 1863, King and J. T. Gardiner—who became a topographer with King and later F. V. Hayden—travelled across the country to California. Both volunteered to assist J. D. Whitney, head of the Geological Survey of California, and eventually were given paid positions with the Survey. Much of King's work was in the Sierra Nevada, where he made a remarkable number of climbs at a time when there was little interest in mountaineering. He was an early investigator of

Yosemite, though he did not at first ascribe the landscape to glacial sculpturing.

In 1867, the twenty-five-year-old King returned east. He persuaded General A. A. Humphreys, Chief of the Army Corps of Engineers, of the merits of an exploration along the fortieth parallel from the California-Nevada boundary eastward, eventually occupying a strip about 160 km wide and 1,300 km long along latitude 40° north. Organizing topographic and geologic work simultaneously, King showed remarkable skill. By 1870, the first publication on mining had been produced by J. D. Hague. Fieldwork was later extended to the east into Colorado and was completed in 1872. The unexpected grand finale was King's investigation of the supposed diamond field in northeastern Colorado and his exposure of the associated swindle; it was the high point of his career.

The publications of the fortieth parallel survey were impressive and the maps, prepared by careful triangulation, even more so. King's volume titled *Systematic Geology* was published in 1878. In it, he emphasized that cataclysmic changes from earthquakes, volcanoes, and climatic shifts were significant in geology, in contrast to the concept of uniformitarianism, which prevailed at the time.

Clarence King had a keen wit and a facile pen. While still engaged in fieldwork on the fortieth parallel survey, he wrote the essays that were published as *Mountaineering in the Sierra Nevada* (1872), considered a classic of American frontier literature. Indeed, his literary reputation, based on this single book, and his friendship with Henry Adams, are better known than his efforts in geology.

More or less contemporary with the completion of King's survey, congressional concern developed over duplication of geological surveys, engendered by arguments between G. M. Wheeler and F. V. Hayden and members of their respective western surveys. Congressman Abram S. Hewitt was responsible for asking the National Academy of Sciences for advice; King reputedly suggested this action to him.

Ultimately stemming from the Academy report, on 3 March 1879 the U.S. Geological Survey (USGS) emerged. Legislation effectively placed federal science under civilian rather than military control. JOHN WESLEY POWELL is credited with steering the legislation through Congress, though some historians credit King with a great deal of effort behind the scenes. Once the agency was created, a determined effort was made to ensure that King rather than Hayden become director. King did play a major public role in the political maneuvering for appointment, and he was fully supported by Powell.

King began by dividing the west into four regions and proposed to have geologists headquartered in Denver, Salt Lake City, and San Francisco, in part to save the cost and time of travel to the field. He also closely allied the early efforts of the organization to the concerns of the mining industry.

At the time the USGS was founded, plans for the 1880 U.S. Census were also being prepared. King proposed that the census include a study of mining in America, with an emphasis on the mining of iron and gold. Raphael Pumpelly headed a census corps that eventually employed fifty-seven men. Some special agents sampled iron ores, and others analyzed them and compiled statistics on production. In 1880, King instructed G. F. Becker and S. F. Emmons to compile information on gold. Meanwhile, King undertook a study of mining laws and customs. This project was finally completed in 1882. As a result of King's efforts and organization, the Tenth U.S. Census is unique in the information provided on mineral natural resources.

Less than two years after his appointment, King resigned as director of the USGS. One reason cited is that he was unable to extend the efforts of the Geological Survey beyond the areas of public lands, but in fact he was bored with administration and desired to make a fortune. Powell became the second director of the USGS, and King's own impact on the subsequent development of the agency has been argued for a century.

After leaving the government, King devoted some of his effort to cattle ranching, but most of his energy and enthusiasm was directed toward silver mining in Mexico; at one time he was actively engaged with three major properties. In the spring of 1882, King began a European tour that lasted more than two years. Although his prime aim was to sell the mines to new investors, he devoted much of his time to travel, art collection, and good living.

Shortly after his return to the United States, King's fortune began to decline, both monetarily and in a scientific sense. In 1892, John Wesley Powell had political difficulties with Congress and the USGS suffered. Powell was urged to resign and serious consideration was given to King as a possible successor, but Powell did not retire until 1894, in part to block King's return, and he was succeeded as director by CHARLES WALCOTT.

King's last significant scientific paper appeared in 1893. In this study he used the laboratory information of Carl Barus on the melting of igneous rocks to calculate the age of Earth. His figure of about 20 million years (Ma) accorded with that of Lord Kelvin, but most geologists at the time regarded Earth as at least five times older (see GEOLOGIC TIME; OLDEST ROCKS IN THE SOLAR SYSTEM). Thereafter King's health declined, his mining consultations dwindled, and his scientific reputation was eclipsed. He died in Phoenix, Arizona, on 24 December 1901.

Bibliography

RABBITT, M. C. *Minerals, Lands, and Geology for the Common Defence and General Welfare.* Vol. 1, *Before 1879.* U.S. Geological Survey. Washington, DC, 1979.

———. *Minerals, Lands, and Geology for the Common Defense and General Welfare.* Vol. 2, *1897–1904.* Washington, DC, 1980.

WILKENS, T. *Clarence King: A Biography.* Albuquerque, NM, 1988.

ELLIS L. YOCHELSON

KUIPER BELT

See Comets

KUIPER, GERARD

Gerard Peter Kuiper (originally Gerrit Pieter Kuiper) was born in The Netherlands on 7 December 1905. He died in Mexico City on 24 December 1973. Educated at Leiden University as an astronomer, Kuiper's earliest research was on the frequency of occurrence of double star systems, a topic that soon led him to the question of planet formation around the Sun and other stars. He emigrated to the United States in 1933, working first at Lick Observatory (San Jose, California) and then at Harvard University and the University of Chicago. In 1961, he moved to the University of Arizona, where he established the Lunar and Planetary Laboratory, a research and educational unit at which many planetary scientists have since been trained.

Kuiper's observational work on the solar system began with his discovery by spectroscopy of the atmosphere of Titan in 1944, in which he noted the importance of finding gas rich in hydrogen atoms that had previously been associated with much larger planets such as Jupiter and Saturn. Kuiper continued observational studies of the planets and satellites until the end of his life, during which he pioneered the use of infrared detectors in astronomical spectroscopy. During World War II, his work in military intelligence positioned him to take advantage of technological developments in infrared detection as this information was declassified. Kuiper's war work on the Alsos mission is described by DeVorkin (1992).

Parallel with his observational work, Kuiper developed a theory of the origin of the solar system, describing the way in which large-scale gravitational instabilities could permit planet formation in a rotating nebula. His theoretical work included insight into the origin of Pluto, and in an explanation of the origin of the short-period comets, he postulated a reservoir of icy planetesimals just beyond the orbit of Pluto. Discoveries made since 1991 appear to corroborate the existence of this reservoir, now called the Kuiper Belt.

Kuiper studied the Moon's surface with telescopes and photographs, recognizing the information it contains for the dynamic history of the terrestrial group of planets. As U.S. government interests turned toward the Moon and near-Earth space as potential military resources, Kuiper garnered NASA and air force contracts and grants to produce atlases of the Moon; the first was *Atlas of the Moon* (1959), reproducing the best telescopic photographs and providing the basis for early planning for the Apollo manned landings. Other atlases of the Moon produced under his supervision were *Orthographic Atlas of the Moon* (1961), *Rectified Lunar Atlas* (1963), *Consolidated Lunar Atlas* (1967), and five collections of photographs obtained by three Ranger probes to the Moon, a project on which he served as the principal scientific investigator. Kuiper's research on the Moon in the 1950s and 1960s provided strong support for the impact theory of crater formation at a time when many astronomers and geologists thought that the craters were volcanic in origin. He did, however, recognize that the lunar maria were modified by volcanism following the intense early bombardment phase of the Moon's history. Together with W. K. Hartmann, Kuiper first recognized the large-scale structure of lunar impact basins. Kuiper's role in evaluating Soviet lunar science is described by Doel (1992).

In establishing the Lunar and Planetary Laboratory (LPL) at the University of Arizona in 1960, Kuiper brought together educators, researchers, and students interested in astronomy, geology, meteorology, chemistry, and physics to forge a multidisciplinary approach to the study of the bodies of the solar system through observations, theory, and laboratory investigations. This early recognition of the many facets of the study of the other planets, and the training of students in what has become known broadly as planetary science, is one of Kuiper's greatest legacies.

Kuiper was a leader in finding superlative sites for ground-based astronomical observatories, particularly those suitable for infrared work, which blossomed in the 1960s and has grown steadily since. He developed the Mount Lemmon site near Tucson, and his enthusiastic testing and promotion of Mauna Kea in Hawaii eventually resulted in the worldwide recognition of Mauna Kea as the premier site for observations in the northern hemisphere. It is now the location of the two largest optical telescopes in the world and the large concentration of other large telescopes operated by many nations and U.S. scientific establishments. In addition, he surveyed potential sites in the desert southwest of the United States and in Mexico.

Seeking better ways to study the infrared radiation from the planets and stars, Kuiper played an influential role in the development of infrared airborne astronomy in the 1960s and 1970s. In 1967 the NASA Convair 990 aircraft with a telescope aboard became available for infrared studies at an altitude of 12,000 m and Kuiper used it extensively for spectroscopy of the Sun, stars, and planets. When the Convair 990 *Galileo* was replaced in 1975 by a bigger plane with a much larger telescope, the new facility was dedicated as the Kuiper Airborne Observatory.

Kuiper undertook major editorial projects beginning with a conference volume, *The Atmospheres of the Earth and Planets,* published in two editions in 1949 and 1952, followed by the four-volume *The Solar System,* published between 1953 and 1963, and the nine-volume compendium *Stars and Stellar Systems* (with B. M. Middlehurst) beginning in 1960.

During his years as director of LPL, Kuiper served as editor of his own journal of planetary and astronomical research, *The Communications of the Lunar and Planetary Laboratory* (published in 195 numbers between 1962 and 1973); the majority of his own research on the Moon, many papers on infrared spectroscopy of the planets and stars, and a major atlas of the spectrum of the Sun, appeared in *Communications.*

As the most senior planetary astronomer in the United States, Kuiper had an important influence on NASA's program of lunar and planetary exploration by spacecraft. He advised NASA in many capacities and participated in several of the missions of the 1960s and early 1970s as a scientist. A full appraisal of his role is described by Tatarewicz (1990).

Kuiper was a demanding individual who thrived on a daily routine of hard work and long hours, and he expected the same of his students and colleagues. As the individual who initiated and promoted physical studies of the bodies of the solar system in the modern era, Gerard Peter Kuiper can justly be considered the father of modern planetary astronomy. More details of his life, career, and influence on the U.S.'s planetary exploration program can be found in the memoir by Cruikshank (1993).

Bibliography

CRUIKSHANK, D. P. "Gerard Peter Kuiper." *Memoirs of the National Academy of Sciences* 62 (1993):259–295.

DEVORKIN, D. H. *Science with a Vengeance.* New York, 1992.

DOEL, R. E. "Evaluating Soviet Lunar Science in Cold War America." *Osiris* 7 (1992):44–70.

KUIPER, G. P., ed. *Atmospheres of the Earth and Planets.* Chicago, 1949, 1952.

———. *The Solar System:* Vol. 1, *The Sun;* Vol. 2, *The Earth as a Planet;* Vol. 3 (with B. M. Middlehurst), *Planets and Satellites,* 1961; Vol. 4 (with B. M. Middlehurst), *The Moon, Meteorites, and Comets.* Chicago, 1953, 1954, 1961, 1963.

———. *Stars and Stellar Systems.* 9 vols. Chicago, 1960.

KUIPER, G. P.; D. W. G. ARTHUR; E. MOORE; J. W. TAPSCOTT; and E. A. WHITAKER. *Photographic Atlas of the Moon.* Chicago, 1959.

KUIPER, G. P.; D. W. G. ARTHUR; and E. A. WHITAKER. *Orthographic Atlas of the Moon.* Tucson, 1961.

KUIPER, G. P.; E. A. WHITAKER; W. K. HARTMANN; and L. H. SPRADLEY. *Rectified Lunar Atlas.* Supp. 2 of *Atlas of the Moon.* Tucson, 1963.

KUIPER, G. P.; E. A. WHITAKER; R. G. STROM; J. W. FOUNTAIN; and S. M. LARSON. *Consolidated Lunar Atlas.* Tucson, 1967.

TATAREWICZ, J. N. *Space Technology and Planetary Astronomy.* Bloomington, IN, 1990.

DALE CRUIKSHANK

L

LAKES

Although lakes cover only about 1 percent of Earth's continental surface, and contain less than 0.02 percent of the world's water, they are one of the most interesting and varied environments. Lakes vary widely in physical, chemical, and biological features, such as size, shape, water depth, circulation patterns, sediment characteristics, water chemistry, and types of water-dwelling organisms. Geologists use the sedimentary records of modern lakes to help understand ancient lake deposits.

Sedimentary deposits of ancient lakes (called lacustrine deposits) comprise only a tiny fraction of the geologic column, but they are extremely important because of the information and detail they contain about ancient environments and climates, and because they may contain economically valuable minerals or fossil fuels. Lacustrine deposits are extremely variable in rock type, mineral content, fossils, thickness, and vertical changes in sedimentary features. Lacustrine deposition is controlled by climate, tectonic setting, and the rock types in the source area.

Origin and Major Types of Lakes

Hutchinson (1957) developed a detailed lake classification in defining seventy-six different types of lakes. In general, most lakes fall into the following categories: tectonic, glacial, volcanic, dissolution, landslide-influenced, fluvial-influenced, wind-formed, coastal, and meteorite impacts.

Many of the world's largest lakes are tectonic in origin, forming in response to large-scale movements of Earth's crust. The three largest lakes in terms of volume of water (Caspian Sea, Lake Baikal, and Lake Tanganyika) are tectonic in origin. Crustal movements influencing lake formation include uplift of the seafloor (Caspian Sea, Aral Sea), uplift around a central basin (Lake Victoria), and rifting. On a much smaller scale, "sag ponds" occur along some faults, due to localized subsidence. Most of the world's tectonic lakes are concentrated along rift zones where Earth's continental crust currently is being torn apart by plate tectonic forces. The largest of these continental rifts, the East African Rift Valley, contains numerous lakes, including Lakes Tanganyika, Malawi, Kivu, Turkana, Mobutu, Magadi, Naivasha, and Natron. The world's deepest lake, Lake Baikal, Russia (1,700 m), is located in the Baikal Rift Zone of the Siberian Platform. The Basin and Range Province of the western United States also contains tectonic lakes in the valleys between fault-block mountains (e.g., Lake Tahoe on the California–Nevada border, and Pyramid Lake, Nevada). Most of the Basin and Range lakes are small remnants of much larger lakes that existed during the latter part of the Pleistocene epoch, 10,000 to 30,000 years ago, such as Lake Bonneville (which covered about 50,000 km² and was about 300 m deep). The Great Salt Lake, Utah, now only about 10 m deep and covering 2,500–6,000 km², is a remnant of Lake Bonneville (Picard and High, 1985). The range in size of the lake is the result of relatively rapid expansion and contraction caused by changes in water level, which are related to varying precipitation.

Many lakes were formed by glacial erosion and deposition during the Pleistocene, when glaciers blanketed the continents north of 40°N latitude. The most prominent glacial lakes are the Great Lakes (Superior, Huron, Michigan, Erie, and Ontario) along the U.S.–Canada border, and the Great Slave and Great Bear lakes in Canada. Lake Superior is the fourth largest lake in the world, in terms of water volume. Other glacial lakes formed as a result of glacial sediment damming valleys carved by glaciers (such as the lakes of the European Alps—Lakes Geneva, Lucerne, Zurich, and Constance), glacial sediment damming rivers and streams (Finger Lakes in New York State), ice carving small basins into the bedrock (some lakes in Norway, southern Sweden, Finland, Scotland, and Ireland), fjord lakes (Norway and British Columbia, Canada), or melting of ice blocks in glacial sediment (such as kettle lakes in Minnesota).

Lakes associated with volcanic activity are commonly small and deep. They may be in the crater of an active but quiet volcano (Volcan Poas, Costa Rica), in a caldera or large basin produced by the explosion and collapse of an underground magma chamber (Crater Lake, Oregon), on collapsed lava flows (Yellowstone Lake), and in valleys dammed by volcanic deposits (Sea of Galilee, and Snag Lake in Mount Lassen National Park, California).

In areas underlain by soluble rock such as limestone, lakes may occupy sinkholes formed by dissolution of the bedrock. Many dissolution lakes are present in Florida, which is largely underlain by limestone.

Many other types of lakes exist. Lakes can form where landslides block a valley and cause damming of drainage. Fluvial-influenced lakes form because of the erosional and depositional action of rivers (such as oxbow lakes along the Mississippi River), and may be associated with levees on river deltas (Lake Pontchartrain, Louisiana). Lakes may occupy wind-formed or deflation basins (in northern Texas, New Mexico, and parts of Australia), basins dammed by windblown sand, and basins between sand dunes (as in the Sand Hills of Nebraska).

Coastal lakes form by sediment deposition that isolates an embayment or irregularity of the coastline (such as the Coorong lakes along the southeastern coast of Australia). The least common lake is that formed by meteorite impacts. Lakes occupy some of the 60–120 meteorite impact craters recognized on Earth (such as Lake Bosumtwi, Ghana, Deep Bay Crater in the Southeast corner of Reindeer Lake in Saskatchewan, and Manicouagan Lake and Ungava or Chubb Lake in Quebec). The origin of some lakes is unresolved, but much debated, such as the ellipsoidal "Carolina bays" of the coastal plain of the southeastern United States.

Lake Structure—The Character of the Water Column

The circulation of lake water and the character of the water column are controlled by differences in water density. The density of water depends primarily on temperature, but it is also controlled by salinity and the amount of sediment in suspension. Water is at its maximum density at a temperature slightly above the freezing point (about 4°C or 39.2°F). Lake water is warmed by solar radiation, but sunlight generally does not penetrate to the bottom, so water temperature decreases with depth. Lake water is commonly thermally stratified into an upper layer with good circulation called the epilimnion, which is warm and well oxygenated, and a lower layer called the hypolimnion, which is cold and relatively still. The hypolimnion is commonly anoxic (lacks oxygen), because decay of organic matter uses up the oxygen in the water. After all of the oxygen is depleted from the bottom waters, additional organic matter that is introduced by settling from the surface will be preserved in the sediments on the bottom. In some lakes, water temperature decreases abruptly from the epilimnion to the hypolimnion. The depth at which water temperature changes most rapidly is called the thermocline. Not all lakes are stratified; lakes in which the entire water column circulates are called holomictic.

Seasonal temperature changes sometimes induce the waters in a lake to overturn, leading to mixing of the upper and lower water masses. When decreasing air temperatures cool the surface water (epilimnion) until it is similar in temperature (and density) to the bottom water (hypolimnion), wind can cause turbulence that leads to a breakdown in stratification and overturning of the water masses. Lakes that overturn yearly are monomictic; those which overturn twice a year (spring and fall) are dimictic. Other lakes may overturn only rarely (oligomictic), or not at all (meromictic). Lakes in temperate regions are more likely to overturn, whereas lakes in tropical regions are likely to overturn only rarely, if at all (oligomictic or meromictic). Air temperatures remain warm all year in the tropics and seldom drop to the point where overturning could occur.

The hydrologic character of a lake (whether or not a lake has outflow) controls the type and arrangement of sedimentary deposits. Hydrologically open lakes (with outflow) have low salinity and relatively stable shorelines, whereas many hydrologically closed lakes (no outflow) have high salinity and shorelines that fluctuate in response to climatic change (Allen and Collinson, 1986). Some lakes pass through open and closed phases during their histories as a result of climatic changes, tectonic events, or changes in drainage patterns.

Chemical Composition of Lake Waters

Lakes are much more variable in salinity and water composition than are ocean waters. Salinity is measured in parts per thousand (ppt) total dissolved solids. For comparison, the salinity of ocean water averages about 34.7 ppt. This means that 1,000 g of water would contain 34.7 g of dissolved solids. Lake waters range from fresh (less than 1 ppt) to saline or hypersaline (more than 250 ppt). For example, Lake Chad in Africa (freshwater) has a salinity of 0.165 ppt; Lake Bogoria, Kenya, has a salinity of 40 ppt; the Great Salt Lake in Utah has a salinity of 250 ppt (hypersaline). Saline or hypersaline lakes form in arid to semiarid areas if evaporation exceeds inflow, and the basin is hydrologically closed. If inflow to the lake is intermittent, the lake may be ephemeral, occasionally drying up (called a playa lake). Constant inflow will result in a perennial saline lake. Ephemeral lakes may undergo rapid salinity changes as they dry up due to evaporitic concentration (such as Lake Eyre, Australia).

The chemistry of lake water depends on the chemistry of the surrounding rocks of the source area. The "typical" lake in a humid region is a dilute freshwater "calcium carbonate lake," in which the principal anion is CO_3. In some arid regions

there are lakes rich in sodium carbonates, called alkali lakes or soda lakes. Other lake waters are enriched in potassium (such as those in the sand hills of Nebraska), or in sulfate (as in southern and central Saskatchewan), or in sodium borate or borax (some desert lakes in the southwestern United States). Other lakes, like those associated with volcanoes in Japan, are acid lakes with pH as low as 1.5 to 2.0.

Unique Sedimentological and Biological/Paleobiological Aspects

The sedimentary records of lakes are among our best sources of data on ancient climates because the structure of the water column, hydrology, and water chemistry are related to climate. Each of these factors influences sediment deposition and the types of organisms that inhabit the lake.

The thickness of lacustrine sedimentary deposits depends on climate and tectonic setting, which influence sedimentation rates and the lifespan of the lake. Tectonic lakes tend to have thick sedimentary deposits because of rapid subsidence and high sediment supply (more than 3 km of sediment in Lake Tanganyika, and more than 5 km in Lake Baikal). Vertical changes in lake deposits are caused by changes in the biology and chemistry of lake waters, shoreline movements, and changes in sediment influx. Changing lake levels produce a record of high and low stands marked by changes in sediment character and fossils. Former low stands are recognized by evaporite minerals, soil horizons, erosion surfaces, or shell beds. Former high stands are recognized by abandoned beaches or wave-cut terraces above present lake shorelines.

Lakes that overturn may produce detailed sedimentary records of these events. Glacial lakes with seasonal overturn commonly deposit varves, or thin layers of sediments, which can be counted and correlated to indicate the age of the sediments.

Fossils preserved in lake sediments are useful in environmental interpretation. Spores and pollen preserved in lake sediments indicate the types of plants that inhabited an area, which are related to climate. Other fossils useful in interpreting ancient lakes include diatoms, ostracodes, mollusks, and fish.

Some lake deposits have economic importance because they contain evaporite minerals (such as borax, trona, gypsum, Epsom salt, and Glauber salt), zeolites, lithium, organic-rich deposits used as fossil fuels (oil shales, petroleum source-rocks, or coal), clays with industrial uses, or because they may be sites of uranium accumulation.

A few examples of well-studied ancient lake sequences include Eocene deposits of Utah, Wyoming, and Colorado (Green River Formation), Triassic and Jurassic lake deposits of the Newark Supergroup in eastern North America, and Devonian lake deposits of the Orcadian Basin in Scotland.

Bibliography

ALLEN, P. A., and J. D. COLLINSON. "Lakes." In *Sedimentary Environments and Facies*, ed. H. G. Reading. Oxford, 1986.

EUGSTER, H. P., and K. KELTS. "Lacustrine Chemical Sediments." In *Chemical Sediments and Geomorphology*, ed. A. Goudie and K. Pye. London, 1983.

FROSTICK, L. E.; R. W. RENAUT; I. REID; and J. J. TIERCELIN, eds. *Sedimentation in the African Rifts*. Geological Society Special Publication No. 25. Oxford, 1986.

HUTCHINSON, G. E. *A Treatise on Limnology. Vol. 1, Geography, Physics, Chemistry.* New York, 1957.

JOHNSON, T. C. "Sedimentation in Large Lakes." *Annual Reviews of Earth and Planetary Sciences* 12 (1984):179–204.

MATTER, A., and M. E. TUCKER, eds. *Modern and Ancient Lake Sediments*. Special Publication 2, International Association of Sedimentologists. Oxford, 1978.

PICARD, M. D., and L. R. HIGH, JR. "Criteria for Recognizing Lacustrine Rocks." In *Recognition of Ancient Sedimentary Environments*, eds. J. K. Rigby and W. K. Hamblin. Society of Economic Paleontologists and Mineralogists Special Publication No. 16. Tulsa, OK, 1972.

PAMELA GORE

LAMARCK, JEAN-BAPTISTE

Jean-Baptiste Lamarck was born in Bazentin, Picardy, France, on 1 August 1744, the eleventh child of a poor noble family. Originally his father thought he should study theology and enter the church, and he was enrolled (against his will) in the Jesuit college at Amiens. Lamarck's father died in 1759, however, and this freed Lamarck to pursue a military career. He fought bravely in the Seven Years' War, being promoted to a lieutenant, but then an injury and ill health forced him to return to civilian life in 1768. Lamarck became interested in natural history, particularly botany, while still in the army. From the military he went to Paris and first tried his hand at banking, then finally studied medicine. In the late 1760s Lamarck became involved with the circle of botanists and their students associated with the Jardin du Roi (the King's Garden). Lamarck's early botanical researches resulted in his publication of the well-received, three-volume *Flore françoise* (1779).

In 1779 Lamarck was made a member of the French Académie des Sciences. In the early 1780s Lamarck traveled through Europe to visit various herbaria, accompanied by the son of the eminent French naturalist Count Georges Louis de Buffon, who was the head of the Jardin du Roi. Lamarck in essence served as a tutor for Buffon's son. In 1788–1789 Lamarck was appointed to a chair of botany at the Jardin du Roi. In 1793, during the French Revolution, the old Jardin du Roi and royal natural history cabinet were reorganized as the Muséum National d'Histoire Naturelle, and Lamarck was made a professor of zoology in charge of "lower animals" such as insects, worms, mollusks, and microscopic animals. Apparently Lamarck was assigned the position that no one else wanted: there were already more prominent botanists at the museum and "higher animals" (such as fishes, amphibians, reptiles, birds, and mammals) were seen as more attractive subjects for research. At the time Lamarck knew little about the subject to which he was assigned, but he dove into it wholeheartedly. Lamarck remained at the museum for the remainder of his career. His health began to fail in the early nineteenth century, and by about 1818 he was blind and could no longer teach (the teaching duties were taken over by his assistant, P.-A. Latreille). Impoverished, and with his reputation at a low ebb, he died in 1829. It was not until fifty or so years later that Lamarck's genius was really acknowledged.

Lamarck took an interest in all of the natural sciences, although he is now best known for his early evolutionary theory. Lamarck studied and reclassified the "lower animals," and it was Lamarck who first made the distinction between "vertebrates" and "invertebrates." Lamarck also first popularized the term "biology" for the study of all living organisms. Among his more important publications along these lines are *Système des animaux sans vertèbres . . .* (1801) and the seven-volume *Histoire naturelle des animaux sans vertèbres* (1815–1822). Lamarck also published major works on physics, chemistry, meteorology, and general natural history, including *Recherches sur les causes des principaux faits physiques* (1794), *Réfutation de la théorie pneumatique* (1796), and *Mémoires de physique et d'histoire naturelle* (1797). Unfortunately, many of Lamarck's physicochemical theories were judged to be inferior to the work of various contemporary chemists, such as Antoine-Laurent Lavoisier, and thus Lamarck's writings on these subjects were widely ignored. Lamarck was also very interested in geology and paleontology, and in *Hydrogéologie* (1802) he developed his ideas along the lines of what would later be considered a "uniformitarianist" point of view. Lamarck maintained that Earth was at least hundreds of thousands, and perhaps hundreds of millions, of years old. He explained the present-day features of the surface of Earth as the result of present-day processes operating over enormous periods of time. In many ways Lamarck was a gradualist.

Lamarck first proposed a cogent scientific theory of organic evolution in the year 1800. He elaborated upon his theory in the book *Philosophie zoologique* (1809). Lamarck believed in the gradual change of species over geological time, and he recognized "evolutionary sequences" of invertebrates (such as mollusks) to prove his point. The primary mechanism driving evolutionary change, according to Lamarck, was the inheritance of certain acquired characteristics that were of use in the perpetuation of the species. If an organism used certain portions of its body throughout its life, such as its legs for running, these portions would be strengthened and further developed, this strengthening and development possibly being passed on to its descendants. Likewise, structures or organs that were not regularly used might wither or weaken, and their withered or weakened state would be passed on to the organism's descendants. In this manner, the morphology and characteristics of organisms and species would change, or evolve, over time.

Lamarck's theory of evolution was not widely accepted in his own lifetime, due in large part to the objections of the man often credited with founding vertebrate paleontology, the "catastrophist" GEORGES CUVIER (1769–1832). However, in the decades immediately following the publica-

tion of CHARLES DARWIN's *On the Origin of Species* (1859) many early evolutionists adopted a Lamarckian point of view. Lamarck's last major publication was *Système analytique des connaissances positives de l'homme* (Analytical System of Man's Positive Knowledge, 1820).

Married three times, Lamarck outlived all his wives. By his first two, he had three children; among them, two daughters by his first marriage, Rosalie and Cornélie, never married and looked after their father in his old age. Lamarck died on 28 December 1829, in Paris, France.

Bibliography

BOWLER, P. *Evolution: The History of an Idea.* Berkeley, CA, 1984.

BURKHARDT, R. W., JR. "The Zoological Philosophy of J. B. Lamarck." In Lamarck, J. B. *Zoological Philosophy*, translated from the French by H. Elliot. Chicago, 1984.

———. *Zoological Philosophy*, translated from the French by H. Elliot. Chicago, 1984.

———. *Hydrogeology*, translated from the French by A. V. Carozzi. Urbana, IL, 1964.

PACKARD, A. S. *Lamarck: The Founder of Evolution; His Life and Work.* New York, 1901.

ROBERT M. SCHOCK

LAND DEGRADATION AND DESERTIFICATION

One of the most significant issues of modern times is the production of enough food to satisfy the needs of an ever-increasing world population. Diminishing land productivity of food, whether for animal feed or human consumption, creates serious problems in all the continents of Earth. Hence, land degradation resulting from natural processes or human activities has begun to gain the attention of research scientists, policy makers, and the general public.

The term "desertification" received worldwide attention following the United Nations Conference on Desertification, which was held in 1977 in Nairobi, Kenya, in the wake of the severe drought of 1968–1973 that devastated the Sahel region of

North Africa. The term was defined as the impoverishment of arid, semiarid, and some subhumid ecosystems as a result of land abuse stemming from human activities.

Since then, the term has been used by nonspecialists, and particularly the popular press, to create the misconception that the desert is a result of misuse. For this reason, the United Nations Environment Program (UNEP) announced after a February 1990 meeting in Nairobi, Kenya, that "land degradation" is a more appropriate term to describe processes that result in diminishing land productivity, particularly in areas with limited rainfall.

Arid and semiarid lands of the earth resulted from a lack or scarcity of rain. For example, deserts are defined as regions that receive less than 25 cm of rain per year. However, these regions were not always dry and were hosts to several wet climate episodes lasting for thousands of years during the past 200,000 years.

The term desert itself conveys this meaning. It evolved from an ancient Egyptian hieroglyph, *Tesert*, via the Latin verb desero, to abandon. From the latter came *desertum*, a waste place or wilderness and *desertus*, meaning abandoned or forsaken. These arid lands hosted kinder climates as recently as 5,000 years ago. With the onset of severe dryness they were relinquished by plants, animals, and humans.

Major fluctuations in the amount of rainfall also have milder and more short-term counterparts. For example, records of the Nile floods indicate a one hundred thirty-five year cycle. The sun-spot cycle of twenty-two years is also considered by some researchers to affect the amount of rainfall, particularly in arid and semiarid regions. Similarly, a seven-year cycle is invoked to explain the repeated droughts in the sub-Saharan belt of North Africa.

Drought

Variations in the amount of rainfall in arid and semiarid lands drastically affect land productivity. Sparse natural vegetation disappears quickly after a period of drought. As the roots of plants die, the soil is loosened and becomes more susceptible to erosion by running surface water and particularly by the wind.

This situation was vividly illustrated by the Dust Bowl in the American prairie states. The land cover in these areas was severely damaged by repeated droughts in the 1930s. Further degradation by overgrazing and plowing exposed vast amounts of soil to strong winds that carried the fine particulates into the atmosphere as dust, turning day into night, scarring the terrain and imposing the need for remedial measures to stabilize the soil.

Without any human input, severe degradation of the landscape may be caused by droughts. The first result of a decrease in the amount of rain is the dryness and hence the disappearance of vegetation. This is true not only of desert shrubs but of resistant trees such as tamarisks and acacias. Semiarid zones are particularly sensitive to such events and may be degraded extensively by droughts.

When droughts occur in regions with relatively dense populations, such as the Rajasthan of northwest India, pressure on land productivity increases dramatically. People and their herds satisfy their needs from increasingly smaller plots and forced stresses on the land cause further degradation.

Soil Erosion

Removal of the vegetation cover exposes fine particles of the soil to the agents of erosion. An occasional heavy rainstorm erodes the soil and scars the terrain with deep ravines and gullies. More important, wind acts on the soil and segregates the particles according to size to be carried in suspension as dust, to accumulate as dunes, or to lag behind with larger rock fragments (El-Baz and Hassan, 1986).

As a result of land degradation due to droughts, wind deposits are discretely zoned depending on the potential of particle transport. The finest particles of clay and silt, up to 0.05 mm in diameter, are winnowed out and whirled into the atmosphere as dust. They settle out of suspension beyond the zones of high wind energy, sometimes at great distances. For example, Saharan dust is usually carried at an altitude of between 1 and 4 km in the atmosphere to settle far away in the Caribbean Sea and the western equatorial region of the Atlantic Ocean.

Fine and medium-sized grains, from 0.05 to 0.5 mm in diameter, bounce in the wind and accumulate into dunes of linear, crescentic, or domical shapes. Once formed, a dune will continue to grow or move as long as the wind blows. Mobile dunes have caused enormous damage to lands worldwide as they encroach on farms, settlements, and roads in nearly all the arid and semiarid lands of Earth (Figure 1).

Particles too large to be lifted by the wind, from 0.5 to 2.0 mm in diameter, may gradually and erratically move or roll along the surface. With high winds, the whole surface covered by such grains appears to be creeping slowly along the wind direction. These grains usually form extensive, flat sand sheets and are commonly used in agriculture.

Pebbles larger than 2.0 mm in diameter lag behind with large fragments to form a "desert pavement" that protects underlying soil from further erosion. Disturbance of this armor of large particles exposes fine-grained soil below to the action of wind, repeating the cycle of erosion. This is what happens when recreational vehicles are used extensively over a naturally protected surface, resulting in destruction of the desert pavement and mobilization of the soil.

Land-Use Patterns

Much of the land degradation due to human activities results from changes in land-use patterns that are imposed by political boundaries. Humans who live in arid and semiarid lands have developed a special sensitivity to their environment. For example, nomadism was, until the nineteenth century, the most prevalent way of life in arid environments. Nomads and Bedouin intermittently roamed the land because it was the only way to use rainfall that is erratic in space and time. In the drylands of Earth it rains in one place but not another, and when the rain returns, it does so in yet another place. That is why nomads roamed in patterns of movement based on thousands of years of experience in exploiting the scarce resources of arid environments.

Creation of sovereign states in much of Africa and Asia during the twentieth century resulted in

Figure 1. Sand masses encroaching on fields on a road near Dakhla Oasis, Egypt. (Courtesy of Farouk El-Baz.)

limiting the freedom of movement of nomads. National boundaries were established where there had been none, and each country closed its borders to roaming herders and their animals. One way of dealing with the plight of nomads was to settle them in villages and towns.

When the nomads are settled, as has happened in most countries in the arid zones of Earth, people gather from all directions and, with their animals, densely populate an area. The land cover can no longer be sustained nor can it support their large numbers, thus beginning the degradation of the terrain. In the apparent solution to the problem of nomads, therefore, lie the roots of long-term degradation of the environment.

The settlement of nomads also requires drilling a number of water wells in a small area and operating them with powerful pumps. Such a solution usually turns out to be counterproductive. Continued pumping of water in enormous quantities from the same area, usually from a limited reservoir, results in a severe decrease of water levels. It may take a decade or more to replenish a depleted aquifer in arid terrain. Furthermore, pumping from one area may deplete nearby locations, forcing people to limit their movement; in so doing they no longer can take advantage of the natural environment. Their herds overgraze the immediate surroundings of the settlements, and severe land degradation sets in.

When settlements grow into towns, urban sprawl on fertile land may also exacerbate the problem by (1) loss of highly productive land by urban expansion; (2) mining of clayey soil for the production of redbrick for building construction; and (3) increased pollution of the environment from increased amounts of waste in small areas. Such degradation due to unchecked urban growth has badly affected such countries as Egypt, where it resulted in a serious decrease of the food production capability of the land (Mainguet, 1991). The outcome was increased food imports, which burdened the national economy.

Agricultural Practices

Irrigated agriculture in marginally fertile soil may also worsen land degradation if it is done without detailed knowledge of the local environment, irrigation techniques, and crop rotation patterns. For example, vast tracts of land in China were de-graded due to a campaign for increasing food production. This was particularly true in the northwestern territories, where the soil was not porous and therefore required special methods to increase its porosity before raising crops. Extensive regions that were not so treated are now fallow, and their fine-grained soils are exposed to the ravages of wind erosion.

Excessive use of brackish or even potable groundwater may degrade the land further. Imperceptibly thin layers of salt are left behind on the top soil as the irrigation water is drained and the moisture in the soil evaporates. Repeated cropping without attempts to diminish the salt content (by soil washing or raising crops that absorb much of the salt) results in the salinization of the land. This process has resulted in the degradation of vast regions in all the semiarid lands of Earth, particularly in Australia, China, India, the African Sahel, and the American Southwest.

Flood irrigation with poor drainage also causes land degradation by waterlogging of the soil. This is particularly true where agriculture is practiced in enclosed regions, such as oases depressions. Such is the case of the New Valley Province in the Western Desert of Egypt. Here groundwater is used to raise crops that require a great amount of water, such as rice, without a way to drain the used water out of the depressions that enclose the oases. The land becomes water-saturated, and the soil is clogged to the extent that it can no longer be used for raising crops.

Sustainable Development

Use of land for food production will undoubtedly increase in the twenty-first century. To assure that this is done with the least possible harm to the environment we must attain the following: (1) better understanding of the nature of the land and the forces that act upon it; (2) mapping of all segments of the surface of Earth that are prone to land degradation; (3) realization of what needs to be done to improve the health of the soil; and (4) developing the land with long-term sustainability in mind.

Research done since the 1960s has dramatically increased our knowledge of the origin and evolution of arid and semiarid lands. The findings came as a result of scientific investigations prompted by cases of prolonged drought in many parts of the world. Since the advent of space exploration, much new knowledge has also been attained by the study of satellite data. Images acquired by spacecraft add a unique perspective, and their repetitive coverage helps monitor environmental change in space and time. This knowledge represents a base on which to build plans for future development projects.

Through efforts by the United Nations Environment Program, there exists a body of information on the location and extent of land degradation worldwide (UNEP, 1992), as well as on research programs to evaluate how land degradation trends can be reversed (UNEP, 1987). Efforts along these lines are also encouraged by donor organizations in the developed world as well as by local efforts in developing countries.

Emphasis on consideration of environmental consequences of development projects has also increased in the wake of the Earth Summit that was convened in June 1992 by the United Nations in Rio de Janeiro. This emphasis makes it imperative to consider the long-term sustainability of development projects, including those projects that strive for expanded production of food. With this goal in mind, the outlook is encouraging for limiting land degradation in the future.

Bibliography

EL-BAZ, F. and M. H. A. HASSAN, eds. *Physics of Desertification.* Dordrecht, Netherlands, 1986.

MAINGUET, M. *Desertification: Natural Background and Human Mismanagement.* Berlin, 1991.

UNITED NATIONS ENVIRONMENT PROGRAM. *UN Project Compendium on Desertification Control and Dryland Development.* Nairobi, Kenya, 1987.

———. *World Atlas of Desertification.* London, 1992.

FAROUK EL-BAZ

LANDSCAPE EVOLUTION

The concept that Earth's landscapes are not immutable, and might change or evolve (which means to change in a systematic way) through time, is at least as old as the shepherd philosophers of the biblical Old Testament. Everyone who has heard Handel's oratorio *Messiah* is familiar with the text from Isaiah (40:4): "Every valley shall be exalted, and every mountain and hill shall be made low, and the crooked straight, and the rough places a plain." Another philosopher (Psalm 121) noted the futility of seeking strength from the dubious durability of the "everlasting" hills. An early Chinese philosopher noted that water, which he judged to be the softest and weakest of all materials, could wear away the most resistant rocks (LaFargue, 1992). Earth scientists who study such phenomena today are called geomorphologists. They use a variety of modern scientific tools to analyze the processes and rates of landscape evolution and to predict future trends. However, any thoughtful person, on viewing the topography of a mountain range from an airplane, will be aware that all the details of the peaks, valleys, and slopes are the results of erosion. Only a little additional reflection is required to conclude that the processes are continuing, and that landscapes are evolving before our eyes.

All of Earth's rocky surface that is above sea level but is not covered by glacier ice constitutes the landscape. It can be studied on a variety of scales, from an astronaut's view of entire mountain ranges and island archipelagos to that of a soil scientist measuring the creep and wash of soil down a hillside during a rainstorm. The subordinate components of a landscape are called landforms. Traditionally, a landform was considered to be a landscape component that could be viewed in its entirety, but our new ability to view our entire planet from space renders that definition useless. Earth is a whole, made up of landscape components that are measured on many scales.

On the largest scale, Earth is nearly a sphere slightly flattened at the poles and bulging at the equator due to rotational forces. The relief, or roughness, of the landscape is small compared to the radius of the earth. The highest mountain is less than 10 km in height, which is less than 0.2 percent of Earth's radius of 6,400 km. On a very large globe of 25 m radius, most mountainous relief would be correctly scaled at 0.25 cm—rough enough to be felt by fingertips, but barely visible. On the next smaller scale, Earth's relief is divisible into continents (about 29%) and ocean basins (71%), which average about 0.8 km above sea level and 3.7 km below sea level, respectively. On the continents, landscapes are traditionally divided into regions called geomorphic (or physiographic) provinces, each of which is characterized by a coherence or unity of landscape and a common geologic history (*see* CONTINENTS, EVOLUTION OF). For example, the geomorphic provinces of the eastern United States are familiar to travelers by such names as the Atlantic coastal plain, the Piedmont, the Appalachian valley and ridge province, the Appalachian plateaus, and so on (Thornbury, 1965). Scenery is the part of a province that can be viewed from an appropriate vantage point—such as a river valley or a named mountain peak. A nice analogy is that the largest landscape components are like theaters, "permanent" structures "in which geologic dramas are enacted" (von Engeln, 1942, p. 17). The next lower levels are like theatrical stages, which are remodeled from time to time. At the viewing level of the audience is the scenery, backdrops, and sets that change during the course of the drama. Scenery is the pleasure of tourists and the raw material for geomorphic analysis.

Most scenery is the result of erosion, which is the removal of rock mass by a variety of agents such as flowing water, glaciers, wind, and waves (*see* EARTH'S GLACIERS AND FROZEN WATERS; GEOLOGIC WORK BY STREAMS; GEOLOGIC WORK BY WIND). One evaluation (Bloom, 1991) credited rivers with 85 to 90 percent of the total transport of eroded rock debris to the sea, glaciers with about 7 percent, subsurface groundwater solution and waves with 1 or 2 percent each, and wind and volcanoes with less than 1 percent each. Landscapes are constructed by a variety of tectonic processes, powered by internal forces of Earth that raise regions above sea level and deform the rocks within them. But except for a few rare and beautiful constructional landforms, such as volcanic peaks and coral islands, most scenery is erosional. A viewer should learn to see valleys as the dynamic, changing landscape components, with the intervening ridges, hills, and mountains as merely the residual landforms that are in the process of being consumed or destroyed.

The tectonic significance of erosional landscape evolution has only recently become appreciated. For example, if a region is uplifted to become a mountain range or plateau in an arid region, or becomes arid as the result of uplift, rivers are ineffective in eroding the mass. Therefore, it becomes self-limiting in its height and begins to spread laterally, as in the case of the Tibetan Plateau or the Altiplano of the South American Andes. On the other hand, a very rapidly rising mountain belt at the colliding margin of two tectonic plates in the eastern part of Taiwan cannot reach a height of more than 1,500 m because erosion is so rapid in

that humid tropical environment. The result is a "steady state" mountain range, rugged but neither growing nor wearing down. Geomorphologists have long been aware that mountain belts have a profound control on climates, but it is a relatively new idea that climate can control the shape and geologic evolution of a mountain range.

Erosion is strongly dependent on relief—the higher and steeper the topography, the more rapid the erosion. Over a century ago, JOHN WESLEY POWELL, one of the first American geomorphologists, asserted: "We may now conclude that the higher the mountain, the more rapid its degradation; that high mountains cannot live much longer than low mountains, and that mountains cannot remain long as mountains: they are ephemeral topographic forms. Geologically all existing mountains are recent; the ancient mountains are gone" (Powell, 1876, p. 193). Powell's "dictum" is still accepted although qualified by the climatic considerations noted in the preceding paragraph. But in terms of mountain building, what is "recent"? Almost the entire landscape of Japan is less than 2 million years (Ma) old, for example. Even excluding the constructional volcanic peaks such as Mount Fuji, many of which are less than 10,000 years old, the rugged and beautiful erosional landscape of the Japanese islands has been carved by erosion from rocks that rose from below sea level only about 2 Ma ago. One can infer that if the Japanese islands were not continuing to be uplifted, they could be largely worn back down to sea level in another few million years. While there are some very ancient landscapes that are said to have persisted for 10 Ma or even more than 100 Ma in places like the arid interior of Australia, many cited examples of very ancient landscapes were probably buried under sedimentary cover and were preserved, to be exhumed and rejuvenated by later tectonic uplift and erosional uncovering.

Traditional theories, now not widely supported, proposed that eroding landscapes evolve through a regular series of stages. Tectonic uplift was relegated to an initial stage when new land was raised above sea level. After the initial, presumably rapid, pulse of uplift, the landmass was passive and simply wore away by erosion until it was reduced to a low, monotonous surface near sea level, called a peneplain ("almost plain"). In this so-called cycle of erosion, the initial uplift produced some kind of broad dome on the scale of an entire mountain range. Rainfall and river runoff, perhaps aided by glaciers on the highest part of the landscape, eroded networks of valleys over the uplifted terrane. A youthful stage of vigorous canyon cutting was hypothesized, with the original uplifted surface becoming rapidly and deeply dissected by erosion. When the entire original topography was destroyed and replaced by a complete network of river valleys, the landscape was said to be mature. This stage could last for a relatively long time of slowly decreasing relief and altitude, until at some unspecified degree of degeneration the landscape entered old age, or the peneplain stage. The analogy with the life cycle of organisms is obvious. Just as certain events, such as the appearance of molars or wisdom teeth in human beings, or the achievement of reproductive maturity in all organisms, are important in the continuum of the life of an organism, certain events in the evolution of a landscape were used to mark its stage of erosional development. For example, the stage of landscape maturity was defined by the final removal of the initial uplifted surface, and by the beginning of floodplain deposition on the widening valley floors of the trunk rivers. As in living things, the boundary between maturity and old age was based on a vague idea of declining energy in the system and a slowing down of the processes of erosion.

Although this idea of a "cycle of erosion" is still a useful concept for introductory lectures in geomorphology, the interaction of tectonics and erosion is much more complex than it implies. It is probably true that tectonic uplift operates at rates at least ten times faster than most erosional lowering of landscapes, so as long as uplift continues, a landscape will become higher and more rugged. But when the locus of tectonic movement shifts elsewhere, perhaps due to the movement of the great lithospheric plates that form the surface of the earth, weathering and erosion will continue to operate and inevitably remove the mass of rock and lower its landscape surface toward sea level, where subaerial erosion stops.

What evidence can geomorphologists offer in support of the assertion that landscapes evolve? Most of the change is too slow for humans to observe, although the weathering of ancient stone monuments offers eloquent proof that rocks exposed to the atmosphere will decay and crumble. We can make small "sandbox" model landscapes and watch them wear away under a shower spray of water, but everyone recognizes that such experiments do not easily scale to the size of real landscapes. Similarly, the rapid erosion and gully formation within a century or a few decades on landscapes that have been deforested is persuasive, but obviously is due to an abnormal, disturbed, condition. Better evidence is available in the measurable annual sediment load of a river which, when divided by the area of the river's drainage basin, gives a number that represents the average rate of lowering of the entire landscape of the basin by the combined processes of weathering, soil formation, and erosion. Such measurements are in the range of 3–6 cm per thousand years for large, temperate-climate, agricultural, or forested drainage basins. Small drainage basins always show relatively greater soil loss than large ones; this implies that uplands erode faster than large valley floors, and that much of the sediment removed from the uplands as they are eroded is stored for at least thousands of years on the floodplains and deltas of the trunk streams. Rivers from semiarid mountains carry the heaviest sediment loads; presumably these landscapes are erosionally evolving at the most rapid rate. Glaciers in mountain valleys also erode at very rapid rates. However, the great ice sheets that have repeatedly overridden the broad lowlands of midwestern North America and central Europe caused relatively little erosion.

The last 2 Ma of Earth history have been unusual. Ice sheets have grown and shrunk again, perhaps twenty times, on the high-latitude continental areas of the Northern Hemisphere. During all that time, Antarctica has been ice covered. Ocean currents and atmospheric circulation have created a strongly zonal climate, with a steep latitudinal gradient from warm humid tropics to ice-covered polar regions. The waxing and waning of ice sheets has caused sea level to fall and rise through a range of about 120 m, causing rivers to first intrench and then subsequently fill their valleys. All major rivers today enter the sea either through drowned valleys called estuaries, or across vast delta plains. In either case, these modern depositional landscapes are less than 10,000 years old, the result of the end of the most recent ice age. Every landscape, whether erosional or depositional, shows traces of evolution under climatic conditions no longer present in the region. For example, northern Europe and North America south to the Missouri and Ohio rivers display many landforms of glacial origin, now relict in a mid-latitude forest environment. In lower latitudes, fossil sand dunes can be found in the rainforests, and even the harshest deserts of today display patterns of river channels that speak of former flowing rivers. Much of the complexity and beauty of the present landscape is the result of climate changes that are geologically "recent."

An important component of the evolving subaerial landscape is not erosional, but depositional. Most of the human population lives on coastal plains, deltas, and river floodplains, all built by the sedimentary debris eroded from adjacent highlands (*see* RIVERS, GEOMORPHOLOGY OF). The mud, sand, and gravel of these depositional landforms can be traced to the rock types in the eroding source areas. Depositional landforms evolve, too. On the ancient Nile delta, the annual river flood deposited a layer of sediment that maintained the soil fertility. The reservoir upstream from the Aswan dam now traps this sediment, and most of the electrical energy of the dam is used to manufacture synthetic fertilizers to compensate for the sediment loss. Meanwhile, the coast of the delta is eroding at a more rapid rate because the wave energy of the Mediterranean Sea is no longer absorbed by the deltaic sediment entering the sea. The Mississippi delta into the Gulf of Mexico fills a buried valley that is nearly 240 m deep at the Louisiana shoreline. Although sea level during the last ice age was only about 120 m lower than present, as the Mississippi deposited its sediment to build the delta, the ocean floor beneath the delta subsided under the great weight, so that the floor of the glacial-age valley has now been warped downward to twice that depth. In spite of the great volume of the drowned valley that has been filled by the Mississippi River, the river has also managed to build its delta forward across the submerged continental shelf so that it is now depositing mud directly onto the continental slope at the outer limits of the shelf. In itself, this huge volume of deposited sediment offers firm proof of the amount of erosion of the landscape within the Mississippi River basin.

Measured erosion rates clearly demonstrate that if a region is not uplifted by tectonic forces, it is doomed to be eroded to a low plain near sea level within a time interval on the order of 10 Ma. The most impressive fact about erosional evolution of landscapes is not the effectiveness of the processes, but that there is any land remaining above sea level. Obviously, our planet is sufficiently active internally so that mountain-building and continent-rejuvenating processes can maintain a diversified

subaerial landscape in the face of formidable erosional processes (*see* COASTAL PROCESSES). The landscape we see today is evolving as the result of variable and complex interactions between constructional and destructional processes. No one can say for sure which set of processes is winning in any particular region.

Bibliography

BLOOM, A. L. *Geomorphology: A Systematic Analysis of Late Cenozoic Landforms*, 2nd ed. Englewood Cliffs, NJ, 1991.

VON ENGELN, O. D. *Geomorphology*. New York, 1942.

LaFARGUE, M. *The Tao of the Tao Te Ching: A Translation and Commentary*. Albany, NY, 1992.

POWELL, J. W. *Report on the Geology of the Eastern Portion of the Uinta Mountains*. Washington, DC, 1876.

THORNBURY, W. D. *Regional Geomorphology of the United States*. New York, 1965.

ARTHUR L. BLOOM

LANDSLIDES AND ROCKFALLS

Landslides and rockfalls together make up mass wastage, the gravity-driven downslope movement of all types of earth materials. Many rates and processes account for mass wastage (Table 1). Major subdivisions of mass wastage are present on Earth's surface as (1) landslides and (2) rockfalls. Both are caused by routine erosion, mainly by rivers, in their removal of the toes of hill masses, which cause oversteepening and eventual failure. The same downslope movement occurs in (3) soil and (4) rock that results from human activity, especially in building transportation routes in cuts across the toes or faces of hillsides.

Landslides occur in masses of alluvium and other transported soil (as used in the engineering sense; particulate matter not equivalent to the induration of rock), in residual soil, weathered rock, weak rock, and in highly fractured rock masses. Landslides typically have areal outlines similar to that of a spoon; with depth, they have a curved bottom (shear) surface along which they slide against stronger underlying earth material (Figure 1). Landslide volumes comprise whatever mass and

Table 1. An Abbreviation of the Classic Worldwide Classification Scheme*

TYPE OF MOVEMENT			TYPE OF MATERIAL		
			BEDROCK	ENGINEERING SOILS	
				Predominantly coarse	Predominantly fine
FALLS			Rock fall	Debris fall	Earth fall
TOPPLES			Rock topple	Debris topple	Earth topple
SLIDES	ROTATIONAL	FEW UNITS	Rock slump	Debris slump	Earth slump
	TRANSLATIONAL	MANY UNITS	Rock block side Rock slide	Debris block slide Debris slide	Earth block slide Earth slide
LATERAL SPREADS			Rock spread	Debris spread	Earth spread
FLOWS			Rock flow (deep creep)	Debris flow (soil creep)	Earth flow
COMPLEX			Combination of two or more principal types of movement		

* Developed by D. J. Varnes (1958 and 1976).

lateral shape are accordant with available gravitational forces, those activated by the weight of the soil mass plus any internal pore water. Such unstable ground is a dynamic mass, though detectable movement may take many years. Movement, caused by internal shearing at the base and sides of the landslide, comes about from the developing weight of the failure mass, often due to infiltrating precipitation and the removal of restraining support by erosion or human activities at the toe of the hill mass. Mathematically speaking, the failure occurs at whatever time the factor of safety of the hillmass degrades to a ratio slightly less than unity (1.0): shear strength of the hillside particles bound together at the shear surface divided by gravitationally induced weight of hillmass, plus pore water, as downslope shear.

The additional destabilizing effect of water at the shearing boundary of the failure mass stems from natural entrapment, or unplanned collection from human activities, of infiltrated precipitation, which includes also excess landscaping waters and leaking sewers and swimming pools. Such fluid, either as perched water or as ground water, is highly destructive to hillslope stability. In masses of soil, the water is termed "pore water"; in rock slopes it is termed "cleft water" because it acts to open up or spread apart the opposing surfaces of the various rock masses. Gravity-driven shear forces will maximize along fractures, which include joints, bedding planes, foliation, shear planes, and faults (Figure 2).

Slope failures are largely predictable on the basis of regional geologic conditions such as hillside angles of repose greater than the angle of internal friction between soil particles or rock blocks, types of soil such as unstable marine clays, the tropics in general, and in regions of active tectonism where slopes are exacerbated by earthquake-induced ground motion. Evidence of mass wastage has been preserved in sedimentary rock of all ages. The mammoth Blackhawk Canyon slide of southern California (Shreve, 1968) is a well-preserved Pleistocene-aged feature.

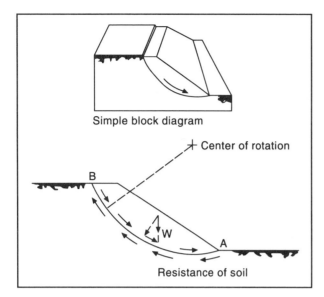

Figure 1. Most typical slope failure in a soil mass. Four forces are shown; (W) representing the gravity-driving weight of the mass (including soil and pore water), a vector promoting stability (acting perpendicular to the failure surface), a shear-force vector acting downslope along the failure plane, and the natural shear resistance on the unfailed side of the failure plane force (Legget & Hatheway, 1988, p. 510).

Instability of rock masses is worst when major rock discontinuities such as bedding or foliation planes are parallel to the bearing of the face of the hillside or the engineered cut, and are inclined into the cut. The opposite condition produces favorable dip, as the ready-made failure surfaces will not be present. When dip into the face is greater than the angle of the natural hillslope, then failure is imminent because there is no rock mass to restrain the rock planes that are totally exposed along their hillside edges.

As there is a full range of particle types involved in typical mass wastage, so is there a range of rates of movement. Relatively dryer and particle-oriented failure masses tend to be slower, while steeper slopes in rock experience failures of full gravity fall. Perhaps at the outer limit of rapidity was the giant debris avalanche of 31 May 1970 at Yungay, Peru, triggered by the great offshore earthquake of magnitude 7.0. The face of the 6,663-m mountain was dislodged and is believed to have arrived at the village of Rhanarica at speeds of up to 320 km/h, destroying the town and killing most of its inhabitants. On the other hand, visible evidence of slow movement rates can be seen in tilted walls, tombstones, and trees that have curved trunks.

Hillsides in tectonically active terrane are especially prone to failure during major earthquakes. The U.S. Geological Survey (USGS) completed a major program of identification and land-use planning for susceptible ground in the San Francisco Bay area in the 1970s (Brabb et al., 1989). Much of the Pacific Rim is plagued with this problem and much of the unsettled debris is volcanic ash, which does not have appreciable shear strength, making the danger all the greater.

Detection and mitigation of slope failures require engineering geologic observation and mapping to determine the obvious or most probable bounds of instability, to deflect incoming water or to remove hillmass water. Crown trenches divert incoming water and sub-horizontal drains are drilled into the hillside to drain water from the hill mass.

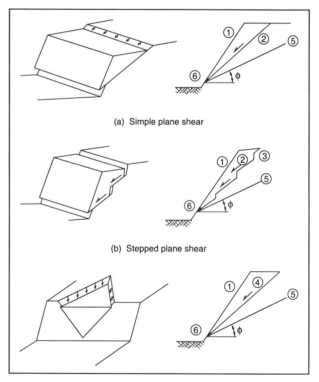

Figure 2. Simplified typical rockslide types; all show an unfavorable geologic structure, dipping into the cut at an angle greater than that of the internal friction ∅. (From *Pit Slope Manual*, Chap. 2, CANMET, 1977, p. 54.)

Simple mathematical representations of the factor of safety are computed in detail along frequent vertical slices of soil masses. The distribution of available shear strength and opposing shear forces are likewise computed along the geologically most-probable failure surfaces in rock. Rock masses also are treatable by the addition of rock bolts or cable tendons drilled into the hillside and resin-cemented to the host rock, deeper than the suspected zone or surface of shear failure. Once in place, the bolts or tendons are stretched into tension by industrial hydraulic jacks and the hillside is tightly compressed so that the otherwise failure-prone rock discontinuities cannot slip into failure.

The geological technology of mass wastage is a rich and well-documented literature. Collin published the first observations of landslides in 1846. The Panama Canal was severely hindered by weak-rock slides (McDonald, 1915; U.S. National Academy of Sciences, 1924), and the USGS and state surveys made major early studies. Frank, Alberta, a town existing solely for the purpose of extracting its nearby Turtle Mountain coals, was destroyed by a 1903 landslide (Daly, 1912) caused by inappropriate mining practices. Heim (1932) sounded the European alarm for the need of comprehensive engineering-oriented geological studies and Hennes (1936) produced the first handbook for that purpose. Sharpe's 1938 book contains the first classification of landslides and Eckel's edited masterpiece handbook of 1958 contains Varnes's expanded and now classic classification.

Beginning in 1950, the Engineering Geology Division of the Geological Society of America began the editing of three outstanding technical treatments of landslides. The 1950 volume contains Terzaghi's compendium of the first mathematical explanations for landslides and the series continues to the present time.

The disastrous spring rains of 1952 in southern California launched the most definitive involvement of geologists in urban hillside control. The California Division of Mines and Geology (1965), as well as many other state geologic surveys, have been providing extensive literature from about that time. Barry Voight (1978) produced a two-volume tome of worldwide coverage of natural and human-induced rockslides and associated avalanches. Brabb and Harrod's (1989) published proceedings of their symposium at the 28th International Geological Congress sets the current worldwide stage for the importance of work to be done to mitigate and control landslides and rockfalls.

By the 1970s, numerous rock engineering works were available and the *Geotechnical Manual for Slopes* (1984), by the Hong Kong government, is highly recommended. The 1976–1981 series of rock slope engineering handbooks by D. R. Piteau & Associates has the greatest integration of geology of all of the engineering manuals. Piteau, a gifted young Canadian geologist, died before his works could be published more widely. Hudson, et al. (1993) have produced a mammoth, five-volume work, including extensive coverage of rockfalls.

Bibliography

BRABB, E. E., and HARROD, B. L., eds. "Landslides; Extent and Economic Significance." *Proceedings, Symposium at the 28th International Geological Congress.* Washington, DC, 1989. Rotterdam, 1989.

CALIFORNIA DIVISION OF MINES AND GEOLOGY. "Landslides and Subsidence; Geologic Hazards Conference." *Proceedings, 2nd Conference on Geologic Hazards.* Los Angeles, 26–27 May, 1965.

CLEAVES, A. B. *Landslide Investigations; A Field Handbook for Use in Highway Location and Design.* Washington, DC, 1961.

COLLIN, A. *Landslides in Clays,* translated by W. R. Schriever. Toronto, 1956. Originally published in 1846.

DALY, R. A. "Report of the Commission Appointed to Investigate Turtle Mountain, Frank, Alberta." Geological Survey of Canada Memoir 27 (1912).

ECKEL, E. B., ed. *Landslides and Engineering Practice.* Transportation Research Board Special Report no. 29, National Research Council. Washington, DC, 1958.

HEIM, A. *Landslides and Human Lives.* Translated from German by N. Skermer. Richmond, British Columbia, 1989.

HENNES, R. G. *Analysis and Control of Landslides.* Engineering Experiment Station Bulletin 91. St. Louis, MO, 1936.

HOEK, E., and J. BRAY. *Rock Slope Engineering,* 3rd ed. London, 1981.

HONG KONG GOVERNMENT. *Geotechnical Manual for Slopes,* 2nd ed. Hong Kong, 1984.

HUDSON, J. A.; E. T. BROWN; C. FAIRHURST; and E. HOEK, eds. *Comprehensive Rock Engineering; Principles, Practice and Projects.* Oxford, Eng., 1993.

LEGGET, R. F., and A. W. HATHEWAY. *Geology and Engineering,* 3rd ed. New York, 1988.

MACDONALD, D. F. *Some Engineering Problems of the Panama Canal in Their Relation to Geology and Topography.* U.S. Bureau of Mines, Bulletin 86, 1915.

PAIGE, S., ed. *Application of Geology to Engineering Practice.* Geological Society of America, Engineering Geology Division. New York, 1950.

PITEAU, D. R., & ASSOCIATES. *Rock Slope Engineering; Planning, Design, Construction and Maintenance of Rock Slopes for Highways and Railways.* Parts A–E. Federal Highway Administration. Washington, DC, 1976–1981.

SHARPE, C. F. S. *Landslides and Related Phenomena: A Study of Mass-Movements of Soil and Rock.* Patterson, NJ, 1962. Reprint of original 1938 edition.

SMITH, D. D. *The Effectiveness of Horizontal Drains.* California Department of Transportation. Sacramento, CA, 1980.

TERZAGHI, K. "Mechanism of Landslides." In *Engineering Geology,* ed. S. Paige. New York, 1950, pp. 83–123.

U.S. NATIONAL ACADEMY OF SCIENCES. *Report of the Committee of The National Academy of Sciences on Panama Canal Slides.* Washington, DC, 1924.

VOIGHT, B., ed. *Rockslides and Avalanches.* Vol. 1, *Natural Phenomena*; Vol. 2, *Engineering Sites.* Amsterdam, 1978.

ALLEN W. HATHEWAY

LAYERED INTRUSIONS

See Mineral Deposits, Igneous

LEHMANN, INGE

Inge Lehmann was born on 13 May 1888 in Copenhagen, Denmark, and died there on 21 February 1993. She is famous for her discovery of Earth's solid inner core inside its fluid core. Her father, Alfred Lehmann, was a scientific pioneer who introduced experimental psychology at the University of Copenhagen. They both loved outdoor activities such as hiking in the mountains.

Inge Lehmann graduated from Hanna Adlers school in 1906. At this school, boys and girls were treated equally, which was unusual at the time. Later in life she found that such equal treatment was not the norm. She was not active in women's liberation, but, busy with her own work and with a fighting spirit, she achieved her own liberation.

In 1920 Lehmann received a Master of Science (M.S.) degree in mathematics from the University of Copenhagen. In 1925 she became attached to the geodetic institution that was in charge of measuring the meridian arc in Denmark. Her first major task was the installation of seismographs for the new seismic station in Copenhagen. She went to Germany in 1927 and was introduced to seismology by Beno Gutenberg during a month stay in Darmstadt.

In 1928 Lehmann received an M.S. degree in geodesy and was appointed chief of the department of seismology at the newly established Geodetic Institute in Copenhagen. The duties in that position were to run seismograph stations in Denmark and Greenland and to report the arrival times of these waves in bulletins. Such bulletins are used to determine earthquake epicenters. She installed seismograph stations at Ivigtut and Scoresbysund in Greenland. Scoresbysund, which started recording in 1928, is situated in such a remote area that ships can reach it only in late summer.

Lehmann became strongly interested in the fact that knowledge of Earth's interior can be obtained from the observations of the earthquake recordings from seismographs. One of the problems in doing that was that the estimates for epicenters were inaccurate because the travel time curves used to calculate them were inaccurate. A travel time curve is a curve showing the travel time of seismic waves from earthquakes as a function of the distance along the surface of the earth. Lehmann found a way to improve the knowledge of travel times because she could determine the slope of the travel time curve by using the European network of stations as an array. For this work she looked at copies of seismograms from these stations and ascertained that she was reading the same wave propagating through Earth to the different stations. These slopes were used by Sir Harold Jeffreys, with whom she corresponded extensively, while he calculated seismic travel time curves in cooperation with Keith Bullen. They used statistical methods to determine travel time curves from many observations of arrival time.

Lehmann's pioneer contribution was the discovery of the inner (solid) core in 1936. She suggested the possibility of an inner core with a wave velocity higher than that of the rest of the core, and she showed that this assumption can explain some seismic phases observed in the shadow zone. This zone is the region from 102 to 142 degrees, where the direct wave from the earthquake is much reduced because Earth's core refracts seismic waves away from it. Only waves that are bent around the core will reach the zone. Such waves are called diffracted waves. They were expected to be weak, whereas she had observed strong phases. The assumption of an inner core was a bold assumption because at the time, it was thought that good evidence showed that Earth's core was a homogenous fluid. The suggestion was immediately accepted by Gutenberg, and a few years later also by Jeffreys. This discovery is one of the most important advances in our knowledge of Earth's interior during the past sixty years.

Lehmann's retirement in 1953 did not mean she stopped work, but she did get relief from her day-to-day duties of station operation. She worked on the structure of the upper mantle, that is, the upper 600–700 km of the earth. She found recordings of nuclear explosions to be very valuable for this work, but earthquake recordings were also used. She worked in her home in Copenhagen as well as at Lamont Geological Observatory of Columbia University (now Lamont Doherty Earth Observatory), at the California Institute of Technology at Pasadena, at the Seismographic Station of the University of California at Berkeley, and at the Dominion Observatory (now Geological Survey of Canada) in Ottawa. She observed that clear differences in upper mantle structure existed between eastern and western North America, and that the upper mantle below Europe was different as well. She was respected for years as a foremost authority on seismic evidence on upper mantle structure. Certain models of Earth used for free oscillation calculations adopted the Lehmann upper mantle structure. (A discontinuity she introduced at 220 km depth is included in the current PREM Earth Model.)

Lehmann's last publication (1987) summarizes in an interesting way works by the geophysicists Gutenberg, Jeffreys, and herself during the period 1925–1936. This paper was first written in Danish for the Danish Geophysical Society, but Bruce Bolt asked for a translation as an introduction to his paper (Bolt, 1987). Lehmann was one of the founders of the Danish Geophysical Society, and she chaired that organization from 1941 to 1944.

In the summer she lived in a small cottage overlooking a quiet lake, where many scientists visited. One of them has described her pleasant memories of visits there (Simon, 1993). Inge Lehmann's last

years were hard. Though blind and weak, her fighting spirit never changed. As late as the summer of 1992 she was able to enjoy a visit to her beloved cottage, located 20 km from her apartment.

Inge Lehmann received numerous awards: the German Geophysical Society's Emil-Wiechert Medal (1964); Dr. of Science, honoris causa, Columbia University of New York (1964); the Royal Danish Academy of Science's Gold Medal (1965); Dr. of Philosophy, honoris causa, University of Copenhagen (1968); and the American Geophysical Union's William Bowie Medal, given for outstanding contributions to fundamental geophysics and unselfish cooperation in research (1971).

A long time ago Lehmann created a fund administered by the Academy of Science in Denmark, and now this fund has been given her entire fortune. It offers a travel award each year, which is given alternatively to psychologists and to geophysicists.

Bibliography

BOLT, B. A. "50 Years of Studies on the Inner Core." *Eos* 68 (1987):73, 80–81.

BOLT, B. A. and E. HJORTENBERG. "Memorial Essay Inge Lehmann (1888–1993)." *Bulletin of the Seismological Society of America* 84 (1994):229–233.

LEHMANN, I. "P." *Publications du Bureau Central Seismologique International: Travaux Scientifiques* Serie A 14 (1936):87–115.

———. "Seismology in the Days of Old." *Eos* 68 (1987): 33–35.

SIMON, R. "A Personal Memoir of Inge Lehmann (1888–1993)." *Eos* 74 (1993):511–512.

ERIK HJORTENBERG

LIBBY, WILLARD

Willard F. Libby was a nuclear chemist, best remembered as the originator and developer of the C^{14} (radiocarbon) dating method, for which he received the Nobel Prize in Chemistry in 1960.

Libby was born on 17 December 1908 in Grand Valley, Colorado, and grew up in a farming community in northern California. He entered the University of California, Berkeley, in 1927 intending, as he often said, to learn enough science to be a modern farmer. He was diverted from this goal by the chemistry faculty, however. As a result he went straight from undergraduate studies to graduate work, receiving his Ph.D. degree in chemistry in 1933, during the Great Depression. He gladly accepted an instructorship in the department. In the following years he did research in several areas of physical and nuclear chemistry, gaining a reputation for originality and daring. In 1941, still an assistant professor, he received a Guggenheim fellowship at Princeton University. While he was there the United States entered World War II. Libby was one of the first two people invited by Professor Harold Urey to join his project at Columbia University to separate the light fissionable isotope U^{235} by a diffusion method, as part of what became the Manhattan Project, the U.S. program to build an atomic bomb (see UREY, HAROLD). Libby accepted and became the head of the chemistry division of that very large enterprise.

The separation of an uncommon isotope of a heavy element in the kilogram quantities needed for nuclear weapons was an unprecedented undertaking. A number of methods were developed at least to a pilot plant stage. It was clear that in principle diffusion of a uranium-containing gas was the most straightforward approach to large-scale production. For several reasons the fluoride UF_6 was the gas to use. There were two major obstacles: (1) the problem of large-scale production of the fine-grained, thin membranes or diffusion barriers and their assembly to perform the multi-stage separation, and (2) the extreme corrosiveness of the gas. The key invention was made by Libby and Anthony Turkevich, who were issued a secret patent for the design and manufacturing method. Libby led the research needed to enable production of the barriers, which was completed while the huge separation plant was under construction at Oak Ridge, Tennessee. The barriers were delivered in time to permit the plant to begin functioning on schedule. The first bomb was exploded over Japan in August 1945.

After the war Libby, along with Urey, Enrico Fermi, Edward Teller, Leo Szilard, and other famous physicists and chemists, accepted professorial positions in newly established science institutes at the University of Chicago, where Fermi and others had earlier achieved the first self-sustained release of nuclear energy in 1942. The Institute

for Nuclear Studies there, now the Fermi Institute, housed an extraordinary concentration of scientific talent. Confidence was high. Discipline boundaries were to a great extent ignored, providing fertile soil for original projects. It was in this environment that Libby's C^{14} dating project began to take shape.

The method was based on a few critical assumptions, which had to be verified at the start. The first was that the radioactive isotope of carbon, C^{14}, is made naturally in the earth's atmosphere by cosmic rays (already suggested by Serge Korff a few years earlier), and that it quickly oxidizes to carbon dioxide, CO_2. Next, that this material, combined with the much more abundant nonradioactive CO_2, quickly mixes uniformly throughout the earth's atmosphere and through carbonate dissolved in the ocean as well. Next, that it quickly enters the "carbon cycle" in which living plants incorporate it, and animals take it in from their food, and that the death and decay of organisms liberates it for the next cycle. Finally, it must be assumed that materials formerly living and preserved, like wood or textiles, no longer are exchanging material with the atmosphere or biosphere, and that, when life stops, C^{14} begins to decay (to nonradioactive N^{14}) at a rate set by its half-life (see GEOLOGIC TIME, MEASUREMENT OF). With these assumptions and appropriate measurements an age (or time of death) can be calculated.

When Libby began his studies in 1946, none of these assumptions had been tested. Even the half-life of C^{14} was extremely uncertain. Moreover, at that time there was no way with existing equipment to measure the C^{14} content or decay rate at the levels that occur in nature without expensive, difficult isotopic enrichment. In the next few years he and his junior collaborators provided proof, at the working level of precision, of all the assumptions listed, measured the half-life still in use in the field today, and built counting equipment capable of measuring the effect. This research was one of the last examples of a major scientific development designed and led to a successful outcome by a single individual.

The first direct comparison of ages calculated from C^{14} with known (historical) ages appeared in 1949. Since the measured half-life is 5,568 years, the method is especially suited to studies of human history and prehistory. When the first "date list" of unknowns was published in 1951, the statistical error of the measurements (one standard deviation) was about 200 years for samples younger than a few thousand years. The range of applicability was about 20,000 years. Since then both measuring methods and fundamental understanding have improved, accompanied by a great increase in the number of laboratories set up for the purpose. There is a scientific journal, *Radiocarbon*, devoted entirely to this subject. It would be difficult to overstate the achievements made by C^{14} analyses in archeology, earth sciences, and many other disciplines, such as the study of solar changes.

In the next few years Libby and his students discovered and studied the natural distribution of the radioactive isotope of hydrogen, H^3 or tritium, which has a half-life of 12.3 years and is also produced by cosmic rays in the atmosphere. It was fortunate that this work preceded by a few years the large-scale testing of nuclear weapons in the atmosphere. The tritium content of the megaton "hydrogen bombs" tested, distributed first through the northern hemisphere and then around the whole earth, swamped the natural tritium level, which could not be freely studied again for decades afterward. Even the C^{14} content of the atmosphere doubled in the interval before the United States and the Soviet Union stopped major atmospheric testing.

In this same period Libby began a small research program to observe the atmospheric and surface distribution of nuclear fission products, especially long-lived Sr^{90} and Cs^{137}, resulting from atmospheric tests and other releases. When fallout became a major issue after the U.S. H-bomb tests in the Marshall Islands in the Pacific in 1954, this project provided much of the limited scientific understanding then available of the processes involved, and of the potential hazards.

In that year Libby accepted an appointment as one of the five U.S. Atomic Energy Commissioners, and moved to Washington, DC. An early achievement was the release of information possessed by the Atomic Energy Commission (AEC) on the fallout resulting from H-bomb tests. Up to his arrival this information had been classified secret.

In his role as commissioner, and afterward, Libby was a tireless advocate of the peaceful uses of nuclear science and technology. These included the relatively noncontroversial use of radioactive isotopes in medical and industrial research as well as the worldwide generation of nuclear power as a substitute for the use of fossil fuels, and projects for large-scale public works, such as the digging of canals and artificial harbors using nuclear explosives. This point of view involved him in many debates, especially in later years.

After the expiration of his term as commissioner, Libby returned to academic life as professor of chemistry at U.C.L.A. There he took an active role in developing new interdisciplinary programs, first as director of the University of California's multicampus Institute of Geophysics and Planetary Physics, and later, among other things, in the creation of U.C.L.A.'s unique Environmental Doctor degree program. He continued to serve on numerous boards and committees concerned with science policy, nuclear matters, and the environment.

Throughout his academic career Libby had a strong interest in teaching, and particularly in the training of research students. He was not an especially gifted classroom lecturer. At the same time his skill in selecting students capable of first-class research achievement, and in developing their motivation and talents, was remarkable.

As a scientist and as a policymaker, Willard Libby was an important figure. Especially in later years, his originality and fearlessness in crossing disciplinary boundaries sometimes led him into error. Still, his contributions to science were most impressive, both through his own work and through the students he trained. Some of his policy views are now widely seen as incorrect. Others, especially his commitment to openness in government, and his enthusiasm for science as an adventure perennially appealing to the young in spirit, are fortunately still with us. Libby died on 8 September 1980 in Los Angeles, California.

Bibliography

ARNOLD, J. R., and W. F. LIBBY. "Age Determinations by Radiocarbon Content: Checks with Samples of Known Age." *Science* 110 (1994):678–680.

———. "Radiocarbon Dates." *Science* 113 (1951):111–120.

GROSS, A. V.; W. H. JOHNSTON; R. L. WOLFGANG; and W. F. LIBBY. "Tritium in Nature." *Science* 113 (1951):1–2.

LIBBY, W. F. *Radiocarbon Dating*. Chicago, 1952.

JAMES ARNOLD

LIFE, EVOLUTION OF

Biological evolution is the modification of characteristics of organisms over time. These changes are heritable and passed from generation to generation. Evolution literally means "unfolding," and the fossil record is a diary of the unfolding of Earth's life over billions of years. This record clearly demonstrates many changes in organisms and their communities over time. Thus, paleontology, the study of fossils, gives us striking proof of the fact of evolution, and a fascinating, but incomplete, chronicle of evolutionary events that shaped life on Earth. To fully understand evolutionary processes, paleontology is combined with information from the anatomy, genetics, biochemistry, behavior, and other properties of related or analogous living organisms (see AMPHIBIANS AND REPTILES; BIRDS; FOSSILIZATION AND THE FOSSIL RECORD; FOSSIL RECORD OF HUMAN EVOLUTION; INVERTEBRATES; MAMMALS; PALEOBOTANY; PALEOECOLOGY; PALEONTOLOGY; TAPHONOMY; TRACE FOSSILS).

Evolution is the central, unifying principle of modern biology. Theodosius Dobzhansky, a geneticist who contributed greatly to today's prevailing views of evolution, referred to as the "modern synthesis," noted that "Nothing in biology makes sense except in the light of evolution." This entry outlines current knowledge about evolution and major theories pertaining to the evolutionary history of life as reflected in the geological record. We will not deal with the origin of life (see LIFE, ORIGIN OF), but rather with theories that attempt to explain the pattern of change and diversification of life following its origin.

The language of systematics, which is the systematic study of relationships among organisms, is the natural language of evolution. Evolutionary "family trees" portraying these relationships are hierarchical classifications of fossil and living organisms. The hierarchy represents the order of successive branching events of lineages (or lines of descent) as they evolved over time. The theory and practice of classifying organisms is taxonomy. Ascending the hierarchy, organisms are grouped successively into the following taxonomic categories: species, genus, family, order, class, phylum, and kingdom.

Although the species is considered the fundamental evolutionary unit and is generally recognizable in nature, its definition varies. The most commonly used concept for sexually reproducing individuals is the biological species concept, in which a species is defined as a group of actually or potentially interbreeding organisms that is reproductively isolated from other such groups. Species (e.g., humans) are composed of populations of organisms participating in a common reproductive community isolated from other such communities (e.g., chimpanzees). Evolutionary changes occur at the population or species level when genetic novelties (mutations) arise in individual organisms and spread through populations by sexual reproduction over generations.

The complexity of the modern synthesis is difficult to capture in brief, but can be represented in the following outline:

1. Mutations (genetic novelties) arise within organisms by chance, and result in genetic variation among individuals within populations.
2. Mutations accumulate in populations over time if they are not harmful, or if they help the organisms to survive or prosper in their surroundings by modifying their characteristics in useful ways.
3. Disadvantageous mutations are usually quickly eliminated, because individuals possessing them will not survive to reproduce, or will produce fewer offspring than those with beneficial mutations.
4. Over generations, sexual reproduction may spread various combinations of the mutations throughout a population. This opportunity to "try out" different assortments of genetic novelties results in individuals with those sets of characteristics most in tune with prevailing conditions.
5. Over many generations, accumulated genetic changes result in divergence from the original species, giving rise to new species and, ultimately, to higher taxonomic groups.

Species can arise in two fundamental ways, anagenesis and cladogenesis. There is evidence for both in the fossil record. Evolutionary change can transform a single lineage over time, a process called anagenesis. Alternatively, isolated populations of a single species can follow different evolutionary pathways by accumulating different characteristics, and split into two distinct species, a process called cladogenesis. The diversity (variety and number) of living forms can increase only by cladogenesis, and the ten million or so species of organisms alive on Earth attest to its importance.

Modes of Evolution

In all scientific fields, the existence of a phenomenon is commonly recognized before its underlying mechanisms are understood. Although paleontology, systematics, and genetics establish the fact of evolution, the relative importance of factors influencing it are hotly debated. Two major theories about evolutionary mode are microevolution (literally, "small evolution") and macroevolution (literally, "large evolution"), which are described below.

One of CHARLES DARWIN's principal theses was that evolution is driven by natural selection of the fittest hereditary variants of a species in each generation. Through differential reproductive success, natural selection preserves traits maximizing the survival and reproduction of individuals in their existing environments. Microevolutionary theory states that natural selection produces new species better adapted to their surrounding environments by incremental adjustments to their basic characteristics. For example, paleontologists think the evolutionary history of North American horses in the last 25 million years (Ma) tracked the expansion of savanna-like grasslands, and increasing seasonality and aridity. New species of horses progressively adapted to the changing environment by evolving increasingly higher crowned teeth, better for chewing the grittier grass and plants that replaced softer woodland plants as a more arid climate developed. During this time, the legs of successive species of horses became progressively longer, which increased efficiency for running in open terrain and migrating longer distances to find more seasonally available food. The theory of microevolution, involving small, incremental stages, can account for these evolutionary changes.

Macroevolution is a theory of evolutionary mode that addresses the great evolutionary differences seen across all organisms, which some scientists believe cannot be accounted for by microevolutionary processes. All living organisms are classified into approximately one hundred phyla, some bearing little resemblance to the others. Although there are fossils intermediate in appearance between some phyla, large gaps between most phyla have led some scientists to argue that micro-

evolutionary changes are inadequate to explain the origins of major groups. Instead, they support macroevolutionary theory, in which evolutionary changes large enough to distinguish taxa above the species level are produced by processes different from microevolution. Macroevolutionists suggest that large, rapid reorganizations of species' characteristics are required to span the gaps between major types of organisms. However, even a huge reorganization requires an organism's complex traits to evolve together in concert to preserve the functionality of its parts. A possible solution to preserving functionality during major reorganizations is the occurrence of mutations that affect control genes, or "master switches" that influence innumerable characteristics of organisms. This is especially true for genes controlling traits during early growth and development, when changes have a cascading influence throughout an organism's life. Opposing macroevolutionary theory, supporters of the modern synthesis argue that microevolutionary mechanisms occurring over long periods of time (e.g., hundreds of thousands of years) can explain larger, macroevolutionary changes; thus, the general scientific preference for simplest explanations requires no invocation of processes different from microevolution.

Some paleontologists extend the meaning of macroevolution to include large-scale evolutionary trends within higher taxonomic groups, such as rates of species origination and extinction, and trends in body design. Just as the natural selection of individuals within species is the mechanism of microevolution, the analogous mechanism in this view of macroevolution is the selection of species that have some unique feature. This "species selection" is thought to shift the general characteristics of a major group in a particular direction. For example, single lineages of horses did not increase in size through time, but horses generally increased in size due to the differential survival of larger lineages.

The largest recent debate in evolutionary paleobiology is whether evolution proceeds gradually (gradualist model) or sporadically (punctuated equilibrium model). In the gradualist model, a species continually adjusts to fit its environment by constant, usually incremental, change. Alternatively, paleontologists proposed the punctuated equilibrium model, in which relatively long periods of unchanging appearance in a species, termed "stasis," are punctuated by rapid shifts that produce new species. This model was developed from the observations of field biologists and the models of geneticists, which suggested that evolutionary novelties occur more commonly and can spread most quickly in small, localized, peripheral populations. These conditions promote reproductive isolation, which leads to cladogenesis. The new species is thought to expand into the old species' geographic range, replacing it. Evidence in support of the punctuated equilibrium model comes from the observation that many fossil species looked essentially the same over long periods of time (e.g., millions of years), and were then (apparently) rapidly replaced by related taxa. This model is also supported by studies in developmental biology that show that developmental systems of organisms may be buffered to resist change for a period of time (stasis) until they rapidly change to a new stable point. Alternatively, defenders of the gradualist model of evolution point out that, although periods of change seen in the fossil record may be brief relative to periods of stasis within a species, the time intervals for change are often quite sufficient to account for changes between species, using evolutionary rates observed in laboratories and field experiments. In reality, the occurrence of both evolutionary modes is likely, and current studies focus on the relative significance of each type within and between major groups.

Evolutionary Diversification

Regardless of evolutionary mode, the birth of new species has produced considerable biologic diversity. Based on tabulations of the number of fossil taxa in successive geologic time intervals, the diversity of life has increased through time. However, diversification has not been a steady climb from the low levels billions of years ago to today's high levels, but rather, a series of increases punctuated by intervals of extinction and replacement (turnover) of species (see EXTINCTIONS). Extinction is part of the evolution of life; in fact, many studies have found a good correlation between origination and extinction rates within taxa, which has kept diversity approximately constant over shorter intervals of time. For example, rates of extinction and origination for the major adaptive groups of North American land mammals (carnivores, omnivores, and herbivores) were in balance throughout the last 10 Ma. Correlations of extinction and origina-

tion have several possible interpretations: (1) extinction and origination might maintain a diversity level controlled by the resources of the environment; (2) they could both be stimulated by environmental change; or (3) species might have intrinsic traits that affect the possibility of the occurrence of extinction and origination.

The largest turnovers in species occurred at the ends of the Permian and Cretaceous geologic time periods (245 and 66 Ma ago, respectively). These dramatic extinctions, which may have been the result of environmentally catastrophic events such as intense volcanism or extraterrestrial impacts, wiped out vast numbers of species (see IMPACT CRATERING AND THE BOMBARDMENT RECORD). Events such as these are rapid and extreme compared to anything normally experienced in the lifetime of an organism. These unpredictable, catastrophic events contrast with the more regular or uniform changes to which organisms can adapt by natural selection.

The diversifications following the Permian and Cretaceous extinctions produced such distinctly different biotas that geologists use them to divide the last 540 (Ma) into three eras: the Paleozoic ("ancient life") era lasted through the Permian, the Mesozoic ("middle life") through the Cretaceous, and the Cenozoic ("new life") through the present. The earliest occurrences of almost all animal phyla are near the beginning of the Paleozoic (see GEOLOGIC TIME; STRATIGRAPHY). This sudden appearance could represent a real evolutionary explosion of animals in different directions of body design and habitat, or merely the beginning of skeletonization in groups that previously possessed only soft parts, unsuitable for fossilization (see HIGHER LIFE FORMS, EARLIEST EVIDENCE OF). A rise of animal activity in the earliest Paleozoic, evidenced by more abundant and diverse animal burrows, trails, and markings, suggests it was a real evolutionary radiation producing very diverse body plans that occupied many different habitats. In any case, it is remarkable that, although life has existed on Earth for some 3.8 billion years (3.8 Ga) complex multicellular life burst onto the scene with few fossilizable antecedents about 540 Ma.

What causes explosions in the numbers of new species and their body designs, such as occurred in earliest Paleozoic time? The favored, ecologic answer is an availability of previously untapped resources that allows new species to expand their morphologies and life habits from those of their ancestors, to take advantage of space and food in a greater variety of environments. Many paleontologists attribute the dramatic diversity increases at the beginning of the Paleozoic, Mesozoic, and Cenozoic eras to invasions of new ecological spaces. About 95 percent of marine invertebrate species became extinct at the end of the Permian, and about 65 percent of all species died out at the close of the Cretaceous. Succeeding rates of origination in the Mesozoic and Cenozoic eras were relatively rapid, and species radiated evolutionarily into new adaptive zones, major new ecological spaces unoccupied by other organisms. For example, Paleozoic clams and brachiopods had similar gross morphologies and occupied similar habitats, and the decline in brachiopod diversity at the end of the Paleozoic is thought to have stimulated a rapid rise in Mesozoic clam diversity (see BRACHIOPODS; MOLLUSKS).

The rapid radiation of mammalian orders into many different habitats during the first 12 Ma of the Cenozoic era is another example of an evolutionary radiation into new adaptive zones. Paleontologists think the evolutionary radiation of mammals resulted from the Cretaceous extinction of dinosaurs, which were the main competitors of mammals. Removing the dinosaurs allowed the smaller, less diverse mammals to invade environments they had not previously occupied, and diversify into a greater variety of forms (see DINOSAURS; MAMMALS).

Lineages may also enter new adaptive zones when they evolve new adaptations for exploiting resources. For instance, the evolution of a vascular system in plants was a key innovation that allowed their evolutionary radiation in the terrestrial realm. Vascular systems transport water in plants on land, which is water-deficient in comparison with the aquatic environment of their ancestors. Vascular systems also act as support systems in air, which is less buoyant, or "thinner," than water (see ANGIOSPERMS; GYMNOSPERMS; PTERIDOPHYTES).

The evolution of life on Earth has resulted in a rich variety of forms adapted to extremely different environments. At this point, life has successfully invaded the sea, land, and air, and scientists even report living bacteria in production fluids from oil reservoirs 3 km beneath Earth's surface. Paleontologists and biologists unravel the evolutionary pattern of these many forms from several records, including sequences of fossils buried in sediments and sequences of protein molecules in living organisms, both of which have lost pieces of the puzzle through time. However, we reconstruct life's evolutionary history not only to discover its pattern, but also to fully understand the processes of evolution, which happens so slowly relative to a human lifetime.

The realization of the fact of evolution is one of the most profound intellectual events in modern times, and has irrevocably altered our view of the human species and its relationship to the rest of the living world. Recognition of our direct kinship and intimate connection to all life makes clear our responsibility to preserve our fellow creatures and their homes on Earth. When Charles Darwin first proposed the theory of evolution, it was threatening to the worldview of many, who found it sterile and mechanistic, with no special or unique status for humankind. There is no reason why the recognition of our status as part of life on Earth should lessen our pride in human achievement, or our wonder as we unravel the complex history of life. Charles Darwin expressed the sentiment best at the end of *The Origin of Species*, the book that first brought the theory of evolution to public attention: "There is grandeur in this view of life, with its several powers, having been originally breathed into a few forms, or into one; and that, whilst this planet has gone cycling on according to the fixed law of gravity, from so simple a beginning endless forms most beautiful and most wonderful have been, and are being, evolved."

Bibliography

GAMLIN, L., and G. VINES. *The Evolution of Life*. New York, 1987.

GOULD, S. J. *Ever Since Darwin*. New York, 1977.

———. *The Panda's Thumb: More Reflections in Natural History*. New York, 1980.

GOULD, S. J., ed. *The Book of Life*. New York, 1993.

SIMPSON, G. G. *The Major Features of Evolution*. New York, 1953.

LAUREL S. COLLINS
TIMOTHY M. COLLINS

LIFE, ORIGIN OF

From the myths of antiquity to the era of modern science, questions concerning origins have a special fascination. How did the universe come into being? Earth and the solar system? Life itself? While the origin of life on Earth is an obvious fact, how life came to be is critical to any speculations regarding the probability of life on other worlds. Studies of the genesis of life involve direct observation and documentation of the occurrence of fossils of the earliest life forms. These data provide the basis for reconstructing the early evolutionary history of life on Earth.

Evidence from the fields of cosmology, astronomy, chemistry, physics, and geology all converge to indicate that the primordial crust of the planet was formed between 4.6 and 4.5 billion years ago (4.6 to 4.5 Ga). The most critical events leading to the origin and early diversification of life fall in the first half of this long history, resulting in a challenge to investigators seeking direct fossil evidence for these events. Rocks of this great age are rare, confined to limited exposures in the interior of ancient continental interiors or shields. Much of the rock from this remote period is igneous in origin or has been subject to such intense metamorphism that any fossils that once might have been present are absent or altered beyond recognition (*see* FOSSILIZATION AND THE FOSSIL RECORD). With the exception of layered columnar fossils called stromatolites, most of the fossils are microscopic cells about the size of modern bacteria, ranging from isolated cells to simple chains or filaments. Most are studied by microscopic observation of thin sections of promising rock samples. The major challenge in studies of this type is differentiating actual microscopic fossils from inclusions or other small structures in the rock that might resemble simple cells (see Schopf and Walter, 1983).

The earliest known cellular remains date from approximately 3.6 to 3.5 Ga, and include material from the North Pole Dome area of Australia (3.56 Ga) and the Onverwacht Group of South Africa (3.54 Ga). All the microfossils in these deposits appear to represent simple prokaryotic organisms—cells characterized by very small size and relatively simple structure, lacking an organized nucleus or other complex internal structure. In form, the fossils include isolated cells, small clumps or aggregates of cells, simple filaments, and laminated

sheets of cells, each layer being one cell thick, accumulating in multiple layers that result in the characteristic columnar fossils known as stromatolites. The fossils appear to document a moderately diverse array of organisms including bacteria that obtain their energy from the breakdown of organic molecules (heterotrophic metabolism) to photosynthetic cyanobacteria that utilized solar energy to synthesize complex organic molecules (autotrophic metabolism). The degree of diversity represented by the earliest known microfossil assemblages would suggest that the actual origin of life occurred distinctly prior to 3.6 Ga.

The diversity of both heterotrophic and autotrophic prokaryotic organisms increases steadily in somewhat younger rocks, with a corresponding increase in the diversity of filamentous and colonial growth forms. Stromatolites become particularly abundant beginning about 2.8 Ga. At approximately 1.5 Ga there is an increase in the size and complexity of cell fossils, marking the appearance of eukaryotic cells (cells with a true nucleus and relatively complex internal structure) of the type found in most organisms today. By perhaps 1 Ga these more complex cells begin to become organized and specialized in true multicellular plants and animals. At the same time, stromatolites, which had dominated the macrofossil record for almost 2 Ga (and still occur today), begin a precipitous decline in abundance, possibly as a consequence of predation by increasing complex animal-like organisms. While direct observation of microscopic fossil remains can provide a basic time frame for the earliest stages in the history of life, data from a diverse array of disciplines are required to reconstruct how the first organisms came to be and what factors impacted the course of their evolution.

Three attributes of the pre-biotic Earth were critical in the chain of events that would lead to the appearance of the first life-forms. These included the composition of the ancient atmosphere, the planetary temperature regime, and the presence of oceans comprised of liquid water. The earliest or primary atmosphere was undoubtedly like that of the outer "gas giants" of the solar system, consisting primarily of hydrogen and helium, with lesser quantities of methane, ammonia, and water vapor (*see* ATMOSPHERES, PLANETARY). Early attempts to reconstruct the origin of life assumed that it was this type of atmosphere that was present on Earth when life appeared. More recent work suggests that much of this primary atmosphere may have been lost to space during the short molten period following the initial accretion of the planet itself. According to this view, the primary atmosphere was replaced by outgassing from the interior of the planet as it cooled. Based on the analysis of present-day outgassing from volcanoes, the secondary atmosphere was made up primarily of nitrogen, carbon dioxide, and water, with relatively little hydrogen compared to the primary atmosphere. Secondary reactions would have added gases such as ammonia and methane to this atmospheric mix. The actual ratio of the various gases that made up the atmosphere at the time of the origin or life may never be known, but the proportions of the various gases do not seem to be critical to the events that followed. What is critical is that none of the models for the atmosphere include free oxygen. A lack of oxygen is consistent with the known atmospheres of any of the planets in the solar system.

As the crust formed with the initial cooling of Earth, much of the water that would make up the oceans was resident in the atmosphere as water vapor. Astronomical evidence suggests that the Sun was sufficiently less energetic in terms of output during this period when as the surface cooled, much of this water might have accumulated as ice. Fortunately, both methane and carbon dioxide are "greenhouse gases," which helped to retain solar energy as the planet cooled, resulting in the formation of liquid oceans as water vapor condensed from the atmosphere (*see* GLOBAL ENVIRONMENTAL CHANGES, NATURAL). The mode of ocean formation (a single sustained condensation event or multiple events) and timing of ocean formation are conjectural. The oldest sedimentary rock record (metamorphosed) from the Ishua Series in Greenland suggests that significant liquid water was present by 3.8 Ga.

Two primary models for the origin of life have been proposed. The most popular assumes that life appeared as the final consequence of a period of prebiotic chemical and biochemical evolution, based on the reaction of elements and compounds in the early environment. In contrast, a second possibility, generally known as panspermia, suggests that the early Earth may have been "seeded" by viable spore-like bodies drifting through space, much as spores of fungi and bacteria will contaminate a culture dish that is opened in the laboratory.

As late as the 1970s, most paleobiologists assumed that the origin of life might have required a billion years or more, following the initial formation of the oceans. With the actual fossil evidence of the earliest cells (3.6 Ga) gradually being pushed back in time toward the origin of the oceans (3.8 Ga), whatever the mechanism for life's origin, it is clear that the process may have occurred in as little as 200 to 300 million years (Ma). This narrowing of the time window for the origin of life might appear to support the panspermia model, but the extreme conditions of open space make it unlikely that living cells could retain their viability for the long ages required for such particles to drift between solar systems. While extraterrestrial sources for living cells seem unlikely, studies of meteorites and comets confirm that a wide range of simple to moderately complex organic molecules do arrive from space (see Chyba et al., 1990). Such chemical seeding might well have sped up the chemical evolution sequence leading to the first cells.

The first stage in the origin of life involves nonbiological synthesis of simple biochemical "building blocks" (monomers), followed by the linkage of these units into more complex biochemical molecules called polymers. Numerous experiments, involving water, gas mixtures analogous to the early atmosphere, and various energy sources (electrical discharge, heat, ultraviolet radiation, etc.), suggest that a diverse array of monomers can be spontaneously formed by reactions between elements and simple compounds that were present in the prebiotic Earth. These experiments, often called "Miller Experiments" in recognition of the pioneering work of Stanley Miller in 1953, produce varying mixtures of amino acids, nucleotides, simple carbohydrates and fatty acids, and critical energy-rich molecules such as adenosine tri-phosphate (ATP). The basis of the energy metabolism of all living cells is the breakdown of ATP molecules.

In living things, simple monomers are linked to form larger polymers, made up of repetitive sequences of monomer subunits. Amino acids are linked to form proteins, simple carbohydrates polymerize to form complex sugars and starches, fatty acids make up lipids, and nucleotides link to form nucleic acids such as ribonucleic acid (RNA) and deoxyribonucleic acid (DNA). Simple monomers, in a dilute biochemical "soup," are unlikely to polymerize effectively without additional concentration and precise spatial orientation. Recent studies suggest that clay minerals may have played a critical role in polymerizing more complex molecules. Charge distribution on the surface of exposed clay minerals can attract and concentrate monomers from the surrounding water, simultaneously orienting the molecular subunits and catalyzing their linkage. Such linkages would be accelerated if ATP molecules were also present to serve as an energy source. None of these processes would have occurred at significant rates on land or in shallow water due to the disruptive effects of intense ultraviolet radiation from the Sun (with no oxygen, there was no protective ozone shield in the upper atmosphere). Submarine vents have been suggested as one possible site that would have provided protection from ultraviolet (UV) exposure while providing a rich chemical environment.

A living cell is more than a simple collection of polymers, and additional processes must have facilitated the formation of structures analogous to cell walls and membranes. Heating and cooling and wetting and drying of amino acid mixtures have been shown to result in the formation of small cell-like structures called proteinoid microspheres. When lipids are present, the walls of these protocells consist of a proteinoid-lipid complex with some of the differential permeability found in cell membranes. These microspheres can differentially accumulate various biochemical molecules, leading to the initial stages in the evolution of membrane-bound chemical systems. Although these microspheres can "grow" and even bud off new microspheres, they are not alive since they cannot replicate with precision. There were probably at least two chemical steps in the evolutionary transition from nonliving microspheres to the simplest prokaryotic cells. The first was the evolution of protein catalysts (enzymes) that would facilitate a wider range of chemical reactions, followed by the evolution of a simple genetic system to assure reliable duplication of these enzymes.

In all modern cells, genetic coding is a three-element process. The primary coding is in the form of double-stranded DNA. Simpler, single-strand RNA molecules are constructed from the DNA template, and the RNA units serve as the pattern for protein enzyme synthesis. The recent discovery of simple RNA sequences that will catalyze their own replication suggests that the earliest genetic code may have used DNA as the primary information storage with direct synthesis of proteins from RNA strands. DNA molecules are more stable than RNA, however, and selection may well

have favored the synthesis of DNA copies of the RNA units, resulting in DNA assuming the primary information storage function.

The development of a stable, DNA-based genetic system was probably the critical event in the transition from nonliving membrane-bound chemical entities to the simplest prokaryotic cells. While multiple life origin events are a theoretical possibility, the universal nature of the DNA genetic code and the extreme specificity of living cells with respect to chemical isomers suggest that all living cells today are probably descended from a single primordial cell type. Other protocell types may have evolved, but, if they did, they were ultimately displaced by the descendants of the single cell lineage.

Structurally and chemically, the simplest living cells are heterotrophic, bacteria-like cells that synthesize ATP (the universal energy "currency" of life) using energy obtained from the breakdown of carbohydrates and other organic polymers. The first cells probably functioned in this manner and used molecules from the surrounding organic "soup" as their food source. In the absence of oxygen, such breakdown reactions are inefficient, in that the "food" molecules are only partially disassembled and the ATP yield is minimal, but the process does work. This simple reaction system, known as glycolysis or anaerobic respiration, is still used as the preliminary step in cell respiration in all living prokaryotic and eukaryotic cells.

Limitations in the availability of "food" molecules might well have resulted in the eventual extinction of these first life forms were it not for the appearance of autotrophic prokaryotes that could use an outside energy source to synthesize energy-rich molecules. The most successful of these early autotrophs were the cyanobacteria, which used light (photosynthesis) as their energy source. The evolution of the cyanobacteria may well be linked to the intense UV levels near the ocean surface as a result of the lack of oxygen and an ozone shield. If cells were to survive in shallower water, one solution would be internal shields that would protect the cell from UV exposure. Chlorophyll is a pigment that absorbs blue and UV light, and chlorophyll pigments, incorporated into membranes, would make an effective UV shield. Early cyanobacteria with such shields could survive in sunlit waters. Chlorophyll, when excited by UV exposure, generates high-energy electrons. The development of very simple electron transport systems would allow some of this energy to be channeled into ATP synthesis. With ample supplies of ATP synthesized from solar energy cyanobacteria could essentially operate anaerobic respiration reactions "in reverse," producing energy-rich molecules like sugars from simple molecules such as carbon dioxide. This is a very simple form of photosynthesis that does not produce oxygen and is essentially identical to a little-used form of photosynthesis (cyclic photo-phosphorylation) found in all plant-like cells. The early appearance of photosynthetic cyanobacteria represented the first stable balance between photosynthetic "producers" and heterotrophic "consumers" that has been the basis for ecosystem structure since that time.

The most successful cyanobacteria were the stromatolites, and their photosynthetic activity provided the resource base for diversification of heterotrophic prokaryotes. Their photosynthesis did require sources of hydrogen that initially would have been available directly from the atmosphere either as hydrogen gas or hydrogen sulfide. Hydrogen, the lightest of the elements, was the gas most prone to leakage into space and, by about 2.5 Ga, cyanobacteria faced a hydrogen crisis. The solution was to adapt to the use of a new hydrogen source—water. Elaboration of the light reactions of photosynthesis led to a new reaction (photolysis) that split water into its component hydrogen and oxygen. The oxygen was an unneeded (and toxic) by-product that was released into the surrounding environment.

As a result of this new variant of photosynthesis, oxygen gradually began to be added to the atmosphere between 2.5 and 2.0 Ga, creating a serious biological crisis. Oxygen is a reactive gas that is toxic to cells; therefore as oxygen levels increased (essentially oxygen pollution of the atmosphere), all cells were increasingly at risk. The solution to this crisis was the evolution of new variations on respiratory metabolic pathways that stripped hydrogen from "food" molecules and linked them up with any oxygen molecules entering the cells. The evolution of this new metabolic variant, known as aerobic respiration, had two consequences. The first was the protection of the cell from oxygen (the water produced when hydrogen was combined with oxygen was harmless), while the second was a huge increase in metabolic efficiency. Being able to strip hydrogens from complex molecules meant that food molecules such as sugar could be broken down completely to carbon dioxide and water with

an ATP yield that was ten to fifteen times greater than that which could be accomplished through anaerobic metabolism! Aerobic cells that could metabolize oxygen were thus not only protected from its toxic effects, they also enjoyed a ten-fold increase in available energy with the same food intake.

The large increase in energy yields brought about by aerobic metabolism provided the energy resources for rapid evolutionary innovation between 2.0 and 1.5 Ga, leading to the appearance of larger and more complex eukaryotic cells and, after 1 Ga, the evolution of complex multicellular life-forms (see LIFE, EVOLUTION OF). In effect, photolysis created and maintains Earth's oxygen rich atmosphere, a chemical anomaly with respect to all other atmospheres in our solar system. Aerobic respiration, a metabolic solution to the problem of increasing oxygen levels in the atmosphere, led to the evolution of complex, multicellular life-forms. Together, the two evolutionary innovations have led to the paradox where almost all organisms in the biosphere require constant supplies of a biochemically toxic gas to maintain their sophisticated cell structure. The descendants of prokaryotic bacteria that did not evolve pathways to disable oxygen within their cells became progressively less common as oxygen levels continued to rise through the late Precambrian and Paleozoic and are now confined to rare environments where oxygen is essentially absent. In effect, the entire fabric of life reflects events and innovations dating back to the dawn of life itself.

Bibliography

Chyba, C. F., P. J. Thomas, L. Brookshaw, and C. Sagan. "Cometary Delivery of Organic Molecules to the Early Earth." *Science* 249 (1990):366–373.

Schopf, J. W., and M. R. Walter "Archean Microfossils: New Evidence of Ancient Microbes." In *Earth's Earliest Biosphere: Its Origin and Evolution*, ed. J. W. Schopf. Princeton, NJ, 1983.

RALPH E. TAGGART

LIMNOLOGY

See Lakes

LINDGREN, WALDEMAR

Waldemar Lindgren's ideas guided, even dominated, the study of mineral deposits throughout the first half of the twentieth century. Born on 14 February 1860 in Kalmar, Sweden, and educated in both Sweden and Germany, Lindgren arrived in the United States in 1883. His training in mineralogy, chemistry, and geology at the renowned Royal Mining Academy at Freiberg in Germany, and letter of introduction from some American friends he had met in Freiberg, helped to secure Lindgren a position as geologist on the Northern Transcontinental Survey being carried out for the Northern Pacific Railroad under the direction of Raphael Pumpelly. The survey was completed early in 1884 and Lindgren then found short-term employment as an assayer and draftsman in the thriving mining camp at Butte, Montana. His earlier training at Freiberg and the recommendation of Pumpelly soon brought Lindgren a post with the U.S. Geological Survey (USGS). In November 1884 Lindgren was assigned to the San Francisco office of the USGS under the direction of George F. Becker. Thus began many years of employment with the USGS and the opportunity to map and study the great mineral districts of western North America. He published many geological papers on the mining districts of California, Nevada, Arizona, Colorado, Wyoming, Montana, New Mexico, and Washington. Some of the papers, such as USGS Professional Paper 43 (1905) on the Clifton-Morenci district in Arizona, and Professional Paper 54 (1906), with F. L. Ransome, on the gold deposits of Cripple Creek, Colorado, quickly became classics and are still standard references.

Invited as visiting professor at Stanford University for the academic year 1897–1898, Lindgren accepted and took a one-year leave from the USGS. The Stanford experience had two important effects—Lindgren discovered that he enjoyed teaching, and he met among his students Herbert C. Hoover, mining engineer and later president of the United States, who remained a close, lifelong friend.

Lindgren was fluent in many languages and his mastery of English was exemplary. He was able to start publishing papers on the mining districts of western North America as soon as he joined the USGS. His papers soon started to draw the attention of influential readers. Near the end of the nineteenth century many American geologists considered that the downward circulation of meteoric waters was probably the most important agent in the formation of mineral deposits. Foremost among those espousing this concept were C. D. Van Hise and S. F. Emmons. Lindgren pointed out that much of the evidence he saw in the field supported the upward flow of mineralizing solutions associated with igneous intrusions, a concept originally championed many years before by the French geologist Elie de Beaumont. Lindgren had been well trained in chemistry at Freiberg and he successfully introduced chemistry to the study of ores. From 1900 onward Lindgren's careful marshaling of the mineralogical, chemical, and field evidence in support of a genetic relation between mineralized veins and bodies of intrusive igneous rock began to convince more and more people of the correctness of his ideas.

Lindgren's growing scientific stature, as discussed by L. C. Graton (1933), led to his promotion in the USGS; in 1908 he became chief of the Division of Metalliferous Geology (succeeding S. F. Emmons), and in 1911 Chief Geologist. His scientific stature was also recognized outside the USGS: in 1908 Lindgren gave a series of invited lectures at the Massachusetts Institute of Technology (MIT) and they were so successful that he was invited to visit and repeat them each year until 1912 when he was offered, and accepted, the post of William Barton Rogers Professor of Geology at MIT. Lindgren's years at MIT allowed him to travel, consult, and work in other countries of the world, such as Cuba, Chile, Bolivia, Canada, and Mexico. Most important, however, he had time to write and publish *Mineral Deposits* (1913), a book containing the substance of his MIT lectures and his most important legacy to geology. *Mineral Deposits* brought together all of Lindgren's incisive thinking and extensive experience. It presented his hypothesis that the form, shape, and pattern of mineralization is controlled by the depth and temperature at which the minerals are emplaced. Many aspects of Lindgren's depth-temperature classification, and even the terms that he introduced—such as "epithermal," "mesothermal," and "hypothermal"—remain current today, although some aspects of his hypothesis have been modified. Later editions of *Mineral Deposits* (1919, 1928, and 1932) kept the text current so that for more than forty years it was the world's leading text on the origin and characteristics of mineral deposits.

Lindgren was a modest, softly spoken man who was greatly revered by those who worked with or for him. He was one of the small group of people who, in 1905, formed the not-for-profit Economic Geology Publishing Company in order to publish *Economic Geology*, the leading journal in the study of mineral deposits. From 1905 to his death in Brookline, Massachusetts, on 3 November 1939, he served as one of the associate editors of *Economic Geology*.

By the time Lindgren died his name was one of the most widely recognized in science. He had been awarded many honors. Three generations of geologists have been trained on Lindgren's writings, but more important he left a legacy of how to study mineral deposits. Lindgren stressed that studies should proceed by careful field examination of evidence, detailed investigation of the microscopic characters of the ore, and only then in forward thinking about the chemistry of deposit-forming processes. This continues to guide geologists in their studies of deposits today.

Bibliography

GRATON, L. C. "Life and Scientific Works of Waldemar Lindgren." In *Ore Deposits of the Western States: Lindgren Volume*. New York, 1933.

LINDGREN, W. "The Copper Deposits of the Clifton-Morenci District, Arizona." U.S. Geological Survey, Professional Paper 43. Washington, DC, 1905.

———. *Mineral Deposits*, 1st ed. New York, 1913. Rev. eds., 1919, 1928, 1932.

LINDGREN, W., and F. L. RANSOME. "Geology and Gold Deposits of the Cripple Creek District, Colorado." U.S. Geological Survey, Professional Paper 54. Washington, DC, 1906.

BRIAN J. SKINNER

LIZARDS, SNAKES, AND TURTLES

See Reptiles

LYELL, CHARLES

Charles Lyell was born at Kirriemuir, Angus, Scotland, on 14 November 1797, the eldest child of Charles Lyell and Frances (Smith) Lyell. In 1798 his father moved to Hampshire in southern England, where Lyell grew up. After attending school at Midhurst in Sussex, Lyell entered Exeter College, Oxford, where he received the usual classical education and graduated with a Bachelor of Arts degree in 1819. At Oxford, William Buckland's lectures on mineralogy and geology aroused Lyell's interest in geology. In 1817 he visited the island of Staffa to study its columnar basalt and in 1818, while traveling with his family through Switzerland, Lyell observed the effects of glaciers in the Alps.

In 1819 Lyell was elected to the Geological Society of London and entered Lincoln's Inn to study law. In 1822 he was admitted to the bar, but defective eyesight (he was shortsighted) made reading for the law difficult. He turned to geology and in 1823 spent two months in Paris, where he met leading naturalists, including the director of the Museum of Natural History, GEORGES CUVIER, and the famous traveler and geologist, Alexander von Humboldt. The geologist Constant Prevost showed him that the alternation of freshwater and marine formations in the Paris basin did not require the geological catastrophes postulated by Cuvier. In 1824 Lyell found that modern limestones formed in Scottish lakes were exactly like ancient freshwater limestones of the Paris basin, suggesting a close analogy between ancient and modern conditions.

In 1828 a geological tour through France and Italy convinced Lyell that the elevation and disturbance of sedimentary rocks had occurred gradually as a result of earthquakes and volcanic activity occurring on the same scale as in the present. He found analogies among rocks formed at widely different geological periods and decided that in the past, geological processes were identical to those of the present. In his *Principles of Geology*, published in three volumes from 1830 to 1833, Lyell argued that the geological past was uniform with the present. During his travels, Lyell recognized that the Tertiary formations of France and Italy were of various ages and might be classified by the proportion of living species among their fossil shells. Lyell noted that throughout the Tertiary period species of shells gradually became extinct to be replaced by new species, their extinction brought about by geological change. According to the proportion of living species among their fossil shells, a seminal observation in the development of historical geology, Lyell classified Tertiary formations into four epochs: Eocene, Miocene, and the older and newer Pliocene.

On 12 July 1832 Lyell married Mary Elizabeth Horner, daughter of Leonard Horner, geologist and educator. The Lyells then took a house in London, where they resided except when on their frequent and often extended travels. Fluent in French and German, Mary Lyell assisted Lyell in his scientific work and usually accompanied him on his travels. Through the 1830s Lyell was occupied continually with revision of the *Principles of Geology*, which was published in a fifth edition in 1837 and exerted a profound influence, especially on young geologists, including Charles Darwin. Lyell showed how the detailed study of rocks, and especially of fossils, could be used to reveal the history of the earth. In 1838 he published the *Elements of Geology*, an introductory work describing the classes of rocks and the succession of geological formations then known.

In July 1841 the Lyells sailed for the United States, where that autumn Lyell delivered the Lowell lectures at Boston. The Lyells remained in America until August 1842, traveling widely to examine the geology of New York State and Pennsylvania, the Atlantic coastal plain as far south as Georgia, and Nova Scotia. In 1845 Lyell published his *Travels in America* and the same year returned to America to deliver the Lowell lectures and to travel widely through the South and along the Mississippi River, travels described in 1849 in his *Second Visit to the United States*. In 1852 he made a third visit to the States to deliver the Lowell lectures, and in 1853 returned a fourth time as a representative of Great Britain at the New York Industrial Exhibition.

During 1853–1854 the Lyells visited Madeira and the Canary Islands to examine the evidence for Leopold von Buch's theory of craters of elevation, which postulated the sudden catastrophic elevation of volcanic mountains. Finding that Madeira and the Canary Islands had been formed by ordinary volcanic eruptions, in 1857 Lyell revisited Sicily to learn whether Elie de Beaumont was cor-

rect that, when lavas were found steeply inclined, they must have been poured out originally on a horizontal surface and uplifted later, an assumption necessary to the theory of craters of elevation. On Mount Etna and Vesuvius, Lyell found that lavas could harden into thick sheets of compact rock on steep slopes and that the mountains had been built up gradually by a long succession of ordinary volcanic eruptions. Lyell's 1858 paper, "On the Structure of Lavas Which Have Consolidated on Steep Slopes. . . ," effectively destroyed the theory of craters of elevation and with it the scientific basis for catastrophism.

In 1859, when Lyell was preparing the sixth edition of the *Elements of Geology*, the discovery of flint implements at Brixham in Devonshire and in the Somme Valley in France, coupled with the recent discovery of a fossil human skeleton at Neanderthal in Germany, suggested that humans had lived in Europe much earlier than was thought previously. As Lyell compared the evidence for early humans to that for successive glacial periods, he found the subject too large to include within the *Elements*, so in 1863 he published a separate work titled *Antiquity of Man*. In *Antiquity of Man* Lyell argued that immense periods of time had been necessary for the differentiation of the various human races and human languages. If the various human races were descended from common ancestors, related species of animals might likewise share a common ancestry. Lyell laid out a broad array of evidence to indicate that species had changed over time, that such change may have occurred as a result of natural selection, and that humankind had evolved from lower animals. Lyell did not draw conclusions from such evidence, thereby inducing his readers to decide its meaning for themselves.

Antiquity of Man enjoyed a large sale, the first edition of four thousand copies disappearing almost immediately and the book moving rapidly through second and third editions. In the second edition, at Charles Darwin's urging, Lyell expressed his opinion that scientists would ultimately agree that change of species had been brought about by variation and natural selection (*see* DARWIN, CHARLES). In 1864, Lyell publicly declared his faith in Darwin's theory and he thoroughly revised the tenth edition of the *Principles of Geology* (1867–1868) to conform to it.

On 25 April 1873 Lady Lyell died of typhoid fever. Her death was a severe blow to Lyell and, after some months of illness, he died on 22 February 1875. Lyell exerted a profound influence on geologists both during his lifetime and later, especially by his insistence that geological phenomena are explicable by natural causes. His objection to catastrophism was that, once a catastrophe was postulated, geological inquiry ceased. Lyell's uniformitarian geology could not be fitted within the brief age of the earth posited by Lord Kelvin and widely accepted in the late nineteenth century, but in the twentieth century the much greater age of Earth determined by radioactive methods and the slow and extremely steady rates of geological change suggested by the theories of continental drift and plate tectonics have vindicated Lyell's confidence in the uniformity of Earth history.

Bibliography

LYELL, C. *Principles of Geology Being an Attempt to Explain the Former Changes of the Earth's Surface by Reference to Causes Now in Operation.* London, Eng., 1830–1833.

———. *Elements of Geology.* London, Eng., 1838.

———. *Travels in North America, in the Years 1841–2; with Geological Observations on the United States, Canada, and Nova Scotia.* London, Eng., 1845.

———. *A Second Visit to the United States of North America.* London, Eng., 1849.

———. *The Geological Evidences of the Antiquity of Man with Remarks on Theories of the Origin of Species by Variation.* London, Eng., 1863.

WILSON, L. G., ed. *Sir Charles Lyell's Scientific Journals on the Species Question.* New Haven, CT, 1970.

———. *Charles Lyell, the Years to 1841.* New Haven, CT, 1972.

LEONARD WILSON